Engineering Thermodynamics

Engineering Thermodynamics

(For students of B.E./B.Tech. in all disciplines of Engineering and Technology all Institutions and Universities)

Onkar Singh

Mechanical Engineering Department

Harcourt Butler Technological Institute, Kanpur (U.P.)

PUBLISHING FOR ONE WORLD

NEW AGE INTERNATIONAL (P) LIMITED, PUBLISHE

(formerly Wiley Eastern Limited)

New Delhi • Bangalore • Chennai • Cochin • Guwahati • Hyderabad
Jalandhar • Kolkata • Lucknow • Mumbai • Ranchi

Visit us at www.newagepublishers.com

Published by New Age International (P) Ltd., Publishers
First Edition : 2006

ISBN : 81-224-1750-7

Rs. 250.00

C-05-09-405

Printed in India at Ajit Printers, Delhi.
Typesetter : Nagri Printer, Delhi.

PUBLISHING FOR ONE WORLD
NEW AGE INTERNATIONAL (P) LIMITED, PUBLISHERS
(formerly Wiley Eastern Limited)
4835/24, Ansari Road, Daryaganj, New Delhi - 110002
Visit us at **www.newagepublishers.com**

Preface

During teaching of the course of engineering thermodynamics I have felt that the students at the undergraduate level of engineering & technology face difficulty in understanding the concepts of engineering thermodynamics and their applications. Also, the students face great difficulty in referring to the number of text books for different topics. The present book is an effort in the direction of presenting the concepts of engineering thermodynamics and their applications in clear, concise and systematic manner at one place. Presentation is made in very simple and easily understandable language and well supported with wide ranging illustrations and solved numerical problems and unsolved numerical problems with answers.

The subject matter in this book covers the syllabus of the basic course on engineering thermodynamics being taught in different institutions and universities across the country. There are total 12 chapters in this text book. These deal with "Fundamental concepts and definitions", "Zeroth law of thermodynamics", "First law of thermodynamics", "Second law of thermodynamics", "Entropy", "Thermodynamic properties of pure substance", "Availability and general thermodynamic relations", "Vapour power cycles", "Gas power cycles", "Introduction to internal combustion engines", "Introduction to refrigeration and air conditioning", "Introduction to heat transfer", and "One dimensional compressible fluid flow". Each chapter has been provided with sufficient number of typical numerical problems of such as steam tables, refrigeration tables, mollier chart, psychrometry chart etc. are also provided at the end of the book for quick reference. I hope that the students and teachers referring to this book will find it useful.

I am highly indebted to my parents and family members for their continuous encouragement and cooperation during the preparation of manuscript. I would like to place on record my gratitude and apologies to my wife Parvin and kids Sneha & Prateek who patiently endured certain neglect and hardships due to my preoccupation with the preparation of this manuscript.

I am thankful to AICTE, New Delhi for the financial support provided to me in the Young Teacher Career Award.

I am also thankful to Mr. L. N. Mishra and other staff members of M/s New Age International for their cooperation throughout the preparation of the text book. At the end I thank to all those who supported directly or indirectly in the preparation of this text book.

I shall be extremely grateful to all the readers of this text book for their constructive criticism, indicating any errors and omissions etc. for improving its quality and form.

Onkar Singh

CONTENTS

CONTENTS

1

Fundamental Concepts and Definitions

1.1 INTRODUCTION AND DEFINITION OF THERMODYNAMICS

Thermodynamics is a branch of science which deals with energy. Engineering thermodynamics is modified name of this science when applied to design and analysis of various energy conversion systems. Thermodynamics has basically a few fundamental laws and principles applied to a wide range of problems. Thermodynamics is core to engineering and allows understanding of the mechanism of energy conversion. It is really very difficult to identify any area where there is no interaction in terms of energy and matter. It is a science having its relevance in every walk of life. Thermodynamics can be classified as 'Classical thermodynamics' and 'Statistical thermodynamics'. Here in engineering systems analysis the classical thermodynamics is employed.

"Thermodynamics is the branch of physical science that deals with the various phenomena of energy and related properties of matter, especially of the laws of transformations of heat into other forms of energy and vice-versa."

Internal combustion engines employed in automobiles are a good example of the energy conversion equipments where fuel is being burnt inside the piston cylinder arrangement and chemical energy liberated by the fuel is used for getting the shaft work from crankshaft. Thermodynamics lets one know the answer for the questions as, what shall be the amount of work available from engine?, what shall be the efficiency of engine?, etc.

For analysing any system there are basically two approaches available in engineering thermodynamics. Approach of thermodynamic analysis means how the analyser considers the system. Macroscopic approach is the one in which complete system as a whole is considered and studied without caring for what is there constituting the system at microscopic level.

Contrary to this the microscopic approach is one having fragmented the system under consideration upto microscopic level and analysing the different constituent subsystems/microsystems. In this approach study is made at the microscopic level. For studying the system the microlevel studies are put together to see the influences on overall system. Thus, the statistical techniques are used for integrating the studies made at microscopic level. This is how the studies are taken up in statistical thermodynamics. In general it can be said that, Macroscopic approach analysis = Σ (Microscopic approach analysis).

1.2 DIMENSIONS AND UNITS

"Dimension" refers to certain fundamental physical concepts that are involved in the process of nature and are more or less directly evident to our physical senses, thus dimension is used for characterizing any physical quantity. Dimensions can be broadly classified as "primary dimensions" and "secondary or derived dimensions". "Basic dimensions such as mass 'M', length 'L', time 't' and temperature 'T' are called primary dimensions, while quantities which are described using primary dimensions are called secondary dimensions such as for energy, velocity, force, volume, etc".

"Units" are the magnitudes assigned to the dimensions. Units assigned to "primary dimensions" are called "basic units" whereas units assigned to "secondary dimensions" are called "derived units". Various systems of units have prevailed in the past such as FPS (Foot-Pound-Second), CGS (Centimetre-Gram-Second), MKS (Metre-Kilogram-Second) etc. but at present SI system (System-International) of units has been accepted world wide. Here in the present text also SI system of units has been used. Following table gives the basic and derived units in SI system of units.

Table 1.1 *SI system of units*

Quantity	*Unit*	*Symbol*
Basic Units		
Length (L)	Metre	m
Mass (M)	Kilogram	kg
Time (t)	Second	s
Temperature (T)	Kelvin	K
Plane angle	Radian	rad
Solid angle	Steradian	sr
Luminous intensity	Candela	cd
Molecular substance	Mole	mol.
Electric Current	Ampere	A
Derived Units		
Force (F)	Newton	N {kg.m/s^2}
Energy (E)	Joule	J {N.m = kg. m^2/s^2}
Power	Watt	W {J/s = kg. m^2/s^3}
Pressure	Pascal	Pa {N/m^2 = kg/(ms^2)}

Equivalence amongst the various systems of unit for basic units is shown in table 1.2.

Table 1. 2 *Various systems of units*

Quantity	Unit-(Symbol) *SI*	*MKS*	*CGS*	*FPS*
Length	Metre (m)	Metre (m)	Centimetre (cm)	Foot (ft)
Mass	Kilogram (kg)	Kilogram (kg)	Gram (gm)	Pound (lb)
Time	Second (s)	Second (s)	Second (s)	Second (s)
Temperature	Kelvin (K)	Centigrade (°C)	Centigrade (°C)	Fahrenheit (°F)

The various prefixes used with SI units are given as under :

Prefix	*Factor*	*Symbol*	*Prefix*	*Factor*	*Symbol*
deca	10	da	deci	10^{-1}	d
hecto	10^2	h	centi	10^{-2}	c
kilo	10^3	k	milli	10^{-3}	m
mega	10^6	M	micro	10^{-6}	μ
giga	10^9	G	nano	10^{-9}	n
tera	10^{12}	T	pico	10^{-12}	p
peta	10^{15}	P	femto	10^{-15}	f
exa	10^{18}	E	atto	10^{-18}	a

The conversion table for one unit into the other is given in table 1.3.

Table 1.3 *Unit Conversion Table*

1 ft = 0.3048 m	1 ft^2 = 0.09290 m^2
1 in = 0.0254 m	1 in^2 = 6.45 cm^2
1 lb = 453.6 gm	1 lb = 0.4536 kg
1 lbf = 4.45 N	1 kgf = 9.81 N
1 lbf/in^2 = 6.989 kN/m^2 = 0.0689 bar = 703 kgf/m^2	
1 bar = 10^5 N/m^2 = 14.5038 1bf/in^2 = 0.9869 atm	
= 1.0197 kgf/cm^2	
1 ft. lbf = 1.356 Joules	
1 Btu = 778.16 ft. lbf = 1.055 kJ	
1Btu/lb = 2.326 kJ/kg	
1 ft^3/lb = 0.0624 m^3/kg,	1 Cal = 4.18 J

1.3 CONCEPT OF CONTINUUM

In Macroscopic approach of thermodynamics the substance is considered to be continuous whereas every matter actually comprises of myriads of molecules with intermolecular spacing amongst them. For analyzing a substance in aggregate it shall be desired to use laws of motion for individual molecules and study at molecular level be put together statistically to get the influence upon aggregate. In statistical thermodynamics this microscopic approach is followed, although it is often too cumbersome for practical calculations.

In engineering thermodynamics where focus lies upon the gross behaviour of the system and substance in it, the statistical approach is to be kept aside and classical thermodynamics approach be followed. In classical thermodynamics, for analysis the atomic structure of substance is considered to be continuous. For facilitating the analysis this concept of continuum is used in which the substance is treated free from any kind of discontinuity. As this is an assumed state of continuum in substance so the order of analysis or scale of analysis becomes very important. Thus, in case the scale of analysis is large enough and the discontinuities are of the order of intermolecular spacing or mean free path then due to relative order of discontinuity being negligible it may be treated continuous.

In the situations when scale of analysis is too small such that even the intermolecular spacing or mean free path are not negligible i.e. the mean free path is of comparable size with smallest significant dimension in analysis then it can not be considered continuous and the microscopic approach for analysis should be followed. For example, whenever one deals with highly rarefied gases such as in rocket flight at very high altitudes or electron tubes, the concept of continuum of classical thermodynamics should be dropped and statistical thermodynamics using microscopic approach should be followed. Thus, in general it can be said that the assumption of continuum is well suited for macroscopic approach where discontinuity at molecular level can be easily ignored as the scale of analysis is quite large. The concept of continuum is thus a convenient fiction which remains valid for most of engineering problems where only macroscopic or phenomenological informations are desired.

For example, let us see density at a point as a property of continuum. Let us take some mass of fluid Δm in some volume ΔV enveloping a point 'P' in the continuous fluid. Average mass density of fluid within volume ΔV shall be the ratio $(\Delta m/\Delta V)$. Now let us shrink the volume ΔV enveloping the point to volume $\Delta V'$. It could be seen that upon reducing the volume, $\Delta V'$ may be so small as to contain relatively few molecules which may also keep on moving in and out of the considered very small volume, thus average density keeps on fluctuating with time. For such a situation the definite value of density can not be given.

Therefore, we may consider some limiting volume ΔV_{limit} such that the fluid around the point may be treated continuous and the average density at the point may be given by the ratio $\left(\dfrac{\Delta m}{\Delta V_{\text{limit}}}\right)$.

Thus, it shows how the concept of continuum although fictitious is used for defining density at a point as given below,

$$\text{Average density at the point} = \lim_{\Delta V \to \Delta V_{\text{limit}}} \left(\frac{\Delta m}{\Delta V}\right)$$

1.4 SYSTEMS, SURROUNDINGS AND UNIVERSE

In thermodynamics the 'system' is defined as the quantity of matter or region in space upon which the attention is concentrated for the sake of analysis. These systems are also referred to as thermodynamic systems. For the study these systems are to be clearly defined using a real or hypothetical boundary. Every thing outside this real/hypothetical boundary is termed as the 'surroundings'. Thus, the surroundings may be defined as every thing surrounding the system. System and surroundings when put together result in universe.

Universe = System + Surroundings

The system is also some times defined as the control system and the boundary defined for separating it from surroundings is called control boundary, the volume enclosed within the boundary is control volume and the space enclosed within the boundary is called control space.

Based on the energy and mass interactions of the systems with surroundings/other systems across the boundary the system can be further classified as the open, close, and isolated system. The 'open system' is one in which the energy and mass interactions take place at the system boundary, for example automobile engine etc.

'Closed system' is the system having only energy interactions at its boundary, for example, boiling water in a closed pan etc. The mass interactions in such system are absent. 'Isolated system' refers to the system which neither has mass interaction nor energy interaction across system boundary, for example Thermos Flask etc. Thus, the isolated system does not interact with the surroundings/systems in any way.

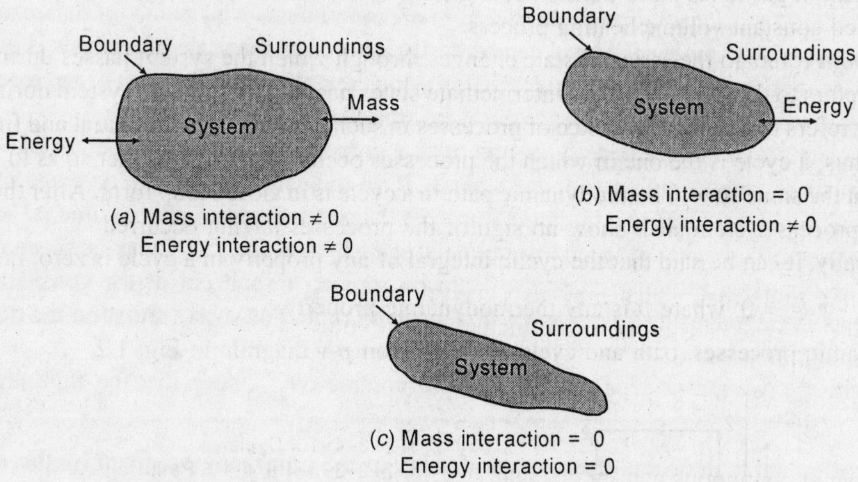

Fig. 1.1 (a) Open system (b) Closed system (c) Isolated system

1.5 PROPERTIES AND STATE

For defining any system certain parameters are needed. 'Properties' are those observable characteristics of the system which can be used for defining it. Thermodynamic properties are observable characteristics of the thermodynamic system. Pressure, temperature, volume, viscosity, modulus of elasticity etc. are the examples of property. These properties are some times observable directly and some times indirectly. Properties can be further classified as the 'intensive property' and 'extensive property'. The *intensive properties* are those properties which have same value for any part of the system or the properties that are independent of the mass of system are called intensive properties, e.g. pressure, temperature etc. *Extensive properties* on the other hand are those which depend upon the mass of system and do not maintain the same value for any path of the system. e.g. mass, volume, energy, enthalpy etc. These extensive properties when estimated on the unit mass basis result in intensive property which is also known as *specific property*, e.g. specific heat, specific volume, specific enthalpy etc.

'State' of a system indicates the specific condition of the system. To know the characteristics of the system quantitatively refers to knowing the state of system. Thus, when the properties of system are quantitatively defined then it refers to the 'state'. For completely specifying the state of a system number of properties may be required which depends upon the complexity of the system. Thermodynamic state in the same way refers to the quantitative definition of the thermodynamic properties of a thermodynamic system e.g. for defining a gas inside the cylinder one may have to define the state using pressure and temperature as 12 bar, 298 K. When the thermodynamic properties defining a state undergo a change in their values it is said to be the 'change of state'.

1.6 THERMODYNAMIC PATH, PROCESS AND CYCLE

Thermodynamic system undergoes changes due to the energy and mass interactions. Thermodynamic state of the system changes due to these interactions. The mode in which the change of state of a system takes place is termed as the *process* such as constant pressure, constant volume process etc. Let us take gas contained in a cylinder and being heated up. The heating of gas in the cylinder shall result in change in state of gas as its pressure, temperature etc. shall increase. However, the mode in which this change of state in gas takes place during heating shall be constant volume mode and hence the process shall be called constant volume heating process.

The *path* refers to the series of state changes through which the system passes during a process. Thus, path refers to the locii of various intermediate states passed through by a system during a process.

Cycle refers to a typical sequence of processes in such a fashion that the initial and final states are identical. Thus, a cycle is the one in which the processes occur one after the other so as to finally bring the system at the same state. Thermodynamic path in a cycle is in closed loop form. After the occurrence of a cyclic process system shall show no sign of the processes having occurred.
Mathematically, it can be said that the cyclic integral of any property in a cycle is zero, i.e.,

$$\oint dp = 0, \text{ where } p \text{ is any thermodynamic property.}$$

Thermodynamic processes, path and cycle are shown on *p-v* diagram in Fig. 1.2

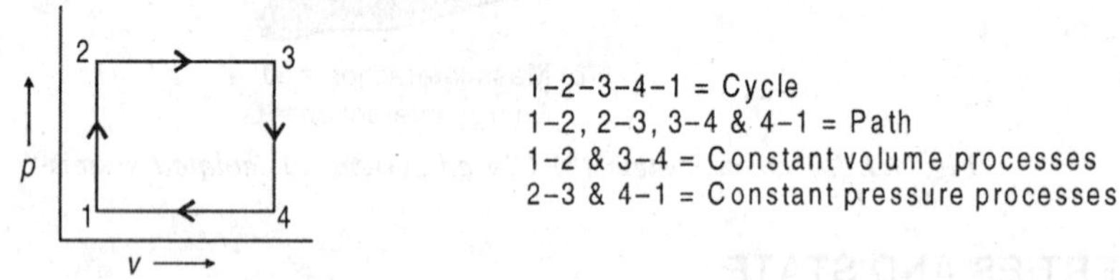

Fig. 1.2 Thermodynamic process, path and cycle.

1.7 THERMODYNAMIC EQUILIBRIUM

Equilibrium is that state of a system in which the state does not undergo any change in itself with passage of time without the aid of any external agent. Equilibrium state of a system can be examined by observing whether the change in state of the system occurs or not. If no change in state of system occurs then the system can be said in equilibrium. Thermodynamic equilibrium is a situation in which thermodynamic system does not undergo any change in its state. Let us consider a steel glass full of hot milk kept in open atmosphere. It is quite obvious that the heat from the milk shall be continuously transferred to atmosphere till the temperature of milk, glass and atmosphere are not alike. During the transfer of heat from milk the temperature of milk could be seen to decrease continually. Temperature attains some final value and does not change any more. This is the equilibrium state at which the properties stop showing any change in themselves.

Generally, thermodynamic equilibrium of a system may be ensured by ensuring the mechanical, thermal, chemical and electrical equilibriums of the system. 'Mechanical equilibrium' of the system can be well understood from the principles of applied mechanics which say that the net force and moment shall be zero in case of such equilibrium. Thus, in the state of mechanical equilibrium the system does not have any tendency to change mechanical state as it is the state at which the applied forces and developed stresses are fully balanced.

'Thermal equilibrium' is that equilibrium which can be stated to be achieved in the absence of any heat interactions. Thus, if the temperature states of the system do not change then thermal equilibrium is said to be attained. Equality of temperature of the two systems interacting with each other shall ensure thermal equilibrium.

'Chemical equilibrium' is the one which can be realized if the chemical potential of the systems interacting are same. The equality of forward rate of chemical reaction and backward rate of chemical reaction can be taken as criterion for ensuring the chemical equilibrium. Similar to this, in case the electrical potential of the systems interacting are same, the 'electrical equilibrium' is said be attained.

Thus, a system can be said to be in thermodynamic equilibrium if it is in mechanical, thermal, chemical and electrical equilibrium.

1.8 REVERSIBILITY AND IRREVERSIBILITY

Thermodynamic processes may have the change of state occuring in two ways. One is the change of state occuring so that if the system is to restore its original state, it can be had by reversing the factors responsible for occurrence of the process. Other change of state may occur such that the above restoration of original state is not possible. Thermodynamic system that is capable of restoring its original state by reversing the factors responsible for occurrence of the process is called reversible system and the thermodynamic process involved is called *reversible process*. Thus, upon reversal of a process there shall be no trace of the process being ocurred, i.e. state changes during the forward direction of occurrence of a process are exactly similar to the states passed through by the system during the reversed direction of the process. It is quite obvious that the such reversibility can be realised only if the system maintains its thermodynamic equilibrium throughout the occurrence of process.

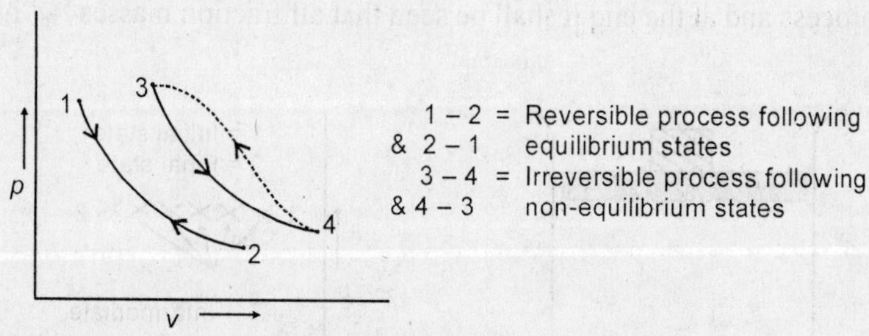

Fig. 1.3 Reversible and Irreversible processes

The *irreversibility* is the characteristics of the system which forbids system from retracing the same path upon reversal of the factors causing the state change. Thus, irreversible systems are those which do not maintain equilibrium during the occurrence of a process. Various factors responsible for the nonattainment of equilibrium are generally the reasons responsible for irreversibility. Presence of friction, dissipative effects etc. have been identified as a few of the prominent reasons for irreversibility. The reversible and irreversible processes are shown on *p-v* diagram in Fig. 1.3 by '1–2 and 2.1' and '3–4 and 4–3' respectively.

1.9 QUASI-STATIC PROCESS

Thermodynamic equilibrium of a system is very difficult to be realised during the occurrence of a thermodynamic process. It may be understood that this kind of equilibrium is rather practically impossible. In case such equilibrium could not be attained then the thermodynamic analysis cannot be done, as the exact analysis of a system not in equilibrium is impossible. 'Quasi-static' consideration is one of the ways to consider the real system as if it is behaving in thermodynamic equilibrium and thus permitting the thermodynamic study. Actually, system does not attain thermodynamic equilibrium only certain assumptions make it akin to a system in equilibrium, for the sake of study and analysis.

Quasi-static literally refers to "almost static" and the infinite slowness of the occurrence of a process is considered as the basic premise for attaining near equilibrium in the system. Here it is considered that the change in state of a system occurs at infinitely slow pace, thus consuming very large time for completion of the process. During the dead slow rate of state change the magnitude of change in a state shall also be infinitely small. This infinitely small change in state when repeatedly undertaken one after the other results in overall state change. Quasi-static process is considered to remain in thermodynamic equilibrium just because of infinitesimal state changes taking place during the occurrence of the process. Quasi static process can be understood from the following example.

Let us consider the heating of gas in a container with certain mass '*W*' kept on the top lid (lid is such that it does not permit leakage across its interface with vessel wall) of the vessel as shown in Fig. 1.4. After certain amount of heat being added to the gas it is found that the lid gets raised up. Thermodynamic state change is shown in figure. The "change in state" is significant. During the change of state since the states could not be considered to be in equilibrium, hence for unsteady state of system, thermodynamic analysis could not be extended. Let us now assume that the total mass comprises of infinitesimal small masses of '*w*' such that all '*w*' masses put together become equal to *W*. Now let us start heat addition to vessel and as soon as the lifting of lid is observed put first fraction mass '*w*' over the lid so as to counter the lifting and estimate the state change. During this process it is found that the state change is negligible. Let us further add heat to the vessel and again put the second fraction mass '*w*' as soon as the lift is felt so as to counter it. Again the state change is seen to be negligible. Continue with the above process and at the end it shall be seen that all fraction masses '*w*' have been put over the lid, thus

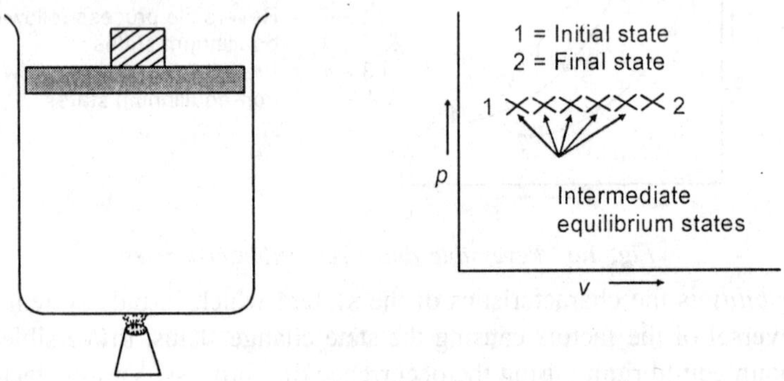

Fig. 1.4 Quasi static process

amounting to mass '*W*' kept over the lid of vessel and the state change occurred is exactly similar to the one which occurred when the mass kept over the lid was '*W*'. In this way the equilibrium nature of system can be maintained and the thermodynamic analysis can be carried out. *p-v* representation for the series of infinitesimal state changes occuring between states 1 & 2 is shown in Fig. 1.4.

1.10 SOME THERMODYNAMIC PROPERTIES

Pressure, temperature, density, volume etc. are some of the thermodynamic properties frequently used. Pressure is defined as the force per unit area. Mathematically, it can be given by the ratio of force applied on a area (say *F*) divided by the area (say *A*) as ;

$$p = F/A, \text{ (N/m}^2\text{)}.$$

In general during analysis one comes across the following four types of pressure,

(*i*) Atmospheric pressure (*ii*) Absolute pressure

(*iii*) Gauge pressure (*iv*) Vacuum pressure

Atmospheric pressure is defined as the pressure exerted by the atmosphere. It is said to be equal to 760 mm of mercury column at 0°C for mercury density of 0.0135951 kg/cm^3, gravitational acceleration of 9.80665 m/s^2 and has magnitude of 1.013 bar (= 1.013×10^5 N/m^2). The instrument used for measuring this pressure is called *barometer.* Italian scientist Torricelli was first to construct the barometer to measure the pressure. In his honour the pressure exerted by one millimeter column of mercury under atmospheric conditions is known as 'Torr' (1 atm = 760 Torr).

Absolute pressure of gas refers to the actual pressure of the gas. Let us consider a U-tube manometer as shown in Fig. 1.5. It shows the manometer with its one limb connected to bulb containing the gas while other limb is open to atmosphere. Fig. 1.5*a* describes a special case in which the pressure of the gas is more than the atmospheric pressure and it is the reason for the rise in level of mercury in the open limb. The difference in the pressure of fluid and atmosphere which is measurable by the rise of mercury column (= *h.d.g.* where *h* is the rise in mercury column, *d* is the density of mercury, *g* is the gravitational acceleration) is known as the *Gauge pressure.* Mathematically, it can be shown that,

Absolute pressure = Atmospheric pressure + Gauge pressure

Figure 1.5*b* shows another typical case in which the pressure of gas is less than the atmospheric pressure and that is why the mercury column is depressed in the open limb. The amount by which the pressure of gas is less than the atmospheric pressure is called *Vacuum pressure.* Thus, the vacuum pressure is the negative gauge pressure. Mathematically it can be shown by,

Absolute pressure = Atmospheric pressure – Vacuum pressure

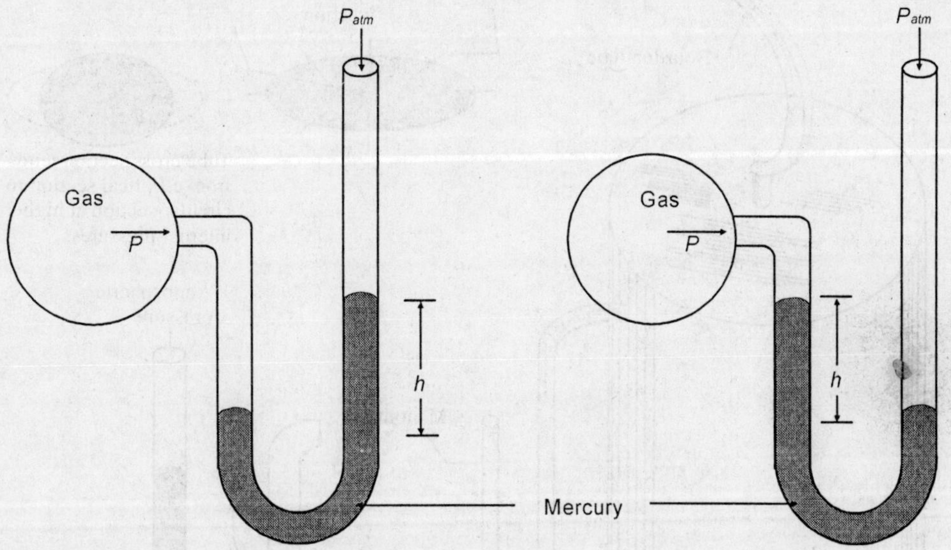

(*a*) Gauge pressure (*b*) vacuum pressure

Fig. 1.5 U-tube manometer

The bar chart shown in Fig. 1.6 further clarifies the interrelationship amongst the different pressures.

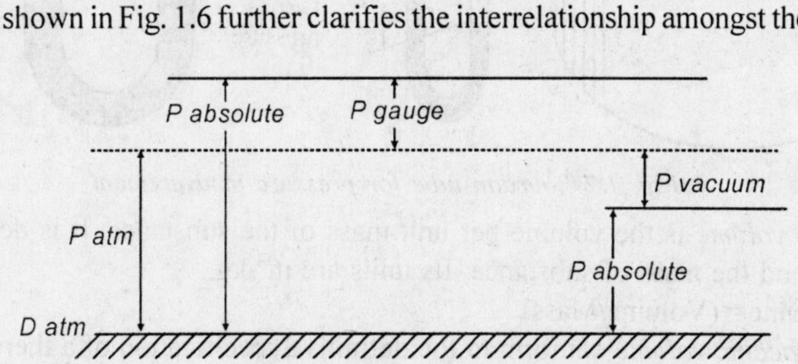

Fig. 1.6 Different pressures

Pressure could also be measured by a Bourdan tube. Bourdan tube has a flattened cross section (oval) closed at one end. Other end of tube connects to the region whose pressure is to be measured. Gas whose pressure is being measured acts on inside of tube surface, thus causing it to change its section from oval section to circular section. Pressure exerted by gas works against tube stresses and air pressure. This change in cross-section from elliptical to circular causes straightening of tube and thus deflecting free end of tube through some distance '*d*' as shown in figure 1.7. This deflection in free end of tube measures the pressure difference between gas pressure and atmospheric pressure. Generally this free end of tube is connected to an indicating hand sweeping over a graduated dial showing the gauge pressure directly.

Temperature is another thermodynamic property which is normally used in Kelvin scale in engineering thermodynamic analysis. It is dealt in detail in subsequent chapter.

Density which refers to the mass per unit volume is the ratio of mass and volume occupied. Its units are kg/m^3.

Density = (Mass/Volume)

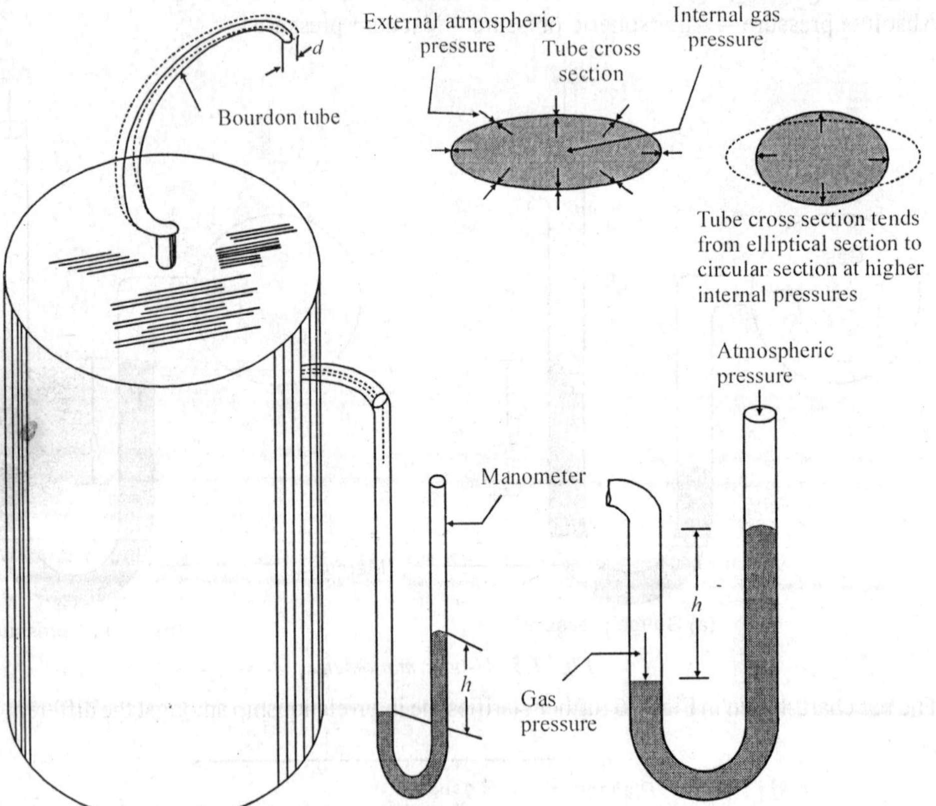

Fig. 1.7 Bourdan tube for pressure measurement

The *specific volume* is the volume per unit mass of the substance. It is defined by ratio of the volume occupied and the mass of substance. Its units are m^3/kg.

Specific volume = (Volume/Mass)

Density or specific volume conform to the definitive specification of a thermodynamic property and are capable of getting associated with other properties such as temperature, pressure and internal energy. Also, the volume occupied by a material is a measure of distance between molecules and thus indicates their molecular energy.

Weight is actually the force due to gravity acting on any substance. Thus, it is the product of mass and gravitational acceleration. Its units are Newtons.

$$\text{Weight} = (\text{mass} \times \text{gravitational acceleration})$$

Specific weight of a substance is the ratio of weight of substance and volume of substance.

$$\text{Specific weight} = (\text{Weight/Volume}) = (\text{density} \times \text{gravitational acceleration})$$

Specific gravity is defined as the ratio of the density of any substance and standard density of some reference substance. For solids and liquids the water density at some specified temperature say 0°C or 4°C is taken as standard density.

1.11 ENERGY AND ITS FORMS

"*Energy* is usually defined as the ability to do mechanical work". It is indeed quite difficult to precisely define the "energy". We feel energy at every moment and can sense it very oftenly.

Another broader definition of energy says that "energy refers to the capacity for producing effects."

Total energy at any moment may be the algebraic summation of the different forms of energy. Conversion of energy from one to other is also possible. In thermodynamics we are primarily interested in studying the change in total energy of a system. Thus, for analysis relative value of energy is considered instead of absolute value.

Energy can be classified in the following general categories;

(*a*) *Energy in transition*: It refers to the energy that is in process of transition between substances or regions because of some driving potential, such as difference or gradient of force, or of temperature, or of electrical potential etc. For example heat, work etc.

(*b*) *Energy stored in particular mass*: It refers to the potential and kinetic energy associated with masses that are elevated or moving with respect to the earth.

Apart from above broad classification the energy can also be categorised into various forms.

(*i*) *Macroscopic energy*: It refers to the energy possessed by a system considered at macroscopic level such as kinetic energy, potential energy etc.

(*ii*) *Microscopic energy*: It refers to the energy defined at molecular level. Summation of energy at molecular level or microscopic energy results in internal energy.

Some of the popular forms of energy are described below :

Potential energy: This type of energy is based on relative position of bodies in a system, i.e. elevation in a gravitational field.

Potential energy for a mass m at elevation z is given as :

$$\text{P.E.} = m.g.z$$

Here g is the gravitational acceleration and elevation is measured from some reference point.

Kinetic energy: It is based on the relative movement of bodies. For a mass m moving with certain velocity c it could be mathematically expressed as;

$$\text{K.E.} = (1/2)\, m.c^2$$

Internal energy: Internal energy of a system is the energy associated with the molecular structure at molecular level.

Let us study fall of a 'weight' from certain height on the floor. Upon hitting the floor, 'weight' comes to dead stop and its potential energy and kinetic energy both reduce to zero. Question arises, where does the vanishing potential energy and kinetic energy go upon 'weight' touching the floor. If we touch the point of contact between 'weight' and floor, it is observed that both these points are slightly hotter than before impact. Thus, it is obvious that the energy has changed its form from potential and kinetic to internal energy and causes rise in temperature of 'weight' and floor at the points of contact.

Internal energy is sum of energy associated with molecules which may have translational, vibrational and rotational motions etc. and respective energies causing these motions.

Internal energy may be thus comprising of sensible energy, latent energy, chemical energy, nuclear energy etc. '*Sensible energy*' refers to the kinetic energy associated with molecules. '*Latent energy*' refers to the energy associated with phase of a substance.

'*Chemical energy*' refers to the energy associated with molecular bonds. '*Nuclear energy*' refers to the energy associated with the bonds within nucleus of atom itself.

Total energy of a system may be given as summation of different forms of energy at a moment. Mathematically;

T.E (Total energy) = K.E + P.E + I.E

where K.E = Kinetic energy

P.E = Potential energy

I.E = Internal energy

Some different forms of energy interaction and associated work interactions with block diagram are given in table 1.4.

Table 1.4 *Some forms of energy and the associated work interactions*

S. No.	*Macroscopic form of energy*	*Governing equation*	*Energy interaction*	*Work interaction*	*Block diagram*
1.	Kinetic energy (translation)	$F = m \cdot \frac{dV}{dt}$	$\Delta E = \frac{1}{2} m \cdot (V_2^2 - V_1^2)$	$= - F \cdot dx$	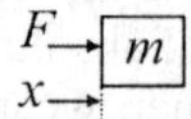
2.	Kinetic energy (rotational)	$T = J \cdot \frac{d\omega}{dt}$	$\Delta E = \frac{1}{2} J \cdot (\omega_2^2 - \omega_1^2)$	$= - T \cdot d\theta$	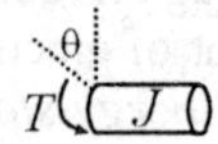
3.	Spring stored energy (translational)	$F = kx$	$\Delta E = \frac{1}{2} k \cdot (x_2^2 - x_1^2)$	$= - F \cdot dx$	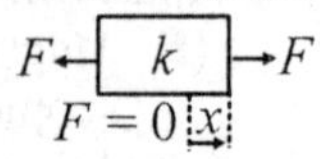
4.	Spring stored energy (rotational)	$T = K \cdot \theta$	$\Delta E = \frac{1}{2} K \cdot (\theta_2^2 - \theta_1^2)$	$= - T \cdot d\theta$	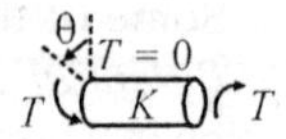
5.	Gravitational energy	$F = mg$	$\Delta E = mg \cdot (Z_2 - Z_1)$	$= - F \cdot dz$	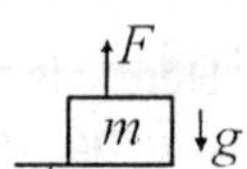
6.	Electrical energy (capacitance)	$u = \frac{q}{c}$	$\Delta E = \frac{1}{2} \frac{q^2}{c} = \frac{1}{2} cu^2$	$= - u \cdot dq$	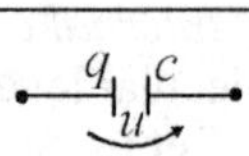
7.	Electrical energy (inductance)	$\phi = L \cdot i$	$\Delta E = \frac{1}{2} Li^2 = \frac{1}{2} \frac{\phi^2}{L}$	$= - i \cdot d\phi$	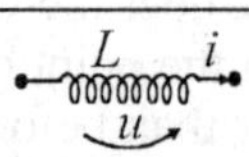

1.12. HEAT AND WORK

When two systems at different temperatures are brought into contact there are observable changes in some of their properties and changes continue till the two don't attain the same temperature if contact is prolonged. Thus, there is some kind of energy interaction at the boundary which causes change in temperatures. This form of energy interaction is called heat. *Thus 'heat' may be termed as the energy interaction at the system boundary which occurs due to temperature difference only.* Heat is observable in transit at the interface i.e. boundary, it can not be contained in a system. In general the heat transfer to the system is assigned with positive (+) sign while the heat transfer from the system is assigned with negative (–) sign. Its units are Calories.

In thermodynamics the work can be defined as follows:

"*Work shall be done by the system if the total effect outside the system is equivalent to the raising of weight and this work shall be positive work*".

In above definition the work has been defined as positive work and says that there need not be actual raising of weight but the effect of the system behaviour must be reducible to the raising of a weight and nothing else. Its units are N. m or Joule. Heat and work are two transient forms of energy.

Let us look at a piston cylinder mechanism (closed system), where high pressure air is filled inside the cylinder fitted with a piston exerting force against some resistance. As the piston moves a distance say 'l', the work would be done. It can be reduced to the raising of weight by replacing this resisting system by a frictionless pulley and lever such that a weight W is raised, Fig. 1.8.

For example, if an electrical battery and resistance is considered as a system, then this system shall do work when electric current flows in an external resistance as this resistance could be replaced by an ideal frictionless motor driving a frictionless pulley and raising a weight.

Here, also in reference to work it is obvious that the work is the entity which is available at the boundary of system, thus work can not be stored rather it is energy interaction in transit at the boundary. From the thermodynamic definition of work the sign convention established as positive work shall be the one which is done by the system while the negative work shall be the one that is done upon the system.

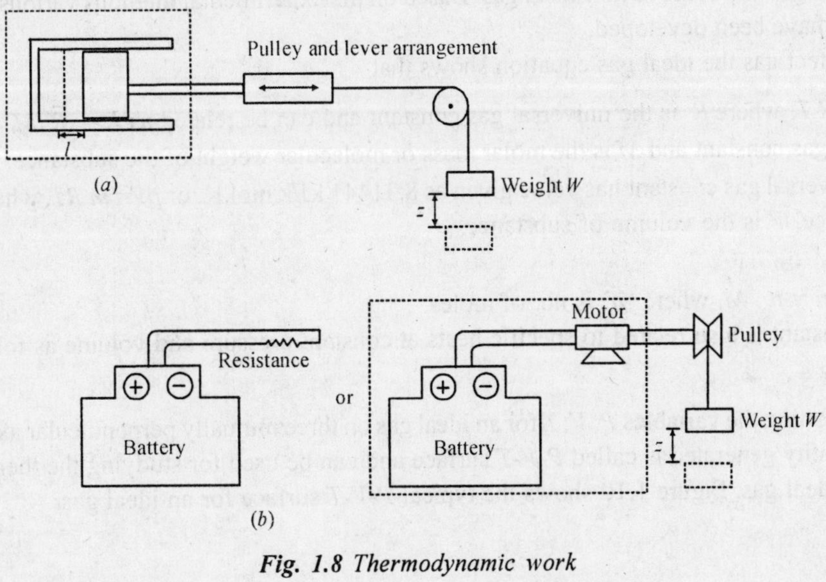

Fig. 1.8 Thermodynamic work

1.13 GAS LAWS

Thermodynamic analysis relies largely upon the gas laws, which are known as Boyle's law (1662) and Charle's law (1787). Boyle's law says that if temperature of a gas is held constant then its molar volume is inversely proportional to the pressure. Mathematically it can be related as $p\bar{v}$ = constant. Here p is the pressure and $\bar{v}$ is the molar volume of gas, i.e. volume per mole.

Charle's law says that for the pressure of gas held constant the volume of gas is directly proportional to the temperature of gas. Mathematically it can be given as $\bar{v}/T$ = constant, where T is the temperature of the gas. It also says that if the molar volume of gas is held constant, the pressure of gas is directly proportional to temperature, i.e. p/T = constant. Figure 1.9 shows the graphical representation.

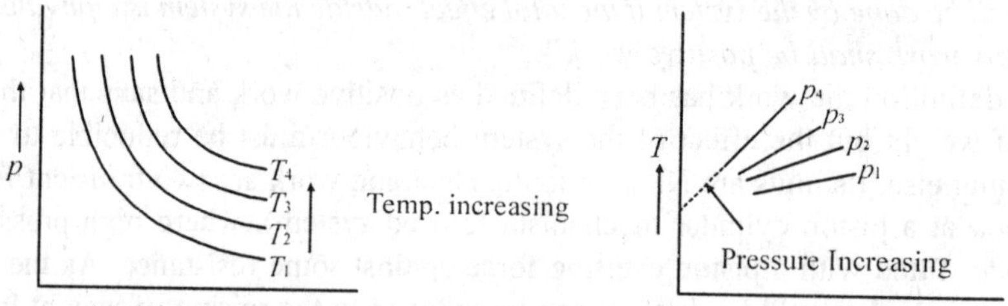

(a) Boyle's law (b) Charle's law

Fig 1.9 Graphical representations of gas laws at constant temperature and at constant pressure

Boyle's and Charle's law when combined together result in,

$$p\bar{v}/T = \text{constant}$$

or $$p\bar{v} = \bar{R}\,T, \text{ where } \bar{R} \text{ is the universal gas constant.}$$

1.14 IDEAL GAS

Engineering thermodynamics deals with different systems having gaseous working fluids. Some gases behave as ideal gas and some as non-ideal gas. Based on the experimental methods various equations of state of gases have been developed.

For perfect gas the ideal gas equation shows that

$p\bar{v} = \bar{R}\,T$, where $\bar{R}$ is the universal gas constant and can be related as $R = \bar{R}/M$, here R is the characteristic gas constant and M is the molar mass or molecular weight of the substance, $\bar{v}$ is volume per mole. Universal gas constant has value given as 8.31441 kJ/k mol.K. or $pV = m\,RT$, where w is mass of the substance, V is the volume of substance,

i.e. $V = n \cdot \bar{v}$

$m = n \cdot M$, where 'n' is no. of moles.

Gas constant is also related to specific heats at constant pressure and volume as follows,

$$R = c_p - c_v$$

Upon plotting the variables P, V, T for an ideal gas on three mutually perpendicular axes, the three dimensional entity generated is called P-V-T surface and can be used for studying the thermodynamic properties of ideal gas. Figure 1.10 shows the typical P-V-T surface for an ideal gas.

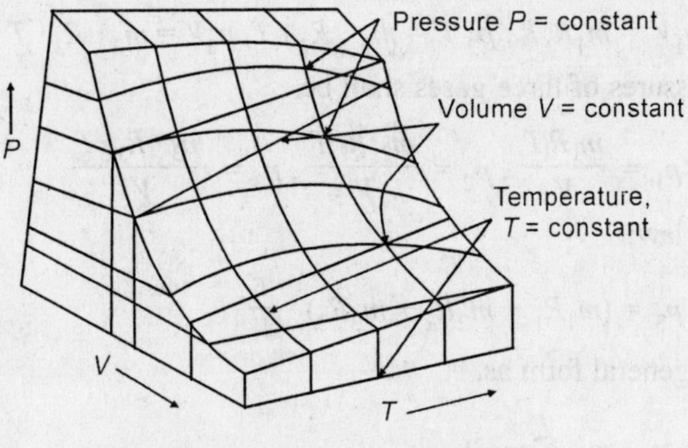

Fig. 1.10 P-V-T surface for ideal gas

For certain gases the molecular weight and gas constant are given in table 1.5.

Table 1.5

Gas	*Molecular Weight, kg/kmol*	*Gas Constant, kJ/kg.K*
Air	28.97	0.287
Carbon dioxide	44.01	0.189
Hydrogen	2.016	4.124
Helium	4.004	2.077
Nitrogen	28.01	0.297
Oxygen	32.00	0.260
Steam	18.02	0.461

1.15 DALTON'S LAW, AMAGAT'S LAW AND PROPERTY OF MIXTURE OF GASES

Dalton's law of partial pressures states that the "total pressure of a mixture of gases is equal to the sum of partial pressures of constituent gases." Partial pressure of each constituent can be defined as the pressure exerted by the gas if it alone occupied the volume at the same temperature.

Thus, for any mixture of gases having '*j*' gases in it, the mathematical statement of Dalton's law says,

$$p = p_1 + p_2 + p_3 + ... + p_j$$

if $\quad V = V_1 = V_2 = V_3 = ... = V_j$

and $\quad T = T_1 = T_2 = T_3 = ... T_j$

Dalton's law can be applied to both mixture of real gases and ideal gases.

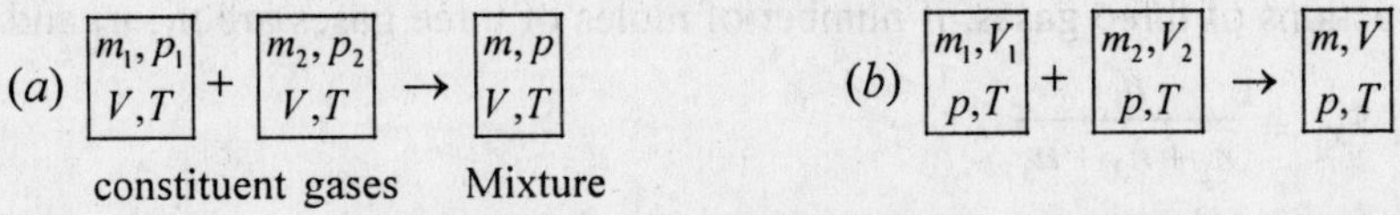

Fig. 1.11 (a) Dalton's law of partial pressures, (b) Amagat's law

Let us take mixture of any three, perfect gases, say, 1, 2, 3 in a container having volume '*V*' and temperature *T*.

Equation of state for these gases shall be,

$$p_1V = m_1R_1T\,;\ p_2V = m_2\cdot R_2\cdot T,\ p_3V = m_3\cdot R_3\cdot T$$

The partial pressures of three gases shall be,

$$p_1 = \frac{m_1R_1T}{V},\ p_2 = \frac{m_2\cdot R_2\cdot T}{V},\ p_3 = \frac{m_3\cdot R_3\cdot T}{V}$$

From Dalton's law;

$$p = p_1 + p_2 + p_3 = (m_1R_1 + m_2R_2 + m_3R_3)\cdot\frac{T}{V}$$

or, it can be given in general form as,

$$pV = T\cdot\sum_{i=1}^{j} m_i\cdot R_i$$

where i refers to constituent gases

Amagat's law of additive volumes states that volume of a gas mixture is equal to the sum of volumes each gas would occupy at the mixture pressure and temperature.

$$V = V_1 + V_2 + V_3 \,..........\, + V_j$$

$$p = p_1 = p_2 = p_3 \,........\, p_j$$

$$T = T_1 = T_2 = T_3 = \,........\, T_j$$

Mass balance upon mixture yields $m = m_1 + m_2 + m_3$

or

$$m = \sum_{i=1}^{j} m_i$$

From above the gas constant for the mixture can be given as;

$$R = \frac{m_1R_1 + m_2R_2 + m_3R_3}{(m_1 + m_2 + m_3)}$$

or, in general form,

$$R = \frac{\sum_{i=1}^{j} m_i\cdot R_i}{\sum_{i=1}^{j} m_i}$$

Mole fraction x_i of a constituent gas can be defined as the ratio of number of moles of that constituent to the total number of moles of all constituents.

Thus mole fractions of three gases, if number of moles of three gases are n_1, n_2 and n_3;

$$x_1 = \frac{n_1}{n_1 + n_2 + n_3}$$

$$x_2 = \frac{n_2}{n_1 + n_2 + n_3}$$

$$x_3 = \frac{n_3}{n_1 + n_2 + n_3}$$

or, in general $x_i = \frac{n_i}{\sum n_i}$

Total no. of moles,

$$n = n_1 + n_2 + n_3 \text{ or, } n = \sum_{i=1}^{j} n_i$$

Sum of mole fractions of all constituent equals to 1,

$$\sum x_i = \frac{\sum n_i}{n} = 1$$

Number of moles of any constituent gas,

$$n_i = n \cdot x_i$$

For M_i being the molecular weight of a constituent gas, the mass m_i of that constituent shall be

$$m_i = n_i \cdot M_i$$

or, $$m_i = n \cdot x_i \cdot M_i$$

and the total mass m, shall be

$$m = \sum m_i = n. \sum x_i \cdot M_i$$

Molecular weight of mixture shall be:

$$M = \frac{m}{n} = \sum x_i \cdot M_i$$

1.16 REAL GAS

When a gas is found to disobey the perfect gas law, i.e. the equation of state for ideal gas, then it is called 'real gas'. Real gas behaviour can also be shown by a perfect gas at the changed thermodynamic states such as high pressure etc.

Deviation of real gas from ideal gas necessitates the suitable equation of state which can be used for interrelating the thermodynamic properties P, V, and T.

From the kinetic theroy of gases it is obvious that the ideal gas equation of state suits the gas behaviour when intermolecular attraction and volume occupied by the molecules themselves is negligibly small in reference to gas volume. At high pressures intermolecular forces and volume of molecules both increase and so the gas behaviour deviates from ideal gas to real gas.

A number of mathematical equations of state suggested by Van der-Waals, Berthelot, Dieterici, Redlich-Kwong, Beattie-Bridgeman and Martin-Hou etc. are available for analysing the real gas behaviour.

Dalton's law and Amagat's law can also be used for real gases with reasonable accuracy in conjunction with modified equations of state.

As the ideal gas equation does not conform to the real gas behaviour in certain ranges of pressures and temperatures, so the perfect gas equation should be modified using compressibility factor for the gas at given pressure and temperature.

Such modified form of equations shall be;

$$P\bar{v} = Z \cdot \bar{R} \cdot T$$

Here Z is the compressibility factor, a function of pressure and temperature.

Thus, compressibility factor is like a correction factor introduced in ideal equation of state for suiting the real gas behaviour. Compressibility factor is an indication of deviation of a gas from ideal gas behaviour and can be mathematically given as;

$$Z = f(P, T)$$

or $$Z = \frac{v_{\text{actual}}}{v_{\text{ideal}}}$$

Here, $v_{\text{ideal}} = \frac{RT}{P}$ i.e. $Z = 1$ for ideal gases while Z can be greater than or less than unity.

Individual graphical representations are available for getting the compressibility factor as shown in Fig 1.12. Compressibility factor charts are available for different substances. Compressibility factors for various substances can also be shown on a generalized compressibility chart using reduced properties. Reduced properties are non-dimensional properties given as ratio of existing property to critical property of substance. Such as reduced pressure is ratio of pressure of gas to critical pressure of gas. Similarly, reduced temperature can be given by ratio of temperature of gas to critical temperature of gas.

Reduced pressure, $$p_R = \frac{p}{p_c}$$

Reduced temperature, $$T_R = \frac{T}{T_c}$$

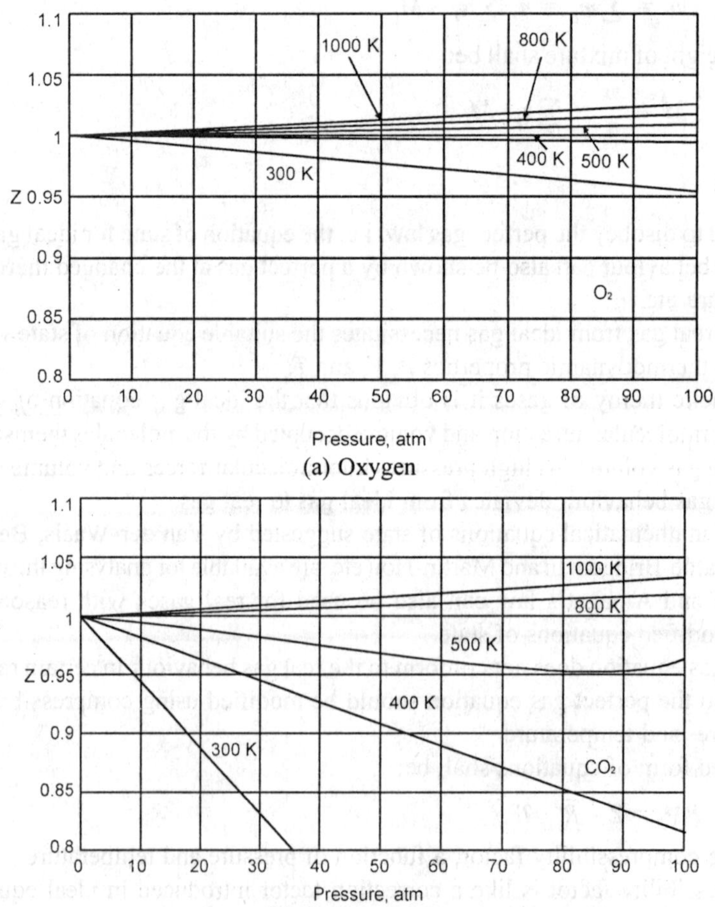

(a) Oxygen

(b) Carbon dioxide

Fig. 1.12 Compressibility factors, Z

where P_c and T_c denote critical pressure and critical temperature respectively. These reduced pressure and reduced temperature are used for getting the generalized compressibility chart of the form, $Z = f(p_R, T_R)$ where Z for all gases is approximately same. This similar behaviour of compressibility correction factor for different gases in respect to reduced pressures and temperatures is called "principle of corresponding states." Fig. 1.13 shows a generalized compressibility chart. In generalized compressibility chart a set of curves is fitted through a set of experimentally determined Z values plotted against reduced

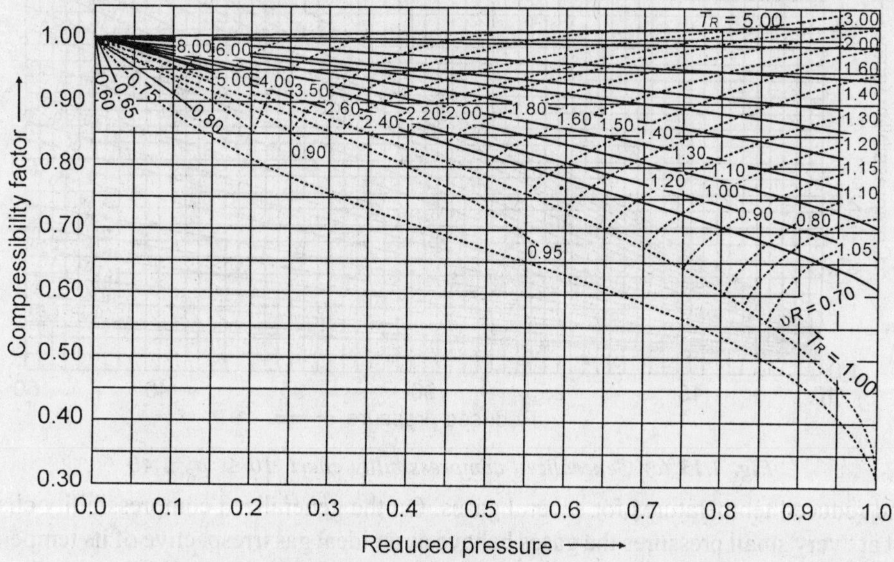

Fig. 1.13 (a) Generalized compressibility chart, $p_R \leq 1.0$

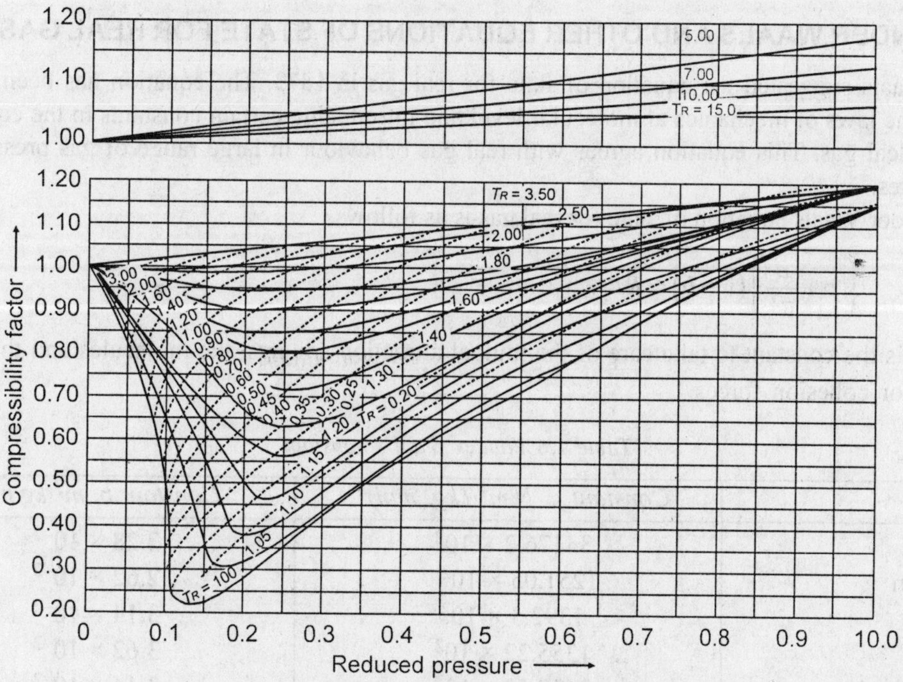

Fig. 1.13 (b) Generalized compressibility chart, $p_R \leq 10.0$

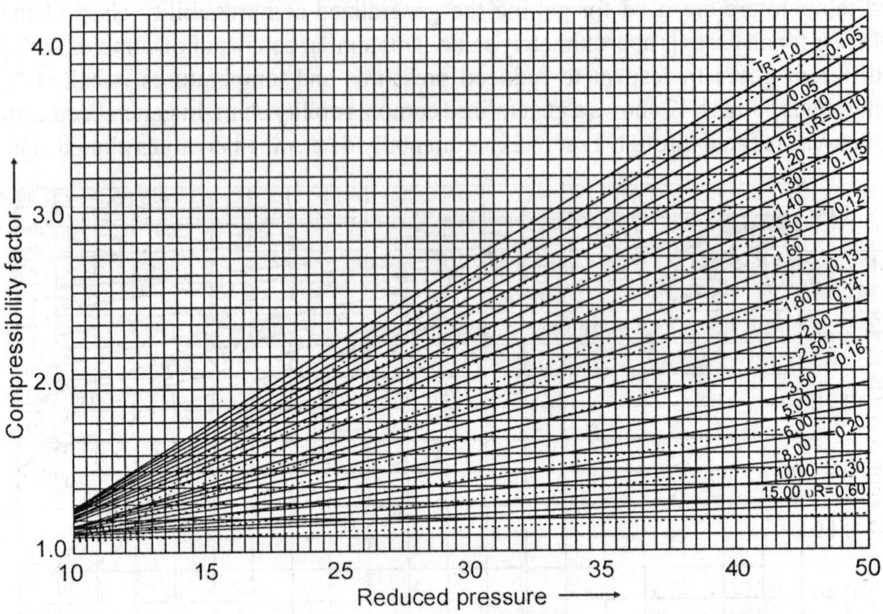

Fig. 1.13 (c) Generalized compressibility chart, $10 \leq p_R \leq 40$

pressure and reduced temperatures for several gases. On the generalized compressibility chart it could be seen that at "very small pressures the gases behave as an ideal gas irrespective of its temperature" and also at "very high temperatures the gases behave as ideal gas irrespective of its pressure".

1.17 VANDER WAALS AND OTHER EQUATIONS OF STATE FOR REAL GAS

Vander Waals suggested the equation of state for real gas in 1873. The equation has been obtained applying the laws of mechanics at molecular level and introducing certain constants in the equation of state for ideal gas. This equation agrees with real gas behaviour in large range of gas pressures and temperatures.

Vander Waals equation of state for real gas is as follows,

$$\left(p+\frac{a}{\bar{v}^2}\right)(\bar{v}-b)=\bar{R}T$$

where 'a' is the constant to take care of the mutual attraction amongst the molecules and thus $\left(\frac{a}{\bar{v}^2}\right)$ accounts for cohesion forces.

***Table 1.6** Vander Waal's constant*

Gas	*Constant a, $N.m^4/(kg.\ mol)^2$*	*Constant b, $m^3/kg.mol$*
Helium	34176.2×10^2	2.28×10^{-2}
Hydrogen	251.05×10^2	2.62×10^{-2}
Oxygen	1392.5×10^2	3.14×10^{-2}
Air	1355.22×10^2	3.62×10^{-2}
Carbon dioxide	3628.50×10^2	3.14×10^{-2}

Constant '*b*' accounts for the volumes of molecules which are neglected in perfect gas equation, thus it denotes "co-volume", Mathematically,

$$a = \frac{27\bar{R}^2 T_c^2}{64 p_c}, \quad b = \left(\bar{R} T_c\right) / \left(8 p_c\right)$$

Here, p_c, T_c are critical point pressures and temperatures having values as given in appendix.

Thus these constants '*a*' & '*b*' are determined from behaviour of substance at the critical point.

In general it is not possible to have a single equation of state which conforms to the real gas behaviour at all pressures and temperatures.

A few more equations of state for real gas as suggested by various researchers are as follows.

Redlich-Kwong equation of state for real gas,

$$p = \frac{\bar{R}T}{(\bar{v} - b)} - \frac{a}{\bar{v} \cdot (\bar{v} + b) \cdot \sqrt{T}}$$

where $\quad a = 0.4278 \left(\frac{\bar{R}^2 \cdot T_c^{2.5}}{p_c} \right)$ and $b = 0.08664 \left(\frac{\bar{R} \cdot T_c}{p_c} \right)$

Berthelot equation of state for real gas,

$$p = \frac{\bar{R}T}{(\bar{v} - b)} - \frac{a}{T \cdot \bar{v}^2},$$

where $\quad a = \left(\frac{27 \cdot \bar{R}^2 \cdot T_c^3}{64 \cdot p_c} \right)$ and $b = \left(\frac{\bar{R} \cdot T_c}{8\, p_c} \right)$

Here *a* and *b* refer to the constants as suggested in respective equations.

Beattie-Bridgeman equation of state given in 1928, for real gas has five constants determined experimentally. It is,

$$p = \frac{\bar{R} \cdot T}{(\bar{v})^2} \left(1 - \frac{C}{\bar{v} \cdot T^3} \right) (\bar{v} + B) - \frac{A}{(\bar{v})^2}$$

where, $\quad A = A_0 \left(1 - \frac{a}{\bar{v}} \right)$ and $B = B_0 \left(1 - \frac{b}{\bar{v}} \right)$

Constants used in Beattie – Bridgeman equation are given in Table 1.7 when p is in $k\,P_a$, $\bar{v}$ is in m^3/k mol, T is in K, and $\bar{R} = 8.314\ k\,P_a m^3/k\ mol.K$.

Table 1.7. *Beattie -Bridgeman constants*

Gas	A_0	*a*	B_0	*b*	*c*
Helium	2.1886	0.05984	0.01400	0.0	40
Hydrogen	20.0117	–0.00506	0.02096	–0.04359	504
Oxygen	151.0857	0.02562	0.04624	0.004208	4.80×10^4
Air	131.8441	0.01931	0.04611	–0.001101	4.34×10^4
Carbon dioxide	507.2836	0.07132	0.10476	0.07235	6.60×10^5

EXAMPLES

1. Find out the pressure difference shown by the manometer deflection of 30 cm of Mercury. Take local acceleration of gravity as 9.78 m/s^2 and density of mercury at room temperature as 13,550 kg/m^3.

Solution:

From the basic principles of fluid statics,

$$\text{Pressure difference} = \rho \cdot gh$$
$$= 13550 \times 30 \times 10^{-2} \times 9.78$$
$$= \mathbf{39755.70\ Pa} \quad \textbf{Ans.}$$

2. An evacuated cylindrical vessel of 30 cm diameter is closed with a circular lid. Estimate the effort required for lifting the lid, if the atmospheric pressure is 76 cm of mercury column (Take g = 9.78 m/s^2)

Solution:

Effort required for lifting the lid shall be equal to the force acting upon the lid. Thus, effort required = Pressure × Area

$$= (76 \times 10^{-2} \times 13550 \times 9.78) \times (3.14 \times (30 \times 10^{-2})^2/4)$$
$$= \mathbf{7115.48\ N} \quad \textbf{Ans.}$$

3. Calculate the actual pressure of air in the tank if the pressure of compressed air measured by manometer is 30 cm of mercury and atmospheric pressure is 101 kPa. (Take g = 9.78 m/s^2)

Solution:

Pressure measured by manometer on the tank is gauge pressure, which shall be

$$= \rho . g . h$$
$$= (13550 \times 9.78 \times 30 \times 10^{-2})$$
$$= 39755.70 \text{ Pa}$$
$$= 39.76 \text{ kPa}$$

$$\text{Actual pressure of air} = \text{Gauge pressure} + \text{atmospheric pressure}$$
$$= 39.76 + 101$$
$$= \mathbf{140.76\ kPa} \quad \textbf{Ans.}$$

4. Determine gauge pressure at a depth of 1 m in a tank filled with oil of specific gravity 0.8. Take density of water as 1000 kg/m^3 and g = 9.81 m/s^2.

Solution:

$$\text{Density of oil} = \text{Specific gravity} \times \text{Density of water}$$
$$\rho_{oil} = 0.8 \times 1000$$
$$\rho_{oil} = 800 \text{ kg/m}^3$$
$$\text{Gauge pressure} = (\rho_{oil} \times g \times h)$$
$$= 800 \times 9.81 \times 1$$
$$= 7848 \text{ N/m}^2$$

or

$$\text{Gauge pressure} = \mathbf{7.848\ kPa.} \quad \textbf{Ans.}$$

5. Calculate the gas pressure using a mercury manometer with one limb open to atmosphere as shown in Fig. 1.14. Barometer reading is 76 cm and density of mercury is 13.6 × 10^3 kg/m^3. Take g = 9.81 m/s^2.

Solution:

Figure shows that the difference of height in mercury columns is 40 cm.

In reference to level *AB* pressure exerted by gas, p_{gas} can be written as sum of atmospheric pressure and pressure due to mercury column at *AB*

$$P_{gas} = (\rho_{mercury} \times 9.81 \times 40 \times 10^{-2}) + \text{Atmospheric pressure}$$
$$= (13.6 \times 10^3 \times 9.81 \times 40 \times 10^{-2}) + (13.6 \times 10^3 \times 9.81 \times 76 \times 10^{-2})$$
$$= 154762.56 \text{ N/m}^2$$

or P_{gas} = **154.76 kPa Ans.**

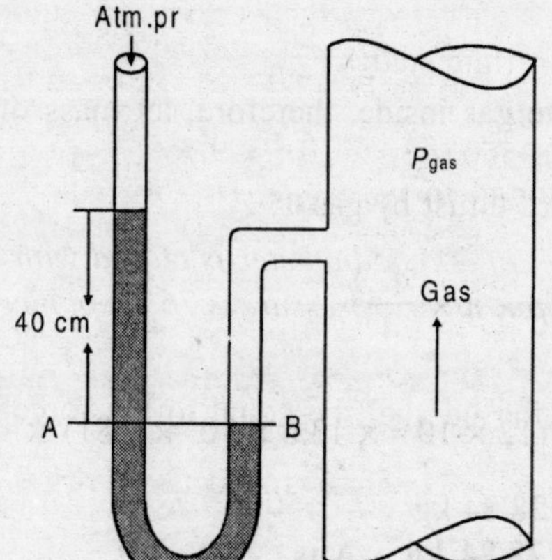

Fig. 1.14

6. 1 kg of water falls from an altitude of 1000 m above ground level. What will be change in the temperature of water at the foot of fall, if there are no losses during the fall. Take specific heat of water as 1 kcal/kg·K

Solution:

Initially when water is at 1000 m, it shall have potential energy in it. This potential energy shall get transformed upon its fall and change the temperature of water.

By law of energy conservation

Potential energy = Heat required for heating water

$$= \frac{1 \times 9.81 \times 1000}{4.18} = 1 \times 1 \times 10^3 \times \Delta T$$

or $\Delta T = 2.35°\text{C}$

Change in temperature of water = **2.35°C Ans.**

7. A spring balance is used for measurement of weight. At standard gravitational acceleration it gives weight of an object as 100 N. Determine the spring balance reading for the same object when measured at a location having gravitational acceleration as 8.5 m/s².

Solution:

At standard gravitational acceleration, mass of object = $\frac{100}{9.81}$

= 10.194 kg

Spring balance reading = Gravitational force in mass

= 10.194 × 8.5

= 86.649 N

= **86.65 N Ans.**

8. An incompressible gas in the cylinder of 15 cm diameter is used to support a piston, as shown. Manometer indicates a difference of 12 cm of Hg column for the gas in cylinder. Estimate the mass of piston that can be supported by the gas. Take density of mercury as 13.6 × 10³ kg/m³.

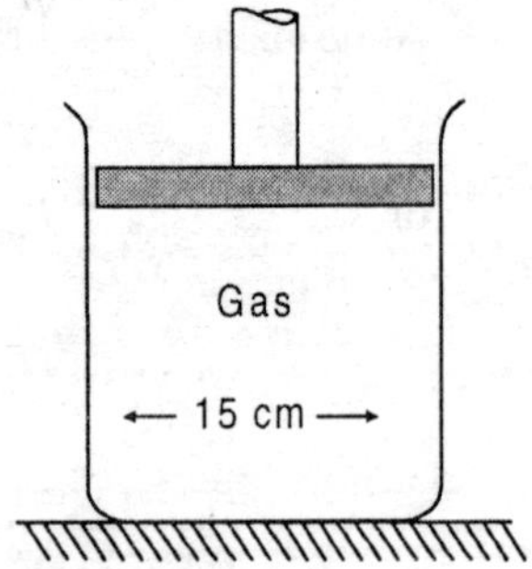

Fig. 1.15

Solution:

Piston shall be supported by the gas inside, therefore, let mass of piston be 'm' kg.

Weight of piston = Upward thrust by gas

$$m.g = p \times \pi \times \frac{d^2}{4}$$

$$m \times 9.81 = (12 \times 10^{-2} \times 13.6 \times 10^3 \times 9.81) \times \frac{\pi}{4} \times (15 \times 10^{-2})^2$$

$$m = 28.84 \text{ kg}$$

Mass of piston = **28.84 kg. Ans.**

9. Determine pressure of steam flowing through a steam pipe when the U-tube manometer connected to it indicates as shown in figure 1.16. During pressure measurement some steam gets condensed in manometer tube and occupies a column of height 2 cm (AB) while mercury gets raised by 10 cm (CD) in open limb. Consider barometer reading as 76 cm of Hg, density of mercury and water as 13.6 × 10³ kg/m³ and 1000 kg/m³ respectively.

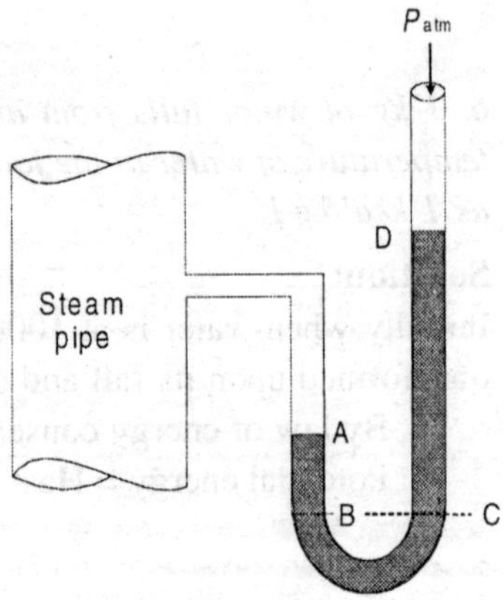

Fig. 1.16

Solution:

Let us make pressure balance at plane BC.

$$p_{steam} + p_{water,\,AB} = p_{atm} + p_{Hg,\,CD}$$

$$p_{steam} = p_{atm} + p_{Hg,\,CD} - p_{water,\,AB}$$

$$p_{atm} = (13.6 \times 10^3 \times 76 \times 10^{-2} \times 9.81)$$

$$p_{atm} = 101396.16 \text{ N/m}^2$$

$$p_{water,\,AB} = (1000 \times 2 \times 10^{-2} \times 9.81)$$

$$p_{water,\,AB} = 196.2 \text{ N/m}^2$$

$$p_{Hg,\,CD} = (13.6 \times 10^3 \times 10 \times 10^{-2} \times 9.81)$$

$$p_{Hg,\,CD} = 13341.6 \text{ N/m}^2$$

Substituting for getting steam pressure,

$$p_{steam} = 101396.16 + 13341.6 - 196.2$$

$$p_{steam} = 114541.56 \text{ N/m}^2$$

or $\quad p_{steam}$ = **114.54 kPa Ans.**

10. A vessel has two compartments 'A' and 'B' as shown with pressure gauges mounted on each compartment. Pressure gauges of A and B read 400 kPa and 150 kPa respectively. Determine the absolute pressures existing in each compartment if the local barometer reads 720 mm Hg.

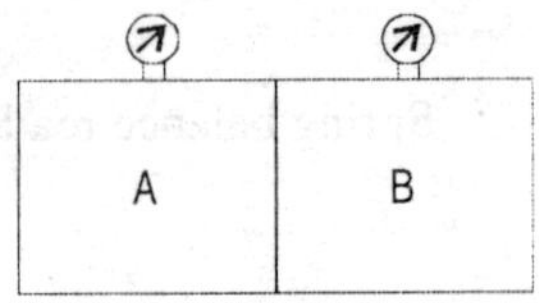

Fig. 1.17

Solution:

Atmospheric pressure from barometer

$= (9810) \times (13.6) \times (0.720)$

$= 96060$ Pa

$= 96.06$ kPa

Absolute pressure in compartment *A*,

$P_{abs,\ A} = P_{gauge,\ A} + P_{atm}$

$= 400 + 96.06 = 496.06$ kPa

Absolute pressure in compartment *B*,

$P_{abs,\ B} = P_{gauge,\ B} + P_{atm}$

$= 150 + 96.06 = 246.06$ kPa

Absolute pressure in compartments *A* & *B*

= 496.06 kPa & 246.06 kPa Ans.

11. Determine the air pressure in a tank having multifluid manometer connected to it, with the tube open to atmosphere as shown in figure. Tank is at an altitude where atmospheric pressure is 90 kPa. Take densities of water, oil and mercury as 1000 kg/m³, 850 kg/m³ and 13600 kg/m³ respectively.

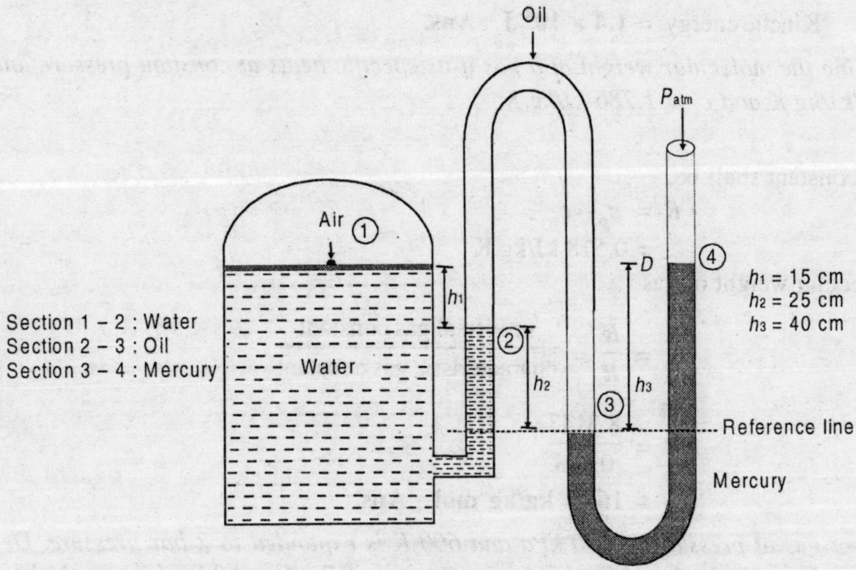

Fig. 1.18

Solution:

It is obvious that the lengths of different fluids in U-tube are due to the air pressure and the pressure of air in tank can be obtained by equalizing pressures at some reference line. Strating from point (1) the pressure can be given as under for equilibrium,

$$p_1 + \rho_{water} \cdot g \cdot h_1 + \rho_{oil} \cdot g \cdot h_2 = p_{atm} + \rho_{mercury} \cdot g \cdot h_3$$

Given : $\rho_{water} = 1000$ kg/m³, $\rho_{oil} = 850$ kg/m³, $\rho_{mercury} = 13600$ kg/m³

$h_1 = 0.15$ m,

$h_2 = 0.25$ m,

$h_3 = 0.40$ m,

$p_{atm} = 90$ kPa

Substituting we get $p_1 = 139.81$ kPa

Air pressure = **139.81 kPa Ans.**

12. Estimate the kinetic energy associated with space object revolving around earth with a relative velocity of 750 m/s and subjected to gravitational force of 4000 N. Gravitational acceleration may be taken as 8 m/s².

Solution:

$$\text{Mass of object} = \frac{\text{Gravitational force}}{\text{Gravitational acceleration}}$$

$$= \frac{4000}{8}$$

$$= 500 \text{ kg}$$

$$\text{Kinetic energy} = \frac{1}{2} \times 500 \times (750)^2$$

$$= 140625000 \text{ J}$$

Kinetic energy = **1.4×10^8 J Ans.**

13. Determine the molecular weight of a gas if its specific heats at constant pressure and volume are $c_p = 2.286$ kJ/kg K and $c_v = 1.786$ kJ/kg K.

Solution:

Gas constant shall be,

$$R = c_p - c_v$$

$$= 0.518 \text{ kJ/kg.K}$$

Molecular weight of gas

$$= \frac{\bar{R}}{R} = \frac{\text{universal gas constant}}{\text{characteristic gas constant}}$$

$$= \frac{8.3143}{0.518}$$

= **16.05 kg/kg mol Ans.**

14. A perfect gas at pressure of 750 kPa and 600 K is expanded to 2 bar pressure. Determine final temperature of gas if initial and final volume of gas are 0.2 m³ and 0.5 m³ respectively.

Solution:

Initial states = 750×10^3 Pa, 600 K, 0.2 m³

Final states = 2 bar or 2×10^5 pa, 0.5 m³.

Using perfect gas equation,

$$\frac{p_1 V_1}{T_1} = \frac{p_2 V_2}{T_2}$$

$$\frac{750 \times 10^3 \times 0.2}{600} = \frac{2 \times 10^5 \times 0.5}{T_2}$$

$$T_2 = 400 \text{ K}$$

Final temperature = **400 K or 127° C Ans.**

15. A vessel of 5 m^3 capacity contains air at 100 kPa and temperature of 300K. Some air is removed from vessel so as to reduce pressure and temperature to 50 kPa and 7°C respectively. Find the amount of of air removed and volume of this mass of air at initial states of air. Take R = 287 J/kg.K for air.

Solution:

Initial states : 100×10^3 Pa, 300 K, 5 m^3

Final states : 50×10^3 Pa, 280 K, 5 m^3

Let initial and final mass of air be m_1 and m_2. From perfect gas equation of air,

$$m_1 = \frac{p_1 V_1}{RT_1};\ m_2 = \frac{p_2 V_2}{RT_2}$$

$$m_1 = \frac{100\times10^3\times5}{287\times300}\ ;\ m_2 = \frac{50\times10^3\times5}{287\times280}$$

Mass of removed, $(m_1 - m_2)$

$$= \left(\frac{100\times10^3\times5}{287\times300}\right) - \left(\frac{50\times10^3\times5}{287\times280}\right)$$

$$m_1 - m_2 = 2.696 \text{ kg}$$

Volume of this mass of air at initial states i.e 100 kPa and 300 K;

$$V = \frac{(m_1 - m_2) . RT_1}{p_1}$$

$$= \frac{2.696\times287\times300}{100\times10^3}$$

$$\text{Volume} = 2.32 \text{ m}^3$$

Mass of air removed = **2.696 kg**

Volume of air at initial states = **2.32 m³** **Ans.**

16. A cylindrical vessel of 1 m diameter and 4 m length has hydrogen gas at pressure of 100 kPa and 27°C. Determine the amount of heat to be supplied so as to increase gas pressure to 125 kPa. For hydrogen take C_p = 14.307 kJ/kg.K, C_v = 10.183 kJ/kg K.

Solution:

Assuming hydrogen to be perfect gas let initial and final states be denoted by subscript 1 and 2.

$$\frac{p_1 V_1}{T_1} = \frac{p_2 V_2}{T_2}, \text{ Here } V_2 = V_1$$

$$T_2 = \frac{p_2 V_2 \cdot T_1}{p_1 V_1} = \frac{p_2\ T_1}{p_1} = \frac{125\times10^3\times300}{100\times10^3}$$

$$T_2 = 375 \text{ K}$$

As it is constant volume heating so, heat supplied,

$$Q = m \cdot C_v\,(T_2 - T_1)$$

From perfect gas characteristics, $R = C_p - C_v$

$$R = 4.124 \text{ kJ/kg} \cdot K$$

Mass of hydrogen, $m = \dfrac{p_1 V_1}{RT_1} = \dfrac{100\times10^3\times\pi\times(0.5)^2\times4}{4.124\times10^3\times300}$

$m = 0.254$ kg

Heat added, $Q = 0.254 \times 10.183 \times (375 - 300)$

Heat to be supplied $= \mathbf{193.99\ kJ}$ **Ans.**

17. Two cylindrical vessels of 2 m^3 each are inter connected through a pipe with valve in-between. Initially valve is closed and one vessel has 20 kg air while 4 kg of air is there in second vessel. Assuming the system to be at 27°C temperature initially and perfectly insulated, determine final pressure in vessels after the valve is opened to attain equilibrium.

Solution:

When the valve is opened then the two vessels shall be connected through pipe and transfer of air shall take place in order to attain equilibrium state. After attainment of equilibrium total mass of air shall be 24 kg.

Final total volume $= 2 \times 2 = 4\ m^3$

Using perfect gas equation.

$$pV = mRT$$

$$p = \frac{mRT}{V}$$

For air, $R = 287$ J/kg K

Substituting values, $p = \dfrac{24\times287\times300}{4}$

$= 516600\ N/m^2$

Final pressure $= \mathbf{516.6\ kPa}$ **Ans.**

18. Determine the pressure of 5 kg carbon dixoide contained in a vessel of 2 m^3 capacity at 27° C, considering it as

(i) perfect gas

(ii) real gas.

Solution:

Given : Volume, $V = 2\ m^3$, Universal gas constt. = 8.314 kJ/kg . K

Temperature, $T = 27°C$

$= (273 + 27)$ K

$T = 300$ K

Mass, $m = 5$ kg

Let pressure exerted be 'p'.

(*i*) Considering it as perfect gas,

$$pV = mR_{CO_2}T$$

$$R_{CO_2} = \frac{\text{Universal gas constt.}}{\text{Molecular weight of } CO_2}$$

$$R_{CO_2} = \frac{8.314\times10^3}{44.01}$$

$$R_{CO_2} = 188.9 \text{ J/kg} \cdot \text{K}$$

Substituting in perfect gas equation,

$$p = \frac{5 \times 188.9 \times 300}{2} = 141675 \text{ N/m}^2$$

Pressure = $\mathbf{1.417 \times 10^5}$ **N/m²   Ans.**

(*ii*) Considering it as real gas let us use Vander-Waals equation;

$$\left(p + \frac{a}{\bar{v}^2}\right)(\bar{v} - b) = \bar{R}T$$

where '$\bar{v}$' is molar specific volume and constants '*a*' and '*b*' can be seen from Table 1.6.

$$\bar{R} = 8.314 \times 10^3$$

Molar specific volume, $\bar{v} = \dfrac{2 \times 44.01}{5}$

$$\bar{v} = 17.604 \text{ m}^3/\text{kg} \cdot \text{mol}$$

Vander-Waals Constant,

$$a = 3628.5 \times 10^2 \text{ N . m}^4/(\text{kg} \cdot \text{mol})^2$$

$$b = 3.14 \times 10^{-2} \text{ m}^3/\text{kg} \cdot \text{mol}$$

Substituting values in Vander-Waals equation,

$$\left(p + \frac{3628.5 \times 10^2}{(17.604)^2}\right)(17.604 - 3.14 \times 10^{-2}) = (8.314 \times 10^3 \times 300)$$

$$p + 1170.86 = 141936.879$$

$$p = 140766.019 \text{ N/m}^2$$

$$\text{Pressure} = 1.408 \times 10^5 \text{ N/m}^2$$

For CO_2 as perfect gas = 1.417×10^5 N/m²

For CO_2 as real gas = $\mathbf{1.408 \times 10^5}$ **N/m²   Ans.**

(using Vander-Waals equation)

***19.** Determine the specific volume of steam at 17672 kPa and 712 K considering it as (a) perfect gas, (b) considering compressibility effects. Take critical pressure = 22.09 MPa, critical temperature =647.3 K,* R_{steam} *= 0.4615 kJ/kg·K.*

Solution:

(*a*) Considering steam as perfect gas,

$$\text{Sp. volume} = \frac{R_{\text{steam}} \cdot T}{p}$$

$$= \frac{0.4615 \times 712}{17672}$$

Specific volume = **0.0186 m³/kg   Ans.**

(*b*) Considering compressibility effects, the specific volume can be given by product of compressibility factor '*Z*' and "specific volume when perfect gas".

$$\text{Reduced pressure} = \frac{p}{\text{Critical pressure}}$$

$$= \frac{17672}{22.09 \times 10^3}$$

Reduced pressure = 0.8

$$\text{Reduced temperature} = \frac{T}{\text{Critical temperature}}$$

$$= \frac{712}{647.3}$$

Reduced temperature = 1.1

From generalized compressibility chart compressibility factor 'Z' can be seen for reduced pressure and reduced temperatures of 0.8 and 1.1. We get,

$$Z = 0.785$$

Actual specific volume = 0.785 × 0.0186

= **0.0146 m³/kg. Ans.**

20. A spherical balloon of 5 m diameter is filled with Hydrogen at 27°C and atmospheric pressure of 1.013 bar. It is supposed to lift some load if the surrounding air is at 17°C. Estimate the maximum load that can be lifted.

Solution:

Balloon filled with H_2 shall be capable of lifting some load due to buoyant forces.

$$\text{Volume of balloon} = \frac{4}{3} \cdot \pi \cdot \left(\frac{5}{2}\right)^3 = 65.45 \text{ m}^3$$

Mass of H_2 in balloon can be estimated considering it as perfect gas.

$$\text{Gas constant for } H_2 = \frac{8.314 \times 10^3}{2} = 4.157 \times 10^3 \text{ J/kg} \cdot \text{K}$$

$$\text{Mass of } H_2 \text{ in balloon} = \frac{P_{\text{balloon}} \cdot V_{\text{balloon}}}{R_{H_2} \cdot T_{H_2}}$$

$$= \frac{1.013 \times 10^5 \times 65.45}{4.157 \times 10^3 \times 300}$$

$$m_{H_2} = 5.316 \text{ kg}$$

Volume of air displaced = Volume of balloon

= 65.45 m³

$$\text{Mass of air displaced} = \frac{1.013 \times 10^5 \times 65.45}{R_{\text{air}} \times (17 + 273)}$$

$$R_{\text{air}} = 0.287 \text{ kJ/kg . K}$$

$$m_{air} = \frac{1.013\times10^5\times65.45}{0.287\times10^3\times290}$$

$$m_{air} = 79.66 \text{ kg}$$

Load lifting capacity due to buoyant force $= m_{air} - m_{H_2}$

$$= 79.66 - 5.316$$

$$= \mathbf{74.344\ kg}$$ **Ans.**

21. A pump draws air from large air vessel of 20 m³ at the rate of 0.25 m³/min. If air is initially at atmospheric pressure and temperature inside receiver remains constant then determine time required to reduce the receiver pressure to $\frac{1}{4}$ th of its original value.

Solution:

Let volume of receiver be V, m^3 and volume sucking rate of pump be v m^3/min, then theoretically problem can be modelled using perfect gas equation. Here p is pressure in receiver and T is temperature in vessel.

$$pVm = mRT$$

Here pressure 'p' and temperature 'T' shall change with respect to time t. Differentiating perfect gas equation with respect to time.

$$V\cdot\frac{dp}{dt} = RT\frac{dm}{dt}$$

Here $\left(\frac{dm}{dt}\right)$ is mass extraction rate from receiver i.e. kg/min. This mass rate can be given using perfect gas equation when volume flow rate (m³/min) is given as v. So.

$$\frac{dp}{dt} = -\frac{pv}{RT} \text{ (–ve as mass gets reduced with time)}$$

Substituting,

$$V\cdot\frac{dp}{dt} = -RT\cdot\frac{pv}{RT}$$

$$V\cdot\frac{dp}{dt} = -pv$$

$$\int_0^t dt = -\frac{V}{v}\int_0^t\frac{dp}{p}$$

$$t = -\frac{V}{v}\ln\left(\frac{p_2}{p_1}\right)$$

Here final pressure, $p_2 = \frac{p_1}{4}$, $V = 20\text{ m}^2$, $v = 0.25\text{ m}^3/\text{min}$

So time, $$t = \frac{V}{v} \ln\left(\frac{1}{4}\right)$$

$$\text{time} = \frac{20}{0.25} \ln(4) = 110.9 \text{ minutes}$$

$$= \mathbf{110.9 \text{ minutes}} \quad \textbf{Ans.}$$

22. In 5 kg mixture of gases at 1.013 bar and 300 K the various constituent gases are as follows, 80% N_2, 18% O_2, 2% CO_2.

Determine the specific heat at constant pressure, gas constant for the constituents and mixture and also molar mass of mixture taking $\gamma = 1.4$ for N_2 and O_2 and $\gamma = 1.3$ for CO_2.

Universal gas constant = 8314 J/kg · K

Solution:

Gas constants for constituent gases shall be,

$$R_{N_2} = \frac{8314}{\text{mol. wt. of } N_2} = \frac{8314}{28} = \mathbf{296.9 \text{ J/kg} \cdot \text{K}}$$

$$R_{O_2} = \frac{8314}{\text{mol. wt. of } O_2} = \frac{8314}{32} = \mathbf{259.8 \text{ J/kg} \cdot \text{K}}$$

$$R_{CO_2} = \frac{8314}{\text{mol. wt. of } CO_2} = \frac{8314}{44} = \mathbf{188.9 \text{ J/kg . K}}$$

Gas constant for mixture, R_{mixture}

$$= \left(\frac{m_{N_2}}{M}\right).R_{N_2} + \left(\frac{m_{O_2}}{M}\right).R_{O_2} + \left(\frac{m_{CO_2}}{M}\right).R_{CO_2}$$

$$= (0.80 \times 296.9) + (0.18 \times 259.8) + (0.02 \times 188.9)$$

$$R_{\text{mixture}} = \mathbf{288.06 \text{ J/kg . K}}$$

Specific heat at constant pressure for constituent gases.

$$C_{p,N_2} = \left(\frac{\gamma}{\gamma - 1}\right). R_{N_2} = \left(\frac{1.4}{0.4}\right) \times 296.9 = \mathbf{1.039 \text{ kJ/kg . K}}$$

$$C_{p,O_2} = \left(\frac{\gamma}{\gamma - 1}\right). R_{O_2} = \left(\frac{1.4}{0.4}\right) \times 259.8 = \mathbf{0.909 \text{ kJ/kg . K}}$$

$$C_{p,CO_2} = \left(\frac{\gamma}{\gamma - 1}\right). R_{CO_2} = \left(\frac{1.3}{0.3}\right) \times 188.9 = \mathbf{0.819 \text{ kJ/kg . K}}$$

$$C_{p,\text{mixture}} = \left(\frac{m_{N_2}}{M}\right). C_{P,N_2} + \left(\frac{m_{O_2}}{M}\right). C_{P,O_2} + \left(\frac{m_{CO_2}}{M}\right). C_{P,CO_2}$$

$$C_{p,\text{mixture}} = (0.80 \times 1.039) + (0.18 \times 0.909) + (0.02 \times 0.819)$$

$$= \mathbf{1.0276 \text{ kJ/kg . K}}$$

$$\text{Molar mass of mixture} = \sum x_i \,.\, M_i = \frac{\text{Total mass of mixture}}{\text{Total no. of moles}}$$

$$x_i = \frac{n_i}{\sum n_i}\text{, here } M_i = \text{mol. wt. of } i\text{th online constituent.}$$

No. of moles of constituent gases,

$$n_{N_2} = \frac{m_{N_2}}{\text{Mol. wt. } N_2} = \frac{0.8 \times 5}{28} = 0.143$$

$$n_{O_2} = \frac{m_{O_2}}{\text{Mol. wt. } O_2} = \frac{0.18 \times 5}{32} = 0.028$$

$$n_{CO_2} = \frac{m_{CO_2}}{\text{Mol. wt. } CO_2} = \frac{0.02 \times 5}{44} = 0.0023$$

$$\text{Total mole numbers in mixture} = n_{N_2} + n_{O_2} + n_{CO_2}$$
$$= (0.143 + 0.028 + 0.0023)$$
$$\sum n_i = 0.1733$$

Mole fractions of constituent gases,

$$x_{N_2} = \frac{n_{N_2}}{\sum n_i} = \frac{0.143}{0.1733} = 0.825$$

$$x_{O_2} = \frac{n_{O_2}}{\sum n_i} = \frac{0.028}{0.1733} = 0.162$$

$$x_{CO_2} = \frac{n_{CO_2}}{\sum n_i} = \frac{0.0023}{0.1733} = 0.0133$$

$$\text{Molecular wt. of mixture} = \text{Molar mass of mixture} = \sum x_i \,.\, M_i$$
$$= (0.825 \times 28) + (0.162 \times 32) + (0.0023 \times 44)$$
$$= \mathbf{28.87 \text{ kg/kmol. Ans.}}$$

23. *A gas mixture comprises of 18% O_2, 75% N_2 and 7% CO_2 by volume at 0.5 MPa and 107°C. For 5 kg mass of mixture carry out gravimetric analysis and determine the partial pressure of gases in mixture.*

Solution:

$$\text{Mole fraction of constituents} \Rightarrow x_i = \frac{n_i}{n} = \frac{V_i}{V}$$

where 'n_i' and 'V_i' are no. of moles and volume fraction of constituent while 'n' and V are total no. of moles and total volume of mixture.

$$x_{O_2} = \frac{0.18}{1} = 0.18$$

$$x_{N_2} = \frac{0.75}{1} = 0.75$$

$$x_{CO_2} = \frac{0.07}{1} = 0.07$$

Molecular weight of mixture = Molar mass

$$= (0.18 \times 32) + (0.75 \times 28) + (0.07 \times 44)$$
$$= 29.84$$

Gravimetric analysis refers to the mass fraction analysis.

Mass fraction of constituents

$$= \frac{m_i}{m} = \frac{(V_i / V) \times \text{Mol. wt. of constituent}}{\text{Mol. wt. of mixture}}$$

$$\left.\begin{aligned} \text{Mole fraction of } O_2 &= \frac{0.18 \times 32}{29.84} = \mathbf{0.193} \\ \text{Mole fraction of } N_2 &= \frac{0.75 \times 28}{29.84} = \mathbf{0.704} \\ \text{Mole fraction of } CO_2 &= \frac{0.07 \times 44}{29.84} = \mathbf{0.104} \end{aligned}\right\} \textbf{Ans.}$$

Partial pressures of constituents = Volume fraction × Pressure of mixture

$$\left.\begin{aligned} \text{Partial pressure of } O_2 &= 0.18 \times 0.5 = \mathbf{0.09\ MPa} \\ \text{Partial pressure of } N_2 &= 0.75 \times 0.5 = \mathbf{0.375\ MPa} \\ \text{Partial pressure of } CO_2 &= 0.07 \times 0.5 = \mathbf{0.35\ MPa} \end{aligned}\right\} \textbf{Ans.}$$

24. A steel insulated tank of 6 m^3 volume is equally divided into two chambers using a partition. The two portions of tank contain N_2 gas at 800 kPa and 480 K and CO_2 gas at 400 kPa and 390 K. Determine the equilibrium temperature and pressure of mixture after removing the partition. Use $\gamma = 1.4$ for N_2, $\gamma = 1.3$ for CO_2.

Solution:

Since tank is insulated so adiabatic mixing can be considered. Let us consider N_2 and CO_2 to behave as perfect gas.

No. of moles of N_2

$$n_{N_2} = \frac{p_{N_2} \cdot V_{N_2}}{\bar{R} \cdot T_{N_2}} = \frac{800 \times 10^3 \times 3}{8314 \times 480} = 0.601$$

No of moles of CO_2

$$n_{CO_2} = \frac{p_{CO_2} \cdot V_{CO_2}}{\bar{R} \cdot T_{CO_2}} = \frac{400 \times 10^3 \times 3}{8314 \times 390} = 0.370$$

Total no. of moles of mixture,

$$n = n_{N_2} + n_{CO_2}$$
$$= 0.601 + 0.370$$
$$= 0.971$$

Specific heat for N_2 at constant volume,

$$C_{v,N_2} = \frac{R_{N_2}}{(\gamma_{N_2}-1)} = \frac{(8314/28)}{(1.4-1)}$$

$$C_{v,N_2} = 742.32 \text{ J/kg . K}$$

Specific heat for CO_2 at constant volume,

$$C_{v,CO_2} = \frac{R_{CO_2}}{(\gamma_{CO_2}-1)} = \frac{(8314/44)}{(1.3-1)}$$

$$C_{v,CO_2} = 629.85 \text{ J/kg . K}$$

Mass of N_2 = n_{N_2} × Mol. wt. of N_2 = 0.601 × 28 = 16.828 kg

Mass of CO_2 = n_{CO_2} × Mol. wt. of CO_2 = 0.370 × 44 = 16.28 kg.

Let us consider the equilibrium temperature of mixture after adiabatic mixing at T. Applying energy conservation principle :

$m_{N_2} . C_{v,N_2} . (T - T_{N_2}) + m_{CO_2} . C_{v,CO_2} . (T - T_{CO_2}) = 0$

$\{16.828 \times 742.32\,(T-480)\} + \{16.28 \times 629.85\,(T-390)\} = 0$

$22745.7 \,.\, T = 9995088.881$

Equilibrium temperature, **T = 439.4 K Ans.**

Equilibrium pressure of mixture, $T_{\text{mixture}} = 439.4$ K, $V_{\text{mixture}} = 6$ m^3

$$p_{\text{mixture}} = \frac{n.\bar{R}.T_{\text{mixture}}}{V_{\text{mixture}}} = \frac{0.971 \times 8314 \times 439.4}{6}$$

Equilibrium pressure = **591.205 kPa Ans.**

25. *2 kg of Hydrogen and 3 kg of Helium are mixed together in an insulated container at atmospheric pressure and 100 K temperature. Determine the specific heat of final mixture if specific heat at constant pressure is 11.23 kJ/kg. K and 5.193 kJ/kg . K for H_2 and He respectively.*

Solution:

Two gases are non reacting, therefore, specific heat of final mixture can be obtained by following for adiabatic mixing.

$$C_{p,\text{mixture}} = \frac{c_{p,H_2} . m_{H_2} + c_{p,He} . m_{He}}{(m_{H_2} + m_{He})}$$

Substituting values,

$$= \frac{(2 \times 11.23) + (3 \times 5.193)}{(3+2)}$$

$$C_{p,\text{mixture}} = \textbf{7.608 kJ/kg . K} \quad \textbf{Ans.}$$

26. *A mixture of 18 kg hydrogen, 10 kg nitrogen and 2 kg of carbon dioxide is put into a vessel at atmospheric conditions. Determine the capacity of vessel and the pressure in vessel if it is heated upto twice of initial temperature. Take ambient temperature as 27ºC.*

Solution:

Gas constant for mixture can be obtained as;

$$R_{\text{mixture}} = \frac{(m_{H_2} . R_{H_2} + m_{N_2} . R_{N_2} + m_{CO_2} . R_{CO_2})}{(m_{H_2} + m_{N_2} + m_{CO_2})}$$

$$R_{H_2} = \frac{8.314}{2} \text{ kJ/kg . K}$$

$$R_{N_2} = \frac{8.314}{28} \text{ kJ/kg . K}$$

$$R_{CO_2} = \frac{8.314}{44} \text{ kJ/kg . K}$$

$$R_{H_2} = 4.15 \text{ kJ/kg . K}$$

$$R_{N_2} = 0.297 \text{ kJ/kg . K}$$

$$R_{CO_2} = 0.189 \text{ kJ/kg . K}$$

$$R_{\text{mixture}} = \frac{(18 \times 4.15 + 10 \times 0.297 + 2 \times 0.189)}{30}$$

$$R_{\text{mixture}} = 2.606 \text{ kJ/kg . K}$$

Considering mixture to be perfect gas;

$$\text{Capacity of vessel } V_{\text{mixture}} = \frac{m_{\text{mixture}} . R_{\text{mixture}} . T}{p}$$

Here, $p = 101.325$ kPa

$$V_{\text{mixture}} = \frac{30 \times 2.606 \times 300.15}{101.325}$$

Capacity of vessel = **231.58 m³ Ans.**

For constant volume heating, final pressure shall be,

$$p_{\text{final}} = p_{\text{initial}} \times \frac{T_{\text{final}}}{T_{\text{initial}}}$$

$p_{\text{final}} = 101.325 \times 2 =$ **202.65 kPa Ans.**

27. Determine the ratio of exit to inlet diameter of a duct having heating element in the mid of duct. Atmospheric air enters the duct at 27°C and gets heated up to 500 K before leaving the duct. The kinetic and potential energy changes during passage of air through duct are negligible.

Solution:

Said air heating process is the case of constant pressure process. Let inlet state be '1' and exit state '2'.

Therefore, by Charle's law volume and temperature can be related as;

$$\frac{V_1}{T_1} = \frac{V_2}{T_2}$$

$$\frac{V_2}{V_1} = \frac{T_2}{T_1}$$

or
$$\frac{\left(\frac{\pi}{4} \times d_2^2\right) \times \text{Velocity at 2}}{\left(\frac{\pi}{4} \times d_1^2\right) \times \text{Velocity at 1}} = \frac{T_2}{T_1}$$

Since
$$\Delta K.E = 0 \text{ , so } \frac{d_2^2}{d_1^2} = \frac{T_2}{T_1}$$

or
$$\frac{d_2}{d_1} = \sqrt{\frac{T_2}{T_1}}$$

$$\text{Exit to inlet diameter ratio} = \sqrt{\frac{500}{300.15}} = 1.29 = \mathbf{1.29} \textbf{ Ans.}$$

28. *A vessel of 2 m³ volume contains hydrogen at atmospheric pressure and 27°C temperature. An evacuating pump is connected to vessel and the evacuation process is continued till its pressure becomes 70 cm of Hg vacuum.*

Estimate the mass of hydrogen pumped out. Also determine the final pressure in vessel if cooling is carried up to 10°C. Take atmospheric pressure as 76 cm of Hg and universal gas constant as 8.314 kJ/kg. K

Solution:

$$\text{For hydrogen, gas constant, } R = \frac{8.314}{2}$$

$$R = 4.157 \text{ kJ/kg . K}$$

Say initial and final states are given by '1' and '2'.

Mass of hydrogen pumped out shall be difference of initial and final mass inside vessel.

Final pressure of hydrogen = Atm. pr. – Vacuum pr.

$$= 76 - 70$$

$$= 6 \text{ cm of Hg.}$$

Therefore, pressure difference = 76 – 6

$$= 70 \text{ cm of Hg.}$$

$$= \frac{70}{76} \times 101.325 \text{ kPa}$$

$$= 93.33 \text{ kPa}$$

$$\text{Mass pumped out} = \frac{p_1V_1}{RT_1} - \frac{p_2V_2}{RT_2}\text{; here } V_1 = V_2 = V \text{ and } T_1 = T_2 = T.$$

$$= \frac{V}{RT}(p_1 - p_2)$$

$$= \frac{2 \times 93.33 \times 10^3}{4.157 \times 300.15 \times 10^3}$$

$$= \mathbf{0.15\ kg.} \quad \mathbf{Ans.}$$

During cooling upto 10°C, the process may be considered as constant volume process. Say the state before and after cooling are denoted by suffix 2 and 3.

Therefore, $$p_3 = \frac{T_3}{T_2} \cdot p_2$$

$$= \frac{283.15}{300.15} \times \frac{6 \times 101.325}{76}$$

Final pressure after cooling $= \mathbf{7.546\ kPa.}$ **Ans.**

EXERCISE

1.1 Define thermodynamics and discuss different approaches to study of thermodynamics.

1.2 Write short notes on the following:
Thermodynamic properties, state, path, process, closed system, isolated system, open system, extensive and intensive properties.

1.3 What is meant by quasi-static process? Also discuss its physical significance.

1.4 Describe thermodynamic equilibrium of a system.

1.5 State thermodynamic definition of work. Also differentiate between heat and work.

1.6 What is energy? What are different forms of it?

1.7 Explain the concept of continuum.

1.8 Define perfect gas.

1.9 Differentiate between characteristic gas constant and universal gas constant.

1.10 State the Dalton's law of partial pressures and assumptions for it.

1.11 What is meant by real gas? Why ideal equation of state cannot be used for it?

1.12 Write equations of state for real gas.

1.13 Define conpressibility factor.

1.14 Write Boyle's law and Charle's law.

1.15 Determine the absolute pressure of gas in a tank if the pressure gauge mounted on tank reads 120 kPa pressure. [221.3 kPa]

1.16 What shall be the volume of a fluid having its specific gravity as 0.0006 and mass as 10 kg? [16.67 m^3]

1.17 Determine the pressure of compressed air in an air vessel, if the manometer mounted on it shows a pressure of 3 m of mercury. Assume density of mercury to be 13.6×10^3 kg/m^3 and atmospheric pressure as 101 kPa. [501.25 kPa]

1.18 Calculate the kinetic energy of a satellite revolving around the earth with a speed of 1 *km/s*. Assume acceleration due to gravity as 9.91 m/s^2 and gravitational force of 5 kN. [254.8 MJ]

1.19 If the gauge pressure of oil in a tube is 6.275 kPa and oil's specific gravity is 0.8, then determine depth of oil inside tube. [80 cm]

1.20 Determine the work required for displacing a block by 50 m and a force of 5 kN. [250 kJ]

1.21 Determine the barometer reading in millimetres of Hg if the vacuum measured on a condenser is 74.5 cm of Hg and absolute pressure is 2.262 kPa. [760 mm]

1.22 Determine the absolute pressures for the following;
(i) Gauge pressure of 1.4 MPa
(ii) Vacuum pressure of 94.7 kPa

Take barometric pressure as 77.2 cm of Hg and density of mercury as 13.6×10^3 kg/m^3.
[1.5 MPa, 8.3 kPa]

1.23 Determine the pressure acting upon surface of a vessel at 200 m deep from surface of sea. Take barometric pressure as 101 kPa and specific gravity of sea water as 1.025. [2.11 MPa]

1.24 A vacuum gauge gives pressure in a vessel as 0.1 bar, vacuum. Find absolute pressure within vessel in bars. Take atmospheric pressure as 76 cm of mercury column, g = 9.8 m/s^2, density of mercury = 13.6 g/cm^3. [0.91 bar]

1.25 Determine the work done upon a spring having spring constant of 50 kN/m. Spring is stretched to 0.1 m from its unstretched length of 0.05 m. [0.0625 kJ]

1.26 Determine the mass of oxygen contained in a tank of 0.042 m^3 at 298 K and 1.5×10^7 Pa considering it as perfect gas. Also determine the mass using compressibility charts. [8.25, 8.84]

1.27 What will be specific volume of water vapour at 1 MPa and 523 K, if it behaves as ideal gas? Also determine the same considering generalized compressibility chart. [0.241 m^3/kg, 0.234 m^3/kg]

1.28 Calculate the pressure of CO_2 gas at 27°C and 0.004 m^3/kg treating it as ideal gas. Also determine the pressure using Van der Waals equation of state. [14.17 MPa, 6.9 MPa]

1.29 Determine molecular weight and gas constant for a mixture of gases having 65% N_2, 35% CO_2 by mole.
[33.6 kg/k mol. 0.247 kJ/kg . K]

1.30 Considering air as a mixture of 78% N_2, 22% O_2 by volume determine gas constant, molecular weight, C_p and C_v for air at 25°C. [0.2879 kJ/kg . K, 28.88 kg/K mol, 1.0106 kJ/kg . K, 0.722 kJ/kg . K]

1.31 What minimum volume of tank shall be required to store 8 kmol and 4 kmol of O_2 and CO_2 respectively at 0.2 MPa, 27°C ? [149.7 m^3]

1.32 Two tanks *A* and *B* containing O_2 and CO_2 have volumes of 2 m^3 and 4 m^3 respectively. Tank *A* is at 0.6 MPa, 37°C and tank *B* is at 0.1 MPa and 17°C. Two tanks are connected through some pipe so as to allow for adiabatic mixing of two gases. Determine final pressure and temperature of mixture.
[0.266 MPa, 30.6°C]

1.33 Determine the molecular weight and gas constant for some gas having C_p = 1.969 kJ/kg . K, C_v = 1.507 kJ/kg . K. [18.04 kg/kmol, 0.461 kJ/kg . K]

2

Zeroth Law of Thermodynamics

2.1 INTRODUCTION

Thermodynamics is the branch of science which deals with the energy interactions. In order to find whether energy interactions are taking place or not some measurable mathematical parameters are needed. These parameters are called thermodynamic properties. Out of number of thermodynamic properties discussed earlier the 'temperature' is one property.

One is well familiar with the qualitative statement of the state of a system such as cold, hot, too cold, too hot etc. based on the day to day experience. The degree of hotness or coldness is relative to the state of observer. For example, let us take an iron bar. Obviously the bar shall have intial temperature equal to the room temperature. Now let us heat this metal bar. Observations at the molecular level show that upon heating the molecular activity inside the bar gets increased. This may be attributed to the more agitated state of molecules as energy is given to them in the form of heating of the bar. From the physiological sensations it can be felt that this has resulted in increase in the degree of hotness of the bar. This qualitative indication of the relative hotness can be exactly defined by using thermodynamic property known as *temperature*. If this hot bar is brought in contact with another bar at room temperature one can feel that after some time the two bars which were initially at high and low temperatures attain the same temperature which is lying between the two temperatures. It is indicative of the fact that there has been exchange of some entity between two bars resulting in the attainment of final equilibrium temperature. This state of attainment of common equilibrium temperature is also termed as the state of thermal equilibrium. Thus, the temperature becomes a potential indicator of the energy interactions in the systems.

A look at the history shows that for quantitative estimation of temperature a German instrument maker Mr. Gabriel Daniel Fahrenheit (1686-1736) came up with idea of instrument like thermometer and developed mercury in glass thermometer. Here he observed that height of mercury column used to change as the bulb of thermometer was brought in the environments having different degrees of hotness. In the year 1742, a Swedish astronomer Mr. Anders Celsius described a scale for temperature measurement. This scale later on became very popular and is known as *Centigrade Scale*. For caliberation of these measuring instruments some reference states of different substances were used initially and the relative state of temperature of the substance could be quantified. Later on with the passage of time things were standardised and internationally acceptable temperature scales and instruments were developed.

2.2 PRINCIPLE OF TEMPERATURE MEASUREMENT AND ZEROTH LAW OF THERMODYNAMICS

After the identification of 'Temperature' as a thermodynamic property for quantification of the energy interactions the big question was its estimation. Based on the relative degree of coldness/hotness concept it was concluded that the absolute value of temperature is difficult to be described. Hence it was mooted to make temperature estimations in reference to certain widely acceptable known thermal states of the substances. Temperature is thus the intensive parameter and requires reference states. These acceptable known thermal states were such as the boiling point of water commonly called steam point, freezing point of water commonly called ice point etc. These easily reproducible and universally acceptable states of the substance are known as *reference states* and the temperature values assigned to them are called *reference temperatures.* Since these reference points and reference temperatures maintain their constant value therefore these are also called *fixed points* and *fixed temperatures* respectively. A list of these fixed points is given in Table 2.1.

Table 2.1 *Some fixed points used for International Practical Temperature Scale*

Sl. No.	*State*	*Temperature °C*
1.	Ice point	0
2.	Steam point	100
3.	Triple point of water	0.010
4.	Triple point of hydrogen	–259.34
5.	Triple point of oxygen	–218.79
6.	Oxygen point (normal boiling point)	–182.96
7.	Silver point (normal freezing point)	961.93
8.	Gold point (normal freezing point)	1064.43
9.	Zinc point (normal freezing point)	419.58
10.	Neon point (normal boiling point)	–246.05
11.	Sulphur point (normal boiling point)	444.60

The methodology adopted was to first develop a temperature measurement system which could show some change in its characteristics (property) due to heat interactions taking place with it. Such systems are called *thermometers*, the characteristics of property which shows change in its value is termed *thermometric property* and the substance which shows change in its thermometric property is called *thermometric substance.* Science that deals with the temperature and its measurement is called *thermometry.* For example in case of *clinical thermometer* the *mercury* in glass is the *thermometric substance* and since there is change in length of mercury column due to the heat interactions taking place between the thermometer and the body whose temperature is to be measured, therefore the *length* is the *thermometric property*. Thus, the underlying principle of temperature measurement is to bring the thermometer in thermal equilibrium with the body whose temperature is to be measured, i.e. when there is no heat interaction or the state when two (thermometer and body) attain same temperature. In this process it is to be noted that thermometer is already caliberated using some standard reference points by bringing thermometer in thermal equilibrium with reference states of the substance.

Zeroth law of thermodynamics states that if the bodies *A* and *B* are in thermal equilibrium with a third body *C* separately then the two bodies *A* and *B* shall also be in thermal equilibrium with each other. This is the principle of temperature measurement. Block diagram shown in Fig. 2.1*a* and 2.1*b* show the zeroth law of thermodynamics and its application for temperature measurement respectively.

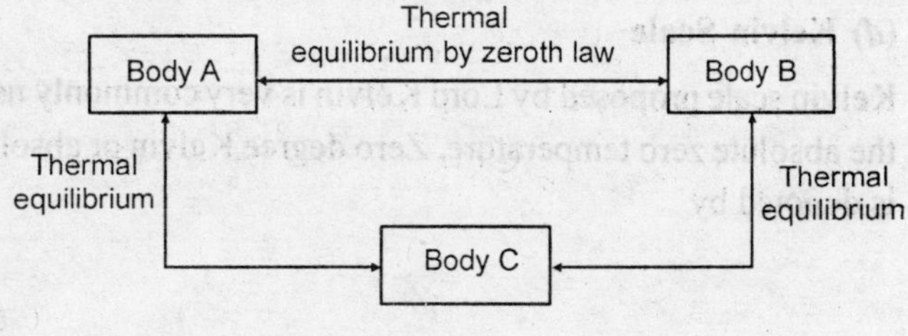

Fig. 2.1a Zeroth law of thermodynamics

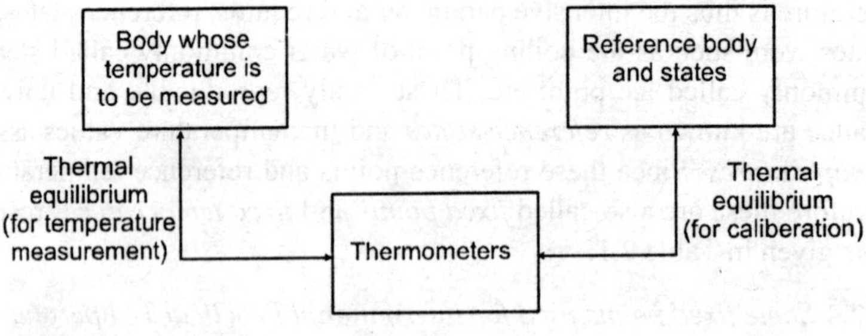

Fig. 2.1b Application of Zeroth law for temperature measurement

2.3 TEMPERATURE SCALES

Number of temperature measuring scales came up from time to time. The text ahead gives a brief idea of the different temperature scales used in thermometry. Different temperature scales have different names based on the names of persons who originated them and have different numerical values assigned to the reference states.

(*a*) Celsius Scale or Centigrade Scale

Anders Celsius gave this Celsius or Centigrade scale using ice point of 0°C as the lower fixed point and steam point of 100°C as upper fixed point for developing the scale. It is denoted by letter *C*. Ice point refers to the temperature at which freezing of water takes place at standard atmospheric pressure. Steam point refers to the temperature of water at which its vaporization takes place at standard atmospheric pressure. The interval between the two fixed points was equally divided into 100 equal parts and each part represented 1°C or 1 degree celsius.

(*b*) Fahrenheit Scale

Fahrenheit gave another temperature scale known as Fahrenheit scale and has the lower fixed point as 32 F and the upper fixed point as 212 F. The interval between these two is equally divided into 180 part. It is denoted by letter *F*. Each part represents 1 F.

(*c*) Rankine Scale

Rankine scale was developed by William John MacQuorn Rankine, a Scottish engineer. It is denoted by letter *R*. It is related to Fahrenheit scale as given below.

$$T_R = T_F + 459.67$$

(*d*) Kelvin Scale

Kelvin scale proposed by Lord Kelvin is very commonly used in thermodynamic analysis. It also defines the absolute zero temperature. Zero degree Kelvin or absolute zero temperature is taken as –273.15°C. It is denoted by

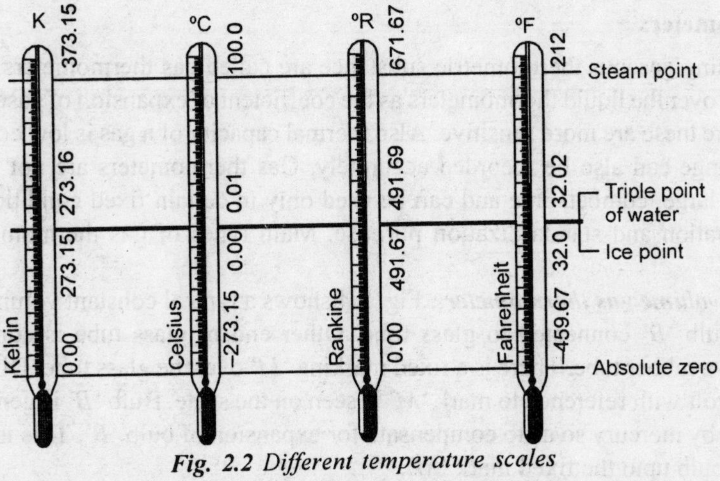

Fig. 2.2 Different temperature scales

letter *K*. Detailed discussion on Kelvin scale has been done in chapter 4 along with absolute thermodynamic temperature scale. Mathematically, it is related to the different temperature scales as follows,

$$\frac{T_C}{100} = \frac{T_F - 32}{180} = \frac{T_K - 273.15}{100} = \frac{T_R - 491.67}{180}$$

$$\frac{T_K}{100} = \frac{T_R}{180}$$

2.4 TEMPERATURE MEASUREMENT

For measurement of temperature number of thermometers are available using different thermometric properties of the thermometric substances. Length, volume, pressure, resistance, e.m.f. etc. are the commonly used thermometric properties for thermometers. Different thermometers developed using these thermometric properties are given below.

(*a*) Liquid Thermometer

Liquid thermometers are those thermometers that employ liquids as the thermometric substance and the change in volume of liquid with heat interaction is the characteristics used for temperature measurement. Commonly used liquids in such thermometers are Mercury and Alcohol. Fig. 2.3 shows the mercury in glass thermometer. In this the change in volume of the mercury results in the rise or fall in the level of mercury column in the glass tube. Out of the two liquids mercury is preferred over alcohol as it has low specific heat and hence absorbs little heat from body. Mercury is comparatively a good conductor of heat. Mercury can be seen in a fine capillary tube conveniently. Mercury does not wet the wall of the tube. Mercury has a uniform coefficient of expansion over a wide range of temperature and remains liquid over a large range as its freezing and boiling points are –39°C and 357°C respectively.

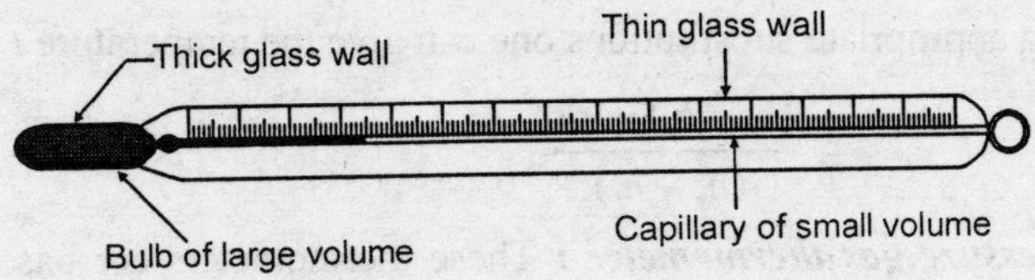

Fig. 2.3 Mercury in glass thermometer

(b) Gas Thermometers

Thermometers using gaseous thermometric substance are called gas thermometers. Gas thermometers are advantageous over the liquid thermometers as the coefficient of expansion of gases is more compared to liquids therefore these are more sensitive. Also thermal capacity of a gas is low compared to liquid so even a small change can also be recorded accurately. Gas thermometers are not suitable for routine work as they are large, cumbersome and can be used only in certain fixed conditions. These are used mainly for calibration and standardization purpose. Main types of gas thermometers are discussed ahead.

(i) Constant volume gas thermometer : Fig. 2.4 shows a typical constant volume gas thermometer having a glass bulb '*B*' connected to glass tube. Other end of glass tube is connected to mercury reservoir through a rubber tube. There is a fixed marking '*M*' over the glass tube. Difference in levels of mercury in reservoir with reference to mark '*M*' is seen on the scale. Bulb '*B*' is generally filled with 1/7th of its volume by mercury so as to compensate for expansion of bulb '*B*'. This is done so as to keep volume of air in bulb upto the fixed mark '*M*'.

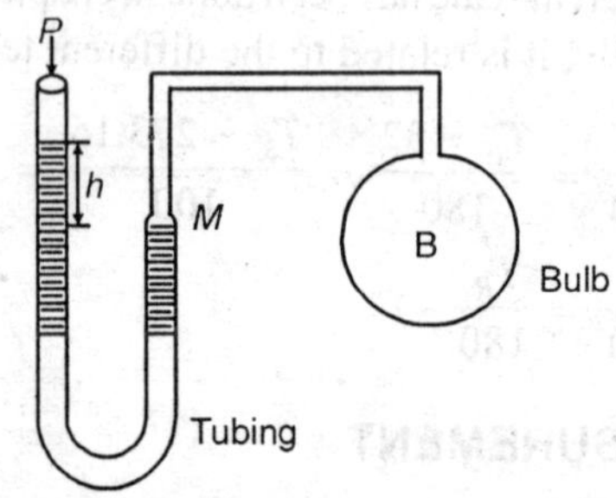

Fig. 2.4 Constant volume air thermometer

Initially the bulb '*B*' is kept in melting ice and reservoir level is suitably adjusted so that mercury level is at mark '*M*'. Corresponding to this difference in level of reservoir and mark '*M*' of h_i height, the ice point pressure shall be,

$$P_i = P + (h_i \cdot \rho \cdot g) = P_0$$

Bulb is kept at the boiling water (steam point) and again the reservoir is adjusted so as to keep mercury at the fixed mark. For difference in mercury levels between mark '*M*' and reservoir level being h_s the pressure corresponding to steam point shall be

$$P_s = P + (h_s \cdot \rho \cdot g) = P_{100}$$

Now for the bulb '*B*' kept in the bath whose temperature is to be measured again the reservoir is to be adjusted so as to keep mercury level at mark '*M*'. At this state if the difference in mercury levels is h_t, then the pressure shall be,

$$P_t = P + (h_t \cdot \rho \cdot g)$$

For a fixed volume, the pressure variation with respect to temperature can be given as,

$$P = P_0 (1 + \alpha \cdot t)$$

Similarly making appropriate substitutions one can give the temperature t as follows,

$$t = \frac{(h_t - h_i) \times 100}{(h_s - h_i)}$$

(ii) Constant pressure gas thermometer : These thermometers are based on the principle that, pressure remaining constant the volume of a given mass of gas is directly proportional to its absolute temperature. Fig. 2.5 shows a constant pressure air thermometer having a silica bulb '*B*' connected to the reservoir '*R*' containing mercury through a *connecting tube* '*A*', *compensating bulb* '*C*' having a compensating tube with volume equal to the connecting tube. Manometer tube contains sulphuric acid.

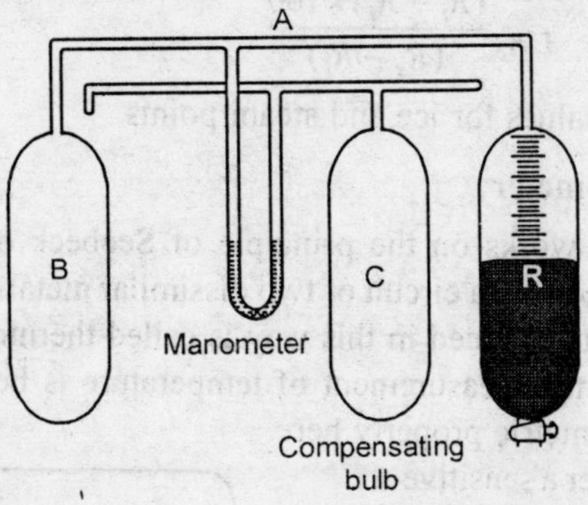

Fig. 2.5 Constant pressure gas thermometer

Initially the reservoir is filled with mercury upto zero marking and the stop cock is closed. The bulbs '*B*' '*R*' and '*C*' are immersed in melting ice. Tubes are sealed when the pressure on the two sides as shown by manometer is the same i.e. the pressure in silica bulb '*B*' and compensating bulb '*C*' are same. When the pressure on two sides of the manometer containing sulphuric acid is same the acid level in two limbs shall be same. This way the pressure of gas and air can be maintained same. Now let us assume the silica bulb '*B*' to have definite number of molecules of air. Also, the compensating bulb and compensating tube contain the same number of molecules of air. If the silica bulb is immersed in the environment whose temperature is to be measured and compensating bulb being kept in melting ice. Both connecting tube and compensating tubes are at the room temperature and the air in silica bulb attains temperature equal to the temperature to be measured.

(*c*) Electrical resistance thermometer

Electrical resistance thermometer first developed by Siemen in 1871, also known as 'Platinum Resistance Thermometer' works on the principle of change in resistance of the thermometric substance (platinum) with temperature. Thus resistance is the thermometric property used in these thermometers. It consists of a pure platinum wire wound in a double spiral on a mica plate. Two ends of the platinum wire are connected to the copper leads (for low temperatures) or platinum leads (for high temperatures). Principle of Wheatstone bridge is employed in these thermometers, as shown in Fig. 2.6. It has a set of compensating leads having exactly similar resistance as leads used. Platinum wire and the compensating leads are enclosed in a sealed glazed porcelain tube having binding terminals at the top. The resistance of wire can be mathematically related as

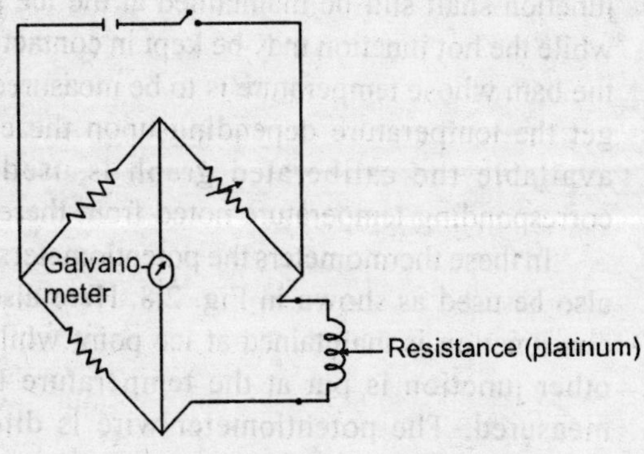

Fig. 2.6 Electrical resistance thermometer, (principle of wheatstone bridge)

$$R_t = R_0 \cdot (1 + a \cdot t + b \cdot t^2)$$

where *a* and *b* are the constants having their values depending upon the nature of material used.

Using fixed points of ice point and steam point the temperature can be mathematically obtained by substituting the different parameters in the following,

$$t = \frac{(R_t - R_i)\times 100}{(R_s - R_i)},$$

where R_i & R_s are resistance values for ice and steam points

(*d*) Thermo Electric Thermometer

Thermo electric thermometer works on the principle of Seebeck effect. *Seebeck effect* says that a current flows or e.m.f. is produced in a circuit of two dissimilar metals having one junction as hot while other as cold junction. Current produced in this way is called thermo electric current while the e.m.f. produced is called thermo e.m.f. Measurement of temperature is being done by knowing the e.m.f. produced which is the thermometric property here.

In such type of thermometer a sensitive galvanometer is connected with thermocouple as shown in Fig. 2.7. One junction is kept at ice point and other in oil bath having any temperature. Upon heating the oil bath it is seen that the thermal e.m.f. is produced by Seebeck effect. Temperature of the oil bath is measured by some calibrated thermometer of any other type. Further the temperature of oil bath is changed to known temperatures and for different temperatures the e.m.f. is noted and a graph is plotted between the temperature of bath and e.m.f.

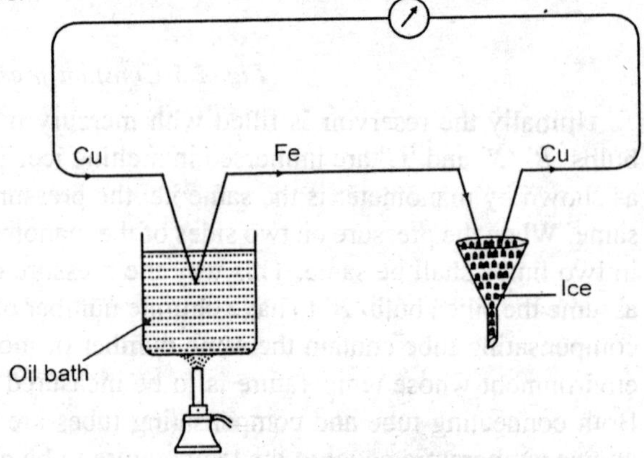

Fig. 2.7 Thermoelectric thermometer using a galvanometer

Now for using this thermocouple the cold junction shall still be maintained at the ice point while the hot junction may be kept in contact with the bath whose temperature is to be measured. To get the temperature depending upon the e.m.f. available the caliberated graph is used and corresponding temperature noted from there.

In these thermometers the potentiometers may also be used as shown in Fig. 2.8. Here also the one junction is maintained at ice point while the other junction is put at the temperature to be measured. The potentiometer wire is directly calibrated to measure temperature.

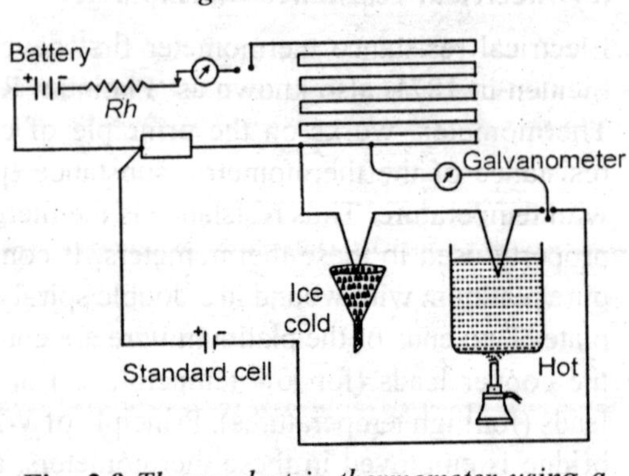

Fig. 2.8 Thermoelectric thermometer using a potentiometer

Here the length of the potentiometer wire at which the balance point is obtained is used for getting temperature.

EXAMPLES

1. Determine the human body temperature in degree celsius (°C) if the temperature in Fahrenheit is 98.6°F.

Solution:

Degree Celsius and Fahrenheit are related as below,

$$T(°C) = \frac{T°(F) - 32}{1.8}$$

Substituting values.

$$T(°C) = \frac{98.6 - 32}{1.8} = 37°C$$

Temperature in degree celsius shall be **37°C. Ans.**

2. A temperature scale is being developed using the following relation.

$$t = a \cdot \ln(p) + \left(\frac{b}{2}\right)$$

where 'p' is thermometric property and 'a' and 'b' are constants. Determine celsius temperature corresponding to thermometric property of 6.5, if ice point and steam point give thermometric property value of 3 and 8.

Solution:

For Ice point; $t = 0°C$ and $p = 3$

For Steam point; $t = 100°C$ and $p = 8$

Using thermometric relation,

$$0 = a\ln(3) + \left(\frac{b}{2}\right)$$

$$100 = a\ln(8) + \frac{b}{2}$$

Solving the above two equations, we get

$a = 101.95$

$b = 224$

Thus, $$t = 101.95.\ \ln(p) + \left(\frac{224}{2}\right)$$

$$t = 101.95\ln(p) + 112$$

For $p = 6.5$, $t =$ **302.83°C Ans.**

3. In a thermoelectric thermometer for t°C temperature, the emf is given as;

$$E = 0.003 \cdot t - 5 \times 10^{-7} \cdot t^2 + 0.5 \times 10^{-3}, \text{ volts}$$

Thermometer is having reference junction at ice point and is calibrated at ice point and steam points. What temperature shall be shown by the thermometer for a substance at 30°C?

Solution:

At ice point; $t = 0°C$, $E_0 = 0.5 \times 10^{-3}$, volts

At steam point, $t = 100°C$, $E_{100} = 0.0265$, volts

When $t = 30°C$

$E_{30} = 9.14 \times 10^{-3}$ volts

Thus temperature shown by this thermometer;

$$t = \left(\frac{E_{30} - E_0}{E_{100} - E_0}\right) \times (T_{100} - T_0)$$

$$= \left(\frac{9.14 \times 10^{-3} - 0.5 \times 10^{-3}}{0.0265 - 0.5 \times 10^{-3}} \right) \times 100$$

$$= \mathbf{33.23°C\ Ans.}$$

4. Estimate the % variation in temperature reading from a thermocouple having its test junction in gas and other reference junction at ice point. The temperature of gas using gas thermometer is found 50°C. Thermocouple is caliberated with emf varying linearly between ice point and steam point. When thermocouple's test junction is kept in gas at t°C and reference junction at ice point, the e.m.f. produced in millivolts is,

$$e = 0.18 \cdot t - 5.2 \times 10^{-4} \times t^2, \text{ millivolts.}$$

Solution:

As ice point and steam points are two reference points, so
at ice point having $t = 0°C$, e.m.f. = 0
at steam point having $t = 100°C$, e.m.f. = 12.8 mV
at gas temperature of 50°C, e.m.f. = 7.7 mV
Since e.m.f. variation is linear so, temperature at e.m.f. of 7.7 mV;

$$= \frac{(100 - 0) \times 7.7}{(12.8 - 0)}$$

$$= 60.16°C$$

Temperature of gas using thermocouple = 60.16°C
% variation in temperature reading with respect to gas thermometer reading of 50°C.

$$= \frac{60.16 - 50}{50} \times 100$$

$$= \mathbf{20.32\%\ Ans.}$$

5. In an unknown temperature scale freezing point of water is 0°X and boiling point of water is 1000°X. Obtain a conversion relation between degrees X and degree celsius. Also determine the absolute zero in degree X.

Solution:

Let the conversion relation be $X = aC + b$
where C is temperature in degree celsius, a & b are constants and X is temperature in °X.
At freezing point, temperature = 0°C, 0°X

or, $0 = a \cdot 0 + b$

$\Rightarrow$ $b = 0$

At boiling point, temperature = 100°C, 1000°X

$1000 = a \cdot 100 + b$

$\Rightarrow$ $a = 10$

Conversion relation

$$\boxed{X = 10 . C} \ \textbf{Ans.}$$

Absolute zero temperature in °C = – 273.15°C
Absolute zero temperature in °X = – 2731.5°X

$$\boxed{-2731.5°X} \ \textbf{Ans.}$$

EXERCISE

2.1 State Zeroth law of thermodynamics.

2.2 Explain, how the zeroth law of thermodynamics can be used for temperature measurement.

2.3 Write short notes on the following:
Thermometry, thermometric substance, thermometric property, Constant volume gas thermometer.

2.4 Sketch and explain the working of constant pressure thermometer.

2.5 Write equivalence amongst different temperature scales. Also write brief note on each of them.

2.6 Obtain triple point of water in Fahrenheit, Rankine and Kelvin scale.

2.7 Heating of a body causes its temperature to change by 30°F. Find out the increase in temperature in °R and °C.

2.8 Temperature of an object changes by 10°C. What is the change in temperature in °R, °F. (–18°R, –18F)

2.9 Prove that the difference between the two temperatures in Celsius scale is same as that in Kelvin scale.

2.10 On some temperature scale 0°C is equivalent to 100°B and 100°C is equivalent to 300°B. Determine the temperature in °C corresponding to 200°B. (50°C)

2.11 During temperature measurement of a body it is seen that the same numerical reading is obtained in Celsius and Fahrenheit scales. What is the temperature in degree Rankine? (419.67°R)

2.12 Write a generic computer program for conversion of temperature in °C, K, °F and °R into one another.

3

First Law of Thermodynamics

3.1 INTRODUCTION

Let us take water in a container and heat it from the bottom. What will happen? Container and the water inside shall start getting heated up. This heating is being sensed by either touching it or by measuring its initial and final temperatures. What has caused it to happen so?

Let us take bicycle wheel and paddle it very fast, Chained wheel starts rotating very fast. Now lèt us apply the brake. The wheel gets stopped. What has made it to stop? Also if we touch the brake shoe and observe its temperature it shall show that the brake shoe has got heated up. Why has it happened so?

Answer for the above question lies in the energy interactions.

The heating up of the container and water has been caused by the heat being added through the burner flame. Heat available in the flame is being transferred to the container resulting into temperature rise of the container and water.

The fast rotation of the bicycle wheel by paddling has been due to work done in the form of paddling and causing wheel to rotate. Subsequent application of the brake has brought wheel to rest as the driving potential with the wheel is gradually dissipated due to the friction between the brake shoe and wheel. Thus the energy of wheel gets transferred to brake shoe bringing it to rest and heating up of the brake shoe (observed by the rise in temperature of brake shoe).

Thus, it is obvious that there is some entity which is responsible for the above phenomenon. This entity is called the energy which is available in different forms, some times causing the wheel rotation, heating up of water etc. Similar to the cases discussed above one may look at other real life systems and understand phenomenon occurring in them. It can be concluded that it is the energy interaction in some form which is responsible for occurrence of such phenomenon. These energy interactions only permit the transformation of energy from one form to other while making the creation and destruction of energy impossible. Hence, it is true that "energy can neither be produced nor destroyed, it can only change it's form". The prevailing law of "energy conservation" also states the same thing.

Here in this chapter we shall look into the first law of thermodynamics, heat, work and its definition, flow and non-flow work, their estimation, steady and unsteady flow processes, their analysis and limitations of the first law of thermodynamics.

3.2 THERMODYNAMIC PROCESSES AND CALCULATION OF WORK

Thermodynamic processes can be precisely categorized as cyclic process and non-cyclic process. The cyclic process is the one in which the initial and final states are identical i.e. system returns to its initial states after occurrence of process. The non cyclic process is the one in which the initial and final states are different i.e. the occurrence of process is accompanied by the state change. Thermodynamic work and its explanation has already been given in Article 1.12.

Let us consider a system consisting of a tank filled with water and fitted with a stirrer at room temperature, Fig. 3.1. Work can be transferred to the system by the stirrer and the temperature of water shall rise. When stirring stops, the system shall cool down till it reaches to the room temperature. Thus, the process is cyclic as the initial and final states are identical.

Let us now take a cylinder having piston and gas filled inside. If the gas is made to expand due to heating, the piston shall undergo displacement and say the piston displacement is dx. If the force exerted by gas on face of piston is F and the cross section area of piston is A, then the displacement work done may be given by :

$$dW = F \cdot dx$$

For the gas pressure being p, the force may be given by $F = p \cdot A$. Substituting for F,

$$dW = p \cdot A \cdot dx$$

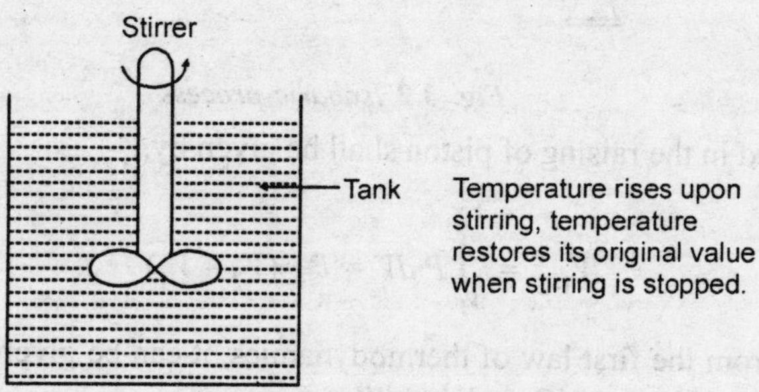

Fig. 3.1 Cyclic process

or, $dW = p \cdot dV$, where dV is the elemental change in volume or the volumetric displacement.

If the total displacement of piston is given by L then the total work can be had by integrating the above dW with respect to x for displacement L, or with respect to volume for volume change.

$$W = \int p{\cdot}dV = \int p \cdot A{\cdot}dx$$

Now, let us examine whether the work estimated above is in conformity to thermodynamic definition of work or not. If the piston displacement is transferred to a suitable link then the weight can be raised, thus it satisfies thermodynamic definition of work. What about the nature of process? cyclic or non cyclic. It is obvious that the initial and final states are not identical therefore, it is a non-cyclic process.

Thus, the work W as defined above refers to thermodynamic work for a non-cyclic process.

Thermodynamic processes can be further classified based on the thermodynamic constraints under which they occur. Different types of thermodynamic processes are as detailed below.

(i) Constant pressure process or isobaric process: It refers to the thermodynamic process in which there is no change in pressure during the process. Such type of processes are also known as isobaric processes. To understand let us take a cylindrical vessel having gas in it. It has a piston above it. Piston is free to reciprocate in the cylinder. Under normal situation piston shall be subjected to atmospheric pressure. Now, let heat be added to cylinder from bottom of cylinder. Due to heat addition, presuming energy transfer taking place reversibly and system always remaining in equilibrium, the gas shall try to expand. Expansion of gas results in raising up of the piston and it attains a new state say 2. Process is shown on p-V diagram in Fig. 3.2.

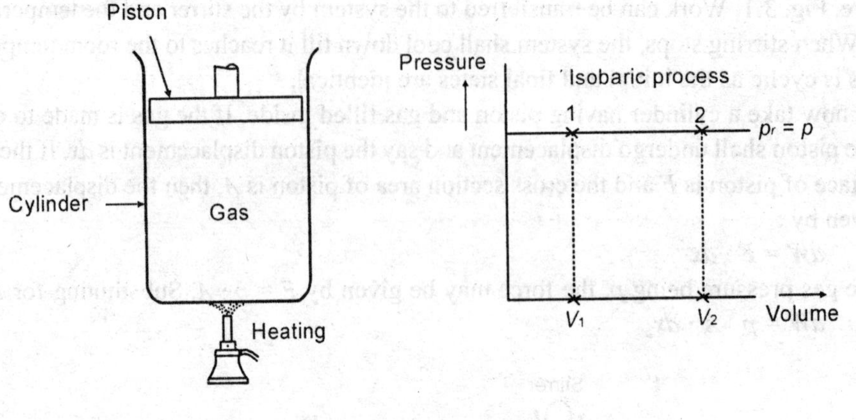

Fig. 3.2 Isobaric process

The work involved in the raising of piston shall be given by,

$$W_{1-2} = \int_{V_1}^{V_2} P \cdot dV = P \cdot (V_2 - V_1)$$

Mathematically from the first law of thermodynamics, it can be given that,

$$dQ = dU + dW$$

$$\int_1^2 dQ = \int_1^2 dU + \int_1^2 dW$$

$$Q_{1-2} = m\, c_v\, (T_2 - T_1) + P(V_2 - V_1)$$

$$= m\, c_v (T_2 - T_1) + mR(T_2 - T_1)$$

Substituting for c_v, i.e. $\quad c_v = \dfrac{R}{(\gamma - 1)}$

$$Q_{1-2} = mR\,(T_2 - T_1) \left\{ \frac{1}{(\gamma - 1)} + 1 \right\}$$

(ii) Constant volume process or isochoric process: When a fluid undergoes a thermodynamic process in a fixed enclosed space such that the process occurs at constant volume, then the process is called constant volume process or isochoric process. Let us consider heating of a gas in fixed enclosure at constant volume. On p–V diagram this process is represented by a vertical line as shown in Fig. 3.3. Area under the process line is zero which indicates that there is rise in pressure but there is no work done as there is no change in volume. Work involved shall be,

$$W_{1-2} = \int_{V_1 = V}^{V_2 = V} P.dV = 0$$

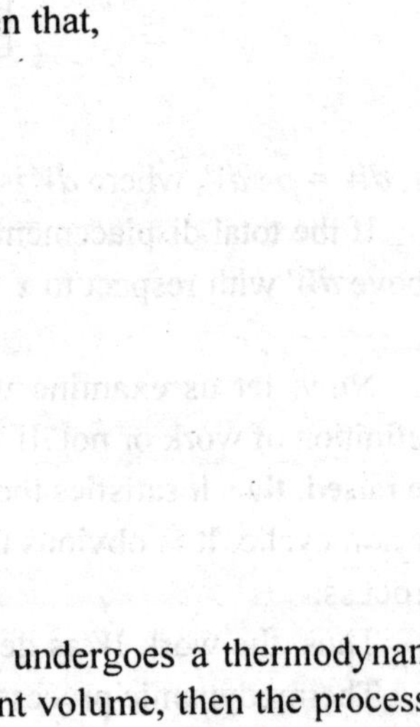

Fig. 3.3 Isochoric process

From first law of thermodynamics,

$$dQ = dU + dW$$

$$\int_1^2 dQ = \int_1^2 dU + \int_1^2 dW = \int_1^2 dU + 0$$

or $$Q_{1-2} = U_2 - U_1 = mc_v\,(T_2 - T_1)$$

Thus, it indicates that the effect of addition of heat in constant volume process is to increase the temperature and consequently the internal energy of system.

(iii) Constant temperature process or isothermal process: Thermodynamic process in which the temperature remains constant is called constant temperature or isothermal process. In this case the gas or vapour may be heated at constant temperature and there shall be no change in internal energy. The work done will be equal to the amount of heat supplied, as shown ahead. For a perfect gas during isothermal process;

$$p_1V_1 = p_2V_2 = \text{Constant, or, } P = \frac{P_1V_1}{V}$$

so work involved, $$W_{1-2} = \int_{V_1}^{V_2} P \cdot dV$$

$$W_{1-2} = \int_{V_1}^{V_2} \frac{p_1V_1}{V}\,dV = P_1V_1 \ln\left(\frac{V_2}{V_1}\right)$$

$$W_{1-2} = P_1V_1 \ln r$$

where r = ratio of final and initial volumes.

By first law of thermodynamics

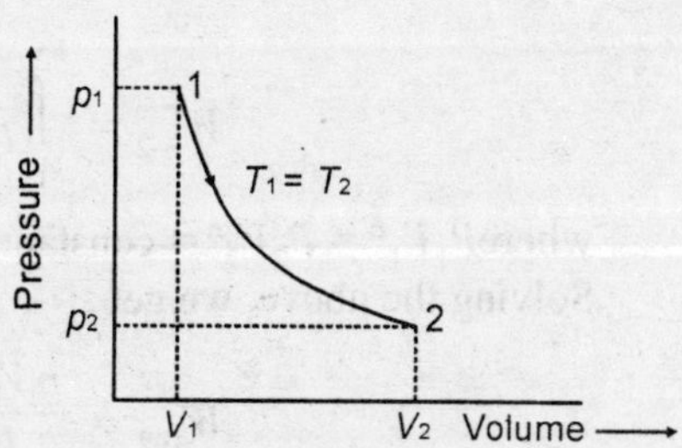

Fig. 3.4 Isothermal process

$$\int_1^2 dQ = \int_1^2 dW + \int_1^2 dU$$

$$Q_{1-2} = W_{1-2} + (U_2 - U_1) = W_{1-2} + 0$$

as $$U_2 - U_1 = mc_v\,(T_2 - T_1), \text{ and } T_1 = T_2$$

(iv) Adiabatic process: An adiabatic process is the thermodynamic process in which there is no heat interaction during the process, i.e. during the process, $Q = 0$. In these processes the work interaction is there at the expense of internal energy. If we talk of adiabatic expansion then it shall mean that work is done at the cost of its own internal energy. The adiabatic process follows the law PV^γ = constant where γ is called adiabatic index and is given by the ratio of two specific heats. Thus, it is obvious that adiabatic expansion shall be accompanied by the fall in temperature while temperature will rise during adiabatic compression. The adiabatic expansion process is shown on Fig. 3.5. Work done during expansion shall be,

$W_{1-2} = \int_{V_1}^{V_2} P \cdot dV$, where PV^γ = constant, therefore solving after substitution. Work shall be,

$$W_{1-2} = \frac{P_1V_1 - P_2V_2}{(\gamma - 1)}$$

From first law of thermodynamics $\int_1^2 dQ = \int_1^2 dU + \int_1^2 dW$

$$Q_{1-2} = (U_2 - U_1) + \frac{p_1V_1 - p_2V_2}{(\gamma - 1)}$$

$$0 = mc_v(T_2 - T_1) + \frac{p_1V_1 - p_2V_2}{(\gamma - 1)}$$

$$\Rightarrow \quad W_{1-2} = mc_v(T_1 - T_2)$$

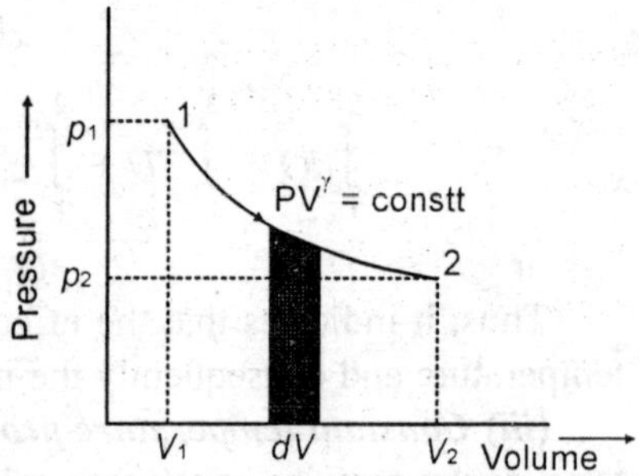

Fig. 3.5 Adiabatic expansion

(v) *Polytropic process*: Polytropic process is the most commonly used process in practice. In this case the thermodynamic process is said to be governed by the law PV^n = constant where n is the index which can vary from $-\infty$ to $+\infty$. Figure 3.6 shows some typical cases in which the value of n is varied and the type of process indicated for different values of n. Thus the various thermodynamics processes discussed above are special cases of polytropic process. Work interaction in case of polytropic process can be given as,

$$W_{1-2} = \int_{V_1}^{V_2} p \cdot dV$$

where $P_1V_1^n = P_2V_2^n$ = constant

Solving the above, we get

$$W_{1-2} = \frac{p_1V_1 - p_2V_2}{(n-1)}$$

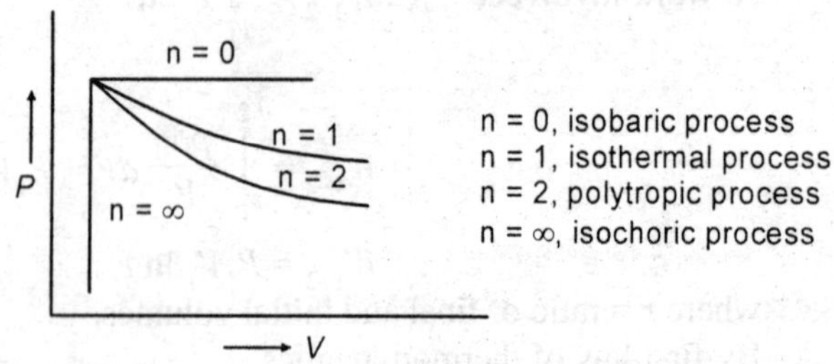

Fig. 3.6 Polytropic process

From first law of thermodynamics,

$$\int_1^2 dQ = \int_1^2 dU + \int_1^2 dW$$

or $$Q_{1-2} = (U_2 - U_1) + \frac{p_1V_1 - p_2V_2}{n-1}$$

or $$Q_{1-2} = mc_v(T_2 - T_1) + \frac{mR(T_1 - T_2)}{(n-1)}$$

also $$c_v = \frac{R}{\gamma - 1} \text{ or } R = c_v(\gamma - 1)$$

$$Q_{1-2} = mc_v\left\{(T_2 - T_1) + \frac{(\gamma - 1)}{(n-1)} \cdot (T_1 - T_2)\right\}$$

Thus heat transfer during a polytropic process for a perfect gas;

$$Q_{1-2} = mc_v(T_2 - T_1) \times \left(\frac{\gamma - n}{1 - n}\right)$$

or also, substituting for c_v

$$Q_{1-2} = \left(\frac{\gamma - n}{\gamma - 1}\right) \times W$$

(vi) Hyperbolic process: Hyperbolic process is the one in which product of pressure and volume remains constant during the process. The curve for such an expansion process is a rectangular hyperbola and hence this is known as hyperbolic expansion.

For a perfect gas $\frac{PV}{T}$ = Constant, if T is also constant then it means that for a perfect gas the hyperbolic process shall also be isothermal process. Figure 3.7 shows hyperbolic expansion process between 1 and 2. Work done during process shall be

$$W_{1-2} = \int_1^2 p \cdot dV \text{ and } p_1V_1 = p_2V_2 = \text{constant}$$

$$W_{1-2} = \int_{V_1}^{V_2} \frac{p_1V_1}{V} dV = p_1V_1 \ln \frac{V_2}{V_1}$$

or, $W_{1-2} = p_1V_1 \ln r$, where $r = \frac{V_2}{V_1}$, ratio of final and initial volumes from first law of thermodynamics,

$$\int_1^2 dQ = \int_1^2 dU + \int_1^2 dW = (U_2 - U_1) + p_1V_1 \ln \left(\frac{V_2}{V_1}\right)$$

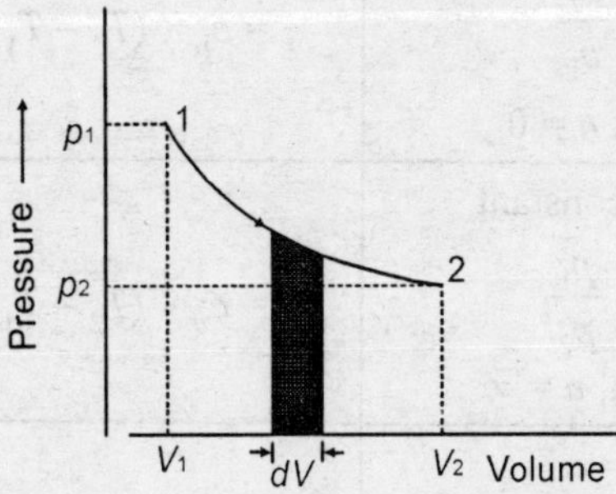

Fig. 3.7 Hyperbolic expansion

(vii) Free Expansion: Free expansion, as the name implies refers to the unrestrained expansion of a gas. Let as take an insulated tank having two compartments separated by a partition, say A and B. Let us assume that compartment A is filled with gas while B is having vacuum. If now the partition is removed and gas allowed to occupy the whole volume of tank, then the gas expands to fill the complete volume space. New pressure of gas will be lesser as compared to initial pressure of gas occupying the compartment A. A close look at the expansion process shows that the expansion due to removal of partition is unresisted expansion due to gas expanding in vacuum. This is also known as free expansion. The reverse of free expansion is impossible and so it is an irreversible process.

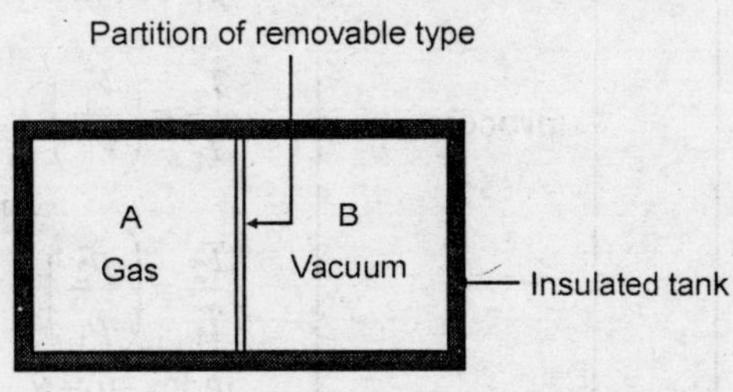

Fig. 3.8 Free expansion

During free expansion no work shall be done by the gas or on the gas due to no boundary displacement in the system.

$$W_{\text{free expansion}} = 0$$

Also in the above there shall be no heat interaction as tank is insulated. From first law of thermodynamics,

$$\Delta Q = \Delta U + \Delta W$$
$$O = \Delta U + O$$

or, $U_{A+B} = U_A$, i.e. initial and final internal energies are same, which means for a perfect gas initial and final temperatures of gas are same.

Table 3.1 *Thermodynamic processes*

Sl. No.	*Process*	*Governing Equations*	*Heat interaction*	*Displacement work or non flow work during state change from 1 to 2*
				$W = \int_1^2 p.dV$
1.	Isobaric process	p = constant $\frac{T_2}{T_1} = \frac{v_2}{v_1}$ index $n = 0$	$q = c_p \times (T_2 - T_1)$	$W = p(V_2 - V_1)$
2.	Isochoric process	V = constant $\frac{T_1}{T_2} = \frac{p_1}{p_2}$ index, $n = \infty$	$q = c_v \times (T_2 - T_1)$	$W = 0$
3.	Isothermal process	T = constant $p_1V_1 = p_2V_2$ index, $n = 1$	$q = p_1V_1 \times \ln\left(\frac{V_2}{V_1}\right)$	$W = P_1V_1 \ln \frac{V_2}{V_1}$
4.	Adiabatic process	$p_1V_1^\gamma = p_2V_2^\gamma$ $\frac{T_2}{T_1} = \left(\frac{V_1}{V_2}\right)^{\gamma-1}$ $\frac{T_2}{T_1} = \left(\frac{p_2}{p_1}\right)^{\frac{\gamma-1}{\gamma}}$ index, $n = \gamma$	$q = 0$	$W = \frac{p_1V_1 - p_2V_2}{\gamma - 1}$

5.	Polytropic process	$p_1V_1^n = p_2V_2^n$ $\frac{T_1}{T_2} = \left(\frac{V_2}{V_1}\right)^{n-1}$ $\frac{T_2}{T_1} = \left(\frac{p_2}{p_1}\right)^{\frac{n-1}{n}}$ $C_n = c_v\left(\frac{\gamma - n}{1-n}\right)$ where C_n is specific heat for polytropic process.	$q = c_v\left(\frac{\gamma - n}{n-1}\right)$ $\times (T_2 - T_1)$ or, $q = \left(\frac{\gamma - n}{\gamma - 1}\right)$ $\times$ work	$W = \frac{p_1V_1 - p_2V_2}{n-1}$
6.	Hyperbolic process	$pV =$ constant but not necessarily $T =$ constant	$q = c_v (T_2 - T_1)$ $+ RT_1 \ln\left(\frac{V_2}{V_1}\right)$	$W = RT_1 \ln\left(\frac{V_2}{V_1}\right)$
7.	Free expansion in adiabatic conditions	Unresisted expansion	$q = 0$	$W = 0$

3.3 NON-FLOW WORK AND FLOW WORK

Work interaction taking place in a system can be classified as flow work or non-flow work based on the nature of process.

Two basic types of processes are

(*i*) Flow process

(*ii*) Non flow process

3.3.1 Flow Process

Flow process is the one in which fluid enters the system and leaves it after work interaction, which means that such processes occur in the systems having open boundary permitting mass interaction across the system boundary.

Figure 3.9 shows the flow process occurring in a system. Flow processes can be further classified into steady flow and non-steady flow processes. Examples of engineering systems having steady flow processes are flow through nozzles, turbines, compressors etc. and the examples of nonsteady flow processes are the filling or emptying of any vessel. Flow process shown indicates various energy and mass interactions taking place across the system boundary.

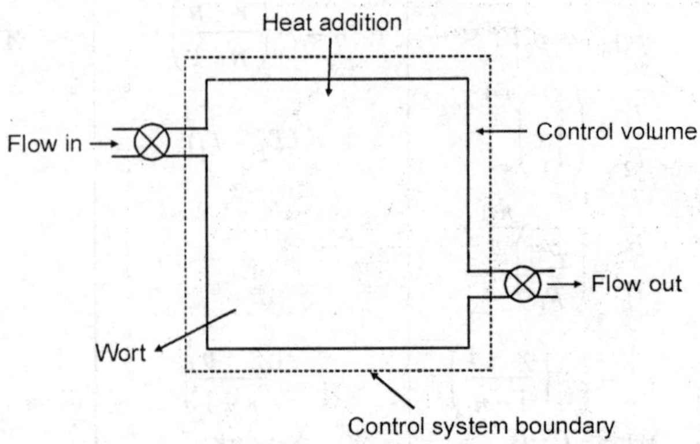

Fig. 3.9 Flow process

As the mass interaction takes place here so for every mass fraction crossing the boundary, work is done upon or by the system. Thus, a portion of work is always required to push the fluid mass into or out of the system. This amount of work is called flow work, or, "work required for causing flow of fluid to or from the system is called flow work".

Here in the control volume shown say, some mass of fluid element is to be pushed into the control volume.

Fluid mass can be injected into the control volume with certain force, say F. The force required for pushing (F) owing to the pressure P of fluid element may be quantified as; $F = P.A$, where A is cross-section area of the passage. For injecting entire mass of fluid, the force F must act through a distance L. Thus, work done in injecting the fluid element across the boundary will be,

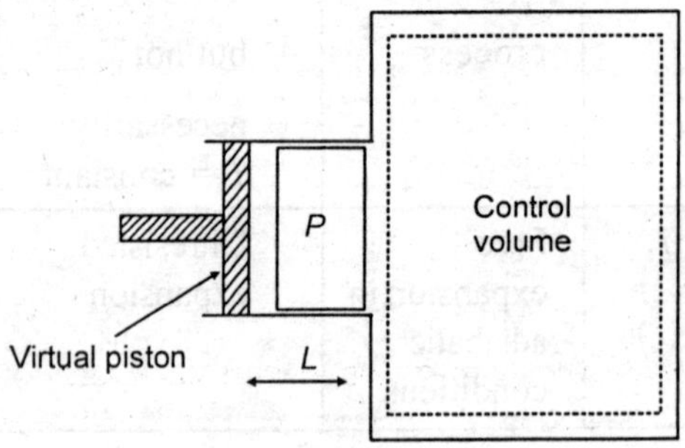

Fig. 3.10 Flow work

$$W = F{\cdot}L = P{\cdot}A{\cdot}L, \ (kJ)$$

or

$$w = P{\cdot}v, \ (kJ/kg)$$

This work is the flow work. Thus, flow work per unit mass can be given as the product of pressure and specific volume. It is also referred to as flow energy or convected energy or transport energy.

3.3.2 Non-Flow Processes

Non-flow process is the one in which there is no mass interaction across the system boundaries during the occurrence of the process.

Figure 3.11 shows block diagram of a piston-cylinder arrangement in which a constant mass of fluid inside the cylinder is shown to undergo compression. Thus, during compression the type of process shall be non-flow process and the work interaction shall be non-flow work.

Say, the force exerted by piston is F, and cross-section area of piston being A, the elemental work done in compressing along the length dL shall be

$$dW = F \cdot dL$$

If pressure of fluid is P then $F = P \cdot A$.

so

$$dW = P \cdot A \cdot dL$$

The total work done in piston displacement, from 1 to 2 shall be,

$$\int_1^2 dW = \int_1^2 P \cdot A \cdot dL$$

$$\int_1^2 dW = \int_1^2 P \cdot dV$$

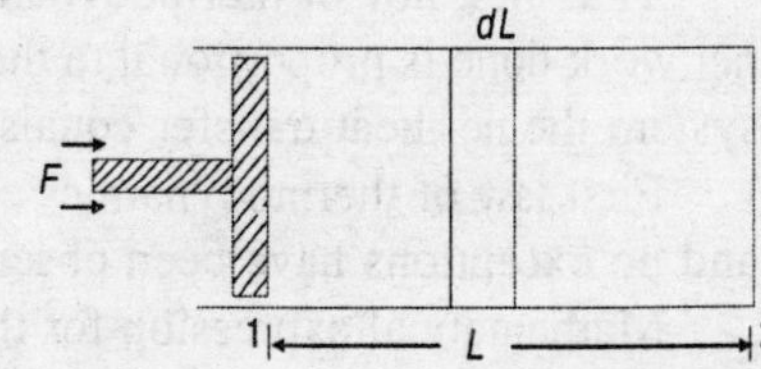

Fig. 3.11 Non-flow process

or $$W_{1-2} = \int_1^2 P \cdot dV$$

Thus, this is called the non-flow work or displacement work.

3.4 FIRST LAW OF THERMODYNAMICS

Benjamin Thompson (Count Runsford) 1753-1814 discovered the equivalence of work and heat in the course of manufacturing canon (1797) by boring solid metal submerged in the water. He was intrigued by the water boiling because of mechanical work of boring, as no heat had been added to the water. In his words, "is it possible that such a quantity of heat as would have caused five pounds of ice cold water to boil could have been furnished by so inconsiderable a quantity of metallic dust merely in consequence of a change in its capacity for heat?" Other experiments later discovered more evidence until some fifty years after the above experiment.

Let as take a bicycle, tyre pump and use it for inflating the bicycle tyre. It is observed that the pump becomes hotter during use. This phenomenon of heating of pump is obviously not from heat transfer but because of the work done. Although the heating of pump could also be realized by heat transfer. It indicates that some effects can be caused equivalently by heat or work and that there exist some relationship between heat and work.

James Prescott Joule (1818-1889) an English scientist and one time student assistant to John Dalton (1766-1844) with assistance from Lord Kelvin showed conclusively that mechanical work and heat are equivalent.

For example, let us take a closed system which permits work interaction and heat interaction both, as in case of stirring in a container, fig. 3.12. As a result of stirring it is seen that the temperature of water gets raised up. This rise in temperature can be accounted by quantifying the amount of heat supplied for raising this temperature. Thus, it is obvious that for any closed system undergoing a cycle $\oint W = J \cdot \oint Q$, where J is Joule's constant.

i.e., the net heat interaction is proportional to the work interaction. Also the constant is known as "Joule's mechanical equivalent of heat". Joule's constant is described as;

$$\frac{W}{Q} = J = 4.18 \ \frac{\text{Joules}}{\text{Calories}}$$

Thus, J is a numerical conversion factor which could be unity if the heat is also given in joules.

For any cyclic process in the closed system the relationship between heat and work shall be, (if the consistent units are used)

$$\oint \delta q = \oint \delta W.$$

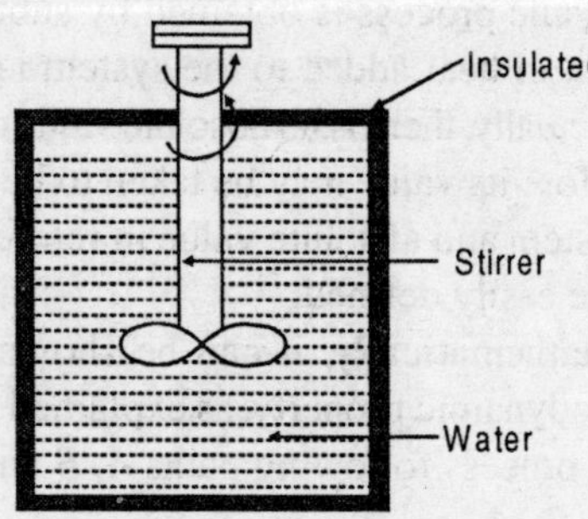

Fig. 3.12 Closed system

Thus first law of thermodynamics states that "in a closed system undergoing a cyclic process, the net work done is proportional to the net heat taken from the surroundings" or "for any cycle of a closed system the net heat transfer equals the net work".

First law of thermodynamics can't be proved but it is supported by a large number of experiments and no exceptions have been observed. It is therefore termed as the law of nature.

Mathematical expression for the first law of thermodynamics can be rearranged and it shall be,

$$\oint (\delta q - \delta W) = 0$$

which shows that the quantity $(\delta q - \delta W)$ is a thermodynamic property.

For non-cyclic process: Let us now take up a system undergoing a non-cyclic process where transfer of heat and work take place and there is some change in the state of system i.e. initial and final states are different.

Figure 3.13 shows the non-cyclic process occurring between states 1 and 2. The change in state is accomplished by the energy interactions. If we assume the system to have the heat interaction ΔQ and work interaction ΔW, then from the basic principles it can be said that :

Energy lost = Energy gained

as the energy can neither be created nor destroyed.

Therefore, between states 1–2 one can write energy balance as,

$$Q_{1-2} - W_{1-2} = U_{1-2}$$

where Q_{1-2}, W_{1-2} and U_{1-2} are the heat, work and stored energy values. *This stored energy is called as internal energy for a system having negligible electrical, magnetic, solid distortion and surface tension effects.*

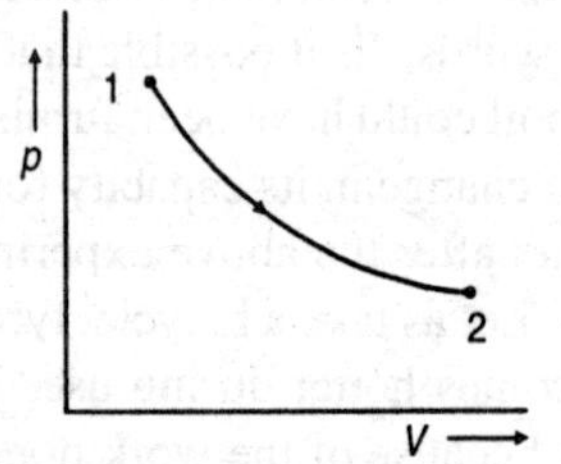

Fig. 3.13 *Non cyclic process*

General expression based on above can be given as follows :

$$\Delta Q - \Delta W = \Delta U$$

or $$\int_1^2 dQ - \int_1^2 dW = \int_1^2 dU$$

or, for elemental interactions; $dQ - dW = dU$

$$dQ = dU + dW$$

Thus, the first law of thermodynamics for non-cyclic processes can be given by

$$\int dQ = \int dU + \int dW$$

Above equations make it obvious that the internal energy change in the closed system during any non-cyclic process is obtained by subtracting the net amount of work done by the system from the net amount of heat added to the system i.e. $\Delta U = Q - W$.

Actually, there is no absolute value of internal energy of any system. Therefore its value may be taken to be zero for any particular state of the system and absolute value in reference to arbitarily assumed state may be easily defined.

Mathematically, it can be shown that the internal energy is a thermodynamic property, as explained ahead. Let us consider the non-cyclic process following paths *A*, *B* and *C* in the directions as shown in Fig. 3.14.

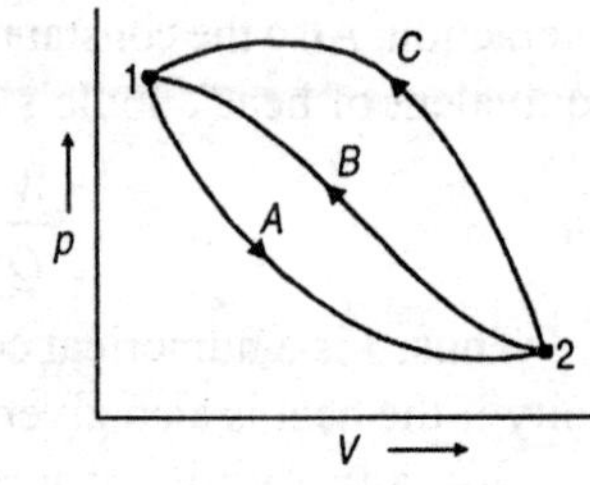

Fig. 3.14 *Two different thermodynamic cycles.*

As the processes *A* & *B* and *A* & *C* constitute a thermodynamic cycle starting and finishing at state 1, the first law of thermodynamics for cyclic process can be employed,

$$\oint (\delta Q - \delta W) = 0$$

For the cycle following path 1–A–2–B–1, the first law of thermodynamics says,

$$\oint_{1-A-2-B-1} (\delta Q - \delta W) = 0$$

or $$\int_{1,A}^{2} (\delta Q - \delta W) + \int_{2,B}^{1} (\delta Q - \delta W) = 0$$

which can be rewritten as,

$$\int_{1,A}^{2} (\delta Q - \delta W) = -\int_{2,B}^{1} (\delta Q - \delta W)$$

or $$\int_{1,A}^{2} (\delta Q - \delta W) = \int_{1,B}^{2} (\delta Q - \delta W) \quad \text{(i)}$$

Also, for the cycle following path 1–A–2–C–1, the first law of thermodynamics can be applied as,

$$\oint_{1-A-2-C-1} (\delta Q - \delta W) = 0$$

or $$\int_{1,A}^{2} (\delta Q - \delta W) + \int_{2,C}^{1} (\delta Q - \delta W) = 0$$

or $$\int_{1,A}^{2} (\delta Q - \delta W) = -\int_{2,C}^{1} (\delta Q - \delta W)$$

or $$\int_{1,A}^{2} (\delta Q - \delta W) = \int_{1,C}^{2} (\delta Q - \delta W) \quad \text{(ii)}$$

From equations (*i*) & (*ii*) it is obvious that

$$\int_{1,A}^{2} (\delta Q - \delta W) = \int_{1,B}^{2} (\delta Q - \delta W) = \int_{1,C}^{2} (\delta Q - \delta W)$$

which shows that $(\delta Q - \delta W)$ is some property as it is independent of the path being followed.

Also, it can be rewritten as,

$$\int_{1,A}^{2} \delta U = \int_{1,B}^{2} \delta U = \int_{1,C}^{2} \delta U$$

or $$\Delta U_{1-2,\,A} = \Delta U_{1-2,\,B} = \Delta U_{1-2,\,C}$$

which means the change in internal energy is independent of the path followed and therefore *internal energy is a thermodynamic property*.

3.5 INTERNAL ENERGY AND ENTHALPY

Let us take a mass at certain elevation in earth's gravitational field and make it move with certain velocity. Energy considerations say that the mass shall have the potential energy (P.E = mgz) and kinetic energy (K.E = (1/2) . mC^2) stored in it. Similarly, several other forms of energy such as due to magnetic, electrical, solid distortion and surface tension effects can be estimated as the contributory components of stored energy.

Difference of heat and work interactions yield the stored energy as given below; $E = Q - W$.

If the energy at macroscopic level as discussed above could be separated from the total stored energy E, then the amount of energy left shall be called internal energy.

Mathematically,

Internal energy, U = (Stored energy) – (Kinetic energy) – (Potential energy) – (Magnetic energy) – (Electrical energy) – (Surface tension energy) – (Solid distortion energy).

Therefore, stored energy is summation of internal energy, potential energy, kinetic energy, magnetic, electrical, surface tension, solid distortion etc. types of energy.

For the situation when magnetic, electric, surface tension, solid distortion effects are negligible, the stored energy shall be;

$$E = U + \text{KE} + \text{PE}$$

or,
$$E = U + \frac{mC^2}{2} + mgz$$

or, on unit mass basis;
$$\boxed{e = u + \frac{C^2}{2} + gz}$$

and the change in stored energy relative to some reference state shall be given as,

$$\Delta E = \Delta U + \Delta\text{KE} + \Delta\text{PE}.$$

Enthalpy (H) of a substance at any point is quantification of energy content in it, which could be given by summation of internal energy and flow energy. Enthalpy is very useful thermodynamic property for the analysis of engineering systems.

Mathematically, it is given as,

$$\boxed{H = U + PV}$$

On unit mass basis, the specific enthalpy could be given as,

$$\boxed{h = u + pv}$$

A look at expression of enthalpy shows that as we can't have absolute value of internal energy, the absolute value of enthalpy can not be obtained. Therefore only change in enthalpy of substance is considered. For certain frequently used substances such as steam, the enthalpy values of steam are available in tabulated form in Steam Tables at different thermodynamic states.

From the definition of enthalpy;

$$h = u + pv$$

or
$$dh = du + p \cdot dv + v \cdot dp.$$

For a constant pressure process, $dp = 0$.

$$dh = du + pdv$$

or,
$$dh = dq_{p\,=\,\text{constt}} \text{ (From first law of thermodynamics)}$$

3.6 SPECIFIC HEATS AND THEIR RELATION WITH INTERNAL ENERGY AND ENTHALPY

Specific heats of the substance refer to the amount of heat interaction required for causing unit change in temperature of the unit mass of substance. This unit change in temperature may be realized under constant volume and constant pressure conditions separately.

Therefore, the above heat value obtained with heat interaction occurring under constant volume conditions is called specific heat at constant volume, denoted as c_v. Whereas the above heat value obtained with heat interaction occurring under constant pressure conditions is called specific heat at constant pressure, denoted as c_p.

Mathematically, the heat interaction causing ΔT temperature change in m mass of substance can be given as,

For isochoric conditions;

$$Q_v = m \cdot c_v \cdot \Delta T \text{ and for isobaric conditions}$$

$$Q_p = m \cdot c_p \cdot \Delta T$$

or

$$c_v = \frac{Q_v}{m \cdot \Delta T} \text{ or, } c_p = \frac{Q_p}{m \cdot \Delta T}$$

For getting the specific heat values, substituting $m = 1$, $\Delta T = 1$,

$$\boxed{c_v = Q_v} \text{ and } \boxed{c_p = Q_p}$$

The specific heat at constant volume can also be given as the partial derivative of internal energy with respect to temperature at constant volume.

Thus

$$c_v = \left(\frac{\partial u}{\partial T}\right)_v$$

or,

$$c_v = \left(\frac{du}{dT}\right)_v$$

Also from first law of thermodynamics, on unit mass basis

$$dq = du + pdv$$

at constant volume, $dv = 0$

$$dq = du$$

or

$$dq = c_v \cdot dT = du$$

$$\boxed{c_v = \frac{dq}{dT}}, \text{ for } v = \text{constant}$$

Specific heat at constant pressure can be given as the partial derivative of enthalpy with respect to temperature at constant pressure.

Mathematically:

$$c_p = \left(\frac{\partial h}{\partial T}\right)_p$$

or

$$c_p = \left(\frac{dh}{dT}\right)_p$$

From definition of enthalpy, at unit mass basis.

$$h = u + pv$$

or $$dh = du + pdv + vdp$$

at constant pressure, $dp = 0$

$$dh = du + pdv$$

substituting from first law of thermodynamics $dq = du + pdv$

$$dh = dq$$

or $$dq = c_p \cdot dT = dh$$

$$\boxed{c_p = \frac{dq}{dT}}, \text{ for } p = \text{constant}$$

Let us try to establish relationship between c_p and c_v.
From enthalpy definition, at unit mass basis

$$h = u + pv$$

or $$h = u + RT \text{ \{for ideal gas\}}$$

Taking partial derivative,

$$dh = du + RdT$$

Also we know for an ideal gas, c_p

$$dh = c_p \cdot dT;\ du = c_v \cdot dT$$

Substituting *dh* and *du*

$$c_p \cdot dT = c_v \cdot dT + R \cdot dT$$

or $$c_p = c_v + R$$

or $$\boxed{c_p - c_v = R}$$

Difference of specific heats at constant pressure and volume is equal to the gas constant for an ideal gas.

Also the ratio of specific heats at constant pressure and volume could be given as γ,

$$\frac{c_p}{c_v} = \gamma$$

Combining above two relations of c_p and c_v we get,

$$c_p = \frac{\gamma . R}{(\gamma . 1)} \text{ and } c_v = \frac{R}{(\gamma . 1)}$$

3.7 FIRST LAW OF THERMODYNAMICS APPLIED TO OPEN SYSTEMS

Let us consider an open system as shown in Fig. 3.15 having inlet at section 1–1 and outlet at section 2–2. The cross-section area, pressure, specific volume, mass flow rate, energy at section 1–1 and 2–2 are

Section 1–1 = A_1, p_1, v_1, m_1, e_1
Section 2–2 = A_2, p_2, v_2, m_2, e_2

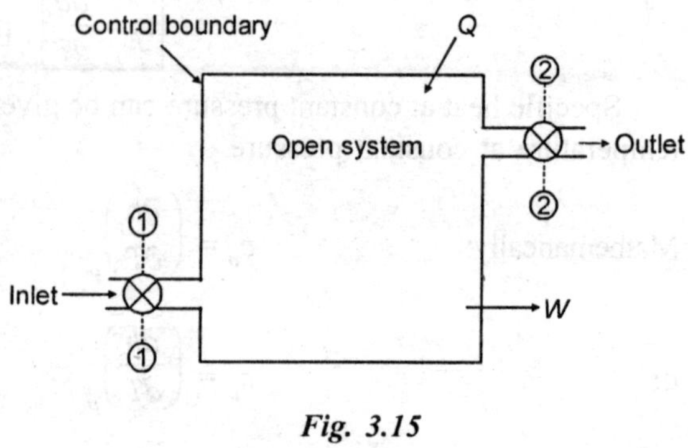

Fig. 3.15

Open system is also having heat and work interactions Q, W as shown in figure above.

Applying the energy balance at the two sections, it can be given as,

Energy added to the system + Stored energy of the fluid at inlet

= Stored energy of the fluid at outlet

Quantifying the various energies;

Energy of fluid at inlet shall comprise of stored energy and flow energy as given here.

$$= m_1(e_1 + p_1v_1)$$

Similarly energy of fluid at outlet shall comprise of stored energy and flow energy,

= Stored energy + Flow energy

$$= m_2(e_2 + p_2v_2)$$

The energy added to the system shall be the net energy interaction due to heat and work interactions.

$$= Q - W$$

Writing energy balance, mathematically;

$$Q - W + m_1(e_1 + p_1v_1) = m_2(e_2 + p_2v_2)$$

or

$$Q + m_1(e_1 + p_1v_1) = W + m_2(e_2 + p_2v_2)$$

If the mass flow rates at inlet and exit are same, then

$$Q + m(e_1 + p_1v_1) = W + m(e_2 + p_2v_2)$$

On unit mass basis

$$q + e_1 + p_1v_1 = w + e_2 + p_2v_2$$

Thus,

Heat + (Stored energy + Flow energy)$_1$ = Work + (Stored energy + Flow energy)$_2$

Stored energy at inlet and outlet can be mathematically given as,

$$e_1 = u_1 + \frac{C_1^2}{2} + gz_1$$

and

$$e_2 = u_2 + \frac{C_2^2}{2} + gz_2$$

where C_1 and C_2 are velocities at inlet and exit, u_1 and u_2 are internal energy at inlet and outlet, z_1 and z_2 are elevations of inlet and exit.

3.8 STEADY FLOW SYSTEMS AND THEIR ANALYSIS

Steady flow refers to the flow in which its properties at any point remain constant with respect to time. Steady system is the system whose properties are independent of time, i.e. any property at a point in system shall not change with time.

Let us take an open system having steady flow. Figure 3.16 shows steady flow system having inlet at section 1–1, outlet at section 2–2, heat addition Q and work done by the system W.

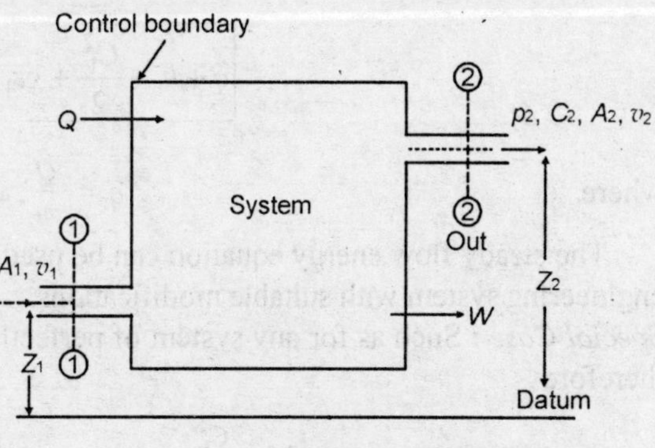

Fig. 3.16 Steady flow system

	At section 1–1	At section 2–2
Pressure, (N/m²)	p_1	p_2
Sp volume, (m³/kg)	v_1	v_2
Velocity, (m/s)	C_1	C_2
Elevation, (m)	z_1	z_2
Cross-section area, (m²)	A_1	A_2
Mass flow rate, (kg/s)	m_1	m_2
Internal energy, (J/kg)	u_1	u_2

As described in earlier article the energy balance when applied to open system results in

$$Q + m_1(e_1 + p_1v_1) = W + m_2(e_2 + p_2v_2)$$

Substituting for e_1 and e_2

$$Q + m_1\left(u_1 + \frac{C_1^2}{2} + gz_1 + p_1v_1\right) = W + m_2\left(u_2 + \frac{C_2^2}{2} + gz_2 + p_2v_2\right)$$

and from definition of enthalpy,

$$h_1 = u_1 + p_1v_1$$
$$h_2 = u_2 + p_2v_2$$

therefore

$$\boxed{Q + m_1\left(h_1 + \frac{C_1^2}{2} + gz_1\right) = W + m_2\left(h_2 + \frac{C_2^2}{2} + gz_2\right)}$$

Above equation is known as steady flow energy equation (S.F.E.E.). If the mass flow rates at inlet and exit are same, i.e. $m_1 = m_2 = m$

then,
$$Q + m\left(h_1 + \frac{C_1^2}{2} + gz_1\right) = W + m\left(h_2 + \frac{C_2^2}{2} + gz_2\right)$$

or, on unit mass basis the S.F.E.E. shall be;

$$\boxed{q + h_1 + \frac{C_1^2}{2} + gz_1 = w + h_2 + \frac{C_2^2}{2} + gz_2}$$

where
$$q = \frac{Q}{m},\ w = \frac{W}{m}$$

The steady flow energy equation can be used as a tool for carrying out thermodynamic analysis of engineering system with suitable modifications.

Special Case : Such as for any system of perfectly insulated type, $Q = 0$

therefore

$$h_1 + \frac{C_1^2}{2} + gz_1 = w + h_2 + \frac{C_2^2}{2} + gz_2$$

Continuity equation applied results in

$$m_1 = m_2$$

or,
$$\frac{A_1 C_1}{v_1} = \frac{A_2 C_2}{v_2}$$

For any system having more than one inlets, outlets and energy, interactions are shown below.

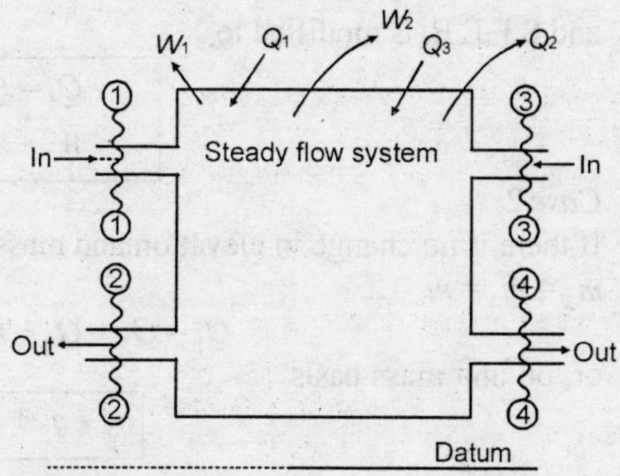

Fig. 3.17 Example of steady flow system

Salient properties at different sections are tabulated as under

	Section 1–1	*Section* 2–2	*Section* 3–3	*Section* 4–4
Pressure, (N/m^2)	p_1	p_2	p_3	p_4
Sp. volume, (m^3/kg)	v_1	v_2	v_3	v_4
Mass flow rate, (kg/s)	m_1	m_2	m_3	m_4
Internal energy, (J/kg)	u_1	u_2	u_3	u_4
Velocity, (m/s)	C_1	C_2	C_3	C_4
Elevation, (m)	z_1	z_2	z_3	z_4
Cross-section area, (m^2)	A_1	A_2	A_3	A_4

Net heat added, $Q = Q_1 - Q_2 + Q_3$

Net work done, $W = W_1 + W_2$

Applying steady flow energy equation on the system shown;

$$Q + m_1\left(u_1 + \frac{C_1^2}{2} + gz_1 + p_1 v_1\right) + m_3\left(u_3 + \frac{C_3^2}{2} + gz_3 + p_3 v_3\right)$$

$$= W + m_2\left(u_2 + \frac{C_2^2}{2} + gz_2 + p_2 v_2\right) + m_4\left(u_4 + \frac{C_4^2}{2} + gz_4 + p_4 v_4\right)$$

Substituting enthalpy values, h_1, h_2, h_3, h_4 and for Q and W;

$$\boxed{\begin{aligned} &Q_1 - Q_2 + Q_3 + m_1\left(h_1 + \frac{C_1^2}{2} + gz_1\right) + m_3\left(h_3 + \frac{C_3^2}{2} + gz_3\right) \\ &= (W_1 + W_2) + m_2\left(h_2 + \frac{C_2^2}{2} + gz_2\right) + m_4\left(h_4 + \frac{C_4^2}{2} + gz_4\right) \end{aligned}}$$

Case 1

If the inlet and exit velocities are negligible, then

$$KE_1 = KE_2 = KE_3 = KE_4 = 0$$

and S.F.E.E. is modified to

$$Q_1 - Q_2 + Q_3 + m_1(h_1 + gz_1) + m_3(h_3 + gz_3) = W_1 + W_2 + m_2(h_2 + gz_2) + m_4(h_4 + gz_4)$$

Case 2

If there is no change in elevation and mass flow rates at all inlets and outlets are same, then, $m_1 = m_2 = m_3 = m_4 = m$

$$Q_1 - Q_2 + Q_3 + m \,.\, h_1 + m\,h_3 = W_1 + W_2 + m\,h_2 + m\,h_4$$

or, on unit mass basis

$$q_1 - q_2 + q_3 + h_1 + h_3 = w_1 + w_2 + h_2 + h_4$$

3.9 FIRST LAW APPLIED TO ENGINEERING SYSTEMS

Here the first law of thermodynamics applied to different engineering systems is discussed. It is assumed in general that the processes are of steady flow type and so the steady flow energy equation can be directly used with modifications in it.

(*a*) *Turbine*: It is the device in which the high temperature and high pressure fluid is expanded to low temperature and pressure resulting in generation of positive work at turbine shaft. Thus, turbine is a work producing device.

Turbines using gas as working fluid are called gas turbine where as turbines using steam are called steam turbines.

Expansion in turbine is assumed to be of adiabatic type so that the maximum amount of work is produced.

Assuming change in kinetic energy, potential energy to be negligible, the steady flow energy equation can be modified and written between 1 and 2 as,

$$O + mh_1 = W_T + mh_2$$

$$W_T = m(h_1 - h_2)$$

i.e., $Q = 0$ and total energy interaction is available in the form of work

Turbine work $= m(h_1 - h_2) = m\,c_p(T_1 - T_2)$

Here m is mass flow rate and T_1, T_2 are temperatures at inlet and outlet.

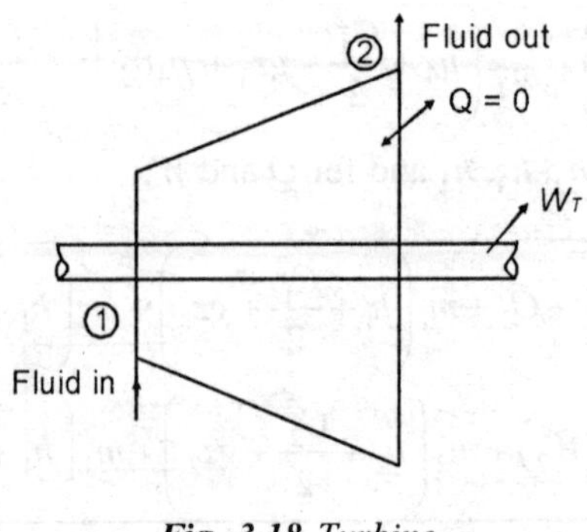

Fig. 3.18 Turbine

(*b*) *Compressor*: Compressor is a work absorbing device used for increasing the pressure of a fluid. Pressure of a fluid is increased by doing work upon it, which is accompanied by increase in temperature depending on the gas properties.

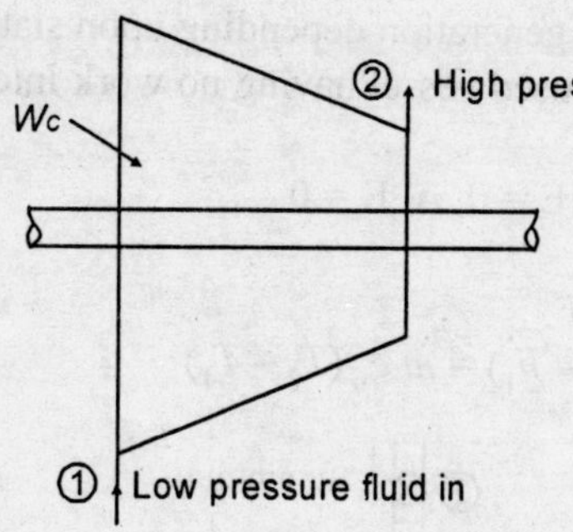

Fig. 3.19 Compressor

For compression of a gas adiabatic process is used as in this there is no heat loss and so minimum work requirement.

Let us assume change in kinetic energy and potential energy to be negligible between 1 & 2 and also flow to be of steady type.

Applying steady flow energy equation in modified form:

$$Q = 0$$
$$\Delta KE = 0$$
$$\Delta PE = 0$$
$$W_c = (-ve) \text{ work for compression}$$
$$mh_1 = -W_c + mh_2$$

or
$$W_c = m(h_2 - h_1)$$

Adiabatic compression work $= m(h_2 - h_1) = mc_p(T_2 - T_1)$

Here T_1, T_2 are temperatures at inlet and outlet and m is mass flow rate.

(*c*) *Pump*: A pump is used for pumping liquid or suction of liquid. In case of pump the following assumptions can be made for using S.F.E.E.

(*i*) Heat transfer is zero, $Q = 0$

(*ii*) Change in internal energy is zero, $\Delta U = 0$

Therefore

$$m\left(p_1v_1 + \frac{C_1^2}{2} + gz_1\right) = m\left(p_2v_2 + \frac{C_2^2}{2} + gz_2\right) - W_{pump}$$

or,
$$W_{pump} = m\{(p_2v_2 - p_1v_1) + \frac{C_2^2 - C_1^2}{2} + g(z_2 - z_1)\}$$

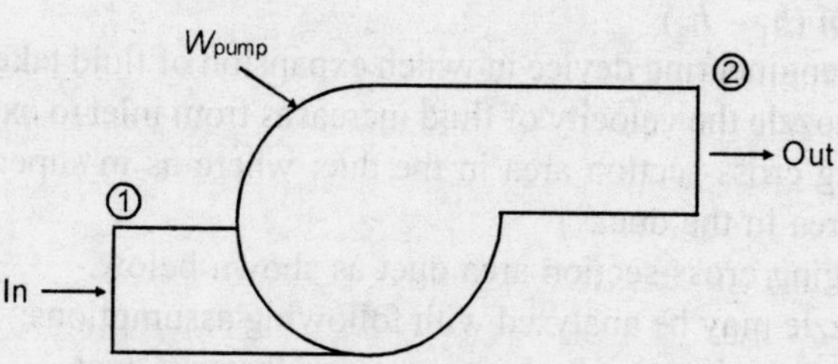

Fig. 3.20 Pump

(*d*) *Boiler*: Boiler is the engineering device used for steam generation at constant pressure. Heat is supplied externally to the boiler for steam generation depending upon state of steam desired.

Boiler may be assumed similar to a closed vessel having no work interaction, no change in kinetic energy, no change in potential energy.

i.e. $W = 0, \Delta KE = 0, \Delta PE = 0.$

Applying steady flow energy equation

$$Q_{boiler} + m(h_1) = m(h_2)$$

or

$$Q_{boiler} = m(h_2 - h_1) = m\, c_p\,(T_2 - T_1)$$

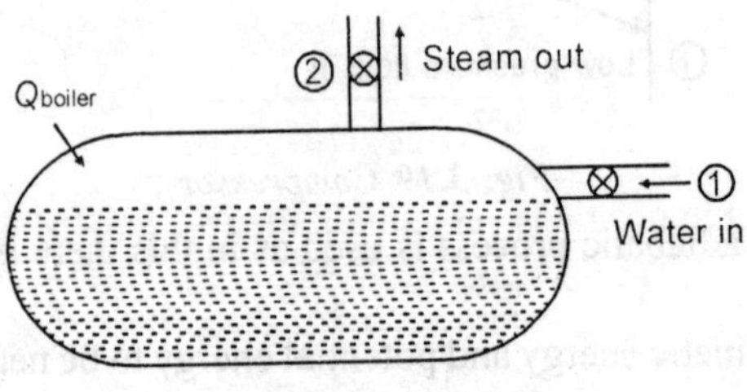

Fig. 3.21 Boiler

(*e*) *Condenser*: Condenser is the device used for condensing vapour into liquid at constant pressure. It is a type of heat exchanger in which another cool fluid is used for condensing the vapours into liquid. Heat exchange between the hot fluid and cold fluid takes place indirectly as cold fluid passes through the tubes and hot vapours are outside tubes in the shell.

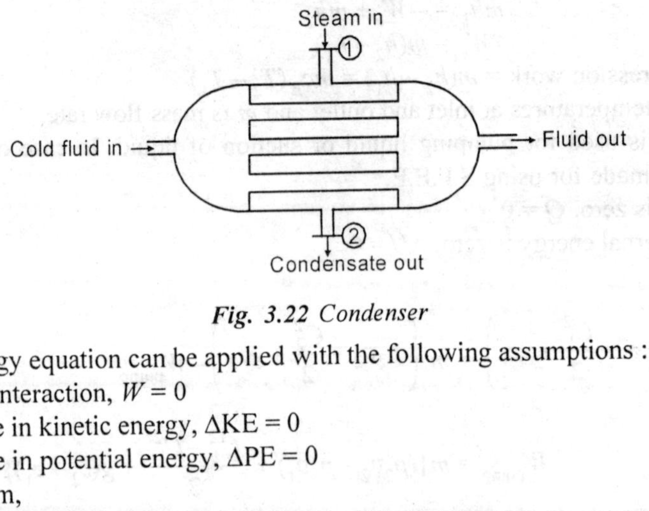

Fig. 3.22 Condenser

Steady flow energy equation can be applied with the following assumptions :

(*i*) No work interaction, $W = 0$

(*ii*) No change in kinetic energy, $\Delta KE = 0$

(*iii*) No change in potential energy, $\Delta PE = 0$

Heat lost by steam,

$$Q = m\,(h_1 - h_2)$$

(*f*) *Nozzle*: Nozzle is the engineering device in which expansion of fluid takes place and pressure drops simultaneously. Thus in nozzle the velocity of fluid increases from inlet to exit. In case of subsonic flow the nozzle has converging cross-section area in the duct where as in supersonic flow the nozzle has diverging cross-section area in the duct.

Let us take a converging cross-section area duct as shown below.

Flow through the nozzle may be analysed with following assumptions:

(*i*) No heat interaction, i.e. $Q = 0$, during passage through duct.

(*ii*) No work interaction, i.e. $W = 0$, during passage through duct.

(*iii*) No change in elevation from 1 to 2, i.e. $\Delta PE = 0$.

Applying S.F.E.E on nozzle,

$$h_1 + \frac{C_1^2}{2} = h_2 + \frac{C_2^2}{2}$$

or,
$$\frac{C_2^2 - C_1^2}{2} = h_1 - h_2$$

$$C_2 = \sqrt{C_1^2 + 2(h_1 - h_2)}$$

or,
$$C_2 = \sqrt{C_1^2 + 2c_p(T_1 - T_2)}$$

In case, the velocity at inlet to nozzle is very small, then C_1 may be neglected and velocity at nozzle exit shall be:

$$C_2 = \sqrt{2c_p(T_1 - T_2)}$$

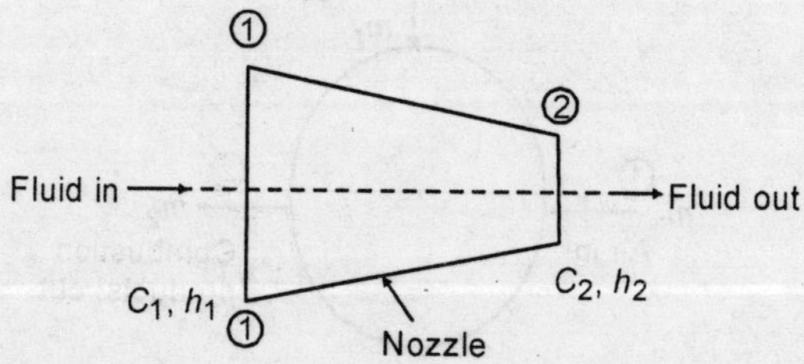

***Fig. 3.23** Nozzle*

(*g*) *Throttling*: Throttling refers to passage of a fluid through some restricted opening under isenthalpic conditions. Thus in the figure shown below the fluid passes through a restriction from section 1 to 2 and undergoes drop in its pressure and increase in volume, but during this passage enthalpy remains constant, such that $h_1 = h_2$.

Based on above throttling process the device called "throttle valve" has been developed in which pressure drop is realized without involving any work and heat interaction, change in kinetic energy and potential energy. Temperature may drop or increase during the throttling process and shall depend upon the Joule-Thomson coefficient, a property based on characteristic of substance.

Joule-Thomson coefficient $\mu = \left(\frac{\partial T}{\partial p}\right)_{h=\text{constt.}}$

and if $\mu = 0$, Temperature remains constant

$\mu > 0$, Temperature decreases.

$\mu < 0$ Temperature increases.

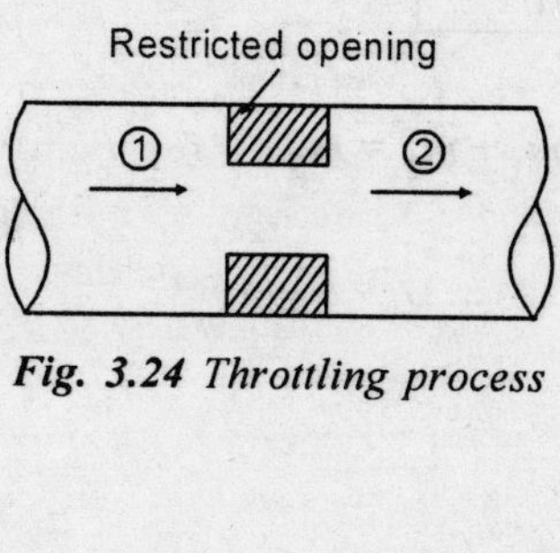

***Fig. 3.24** Throttling process*

(*h*) *Combustion chamber*: Combustion chambers are commonly used in gas turbine installations, in which fuel is injected at high pressure into a chamber having high pressure, high temperature air in it and ignited for heat release at constant pressure.

Mass balance yields:

$$m_2 = m_1 + m_f$$

where m_1, m_2, m_f are mass flow rates at 1, 2, and f shown in figure.

S.F.E.E. may be applied with assumptions of $\Delta KE = 0$,

$$\Delta PE = 0,$$

$$W = 0$$

Here, $Q = m_f \times$ Calorific value of fuel

or

$$Q = m_f \times CV$$

$$\boxed{Q + m_1 h_1 + m_f h_f = m_2 h_2}$$

substituting for Q

$$m_f CV + m_1 h_1 + m_f h_f = m_2 h_2$$

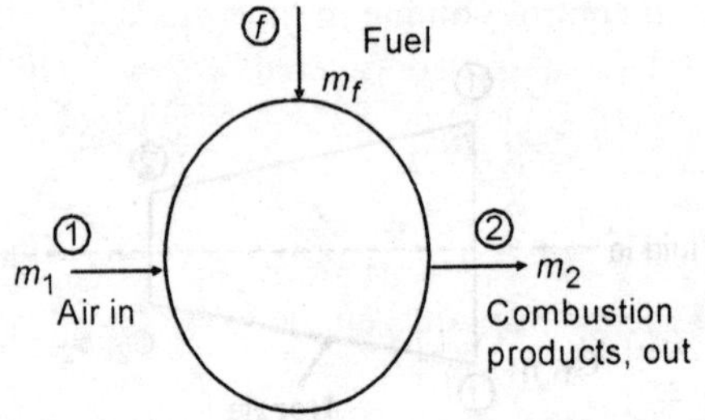

Fig. 3.25 Combustion chamber

(*i*) *Adiabatic mixing*: Adiabatic mixing refers to mixing of two or more streams of same or different fluids under adiabatic conditions.

Let us consider two streams of same fluid with mass flow rates m_1 and m_2 to get mixed together adiabatically.

Assumptions for applying S.F.E.E shall be;

(*i*) No heat interaction, $Q = 0$

(*ii*) No work interaction, $W = 0$

(*iii*) No change in kinetic energy, $\Delta KE = 0$

(*iv*) No change in potential energy, $\Delta PE = 0$.

Thus,

$$m_1 \cdot h_1 + m_2 \cdot h_2 = m_3 \cdot h_3.$$

or

$$m_1 \cdot c_p \cdot T_1 + m_2 \cdot c_p \cdot T_2 = m_3 \cdot c_p \cdot T_3$$

or

$$\boxed{T_3 = \frac{m_1 T_1 + m_2 T_2}{m_3}}$$

By mass balance,

$$m_1 + m_2 = m_3$$

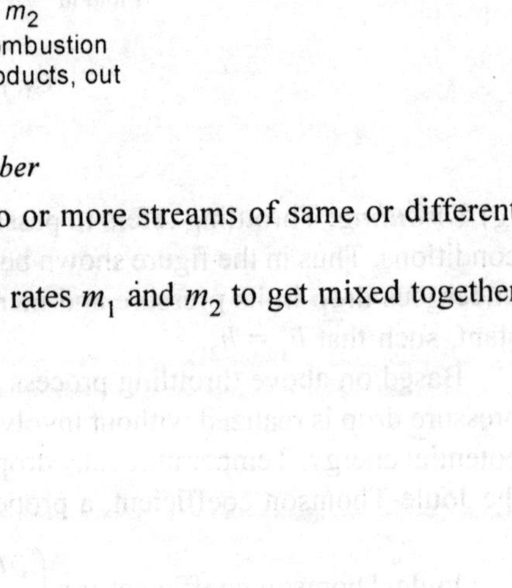

Fig. 3.26 Adiabatic mixing

3.10 UNSTEADY FLOW SYSTEMS AND THEIR ANALYSIS

In earlier discussions, for a steady flow system, it has been assumed that the properties do not change with time.

However, there exist a number of systems such as filling up of a bottle or emptying of a vessel etc. in which properties change continuously as the process proceeds. Such systems can not be analysed with the steady state assumptions. Unsteady flow processes are also known as transient flow processes or variable flow processes.

Let us take example of filling up of the bottle.

The bottle is filled up gradually, therefore it is case of an unsteady system.

By conservation of mass, the unsteady process over a period of time '*dt*' can be expressed as following in generic form.

(Mass entering the control volume in time *dt*)

– (Mass leaving the control volume in time *dt*)

= Net change in mass in control volume in time *dt*.

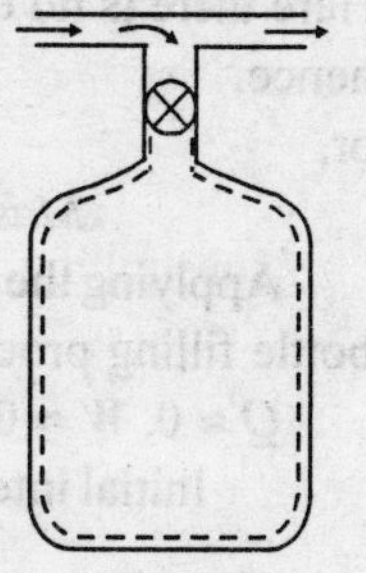

Fig. 3.27 Filling of the bottle

If the mass flow rate at inlet and exit are given as m_i, m_e then

$$\frac{dm_i}{dt} - \frac{dm_e}{dt} = \frac{dm_{cv}}{dt}$$

and also, $\Sigma m_i - \Sigma m_e = (m_{\text{final}} - m_{\text{initial}})_{cv}$

By the conservation of energy principle applied on control volume for time '*t*', energy balance yields;

Net energy interaction across the boundary in time *dt*

+ Energy entering into control volume in time *dt*

– Energy leaving out of control volume in time *dt*

= Change in energy in control volume in time *dt*

Mathematically, it can be given as:

$$\boxed{(Q - W) + \Sigma E_i - \Sigma E_e = \Delta E_{cv}}$$

where
$$E_i = \int_0^t m_i\left(h_i + \frac{C_i^2}{2} + gz_i\right) . dt$$

$$E_e = \int_0^t m_e\left(h_e + \frac{C_e^2}{2} + gz_e\right) . dt$$

Thus, the above mass balance and energy balance can be used for analysing the unsteady flow systems with suitable assumptions.

It may be assumed that the control volume state is uniform and fluid properties are uniform and steady at inlet and exit.

Simplified form of energy balance written above can be given as;

$$Q - W + \Sigma\, m_i\left(h_i + \frac{C_i^2}{2} + gz_i\right) - \Sigma\, m_e\left(h_e + \frac{C_e^2}{2} + gz_e\right)$$

$$= (m_{\text{final}} \cdot u_{\text{final}} - m_{\text{initial}} \cdot u_{\text{initial}})_{cv}$$

If the changes in kinetic energy and potential energy are negligible, then energy balance gets modified as;

$$Q - W + \Sigma\, m_i \cdot h_i - \Sigma\, m_e \cdot h_e = (m_{\text{final}} \cdot u_{\text{final}} - m_{\text{initial}} \cdot u_{\text{initial}})_{cv}$$

Case 1: Let us now use the energy and mass balance to the unsteady flow process of filling up a bottle as shown in Figure 3.27. Bottle is initially empty and connected to a pipe line through valve for being filled.

Let us denote initial state of system by subscript 1 and final state by 2.
Initially as bottle is empty, so $m_1 = 0$
From mass balance

$$\Sigma m_i - \Sigma m_e = (m_2 - 0)_{cv}$$

Here there is no exit from the bottle so $m_e = 0$

hence, $$\Sigma m_i = m_2$$

or, $$m_i = m_2$$

Mass entered into bottle = Final mass inside the bottle

Applying the energy balance assuming change in kinetic and potential energy to be negligible, treating bottle filling process to be occurring in insulated environment, and no work interaction, we get

$Q \approx 0$, $W \approx 0$, $\Delta KE \approx 0$, $\Delta PE \approx 0$,

Initial internal energy in bottle = 0

Mass leaving = 0

$$0 = -\Sigma m_i \cdot h_i + (m_2 \cdot u_2)_{cv}$$

or $$m_i \cdot h_i = m_2 u_2$$

also $$\boxed{h_i = u_2 \text{ as } m_i = m_2}$$

Enthalpy of fluid entering bottle = Final internal energy of fluid in bottle.

If fluid is ideal gas, then $c_p \cdot T_i = c_v \cdot T_2$

or $$T_2 = \gamma \cdot T_i$$

where $$\frac{c_p}{c_v} = \gamma$$

Case 2: Let us now take a case of emptying of bottle. Arrangement is shown in Fig. 3.28.

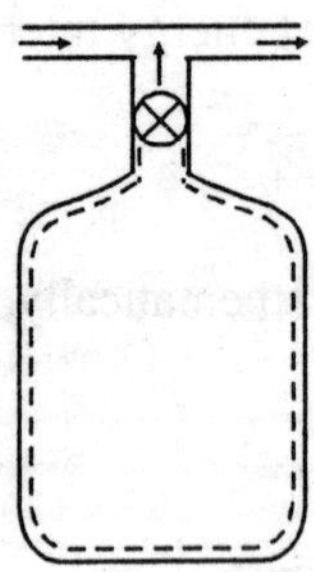

***Fig. 3.28** Emptying of bottle*

Initially bottle has mass m_1 and finally as a result of emptying, say mass left is m_2 after some time.

Applying mass balance, (as mass entering is zero),

or, $$0 - \Sigma m_e = (m_2 - m_1)_{cv}$$

or $$\Sigma m_e = (m_1 - m_2)_{cv}$$

or $$\boxed{m_e = (m_1 - m_2)_{cv}}$$

Total mass leaving the bottle = (Mass reduced in bottle)

Applying energy balance, with the assumptions given below;

(*i*) No heat interaction i.e. $Q = 0$

(*ii*) No work interaction i.e. $W = 0$

(*iii*) No change in kinetic energy i.e. $\Delta KE = 0$

(*iv*) No change in potential energy i.e. $\Delta PE = 0$

$$-\Sigma m_e \cdot h_e = (m_2 u_2 - m_1 u_1)_{cv}$$

or, $$(-m_e \cdot h_e) = (m_2 u_2 - m_1 u_1)_{cv}$$

Substituting for 'm_e' we get $(m_2 - m_1)_{cv} \cdot h_e = (m_2 u_2 - m_1 u_1)_{cv}$

In case of complete emptying, $m_2 = 0$

and so, $\boxed{h_e = u_1}$

3.11 LIMITATIONS OF FIRST LAW OF THERMODYNAMICS

First law of thermodynamics based on law of energy conservation has proved to be a powerful tool for thermodynamic analysis. But over the period of time when it was applied to some real systems, it was observed that theoretically first law stands valid for the processes which are not realizable practically. It was then thought that there exist certain flaws in first law of thermodynamics and it should be used with certain limitations.

Say for example let us take a bicycle wheel and paddle it to rotate. Now apply brake to it. As a result of braking wheel comes to rest upon coming in contact with brake shoe. Stopping of wheei is accompanied by heating of brake shoe. Examining the situation from Ist law of thermodynamics point of view it is quite satisfying that rotational energy in wheel has been transformed into heat energy with shoe, thus causing rise in its temperature:

Now, if we wish to introduce the same quantity of heat into brake shoe and wish to restore wheel motion then it is not possible simply, whereas theoretically first law permits the conversion from heat to work (rotation of wheel in this case) as well.

Therefore, it is obvious that Ist law of thermodynamics has certain limitations as given below:

(*i*) First law of thermodynamics does not differentiate between heat and work and assures full convertibility of one into other whereas full conversion of work into heat is possible but the vice-versa is not possible.

(*ii*) First law of thermodynamics does not explain the direction of a process. Such as theoretically it shall permit even heat transfer from low temperature body to high temperature body which is not practically feasible. Spontaneity of the process is not taken care of by the first law of thermodynamics.

Perpetual motion machine of the first kind (PMM-I) is a hypothetical device conceived, based on violation of First law of thermodynamics. Let us think of a system which can create energy as shown below.

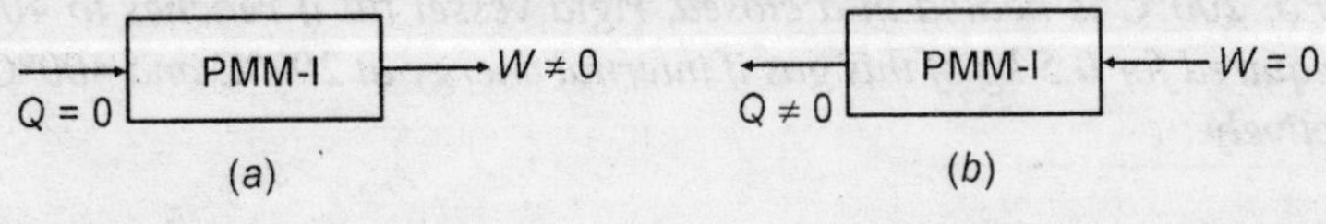

Fig. 3.29 PMM-I, based on violation of Ist law of thermodynamics

Here a device which is continuously producing work without any other form of energy supplied to it has been shown in (*a*), which is not feasible.

Similarly a device which is continuously emitting heat without any other form of energy supplied to it has been shown in (*b*), which is again not feasible.

Above two imaginary machines are called Perpetual Motion Machines of 1st kind.

EXAMPLES

1. Figure shows a system comprising of gas in cylinder at pressure of 689 kPa.

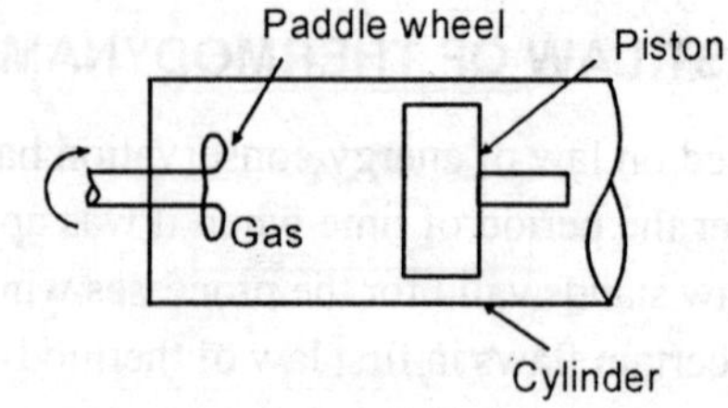

Fig. 3.30

Fluid expands from a volume of 0.04 m³ to 0.045 m³ while pressure remains constant. Paddle wheel in the system does a work of 4.88 kJ on the system. Determine (a) work done by system on the piston (b) the net amount of work done on or by the system.

Solution:

(*a*) It is a closed system. If the pressure on face of piston is uniform, then the work done on piston can be obtained as,

$$W = p \int_1^2 dV$$

$$= 689 \times 10^3 (0.045 - 0.04)$$

Work done on piston, $W = 3445$ J or 3.445 kJ

Work done on piston = **3.445 kJ Ans.**

(*b*) Paddle work done on the system = – 4.88 kJ

Net work of system ⇒ $W_{net} = W_{piston} + W_{paddle}$

$$= 3445 - 4880$$

$$W_{net} = -1435 \text{ J}$$

Work done on system = 1435 J or 1.435 kJ.

Work done on system = **1.435 kJ Ans.**

2. A gas at 65 kPa, 200°C is heated in a closed, rigid vessel till it reaches to 400°C. Determine the amount of heat required for 0.5 kg of this gas if internal energy at 200°C and 400°C are 26.6 kJ/kg and 37.8 kJ/kg respectively.

Solution:

Given $m = 0.5$ kg

$u_1 = 26.6$ kJ/kg

$u_2 = 37.8$ kJ/kg

As the vessel is rigid therefore work done shall be zero.

$$W = 0$$

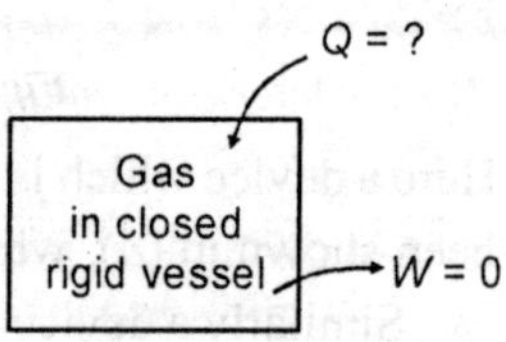

Fig. 3.31

From first law of thermodynamics;

$$Q = U_2 - U_1 + W = m(u_2 - u_1) + 0$$

$$Q = 0.5\,(37.8 - 26.6)$$

$$Q = 5.6 \text{ kJ}$$

Heat required = **5.6 kJ Ans.**

3. Carbon dioxide passing through a heat exchanger at a rate of 50 kg/hr is to be cooled down from 800°C to 50°C. Determine the rate of heat removal assuming flow of gas to be of steady and constant pressure type. Take c_p = 1.08 kJ/kg K.

Solution:

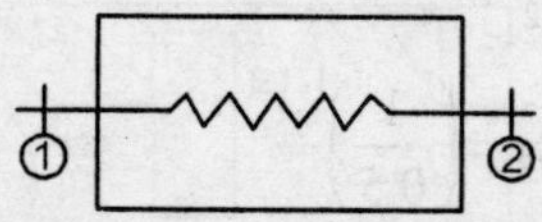

T_1 = 800 °C T_2 = 50 °C

Fig. 3.32

Given, $m = 50$ kg/hr

Writing down the steady flow energy equation.

$$q + h_1 + \frac{C_1^2}{2} + gz_1 = h_2 + \frac{C_2^2}{2} + gz_2 + w$$

Here let us assume changes in kinetic and potential energy to be negligible. During flow the work interaction shall also be zero.

Hence $q = h_2 - h_1$

or $Q = m\,(h_2 - h_1)$

$= m \cdot c_p \cdot (T_2 - T_1)$

$= 50 \times 1.08 \times (750)$

$= 40500$ kJ/hr

Heat should be removed at the rate of **40500 kJ/hr. Ans.**

4. A completely evacuated cylinder of 0.78 m³ volume is filled by opening its valve to atmosphere and air rushing into it. Determine the work done by the air and by surroundings on system.

Solution : Total work done by the air at atmospheric pressure of 101.325 kPa,

$$W = \int_{\text{cylinder}} p.dv + \int_{\text{air}} p.dv$$

$= 0 + p.\Delta v$, it is –ve work as air boundary shall contract

Work done by air = – 101.325 × 0.78 = **– 79.03 kJ**

Work done by surroundings on system = **+ 79.03 kJ** } **Ans.**

5. A system comprising of a gas of 5 kg mass undergoes expansion process from 1 MPa and 0.5 m³ to 0.5 MPa. Expansion process is governed by, $p.v^{1.3}$ = constant. The internal energy of gas is given by, u = 1.8 pv + 85, kJ/kg. Here 'u' is specific internal energy, 'p' is pressure in kPa, 'v' is specific volume in m³/kg. Determine heat and work interaction and change in internal energy.

Solution:

Given mass of gas, $m = 5$ kg, $pv^{1.3}$ = constant

Assuming expansion to be quasi-static, the work may be given as,

$$W = m\int p.dv$$

$$= \frac{p_2 V_2 - p_1 V_1}{(1-n)}$$

From internal energy relation, change in specific internal energy,

$$\Delta u = u_2 - u_1 = 1.8\,(p_2 v_2 - p_1 v_1), \text{ kJ/kg}$$

Total change, $\Delta U = 1.8 \times m \times (p_2v_2 - p_1v_1)$, kJ

$$\Delta U = 1.8 \times (p_2V_2 - p_1V_1)$$

Between states 1 and 2,

$$p_1V_1^{1.3} = p_2V_2^{1.3}$$

or

$$p_1V_1^{1.3} = p_2V_2^{1.3}$$

$$\Rightarrow \quad V_2 = (0.5) \cdot \left(\frac{1}{0.5}\right)^{1/1.3}$$

$$V_2 = 0.852 \text{ m}^3$$

Total change in internal energy, $\Delta U = -133.2$ kJ

Work,

$$W = \frac{(0.5 \times 0.852 - 1 \times 0.5) \times 10^3}{(1 - 1.3)}$$

$$W = 246.67 \text{ kJ}$$

From first law,

$$\Delta Q = \Delta U + W$$

$$= -133.2 + 246.7$$

$$\Delta Q = 113.5 \text{ kJ}$$

Heat interaction = **113.5 kJ**
Work interaction = **246.7 kJ**
Change in internal energy = **–133.2 kJ** **Ans.**

6. A gas contained in a cylinder is compressed from 1 MPa and 0.05 m³ to 2 MPa. Compression is governed by $pV^{1.4}$ = constant. Internal energy of gas is given by;

$$U = 7.5\, pV - 425, \text{ kJ.}$$

where p is pressure in kPa and V is volume in m³.

Determine heat, work and change in internal energy assuming compression process to be quasi-static.

Also find out work interaction, if the 180 kJ of heat is transferred to system between same states. Also explain, why is it different from above.

Solution:

Final state, volume

$$V_2 = \left(\frac{p_1}{p_2}\right)^{1/1.4} \cdot V_1$$

$$= \left(\frac{1}{2}\right)^{1/1.4} \cdot 0.05$$

$$V_2 = 0.03 \text{ m}^3$$

Change in internal energy,

$$\Delta U = U_2 - U_1 = (7.5\, p_2V_2 - 7.5\, p_1V_1)$$

$$= 7.5 \times 10^3 (2 \times 0.03 - 1 \times 0.05)$$

$$\Delta U = 75 \text{ kJ}$$

For quasi-static process,

Work, $$W = \int_1^2 p \cdot dV$$

$$= \frac{p_2V_2 - p_1V_1}{1-n}$$

$$= \frac{(2\times 0.03 - 1\times 0.05)\times 10^3}{(1-1.4)}$$

$$W = 25 \text{ kJ, (–ve)}$$

From first law of thermodynamics,

Heat interaction $\Delta Q = \Delta U + W$

$= 75 + (-25)$

$= 50$ kJ

Heat = **50 kJ**

Work = **25 kJ (–ve)** **Ans.**

Internal energy change = **75 kJ**

If 180 kJ heat transfer takes place, then from Ist law,

$$\Delta Q = \Delta U + W$$

Since end states remain same, therefore ΔU, i.e. change in internal energy remains unaltered.

$$180 = 75 + W$$

or $W =$ **105 kJ**

This work is different from previous work because the process is not quasi-static in this case. **Ans.**

7. Determine the heat transfer and its direction for a system in which a perfect gas having molecular weight of 16 is compressed from 101.3 kPa, 20°C to a pressure of 600 kPa following the law $pV^{1.3}$ = constant. Take specific heat at constant pressure of gas as 1.7 kJ/kg.K.

Solution:

Characteristic gas constant, $$R = \frac{\text{Universal gas constant}}{\text{Molecular weight}}$$

$$= \frac{8.3143\times 10^3}{16}, \text{ J/kg.K}$$

$= 519.64$, J/kg.K

or $= 0.51964$, kJ/kg.K

$R = 0.520$, kJ/kg.K

$C_v = C_p - R$

$= 1.7 - 0.520$

$C_v = 1.18$, kJ/kg.K

or $$\gamma = \frac{C_p}{C_v} = \frac{1.7}{1.18} = 1.44$$

For polytropic process, $$V_2 = \left(\frac{p_1}{p_2}\right)^{1/1.3} \cdot V_1$$

or $$T_2 = T_1\left(\frac{p_2}{p_1}\right)^{\frac{1.3-1}{1.3}}$$

$$T_2 = 293\cdot\left(\frac{600}{101.3}\right)^{0.231}$$

$$T_2 = 441.9 \text{ K}$$

Work, $$W = \frac{R(T_1 - T_2)}{(1.3-1)}$$

$$W = 258.1 \text{ kJ/kg}$$

For polytropic process

Heat, $$Q = \left(\frac{\gamma - n}{\gamma - 1}\right)\cdot W = \left(\frac{1.44-1.3}{1.44-1}\right) \times 258.1$$

$$= \mathbf{82.12, \text{ kJ/kg (+ve) Ans.}}$$

8. In a nozzle air at 627°C and twice atmospheric pressure enters with negligible velocity and leaves at a temperature of 27°C. Determine velocity of air at exit, assuming no heat loss and nozzle being horizontal. Take $C_p = 1.005$ kJ/kg.K for air.

Solution: Applying steady flow energy equation with inlet and exit states as 1, 2 with no heat and work interaction and no change in potential energy.

$$h_1 + \frac{C_1^2}{2} = h_2 + \frac{C_2^2}{2}$$

Given that, $C_1 \approx 0$, negligible inlet velocity

$$C_2 = \sqrt{2(h_1 - h_2)}$$

Exit velocity, $$C_2 = \sqrt{2\cdot C_p\cdot(T_1 - T_2)}$$

Given, $T_1 = 900$ K, $T_2 = 300$ K

or $$C_2 = \sqrt{2\times 1.005\times 10^3 (900-300)}$$

$$C_2 = 1098.2 \text{ m/s}$$

Exit velocity = **1098.2 m/s. Ans.**

9. An air compressor requires shaft work of 200 kJ/kg of air and the compression of air causes increase in enthalpy of air by 100 kJ/kg of air. Cooling water required for cooling the compressor picks up heat of 90 kJ/kg of air. Determine the heat transferred from compressor to atmosphere.

Solution:

Work interaction, $W = -200$ kJ/kg of air

Increase in enthalpy of air = 100 kJ/kg of air

Total heat interaction, Q = Heat transferred to water + Heat transferred to atmosphere.

Writing steady flow energy equation on compressor, for unit mass of air entering at 1 and leaving at 2.

$$h_1 + \frac{C_1^2}{2} + gZ_1 + Q = h_2 + \frac{C_2^2}{2} + gZ_2 + W$$

Assuming no change in potential energy and kinetic energy;

$$\Delta KE = \Delta PE = 0$$

$$Q = (h_2 - h_1) + W$$

$$= 100 + (-200)$$

$$Q = -100 \text{ kJ/kg of air}$$

Also

$$Q = -100 = \text{Heat transferred to water} + \text{Heat transferred to atmosphere}$$

or

$$-100 = (-90) + \text{Heat transferred to atmosphere}$$

Heat transferred to atmosphere = **– 10 kJ/kg** of air **Ans.**

10. *In a cinema hall with seating capacity of 500 persons the comfort conditions are created by circulating hot water through pipes in winter season. Hot water enters the pipe with enthalpy of 80 kcal/kg and leaves the pipe with enthalpy of 45 kcal/kg. The difference in elevation of inlet pipe and exit pipe is 10 m with exit pipe being higher than inlet pipe. Heat requirement per person is 50 kcal/hr. Estimate the quantity of water circulated per minute, neglecting changes in velocity.*

Solution:

Above problem can be solved using steady flow energy equations upon hot water-flow,

$$Q + m_1\left(h_1 + \frac{C_1^2}{2} + gz_1\right) = W + m_2.\left(h_2 + \frac{C_2^2}{2} + gz_2\right)$$

Here total heat to be supplied $= 500 \times 50$

Heat lost by water (–ve), $Q = -25000$ kcal/hr.

There shall be no work interaction and change in kinetic energy, so, steady flow energy equation shall be,

$$Q + m_1(h_1 + gz_1) = m_2(h_2 + gz_2)$$

Here $m_1 = m_2 = m$, $h_1 = 80$ kcal/kg, $h_2 = 45$ kcal/kg

$$Q + m(h_1 - h_2) = m(gz_2 - gz_1)$$

$$-(25000 \times 10^3 \times 4.18) + m(80 - 45) \times 10^3 \times 4.18$$

$$= m \times (9.81 \times 10)$$

$$m = 714.76 \text{ kg/hr}$$

or, Water circulation rate = **11.91 kg/min. Ans.**

11. *Figure 3.33 shows steam injector for lifting water from a depth of 2 m from axis of injector. Determine the rate at which steam should be supplied for pumping unit mass of water. Enthalpy of steam entering and water entering injector are 72.0 kcal/kg and 24.6 kcal/kg respectively.*

Enthalpy of steam and water mixture leaving injector is 100 kcal/kg. Velocity of steam entering injector is 50 m/s while mixture leaves with 25 m/s. Heat loss from injector to surroundings is 12 kcal/kg of steam sent through injector.

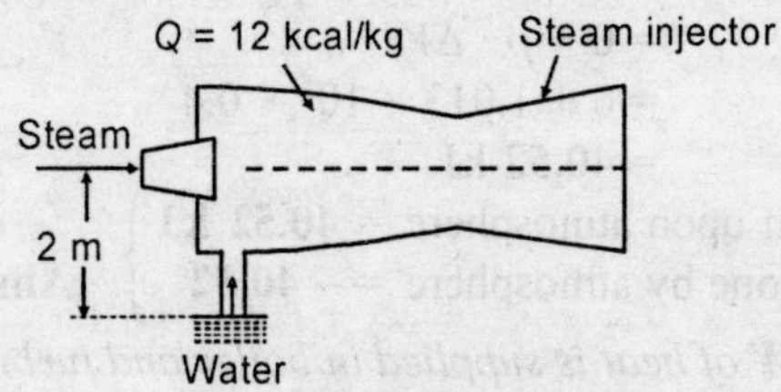

Fig. 3.33

Solution:

Let mass of steam to be supplied per kg of water lifted be 'm' kg. Applying law of energy conservation upon steam injector, for unit mass of water lifted.

Energy with steam entering + Energy with water entering = Energy with mixture leaving + Heat loss to surroundings.

$$\underbrace{m\left[\frac{(50)^2}{2}\right.}_{\text{K.E.}} + \underbrace{\left.(720\times10^2\times4.18)\right]}_{\text{Enthalpy}} + 1 \times \underbrace{[(24.6 \times 10^3 \times 4.18)}_{\text{Enthalpy}}$$

$$\underbrace{+ (9.81 \times 2)]}_{\text{P.E}} = (1 + m)\left[\underbrace{(100\times10^3\times4.18)}_{\text{Enthalpy}} + \underbrace{\frac{(25)^2}{2}}_{\text{K.E}}\right]$$

$$+ \underbrace{[m \times 12 \times 10^3 \times 4.18]}_{\text{Heat loss}}$$

$m\,[3010850] + [102847.62] = (1 + m)\,.\,(418312.5) + m[50160]$

Upon solving, $m = 0.124$ kg steam/kg of water

Steam supply rate = **0.124 kg/s** per kg of water. **Ans.**

12. An inelastic flexible balloon is inflated from initial empty state to a volume of 0.4 m³ with H_2 available from hydrogen cylinder. For atmospheric pressure of 1.0313 bar determine the amount of work done by balloon upon atmosphere and work done by atmosphere.

Solution:

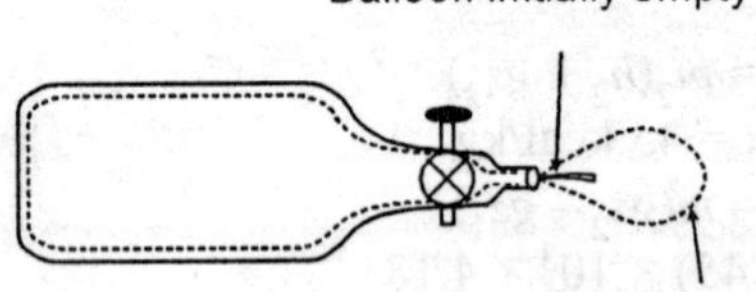

Fig. 3.34

Here let us assume that the pressure is always equal to atmospheric pressure as balloon is flexible, inelastic and unstressed and no work is done for stretching balloon during its filling. Figure 3.34 shows the boundary of system before and after filling balloon by firm line and dotted line respectively.

Displacement work, $W = \int_{\text{cylinder}} p.dV + \int_{\text{balloon}} p.dV$

$\int p.dV = 0$ as cylinder shall be rigid.

$= 0 + p \cdot \Delta V$

$= 0 + 1.013 \times 10^5 \times 0.4$

$= 40.52$ kJ

Work done by system upon atmosphere = **40.52 kJ**
Work done by atmosphere = – **40.52** **Ans.**

13. In a steam power plant 5 kW of heat is supplied in boiler and turbine produces 25% of heat added while 75% of heat added is rejected in condenser. Feed water pump consumes 0.2% of this heat added for pumping condensate to boiler. Determine the capacity of generator which could be used with this plant.

Solution:

Given, $Q_{add} = 5000$ J/s

so, $W_T = 0.25 \times 5000 = 1250$ J/s

$$Q_{rejected} = 0.75 \times 5000 = 3750 \text{ J/s}$$

$$W_p = (-)\ 0.002 \times 5000 = 10 \text{ J/s}$$

$$\text{Capacity of generator} = W_T - W_P$$

$$= 1250 - 10$$

$$= 1240 \text{ J/s or } 1240 \text{ W}$$

$$= \mathbf{1.24\ kW\ Ans.}$$

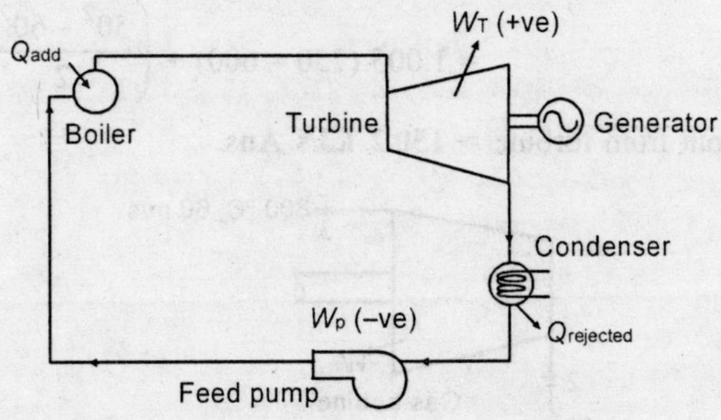

Fig. 3.35

***14.** In a gas turbine installation air is heated inside heat exchanger upto 750°C from ambient temperature of 27°C. Hot air then enters into gas turbine with the velocity of 50 m/s and leaves at 600°C. Air leaving turbine enters a nozzle at 60 m/s velocity and leaves nozzle at temperature of 500°C. For unit mass flow rate or air determine the following assuming adiabatic expansion in turbine and nozzle,*

(a) heat transfer to air in heat exchanger

(b) power output from turbine

(c) velocity at exit of nozzle.

Take c_p for air as 1.005 kJ/kg°K.

Solution:

In heat exchanger upon applying S.F.E.E. with assumptions of no change in kinetic energy, no work interaction, no change in potential energy, for unit mass flow rate of air,

$$h_1 + Q_{1-2} = h_2$$

$$Q_{1-2} = h_2 - h_1$$

$$Q_{1-2} = C_p \cdot (T_2 - T_1)$$

Heat transfer to air in heat exchanger Q_{1-2} = **726.62 kJ Ans.**

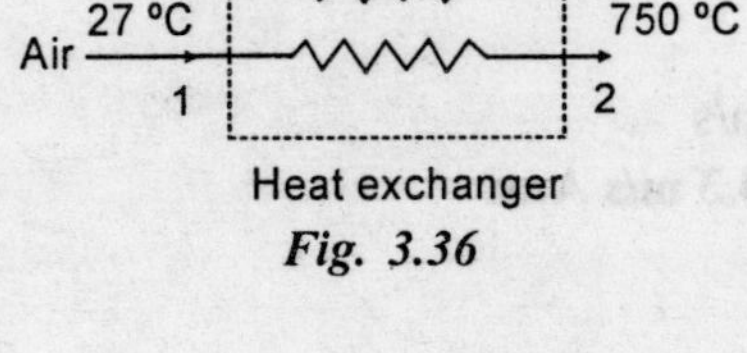

Fig. 3.36

In gas turbine let us use S.F.E.E., assuming no change in potential energy, for unit mass flow rate of air

$$h_2 + \frac{C_2^2}{2} = h_3 + \frac{C_3^2}{2} + W_T$$

$$W_T = (h_2 - h_3) + \left(\frac{C_2^2 - C_3^2}{2}\right)$$

$$= C_p(T_2 - T_3) + \left(\frac{C_2^2 - C_3^2}{2}\right)$$

$$= 1.005\ (750 - 600) + \left(\frac{50^2 - 60^2}{2}\right) \times 10^{-3}$$

Power output from turbine = **150.2 kJ/s Ans.**

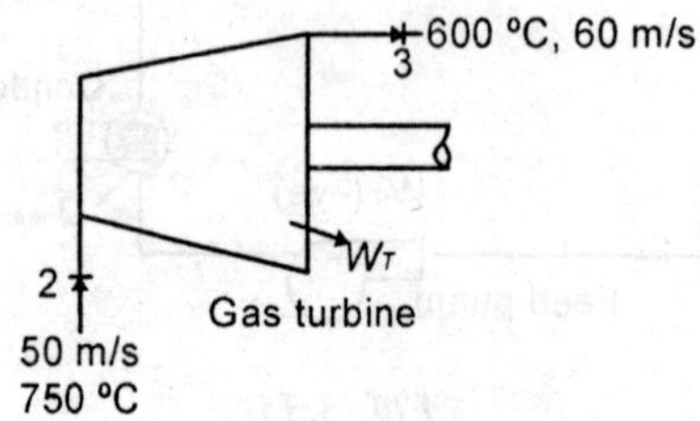

Fig. 3.37

Applying S.F.E.E. upon nozzle assuming no change in potential energy, no work and heat interactions, for unit mass flow rate,

$$h_3 + \frac{C_3^2}{2} = h_4 + \frac{C_4^2}{2}$$

$$\frac{C_4^2}{2} = (h_3 - h_4) + \frac{C_3^2}{2}$$

$$= C_p(T_3 - T_4) + \frac{C_3^2}{2}$$

$$= 1.005\ (600 - 500) + \left(\frac{60^2}{2}\right) \times 10^{-3}$$

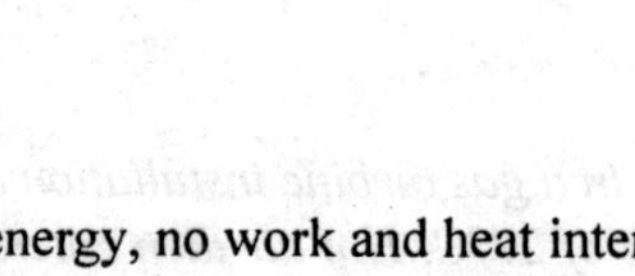

$$\frac{C_4^2}{2} = 102.3$$

$$C_4 = 14.3 \text{ m/s}$$

Velocity at exit of nozzle = **14.3 m/s Ans.**

Fig. 3.38

15. One mol of air at 0.5 MPa and 400 K, initially undergoes following processes, sequentially

(a) heating at constant pressure till the volume gets doubled.

(b) expansion at constant temperature till the volume is six times of initial volume.

Determine the work done by air.

Solution:

For constant pressure heating, say state changes from 1 to 2

$$W_a = \int_1^2 p_1 \, dV$$

$$W_a = p_1 (V_2 - V_1)$$

It is given that $V_2 = 2V_1$

so $W_a = p_1 V_1$

$$W_a = RT_1$$

For subsequent expansion at constant temperature say state changes from 2 to 3.

Also given that $\dfrac{V_3}{V_1} = 6$, so $\dfrac{V_3}{V_2} = 3$

Work, $W_b = \displaystyle\int_2^3 p dV$

$$= \int_2^3 \frac{RT}{V} dV = RT_2 \ln \frac{V_3}{V_2}$$

$$W_b = RT_2 \ln (3)$$

Temperature at 2 can be given by perfect gas considerations as,

$$\frac{T_2}{T_1} = \frac{V_2}{V_1}$$

or $T_2 = 2\,T_1$

Total work done by air, $W = W_a + W_b$

$$= RT_1 + RT_2 \ln (3)$$

$$= RT_1 + 2RT_1 \ln (3)$$

$$= RT_1 (1 + 2 \ln 3) = 8.314 \times 400 \, (1 + 2 \ln 3)$$

Work done = **10632.69 kJ Ans.**

16. Determine the work done by gas for the arrangement shown in Fig. 3.39. Here spring exerts a force upon piston which is proportional to its deformation from equilibrium position. Spring gets deflected due to heating of gas till its volume becomes thrice of original volume. Initial states are 0.5 MPa and 0.5 m^3 while final gas pressure becomes 1 MPa. Atmospheric pressure may be taken as 1.013×10^5 Pa.

Solution:

Let stiffness of spring be k and it undergoes a deflection by 'x' along x-axis. Force balance at any equilibrium position of piston shall be,

$$p \cdot A = p_{atm} \cdot A + kx,$$

here x shall be linear displacement of piston due to expansion of gas. Let volume of gas change from V_0 to some value V. Then, $x = \dfrac{V - V_0}{A}$, V_0 is volume of gas when spring is at its natural length.

$$p \cdot A = p_{atm} \cdot A + k\left(\frac{V - V_0}{A}\right)$$

or $$(p - p_{atm}) = \frac{k(V - V_0)}{A^2}$$

Work done by gas between initial and final states, $W = \int_i^f p.dV$

$$W = p_{atm}(V_f - V_i) + \frac{k}{A^2}\left\{\frac{V^2}{2} - V_0.V\right\}_i^f$$

$$= p_{atm}(V_f - V_i) + \frac{k}{A^2}\left\{\frac{V_f^2 - V_i^2}{2} - V_0 V_f + V_0.V_i\right\}$$

$$= (V_f - V_i)\, p_{atm} + \frac{k}{A^2}\left\{\left(\frac{V_f^2 - V_i^2 - 2V_0 V_f + 2V_0.V_i}{2}\right)\right\}$$

$$= (V_f - V_i)\, p_{atm} + \left\{\frac{k}{2A^2}((V_f - V_0) + (V_i - V_0))(V_f - V_i)\right\}$$

$$= (V_f - V_i)\left\{P_{atm}\,\frac{k}{2A^2}((V_f - V_0) + (V_i - V_0))\right\}$$

from above force balance, $(p_f - p_{atm}) = \frac{k}{A^2}(V_f - V_0)$

$$(p_i - p_{atm}) = \frac{k}{A^2}(V_i - V_0)$$

or $$W = (V_f - V_i)\left\{p_{atm} + \left(\frac{p_f - p_{atm}}{2}\right) + \left(\frac{p_i - p_{atm}}{2}\right)\right\}$$

$$= (V_f - V_i)\left\{\frac{p_i + p_f}{2}\right\}$$, substituting pressure and volume values, $V_f = 3V_i$

$W = 0.75 \times 10^6$ J Ans.

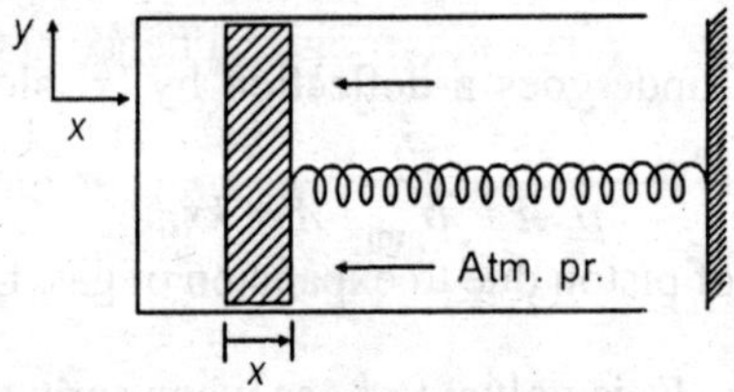

Fig. 3.39

17. A closed insulated container has frictionless and smooth moving insulated partition as shown in Fig. 3.40 such that it equally divides total 1 m³ of volume, when both the gases are at initial pressure of 0.5 MPa and ambient temperature of 27°C. Subsequently the nitrogen is heated using electrical heating element such that volume of N_2 becomes 3/4 of total volume of container. Determine, (i) final pressure of hydrogen, (ii) Work done by partition, (iii) Workdone by N_2 and H_2 (iv) Heat added to N_2 by electric heater.

Take $C_{p, N_2} = 1.039$ kJ/kg. K, $C_{p, H_2} = 14.307$ kJ/kg . K,

$R_{N_2} = 0.2968$ kJ/kg . K $R_{H_2} = 4.1240$ kJ/kg . K

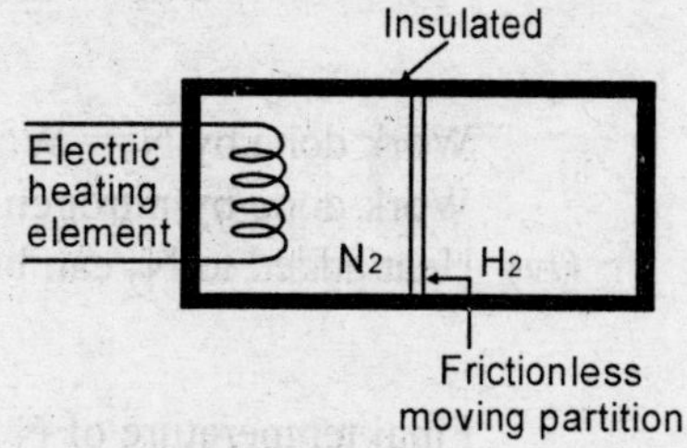

Fig. 3.40

Solution:

With the heating of N_2 it will get expanded while H_2 gets compressed simultaneously. Compression of H_2 in insulated chamber may be considered of adiabatic type.

Adiabatic Index of compression for H_2 can be obtained as,

$$C_{p,H_2} = R_{H_2}\left(\frac{\gamma_{H_2}}{\gamma_{H_2}-1}\right)$$

$$14.307 = 4.124\left(\frac{\gamma_{H_2}}{\gamma_{H_2}-1}\right)$$

$$\gamma_{H_2} = 1.405$$

Adiabatic Index of expansion for N_2, $C_{p,N_2} = R_{N_2}\left(\frac{\gamma_{N_2}}{\gamma_{N_2}-1}\right)$

$$1.039 = 0.2968\left(\frac{\gamma_{N_2}}{\gamma_{N_2}-1}\right)$$

$$\gamma_{N_2} = 1.399$$

(*i*) For hydrogen, $p_1 V_1^\gamma = p_2 V_2^\gamma$

Here $\gamma = \gamma_{H_2} = 1.405$, $V_1 = 0.5$ m³

$p_1 = 0.5 \times 10^6$ Pa, $V_2 = 0.25$ m³

$$\text{Final pressure of } H_2 = 0.5 \times 10^6 \left(\frac{0.5}{0.25}\right)^{1.405}$$

$$= \mathbf{1.324\ MPa\ Ans.}$$

(*ii*) since partition remains in equilibrium throughout hence no work is done by partition. It is a case similar to free expansion.

Partition work = 0. Ans.

(*iii*) Work done upon H_2,

$$W_{H_2} = \frac{P_1V_1 - p_2V_2}{(\gamma_{H_2}-1)}$$

Here $p_1 = 0.5 \times 10^6$ Pa, $p_2 = 1.324 \times 10^6$ Pa, $V_1 = 0.5$ m³, $V_2 = 0.25$ m³.

$$\text{Work done by hydrogen, } W_{H_2} = \frac{(-)0.081\times10^6}{0.405}$$

$$= (-)\ 2 \times 10^5 \text{ J } \textbf{Ans.}$$

Work done by N_2 = Work done upon H_2

Work done by nitrogen = **+ 2 × 10⁵ J Ans.**

(iv) Heat added to N_2 can be obtained using first law of thermodynamics as

$$Q_{N_2} = \Delta U_{N_2} + W_{N_2} \Rightarrow Q_{N_2} = mc_v(T_2 - T_1) + W_{N_2}$$

Final temperature of N_2 can be obtained considering it as perfect gas.

$$\text{Therefore,} \quad T_2 = \frac{p_2 V_2 T_1}{p_1 V_1}$$

p_2 = Final pressure of N_2 which will be equal to that of H_2 as the partition is free and frictionless.

$$p_2 = 1.324 \times 10^6 \text{ Pa.}$$

$$T_2 = \text{Final temperature of } N_2 = \frac{1.324\times10^6\times0.75\times300}{0.5\times10^6\times0.5}$$

$$= 1191.6 \text{ K}$$

$$\text{mass of } N_2,\ m = \frac{p_1V_1}{RT_1} = \frac{0.5\times10^6\times0.5}{0.2968\times10^3\times300} = 2.8 \text{ kg.}$$

Specific heat at constant volume, $C_v = C_p - R \Rightarrow C_{v,N_2} = 0.7422$ kJ/kg . K.

Heat added to N_2, $Q_{N_2} = \{2.8 \times (1191.6 - 300) \times 0.7422 \times 10^3\} + 2 \times 10^5$

= 2052.9 kJ Ans.

18. *A cylinder of 2 m³ has air at 0.5 MPa and temperature of 375°K. Air is released in atmosphere through a valve on cylinder so as to run a frictionless turbine. Find the amount of work available from turbine assuming no heat loss and complete kinetic energy being used for running turbine. Take $C_{p,\,air}$ = 1.003 kJ/kg . K, $C_{v,\,air}$ = 0.716 kJ/kg . K, R_{air} = 0.287 kJ/kg . K.*

Solution:

Let initial states and final states of air inside cylinder be given by m_1, p_1, V_1, T_1, and m_2, p_2, V_2, T_2 respectively. It is a case of emptying of cylinder.

$$\text{Initial mass of air,} \quad m_1 = \frac{p_1V_1}{P_{air}\cdot T_1} = 9.29 \text{ kg.}$$

For adiabatic expansion during release of air through valve from 0.5 MPa to atmospheric pressure.

$$T_2 = T_1\left(\frac{p_2}{p_1}\right)^{\frac{\gamma-1}{\gamma}}$$

$$= 375\left(\frac{1.013\times10^5}{0.5\times10^6}\right)^{\frac{1.4-1}{1.4}}$$

$$T_2 = 237.65 \text{ K}$$

Final mass of air left in tank, $m_2 = \dfrac{p_2V_2}{RT_2}$

$$m_2 = \frac{1.013\times10^5\times2}{(0.287\times10^3\times237.65)} = 2.97 \text{ kg}$$

Writing down energy equation for unsteady flow system

$$(m_1 - m_2)\left(h_2 + \frac{C^2}{2}\right) = m_1u_1 - m_2u_2$$

$$(m_1 - m_2)\frac{C^2}{2} = (m_1\,u_1 - m_2u_2) - (m_1 - m_2)h_2$$

Kinetic energy available for running turbine

$= (m_1\,C_v\,T_1 - m_2\,C_v\,T_2) - (m_1 - m_2)\cdot C_p\cdot T_2$

$= (9.29 \times 0.716 \times 10^3 \times 375) - (2.97 \times 0.716 \times 10^3 \times 237.65) - \{(9.29 - 2.97) \times 1.003 \times 10^3 \times 237.65\}$

$= 482.54 \times 10^3$ J

Amount of work available = **482.54 kJ Ans.**

19. *A rigid and insulated tank of 1 m³ volume is divided by partition into two equal volume chambers having air at 0.5 MPa, 27°C and 1 MPa, 500 K. Determine final pressure and temperature if the partition is removed.*

Solution:

Using perfect gas equation for the two chambers having initial states as 1 and 2 and final state as 3.

$p_1 = 0.5 \times 10^6$ Pa, $V_1 = 0.5$ m³, $T_1 = 300$ K

$$n_1 = \frac{p_1V_1}{\bar{R}T_1}$$

$$= \frac{0.5\times10^6\times0.5}{8314\times300}$$

$$n_1 = 0.1002$$

and $n_2 = \dfrac{p_2V_2}{\bar{R}T_2}$ where $p_2 = 1 \times 10^6$ Pa, $V_2 = 0.5$ m³, $T_2 = 500$ K

$$= \frac{1\times10^6\times0.5}{8314\times500}$$

$$n_2 = 0.1203$$

For tank being insulated and rigid we can assume, $\Delta U = 0$, $W = 0$, $Q = 0$, so writing ΔU,

$$\Delta U = n_1\,C_v(T_3 - T_1) + n_2\,C_v(T_3 - T_2) = 0$$

or $T_3 = 409.11$ K

Using perfect gas equation for final mixture,

$$p_3 = \frac{(n_1 + n_2)\bar{R}T_3}{(V_1 + V_2)} = 0.75 \text{ MPa}$$

Final pressure and temperature = **0.75 MPa, 409.11 K. Ans.**

20. An evacuated bottle of 0.5 m³ volume is slowly filled from atmospheric air at 1.0135 bars until the pressure inside the bottle also becomes 1.0135 bar. Due to heat transfer, the temperature of air inside the bottle after filling is equal to the atmospheric air temperature. Determine the amount of heat transfer.

[U.P.S.C., 1994]

Solution:

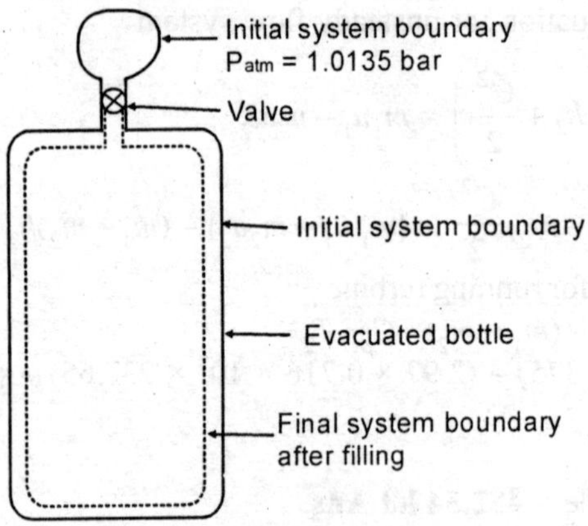

Fig. 3.41

Displacement work; $W = 1.0135 \times 10^5 \times (0 - 0.5)$

$W = -\,0.50675 \times 10^5$ N.m.

Heat transfer, $Q = 0.50675 \times 10^5$ N.m.

Heat transfer = 0.50675×10^5 N.m. **Ans.**

21. A compressed air bottle of 0.3 m³ volume contains air at 35 bar, 40°C. This air is used to drive a turbogenerator sypplying power to a device which consumes 5 W. Calculate the time for which the device can be operated if the actual output of the turbogenerator is 60% of the maximum theoretical output. The ambient pressure to which the tank pressure has fallen is 1 bar. For air, $\frac{C_p}{C_v} = 1.4$.

[U.P.S.C. 1993]

Solution:

Here turbogenerator is fed with compressed air from a compressed air bottle. Pressure inside bottle gradually decreases from 35 bar to 1 bar. Expansion from 35 bar to 1 bar occurs isentropically. Thus, for the initial and final states of pressure, volume, temperature and mass inside bottle being given as P_1, V_1, T_1 & m_1 and P_2, V_2, T_2 & m_2 respectively. It is transient flow process similar to emptying of the bottle.

$$\left(\frac{P_2}{T_1}\right)^{\frac{\gamma-1}{\gamma}} = \frac{T_2}{T_1}, \text{ Given: } P_1 = 35 \text{ bar}, T_1 = 40°\text{C or } 313 \text{ K}$$

$V_1 = 0.3$ m³; $V_2 = 0.3$ m³

$P_2 = 1$ bar.

$$T_2 = T_1 \left(\frac{P_2}{T_1}\right)^{\frac{\gamma-1}{\gamma}}$$

$$T_2 = 113.22 \text{ K}$$

By perfect gas law, initial mass in bottle, $m_1 = \dfrac{P_1V_1}{RT_1} = \dfrac{35\times 10^2 \times 0.3}{0.287\times 313}$

$$m_1 = 11.68 \text{ kg}$$

Final mass in bottle, $m_2 = \dfrac{P_2V_2}{RT_2} = \dfrac{1\times 10^2 \times 0.3}{0.287\times 113.22}$

$$m_2 = 0.923 \text{ kg}$$

Energy available for running turbo generator or work;

$$W + (m_1 - m_2)\, h_2 = m_1\, u_1 - m_2\, u_2$$

$$W = (m_1u_1 - m_2u_2) - (m_1 - m_2)\, h_2$$

$$= (m_1\, c_v\, T_1 - m_2\, c_v\, T_2) - (m_1 - m_2)\cdot c_p \cdot T_2$$

Taking $c_v = 0.718$ kJ/kg . K and $c_p = 1.005$ kJ/kg · K

$$W = \{(11.68 \times 0.718 \times 313) - (0.923 \times 0.718 \times 113.22)\}$$
$$- \{(11.68 - 0.923) \times 1.005 \times 113.22\}$$

$$W = 1325.86 \text{ kJ.}$$

This is the maximum work that can be had from the emptying of compressed air bottle between given pressure limits.

Turbogenerator's actual output = 5 kJ/s

Input to turbogenerator = $\dfrac{5}{0.6} = 8.33$ kJ/s.

Time duration for which turbogenerator can be run;

$$\Delta t = \frac{1325.86}{8.33}$$

$$\Delta t = 159.17 \text{ sec.}$$

Duration ≈ **160 seconds**	**Ans.**

22. 3 kg of air at 1.5 bar pressure and 77°C temperature at state 1 is compressed polytropically to state 2 at pressure 7.5 bar, index of compression being 1.2. It is then cooled at constant temperature to its original state 1. Find the net work done and heat transferred. *[U.P.S.C. 1992]*

Solution:

Different states as described in the problem are denoted as 1, 2 and 3 and shown on p-V diagram.

Process 1-2 is polytropic process with index 1.2

So, $$\frac{T_2}{T_1} = \left(\frac{P_2}{P_1}\right)^{\frac{n-1}{n}}$$

or, $$T_2 = T_1 \left(\frac{P_2}{P_1}\right)^{\frac{n-1}{n}}$$

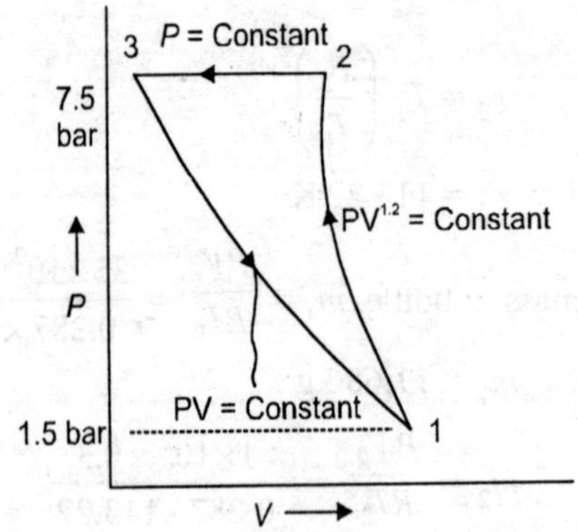

Fig. 3.42

$$= 350 \cdot \left(\frac{7.5}{1.5}\right)^{\frac{1.2-1}{1.2}}$$

$$T_2 = 457.68 \text{ K}$$

At state 1, $$P_1V_1 = mRT_1$$

$$\frac{1.5\times10^5}{10^3} \cdot V_1 = 3 \times 0.287 \times 350$$

or, $$V_1 = 2.009 \approx 2.01 \text{ m}^3$$

For process 1-2, $$V_2^{1.2} = \frac{P_1V_1^{1.2}}{P_2},\ V_2 = \left\{\frac{1.5\times(2.01)^{1.2}\times10^5}{7.5\times10^5}\right\}^{\frac{1}{1.2}}$$

or, $$V_2 = 0.526 \text{ m}^3$$

Process 2-3 is constant pressure process, so $\frac{P_2V_2}{T_2} = \frac{P_3V_3}{T_3}$ gets modified as,

$\Rightarrow$ $$V_3 = \frac{P_2 \cdot T_3}{T_2}$$

Here process 3-1 is isothermal process, so $T_1 = T_3$

or, $$V_3 = \frac{0.526\times350}{457.68}$$

or, $$V_3 = 0.402 \text{ m}^3$$

During process 1-2 the compression work;

$$W_{1-2} = \frac{m.R.(T_1 - T_2)}{1-n}$$

$$= \frac{3\times0.287(457.68-350)}{(1-1.2)}$$

$$W_{1-2} = -463.56 \text{ kJ}$$

Work during process 2-3,

$$W_{2-3} = P_2 (V_3 - V_2)$$
$$= 7.5 \times 10^5 (0.402 - 0.526)$$
$$= -93 \text{ kJ}$$

Work during process 3-1,

$$W_{3-1} = P_3 V_3 \ln \left(\frac{V_1}{V_3}\right) = 7.5 \times 10^5 \times 0.402 \times \ln \left(\frac{2.01}{0.402}\right)$$

$$W_{3-1} = 485.25 \text{ kJ}$$

Net work,
$$W_{net} = W_{1-2} + W_{2-3} + W_{3-1}$$
$$= -463.56 - 93 + 485.25$$

Network = – 71.31 kJ **Ans.**

–ve work shows work done upon the system. Since it is the cycle, so

$$W_{net} = Q_{net}$$

$$\oint dW = \oint dQ = -71.31 \text{ kJ}$$

Heat transferred from system = 71.31 kJ **Ans.**

23. A compressed air bottle of volume 0.15 m^3 contains air at 40 bar and 27°C. It is used to drive a turbine which exhausts to atmosphere at 1 bar. If the pressure in the bottle is allowed to fall to 2 bar, determine the amount of work that could be delivered by the turbine. *[U.P.S.C. 1998]*

Solution:

$$c_p = 1.005 \text{ kJ/kg . K}, c_v = 0.718 \text{ kJ/kg . K}, \gamma = 1.4$$

Initial mass of air in bottle $\Rightarrow m_1 = \dfrac{p_1 V_1}{RT_1} = \dfrac{40 \times 10^2 \times 0.15}{0.287 \times 300}$

$$m_1 = 6.97 \text{ kg}$$

Final mass of air in bottle $\Rightarrow m_2 = \dfrac{p_2 V_2}{RT_2}$

$$\frac{T_2}{T_1} = \left(\frac{P_2}{P_1}\right)^{\frac{\gamma-1}{\gamma}}, \quad m_2 = \frac{2 \times 10^2 \times 0.15}{0.287 \times 127.36}$$

$$= \left(\frac{2}{40}\right)^{\frac{1.4-1}{1.4}}, \quad m_2 = 0.821 \text{ kg.}$$

$$T_2 = 127.36 \text{ K}$$

Energy available for running of turbine due to emptying of bottle,

$$= (m_1 \, c_v \, T_1 - m_2 \, c_v \, T_2) - (m_1 - m_2) \cdot c_p \cdot T_2$$
$$= \{(6.97 \times 0.718 \times 300) - (0.821 \times 0.718 \times 127.36)\}$$
$$- \{(6.97 - 0.821) \times 1.005 \times 127.35\}$$
$$= 639.27 \text{ kJ.}$$

Work available from turbine = 639.27 kJ **Ans.**

EXERCISE

3.1 Define the first law of thermodynamics. Also give supporting mathematical expression for it.
3.2 How the first law of thermodynamics is applied to a closed system undergoing a non-cyclic process?
3.3 Show that internal energy is a property.
3.4 Explain the following :
(*a*) Free expansion
(*b*) Polytropic process
(*c*) Hyperbolic process
Also obtain expressions for work in each case.
3.5 Show that for a polytropic process.

$$Q = \left(\frac{\gamma - n}{\gamma - 1}\right) W$$

where Q and W are heat and work interactions and n is polytropic index.
3.6 Derive the steady flow energy equation.
3.7 Explain a unsteady flow process.
3.8 Show that for a given quantity of air supplied with a definite amount of heat at constant volume, the rise in pressure shall be directly proportional to initial absolute pressure and inversely proportional to initial absolute temperature.
3.9 How much work is done when 0.566 m^3 of air initially at a pressure of 1.0335 bar and temperature of 7°C undergoes an increase in pressure upto 4.13 bar in a closed vessel? [0]
3.10 An ideal gas and a steel block having same volumes at same temperature and pressure. Pressure on both is increased isothermally to five times of its initial value. Show with the help of P–V diagram, whether the quantities of work shall be same in two processes or different. If different then which one is greater. Assume processes to be quasi-static.
3.11 An inventor has developed an engine getting 1055 MJ from fuel and rejecting 26.375 MJ in exhaust and delivering 25 kWh of mechanical work. Is this engine possible? [No]
3.12 For an ideal gas the pressure is increased isothermally to 'n' times its initial value. How high would the gas be raised if the same amount of work were done in lifting it? Assume process to be quasi-static.
3.13 A system's state changes from a to b as shown on P–V diagram

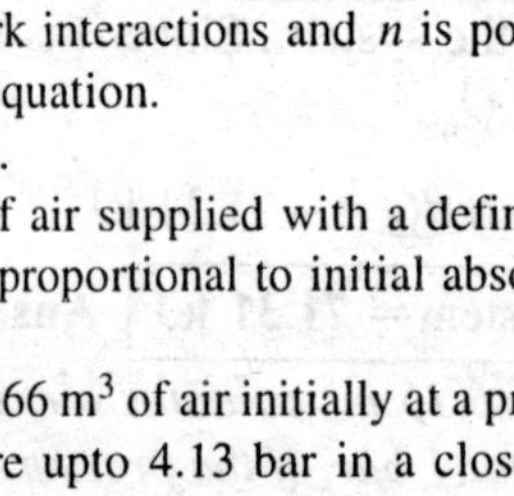

Fig. 3.43

Along path '*acb*' 84.4 kJ of heat flows into the system and system does 31.65 kJ of work. Determine heat flow into the system along path '*adb*' if work done is 10.55 kJ. When system returns from '*b*' to '*a*' following the curved path then work done on system is 21.1 kJ. How much heat is absorbed or rejected? If internal energy at '*a*' and '*d*' are 0 and 42.2 kJ, find the heat absorbed in processes '*ad*' and '*db*'. [63.3 kJ, – 73.85 kJ, 52.75 kJ, 10.55 kJ]
3.14 A tank contains 2.26 m^3 of air at a pressure of 24.12 bar. If air is cooled until its pressure and temperature becomes 13.78 bar and 21.1°C respectively. Determine the decrease of internal energy. [– 5857.36 kJ]

3.15 Water in a rigid, insulating tank is set in rotation and left. Water comes to rest after some time due to viscous forces. Considering the tank and water to constitute the system answer the following.

(*i*) Is any work done during the process of water coming to rest?

(*ii*) Is there a flow of heat?

(*iii*) Is there any change in internal energy (*U*)?

(*iv*) Is there any change in total energy (*E*)? [No, No, Yes, No]

3.16 Fuel-air mixture in a rigid insulated tank is burnt by a spark inside causing increase in both temperature and pressure. Considering the heat energy added by spark to be negligible, answer the following :

(*i*) Is there a flow of heat into the system?

(*ii*) Is there any work done by the system?

(*iii*) Is there any change in internal energy (*U*) of system?

(*iv*) Is there any change in total energy (*E*) of system? [No, No, No, No]

3.17 Calculate the work if in a closed system the pressure changes as per relation $p = 300 \,.\, V + 1000$ and volume changes from 6 to 4 m^3. Here pressure '*p*' is in Pa and volume '*V*' is in m^3. [–5000J]

3.18 Hydrogen from cylinder is used for inflating a balloon to a volume of 35m^3 slowly. Determine the work done by hydrogen if the atmospheric pressure is 101.325 kPa. [3.55 MJ]

3.19 Show that the work done by an ideal gas is mRT_1, if gas is heated from initial temperature T_1 to twice of initial temperature at constant volume and subsequently cooled isobarically to initial state.

3.20 Derive expression for work done by the gas in following system. Piston-cylinder device shown has a gas initially at pressure and volume given by P_1, V_1. Initially the spring does not exert any force on piston. Upon heating the gas, its volume gets doubled and pressure becomes P_2.

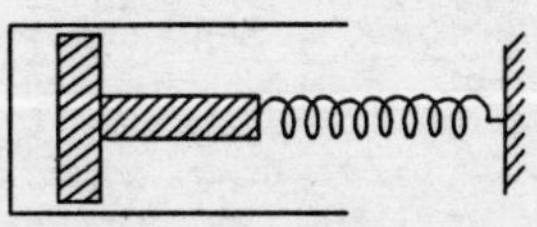

Fig. 3.44 Piston-cylinder arrangement

3.21 An air compressor with pressure ratio of 5, compresses air to $\frac{1}{4}$th of the initial volume. For inlet temperature to be 27°C determine temperature at exit and increase in internal energy per kg of air. [101.83°C, 53.7 kJ/kg]

3.22 In a compressor the air enters at 27°C and 1 atm and leaves at 227°C and 1 MPa. Determine the work done per unit mass of air assuming velocities at entry and exit to be negligible. Also determine the additional work required, if velocities are 10 m/s and 50 m/s at inlet and exit respectively. [200.9 kJ/kg, 202.1 kJ/kg]

3.23 Turbojet engine flies with velocity of 270 m/s at the altitude where ambient temperature is –15°C. Gas temperature at nozzle exit is 873 K and fuel air ratio is 0.019. Corresponding enthalpy values for air and gas at inlet and exit are 260 kJ/kg and 912 kJ/kg respectively. Combustion efficiency is 95% and calorific value of fuel is 44.5 MJ/kg. For the heat losses from engine amounting to 21 kJ/kg of air determine the velocity of gas jet at exit. [613.27 m/s]

3.24 Oxygen at 3MPa and 300°C flowing through a pipe line is tapped out to fill an empty insulated rigid tank. Filling continues till the pressure equilibrium is not attained. What shall be the temperature of the oxygen inside the tank? If $\gamma = 1.39$. [662.5°C]

3.25 Determine work done by fluid in the thermodynamic cycle comprising of following processes :

(*a*) Unit mass of fluid at 20 atm and 0.04 m^3 is expanded by the law $PV^{1.5}$ = constant, till volume gets doubled.

(*b*) Fluid is cooled isobarically to its original volume.

(*c*) Heat is added to fluid till its pressure reaches to its original pressure, isochorically. [18.8 kJ]

3.26 An air vessel has capacity of 10 m^3 and has air at 10 atm and 27°C. Some leakage in the vessel causes air pressure to drop sharply to 5 atm till leak is repaired. Assuming process to be of reversible adiabatic type determine the mass of air leaked. [45.95 kg]

3.27 Atmospheric air leaks into a cylinder having vacuum. Determine the final temperature in cylinder when inside pressure equals to atmospheric pressure, assuming no heat transferred to or from air in cylinder. [144.3°C]

3.28 Determine the power available from a steam turbine with following details;
Steam flow rate = 1 kg/s
Velocity at inlet and exit = 100 m/s and 150 m/s
Enthalpy at inlet and exit = 2900 kJ/kg, 1600 kJ/kg
Change in potential energy may be assumed negligible. [1293.75 kW]

4

Second Law of Thermodynamics

4.1 INTRODUCTION

Earlier discussions in article 3.11 throw some light on the limitations of first law of thermodynamics. A few situations have been explained where first law of thermodynamics fails to mathematically explain non-occurrence of certain processes, direction of process etc. Therefore, need was felt to have some more law of thermodynamics to handle such complex situations. Second law came up as embodiment of real happenings while retaining the basic nature of first law of thermodynamics. Feasibility of process, direction of process and grades of energy such as low and high are the potential answers provided by IInd law. Second law of thermodynamics is capable of indicating the maximum possible efficiencies of heat engines, coefficient of performance of heat pumps and refrigerators, defining a temperature scale independent of physical properties etc.

4.2 HEAT RESERVOIR

Heat reservoir is the system having very large heat capacity i.e. it is a body capable of absorbing or rejecting any finite amount of energy without any appreciable change in its' temperature. Thus in general it may be considered as a system in which any amount of energy may be dumped or extracted out and there shall be no change in its temperature. Such as *atmosphere* to which large amount of heat can be rejected without measurable change in its temperature. Large river, sea etc. can also be considered as reservoir, as dumping of heat to it shall not cause appreciable change in temperature.

Heat reservoirs can be of two types depending upon nature of heat interaction i.e. heat rejection or heat absorption from it. Heat reservoir which rejects heat from it is called *source.* While the heat reservoir which absorbs heat is called *sink.* Some times these heat reservoirs may also be called Thermal Energy Reservoirs (TER).

4.3 HEAT ENGINE

Heat engine is a device used for converting heat into work as it has been seen from nature that conversion from work to heat may take place directly but the vice-versa is not simple to be realized. Heat and work have been categorized as two forms of energy of low grade and high grade type. Conversion of high grade of energy to low grade of energy may be complete (100%), and can occur directly whereas complete conversion of low grade of energy into high grade of energy is not possible. For converting low grade of energy (heat) into high grade of energy (work) some device called heat engine is required.

Thus, heat engine may be precisely defined as "a device operating in cycle between high temperature source and low temperature sink and producing work". Heat engine receives heat from source, transforms some portion of heat into work and rejects balance heat to sink. All the processes occurring in heat engine constitute cycle.

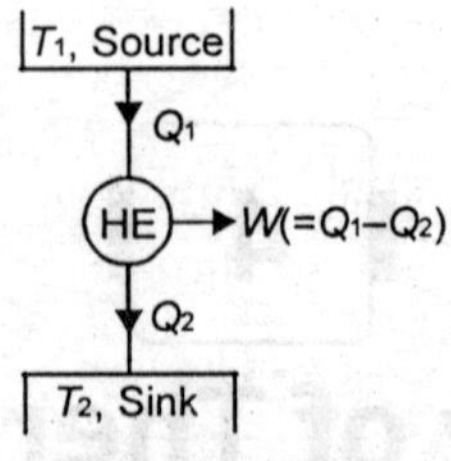

Fig. 4.1 Heat engine

Block diagram representation of a heat engine is shown above. A practical arrangement used in gas turbine plant is also shown for understanding the physical singnificance of heat engine.

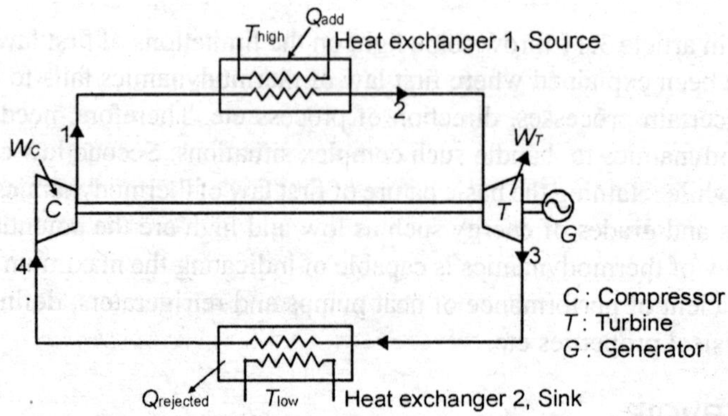

Fig. 4.2 Closed cycle gas turbine power plant

Gas turbine installation shows that heat is added to working fluid from 1–2 in a 'heat exchanger 1' and may be treated as heat supply by source. Working fluid is expanded in turbine from 2–3 and produces positive work. After expansion fluid goes to the 'heat exchanger 2' where it rejects heat from it like heat rejection in sink. Fluid at state 4 is sent to compressor for being compressed to state 1. Work required for compression is quite small as compared to positive work available in turbine and is supplied by turbine itself.

Therefore, heat engine model for it shall be as follows,

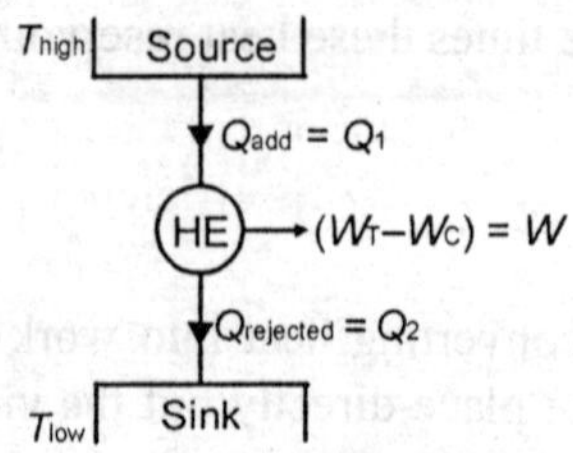

Fig. 4.3 Heat engine representation for gas turbine plant

Efficiency of heat engine can be given by the ratio of net work and heat supplied.

$$\eta_{\text{heat engine}} = \frac{\text{Net work}}{\text{Heat supplied}} = \frac{W}{Q_1}$$

For gas turbine plant shown

$$W = W_T - W_C$$

and
$$Q_1 = Q_{\text{add}}$$

Also since it is operating in cycle,

$$W_T - W_C = Q_{\text{add}} - Q_{\text{rejected}}$$

therefore,

$$\eta_{\text{heat engine}} = \frac{W_T - W_C}{Q_{\text{add}}}$$

$$= \frac{Q_{\text{add}} - Q_{\text{rejected}}}{Q_{\text{add}}}$$

$$\eta_{\text{heat engine}} = 1 - \frac{Q_{\text{rejected}}}{Q_{\text{add}}}$$

4.4 HEAT PUMP AND REFRIGERATOR

Heat pump refers to a device used for extracting heat from a low temperature surroundings and sending it to high temperature body, while operating in a cycle. In other words heat pump maintains a body or system at temperature higher than temperature of surroundings, while operating in cycle. Block diagram representation for a heat pump is given below:

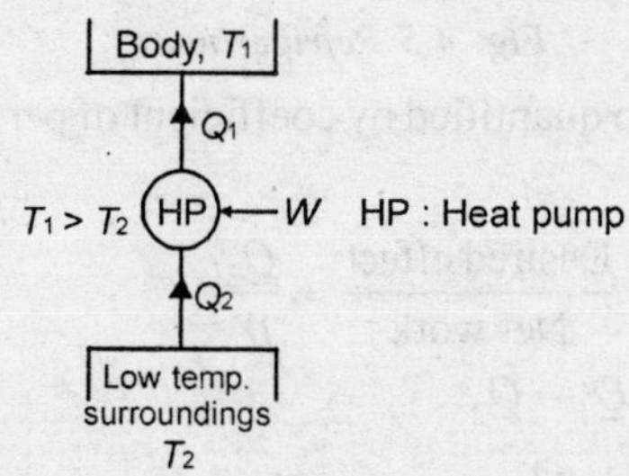

Fig. 4.4 Heat pump

As heat pump transfers heat from low temperature to high temperature, which is non spontaneous process, so external work is required for realizing such heat transfer. Heat pump shown picks up heat Q_2 at temperature T_2 and rejects heat Q_1 for maintaining high temperature body at temperature T_1.

For causing this heat transfer heat pump is supplied with work W as shown.

As heat pump is not a work producing machine and also its objective is to maintain a body at higher temperature, so its performance can't be defined using efficiency as in case of heat engine. Performance of heat pump is quantified through a parameter called coefficient of performance (C.O.P). *Coefficient of performance* is defined by the ratio of desired effect and net work done for getting the desired effect.

$$\text{C.O.P.} = \frac{\text{Desired effect}}{\text{Net work done}}$$

For heat pump :

$$\text{Net work} = W$$

Desired effect = heat transferred Q_1 to high temperature body at temperature, T_1.

$$(COP)_{HP} = \frac{Q_1}{W}$$

also $$W = Q_1 - Q_2$$

so $$\boxed{(COP)_{HP} = \frac{Q_1}{Q_1 - Q_2}}$$

Refrigerator is a device similar to heat pump but with reverse objective. It maintains a body at temperature lower than that of surroundings while operating in a cycle. Block diagram representation of refrigerator is shown in Fig 4.5.

Refrigerator also performs a non spontaneous process of extracting heat from low temperature body for maintaining it cool, therefore external work W is to be done for realizing it.

Block diagram shows how refrigerator extracts heat Q_2 for maintaining body at low temperature T_2 at the expense of work W and rejects heat to high temperature surroundings.

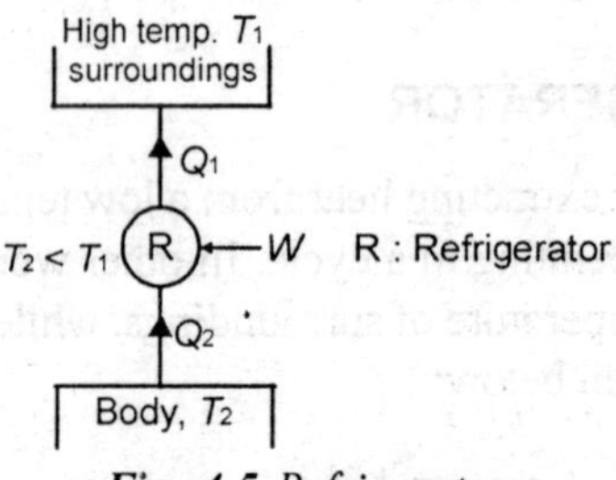

Fig. 4.5 Refrigerator

Performance of refrigerator is also quantified by coefficient of performance, which could be defined as:

$$(COP)_{refrigerator} = \frac{\text{Desired effect}}{\text{Net work}} = \frac{Q_2}{W}$$

Here $$W = Q_1 - Q_2$$

or $$(COP)_{refrigerator} = \frac{Q_2}{Q_1 - Q_2}$$

COP values of heat pump and refrigerator can be interrelated as:

$$(COP)_{HP} = (COP)_{refrigerator} + 1$$

4.5 STATEMENTS FOR IIND LAW OF THERMODYNAMICS

Rudolph Julius Emmanuel Clausius, a German physicist presented a first general statement of second law of thermodynamics in 1850 after studying the work of Sadi Carnot. It was termed as Clausius statement of second law. Lord Kelvin and Max Planck also came up with another statement of second law which was termed as Kelvin-Planck statement for second law of thermodynamics. Thus, there are two statements of second law of thermodynamics, (although they are equivalent as explained ahead).

Clausius statement of second law of thermodynamics: "It is impossible to have a device that while operating in a cycle produces no effect other than transfer of heat from a body at low temperature to a body at higher temperature."

Above statement clearly indicates that if a non spontaneous process such as transferring heat from low temperature body to high temperature body is to be realized then some other effects such as external work requirement is bound to be there. As already seen in case of refrigerator the external work is required for extracting heat from low temperature body and rejecting it to high temperature body.

Kelvin-Planck statement of second law of thermodynamics: "It is impossible for a device operating in a cycle to produce net work while exchanging heat with bodies at single fixed temperature".

It says that in order to get net work from a device operating in cycle (i.e. heat engine) it must have heat interaction at two different temperatures or with body/reservoirs at different temperatures (i.e. source and sink).

Thus, above two statements are referring to feasible operation of heat pump/refrigerator and heat engine respectively. Devices based on violation of IInd law of thermodynamics are called Perpetual motion machines of 2nd kind (PMM-II). Fig 4.6 shows such PMM-II.

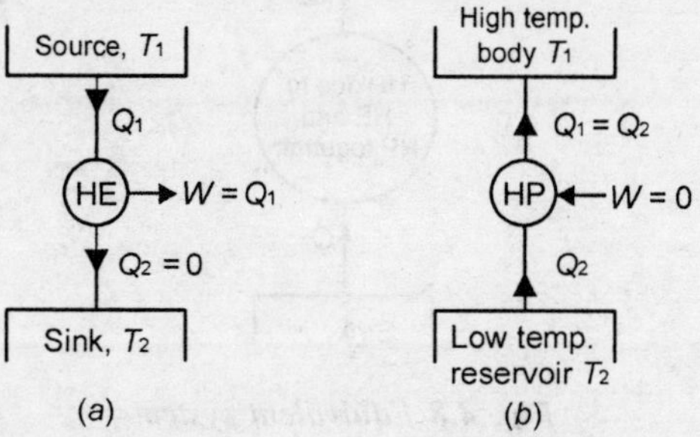

***Fig. 4.6** Perpetual Motion Machine of IInd kind*

PMM-II shown in Fig. 4.6*a*, refers to a heat engine which produces work while interacting with only one reservoir. PMM-II shown in Fig. 4.6*b*, refers to the heat pump which transfers heat from low temperature to high temperature body without spending work.

4.6 EQUIVALENCE OF KELVIN-PLANCK AND CLAUSIUS STATEMENTS OF IIND LAW OF THERMODYNAMICS

Kelvin-Planck and Clausius statements of IInd law of thermodynamics are actually two different interpretations of the same basic fact. Here the equivalence of two statements has been shown. For establishing equivalence following statements may be proved.

(*a*) System based on violation of Kelvin-Planck statement leads to violation of Clausius statement.

(*b*) System based on violation of Clausius statement leads to violation of Kelvin-Planck statement.

(*a*) Let us assume a heat engine producing net work while exchanging heat with only one reservoir at temperature T_1, thus based on violation of Kelvin Planck statement. Let us also have a perfect heat pump operating between two reservoirs at temperatures T_1 and T_2. Work requirement of heat pump may be met from the work available from heat engine. Layout shown ahead explains the proposed arrangement.

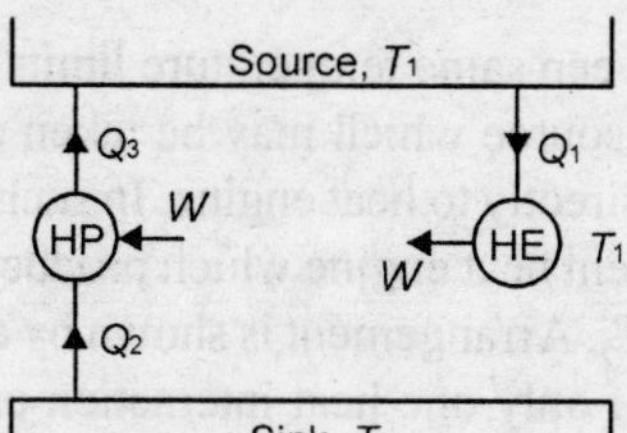

***Fig. 4.7** System based on violation of Kelvin Planck statement*

If heat pump takes input work from output of heat engine then,

$$Q_3 = Q_2 + W$$

and $$W = Q_1$$

or $$Q_3 = Q_1 + Q_2$$

Combination of heat engine and heat pump shall thus result in an equivalent system working as heat pump transferring heat from low temperature T_2 to high temperature T_1 without expense of any external work. This heat pump is based on violation of Clausius statement and therefore not possible.

Hence, it shows that violation of Kelvin Planck statement leads to violation of Clausius statement.

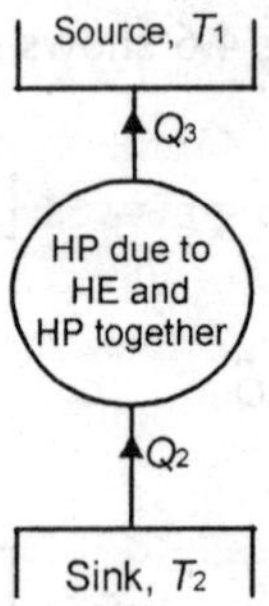

Fig. 4.8 Equivalent system

(*b*) Let us assume a heat pump which operating in cycle transfers heat from low temperature reservoir to high temperature reservoir without expense of any work, thus based on violation of Clausius statement.

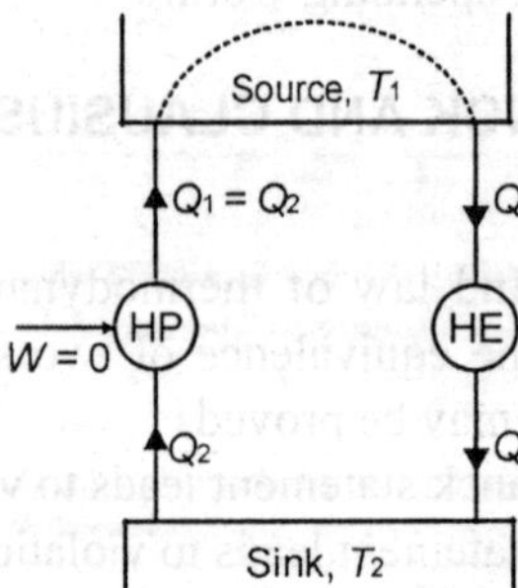

Fig. 4.9 System based on violation of Clausius statement

Heat pump transfers heat Q_1 to high temperature reservoir while extracting heat Q_2 from low temperature reservoir. Mathematically, as no work is done on pump, so

$$Q_2 = Q_1$$

Let us also have a heat engine between same temperature limits of T_1 and T_2 and produce net work W. Heat engine receives heat Q_3 from source which may be taken equal to Q_1. Let us now devise for heat rejected from heat pump be given directly to heat engine. In such a situation the combination of heat pump and heat engine results in equivalent heat engine which produces work '*W*' while exchanging heat with only one reservoir at temperature T_2. Arrangement is shown by dotted lines. This type of equivalent system is producing work as a result of only one heat interaction and thus violation of Kelvin Planck statement.

Thus, it shows that violation of Clausius statement also causes violation of Kelvin Planck statement.

Hence from (*a*) and (*b*) proved above it is obvious that the Clausius and Kelvin-Planck statements are equivalent. Conceptually the two statements explain the basic fact that,

(*i*) net work can't be produced without having heat interactions taking place at two different temperatures.

(*ii*) non spontaneous process such as heat flow from low temperature body to high temperature body is not possible without spending work.

4.7 REVERSIBLE AND IRREVERSIBLE PROCESSES

Reversible processes as described in chapter 1 refer to "the thermodynamic processes occurring in the manner that states passed through are always in thermodynamic equilibrium and no dissipative effects are present." Any reversible process occurring between states 1–2 upon reversal, while occurring from 2–1 shall not leave any mark of process ever occurred as states traced back are exactly similar to those in forward direction. Reversible processes are thus very difficult to be realized and also called ideal processes. All thermodynamic processes are attempted to reach close to the reversible process in order to give best performance.

Thermodynamic process which does not fulfil conditions of a reversible process are termed irreversible processes. Irreversibilities are the reasons causing process to be irreversible. Generally, the irreversibilities can be termed as internal irreversibility and external irreversibility. Internal irreversibility is there because of internal factors whereas external irreversibility is caused by external factors at the system-surrounding interface. Generic types of irreversibilities are due to;

(*i*) Friction,
(*ii*) Electrical resistance,
(*iii*) Inelastic solid deformations,
(*iv*) Free expansion
(*v*) Heat transfer through a finite temperature difference,
(*vi*) Non equilibrium during the process, etc.

(*i*) ***Friction:*** Friction is invariably present in real systems. It causes irreversibility in the process as work done does not show equivalent rise in kinetic or potential energy of the system. Fraction of energy wasted due to frictional effects leads to deviation from reversible states.

(*ii*) ***Electrical resistance***: Electrical resistance in the system also leads to presence of dissipation effects and thus irreversibilities. Due to electric resistance dissipation of electrical work into internal energy or heat takes place. The reverse transformation from heat or internal energy to electrical work is not possible, therefore leads to irreversibility.

(*iii*) ***Inelastic solid deformation:*** Deformation of solids, when of inelastic type is also irreversible and thus causes irreversibility in the process. If deformation occurs within elastic limits then it does not lead to irreversibility as it is of reversible type.

(*iv*) ***Free expansion:*** Free expansion as discussed earlier in chapter 3, refers to the expansion of unresisted type such as expansion in vacuum. During this unresisted expansion the work interaction is zero and without expense of any work it is not possible to restore initial states. Thus, free expansion is irreversible.

(*v*) ***Heat transfer through a finite temperature difference:*** Heat transfer occurs only when there exist temperature difference between bodies undergoing heat transfer. During heat transfer if heat addition is carried out in finite number of steps then after every step the new state shall be a non-equilibrium state. In order to have equilibrium states in between, the heat transfer process may be carried out in infinite number of steps. Thus, infinitesimal heat transfer every time causes infinitesimal temperature variation. These infinitesimal state changes shall require infinite time and process shall be of quasi-static type, therefore reversible. Heat transfer through a finite temperature difference which practically occurs is accompanied by irreversible state changes and thus makes processes irreversible.

(*vi*) ***Non equilibrium during the process:*** Irreversibilities are introduced due to lack of thermodynamic equilibrium during the process. Non equilibrium may be due to mechanical inequilibrium, chemical inequilibrium, thermal inequilibrium, electrical inequilibrium etc. and irreversibility are called mechanical irreversibility, chemical irreversibility, thermal irreversibility, electrical irreversibility respectively. Factors discussed above are also causing non equilibrium during the process and therefore make process irreversible.

Comparative study or reversible and irreversible processes shows the following major differences.

Difference between reversible and irreversible processes

Reversible process	*Irreversible process*
(*i*) Reversible process can not be realized in practice	(*i*) All practical processes occurring are irreversible processes
(*ii*) The process can be carried out in the reverse direction following the same path	(*ii*) Process, when carried out in reverse direction follows the path different from that in forward direction.
(*iii*) A reversible process leaves on trace of occurrence of process upon the system and surroundings after its' reversal.	(*iii*) The evidences of process having occurred are evident even after reversal of irreversible process.
(*iv*) Such processes can occur in either directions without violating second law of thermodynamics.	(*iv*) Occurrence of irreversible processes in either direction is not possible, as in one direction it shall be accompanied with the violation of second law of thermodynamics.
(*v*) A system undergoing reversible processes has maximum efficiency. So the system with reversible processes are considered as reference systems or bench marks.	(*v*) System having irreversible processes do not have maximum efficiency as it is accompanied by the wastage of energy.
(*vi*) Reversible process occurs at infinitesimal rate i.e. quasi-static process.	(*vi*) Irreversible processes occur at finite rate.
(*vii*) System remains throughout in thermodynamic equilibrium during occurrence of such process.	(*vii*) System does not remain in thermodynamic equilibrium during occurrence of irreversible processes.
(*viii*) Examples; Frictionless motion, controlled expansion and compression, Elastic deformations, Electric circuit with no resistance, Electrolysis, Polarization and magnetisation process etc.	(*viii*) Examples; Viscous fluid flow, inelastic deformation and hysteresis effect, Free expansion, Electric circuit with resistance, Mixing of dissimilar gases, Throttling process etc.

4.8 CARNOT CYCLE AND CARNOT ENGINE

Nicholas Leonard Sadi Carnot, an engineer in French army originated use of cycle (Carnot) in thermodynamic analysis in 1824 and these concepts provided basics upon which second law of thermodynamics was stated by Clausius and others.

Carnot cycle is a reversible thermodynamic cycle comprising of four reversible processes.

Thermodynamic processes constituting Carnot cycle are;

(*i*) Reversible isothermal heat addition process, (1–2, Q_{add})
(*ii*) Reversible adiabatic expansion process (2–3, W_{expn} +ve)
(*iii*) Reversible isothermal heat release process (3–4, $Q_{rejected}$)
(*iv*) Reversible adiabatic compression process (4–1, W_{compr} –ve)

Carnot cycle is shown on *P–V* diagram between states 1, 2, 3 4, and 1. A reciprocating piston-cylinder assembly is also shown below *P–V* diagram.

Process 1 –2 is isothermal heat addition process of reversible type in which heat is transferred to system isothermally. In the piston cylinder arrangement heat Q_{add} can be transferred to gas from a constant temperature source T_1 through a cylinder head of conductor type.

First law of thermodynamics applied on 1–2 yields;

$$Q_{add} = U_2 - U_1 + W_{1-2}$$

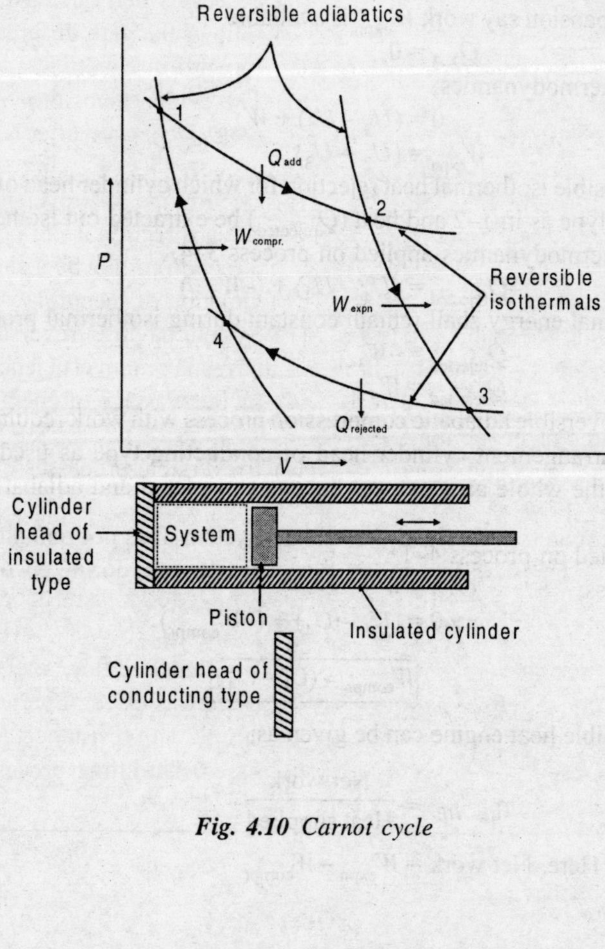

Fig. 4.10 Carnot cycle

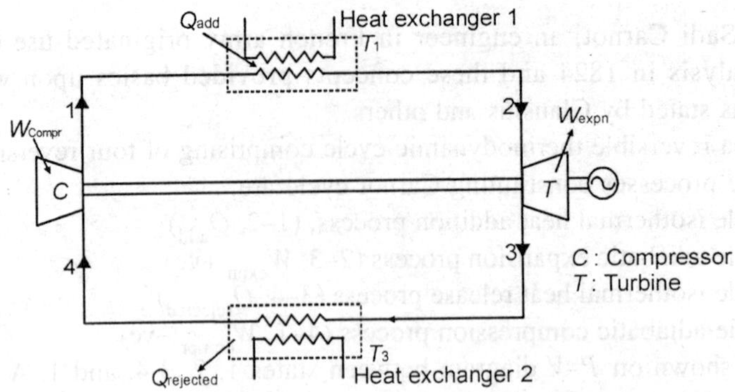

Fig. 4.11 Gas turbine plant: Carnot heat engine

For the perfect gas as working fluid in isothermal process no change in internal energy occurs, therfore $U_2 = U_1$

and $$Q_{add} = W_{1-2}$$

Process 2–3 is reversible adiabatic expansion process which may be had inside cylinder with cylinder head being replaced by insulating type cylinder head so that complete arrangement is insulated and adiabatic expansion carried out.

During adiabatic expansion say work W_{expn} is available,

$$Q_{2-3} = 0$$

From first law of thermodynamics;

$$0 = (U_3 - U_2) + W_{expn}$$

or $$W_{expn} = (U_2 - U_3)$$

Process 3–4 is reversible isothermal heat rejection for which cylinder head of insulating type may be replaced by conducting type as in 1–2 and heat ($Q_{rejected}$) be extracted out isothermally.

From first law of thermodynamics applied on process 3–4,

$$-Q_{rejected} = (U_4 - U_3) + (-W_{3-4})$$

for perfect gas internal energy shall remain constant during isothermal process. Thus, $U_3 = U_4$

$$-Q_{rejected} = -W_{3-4}$$

or $$Q_{rejected} = W_{3-4}$$

Process 4–1 is the reversible adiabatic compression process with work requirement for compression. In the piston cylinder arrangement cylinder head of conducting type as used in 3–4 is replaced by insulating type, so that the whole arrangement becomes insulated and adiabatic compression may be realized,

From first law applied on process 4–1

For adiabatic process; $$Q_{4-1} = 0$$

$$\Rightarrow 0 = (U_1 - U_4) + (-W_{compr})$$

or $$\boxed{W_{compr} = (U_1 - U_4)}$$

Efficiency of reversible heat engine can be given as;

$$\eta_{rev,\,HE} = \frac{\text{Net work}}{\text{Heat supplied}}$$

Here, Net work = $W_{expn} - W_{compr}$

and heat is supplied only during process 1–2, therefore

$$\text{heat supplied} = Q_{add}$$

Substituting in the expression for efficiency.

$$\eta_{rev,\,HE} = \frac{W_{expn} - W_{compr}}{Q_{add}}$$

Also for a cycle

$$\sum_{cycle} W = \sum_{cycle} Q$$

so

$$W_{net} = Q_{add} - Q_{rejected}$$

Hence

$$\eta_{rev,\,HE} = 1 - \frac{Q_{rejected}}{Q_{add}}$$

As the heat addition takes place at high temperature, while heat rejection takes place at low temperature, so writing these heat interactions as Q_{high}, Q_{low} we get,

$$\boxed{\eta_{rev,\,HE} = 1 - \frac{Q_{low}}{Q_{high}}}$$

$$\boxed{\eta_{Carnot} = 1 - \frac{Q_{low}}{Q_{high}}}$$

Piston-cylinder arrangement shown and discussed for realizing Carnot cycle is not practically feasible as;

(*i*) Frequent change of cylinder head i.e. of insulating type and diathermic type for adiabatic and isothermal processes is very difficult.

(*ii*) Isothermal heat addition is practically very difticult to be realized

(*iii*) Reversible adiabatic expansion and compression are not possible.

(*iv*) Even if near reversible isothermal heat addition and rejection is to be achieved then time duration for heat interaction should be very large i.e. infinitesimal heat interaction occurring at dead slow speed. Near reversible adiabatic processes can be achieved by making them to occur fast. In a piston-cylinder reciprocating engine arrangement such speed fluctuation in a single cycle is not possible.

Carnot heat engine arrangement is also shown with turbine, compressor and heat exchangers for adiabatic and isothermal processes. Fluid is compressed in compressor adiabatically, heated in heat exchanger at temperature T_1, expanded in turbine adiabatically, cooled in heat exchanger at temperature T_3 and sent to compressor for compression. Here also following practical difficulties are confronted;

(*i*) Reversible isothermal heat addition and rejection are not possible.

(*ii*) Reversible adiabatic expansion and compression are not possible.

Carnot cycle can also operate reversibly as all processes constituting it are of reversible type. Reversed Carnot cycle is shown below;

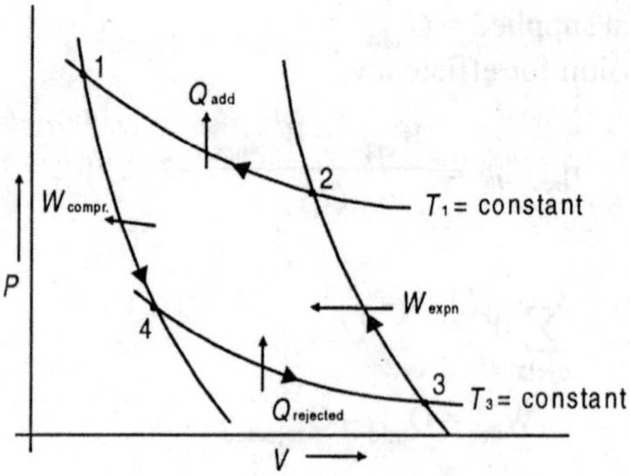

Fig. 4.12 Reversed Carnot cycle

Heat engine cycle in reversed form as shown above is used as ideal cycle for refrigeration and called "Carnot refrigeration cycle".

4.9 CARNOT THEOREM AND ITS COROLLARIES

Carnot theorem states that "any engine cannot have efficiency more than that of reversible engine operating between same temperature limits."

Different corollaries of Carnot theorem are,

(*i*) Efficiency of all reversible engines operating between same temperature limits is same.

(*ii*) Efficiency of a reversible engine does not depend on the working fluid in the cycle.

Using Clausius and Kelvin Planck statements the Carnot theorem can be proved easily. Let us take two heat engines HE_I and HE_{II} operating between same temperature limits T_1, T_2 of source and sink as shown in Fig. 4.13*a*.

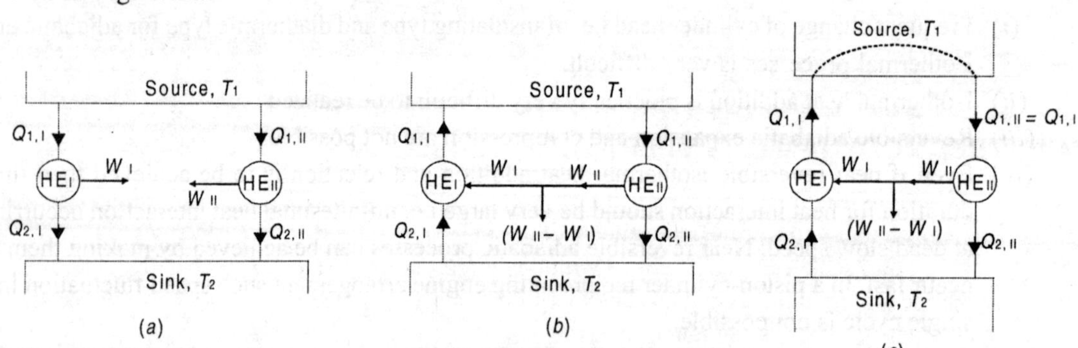

Fig. 4.13 Proof of Carnot's theorem

Arrangement shown has heat engine, HE_I getting $Q_{1,I}$ from source, rejecting $Q_{2,I}$ and producing work W_I. Heat engine, HE_{II} receives $Q_{1,II}$, rejects $Q_{2,II}$ and produces work W_{II}.

$$W_I = Q_{1,I} - Q_{2,I}$$

$$W_{II} = Q_{1,II} - Q_{2,II}$$

Efficiency of engines HE_I, HE_{II}

$$\eta_{HE_I} = \frac{W_I}{Q_{1,I}}$$

$$\eta_{HE_{II}} = \frac{W_{II}}{W_{1,II}}$$

Now let us assume that engine HE_1 is reversible engine while HE_{II} is any engine.

As per Carnot's theorem efficiency of HE_I (reversible engine) is always more than that of HE_{II}. Let us start with violation of above statement, i.e,

$$\eta_{HE_I} < \eta_{HE_{II}}$$

or
$$\frac{W_I}{Q_{1,I}} < \frac{W_{II}}{Q_{1,II}}$$

Let us take the heat addition to each engine to be same i.e.

$$Q_{1,I} = Q_{1,II}$$

Hence
$$W_I < W_{II}$$

Also we have assumed that engine HE_1 is of reversible type, so let us operate it in reversed manner, as shown in Figure 4.13*b*.

Let us also assume that the work requirement of reversed heat engine, HE_1 be fed by work output W_{II} of the heat engine HE_{II}. Since W_{II} is more than W_1, a net work $(W_{II} - W_I)$ shall also be available as output work after driving HE_1. Also since $Q_{1,I}$ and $Q_{1,II}$ are assumed to be same, the heat rejected by reversed HE_1 may be supplied to heat engine, HE_{II} as shown in figure 4.13c by dotted lines. Thus, it results into an equivalent heat engine which produces net work $(W_{II} - W_I)$ while heat interaction takes place with only one reservoir at temperature, T_2. This is a violation of Kelvin Planck statement, so the assumption made in beginning that efficiency of reversible engine is less than that of other engine, is not correct.

Hence, it is established that out of all heat engines operating within same temperature limits, the reversible engine has highest efficiency.

Similarly for showing the correctness of corollaries of Carnot theorem the heat engines and their combinations be considered like above and proved using Kelvin-Planck and Clausius statements.

4.10 THERMODYNAMIC TEMPERATURE SCALE

After the Carnot's theorem and its corollary were stated and verified, it was thought to have a thermodynamic temperature scale, independent of thermometric substance and principles of thermometry. Such a temperature scale can be developed with the help of reversible heat engine concept and is called thermodynamic temperature scale. Defining thermodynamic temperature scale refers to the assigning of numerical values to different temperatures using reversible heat engines.

From the previous discussions on heat engines it is obvious that the efficiency of a reversible heat engine depends on the temperatures of reservoir with which heat interaction takes place.

Mathematically, it can be easily given by any unknown function '*f*';

$$\eta_{rev, HE} = f\,(T_{high},\ T_{low})$$

where T_{high} and T_{low} are the two temperatures of high temperature source and low temperature sink.

or
$$\eta_{rev, HE} = 1 - \frac{Q_{low}}{Q_{high}} = f\,(T_{high},\ T_{low})$$

Unknown function '*f*' may be substituted by another unknown function, say ϕ

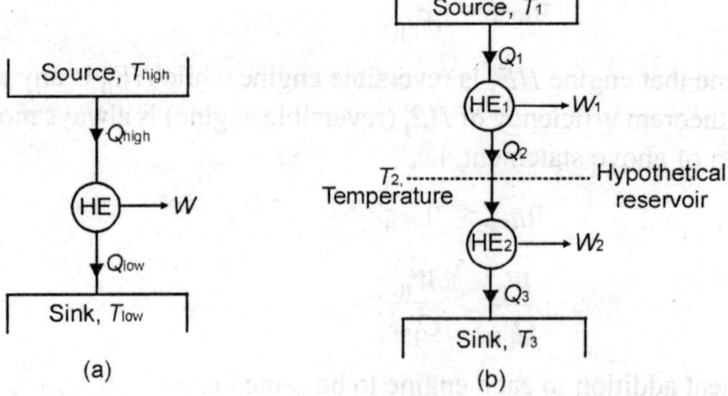

Fig. 4.14 Reversible heat engine and its combinations

$$\frac{Q_{high}}{Q_{low}} = \phi\,(T_{high}, T_{low})$$

Thus, some functional relationship as defined by 'ϕ' is established between heat interactions and temperatures.

Let us now have more than one reversible heat engines operating in series as shown in Figure 4.14b, between source and sink having T_1 and T_3 temperatures. In between an imaginary reservoir at temperature T_2 may be considered.

From the above for two reversible heat engines;

$$\frac{Q_1}{Q_2} = \phi\,(T_1, T_2)$$

and
$$\frac{Q_2}{Q_3} = \phi\,(T_2, T_3)$$

Combination of two heat engines may be given as shown here,

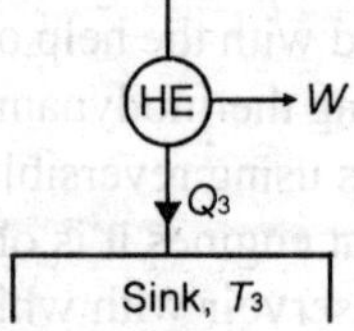

Fig. 4.15 Equivalent heat engine for two reversible heat engines operating in series.

$$\frac{Q_1}{Q_3} = \frac{Q_1/Q_2}{Q_2/Q_3}$$

or
$$\frac{Q_1}{Q_3} = \phi\,(T_1, T_3)$$

$$\frac{Q_1}{Q_3} = \frac{Q_1}{Q_2} \cdot \frac{Q_2}{Q_3}$$

or $$\phi(T_1, T_3) = \phi(T_1, T_2) \cdot \phi(T_2, T_3)$$

Above functional relation is possible only if it is given by another function ψ as follows.

$$\phi(T_1, T_2) = \frac{\psi(T_1)}{\psi(T_2)}$$

$$\phi(T_2, T_3) = \frac{\psi(T_2)}{\psi(T_3)}$$

$$\phi(T_1, T_3) = \frac{\psi(T_1)}{\psi(T_3)}$$

Thus,

$$\frac{Q_1}{Q_2} = \frac{\psi(T_1)}{\psi(T_2)}$$

$$\frac{Q_2}{Q_3} = \frac{\psi(T_2)}{\psi(T_3)}$$

$$\frac{Q_1}{Q_3} = \frac{\psi(T_1)}{\psi(T_3)}$$

Lord Kelvin based upon his observations proposed that the function $\psi(T)$ can be arbitrarily chosen based on Kelvin scale or absolute thermodynamic temperature scale as;

$$\psi(T) = \text{Temperature } T \text{ in Kelvin Scale}$$

Therefore,

$$\frac{Q_1}{Q_2} = \frac{\psi(T_1)}{\psi(T_2)} = \frac{T_1}{T_2}$$

$$\frac{Q_2}{Q_3} = \frac{\psi(T_2)}{\psi(T_3)} = \frac{T_2}{T_3}$$

$$\frac{Q_1}{Q_3} = \frac{\psi(T_1)}{\psi(T_3)} = \frac{T_1}{T_3}$$

where T_1, T_2, T_3 are temperatures in *absolute thermodynamic scale.*

Here heat absorbed and heat rejected is directly proportional to temperatures of reservoirs supplying and accepting heats to heat engine. For a Carnot heat engine or reversible heat engine operating between reservoirs at temperature T and triple point of water, T_t;

$$\frac{Q}{Q_t} = \frac{T}{T_t} = \frac{T}{273.16}$$

or $$\boxed{T = 273.16 \cdot \frac{Q}{Q_t}}$$

Here for a known Q and Q_t values the temperature T can be defined. Thus, heat interaction acts as thermometric property in thermodynamic temperature scale, which is independent of thermometric substance. It may be noted that negative temperatures cannot exist on thermodynamic temperature scale.

Let us now have a large number of reversible heat engines (Carnot engines) operating in series as shown in Figure 4.16.

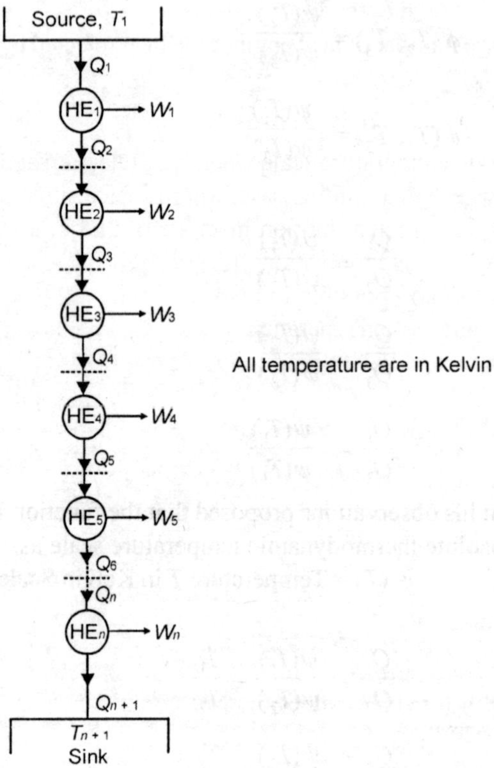

Fig. 4.16 Series of reversible heat engines

From thermodynamic temperature scale for different engines,

$$\frac{Q_1}{Q_2} = \frac{T_1}{T_2}$$

$$\frac{Q_2}{Q_3} = \frac{T_2}{T_3}$$

$$\frac{Q_3}{Q_4} = \frac{T_3}{T_4}$$

For nth engine

$$\frac{Q_n}{Q_{n+1}} = \frac{T_n}{T_{n+1}}$$

Here work output from each engine shall continuously diminish the heat supplied to subsequent heat engine. Let us assume work outputs from '*n*' engines to be same; i.e.

$$W_1 = W_2 = W_3 = W_4 = \dots = W_n$$

or $$(Q_1 - Q_2) = (Q_2 - Q_3) = (Q_3 - Q_4) = \dots = (Q_n - Q_{n+1})$$

or $$(T_1 - T_2) = (T_2 - T_3) = (T_3 - T_4) = \dots = (T_n - T_{n+1})$$

It is obvious that for a large number of heat engines the heat rejected by *n*th engine shall be negligible i.e for very large value of *n*, $Q_{n+1} \to 0$

or for $\text{Lim } n \to \infty$, $\text{Lim } Q_{n+1} \to 0$

Thus, from thermodynamic temperature scale when heat rejection approaches zero, the temperature of heat rejection also tends to zero as a limiting case. But in such a situation when heat rejection is zero, the heat engine takes form of a perpetual motion machine of 2nd kind, where work is produced with only heat supplied to it. Thus, it leads to violation of Kelvin-Planck statement. Hence it is not possible.

Also it can be said that "it is impossible to attain absolute zero temperature in finite number of operations." There exists absolute zero temperature on thermodynamic temperature scale, but cannot be attained without violation of second law of thermodynamics. This fact is popularly explained by *third law of thermodynamics.*

Carnot cycle efficiency can now be precisely defined as function of source and sink temperatures.

$$\eta_{\text{carnot}} = 1 - \frac{Q_{\text{low}}}{Q_{\text{high}}}$$

$$\eta_{\text{carnot}} = 1 - \frac{T_{\text{low}}}{T_{\text{high}}}$$

Thus, it is seen that Carnot cycle efficiency depends only upon lower and higher temperatures. Carnot cycle efficiency is high for small values of sink temperature (T_{low}) and larger values of source temperature (T_{high}).

Therefore for maximum efficiency, Carnot cycle must operate between maximum possible source and minimum possible sink temperatures.

EXAMPLES

1. Using IInd law of thermodynamics show that the following are irreversible

(i) Free expansion.

(ii) Heat transfer through finite temperature difference.

Solution:

(*i*) Let us consider a perfectly insulated tank having two compartments divided by thin wall. Compartment I has gas while II has vacuum. When wall is punctured then gas in I expands till pressure in I and II gets equalised. Let us assume that free expansion is reversible i.e. the gas in II returns into I and original states are restored.

When gas is allowed to expand, say it produces work *W* from a device *D* due to expansion. This work *W* is available due to change in internal energy of gas. Internal energy of gas can be restored by adding equivalent heat *Q* to it from a source as shown. This whole arrangement if consolidated can be treated as a device which is producing work while exchanging heat with single body. Thus, it is violation of IInd law of thermodynamics, therefore the assumption that free expansion is reversible is incorrect. Free expansion is irreversible.

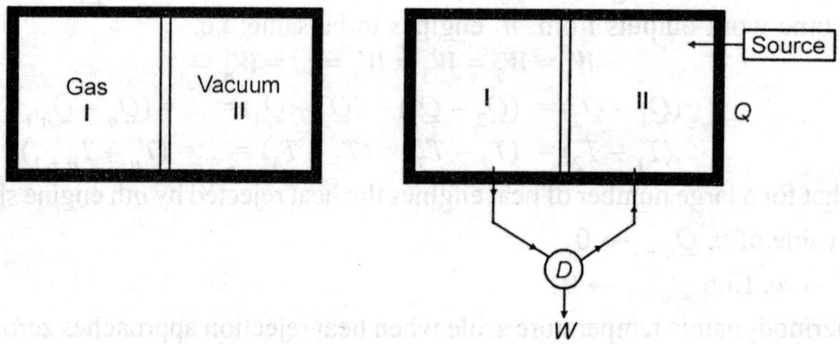

Fig. 4.17 Free expansion

(*ii*) For showing that the heat transfer through finite temperature difference is irreversible, let us start with the fact that such heat transfer is reversible. Let us take a heat source (T_1) and sink (T_2) and assume that a heat Q_{1-2} flows from T_1 to T_2.

Let us have a heat engine operating between T_1 and T_2 as shown and producing work W. Let us reverse heat transfer process from T_2 to T_1 i.e. Q_{2-1}, as assumed. Let us assume $Q_2 = Q_{2-1}$.

This assumption paves the way for eliminating sink. Let us now remove sink and directly supply Q_2 as Q_{2-1} (= Q_2). This results in formation of a heat engine which produces work while exchanging heat with single reservoir, the violation of IInd law of thermodynamics. (Kelvin Planck statement).

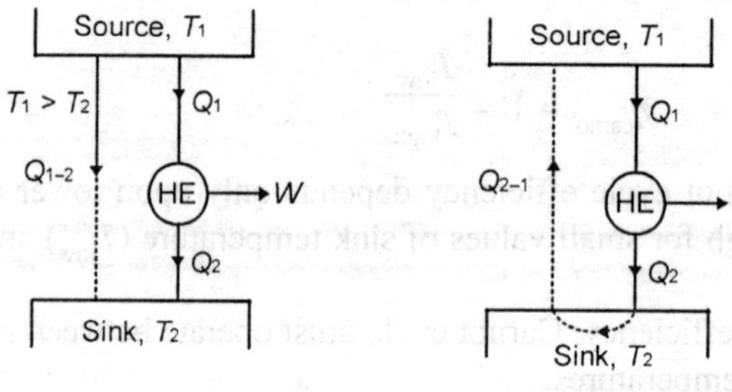

Fig. 4.18 Heat transfer through a finite temperature difference

Hence, assumption that heat transfer through finite temperature is reversible, stands incorrect. Therefore, heat transfer through finite temperature difference is irreversible.

2. Determine the heat to be supplied to a Carnot engine operating between 400°C and 15°C and producing 200 kJ of work.

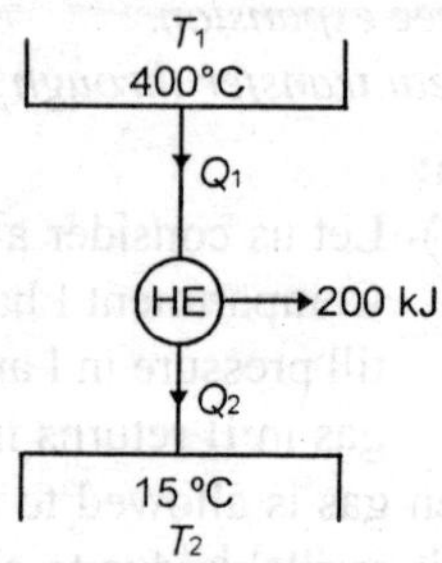

Fig. 4.19

Solution :

To find out $Q_1 = ?$

In Carnot engine from thermodynamic temperature scale;

$$\frac{Q_1}{Q_2} = \frac{T_1}{T_2}$$

and work $W = Q_1 - Q_2$

Thus $$\frac{Q_1}{Q_2} = \frac{673}{288} \quad (1)$$

and $$Q_1 - Q_2 = 200 \text{ kJ} \quad (2)$$

From eqn. 1 and 2, upon solving

$$Q_1 = 349.6 \text{ kJ}$$

and $$Q_2 = 149.6 \text{ kJ}$$

Heat to be supplied = **349.6 kJ Ans.**

3. *A refrigerator operates on reversed carnot cycle. Determine the power required to drive refrigerator between temperatures of 42°C and 4°C if heat at the rate of 2 kJ/s is extracted from the low temperature region.*

Solution:

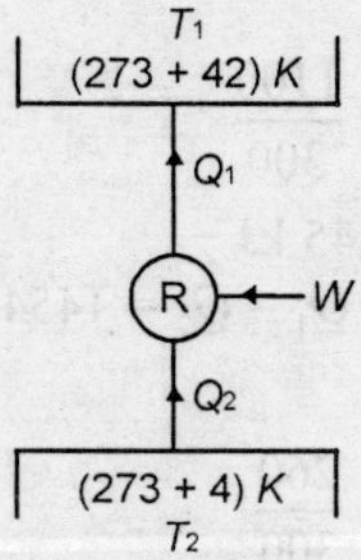

Fig. 4.20

To find out, $W = ?$

Given: $T_1 = 315 \text{ K}, T_2 = 277 \text{ K}$

and $Q_2 = 2 \text{ kJ/s}$

From thermodynamic temperature scale;

$$\frac{Q_1}{Q_2} = \frac{T_1}{T_2}$$

or $$\frac{Q_1}{2} = \frac{315}{277}$$

or $$Q_1 = 2.274 \text{ kJ/s}$$

$$\text{Power/Work input required} = Q_1 - Q_2$$
$$= 2.274 - 2$$
$$\text{Power required} = 0.274 \text{ kJ/s}$$

Power required for driving refrigerator = **0.274 kW Ans.**

4. *A reversible heat engine operates between two reservoirs at 827°C and 27°C. Engine drives a Carnot refrigerator maintaining –13°C and rejecting heat to reservoir at 27°C. Heat input to the engine is 2000 kJ and the net work available is 300 kJ. How much heat is transferred to refrigerant and total heat rejected to reservoir at 27°C?*

Solution:

Block diagram based on the arrangement stated;

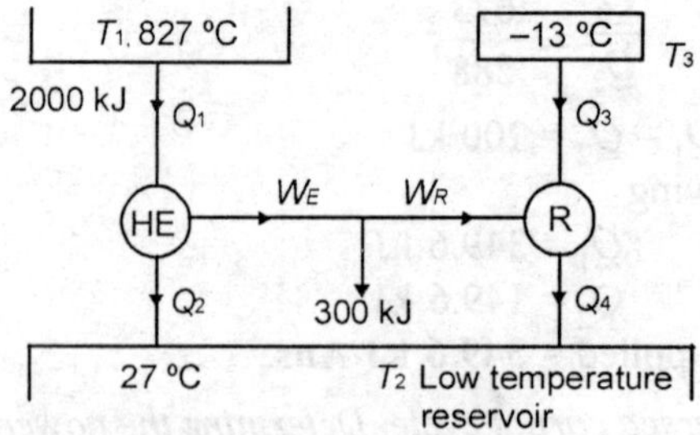

Fig. 4.21

We can write, for heat engine, $$\frac{Q_1}{Q_2} = \frac{T_1}{T_2}$$

$$\frac{Q_1}{Q_2} = \frac{1100}{300}$$

Substituting $Q_1 = 2000$ kJ, we get $Q_2 = 545.45$ kJ

Also $$W_E = Q_1 - Q_2 = 1454.55 \text{ kJ}$$

For refrigerator,

$$\frac{Q_3}{Q_4} = \frac{260}{300} \qquad (1)$$

Also, $$W_R = Q_4 - Q_3 \qquad (2)$$

and $$W_E - W_R = 300$$

or $$W_R = 1154.55 \text{ kJ}$$

Eqn. (1) & (2) result in,

$$Q_4 - Q_3 = 1154.55 \qquad (3)$$

From eqn. (1) & (3),

$$Q_3 = 7504.58 \text{ kJ}$$

$$Q_4 = 8659.13 \text{ kJ}$$

Total heat transferred to low temperature reservoir

$$= Q_2 + Q_4 = 9204.68 \text{ kJ}$$

Heat transferred to refrigerant = **7504.58 kJ**

Total heat transferred to low temperature reservoir = **9204.68 kJ** **Ans.**

5. In a winter season when outside temperature is –1°C, the inside of house is to be maintained at 25°C. Estimate the minimum power required to run the heat pump of maintaining the temperature. Assume heating load as 125 MJ/h.

Solution:

$$COP_{HP} = \frac{Q_1}{W} = \frac{Q_1}{Q_1 - Q_2} = \frac{1}{\left(1 - \dfrac{Q_2}{Q_1}\right)}$$

Also we know

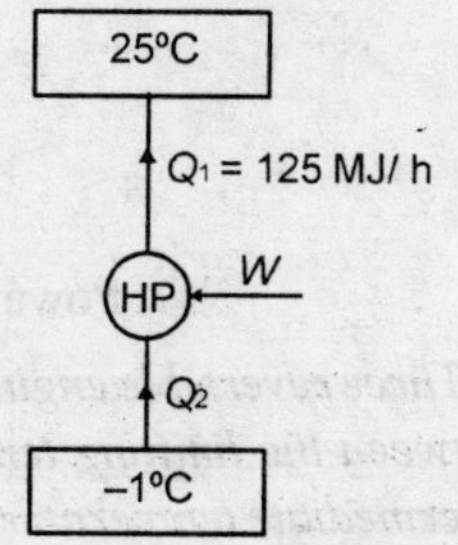

Fig. 4.22

$$\frac{Q_1}{Q_2} = \frac{298.15}{272.15}$$

Thus $$COP_{HP} = 11.47$$

Also $$COP_{HP} = \frac{Q_1}{W}, \text{ Substituting } Q_1$$

therefore $$W = 10.89 \text{ MJ/h}$$

or, $$3.02 \text{ kW}$$

Minimum power required = **3.02 kW Ans.**

6. A cold storage plant of 40 tonnes of refrigeration capacity runs with its performance just $\frac{1}{4}$ th of its Carnot COP. Inside temperature is –15°C and atmospheric temperature is 35°C. Determine the power required to run the plant. [Take : One ton of refrigeration as 3.52 kW]

Solution:

Cold storage plant can be considered as a refrigerator operating between given temperatures limits.

Capacity of plant = Heat to be extracted = 140.8 kW

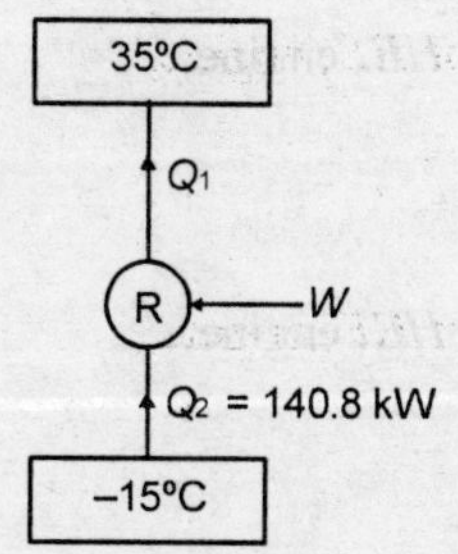

Fig. 4.23

$$\text{Carnot COP of plant} = \frac{1}{\left(^{308}/_{258.15}\right) - 1}$$

$$= 5.18$$

Actual $$\text{COP} = \frac{5.18}{4} = 1.295$$

Also actual $$\text{COP} = \frac{Q_2}{W}, \text{ hence W} = 108.73 \text{ kW.}$$

Power required = **108.73 kW Ans.**

7. What would be maximum efficiency of engine that can be had between the temperatures of 1150°C and 27°C ?

Solution:

Highest efficiency is that of Carnot engine, so let us find the Carnot cycle efficiency for given temperature limits.

$$\eta = 1 - \left(\frac{273 + 27}{273 + 1150}\right)$$

η = **0.7891 or 78.91% Ans.**

8. A domestic refrigerator maintains temperature of – 8°C when the atmospheric air temperature is 27°C. Assuming the leakage of 7.5 kJ/min from outside to refrigerator determine power required to run this refrigerator. Consider refrigerator as Carnot refrigerator.

Solution:

Here heat to be removed continuously from refrigerated space shall be 7.5 kJ/min or 0.125 kJ/s.

For refrigerator, C.O.P. shall be,

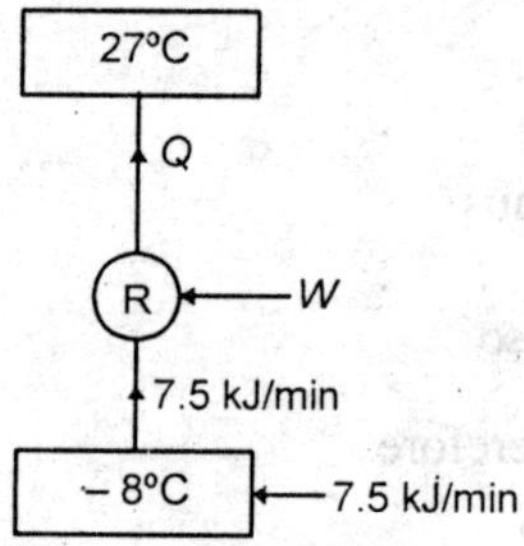

Fig. 4.23

$$\frac{0.125}{W} = \frac{265}{(300-265)}$$

or $$W = 0.0165 \text{ kJ/s.}$$

Power required = **0.0165 kW Ans.**

9. Three reversible engines of Carnot type are operating in series as shown between the limiting temperatures of 1100 K and 300 K. Determine the intermediate temperatures if the work output from engines is in proportion of 3 : 2 : 1.

Solution:

Here, $W_1 : W_2 : W_3 = 3 : 2 : 1$

Efficiency of engine, HE_1,

$$\frac{W_1}{Q_1} = \left(1 - \frac{T_2}{1100}\right) \Rightarrow Q_1 = \frac{1100.W_1}{(1100 - T_2)}$$

for HE_2 engine,

$$\frac{W_2}{Q_2} = \left(1 - \frac{T_3}{T_2}\right)$$

for HE_3 engine,

$$\frac{W_3}{Q_3} = \left(1 - \frac{300}{T_3}\right)$$

From energy balance on engine, HE_1

$$Q_1 = W_1 + Q_2 \Rightarrow Q_2 = Q_1 - W_1$$

Above gives,

$$Q_2 = \left\{\frac{1100\, W_1}{(1100 - T_2)} - W_1\right\} = W_1\left\{\frac{T_2}{1100 - T_2}\right\}$$

Substituting Q_2 in efficiency of HE_2

$$\frac{W_2}{W_1\left(\frac{T_2}{1100 - T_2}\right)} = \left(1 - \frac{T_3}{T_2}\right)$$

or $$\frac{W_2}{W_1} = \left(\frac{T_2}{1100 - T_2}\right)\left(\frac{T_2 - T_3}{T_2}\right) = \left(\frac{T_2 - T_3}{1100 - T_2}\right)$$

or $$\left\{\frac{2}{3} = \left(\frac{T_2 - T_3}{1100 - T_2}\right)\right\}$$

or $$2200 - 2T_2 = 3T_2 - 3T_3$$

$$\boxed{5T_2 - 3T_3 = 2200}$$

Energy balance on engine HE_2 gives,

$$Q_2 = W_2 + Q_3$$

Substituting in efficiency of HE_2,

$$\frac{W_2}{(W_2 + Q_3)} = \left(\frac{T_2 - T_3}{T_2}\right)$$

or $$W_2 . T_2 = (W_2 + Q_3)(T_2 - T_3)$$

or $$Q_3 = \frac{W_2\ T_3}{(T_2 - T_3)}$$

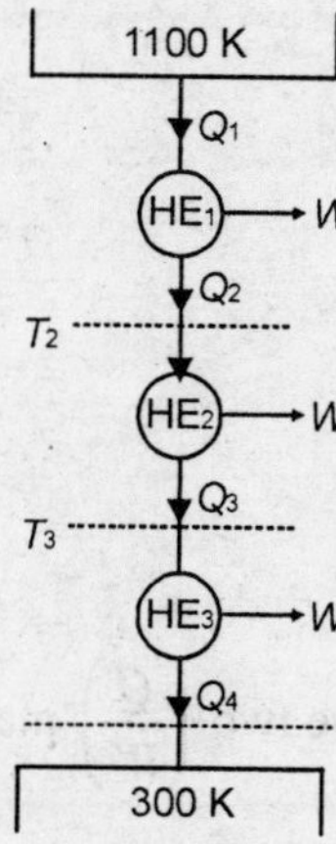

Fig. 4.25

Substituting Q_3 in efficiency of HE_3,

$$\frac{W_3}{\left(\dfrac{W_2 T_3}{T_2 - T_3}\right)} = \frac{T_3 - 300}{T_3}$$

$$\frac{W_3}{W_2} = \left(\frac{T_3}{T_2 - T_3}\right)\left(\frac{T_3 - 300}{T_3}\right)$$

$$\frac{1}{2} = \frac{T_3 - 300}{T_2 - T_3}$$

$$3T_3 - T_2 = 600$$

Solving, equations of T_2 and T_3, $T_3 = 433.33$ K

$$T_2 = 700 \text{ K}$$

Intermediate temperatures: **700 K** and **433.33 K Ans.**

10. *A Carnot engine getting heat at 800 K is used to drive a Carnot refrigerator maintaining 280 K temperature. Both engine and refrigerator reject heat at some temperature, T, when heat given to engine is equal to heat absorbed by refrigerator. Determine efficiency of engine and C.O.P. of refrigerator.*

Solution:

Efficiency of engine,

$$\frac{W}{Q_1} = \left(\frac{800 - T}{800}\right)$$

For refrigerator, COP

$$\frac{Q_3}{W} = \frac{280}{(T - 280)}$$

It is given that $$Q_1 = Q_3 = Q$$

so, from engine

$$\frac{W}{Q} = \left(\frac{800 - T}{800}\right)$$

From refrigerator,

$$\frac{Q}{W} = \frac{280}{T - 280}$$

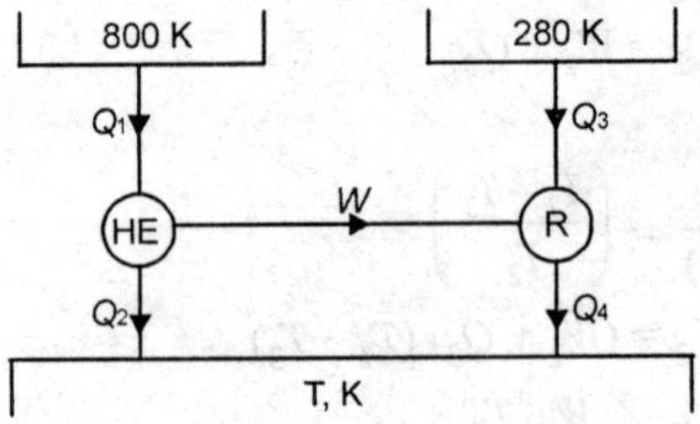

Fig. 4.26

From above two $\left(\frac{Q}{W}\right)$ may be equated,

$$\frac{T-280}{280} = \frac{800-T}{800}$$

Temperature, T = 414.8 K

$$\text{Efficiency of engine} = \left(\frac{800-414.8}{800}\right) = \mathbf{0.4815\ Ans.}$$

$$\text{C.O.P. of refrigerator} = \left(\frac{280}{414.8-280}\right) = \mathbf{2.077\ Ans.}$$

***11.** 0.5 kg of air executes a Carnot power cycle having a thermal efficiency of 50%. The heat transfer to the air during isothermal expansion is 40 kJ. At the beginning of the isothermal expansion the pressure is 7 bar and the volume is 0.12 m³. Determine the maximum and minimum temperatures for the cycle in Kelvin, the volume at the end of isothermal expansion in m³, and the work and heat transfer for each of the four processes in kJ. For air c_p = 1.008 kJ/kg . K, c_v= 0.721 kJ/kg. K. [U.P.S.C. 1993]*

Solution:

Given: $\eta_{carnot} = 0.5$, $m = 0.5$ kg

$P_2 = 2$ bar, $V_2 = 0.12$ m³

Let thermodynamic properties be denoted with respect to salient states;

$$\text{Carnot efficiency } \eta_{carnot} = 1 - \frac{T_1}{T_2}$$

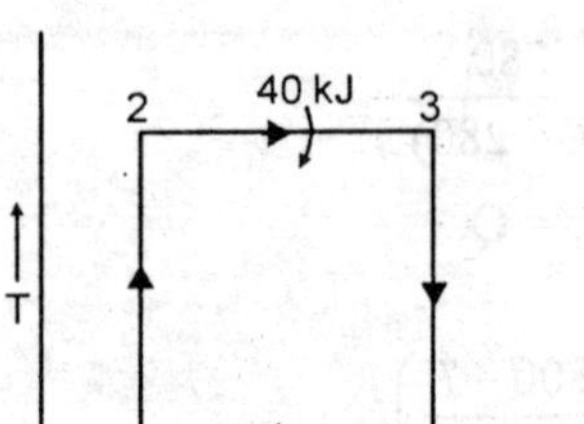

Fig. 4.27

or, $$\frac{T_1}{T_2} = 0.5$$

or, $$T_2 = 2T_1$$

Corresponding to state 2, $P_2\ V_2 = mRT_2$

$$7 \times 10^5 \times 0.12 = 0.5 \times 287 \times T_2$$

$$T_2 = 585.36 \text{ K}$$

Heat transferred during process 2-3 (isothermal expansion), $Q_{23} = 40$ kJ

$$Q_{23} = W_{23} = P_2V_2 \ln\left(\frac{V_3}{V_2}\right)$$

$$40 = mRT_2 \ln \times \left(\frac{V_3}{V_2}\right) = 0.5 \times 0.287 \times 585.36 \ln\left(\frac{V_3}{0.12}\right)$$

$$V_3 = 0.1932 \text{ m}^3$$

Temperature at state 1, $$T_1 = \frac{T_2}{2}$$

$$T_1 = 292.68 \text{ K}$$

During process 1–2,

$$\frac{T_2}{T_1} = \left(\frac{P_2}{P_1}\right)^{\frac{\gamma-1}{\gamma}}$$

$$\gamma = \frac{c_p}{c_v} = \frac{1.008}{0.721},\ \gamma = 1.398$$

Thus, $$P_1 = 0.613 \text{ bar}$$

$$P_1\ V_1 = mRT_1$$

$$0.613 \times 10^5 \times V_1 = 0.5 \times 287 \times 292.68$$

$$V_1 = 6.85 \times 10^{-4} \text{ m}^3$$

Heat transferred during process 4 – 1 (isothermal compression) shall be equal to the heat transferred during process 2 – 3 (isothermal expansion).

For isentropic process, $dQ = 0$

$$dW = dU$$

During process 1 – 2, isentropic process, $W_{12} = -mc_v\ (T_2 - T_1)$

$$Q_{12} = 0,\ W_{12} = -0.5 \times 0.721\ (585.36 - 292.68)$$

$$W_{12} = -\ 105.51 \text{ kJ, (–ve work)}$$

During process 3 – 4, isentropic process, $W_{34} = -mc_v\ (T_4 - T_3)$

$$Q_{34} = 0,\ W_{34} = +\ 0.5 \times 0.721 \times (585.36 - 292.68)$$

$$W_{34} = +\ 105.51 \text{ kJ (+ve work)}$$

Ans.

Process	*Heat transfer*	*Work interaction*
1 – 2	0	– 105.51, kJ
2 – 3	40 kJ	40 kJ
3 – 4	0	+ 105.51, kJ
4 – 1	– 40 kJ	– 40 kJ

Maximum temperature of cycle = 585.36 kJ
Minimum temperature of cycle = 292.68 kJ
Volume at the end of isothermal expansion = 0.1932 m^3

12. A reversible engine as shown in figure during a cycle of operation draws 5 MJ from the 400 K reservoir and does 840 kJ of work. Find the amount and direction of heat interaction with other reservoirs. *[U.P.S.C. 1999]*

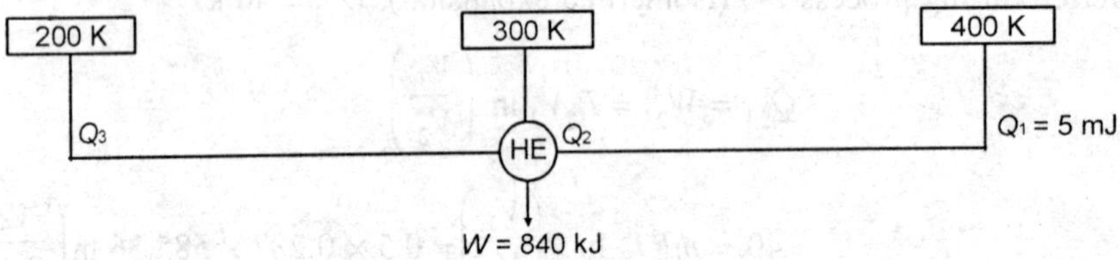

Fig. 4.28

Solution:

Let us assume that heat engine rejects Q_2 and Q_3 heat to reservoir at 300 K and 200 K respectively. Let us assume that there are two heat engines operating between 400 K and 300 K temperature reservoirs and between 400 K and 200 K temperature reservoirs. Let each heat engine receive Q_1' and Q_1'' from reservoir at 400 K as shown below:

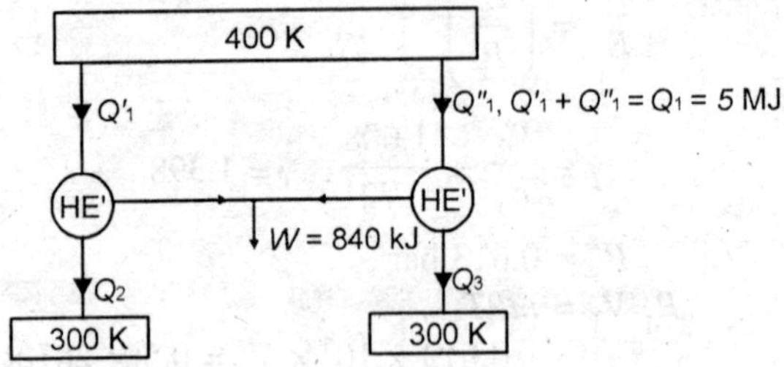

Fig. 4.29 Assumed arrangement

Thus, $Q'_1 + Q''_1 = Q_1 = 5 \times 10^3$ kJ

also, $\dfrac{Q_1'}{Q_2} = \dfrac{400}{300}$, or, $Q'_1 = \dfrac{4}{3}Q_2$

and $\dfrac{Q_1''}{Q_3} = \dfrac{400}{200}$ or, $Q''_1 = 2Q_3$

Substituting Q'_1 and Q''_1

$$\frac{4}{3}Q_2 + 2Q_3 = 5000$$

Also from total work output, $Q'_1 + Q''_1 - Q_2 - Q_3 = W$

$$5000 - Q_2 - Q_3 = 840$$

$$Q_2 + Q_3 = 4160$$

$$Q_3 = 4160 - Q_2$$

Substituting Q_3, $\frac{4}{3}Q_2 + 2(4160 - Q_2) = 5000$

$$\frac{4}{3}Q_2 - 2\,Q_2 = 5000 - 8320$$

$$\frac{-2Q_2}{3} = -\,3320$$

$$Q_2 = 4980 \text{ kJ}$$

and $$Q_3 = -\,820 \text{ kJ}$$

Negative sign with Q_3 shows that the assumed direction of heat Q_3 is not correct and actually Q_3 heat will flow from reservoir to engine. Actual sign of heat transfers and magnitudes are as under:

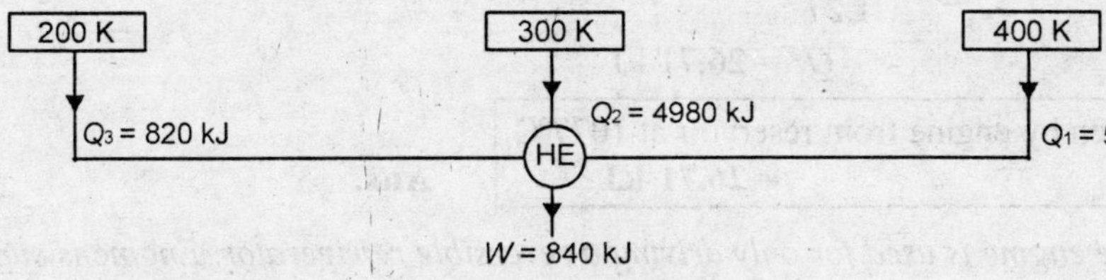

Fig 4.30

Q_2 = 4980 kJ, from heat engine
Q_3 = 820 kJ, to heat engine **Ans.**

13. A heat pump working on a reversed Carnot cycle takes in energy from a reservoir maintained at 3°C and delivers it to another reservoir where temperature is 77°C. The heat pump drives power for it's operation from a reversible engine operating within the higher and lower temperature limits of 1077°C and 77°C. For 100 kJ/s of energy supplied to the reservoir at 77°C, estimate the energy taken from the reservoir at 1077°C. *[U.P.S.C. 1994]*

Solution:

Arrangement for heat pump and heat engine operating together is shown here. Engine and pump both reject heat to the reservoir at 77°C (350 K).

For heat engine.

$$\eta_E = 1 - \frac{350}{1350} = \frac{W}{Q_1}$$

$$0.7407 = \frac{Q_1 - Q_2}{Q_1}$$

$$0.7407 = 1 - \frac{Q_2}{Q_1}$$

$$Q_2 = 0.2593\ \text{Q}_1$$

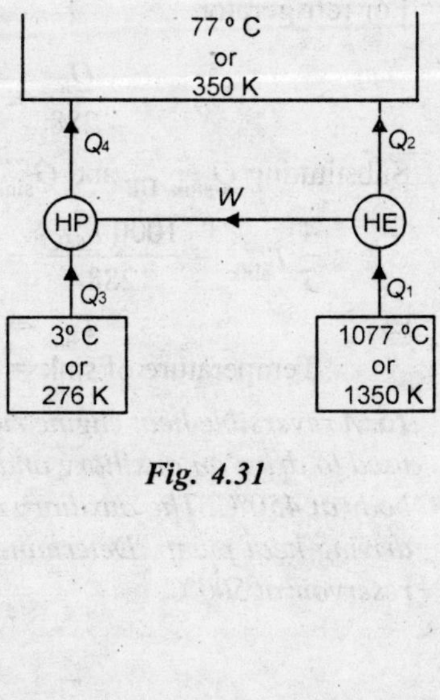

Fig. 4.31

For heat pump

$$COP_{HP} = \frac{Q_4}{Q_4 - Q_3}$$

$$COP_{HP} = \frac{350}{350-276} = \frac{Q_4}{Q_4 - Q_3}$$

$$\Rightarrow \qquad Q_4 = 1.27 Q_3$$

Work output from engine = Work input to pump

$$Q_1 - Q_2 = Q_4 - Q_3 \Rightarrow Q_1 - 0.2593\, Q_1 = Q_4 - \frac{Q_4}{1.27}$$

Also it is given that $Q_2 + Q_4 = 100$

Substituting Q_2 and Q_4 as function of Q_1 in following expression,

$$Q_2 + Q_4 = 100$$

$$0.2539\, Q_1 + \frac{Q_1}{1.27} = 100$$

$$Q_1 = 26.71 \text{ kJ}$$

Energy taken by engine from reservoir at 1077°C = **26.71 kJ**

Ans.

14. A reversible engine is used for only driving a reversible refrigerator. Engine is supplied 2000 kJ/s heat from a source at 1500 K and rejects some energy to a low temperature sink. Refrigerator is desired to maintain the temperature of 15°C while rejecting heat to the same low temperature sink. Determine the temperature of sink if total 3000 kJ/s heat is received by the sink.

Solution:

Let temperature of sink be T_{sink} K.

Given: $Q_{sink,\,HE} + Q_{sink,\,R} = 3000$ kJ/s

Since complete work output from engine is used to run refrigerator so,

$$2000 - Q_{sink,\,HE} = Q_{sink,\,R} - Q_R$$

$$Q_R = 3000 - 2000 = 1000 \text{ kJ/s}$$

Also for engine,

$$\frac{2000}{1500} = \frac{Q_{sink,\,HE}}{T_{sink}} \Rightarrow Q_{sink,\,HE} = \frac{4}{3} T_{sink},$$

For refrigerator,

$$\frac{Q_R}{288} = \frac{Q_{sink,R}}{T_{sink}} \Rightarrow Q_{sink,\,R} = \frac{1000\, T_{sink}}{288}$$

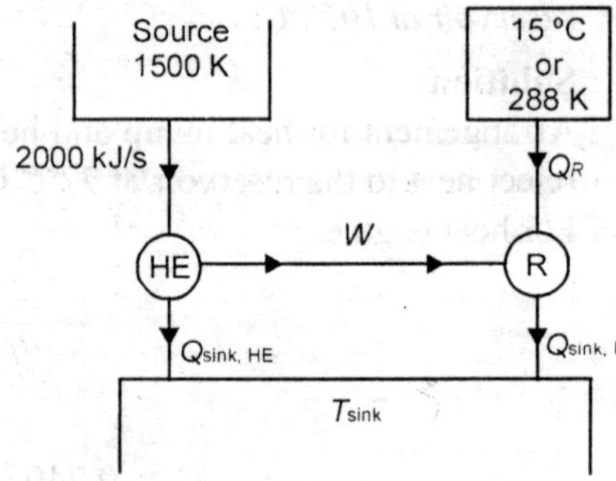

Fig. 4.32

Substituting $Q_{sink,\,HE}$ and $Q_{sink,\,R}$ values.

$$\frac{4}{3} T_{sink} + \frac{1000\, T_{sink}}{288} = 3000$$

$$\Rightarrow \qquad T_{sink} = 624.28 \text{ K}$$

Temperature of sink = **351.28°C Ans.**

15. A reversible heat engine runs between 500°C and 200°C temperature reservoirs. This heat engine is used to drive an auxiliary and a reversible heat pump which runs between reservoir at 200°C and the body at 450°C. The auxiliary consumes one third of the engine output and remaining is consumed for driving heat pump. Determine the heat rejected to the body at 450°C as fraction of heat supplied by reservoir at 500°C.

Solution:

Let the output of heat engine be W. So $\frac{W}{3}$ is consumed for driving auxiliary and remaining $\frac{2W}{3}$ is consumed for driving heat pump for heat engine,

$$\eta = \frac{W}{Q_1} = 1 - \frac{473}{773}$$

$$\frac{W}{Q_1} = 0.3881$$

$$\text{COP of heat pump} = \frac{723}{723-473} = \frac{Q_3}{2W/3}$$

$$\Rightarrow \qquad 2.892 = \frac{3Q_3}{2W}$$

Substituting W,

$$\frac{Q_3}{Q_1} = 0.7482$$

Ratio of heat rejected to body at 450°C to the heat supplied by the reservoir = **0.7482** **Ans.**

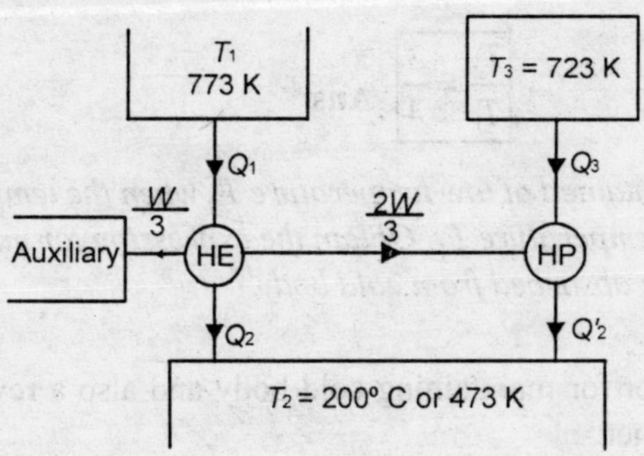

Fig. 4.33

16. A reversible heat engine operates between a hot reservoir at T_1 and a radiating surface at T_2. Heat radiated from the surface is proportional to the surface area and temperature of surface raised to power 4. Determine the condition for minimum surface area for a given work output.

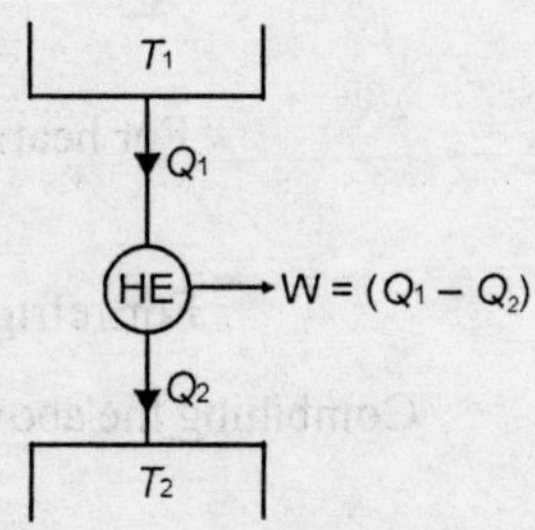

Fig. 4.34

Solution:

Heat rejected = Heat radiated from surface at $T_2 = K \cdot A \cdot T_2^4$, where A is surface area and K is proportionality constant.

$$\frac{Q_2}{W} = \frac{T_2}{T_1 - T_2}$$

$$\frac{K.A.T_2^4}{W} = \frac{T_2}{T_1 - T_2}$$

$$\Rightarrow \qquad A = \frac{W}{T_2^3(T_1 - T_2).K}$$

In order to have minimum surface area the denominator in above expression of A should be maximum i.e. $T_2^3\ (T_1 - T_2)$ should be maximum. Differentiating with respect to T_2.

$$\frac{d}{dT_2}\{(T_2^3(T_1 - T_2))\} = 0$$

$$3T_1.\ T_2^2 - 4T_2^3 = 0$$

$$\Rightarrow \qquad \frac{T_2}{T_1} = \frac{3}{4} \Rightarrow T_2 = T_1 . \frac{3}{4}$$

Taking second differential

$$\frac{d^2}{dT_2^2}\{T_2^3.(T_1 - T_2)\} = 6T_1 . T_2 - 12T_2^2$$

Upon substitution it is –ive so $\left(\frac{T_2}{T_1} = \frac{3}{4}\right)$ is the condition for $\{T_2^3 (T_1 - T_2)\}$ to be maximum and so the minimum surface area

$$\boxed{\frac{T_2}{T_1} = \frac{3}{4}} \quad \textbf{Ans.}$$

17. A cold body is to be maintained at low temperature T_2 when the temperature of surrounding is T_3. A source is available at high temperature T_1. Obtain the expression for minimum theoretical ratio of heat supplied from source to heat absorbed from cold body.

Solution:

Let us consider a refrigerator for maintaining cold body and also a reversible heat engine for driving refrigerator to operate together.

To obtain; $\frac{Q_1}{Q_3}$

For heat engine, $\frac{W}{Q_1} = \frac{T_1 - T_3}{T_1}$

For refrigerator, $\frac{Q_3}{W} = \frac{T_2}{T_3 - T_2}$

Combining the above two:

$$\frac{Q_1}{Q_3} = \frac{T_1 \times (T_3 - T_2)}{T_2 \times (T_1 - T_3)}$$

Ratio of heat supplied from source to heat absorbed from cold body $= \frac{T_1.(T_3 - T_2)}{T_2.(T_1 - T_3)}$ **Ans.**

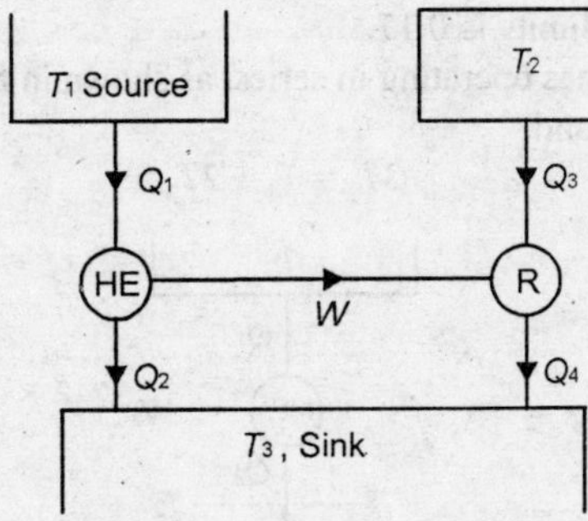

Fig. 4.35

EXERCISE

4.1 State the Kelvin Planck and Clausius statements of 2nd law of thermodynamics.

4.2 Show the equivalence of two statements of 2nd law of thermodynamics.

4.3 Write short notes on the following:
Heat reservoir, Heat engine, Heat pump and refrigerator.

4.4 Explain the reversible and irreversible processes.

4.5 Describe Carnot cycle and obtain expression for its efficiency as applied to a heat engine.

4.6 Why Carnot cycle is a theoretical cycle? Explain.

4.7 Show that coefficient of performance of heat pump and refrigerator can be related as;
$$COP_{Ref} = COP_{HP-1}$$

4.8 State Carnot theorem. Also prove it.

4.9 Show that the efficiencies of all reversible heat engines operating between same temperature limits are same.

4.10 Show that efficiency af an irreversible engine is always less than the efficiency of reversible engine operating between same temperature limits.

4.11 Assume an engine to operate on Carnot cycle with complete reversibility except that 10% of work is required to overcome friction. For the efficiency of reversible cycle being 30% what shall be the efficiency of assumed engine.
If same magnitude of energy is required to overcome friction, if machine operated as heat pump, then what shall be ratio between refrigerating effect and work required. [27%, 2.12]

4.12 A Carnot engine operating between certain temperature limits has an efficiency of 30%. Determine the ratio of refrigerating effect and work required for operating the cycle as a heat pump between the same temperature limits. [2.33]

4.13 An inventor claims to have developed an engine that takes in 1055 MJ at a temperature of 400K and rejects 42.2 MJ at a temperature of 200 K while delivering 15kWh of mechanical work. Check whether engine is feasible or not. [Engine satisfies Ist law but violates 2nd law]

4.14 Determine which of the following is the most effective way to increase Carnot engine efficiency
(i) To increase T_2 while keeping T_1 fixed.
(ii) To decrease T_1 while keeping T_2 fixed. [If T_1 is decreased]

4.15 A refrigerator has COP one half as great as that of a Carnot refrigerator operating between reservoirs at temperatures of 200 K and 400 K, and absorbs 633 KJ from low temperature reservoir. How much heat is rejected to the high temperature reservoir. [1899 kJ]

4.16 Derive a relationship between COP of a Carnot refrigerator and the efficiency of same refrigerator when operated as an engine.
Is a Carnot engine having very high efficiency suited as refrigerator?

4.17 Calculate COP of Carnot refrigerator and Carnot heat pump, if the efficiency of the Carnot engine between same temperature limits is 0.17. [5, 6]

4.18 For the reversible heat engines operating in series, as shown in figure 4.36. Show the following, if work output is twice that of second.

$$3T_2 = T_1 + 2T_3$$

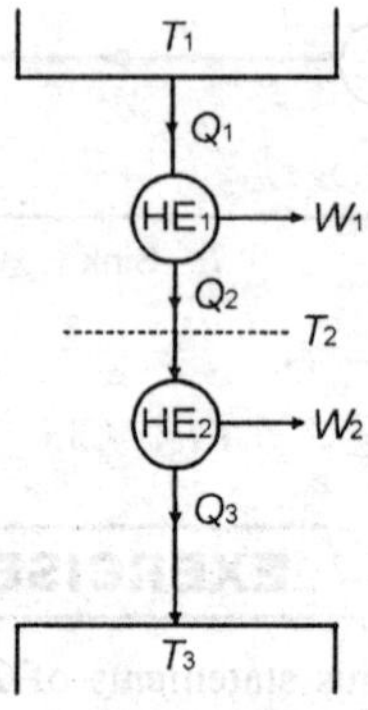

Fig. 4.36

4.19 A domestic refrigerator is intended to freeze water at 0°C while water is available at 20°C. COP of refrigerator is 2.5 and power input to run it is 0.4 kW. Determine capacity of refrigerator if it takes 14 minutes to freeze. Take specific heat of water as 4.2 kJ/kg. °C. [10 kg]

4.20 A cold storage plant of 49.64 hp power rating removes 7.4 MJ/min and discharges heat to atmospheric air at 30°C. Determine the temperature maintained inside the cold storage. [–40°C]

4.21 A house is to be maintained at 21°C from inside during winter season and at 26°C during summer. Heat leakage through the walls, windows and roof is about 3 × 10³ kJ/hr per degree temperature difference between the interior of house and environment temperature. A reversible heat pump is proposed for realizing the desired heating/cooling. What minimum power shall be required to run the heat pump in reversed cycle if outside temperature during summer is 36°C? Also find the lowest environment temperature during winter for which the inside of house can be maintained at 21°C. [0.279 kW, 11°C]

4.22 Estimate the minimum power requirement of a heat pump for maintaining a commercial premises at 22°C when environment temperature is –5°C. The heat load on pump is 1 × 10⁷ kJ/day.

4.23 A reversible engine having 50% thermal efficiency operates between a reservoir at 1527°C and a reservoir at some temperature *T*. Determine temperature *T* in *K*.

4.24 A reversible heat engine cycle gives output of 10 kW when 10 kJ of heat per cycle is supplied from a source at 1227°C. Heat is rejected to cooling water at 27°C. Estimate the minimum theoretical number of cycles required per minute. [75]

4.25 Some heat engine *A* and a reversible heat engine *B* operate between same two heat reservoirs. Engine *A* has thermal efficiency equal to two-third of that of reversible engine *B*. Using second law of thermodynamics show that engine *A* shall be irreversible engine.

4.26 Show that the COP of a refrigeration cycle operating between two reservoirs shall be, $COP_{ref} = \left(\frac{1}{\eta_{max}} - 1\right)$, if η_{max} refers to thermal efficiency of a reversible engine operating between same temperature limits.

4.27 A heat pump is used for maintaining a building at 20°C. Heat loss through roofs and walls is at the rate of 6 × 10⁴ kJ/h. An electric motor of 1 kW rating is used for driving heat pump. On some day when environment temperature is 0°C would it be possible for pump to maintain building at desired temperature? [No]

5 Entropy

5.1 INTRODUCTION

Till now the detailed explanation of zeroth law, first law and second law of thermodynamics have been made. Also we have seen that the first law of thermodynamics defined a very useful property called internal energy. For overcoming the limitations of first law the second law of thermodynamics had been stated. Now we need some mathematical parameter for being used as decision maker in respect of feasibility of process, irreversibility, nature of process etc. Here in this chapter a mathematical function called 'entropy' has been explained. 'Entropy' is the outcome of second law and is a thermodynamic property. Entropy is defined in the form of calculus operation, hence no exact physical description of it can be given. However, it has immense significance in thermodynamic process analysis.

5.2 CLAUSIUS INEQUALITY

Let us take any reversible process 1–2 as shown on *P–V* diagram. Let us also have a reversible adiabatic process 1–1′ followed by reversible isothermal process 1′–2′ and a reversible adiabatic process 2′ – 2, as approximation to the original process 1 – 2 such that area under 1 – 2 equals to that under 1–1′–2′ –2. By first law of thermodynamics on 1–2.

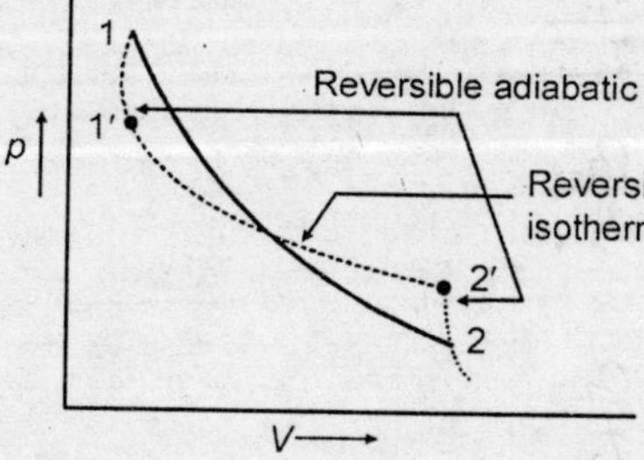

Fig. 5.1 Reversible adiabatic, isothermal and reversible process

$$Q_{1-2} = (U_2 - U_1) + W_{1-2}$$

First law on 1–1′–2′–2 processes; (Heat and work are path functions and internal energy is point function)

$$Q_{1-1'-2'-2} = (U_2 - U_1) + W_{1-1'-2'-2}$$

As already assumed that $W_{1-2} = W_{1-1'-2'-2}$

so $Q_{1-1'-2'-2} = Q_{1-2}$

In the path 1 – 1′ –2′ –2 during adiabatic processes 1 – 1′ & 2 –2′ there is no heat interaction so the total heat interaction in 1 –2 is getting occurred during isothermal process 1′ – 2′ only.

Hence, it is always possible to replace any reversible process by a series of reversible adiabatic, reversible isothermal and reversible adiabatic processes between the same end states provided the heat interaction and work involved remains same.

If the number of reversible adiabatic and reversible isothermal processes is quite large then the series of such processes shall reach close to the original reversible process.

Let us undertake this kind of substitution for the processes in a reversible cycle

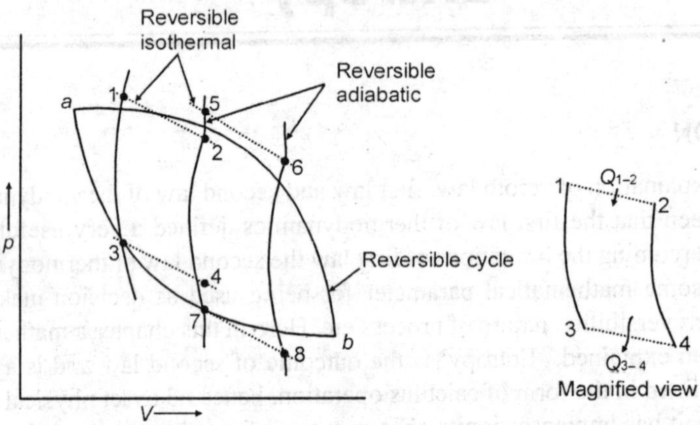

Fig. 5.2 A reversible cycle replaced by reversible adiabatics and reversible isotherms

Figure 5.2 shows replacement of original processes in cycle $a - b - a$ by adiabatic and isothermals. This shall result in a number of Carnot cycles appearing together in place of original cycle. Two Carnot cycles thus formed are shown by 1 –2 – 4 –3 and 5 – 6 – 8 – 7. Magnified view of first Carnot cycle is also shown separately where heat supplied at high temperature is Q_{1-2} and heat rejected at low temperature is Q_{3-4}.

From thermodynamic temperature scale;

For Carnot cycle 1 – 2 – 4 – 3,

$$\frac{Q_{1-2}}{Q_{3-4}} = \frac{T_{1-2}}{T_{3-4}}$$

or

$$\frac{Q_{1-2}}{T_{1-2}} = \frac{Q_{3-4}}{T_{3-4}}$$

For Carnot cycle 5 – 6 – 8 – 7,

$$\frac{Q_{5-6}}{T_{5-6}} = \frac{Q_{7-8}}{T_{7-8}}$$

Now taking sign conventions for heat added and rejected;

$$\frac{Q_{1-2}}{T_{1-2}} + \frac{Q_{3-4}}{T_{3-4}} = 0$$

and

$$\frac{Q_{5-6}}{T_{5-6}} + \frac{Q_{7-8}}{T_{7-8}} = 0$$

Hence, if there are '*n*' number of Carnot cycles replacing the original reversible cycle, then

$$\left(\frac{Q_{1-2}}{T_{1-2}}+\frac{Q_{3-4}}{T_{3-4}}\right)+\left(\frac{Q_{5-6}}{T_{5\text{-}6}}+\frac{Q_{7-8}}{T_{7-8}}\right)+\ldots=0$$

or, it can be given as summation of the ratio of heat interaction (Q) to the temperature (T) at which it occurs, being equal to zero.

$$\sum\frac{Q}{T}=0$$

If number of Carnot cycles is very large, then the zig-zag path formed due to replacing adiabatics and isotherms shall reach very close to original cycle.

In such situation the cyclic integral of $\frac{Q}{T}$ may be given in place of above.

or $$\oint\left(\frac{dQ}{T}\right)_{\text{rev}}=0$$

Here it indicates that $\left(\frac{dQ}{T}\right)$ is some thermodynamic property. Above expression developed for a reversible heat engine cycle also remains valid for internally reversible engines. In case of internally reversible engines T shall be temperature of working fluid at the time of heat interaction.

$$\oint\left(\frac{dQ}{T}\right)_{\text{int, rev}}=0$$

Let us now try to find out what happens to $\oint\frac{dQ}{T}$ when we have an irreversible engine cycle. Let there be a reversible and irreversible heat engine operating between same temperature limits, such that heat added to them is same.

Form Carnot's theorem for both reversible and irreversible heat engine cycles,

$$\eta_{\text{rev}}>\eta_{\text{irrev}}$$

or $$\left(1-\frac{Q_{\text{rejected}}}{Q_{\text{add}}}\right)_{\text{rev}}>\left(1-\frac{Q_{\text{rejected}}}{Q_{\text{add}}}\right)_{\text{irrev}}$$

or $$\left(\frac{Q_{\text{rejected}}}{Q_{\text{add}}}\right)_{\text{irrev}}>\left(\frac{Q_{\text{rejected}}}{Q_{\text{add}}}\right)_{\text{rev}}$$

For same heat added, i.e. $Q_{\text{add, rev}}=Q_{\text{add, irrev}}=Q_{\text{add}}$

$$Q_{\text{rejected, irrev}}>Q_{\text{rejected, rev}}$$

or $$\frac{Q_{\text{rejected, irrev}}}{Q_{\text{rejected, rev}}}>1$$

For absolute thermodynamic temperature scale,

$$\left(\frac{Q_{\text{add}}}{Q_{\text{rejected}}}\right)_{\text{rev}}=\frac{T_{\text{add}}}{T_{\text{rejected}}}$$, upon substitution we get,

$$\frac{Q_{\text{rejected, irrev}}}{Q_{\text{add}}} > \frac{T_{\text{rejected}}}{T_{\text{add}}}$$

or $$\frac{Q_{\text{rejected, irrev}}}{T_{\text{rejected}}} > \frac{Q_{\text{add}}}{T_{\text{add}}}$$

Upon substituting sign convention, we get

$$\frac{Q_{\text{add}}}{T_{\text{add}}} + \frac{Q_{\text{rejected, irrev}}}{T_{\text{rejected}}} < 0$$

or $$\frac{Q_{\text{add, irrev}}}{T_{\text{add}}} + \frac{Q_{\text{rejected, irrev}}}{T_{\text{rejected}}} < 0$$

If it is given in the form of cyclic integral.

or $$\boxed{\oint \left(\frac{dQ}{T}\right)_{\text{irrev}} < 0}$$

Now combining for reversible and irreversible paths it can be given as;

$$\oint \left(\frac{dQ}{T}\right) \le 0$$

This is called *Clausius inequality*.

here, $$\oint \left(\frac{dQ}{T}\right) = 0 \text{ for reversible cycle}$$

$$\oint \left(\frac{dQ}{T}\right) < 0, \text{ for irreversible cycle}$$

$$\oint \left(\frac{dQ}{T}\right) > 0, \text{ for impossible cycle}$$

5.3 ENTROPY–A PROPERTY OF SYSTEM

From Clausius inequality mathematically it is shown that for a reversible cycle.

$$\oint \left(\frac{dQ}{T}\right)_{\text{rev}} = 0$$

Let us take a reversible cycle comprising of two processes A and B as shown and apply Clausius inequality.

$$\int_{a \text{ path } A}^{b} \frac{dQ}{T} + \int_{b \text{ path } B}^{a} \frac{dQ}{T} = 0$$

$$\int_{a \text{ path } A}^{b} \frac{dQ}{T} = -\int_{b \text{ path } B}^{a} \frac{dQ}{T}$$

or

$$\int_{a \text{ path } A}^{b} \frac{dQ}{T} = \int_{a \text{ path } A}^{b} \frac{dQ}{T}$$

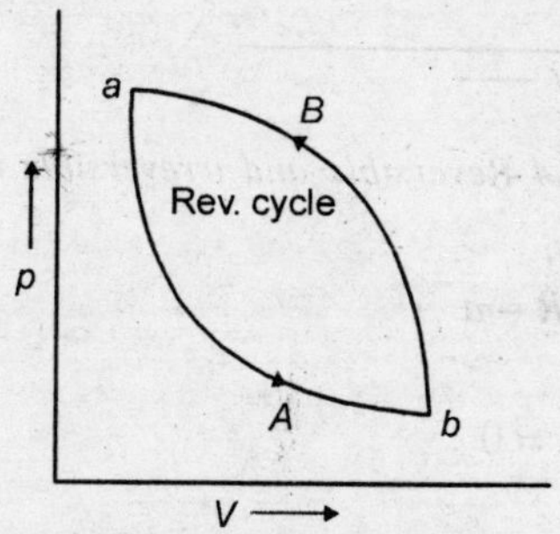

Fig. 5.3 Reversible cycle

Hence, it shows that $\left(\frac{dQ}{T}\right)$ is some property and does not depend upon path followed. This thermodynamic property is called "entropy". Entropy is generally denoted by 'S' or 'ϕ'.

Thus, the energy interactions in the form of heat are only accompanied by entropy changes.

Writing is as function of entropy change.

$$\int_{a \text{ path } A}^{b} \frac{dQ}{T} = \int_{a \text{ path } A}^{b} \frac{dQ}{T} = S_b - S_a$$

or

$$\int \left(\frac{dQ}{T}\right)_{\text{rev}} = \int dS$$

Since entropy is point function and depends only upon end states therefore entropy change for any process following reversible or irreversible path shall be same.

$$(S_b - S_a)_{\text{rev, path}} = (S_b - S_a)_{\text{irrev, path}}$$

$$\boxed{\int_a^b \frac{dQ_{\text{rev}}}{T} = \Delta S_{\text{rev, path}} = \Delta S_{\text{irrev,path}}}$$

Entropy is an extensive property and has units J/K. Specific entropy may be, given on unit mass basis

$$s = \frac{S}{m} \text{ (J/kg.K)}$$

Entropy, as obvious from definition is defined for change in entropy value, therefore absolute value of entropy cannot be defined. Entropy is always given as change, i.e. relative values can be expressed.

Let us now have two thermodynamic cycles $a - b - a$ following paths $(a - R - b,\ b - R - a)$ and $(a - R - b,\ b - I - a)$.

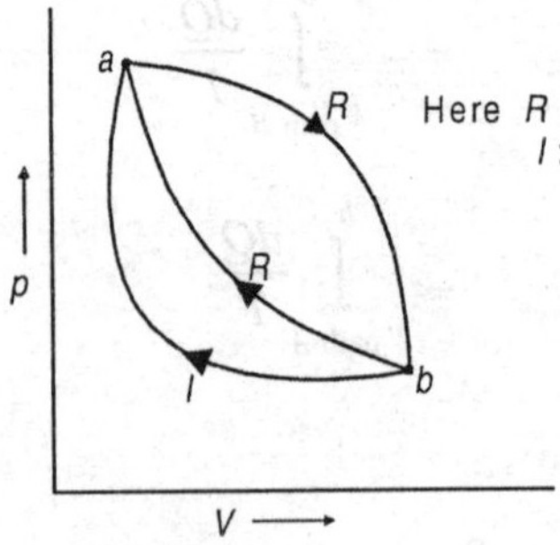

Fig. 5.4 Reversible and irreversible cycle

We have from Clausius inequality.

For reversible cycle $a - R - b - R - a$

$$\oint\left(\frac{dQ}{T}\right)_{\text{rev}} = 0$$

$$\int_{a,R}^{b} \frac{dQ}{T} + \int_{b,R}^{a} \frac{dQ}{T} = 0$$

For irreversible cycle $a - R - b - I - a$

$$\oint\left(\frac{dQ}{T}\right) < 0$$

or
$$\int_{a,R}^{b} \frac{dQ}{T} + \int_{b,I}^{a} \frac{dQ}{T} < 0$$

or
$$\int_{a,R}^{b} \frac{dQ}{T} < -\int_{b,I}^{a} \frac{dQ}{T}$$

Also from definition of entropy.

$$\int\left(\frac{dQ}{T}\right)_{\text{rev}} = \int dS$$

so from above

$$\int_{a,R}^{b} \frac{dQ}{T} = \int_{a}^{b} dS$$

or
$$\left(\frac{dQ}{T}\right)_{\text{rev}} = dS$$

For reversible cycle.

$$\int_{a,R}^{b} \frac{dQ}{T} = -\int_{b,R}^{a} \frac{dQ}{T}$$

Substituting it in expression for irreversible cycle.

$$-\int_{b,R}^{a} \frac{dQ}{T} < -\int_{b,I}^{a} \frac{dQ}{T}$$

also from definition of entropy.

$$-\int_{b}^{a} dS < -\int_{b,I}^{a} \frac{dQ}{T}$$

or $$\int_{b,I}^{a} \frac{dQ}{T} < \int_{b}^{a} dS$$

or, in general it can be given as,

$$\left(\frac{dQ}{T}\right)_{\text{irrev}} < dS$$

Combining the above two we get inequality as following,

$$\boxed{dS \geq \frac{dQ}{T}}$$

where $$dS = \frac{dQ}{T} \quad \text{for reversible process}$$

$$dS > \frac{dQ}{T} \quad \text{for irreversible process}$$

Mathematical formulation for entropy ($dQ_{\text{rev}} = T \cdot dS$) can be used for getting property diagrams between "temperature and entropy" ($T-S$), "enthalpy and entropy" ($h-S$).

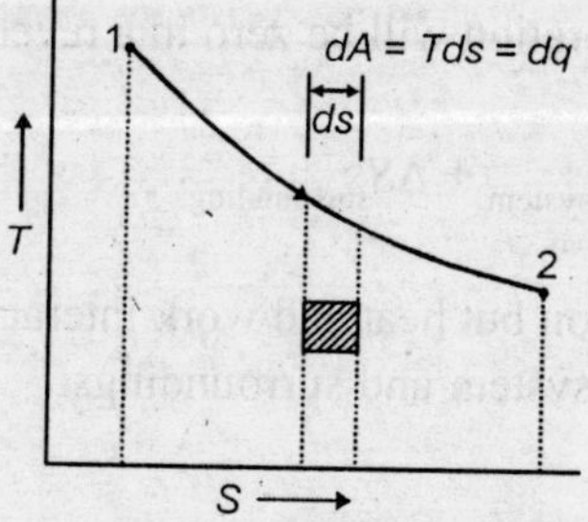

Fig. 5.5 T-S diagram

Area under process curve on T–S diagram gives heat transferred, for internally reversible process

$$dQ_{\text{int, rev}} = T \cdot dS$$

$$\boxed{Q_{\text{int, rev}} = \int_{1}^{2} T \cdot dS}$$

5.4 PRINCIPLE OF ENTROPY INCREASE

By second law, entropy principle has been obtained as,

$$\boxed{dS \geq \frac{dQ}{T}}$$

For an isolated system:

$$dQ = 0, \text{ therefore } dS_{\text{isolated}} \geq 0$$

for a reversible process $dS_{\text{isolated}} = 0$

i.e. $S_{\text{isolated}} = \text{constant}$

for an irreversible process $dS_{\text{isolated}} > 0$

which means the entropy of an isolated system always increases, if it has irreversible processes,

In general form $\boxed{dS_{\text{isolated}} \geq 0}$

It may be concluded here that the entropy of an isolated system always increases or remains constant in case of irreversible and reversible processes respectively. This is known as "Principle of entropy increase" of "entropy principle".

Universe which comprises of system and surroundings may also be treated as isolated system and from entropy principle;

$$dS_{\text{Universe}} \geq 0$$

Which means that entropy of universe either increases continuously or remains constant depending upon whether processes occurring in it are of "irreversible" or "reversible" type respectively.

Since, Universe = System + Surrounding.

therefore

$$dS_{\text{system}} + dS_{\text{surrounding}} \geq 0$$

or

$$\Delta S_{\text{system}} + \Delta S_{\text{surrounding}} \geq 0$$

Since most of the processes occurring generally are of irreversible type in universe, so it can be said that in general entropy of universe keeps on increasing and shall be maximum after attaining the state of equilibrium, which is very difficult to attain.

In the above expression, system and surroundings are treated as two parts of universe (an isolated system). So the total entropy change during a process can be given by sum of "entropy change in system" and "entropy change in surroundings". This total entropy change is also called "entropy generation" or "entropy production". Entropy generation will be zero in a reversible process.

Therefore

$$\Delta S_{\text{total}} = S_{\text{gen}} = \Delta S_{\text{system}} + \Delta S_{\text{surrounding}}$$

For closed systems

In case of closed systems there is no mass interaction but heat and work interactions are there. Entropy change is related to heat interactions occurring in system and surroundings.

Total entropy change.

or

Entropy generated

$$\Delta S_{\text{total}} = S_{\text{gen}} = \Delta S_{\text{system}} + \Delta S_{\text{surrounding}}$$

For system changing it's state from 1 to 2 i.e. initial and final state.

$$\Delta S_{\text{system}} = S_2 - S_1$$

For surroundings entropy change depends upon heat interactions.

$$\Delta S_{\text{surrounding}} = \frac{Q_{\text{surrounding}}}{T_{\text{surrounding}}}$$

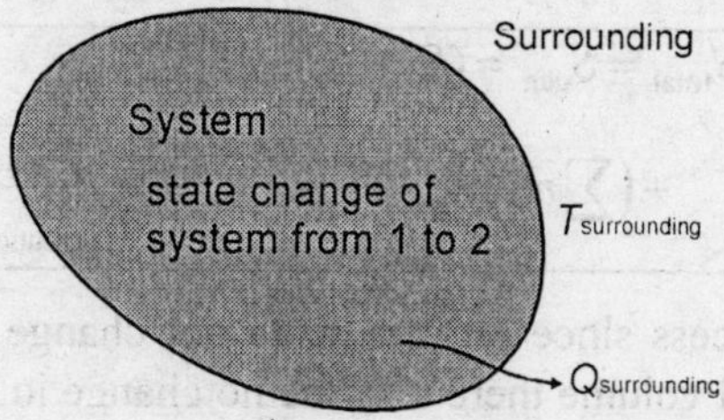

***Fig. 5.6** Closed system*

or $$-\Delta S_{\text{total}} = S_{\text{gen}} = (S_2 - S_1) + \frac{Q_{\text{surrounding}}}{T_{\text{surrounding}}}$$

or $$S_2 - S_1 = m(S_2 - S_1)$$

where m is mass in system and s_1 and s_2 are specific entropy values at initial and final state, then

$$\boxed{S_{\text{gen}} = m(s_2 - s_1) + \frac{Q_{\text{surrounding}}}{T_{\text{surrounding}}}}$$

For open systems

In case of open systems the mass interactions also take place along with energy interactions. Here mass flow into and out of system shall also cause some entropy change, so a control volume as shown in figure is to be considered. Entropy entering and leaving at section $i - i$ and $o - o$ are considered. Mass flow carries both energy and entropy into or out of control volume. Entropy transfer with mass flow is called "entropy transport". It is absent in closed systems.

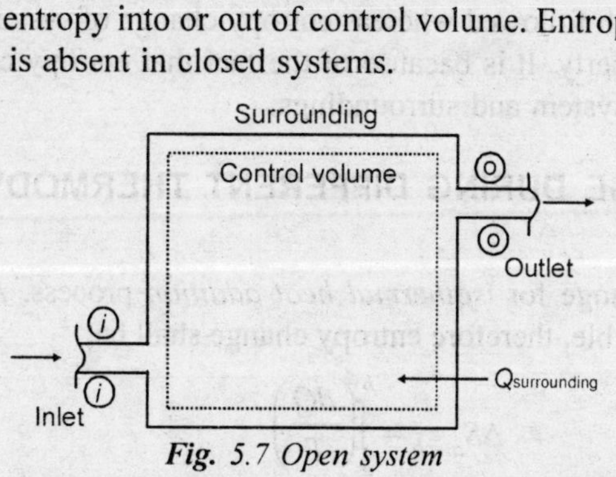

***Fig. 5.7** Open system*

If control volume undergoes state change from 1 to 2, then entropy change in control volume shall be $(S_2 - S_1)$ while entropy entering and leaving out may be given as S_i and S_o respectively.

By principle of entropy increase, total entropy change shall be,

$$\Delta S_{\text{total}} = S_{\text{gen}} = (S_2 - S_1) + (S_o - S_i) + \frac{Q_{\text{surrounding}}}{T_{\text{surrounding}}}$$

Entropy entering and leaving out may be given as sum of entropy of all mass flows into and out of system in case of *uniform flow process*.

Therefore
$$S_i = \sum m_i \,.\, s_i$$
$$S_o = \sum m_o \,.\, s_o$$
where m_i and m_o are mass flows into and out of system, and s_i and s_o are specific entropy associated with mass entering and leaving.

Substituting,

$$\boxed{\begin{aligned}\Delta S_{\text{total}} &= S_{\text{gen}} = (S_2 - S_1) \\ &+ \left(\sum m_o \,.\, s_o - \sum m_i \,.\, s_i\right) + \frac{Q_{\text{surrounding}}}{T_{\text{surrounding}}}\end{aligned}}$$

In case of steady flow process since properties do not change with respect to time during any process, therefore within control volume there shall be no change in entropy.

i.e.
$$S_1 = S_2$$

Total entropy change or entropy generation for this case shall be;

$$\Delta S_{\text{total}} = S_{\text{gen}} = (S_o - S_i) + \frac{Q_{\text{surrounding}}}{T_{\text{surrounding}}}$$

$$\boxed{\Delta S_{\text{total}} = S_{\text{gen}} = \left(\sum m_o \,.\, s_o - \sum m_i \,.\, s_i\right) + \frac{Q_{\text{surrounding}}}{T_{\text{surrounding}}}}$$

In all the cases discussed above $\Delta S_{\text{total}} \geq 0$ or $S_{\text{gen}} \geq 0$.

Entropy generated can be taken as criterion to indicate feasibility or process as follows;

* If S_{gen} or $\Delta S_{\text{total}} = 0$ then process is reversible.

* If S_{gen} or $\Delta S_{\text{total}} > 0$ then process is irreversible.

* If S_{gen} or $\Delta S_{\text{total}} < 0$ then process is impossible.

One thing is very important about entropy generated that S_{gen} is not a thermodynamic property and it's value depends on the path followed whereas entropy change of system $(S_2 - S_1)$ is a point function and so thermodynamic property. It is bacause of the fact that entropy change is cumulative effect of entropy transfer/change in system and surroundings.

5.5 ENTROPY CHANGE DURING DIFFERENT THERMODYNAMIC PROCESSES

Isothermal process

Let us find out entropy change for *isothermal heat addition* process. As isothermal process can be considered internally reversible, therefore entropy change shall be;

$$\Delta S_{a-b} = \int_a^b \left(\frac{dQ}{T}\right)$$

or
$$\Delta S_{a-b} = \frac{1}{T}\int_a^b dQ$$

or
$$\boxed{\Delta S_{a-b} = \frac{Q_{a-b}}{T}}$$

where Q_{a-b} is total heat interaction during $a - b$ at temperature T.

Isentropic process

It is the process during which change in entropy is zero and entropy remains constant during process.

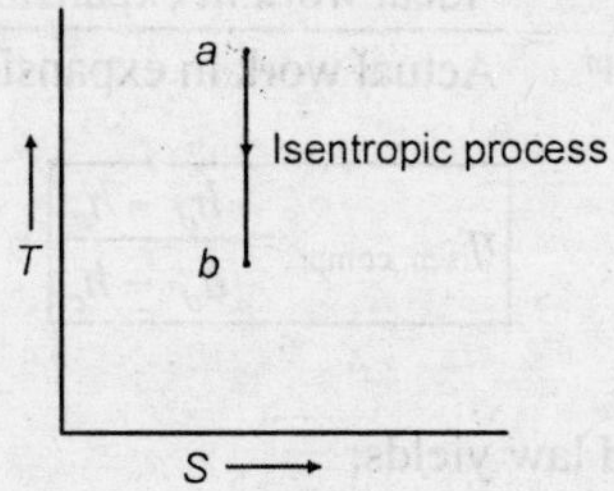

Fig. 5.8 Isentropic process

It indicates that when $\Delta S_{a-b} = 0$.

then $$Q_{a-b} = 0$$

which means there is no heat interaction during such process and these are adiabatic process.

Hence, it can be said that "a reversible isentropic process shall be adiabatic, where as if isentropic process is adiabatic then it may or may not be reversible.

Thus, adiabatic process may or may not be reversible. It means in reversible adiabatic process all states shall be in equilibrium and no dissipative effects are present along with no heat interaction whereas in adiabatic process there is no heat interaction but process may be irreversible.

Finally, it can be concluded that an adiabatic process may or may not be isentropic whereas a reversible adiabatic process is always isentropic.

An adiabatic process of non isentropic type is shown below where irreversibility prevails, say due to internal friction.

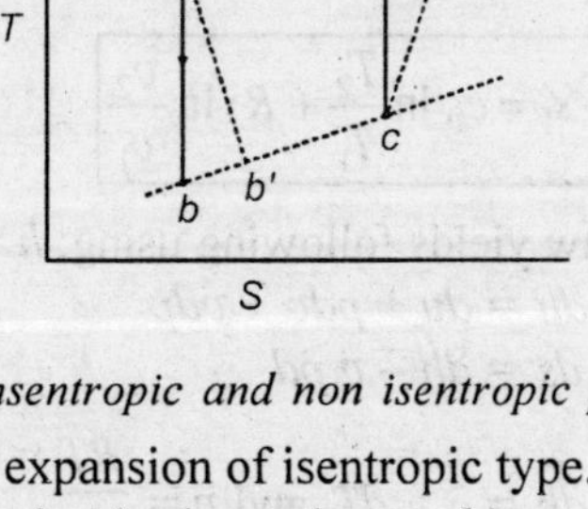

Fig. 5.9 Insentropic and non isentropic processes

Here $a - b$ is reversible adiabatic expansion of isentropic type.

Non isentropic or adiabatic expansion is shown by $a - b'$.

Isentropic expansion efficiency may be defined as ratio of actual work to ideal work available during expansion.

$$\eta_{\text{isen, expn}} = \frac{\text{Actual work in expansion}}{\text{Ideal work in expansion}}$$

$$\boxed{\eta_{\text{isen,expn}} = \frac{h_a - h_{b'}}{h_a - h_b}}$$

Similarly, isentropic and non isentropic compression process are shown as $c-d$ and $c-d'$ respectively. Isentropic compression efficiency can be defined on same lines as,

$$\eta_{\text{isen, compr}} = \frac{\text{Ideal work in expansion}}{\text{Actual work in expansion}}$$

$$\boxed{\eta_{\text{isen, compr}} = \frac{h_d - h_c}{h_{d'} - h_c}}$$

For ideal gases

Combination of first and second law yields;

$$Tds = du + pdv$$

also we know $du = c_v \cdot \text{dT}$ and for perfect gas $p = \frac{RT}{v}$

substituting for du and R

$$T \cdot ds = c_v \cdot dT + \frac{RT}{v} \cdot dv$$

or

$$ds = c_v \cdot \frac{dT}{T} + \frac{Rdv}{v}$$

or

$$s_2 - s_1 = \int_1^2 c_v \cdot \frac{dT}{T} + \int_1^2 \frac{Rdv}{v}$$

If c_v is function of temperature then,

$$\boxed{s_2 - s_1 = \int_1^2 c_v(T) \cdot \frac{dT}{T} + R \cdot \ln \frac{v_2}{v_1}}$$

If specific heat is constant then,

$$\boxed{s_2 - s_1 = c_v \ln \frac{T_2}{T_1} + R \cdot \ln \frac{v_2}{v_1}}$$

Also combination of Ist and IInd law yields following using; $h = u + pv$,

or

$$dh = du + pdv + vdp$$

$$T \cdot ds = dh - v\,pd$$

substituting $dh = c_p \cdot dT$, and $v = \frac{RT}{p}$

$$T \cdot ds = c_p \cdot dT - v \cdot dp$$

or

$$ds = c_p \cdot \frac{dT}{T} - \frac{Rdp}{p}$$

entropy change

$$s_2 - s_1 = \int_1^2 \frac{c_p\,dT}{T} - \int_1^2 \frac{Rdp}{p}$$

If specific heat is function of temperature then

$$\boxed{s_2 - s_1 = \int_1^2 \frac{c_p\,(T)dT}{T} - R \ln \frac{p_2}{p_1}}$$

If specific heat is constant, then

$$\boxed{s_2 - s_1 = c_p \ln \frac{T_2}{T_1} - R \ln \frac{p_2}{p_1}}$$

Above expressions given in enclosed box may be suitable used for getting the change in entropy.

Polytropic Process

Entropy change in a polytropic process having governing equation as pv^n = constant, can be obtained as below,

For polytropic process between 1 and 2, $p_1 v_1^n = p_2 v_2^n$

or
$$\left(\frac{p_1}{p_2}\right) = \left(\frac{v_2}{v_1}\right)^n$$

Also, from gas laws,

$$\frac{p_1 v_1}{T_1} = \frac{p_2 v_2}{T_2}$$

$$\frac{p_1}{p_2} = \left(\frac{v_2}{v_1} \times \frac{T_1}{T_2}\right)$$

Above two pressure ratios give,

$$\frac{v_2}{v_1} = \left(\frac{T_1}{T_2}\right)^{\left(\frac{1}{n-1}\right)}$$

Substituting $\left(\frac{v_2}{v_1}\right)$ in the entropy change relation derived earlier.

$$s_2 - s_1 = c_v \ln \frac{T_2}{T_1} + R \ln \frac{v_2}{v_1}$$

$$= c_v \ln\left(\frac{T_2}{T_1}\right) + R \ln\left(\frac{T_1}{T_2}\right)^{\frac{1}{n-1}}$$

$$s_2 - s_1 = c_v \ln\left(\frac{T_2}{T_1}\right) + \left(\frac{R}{n-1}\right). \ln \frac{T_1}{T_2}$$

For perfect gas
$$R = c_p - c_v$$
$$R = \gamma\ .\ c_v - c_v$$

$$R = c_v (\gamma - 1)$$

Substituting R in entropy change

$$s_2 - s_1 = c_v \ln \left(\frac{T_2}{T_1}\right) + \frac{c_v(\gamma-1)}{(n-1)} \ln \left(\frac{T_1}{T_2}\right)$$

$$= c_v \ln \frac{T_2}{T_1}\left\{1 - \left(\frac{\gamma-1}{n-1}\right)\right\}$$

$$s_2 - s_1 = c_v \ln \frac{T_2}{T_1}\left(\frac{n-\gamma}{n-1}\right)$$

Entropy change in polytropic process.

$$\boxed{s_2 - s_1 = \left(c_v \cdot \ln \frac{T_2}{T_1}\right)\left(\frac{n-\gamma}{n-1}\right)}$$

5.6 ENTROPY AND ITS RELEVANCE

Entropy has been introduced as a property based on the concept of IInd law of thermodynamics and derived from the thermodynamics involved in heat engines. A large number of definitions are available for entropy. To understand entropy let us take some gas in a closed vessel and heat it. Upon heating of gas the motion of gas molecules inside the vessel gets increased. State of molecular motion inside vessel depends upon the quantum of heat supplied. If we measure new kinetic energy of gas molecules, it is found to be larger than that initially. Also, the rate of intermolecular collision and randomness in molecular motion gets increased. In nutshell it could be said that heating has caused increase in energy level of gas molecules and thus resulting in increased disorderness inside the vessel. Higher is the energy with molecules, higher shall be the degree of disorderness. Entropy is closely defined using the degree of disorderness. It is said that greater is the molecular disorderness in the system greater shall be entropy. Mathematically, it can be supported by greater entropy value due to large heat supplied (dQ/T).

Thus "entropy can be defined as a parameter for quantifying the degree of molecular disorderness in the system". "Entropy is a measure of driving potential available for occurrence of a process".

Entropy is also an indicator of the direction of occurrence of any thermodynamic process. Mathematically, it has been seen from second law of thermodynamics that entropy of an isolated system always increases. Therefore, a process shall always occur in such a direction in which either entropy does not change or increases. In general almost all real processes are of irreversible type so entropy tends to increase. As entropy cannot be defined absolutely so the change in entropy should always have a positive or zero value.

5.7 THERMODYNAMIC PROPERTY RELATIONSHIP

Different thermodynamic properties such as P, V, T, U, H, S etc. can be related to one another using the combination of mathematical forms of first law, second law of thermodynamics and definitions of properties. Here specific values of properties are related.

For a non-flow process in closed system.

$$dq = du + dw$$

or

$$dq = du + p \cdot dv$$

Also, for a reversible process from definition of entropy, by second law we can write

$$dq = Tds$$

Combining above two,

$$\boxed{Tds = du + p \cdot dv}$$

From definition of enthalpy, specific enthalpy

$$h = u + pv$$

or $$dh = du + p \cdot dv + v \cdot dp$$

substituting from above

$$\boxed{d = T \cdot pds + vdp}$$

Above relations may be used for getting the variation of one property with the other, such as for constant pressure process, $dh = T \cdot ds$

or $$\left(\frac{dh}{ds}\right)_{p=\text{constt.}} = T$$

which means slope of constant pressure line on enthalpy – entropy diagram $(h-s)$ is given by temperature.

Also from above two relations

$$\boxed{T \cdot ds = c_v \cdot dT + p \cdot dv} \qquad \{\text{as } du = c_v \cdot dT\}$$

Substituting for dh and rearranging, $dh = T \cdot ds + v \cdot dp$ $\{\text{as } dh = c_p \cdot dT\}$

or $$\boxed{Tds = c_p \cdot dT - vdp}$$

For a constant pressure process above yields

$$\left(\frac{dT}{ds}\right)_{p=\text{constt.}} = \frac{T}{c_p}$$

It gives the slope of constant volume line on $T-s$ diagram.

Similarly for a constant volume process,

$$\left(\frac{dT}{ds}\right)_{v=\text{constt.}} = \frac{T}{c_v}$$

It gives the slope of constant volume line on $T-s$ diagram.

It can be concluded from the above mathematical explanations for slope that *slope of constant volume line is more than the slope of constant pressure line as* $c_p > c_v$.

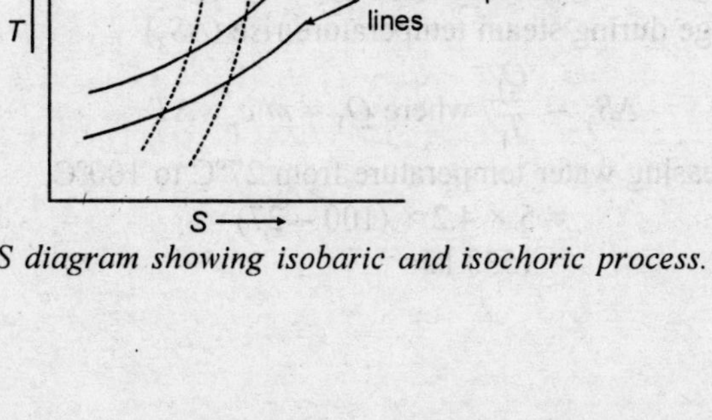

Fig. 5.10 T-S diagram showing isobaric and isochoric process.

5.8 THIRD LAW OF THERMODYNAMICS

'Third law of thermodynamics', an independent principle uncovered by 'Nernst' and formulated by 'Planck', states the "Entropy of a pure substance approaches zero at absolute zero temperature." This fact can also be corroborated by the definition of entropy which says it is a measure of molecular disorderness. At absolute zero temperature substance molecules get frozen and do not have any activity, therefore it may be assigned zero entropy value at crystalline state. Although the attainment of absolute zero temperature is impossible practically, however theoretically it can be used for defining absolute entropy value with respect to zero entropy at absolute zero temperature. Second law of thermodynamics also shows that absolute zero temperature can't be achieved, as proved earlier in article 4.10. Third law of thermodynamics is of high theoretical significance for the sake of absolute property definitions and has found great utility in thermodynamics.

EXAMPLES

1. Calculate the change in entropy of air, if it is throttled from 5 bar, 27°C to 2 bar adiabatically.

Solution:

Here $p_1 = 5$ bar, $T_1 = 300$ K.

$$p_2 = 2 \text{ bar}, c_{p\,\text{air}} = 1.004 \text{ kJ/kg.K}$$
$$R = 0.287 \text{ kJ/kg.K}$$

Entropy change may be given as;

$$s_2 - s_1 = \left\{ c_p \ln \frac{T_2}{T_1} - R \ln \frac{p_2}{p_1} \right\}$$

for throttling process $h_1 = h_2$

i.e. $c_p T_1 = c_p T_2$

or $T_1 = T_2$

Hence,

$$\text{Change in entropy} = 1.004 \ln(1) - 0.287 \ln\left(\frac{2}{5}\right)$$
$$= 0.263 \text{ kJ/kg.K}$$

Change in entropy = **0.263 kJ/kg.K Ans.**

2. Find the change in entropy of steam generated at 400°C from 5 kg of water at 27°C and atmospheric pressure. Take specific heat of water to be 4.2 kJ/kg.K, heat of vaporization at 100°C as 2260 kJ/kg and specific heat for steam given by; $c_p = R\,(3.5 + 1.2T + 0.14T^2)$, J/kg.K

Solution:

Total entropy change = Entropy change during water temperature rise (ΔS_1).

+ Entropy change during water to steam change (ΔS_2)
+ Entropy change during steam temperature rise (ΔS_3)

$$\Delta S_1 = \frac{Q_1}{T_1} \text{ where } Q_1 = m\, c_p \cdot \Delta T$$

Heat added for increasing water temperature from 27°C to 100°C.

$$= 5 \times 4.2 \times (100 - 27)$$
$$= 1533 \text{ kJ}$$

therefore, $$\Delta S_1 = \frac{1533}{300} = 5.11 \text{ kJ/K}$$

Entropy change during phase transformation;

$$\Delta S_2 = \frac{Q_2}{T_2}$$

Here Q_2 = Heat of vaporization = 5 × 2260 = 11300 kJ

Entropy change, $$\Delta S_2 = \frac{11300}{373.15} = 30.28 \text{ kJ/K.}$$

Entropy change during steam temperature rise;

$$\Delta S_3 = \int_{373.15}^{673.15} \frac{dQ}{T}$$

Here $dQ = mc_p \cdot dT$; for steam $R = \frac{8.314}{18} = 0.462$ kJ/kg.K

Therefore, c_p for steam $= 0.462\,(3.5 + 1.2 \cdot T + 0.14T^2) \times 10^{-3}$
$= (1.617 + 0.5544\,T + 0.065\,T^2) \times 10^{-3}$

or $$\Delta S_3 = \int_{373.15}^{673.15} 5 \times 10^{-3} \times \left(\frac{1.617}{T} + 0.5544 + 0.065\,T\right) dT$$

$$= 51843.49 \times 10^{-3} \text{ kJ/K}$$

$$\Delta S_3 = 51.84 \text{ kJ/K}$$

Total entropy change = 5.11 + 30.28 + 51.84
= **87.23 kJ/K Ans.**

3. Oxygen is compressed reversibly and isothermally from 125 kPa and 27°C to a final pressure of 375 kPa. Determine change in entropy of gas?

Solution:

Gas constant for oxygen:

$$R = \frac{8.314}{32} = 0.259 \text{ kJ/kg.K}$$

For reversible process the change in entropy may be given as;

$$\Delta s = c_p \ln \frac{T_2}{T_1} - R \ln \frac{p_2}{p_1}$$

Substituting values of initial & final states

$$\Delta s = -R \ln \left(\frac{375}{125}\right) = -0.285 \text{ kJ/kg.K}$$

Entropy change = – **0.285 kJ/kg.K Ans.**

4. Determine the change in entropy of universe if a copper block of 1 kg at 150°C is placed in a sea water at 25°C. Take heat capacity of copper as 0.393 kJ/kg.K.

Solution:

Entropy change in universe

$$\Delta S_{universe} = \Delta S_{block} + \Delta S_{water}$$

where $\Delta S_{block} = mC.\ \ln \frac{T_2}{T_1}$

Here hot block is put into sea water, so block shall cool down upto sea water at 25°C as sea may be treated as sink.

Therefore, $T_1 = 150°C$ or $423.15\ K$

and $T_2 = 25°C$ or $298.15\ K$

$$\Delta S_{block} = 1 \times 0.393 \times \ln\left(\frac{298.15}{423.15}\right)$$

$$= -0.1376 \text{ kJ/K}$$

Heat lost by block = Heat gained by water

$$= -1 \times 0.393 \times (423.15 - 298.15)$$

$$= -49.125 \text{ kJ}$$

Therefore, $\Delta S_{water} = \frac{49.125}{298.15} = 0.165$ kJ/k

Thus, $\Delta S_{universe} = -0.1376 + 0.165$

$$= 0.0274 \text{ kJ/k or } 27.4 \text{ J/K}$$

Entropy change of universe = **27.4 J/K. Ans.**

5. Determine change in entropy of universe if a copper block of 1 kg at 27°C is dropped from a height of 200 m in the sea water at 27°C. (Heat capacity for copper= 0.393 kJ/kg.K)

Solution:

$$\Delta S_{universe} = \Delta S_{block} + \Delta S_{sea\ water}$$

Since block and sea water both are at the same temperature so,

$$\Delta S_{universe} = \Delta S_{sea\ water}$$

Conservation of energy equation yields;

$$Q - W = \Delta U + \Delta PE + \Delta KE$$

Since in this case, $W = 0,\ \Delta KE = 0,\ \Delta U = 0$

$$Q = \Delta PE$$

Change in potential energy = ΔPE = mgh = 1 × 9.81 × 200

$$= 1962\ J$$

$$Q = 1962\ J$$

$$\Delta S_{universe} = \Delta S_{sea\ water} = \frac{1962}{300} = 6.54 \text{ J/kg.K}$$

Entropy change of universe = **6.54 J/kg.K. Ans.**

6. Determine entropy change of universe, if two copper blocks of 1 kg & 0.5 kg at 150°C and 0°C are joined together. Specific heats for copper at 150°C and 0°C are 0.393 kJ/kg.K and 0.381 kJ/kg.K respectively.

Solution:

Here, $\Delta S_{universe} = \Delta S_{block\ 1} + \Delta S_{block\ 2}$

Two blocks at different temperatures shall first attain equilibrium temperature. Let equilibrium temperature be T_f.

Then from energy conservation.

$$1 \times 0.393 \times (423.15 - T_f) = 0.5 \times 0.381 \times (T_f - 273.15)$$

$$T_f = 374.19\ K$$

Hence, entropy change in block 1, due to temperature changing from 423.15 *K* to 374.19 *K*.

$$\Delta S_1 = 1 \times 0.393 \times \ln\left(\frac{374.19}{423.15}\right) = -0.0483 \text{ kJ/k}$$

Entropy change in block 2

$$\Delta S_2 = 0.5 \times 0.381 \times \ln\left(\frac{374.19}{273.15}\right) = 0.0599 \text{ kJ/k}$$

Entropy change of universe = 0.0599 – 0.0483
= 0.0116 kJ/K

Entropy change of universe = **0.0116 kJ/K. Ans.**

7. A cool body at temperature T_1 is brought in contact with high temperature reservoir at temperature T_2. Body comes in equilibrium with reservoir at constant pressure. Considering heat capacity of body as C, show that entropy change of universe can be given as;

$$C\left[\left(\frac{T_1 - T_2}{T_2}\right) - \ln\frac{T_1}{T_2}\right]$$

Solution:

Since body is brought in contact with reservoir at temperature T_2, the body shall come in equilibrium when it attains temperature equal to that of reservoir, but there shall be no change in temperature of the reservoir.

Entropy change of universe $\Delta S_{\text{universe}} = \Delta S_{\text{body}} + \Delta S_{\text{reservoir}}$

$$\Delta S_{\text{body}} = C \ln \frac{T_2}{T_1}$$

$$\Delta S_{\text{reservoir}} = \frac{-C(T_2 - T_1)}{T_2}$$

as, heat gained by body = Heat lost by reservoir
$= C\,(T_2 - T_1)$

Thus,
$$\Delta S_{\text{universe}} = C \ln \frac{T_2}{T_1} - \frac{C(T_2 - T_1)}{T_2}$$

or, rearranging the terms,

$$\Delta S_{\text{universe}} = \frac{C(T_1 - T_2)}{T_2} - C \ln\left(\frac{T_1}{T_2}\right)$$

Hence proved.

8. Determine the rate of power loss due to irreversibility in a heat engine operating between temperatures of 1800 K and 300 K. Engine delivers 2 MW of power when heat is added at the rate of 5 MW.

Solution:

For irreversible operation of engine

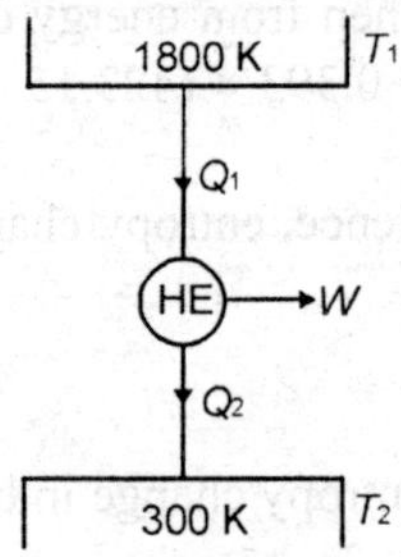

Fig. 5.11

$$\text{Rate of entropy generation} = \frac{Q_1}{T_1} + \frac{Q_2}{T_2}$$

$$= \frac{-5}{1800} + \frac{Q_2}{T_2}$$

Also, $W = Q_1 - Q_2 = 5 \times 10^6 - Q_2$

given $W = 2\text{ MW} = 2 \times 10^6\text{ W}$

so $Q_2 = 3 \times 10^6\text{ W}$

Therefore

$$\text{entropy generated} = \left(\frac{-5}{1800} + \frac{3}{300}\right) \times 10^6$$

$$\Delta S_{gen} = 7222.22\text{ W/K}$$

Work lost $= T_2 \times \Delta S_{gen}$

$= 300 \times 7222.22$

$= 2.16 \times 10^6\text{ W}$

or $= 2.16\text{ MW}.$

Work lost = **2.16 MW. Ans.**

9. A system at 500 K and a heat reservoir at 300 K are available for designing a work producing device. Estimate the maximum work that can be produced by the device if heat capacity of system is given as;

$$C = 0.05\,T^2 + 0.10T + 0.085,\ \text{J/K}$$

Solution:

System and reservoir can be treated as source and sink. Device thought of can be a Carnot engine operating between these two limits. Maximum heat available from system shall be the heat rejected till it's temperature drops from 500 K to 300 K.

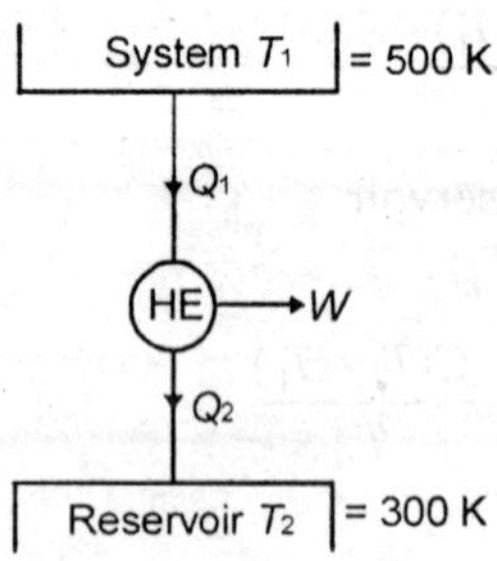

Fig. 5.12

Therefore,

$$\text{Maximum heat} \quad Q_1 = \int_{T_1}^{T_2} C \cdot dT$$

$$Q_1 = \int_{500}^{300} (0.05\, T^2 + 0.10T + 0.085)\, dT$$

$$Q_1 = 1641.35 \times 10^3 \text{ J}$$

Entropy change of system, $\Delta S_{system} = \int_{500}^{300} C \frac{dT}{T}$

$$= -\, 4020.043 \text{ J/K}$$

$$\Delta S_{reservoir} = \frac{Q_2}{T_2} = \frac{Q_1 - W}{T_2}$$

$$= \left(\frac{1641.35 \times 10^3 - W}{300}\right)$$

Also, we know from entropy principle

$$\Delta S_{universe} \geq 0$$

and $\Delta S_{universe} = \Delta S_{system} + \Delta S_{reservoir}$

Thus, upon substituting

$$(\Delta S_{system} + \Delta S_{reservoir}) \geq 0$$

$$\left\{-4020.043 + \left(\frac{1641.35 \times 10^3 - W}{300}\right)\right\} \geq 0$$

or $1451.123 - \frac{W}{300} \geq 0$

$$1451.123 \geq \frac{W}{300}$$

$$W \leq 435337.10$$

or $W \leq 435.34$ kJ

Hence Maximum work = **435.34 kJ**

For the given arrangement, device can produce maximum **435.34 kJ** of work. **Ans.**

10. *Determine the change in enthalpy and entropy if air undergoes reversible adiabatic expansion from 3MPa, 0.05 m^3 to 0.3 m^3.*

Solution:

For reversible adiabatic process governing equation for expansion,

$$PV^{1.4} = \text{Constt.}$$

Also, for such process entropy change = 0.

Initial state : 3MPa, 0.05 m^3

Final state : 0.3 m^3

Using $\frac{p_2}{p_1} = \left(\frac{V_1}{V_2}\right)^{1.4}$ or $V = \left(\frac{p_1 V_1^{1.4}}{p}\right)^{\frac{1}{1.4}}$

we get $p_2 = 0.244\ MPa$

From first law, second law and definition of enthalpy;

$$dH = T{\cdot}dS + Vdp$$

or, for adiabatic process of reversible type, $dS = 0$.

$$dH = V{\cdot}dp$$

$$\int_1^2 dH = \int_1^2 V\cdot dp$$

Substituting V, and actual states

$$H_2 - H_1 = \int_{3000}^{244} \left(\frac{3000 \times 0.05^{1.4}}{p}\right)^{\frac{1}{1.4}} dp$$

or
$$\Delta H = 268.8 \text{ kJ}$$

Enthalpy change = **268.8 kJ.**

Entropy change = **0 Ans.**

11. During a free expansion 2 kg air expands from 1 m³ to 10m³ volume in an insulated vessel. Determine entropy change of (a) the air (b) the surroundings (c) the universe.

Solution:

During free expansion temperature remains same and it is an irreversible process. For getting change in entropy let us approximate this expansion process as a reversible isothermal expansion.

(*a*) Change in entropy of air

$$\Delta S_{\text{air}} = m.R \ln \frac{V_2}{V_1}$$

$$= 2 \times 287 \ln \left(\frac{10}{1}\right)$$

$$\Delta S_{\text{air}} = 1321.68 \text{ J/K}$$

$$= \mathbf{1321.68 \text{ J/K Ans.}}$$

(*b*) During free expansion on heat is gained or lost to surroundings so,

$$\Delta S_{\text{surroundings}} = 0$$

Entropy change of surroundings,

= **0 Ans.**

(*c*) Entropy change of universe

$$\Delta S_{\text{universe}} = \Delta S_{\text{air}} + \Delta S_{\text{surroundings}}$$

$$= 1321.68 \text{ J/k}$$

$$= \mathbf{1321.68 \text{ J/k Ans.}}$$

12. Determine the change in entropy of 0.5 kg of air compressed polytropically from 1.013×10^5 P_a to 0.8 MPa and 800 K following index 1.2. Take $C_v = 0.71$ kJ/kg . K.

Solution:

Let initial and final states be denoted by 1 and 2.

For polytropic process pressure and temperature can be related as,

$$\left(\frac{p_2}{p_1}\right)^{\frac{n-1}{n}} = \frac{T_2}{T_1}$$

or
$$T_2 = 800 \times \left(\frac{0.8\times10^6}{1.013\times10^5}\right)^{\frac{1.2-1}{1.2}}$$

Temperature after compression = 1128.94 K

Substituting in entropy change expression for polytropic process,

$$(s_2 - s_1) = C_v\left(\frac{n-\gamma}{n-1}\right).\ln\frac{T_2}{T_1}$$

$$= 0.71 \times 10^3\left(\frac{1.2-1.4}{1.2-1}\right)\ln\left(\frac{1128.94}{800}\right)$$

$$= -244.54, \text{ kJ/kg . K}$$

$$\text{Total entropy change} = m\,(s_2 - s_1)$$

$$= 0.5 \times 244.54$$

$$\Delta S = \mathbf{122.27\ J/K}$$ **Ans.**

13. *A heat engine is working between the starting temperature limits of T_1 and T_2 of two bodies. Working fluid flows at rate 'm' kg/s and has specific heat at constant pressure as C_p. Determine the maximum obtainable work from engine.*

Solution:

In earlier discussions we have seen that in order to have highest output from engine, it should operate in reversible cycle and satisfy following relation,

$$\oint\frac{dQ}{T} = 0$$

Let us assume that the two bodies shall attain final temperature of T_f and engine shall than get stopped. so,

$$\int_{T_1}^{T_f} mC_p\frac{dT}{T} + \int_{T_2}^{T_f} mC_p\frac{dT}{T} = 0$$

or
$$mC_p\left\{\ln\left(\frac{T_f}{T_1}\right) + \ln\left(\frac{T_f}{T_2}\right)\right\} = 0$$

or
$$mC_p \cdot \ln\left(\frac{T_f^2}{T_1 \cdot T_2}\right) = 0$$

Here, $mC_p \neq 0$, so, $\ln\left(\frac{T_f^2}{T_1 \cdot T_2}\right) = 0$

or
$$T_f = \sqrt{T_1 \cdot T_2}$$

$$\text{Maximum work} = Q_{\text{supplied}} - Q_{\text{rejected}}$$

$$= mC_p(T_1 - T_f) - mC_p(T_f - T_2)$$

$$= mC_p\{T_1 - 2T_f + T_2\}$$

$$= mC_p\{T_1 - 2\cdot\sqrt{T_1\cdot T_2} + T_2\}$$

$$\text{Maximum work} = \boldsymbol{mC_p\left\{\sqrt{T_1}-\sqrt{T_2}\right\}^2} \textbf{ Ans.}$$

14. A heat engine operates between source at 600 K and sink at 300 K. Heat supplied by source is 500 kcal/s. Evaluate feasibility of engine and nature of cycle for the following conditions. (i) Heat rejected being 200 kcal/s, (ii) Heat rejected being 400 kcal/s (iii) Heat rejected being 250 kcal/s.

Solution:

Clausius inequality can be used for cyclic process as given below; consider '1' for source and '2' for sink.

$$\oint\frac{dQ}{T} = \frac{Q_1}{T_1} - \frac{Q_2}{T_2}$$

(*i*) For Q_2 = 200 kcal/s

$$\oint\frac{dQ}{T} = \frac{500}{600} - \frac{200}{300} = 0.1667$$

As $\oint\frac{dQ}{T} \not> 0$, therefore under these conditions engine is *not possible*. **Ans.**

(*ii*) For Q_2 = 400 kcal/s

$$\oint\frac{dQ}{T} = \frac{500}{600} - \frac{400}{300} = -0.5$$

Here $\oint\frac{dQ}{T} < 0$, so engine is *feasible* and cycle is *reversible* **Ans.**

(*iii*) For Q_2 = 250 kcal/s

$$\oint\frac{dQ}{T} = \frac{500}{600} - \frac{250}{300} = 0$$

Here, $\oint\frac{dQ}{T} = 0$, so engine is *feasible* and cycle is *reversible.* **Ans.**

15. Along a horizontal and insulated duct the pressure and temperatures measured at two points are 0.5 MPa, 400 K and 0.3 MPa, 350 K. For air flowing through duct determine the direction of flow.

Solution:

Let the two points be given as states 1 and 2, so,

$$p_1 = 0.5 \text{ MPa}, T_1 = 400 \text{ K}$$
$$p_2 = 0.3 \text{ MPa}, T_2 = 350 \text{ K}$$

Let us assume flow to be from 1 to 2

So entropy change $\qquad \Delta s_{1-2} = s_1 - s_2 = C_p \ln\left(\frac{T_1}{T_2}\right) - R\ln\left(\frac{p_1}{p_2}\right)$

For air, $\qquad R = 0.287$ kJ/kg . K

$C_p = 1.004$ kJ/kg . K

Hence $\qquad s_1 - s_2 = 1.004 \ln\left(\frac{400}{350}\right) - 0.287\ln\left(\frac{0.5}{0.3}\right)$

$$= -0.01254 \text{ kJ/kg . K}$$

or $$s_1 - s_2 = 0.01254 \text{ kJ/kg .K}$$

It means $s_2 > s_1$ hence the assumption that flow is from 1 to 2 is correct as from second law of thermodynamics the entropy increases in a process i.e. $s_2 \geq s_1$.

Hence flow occurs from 1 to 2 i.e. *from* ***0.5 MPa, 400 K*** *to* ***0.3 MPa & 350 K.*** **Ans.**

16. *An ideal gas is heated from temperature T_1 to T_2 by keeping its volume constant. The gas is expanded back to it's initial temperature according to the law p_v^n = constant. If the entropy change in the two processes are equal, find the value of 'n' in terms of adiabatic index γ.* *[U.P.S.C. 1997]*

Solution:

During constant volume process change in entropy $\Delta S_{12} = mc_v \,.\, \ln \dfrac{T_2}{T_1}$

Change in entropy during polytropic process, $\Delta S_{23} = mc_v \left(\dfrac{\gamma - n}{n-1}\right) \ln \dfrac{T_2}{T_1}$

Since the entropy change is same, so

$$\Delta S_{12} = \Delta S_{23}$$

$$mc_v \ln \frac{T_2}{T_1} = mc_v \left(\frac{\gamma - n}{n-1}\right) \ln \frac{T_2}{T_1}$$

or $$\boxed{n = \frac{\gamma + 1}{2}}$$ **Ans.**

17. *A closed system executed a reversible cycle 1–2–3–4–5–6–1 consisting of six processes. During processes 1–2 and 3–4 the system receives 1000 kJ and 800 kJ of heat, respectively at constant temperatures of 500 K and 400 K, respectively. Processes 2–3 and 4–5 are adiabatic expansions in which the steam temperature is reduced from 500 K to 400 K and from 400 K to 300 K respectively. During process 5–6 the system rejects heat at a temperature of 300 K. Process 6–1 is an adiabatic compression process. Determine the work done by the system during the cycle and thermal efficiency of the cycle.*

[U.P.S.C. 1995]

Solution:

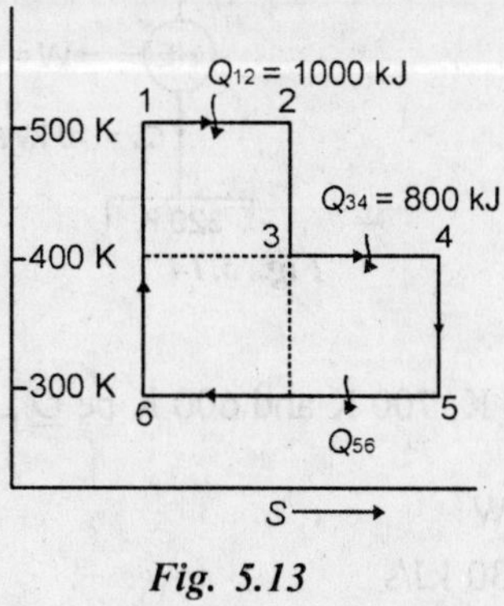

Fig. 5.13

$$\text{Heat added} = Q_{12} + Q_{14}$$

$$\text{Total heat added} = 1800 \text{ kJ}$$

For heat addition process 1–2,

$$Q_{12} = T_1 \cdot (S_2 - S_1)$$
$$1000 = 500 \cdot (S_2 - S_1)$$

or, $$S_2 - S_1 = 2$$

For heat addition process 3–4, $Q_{34} = T_3 \cdot (S_4 - S_3)$

$$800 = 400 \cdot (S_4 - S_3)$$

or, $$S_4 - S_3 = 2$$

Heat rejected in process 5–6

$$Q_{56} = T_5 \cdot (S_5 - S_6)$$
$$= T_5 \cdot \{(S_2 - S_1) + (S_4 - S_3)\}$$
$$Q_{56} = 300 \cdot \{2 + 2\} = 1200 \text{ kJ}$$

Net work done = Net heat

$$= (Q_{12} + Q_{34}) - Q_{56}$$
$$W_{net} = 1800 - 1200$$
$$W_{net} = 600 \text{ kJ}$$

$$\text{Thermal efficiency of cycle} = \frac{W_{net}}{\text{Heat added}}$$
$$= \frac{600}{1800}$$
$$= 0.3333 \text{ or } 33.33\%$$

Work done = **600 kJ**
Thermal efficiency = **33.33%**

Ans.

***18.** A reversible heat engine has heat interaction from three reservoirs at 600 K, 700 K and 800 K. The engine rejects 10 kJ/s to the sink at 320 K after doing 20 kW of work. The heat supplied by reservoir at 800 K is 70% of the heat supplied by reservoir at 700 K then determine the exact amount of heat interaction with each high temperature reservoir.*

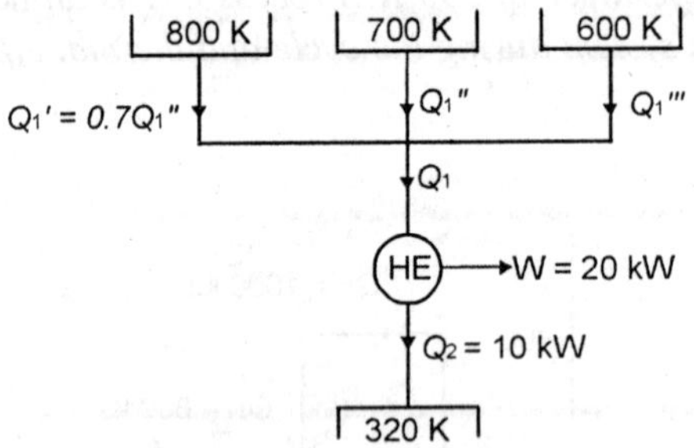

Fig. 5.14

Solution:

Let heat supplied by reservoir at 800 K, 700 K and 600 K be Q'_1, Q''_1, Q'''_1.

Here,

$$Q_1 - Q_2 = W$$

$\Rightarrow$ $$Q_1 = 30 \text{ kJ/s}$$

Also given that, $$Q'_1 = 0.7\, Q''_1$$

$$Q'''_1 = Q_1 - (0.7\, Q''_1 + Q''_1)$$

$$Q'''_1 = Q_1 - 1.7\,Q''_1$$

For reversible engine

$$\Rightarrow \frac{Q'_1}{800} + \frac{Q''_1}{700} + \frac{Q'''_1}{600} - \frac{Q_2}{320} = 0$$

$$\Rightarrow \frac{0.7Q'_1}{800} + \frac{Q''_1}{700} + \frac{(Q_1 - 1.7Q''_1)}{600} - \frac{10}{320} = 0$$

$$\Rightarrow \quad Q''_1 = 88.48 \text{ kJ/s}$$

$$Q'_1 = 61.94 \text{ kJ/s}$$

$$Q''' = -\,120.42 \text{ kJ/s}$$

Heat supplied by reservoir at 800 K = **61.94 kJ/s**

Heat supplied by reservoir at 700 K = **88.48 kJ/s**

Heat supplied by reservoir at 600 K = **120.42 kJ/s Ans.**

EXERCISE

5.1 Discuss the significance of Clausius inequality.

5.2 Define the 'entropy'. Also explain how it is a measure of irreversibility?

5.3 Explain the difference between isentropic process and adiabatic process.

5.4 How does the second law of thermodynamics overcome limitations of first law of thermodynamics?

5.5 Show that entropy of universe is increasing.

5.6 Is the adiabatic mixing of fluids irreversible ? If yes, explain.

5.7 Why does entropy generally increase ? Explain.

5.8 Explain the entropy principle and apply it to a closed system.

5.9 How the feasibility of any process can be ensured?

5.10 Give the third law of thermodynamics.

5.11 Explain why the slope of constant volume line is more than the slope of constant pressure line on *T–S* diagram.

5.12 Explain, whether the arrangement shown below for a reversible engine is feasible. If no then why? Give the correct arrangement.

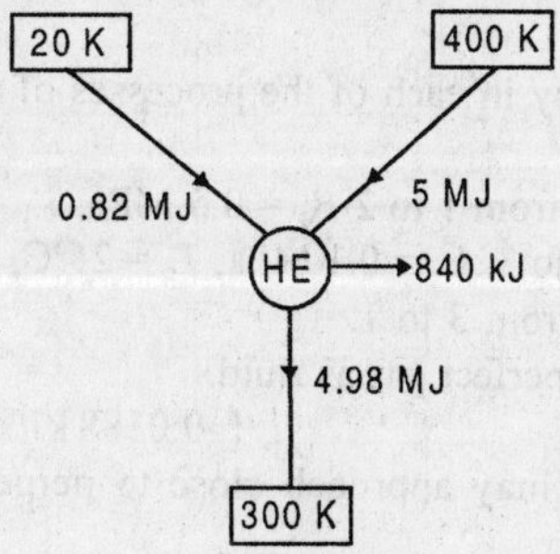

Fig. 5.15

5.13 Using second law of thermodynamics check the following and also indicate nature of cycle.

(*i*) Heat engine receiving 1000 kJ of heat from a reservoir at 500 K and rejecting 700 kJ heat to a sink at 27°C.

(*ii*) Heat engine receiving 1000 kJ of heat from a reservoir at 500 K and rejecting 600 kJ of heat to a sink at 27°C.

(*i*) Possible, irreversible cycle
(*ii*) Possible, reversible cycle

5.14 Determine the change in entropy of air during it's heating in a perfectly insulated rigid tank having 5 kg of air at 2 atm. Air is heated from 40°C to 80°C temperature.

5.15 Calculate change in entropy of air during the process in which a heat engine rejects 1500 kJ of heat to atmosphere at 27°C during its operation. [5 kJ/K]

5.16 Determine the final temperature and total entropy change during a process in which metal piece of 5 kg at 200°C falls into an insulated tank containing 125 kg of water at 20°C. Specific heat of metal = 0.9 kJ/kg.K, Specific heat of water = 4.184 kJ/kg.K. [21.53°C, 0.592 kJ/K]

5.17 Show that for air undergoing isentropic expansion process;

$$ds = c_p \frac{dv}{v} + c_v \frac{dp}{p}$$

5.18 Determine the change in entropy of air, if it is heated in a rigid tank from 27°C to 150°C at low pressure. [246.8 J/kg.K]

5.19 An electrical resistance of 100 ohm is maintained at constant temperature of 27°C by a continuously flowing cooling water. What is the change in entropy of the resistor in a time interval of one minute ? [0]

5.20 A water tank of steel is kept exposed to sun. Tank has capacity of 10 m^3 and is full of water. Mass of steel tank is 50 kg and during bright sun temperature of water is 35°C and by the evening water cools down to 30°C. Estimate the entropy change during this process. Take specific heat for steel as 0.45 kJ/kg.K and water as 4.18 kJ/kg.K. [5.63 kJ/K]

5.21 Heat engine operating on Carnot cycle has a isothermal heat addition process in which 1 MJ heat is supplied from a source at 427°C. Determine change in entropy of (i) working fluid, (ii) source, (iii) total entropy change in process. [1.43 kJ/K, – 1.43 kJ/K, 0]

5.22 A system operating in thermodynamic cycle receives Q_1 heat at T_1 temperature and rejects Q_2 at temperature T_2. Assuming no other heat transfer show that the net work developed per cycle is given as,

$$W_{cycle} = Q_1 \frac{Q_1}{T_1} + \left(1 - \frac{T_2}{T_1}\right) - T_2 \cdot S_{gen}$$

where S_{gen} is amount of entropy produced per cycle due to irreversibilities in the system.

5.23 A rigid tank contains 5 kg of ammonia at 0.2 MPa and 298 K. Ammonia is then cooled until its pressure drops to 80 kPa. Determine the difference in entropy of ammonia between initial and final state. [–14.8 kJ/K]

5.24 Determine the change in entropy in each of the processes of a thermodynamic cycle having following processes;

(*i*) Constant pressure cooling from 1 to 2 P_1 = 0.5 MPa, V_1 = 0.01 m^3

(*ii*) Isothermal heating from 2 to 3, P_3 = 0.1 MPa, T_3 = 25°C, V_3 = 0.01 m^3

(*iii*) Constant volume heating from 3 to 1.

Take C_p = 1 kJ/kg . K for perfect gas as fluid.

[–0.0188 kJ/kg . K, 0.00654 kJ/kg . K, 0.0134 kJ/kg . K]

5.25 Conceptualize some toys that may approach close to perpetual motion machines. Discuss them in detail.

6

Thermodynamic Properties of Pure Substance

6.1 INTRODUCTION

Engineering systems have an inherent requirement of some substance to act as working fluid i.e. transport agent for energy and mass interactions. Number of working fluids are available and are being used in different systems suiting to the system requirements. Steam is also one of such working fluids used exhaustively because of its favourable properties. In thermal power plants steam is being extensively used. Water has capability to retain its chemical composition in all of its' phases i.e. steam and ice, and also it is almost freely available as gift of nature.

Pure substance refers to the "*substance with chemical homogeneity and constant chemical composition.*" H_2O is a pure substance as it meets both the above requirements. Any substance, which undergoes a chemical reaction, cannot be pure substance.

6.2 PROPERTIES AND IMPORTANT DEFINITIONS

Pure substance as defined earlier is used for operating various systems, such as steam is used for power generation in steam power plants. Hence, for thermodynamic analysis thermodynamic properties are required. Pressure and temperature are the properties that can be varied independently over wide range in a particular phase. Therefore, the behaviour of properties of pure substance have to be studied and mathematical formulations be made for their estimation.

Various dependent properties discussed ahead shall be enthalpy, internal energy, specific volume, entropy etc.

Some of terms used in discussion ahead are given as under.

(*a*) *Sensible heating*: It refers to the heating of substance in single phase. It causes rise in temperature of substance. In case of cooling in above conditions it shall be called sensible cooling.

(*b*) *Latent heating*: It is the heating of substance for causing its phase change without any change in it's temperature. If heat is extracted for causing phase change without any change in its temperature it will be called latent cooling.

(*c*) *Normal boiling point*: It is the temperature at which vapour pressure equals to atmospheric pressure and at this temperature phase change from liquid to gas begins.

(*d*) *Melting point*: It is the temperature at which phase change from solid to liquid takes place upon supplying latent heat.

(*e*) *Saturation states*: Saturation state of a substance refers to the state at which its phase transformation takes place without any change in pressure and temperature. These can be saturated solid state, saturated liquid state and saturated vapour state. For example saturated vapour state refers to the state of water at which its phase changes to steam without varying pressure and temperature.

(*f*) *Saturation pressure*: It is the pressure at which substance changes its phase for any given temperature. Such as at any given temperature water shall get converted into steam at a definite pressure only, this pressure is called saturation pressure corresponding to given temperature. For water at 100°C the saturation pressure is 1 atm pressure.

(*g*) *Saturation temperature*: It refers to the temperature at which substance changes its phase for any given pressure. For water at 1 atm pressure the saturation temperature is 100°C.

(*h*) *Triple point*: Triple point of a substance refers to the state at which substance can coexist in solid, liquid and gaseous phase in equilibrium. For water it is 0.01°C i.e. at this temperature ice, water and steam can coexist in equilibrium. Table 6.1 given below gives triple point data for number of substances.

Table 6.1 *Triple point and critical point*

	Triple point		*Critical point*	
Substance	*Pressure, kPa*	*Temperature, °C*	*Pressure, MPa*	*Temperature, °C*
Water	0.611	0.01	22.12	374.15
Helium	5.1	– 271	0.23	– 268
Hydrogen	7.0	– 259	1.3	– 213
Oxygen	0.15	– 219	5.0	– 119
Nitrogen	12.5	– 210	3.4	– 147
Ammonia	6.1	– 78	11.3	132
Carbondioxide	517	– 57	7.39	31
Mercury	1.65×10^{-7}	– 39	18.2	899

(*i*) *Critical states*: "Critical state refers to that state of substance at which liquid and vapour coexist in equilibrium." In case of water at 22.12 MPa, and 374.15°C the water and vapour coexist in equilibrium, thus it is the highest pressure and temperature at which distinguishable water and vapour exist together. Data for critical state of many substances is given in the table 6.1. Specific volume at critical point for water is 0.00317 m^3/kg.

(*j*) *Dryness fraction*: It is the mass fraction of vapour in a mixture of liquid and vapour at any point in liquid-vapour mixture region. It is generally denoted by 'x'. It is also called quality of steam.

(*k*) *Compressed liquid or subcooled liquid*: Liquid at temperature less than saturation temperature corresponding to a given pressure is called compressed liquid or subcooled liquid. Degree of subcooling is given by the temperature difference between liquid temperature and saturation temperature of liquid at given pressure.

Degree of subcooling = Saturation temperature at given pressure – Temperature of liquid.

(*l*) *Superheated steam*: Steam having temperature more than the saturation temperature corresponding to given pressure is called superheated steam. Amount of superheating is quantified by degree of superheating. Degree of superheating is given by difference between temperature of steam and saturation temperature at given pressure.

Degree of superheating = Temperature of steam – Saturation temperature at given pressure.

6.3 PHASE TRANSFORMATION PROCESS

Let us study phase transformation from ice to steam by taking ice at –20°C in an open vessel i.e. at atmospheric pressure, and heat it from bottom. Salient states passed through the phase change are as given under. Melting point of ice is 0°C and boiling point of water is 100°C for water at 1 atmospheric pressure.

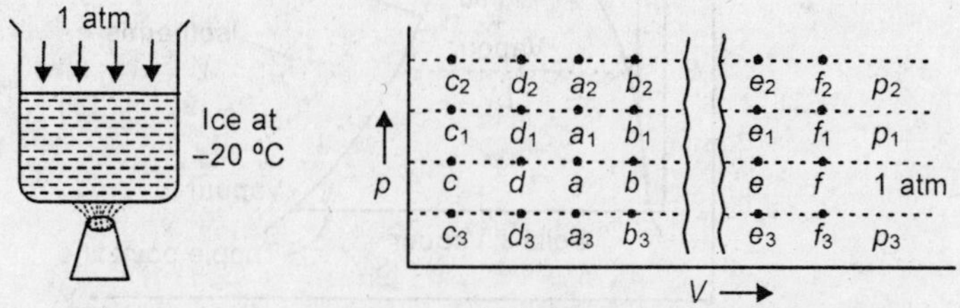

(*a*) (*b*) Volume change with pressure

***Fig. 6.1** Phase transformation process*

Say initial state is given by '*a*' at –20°C and 1 atmospheric pressure.

(*i*) Upon heating the ice its temperature increases from –20°C to 0°C while being in solid phase. Temperature increase is accompanied by increase in volume and new state '*b*' is attained. This heating is sensible heating as heating causes increase in temperature in same phase.

(*ii*) After ice reaches to 0°C, the melting point, it is ready for phase transformation into water. Further heat addition to it causes melting and now water at 0°C is available. This heating is called latent heating and heat added is called latent heat. New state attained is '*c*' and volume gets reduced due to typical characteristic of water. As defined earlier state '*b*' is called *saturation solid state* as phase can change here without any change in pressure and temperature. State '*c*' is called *saturated liquid state* with respect to solidification.

(*iii*) Further heating of water at 0°C shall cause increase in its temperature upto 100°C. This heat addition is accompanied by increase in volume and state changes from '*c*' to '*d*' as shown an *p-V* diagram. Here typical behaviour of water from 0 to 4°C is neglected. This heating is sensible heating in liquid phase. State '*d*' is called saturated liquid state with respect to vaporization. Thus, there are two saturated liquid states '*c*' and '*d*' depending upon direction of transformation.

(*iv*) Water at 100°C and 1 atmosphere is ready for getting vaporized with supply of latent heat of vaporization. Upon adding heat to it the phase transformation begins and complete liquid gradually gets transformed into steam at state '*e*'. This phase change is accompanied by large increase in volume. Heating in this zone is called latent heating. State '*e*' is called *saturated vapour state* or saturated steam state.

(*v*) Steam at 100°C upon heating becomes hotter and its temperature rises. Say, the heating causes temperature rise upto 200°C. This increase in temperature is also accompanied by increase in volume up to state '*f*' as shown on *p-V* diagram. This heating is sensible heating in gaseous phase.

Similar phase transformations can be realized at different pressures and such salient states be identified and marked. Joining all saturated solid states at different pressures results in a locii, which is called "saturated solid line." Similarly, joining all saturated liquid states with respect to solidification and saturated liquid states with respect to vaporization results in two 'saturated liquid lines'. Locii of all saturated vapour states at different pressure generates 'saturated vapour line'. The lines thus obtained are shown in Fig. 6.2 detailing *p-V* diagram for water. Point at which "saturated liquid line" with respect

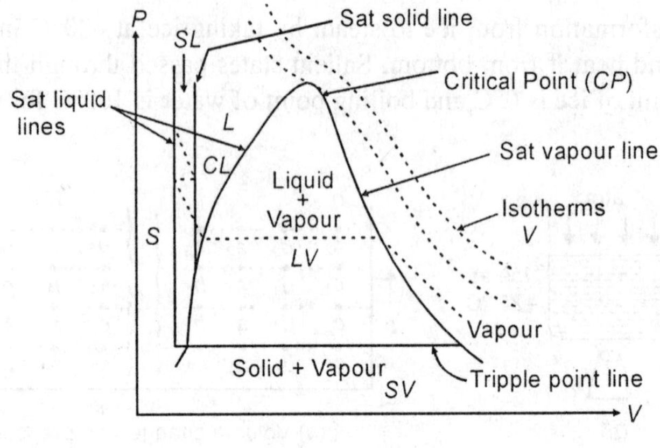

Fig. 6.2 P-V diagram for water

to vaporization meets with "saturated vapour line" is called "Critical point". Critical point is also sometimes referred to as "Critical state".

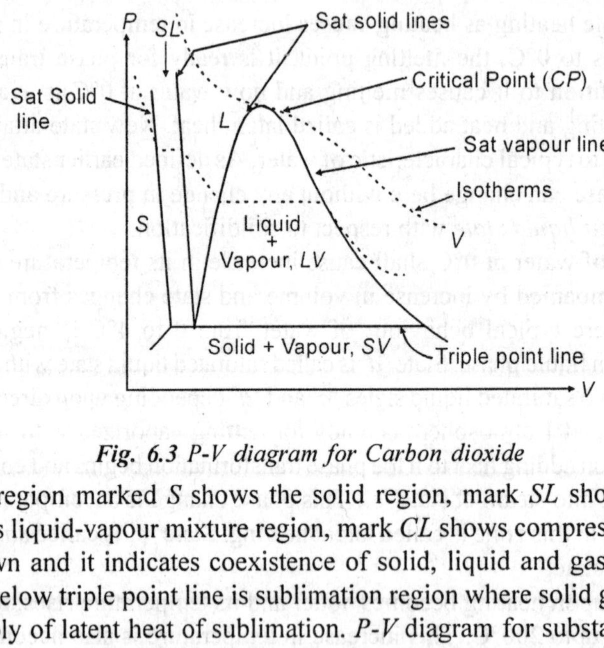

Fig. 6.3 P-V diagram for Carbon dioxide

On *P-V* diagram region marked *S* shows the solid region, mark *SL* shows solid-liquid mixture region, mark *LV* shows liquid-vapour mixture region, mark *CL* shows compressed liquid region. Triple point line is also shown and it indicates coexistence of solid, liquid and gas in equilibrium. Region marked *SV* and lying below triple point line is sublimation region where solid gets transformed directly into vapour upon supply of latent heat of sublimation. *P-V* diagram for substance which has different characteristics of contraction upon freezing is also shown here, Fig. 6.3. Carbon dioxide is such substance. Difference in *p-V* diagrams for two different substances (water and CO_2) may be understood from here.

6.4 GRAPHICAL REPRESENTATION OF PRESSURE, VOLUME AND TEMPERATURE

Graphical representations based on variation of thermodynamic properties can be obtained from the study of actual phase transformation process. In earlier discussion the variation of pressure and volume during phase transformation from ice to steam has been shown and explained. On these lines the standard graphical representations in terms of *p-V*, *T-V*, *p-T* and *p-V-T* for water can be obtained. *p-V* diagram as obtained has already been discussed in article 6.2.

(*i*) ***T-V* diagram:** It gives variation of temperature with volume. Let us look at different steps involved in phase transformation and how are the temperature variations.

a – b: Temperature rises from –20°C to 0°C with volume increase, (ice)

b – c: Temperature remains constant at 0°C due to phase change and volume decreases, (ice to water)

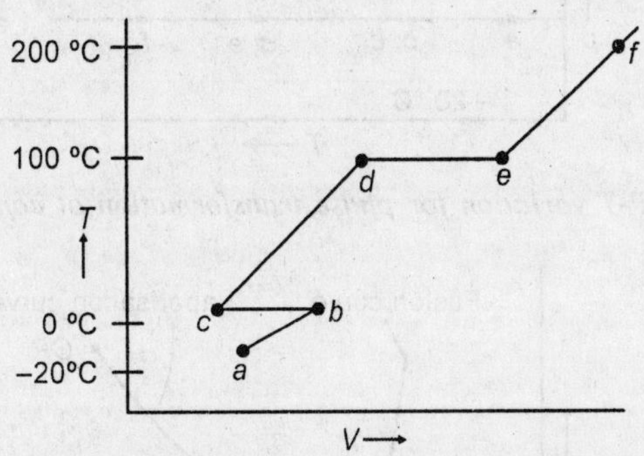

Fig. 6.4 Temperature-Volume variation at 1 atm pressure

c – d: Temperature increases from 0°C to 100°C and volume increases, (water).

d – e: Temperature remains constant at 100°C, phase changes from liquid to vapour and volume increases, (water to steam).

e – f: Temperature increases from 100°C to 200°C and volume increases, (steam).

Similar to above, the *T-V* variations at different pressures can be obtained and identical states i.e. saturation states be joined to get respective saturation lines as shown in Fig. 6.5 ahead.

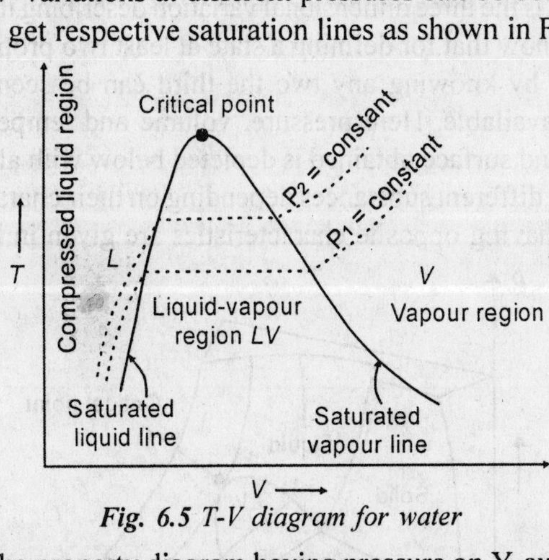

Fig. 6.5 T-V diagram for water

(*ii*) ***P-T* diagram:** It is the property diagram having pressure on Y-axis and temperature on X-axis. This can also be obtained by identifying and marking salient states during phase transformation and subsequently generating locii of identical states at different pressures.

For the phase change discussed in article 6.3, pressure and temperature variation shall be as described ahead. This phase change occurs at constant pressure of 1 atm.

$a-b$: Temperature rises from –20°C to 0°C, phase is solid.
$b-c$: Temperature does not rise, phase changes from ice to water (solid to liquid)
$c-d$: Temperature rises from 0°C to 100°C, phase is liquid
$d-e$: Temperature does not rise, phase changes from liquid to gas (water to steam)
$e-f$: Temperature rises from 100°C to 200°C, phase is gas (steam).

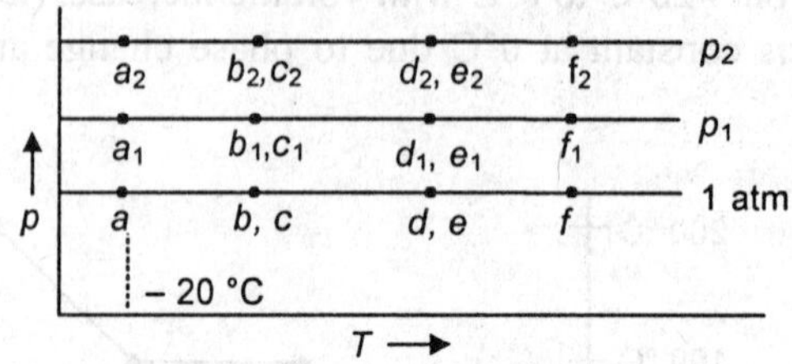

Fig. 6.6 P-T variation for phase transformation at constant pressures

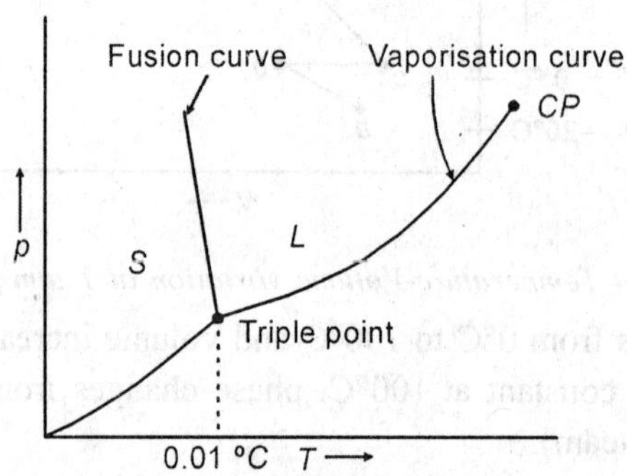

Fig. 6.7 P-T diagram for water

(*iii*) ***P-V-T* surface:** This is the three dimensional variation describing three thermodynamic properties *P*, *V* and *T*. As we know that for defining a state at least two properties are needed therefore on *P-V-T* surface also, by knowing any two the third can be seen and continuous variation of these properties is available. Here pressure, volume and temperature are taken on mutually perpendicular axis and surface obtained is depicted below with all salient points. *P-V-T* surface shall be different for different substances depending on their characteristics. Here *P-V-T* surface for two substances having opposite characteristics are given in Fig. 6.8.

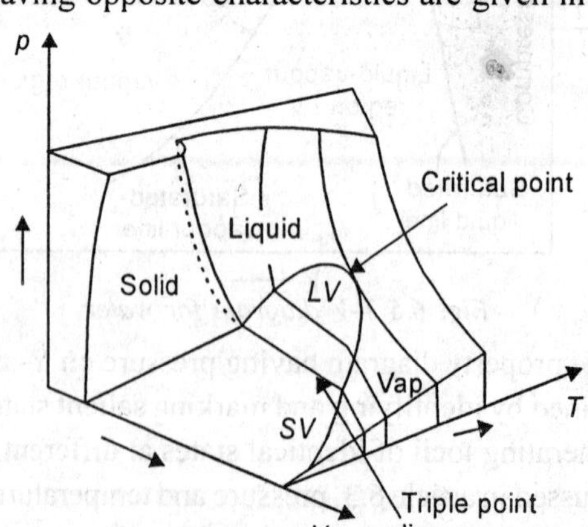

(A) P-V-T surface for water (which expands upon freezing.)

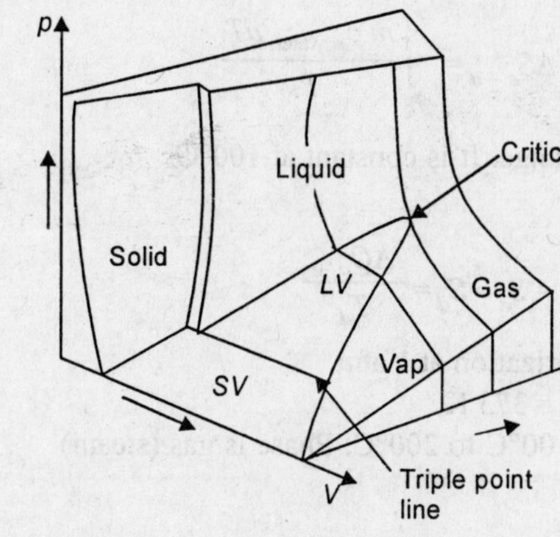

(B) P-V-T surface for Carbon dioxide (which contracts upon freezing.)

***Fig. 6.8** P-V-T surface for water and CO_2*

6.5 THERMODYNAMIC RELATIONS INVOLVING ENTROPY

Entropy change during phase transformation process can be studied with temperature at any given pressure, based on discussions in article 6.3. Entropy changes for every state change are estimated and plotted on *T-S* diagram for 1 atm. pressure and '*m*' mass of ice.

$a - b$: Temperature changes from –20°C to 0°C. Phase is solid

$$\text{Entropy change} \quad \Delta S_{a-b} = S_b - S_a = \int_{T_a}^{T_b} \frac{dQ_{a-b}}{T}$$

Here $\quad dQ_{a-b} = m \cdot c_{p,\text{ ice}} \cdot dT$

and $\quad T_a = 253 \text{ K}, T_b = 273 \text{ K}.$

$$\text{or} \quad \Delta S_{a-b} = \int_{T_a}^{T_b} \frac{m\, c_{p,\text{ice}} . dT}{T}$$

$b - c$: Temperature does not change. It is constant at 0°C.
Entropy change,

$$\Delta S_{b-c} = S_c - S_b = \frac{\Delta Q_{b-c}}{T_b}$$

Here, $\quad \Delta Q_{b-c}$ = Latent heat of fusion/melting of ice at 1 atm

$T_b = 273$ K

$c - d$: Temperature rises from 0°C to 100°C. Phase is liquid.
Entropy change,

$$\Delta S_{c-d} = S_d - S_c = \int_{T_c}^{T_d} \frac{dQ_{c-d}}{T}$$

Here $dQ_{c-d} = m \cdot c_{p,\text{water}} \cdot dT$ and $T_c = 273$ K, $T_d = 373$ K.

or, $$\Delta S_{c-d} = \int_{T_c}^{T_d} \frac{m\, c_{p,\text{water}}\, dT}{T}$$

$d - e$: Temperature does not change. It is constant at 100°C.
Entropy change,

$$\Delta S_{d-e} = S_e - S_d = \frac{\Delta Q_{d-e}}{T_d}$$

Here, ΔQ_{d-e} = latent heat of vaporization at 1 atm.
and $T_d = 373$ K.

$e - f$: Temperature rises from 100°C to 200°C. Phase is gas (steam)
Entropy change,

$$\Delta S_{e-f} = S_f - S_e = \int_{T_e}^{T_f} \frac{dQ_{e-f}}{T}$$

Here, $dQ_{e-f} = m \cdot c_{p,\text{steam}} \cdot dT$ and $T_e = 373\ K$, $T_f = 473$ K

or, $$\Delta S_{e-f} = \int_{T_e}^{T_f} \frac{m\, c_{p,\text{steam}} \cdot dT}{T}$$

Above entropy change when plotted on *T–S axis* result as below.

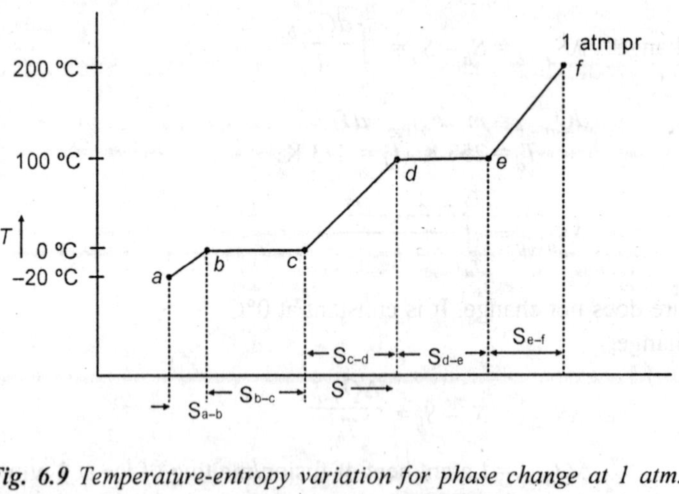

Fig. 6.9 Temperature-entropy variation for phase change at 1 atm.

On the similar lines, (as discussed above) the *T–S* variation may be obtained for water at other pressures. In the *T–S* diagram different important zones and lines are earmarked. Out of whole *T–S* diagram, generally the portion detailing liquid, liquid-vapour zone, vapour zone are of major interest due to steam undergoing processes and subsequently condensed and evaporated. Both *T–S* diagrams are shown here.

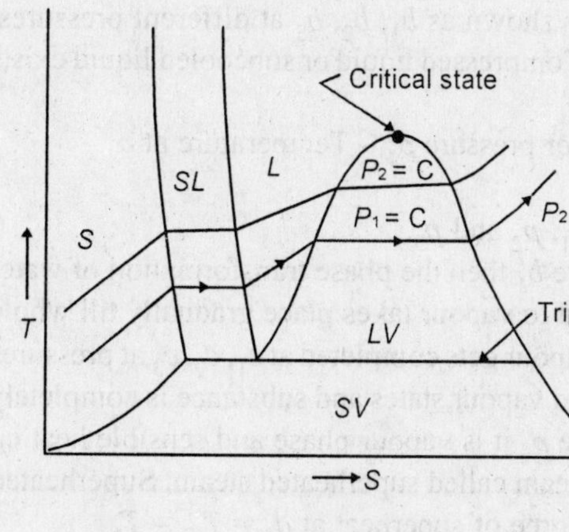

Fig. 6.10 T–S diagram for water

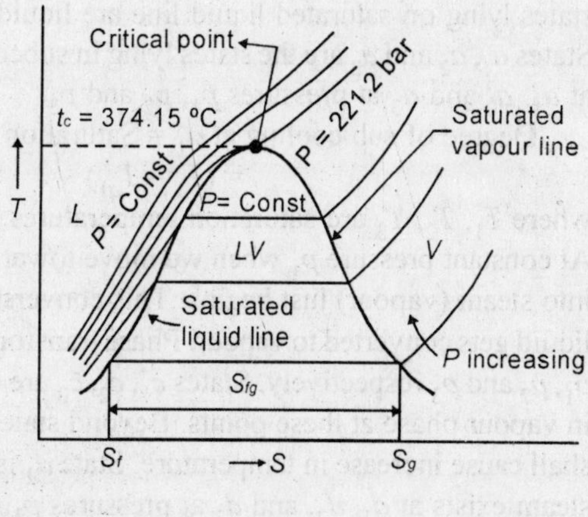

Fig. 6.11 T–S diagram showing liquid and vapour regions

6.6 PROPERTIES OF STEAM

For thermodynamic analysis the following thermodynamic properties of steam are frequently used.

"Saturation temperature (T), saturation pressure (P), specific volume (v), enthalpy (h), entropy (s), internal energy (u)"

Let us look at T–S diagram below.

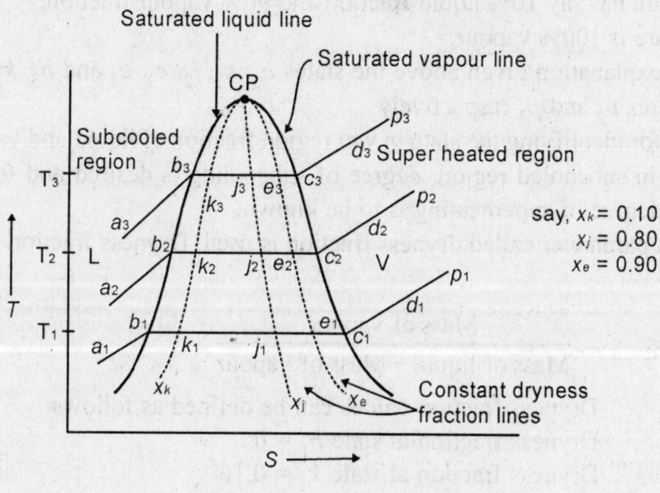

Fig. 6.12 T–S diagram

Thermodynamic properties and nomenclature used is indicated inside bracket along with property. Discussion is based on unit mass, of steam/mixture. T–S diagram for 2-phases i.e. liquid and vapour has saturated liquid line and saturated vapour line meeting at critical point. Three constant pressure lines corresponding to pressure P_1, P_2 and P_3 are shown. Let us take a constant pressure line for pressure p_1 which has states $a_1, b_1, k_1, j_1, e_1, c_1, d_1$ shown upon it. Region on the left of saturated liquid line is liquid region. Region enclosed between saturated liquid line and saturated vapour line is liquid-vapour mixture region or also called wet region. Region on the right of saturated vapour line is vapour region. All the states lying on saturated liquid line are liquid (water) states shown as b_1, b_2, b_3 at different pressures. States a_1, a_2 and a_3 are the states lying in subcooled region. Compressed liquid or subcooled liquid exists at a_1, a_2 and a_3 at pressures p_1, p_2 and p_3.

Degree of sub cooling at a_1 = Saturation temperature for pressure p_1 – Temperature at a_1

$$= (T_1 - T_{a1})$$

where T_1, T_2, T_3 are saturation temperatures at pressures p_1, p_2 and p_3.

At constant pressure p_1 when we move towards right of state b_1 then the phase transformation of water into steam (vapour) just begins. This conversion from liquid to vapour takes place gradually till whole liquid gets converted to vapour. Phase transformation into vapour gets completed at c_1, c_2, c_3 at pressures p_1, p_2 and p_3 respectively. States c_1, c_2, c_3 are called saturated vapour states and substance is completely in vapour phase at these points. Beyond state c_1, at pressure p_1 it is vapour phase and sensible heating shall cause increase in temperature. State d_1 is the state of steam called superheated steam. Superheated steam exists at d_1, d_2, and d_3 at pressures p_1, p_2 and p_3. Degree of superheat at $d_1 = T_{d1} - T_1$

Similarly, degree of superheat at $d_2 = T_{d2} - T_2$,

degree of superheat at $d_3 = T_{d3} - T_3$.

As pressure is increased upto critical pressure then constant pressure line is seen to become tangential to critical point (CP) at which water instantaneously flashes into vapour.

In the wet region states k_1, j_1 and e_1 are shown at pressure p_1.

At state b_1 mixture is 100% liquid

At state k_1 mixture has larger liquid fraction, say 90% liquid fraction and 10% vapour fraction.

At state j_1 mixture has say 20% liquid fraction and 80% vapour fraction.

At state e_1 mixture has say 10% liquid fraction and 90% vapour fraction.

At state c_1 mixture is 100% vapour.

Similarly, from explanation given above the states b_2, k_2, j_2, e_2, c_2 and b_3, k_3, j_3, e_3 and c_3 can be understood at pressures p_2 and p_3 respectively.

At any pressure for identifying the state in wet region fraction of liquid and vapour must be known, for identifying state in subcooled region, degree of subcooling is desired and for identifying state in superheated region, degree of superheating is to be known.

For wet region a parameter called dryness fraction is used. Dryness fraction as defined earlier can be given as, x:

$$\text{Dryness fraction; } x = \frac{\text{Mass of vapour}}{\text{Mass of liquid} + \text{Mass of vapour}}$$

Dryness fraction values can be defined as follows;

Dryness fraction at state b_1 = 0

Dryness fraction at state k_1 = 0.10

$$\left\{\text{as assumed} = \frac{0.10\,(\text{m})}{(0.9 + 0.1)\text{m}}\right\}$$

Dryness fraction at state j_1 = 0.80

Dryness fraction at state e_1 = 0.90

Dryness fraction at state c_1 = 1.00

Thus, saturated liquid line and saturated vapour line are locii of all states having 0 and 1 dryness fraction values.

At critical point dryness fraction is either 0 or 1.

For different pressures the locii of constant dryness fraction points may be obtained and it yields constant dryness fraction lines corresponding to $x_k(0.10)$, $x_j(0.80)$, $x_e(0.90)$, as shown by dotted lines.

Let us use subscript 'f' for liquid states and 'g' for vapour states.

Therefore, enthalpy corresponding to saturated liquid state = h_f,

Enthalpy corresponding to saturated vapour state = h_g.

Similarly, entropy may be given as s_f and s_g.

Specific volume may be given as v_f and v_g.

Internal energy may be given as u_f and u_g.

At any pressure for some dryness fraction x, the total volume of mixture shall comprise of volume occupied by liquid and vapour both.

$$\text{Total volume } V = V_f + V_g.$$

Similarly,

Total mass $m = m_f + m_g$, i.e. mass of liquid and mass of vapour put together substituting for volume,

$$m \cdot v = m_f \cdot v_f + m_g \cdot v_g.$$

where m is total mass and v is specific volume of mixture

$$v = \left(1 - \frac{m_g}{m}\right) \cdot v_f + \left(\frac{m_g}{m}\right) v_g$$

or

$$v = \left(1 - \frac{m_g}{m}\right) \cdot v_f + \left(\frac{m_g}{m}\right) \cdot v_g$$

or

$$v = \left(1 - \frac{m_g}{m}\right) \cdot v_f + \left(\frac{m_g}{m}\right) \cdot v_g$$

From definition,

$$x = \frac{m_g}{m_f + m_g} = \frac{m_g}{m}$$

or,

$$v = (1 - x) \cdot v_f + x \cdot v_g$$

or,

$$v = v_f + x \cdot (v_g - v_f)$$

or,

$$v = v_f + x \cdot v_{fg}.$$

here v_{fg} indicates change in specific volume from liquid to vapour

Similarly, enthalpy, entropy and internal energy may be defined for such states in wet region.

i.e.

$$h = h_f + x\,(h_g - h_f)$$

or

$$h = h_f + x \cdot h_{fg}$$

here h_{fg} is difference in enthalpy between saturated liquid and vapour states. Actually h_{fg} is energy or heat required for vaporization or heat to be extracted for condensation i.e. latent heat.

$$\text{Similarly, } s = s_f + x \cdot s_{fg} \text{ where } s_{fg} = s_g - s_f$$

$$u = u_f + x \cdot u_{fg} \text{ where } u_{fg} = u_g - u_f.$$

6.7 STEAM TABLES AND MOLLIER DIAGRAM

Steam being pure substance has its unique and constant properties at different pressures and temperatures. Therefore, thermodynamic properties can be estimated once and tabulated for future use. Steam table is a tabular presentation of properties such as specific enthalpy, entropy, internal energy and specific volume at different saturation pressures and temperatures. Steam table may be on pressure basis or on temperature basis. The table on pressure basis has continuous variation of pressure and corresponding to it : saturation temperature (T_{sat}), enthalpy of saturated liquid (h_f), enthalpy of saturated vapour (h_g), entropy of saturated liquid (s_f), entropy of saturated vapour (s_g), specific volume of saturated liquid (v_f), specific volume of saturated vapour (v_g), internal energy of saturated liquid (u_f), internal energy of saturated vapour (u_g) are given on unit mass basis, i.e. as shown in table 6.2.

Similar to above the temperature based table which gives continuous variation of temperature and corresponding to it saturation pressure and other properties as h_f, h_g, h_{fg}, s_f, s_g, s_{fg}, v_f, v_g, u_f, u_g and u_{fg} are given.

Similarly, steam properties for superheated steam are also estimated and tabulated at some discrete pressures for varying degree of superheat. Super heated steam table are available for getting enthalpy, entropy, specific volume and internal energy separately. Example of superheated steam table for enthalpy is given here:

Table 6.2 *Pressure based steam table*

Pressure	*Sat. Temp.*	*Enthalpy*			*Entropy*			*Specific volume*		*Internal energy*		
	T_{sat} °C	h_f kJ/kg	h_g kJ/kg	h_{fg} kJ/kg	s_f kJ/kg°K	s_g kJ/kg°K	s_{fg} kJ/kg°K	v_f m³/kg	v_g m³/kg	u_f kJ/kg	u_g kJ/kg	u_{fg} kJ/kg

Table 6.3 *Temperature based steam table*

Tempe-rature	*Sat. Pressure*	*Enthalpy*			*Entropy*			*Specific volume*		*Internal energy*		
°C	kPa	h_f kJ/kg	h_g kJ/kg	h_{fg} kJ/kg	s_f kJ/kg°K	s_g kJ/kg°K	s_{fg} kJ/kg°K	v_f m³/kg	v_g m³/kg	u_f kJ/kg	u_g kJ/kg	u_{fg} kJ/kg

Table 6.4 *Superheated steam table for enthalpy*

Pressure	*Sat. temp.* °C	*Enthalpy values for varying degree of superheat (kJ/kg)*						
kPa	(T_{sat})	T_1	T_2	T_3	T_4	T_5	T_6	T_7

Here T_1, T_2, T_3, T_4, T_5, T_6, T_7 ... are more than T_{sat} and have increasing value of degree of superheat. Steam tables as discussed above are available in appendix in this text book.

Mollier diagram is the enthalpy-entropy (h–s) diagram for steam. This diagram is obtained on the basis of following equation depending upon the phase transformation as discussed earlier.

$Tds = dh - vdp$. (First and second law combined)

For constant pressure $\left(\frac{dh}{ds}\right)_p = T$

Enthalpy entropy diagram as obtained for all phases of water is as given in Figure 6.13 here.

Generally liquid and vapour region is only of interest in engineering systems, so mostly used portion of *h–s* diagram is taken out and shown in Fig. 6.14. It is popularly known as *mollier diagram* or *mollier chart.*

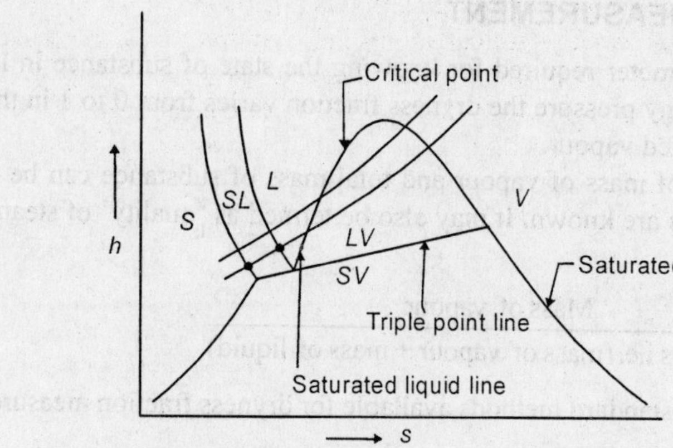

Fig. 6.13 Enthalpy-entropy diagram for all phases

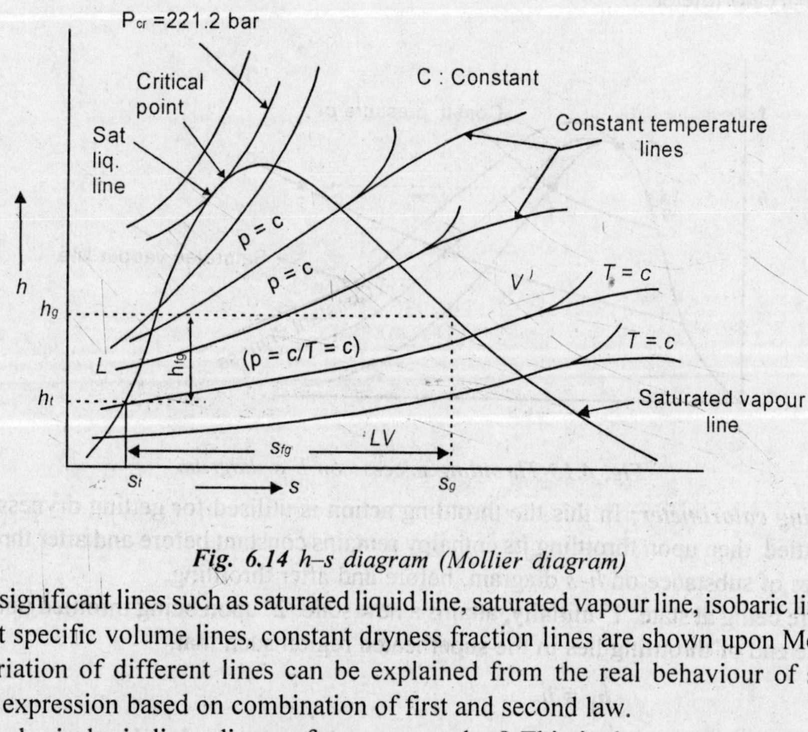

Fig. 6.14 h–s diagram (Mollier diagram)

Different significant lines such as saturated liquid line, saturated vapour line, isobaric lines, isothermal lines, constant specific volume lines, constant dryness fraction lines are shown upon Mollier diagram. Nature of variation of different lines can be explained from the real behaviour of substance and mathematical expression based on combination of first and second law.

Such as, why isobaric lines diverge from one another? This is due to the increase in saturation temperature with increase in pressure. Slope of isobar is equal to saturation temperature as shown in the beginning, therefore it also increases with increasing pressure.

Why isothermal lines are not visible in wet region? It is because constant temperature lines and constant pressure lines coincide upon in wet region. For every pressure there shall be definite saturation temperature which remains constant in wet region.

Mollier chart is also given in appendix, at the end of this book.

6.8 DRYNESS FRACTION MEASUREMENT

Dryness fraction is the basic parameter required for knowing the state of substance in liquid-vapour mixture region (wet region). For any pressure the dryness fraction varies from 0 to 1 in the wet region i.e. from saturated liquid to saturated vapour.

Dryness fraction being ratio of mass of vapour and total mass of substance can be conveniently estimated if these two mass values are known. It may also be termed as 'quality' of steam or 'dryness factor'.

$$\text{Dryness fraction} = \frac{\text{Mass of vapour}}{\text{Total mass i.e. (mass of vapour + mass of liquid)}}$$

Here we shall be looking into standard methods available for dryness fraction measurement. These are;

(*i*) Throttling calorimeter
(*ii*) Separating calorimeter
(*iii*) Separating and throttling calorimeter
(*iv*) Electrical calorimeter.

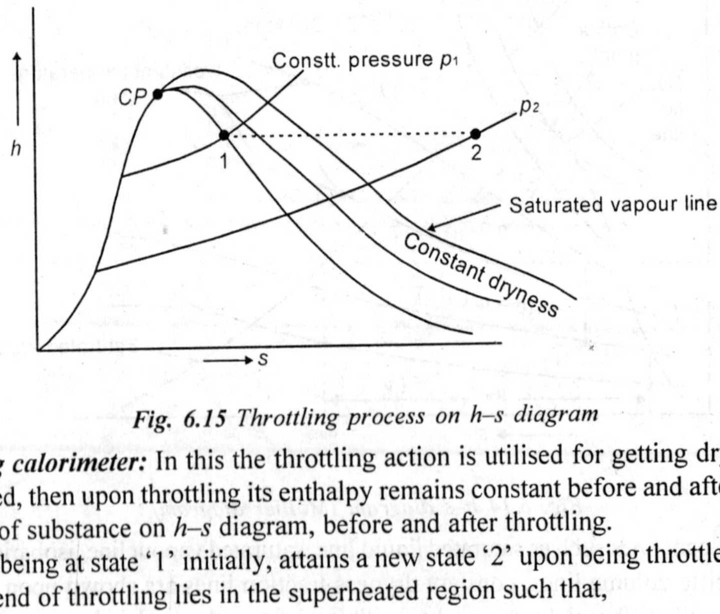

Fig. 6.15 Throttling process on h–s diagram

(*i*) ***Throttling calorimeter:*** In this the throttling action is utilised for getting dryness fraction. If a mixture is throttled, then upon throttling its enthalpy remains constant before and after throttling. Let us look upon states of substance on *h–s* diagram, before and after throttling.

Wet mixture being at state '1' initially, attains a new state '2' upon being throttled upto pressure p_2. This state at the end of throttling lies in the superheated region such that,

$$h_1 = h_2$$

Say, dryness fraction at state 1 is *x*, then enthalpy at this point can be given as

$$h_1 = h_{f\text{ at }p_1} + x \times h_{fg\text{ at }p_1}$$

In the above expression $h_{f \text{ at } p_1}$ and $h_{fg \text{ at } p_1}$ can be seen from steam table if pressure of wet steam is known. Also the enthalpy at state 2 (end of throttling) can be seen from superheated steam table if pressure and temperature at '2' are known.

$$\text{Substituting in, } h_1 = h_2$$

$$h_{f \text{ at } p_1} + x \times h_{fg \text{ at } p_1} = h_2$$

Here h_2, $h_{f \text{ at } p_1}$, $h_{fg \text{ at } p_1}$ are all known as explained above.

$$\text{Therefore, } x = \frac{h_2 - h_{f \text{ at } p_1}}{h_{fg \text{ at } p_1}}$$

Now the arrangements are to be made for, (*a*) measurement of pressure of wet steam in the begining, (*b*) throttling of wet mixture such that state at the end of throttling lies in superheated region, (*c*) measurement of pressure, temperature of throttled steam. Arrangement used in throttling calorimeter is as shown in Fig. 6.16.

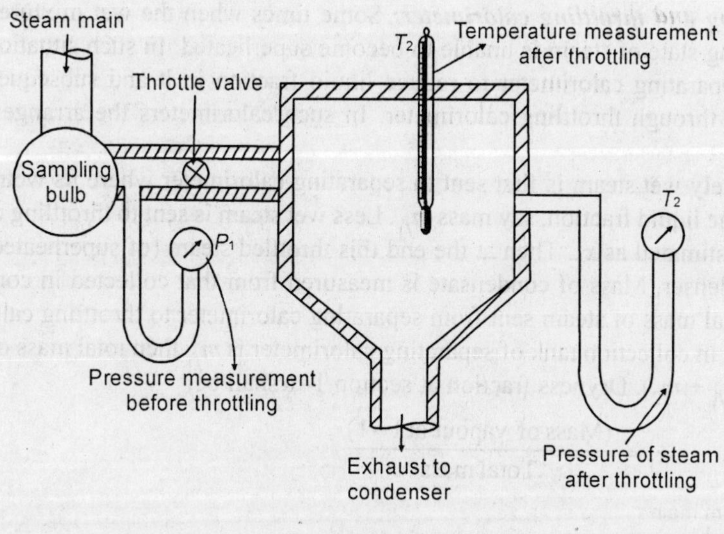

Fig. 6.16 Throttling calorimeter

(*ii*) ***Separating calorimeter:*** In this type of calorimeter the known mass of wet mixture is collected through a sampling bulb and sent to a separating chamber. Separating chamber has the series of obstacles, and zig-zag path inside it so that when mixture passes through them the liquid particles get separated due to sudden change in direction of flow and gravity action. Liquid thus separated out is collected in a collection tank and is measured. Thus by knowing the two mass values dryness fraction can be estimated as;

$$\text{Dryness fraction} = \frac{\text{(Total mass of liquid)}}{\text{Total mass}}$$

Layout of separating calorimeter is given in Fig. 6.17.

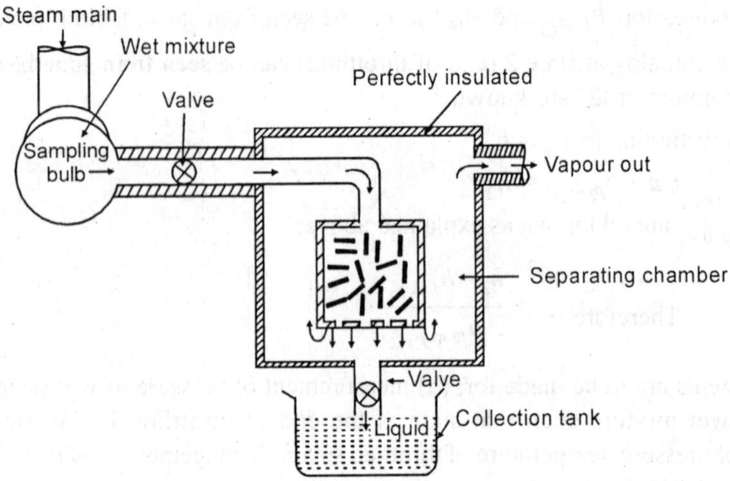

Fig. 6.17 Separating calorimeter

(iii) ***Separating and throttling calorimeter:*** Some times when the wet mixture is extremely wet then upon throttling state of steam is unable to become superheated. In such situations mixture is first passed through separating calorimeter to reduce liquid fraction in it and subsequently this less wet mixture is passed through throttling calorimeter. In such calorimeters the arrangement is as shown below.

Here excessively wet steam is first sent to separating calorimeter where its wetness is reduced by separating out some liquid fraction, say mass m_{f_1}. Less wet steam is sent to throttling calorimeter and its dryness fraction estimated as x_2. Then at the end this throttled steam (of superheated type) is made to pass through condenser. Mass of condensate is measured from that collected in condensate tank, say m_2. Thus m_2 is total mass of steam sent from separating calorimeter to throttling calorimeter. As mass of liquid collected in collection tank of separating calorimeter is m_{f_1} then total mass of wet steam under examination is $(m_{f_1} + m_2)$. Dryness fraction at section 1–1 shall be;

$$x_1 = \frac{(\text{Mass of vapour at } 1-1)}{\text{Total mass}}$$

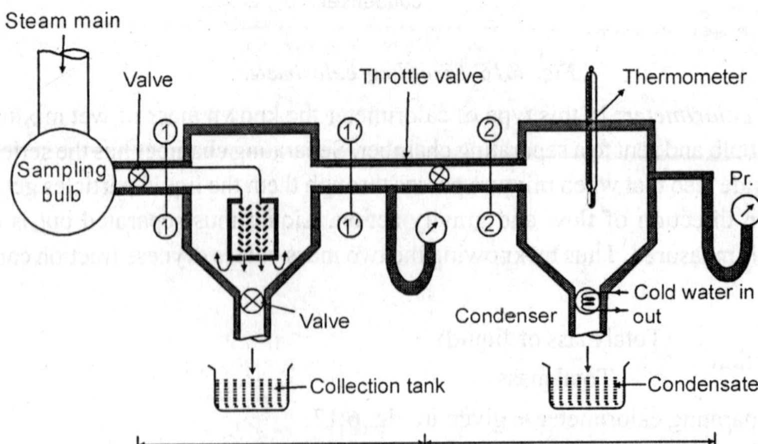

Fig. 6.18 Separating and throttling calorimeter

Mass of vapour at 1–1 shall be similar to mass of vapour entering at 2–2.

Mass of vapour at 2–2 = $x_2 \times m_2$

Hence, dryness fraction at 1–1, $x_1 = \dfrac{x_2 \cdot m_2}{m_{f1} + m_2}$

Separating and throttling processes occurring are also shown on *h–s* diagram in Fig. 6.19.

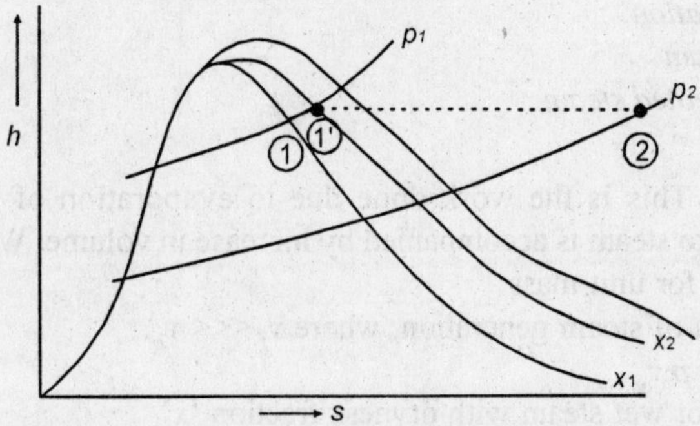

Fig. 6.19 Separating and throttling together on h–s diagram

(iv) ***Electrical Calorimeter:*** In electrical calorimeter too the principle employed is similar to that of throttling calorimeter. Here also wet mixture is brought to the superheated state, by heating and not by throttling. For known amount of heat added and the final enthalpy for superheated steam being known, one can find out the initial enthalpy.

For mass *m* of mixture, heat Q_{add} added by heater, and the enthalpies before and after heating being h_1, h_2, steady flow energy equation may be written as; $mh_1 + Q_{add} = mh_2$.

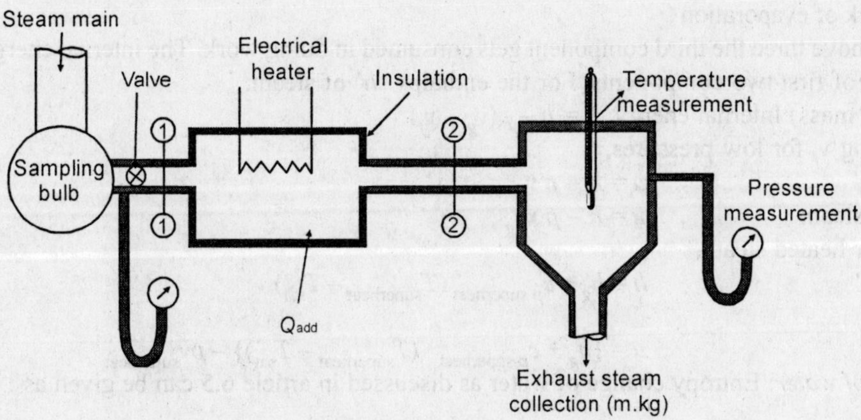

Fig. 6.20 Electrical calorimeter

Here, for electrical heater Q_{add} = V.I, where V and I are voltage and current.

h_2 is known, as mixture is brought to superheated state and pressure and temperature measured indicate enthalpy from superheated steam table, also $h_1 = h_{f\,at\,p1} + x \,.\, h_{fg\,at\,p1}$, here $h_{f\,at\,p1}$ & $h_{fg\,at\,p1}$ can be seen from steam table.

Now h_1 being known, for known *m*, h_2, Q_{add}, dryness fraction '*x*' can be easily obtained.

EXAMPLES

1. Derive the expressions for the following :

(a) Work of evaporation or external work of evaporation
(b) True latent heat
(c) Internal energy of steam
(d) Entropy of water
(e) Entropy of evaporation
(f) Entropy of wet steam
(g) Entropy of superheated steam

Solution:

(*a*) *Work of evaporation*: This is the work done due to evaporation of water to steam as phase transformation from water to steam is accompanied by increase in volume. Work of evaporation can be estimated as; $= p\,(v_g - v_f)$, for unit mass

For very low pressures of steam generation, where $v_f <<< v_g$.

Work of evaporation $= p \cdot v_g$

Work of evaporation for wet steam with dryness fraction '*x*'

$$= p \cdot x \cdot v_g.$$

(*b*) *True latent heat*: Latent heat causing the phase transformation from water to steam is accompanied with change in volume as well. Therefore, latent heat shall have two components i.e. (i) true latent heat causing phase change and (ii) work of evaporation due to volume increase. Mathematically, for unit mass, True latent heat $= h_{fg} - p(v_g - v_f)$.

(*c*) *Internal energy of steam*: For a given mass of steam the total heat energy with steam can be said to comprise of

(*i*) Sensible heat
(*ii*) True latent heat
(*iii*) Work of evaporation

Out of above three the third component gets consumed in doing work. The internal energy of steam shall consist of first two components. For the enthalpy '*h*' of steam,

(for unit mass) Internal energy, $u = h - p(v_g - v_f)$

neglecting v_f for low pressures,

$$u = h - p \cdot v_g$$

for wet steam $\quad u = h - p \cdot x v_g$

for super heated steam,

$$h = h_g + c_{p\,\text{superheat}}\,(T_{\text{superheat}} - T_{\text{sat}})$$

Hence

$$u = \{h_g + c_{p\,\text{superheat}} \cdot (T_{\text{superheat}} - T_{\text{sat}})\} - p v_{\text{superheat}}$$

(*d*) *Entropy of water*: Entropy change of water as discussed in article 6.5 can be given as :

$$\int_1^2 ds = \int_{T_1}^{T_2} c_{p\,\text{water}} \frac{dT}{T}$$

for constant specific heat of water

$$(s_2 - s_1) = c_{p\,\text{water}} \ln\left(\frac{T_2}{T_1}\right)$$

Absolute entropy may be given in reference to absolute zero temperature

$$s = c_{p\,\text{water}} \cdot \ln\left(\frac{T}{273.15}\right)$$

(*e*) *Entropy of evaporation*: During evaporation heat absorbed is equal to latent heat of evaporation. Therefore, for unit mass

$$s_{\text{evaporation}} = \frac{h_{fg}}{T_{\text{sat}}}$$

During incomplete evaporation (for wet steam)

$$s_{\text{evaporation}} = \frac{x \cdot h_{fg}}{T_{\text{sat}}}$$

(*f*) *Entropy of wet steam*:

Entropy of wet steam = Entropy of water + Entropy of evaporation

For unit mass,

$$s_{\text{wet}} = c_{p\,\text{water}} \ln \frac{T_2}{T_1} + x \cdot \frac{h_{fg}}{T_2}$$

(*g*) *Entropy of super heated steam*: Entropy change during constant pressure heating for superheating unit mass of the steam.

$$= c_{p\,\text{steam}} \ln\left(\frac{T_{\text{sup}}}{T_{\text{sat}}}\right)$$

Total entropy of super heated steam, starting with water at temperature T_1.

$$s_{\text{superheat}} = c_{p\,\text{water}} \ln\left(\frac{T_{\text{sat}}}{T_1}\right) + \frac{h_{fg}}{T_{\text{sat}}} + c_{p\,\text{steam}} \cdot \ln\left(\frac{T_{\text{super heat}}}{T_{\text{sat}}}\right).$$

2. Throttling calorimeter has steam entering to it at 10 MPa and coming out of it at 0.05 MPa and 100°C. Determine dryness fraction of steam.

Solution:

During throttling, $h_1 = h_2$

At state 2, enthalpy can be seen for superheated steam at 0.05 MPa and 100°C.

Thus, $h_2 = 2682.5$ kJ/kg

At state 1, before throttling

$$h_{f\,\text{at10MPa}} = 1407.56 \text{ kJ/kg}$$

$$h_{fg\,\text{at10MPa}} = 1317.1 \text{ kJ/kg}$$

$$h_1 = h_{f\,\text{at10MPa}} + x_1\, h_{fg\,\text{at10MPa}} = h_2$$

$$2682.5 = 1407.56 + (x_1 \cdot 1317.1)$$

$$x_1 = 0.968$$

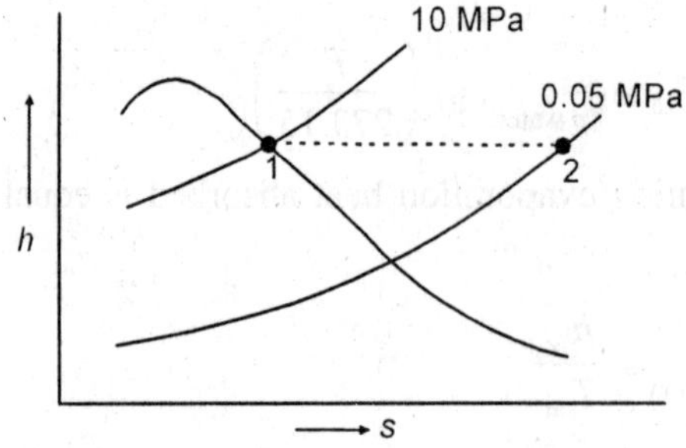

Fig. 6.21

Dryness fraction is **0.968.** **Ans.**

3. Determine internal energy of steam if its enthalpy, pressure and specific volumes are 2848 kJ/kg, 12 MPa and 0.017 m³/kg.

Solution:

Internal energy, $u = h - pv$

$= (2848 - 12 \times 10^3 \times 0.017)$ kJ/kg

$= 2644$ kJ/kg

Internal energy = **2644 kJ/kg** **Ans.**

4. Determine entropy of 5 kg of steam at 2 MPa and 300°C. Take specific heat of super heated steam as 2.1 kJ/kg.K.

Solution:

Steam state 2 MPa and 300°C lies in superheated region as saturation temperature at 2 MPa is 212.42°C and $h_{fg} = 1890.7$ kJ/kg.

Entropy of unit mass of superheated steam with reference to absolute zero.

$$= c_{p\,\text{water}} \ln\left(\frac{T_{\text{sat}}}{273.15}\right) + \frac{h_{fg,2\text{MPa}}}{T_{\text{sat}}} + c_{p\,\text{superheat}} \ln\frac{T_{\text{super heat}}}{T_{\text{sat}}}$$

Substituting values

$$= 4.18 \ln\left(\frac{485.57}{273.15}\right) + \left(\frac{1890.7}{485.57}\right) + \left[2.1 \ln\left(\frac{573.15}{485.57}\right)\right]$$

$= 6.646$ kJ/kg.K.

Entropy of 5 kg of steam = 33.23 kJ/K

Entropy of steam = **33.23 kJ/K.** **Ans.**

5. Water in a pond boils at 110°C at certain depth in water. At what temperature the water shall boil if we intend to boil it at 50 cm depth from above mentioned level.

Solution:

Boiling point = 110°C, pressure at which it boils = 143.27 kPa (from steam table, sat. pressure for 110°C)

At further depth of 50 cm the pressure

$= 143.27 - ((10^3 \times 9.81 \times 0.50) \times 10^{-3})$

$= 138.365$ kPa.

Boiling point at this depth = $T_{sat,\ 138.365}$

From steam table this temperature = 108.866 = 108.87°C

Boiling point = **108.87°C** **Ans.**

6. Water-vapour mixture at 100°C is contained in a rigid vessel of 0.5 m³ capacity. Water is now heated till it reaches critical state. What was the mass and volume of water initially?

Solution:

In a rigid vessel it can be treated as constant volume process.

$$v_1 = v_2$$

Since final state is given to be critical state, then specific volume at critical point,

$$v_2 = 0.003155 \text{ m}^3/\text{kg}$$

At 100°C saturation temperature, from steam table

$$v_f = 0.001044 \text{ m}^3/\text{kg},\ v_g = 1.6729 \text{ m}^3/\text{kg}$$

Thus for initial quality being x_1

$$v_1 = v_{f\,100°C} + x_1 \cdot v_{fg100°C}\ .$$

or

$$0.003155 = 0.001044 + x_1 \times 1.671856$$

$$x_1 = 0.0012627$$

$$\text{Mass of water initially} = \text{Total mass} \cdot (1 - x_1)$$

$$\text{Total mass of fluid} = \frac{V}{v_2} = \frac{0.5}{0.003155} = 158.48 \text{ kg}$$

Mass of water = 158.48 kg

Volume of water = 158.48 × 0.001044

= 0.1652 m³

Mass of water = **158.48 kg,** **Ans.**
Volume of water = **0.1652 m³**

7. Determine slope of an isobar at 2 MPa and 500°C on mollier diagram.

Solution:

On mollier diagram ($h - s$ diagram) the slope of isobaric line may be given as

$$\left(\frac{dh}{ds}\right)_{p\,=\,\text{const}} = \text{Slope of isobar}$$

From Ist and IInd law combined;

$$Tds = dh - vdp$$

for constant pressure

$$\left(\frac{dh}{ds}\right)_{p\,=\,\text{const}} = T$$

Here temperature $T = 773.15$ K

$$\text{hence slope} = \left(\frac{dh}{ds}\right)_{p\,=\,\text{const}} = 773.15$$

= **773.15** **Ans.**

8. Determine enthalpy, specific volume, entropy for mixture of 10% quality at 0.15 MPa.

Solution:

Given, $x = 0.10$

At 0.15 MPa, from steam table;

$h_f = 467.11$ kJ/kg $h_g = 2693.6$ kJ/kg

$v_f = 0.001053$ m³/kg $v_g = 1.1593$ m³/kg

$s_f = 1.4336$ kJ/kg.K $s_g = 7.2233$ kJ/kg.K

Enthalpy at $x = 0.10$

$$h = h_f + x.h_{fg} = 467.11 + \{0.10 \times (2693.6 - 467.11)\}$$

$$h = 689.759 \text{ kJ/kg}$$

Specific volume, $v = v_f + x.v_{fg}$

$$= 0.001053 + \{0.10 \times (1.1593 - 0.001053)\}$$

$$v = 0.116877 \text{ m}^3\text{/kg}$$

Entropy, $s = s_f + x.s_{fg}$

$$= 1.4336 + \{0.10 \times (7.2233 - 1.4336)\}$$

$$s = 2.01257 \text{ kJ/kg.K}$$

$h = 689.759$ kJ/kg
$v = 0.116877$ m³/kg
$s = 2.01257$ kJ/kg.K **Ans.**

9. In a piston-cylinder arrangement the steam at 1.0 MPa, 80% dryness fraction, and 0.05 m³ volume is heated to increase its volume to 0.2 m³. Determine the heat added.

Solution:

Given;

Initial states, 1: $P_1 = 1.0$ MPa, $V_1 = 0.05$ m³, $x_1 = 0.80$

Final state, 2: $V_2 = 0.2$ m³, $P_2 = 1$ MPa

Work done during constant pressure process, $W = P_1(V_2 - V_1)$

$$W = 1000\ (0.2 - 0.05)$$

$$W = 150 \text{ kJ}$$

$$\text{Mass of steam} = \frac{V_1}{v_1}$$

From steam table at P_1; $v_f = 0.001127$ m³/kg

$v_g = 0.19444$ m³/kg

$u_f = 761.68$ kJ/kg

$u_{fg} = 1822$ kJ/kg

so $v_1 = v_f + x_1\, v_{fg}$

$$v_1 = 0.15578 \text{ m}^3\text{/kg}$$

$$\text{Hence, mass of steam} = \frac{0.05}{0.15578} = 0.32097 \text{ kg}$$

$$\text{Specific volume at final state} = \frac{V_2}{\text{mass of steam}}$$

$$v_2 = 0.62311 \text{ m}^3\text{/kg.}$$

Corresponding to this specific volume the final state is to be located for getting the internal energy at final state at 1 MPa

$$v_2 > v_{g1\text{ MPa}}$$

hence state lies in superheated region, from the steam table by interpolation we get temperature as;

State lies between temperature of 1000°C and 1100°C

so exact temperature at final state

$$= 1000 + \frac{100 \times (0.62311 - 0.5871)}{(0.6335 - 0.5871)}$$

$$= 1077.61°\text{C}$$

Thus internal energy at final state, 1 MPa, 1077.61°C;

$$u_2 = 4209.6 \text{ kJ/kg}$$

Internal energy at initial state,

$$u_1 = u_f + x_1.u_{fg} = 761.68 + 0.8 \times 1822$$

$$u_1 = 2219.28 \text{ kJ/kg}$$

From first law of thermodynamics;

$$Q - W = \Delta U$$

$$Q = (U_2 - U_1) + W$$

$$= m(u_2 - u_1) + W$$

$$= 0.32097\ (4209.6 - 2219.28) + 150$$

$$Q = 788.83 \text{ kJ}$$

Heat added = **788.83 kJ** **Ans.**

10. Steam at 800 kPa and 200°C in a rigid vessel is to be condensed by cooling. Determine pressure and temperature corresponding to condensation.

Solution:

Here steam is kept in rigid vessel, therefore its' specific volume shall remain constant.

Specific volume at initial state, 800 kPa, 200°C, v_1,

It is superheated steam as T_{sat} = 170.43°C at 800 kPa.

From superheated steam table; v_1 = 0.2404 m³/kg.

At the begining of condensation specific volume = 0.2404 m³/kg.

$$v_2 = 0.2404 \text{ m}^3/\text{kg}$$

This v_2 shall be specific volume corresponding to saturated vapour state for condensation.

Thus $v_2 = v_g$ = 0.2404 m³/kg

Looking into steam table v_g = 0.2404 m³/kg shall lie between temperatures 175°C (v_g = 0.2168 m³/kg) and 170°C (v_g = 0.2428 m³/kg) and pressures 892 kPa (175°C) and 791.7 kPa (170°C).

By interpolation, temperature at begining of condensation

$$T_2 = 175 - \frac{(175 - 170) \times (0.2404 - 0.2168)}{(0.2428 - 0.2168)}$$

$$T_2 = 170.46°\text{C}$$

$$\text{Similarly, pressure, } p_2 = 892 - \frac{(892 - 791.7) \times (0.2404 - 0.2168)}{(0.2428 - 0.2168)}$$

$$p_2 = 800.96 \text{ kPa}$$

Pressure and temperature at condensation = **800.96 kPa** and **170.46°C** **Ans.**

11. Feed water pump is used for pumping water from 30°C to a pressure of 200 kPa. Determine change in enthalpy assuming water to be incompressible and pumping to be isentropic process.

Solution:

From Ist and IInd law;

$$Tds = dh - vdp$$

for isentropic process, $ds = 0$

hence $$dh = v \cdot dp$$

i.e. $$(h_2 - h_1) = v_1 (p_2 - p_1)$$

Corresponding to initial state of saturated liquid at 30°C; from steam table;

$$p_1 = 4.25 \text{ kPa}, v_f = v_1 = 0.001004 \text{ m}^3/\text{kg}$$

Therefore

$$(h_2 - h_1) = 0.001004 \, (200 - 4.25)$$
$$(h_2 - h_1) = 0.197 \text{ kJ/kg}$$

Enthalpy change = **0.197 kJ/kg** **Ans.**

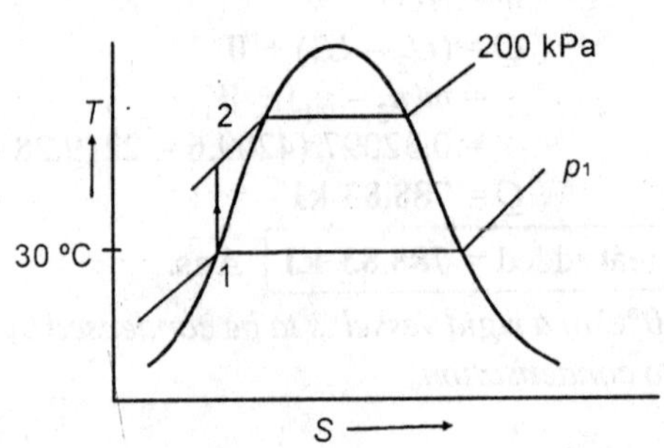

Fig. 6.22

12. A rigid vessel contains liquid-vapour mixture in the ratio of 3:2 by volume. Determine quality of water vapour mixture and total mass of fluid in vessel if the volume of vessel is 2 m³ and initial temperature is 150°C.

Solution:

From steam table at 150°C,

$$v_f = 0.001091 \text{ m}^3/\text{kg}, \; v_g = 0.3928 \text{ m}^3/\text{kg}$$

Volume occupied by water = 1.2 m³

Volume of steam = 0.8 m³

$$\text{Mass of water} \Rightarrow m_f = \frac{1.2}{0.001091} = 1099.91 \text{ kg}$$

$$\text{Mass of steam} \Rightarrow m_g = \frac{0.8}{0.3928} = 2.04 \text{ kg}$$

$$\text{Total mass in tank} = m_f + m_g = 1103.99 \text{ kg}$$

$$\text{Quality or Dryness fraction, } x = \frac{2.04}{1103.99} = 0.001848$$

Mass = **1103.99 kg, Quality = 0.001848** **Ans.**

13. Steam turbine expands steam reversibly and adiabatically from 4 MPa, 300°C to 50°C at turbine exit. Determine the work output per kg of steam.

Solution:

From SFEE on steam turbine;

$$W = (h_1 - h_2)$$

Initially at 4 MPa, 300°C the steam is super heated so enthalpy from superheated steam table or Mollier diagram.

$$h_1 = 2886.2 \text{ kJ/kg}, \; s_1 = 6.2285 \text{ kJ/kg.K}$$

Reversible adiabatic expansion process has entropy remaining constant. On Mollier diagram the state 2 can be simply located at intersection of constant temperature line for 50°C and isentropic expansion line.

Else from steam tables at 50°C saturation temperature;

$$h_f = 209.33 \text{ kJ/kg}, \; s_f = 0.7038 \text{ kJ/kg.K}$$

$$h_{fg} = 2382.7 \text{ kJ/kg}, \; s_{fg} = 7.3725 \text{ kJ/kg.K}$$

Here $s_1 = s_2$, let dryness fraction at 2 be x_2,

$$6.2285 = 0.7038 + x_2 \times 7.3725$$

$$x_2 = 0.7494$$

Hence enthalpy at state 2,

$$h_2 = h_f + x_2 . h_{fg} = 209.33 + 0.7494 \times 2382.7$$

$$h_2 = 1994.93 \text{ kJ/kg}$$

$$\text{Steam turbine work} = (2886.2 - 1994.93)$$

$$= 891.27 \text{ kJ/kg}$$

Turbine output = **891.27 kJ/kg. Ans.**

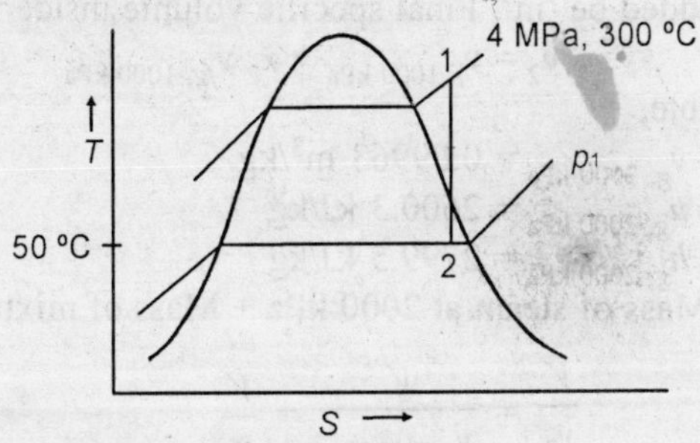

Fig. 6.23

14. In a closed vessel the 100 kg of steam at 100 kPa, 0.5 dry is to be brought to a pressure of 1000 kPa inside vessel. Determine the mass of dry saturated steam admitted at 2000 kPa for raising pressure. Also determine the final quality.

Solution:

It is a constant volume process.

Volume of vessel V = (Mass of vapour) × (Specific volume of vapour)

Initial specific volume, v_1

$$v_1 = v_{f,\,100\text{kPa}} + x_1 . v_{fg,\,100\text{kPa}}$$

at 100 kPa from steam table;

$$h_{f,\,100\text{ kPa}} = 417.46 \text{ kJ/kg}$$

$$u_{f,\ 100\ kPa} = 417.36 \text{ kJ/kg}$$
$$v_{f,\ 100\ kPa} = 0.001043 \text{ m}^3\text{/kg}$$
$$h_{fg,\ 100\ kPa} = 2258 \text{ kJ/kg},$$
$$u_{fg,\ 100\ kPa} = 2088.7 \text{ kJ/kg},$$
$$v_{g,\ 100\ kPa} = 1.6940 \text{ m}^3\text{/kg}$$

given $\quad x_1 = 0.5, v_1 = 0.8475 \text{ m}^3\text{/kg}$

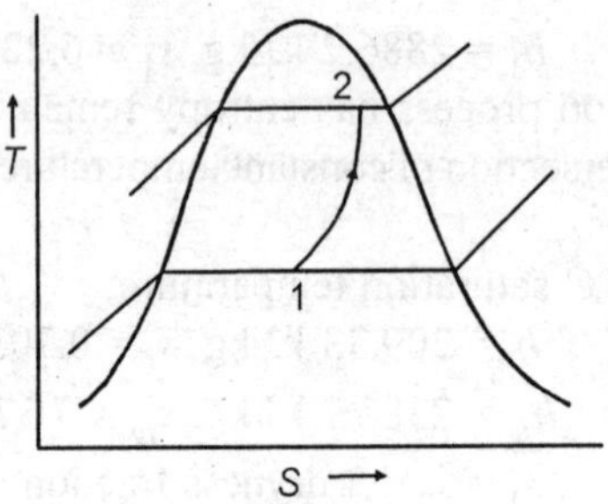

Fig. 6.24

Enthalpy at 1, $h_1 = 417.46 + 0.5 \times 2258 = 1546.46$ kJ/kg

Thus, volume of vessel $= (100 \times 0.5) \times (0.8475)$

$$V = 42.375 \text{ m}^3$$

Internal energy in the beginning $= U_1 = m_1 \times u_1$

$$= 100\ (417.36 + 0.5 \times 2088.7)$$
$$U_1 = 146171 \text{ kJ}$$

Let the mass of dry steam added be 'm', Final specific volume inside vessel, v_2

$$v_2 = v_{f,\ 1000\ kPa} + x_2 \cdot v_{fg,\ 1000\ kPa}$$

At 2000 kPa, from steam table,

$$v_{g,\ 2000\ kPa} = 0.09963 \text{ m}^3\text{/kg}.$$
$$u_{g,\ 2000\ kPa} = 2600.3 \text{ kJ/kg}$$
$$h_{g,\ 2000\ kPa} = 2799.5 \text{ kJ/kg}$$

Total mass inside vessel = Mass of steam at 2000 kPa + Mass of mixture at 100 kPa

$$\frac{V}{v_2} = \frac{V}{v_{g\ 2000\ kPa}} + \frac{V}{v_1}$$

$$\frac{42.375}{v_2} = \frac{42.375}{0.09963} + \frac{42.375}{0.8475}$$

$$v_2 = 0.089149$$

substituting $\quad v_2 = v_{f,\ 1000\ kPa} + x_2 \cdot v_{fg,\ 1000\ kPa}$

At 1000 kPa from steam table,

$$h_{f,\ 1000\ kPa} = 762.81 \text{ kJ/kg},$$
$$h_{fg,\ 1000\ kPa} = 2015.3 \text{ kJ/kg}\ v_{f,\ 1000\ kPa} = 0.001127 \text{ m}^3\text{/kg}$$
$$v_{g,\ 1000\ kPa} = 0.19444 \text{ m}^3\text{/kg}$$

it gives $\quad x_2 = 0.455$

For adiabatic mixing

$$(100 + m) \cdot h_2 = 100 \times h_1 + m \times h_{g,\,2000\text{ kPa}}$$

$$(100 + m) \cdot (762.81 + 0.455 \times 2015.3) = \{100 \times (1546.46)\} + \{m \times 2799.5\}$$

It gives upon solving;

$$(100 + m)\,(1679.77) = 154646 + (2799.5)m$$

$$m = 11.91 \text{ kg}$$

Mass of dry steam at 2000 kPa to be added = **11.91 kg** Quality of final mixture = **0.455**

Ans.

15. In a condenser the following derivations were made,
Recorded condenser vacuum = 71.5 cm of Mercury
Barometer reading = 76.8 cm of Mercury
Temperature of condensation = 35°C
Temperature of hot well = 27.6°C
Mass of condensate per hour = 1930 kg
Mass of cooling water per hour = 62000 kg
Inlet temperature = 8.51°C
Outlet temperature 26.24°C
Determine the state of steam entering condenser.

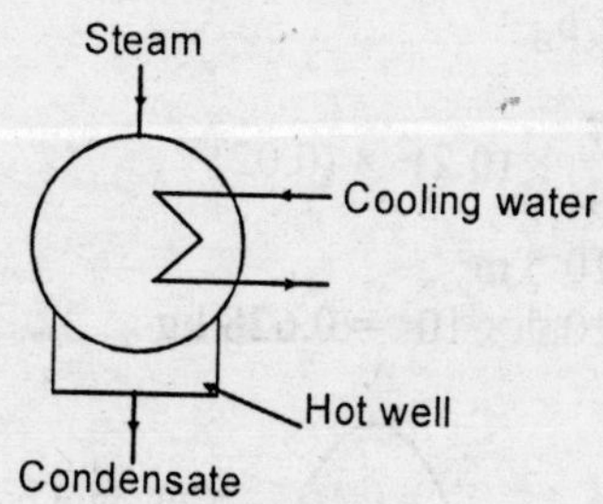

Fig. 6.25

Solution:

From Dalton's law of partial pressure the total pressure inside condenser will be sum of partial pressures of vapour and liquid inside.

$$\text{Condenser pressure} = \left\{\frac{76.8 - 71.5}{(73.55)} \times 101.325\right\} \text{kPa}$$

$$= 7.30 \text{ kPa}$$

Partial pressure of steam corresponding to 35°C from steam table;

$$= 5.628 \text{ kPa}$$

Enthalpy corresponding to 35°C from steam table,

$$h_f = 146.68 \text{ kJ/kg}$$

$$h_{fg} = 2418.6 \text{ kJ/kg}$$

Let quality of steam entering be 'x'.

From energy balance;

$$m_w(26.24 - 8.51) \times 4.18 = 1930\,(146.68 + x \times 2418.6 - 4.18 \times 27.6)$$

or $\qquad 62000\ (17.73 \times 4.18) = 1930\ (31.31 + x \times 2418.6)$

which gives $\qquad x = 0.97$

Dryness fraction of steam entering = 0.97 Ans.

16. In a vertical vessel of circular cross section having diameter of 20 cm water is filled upto a depth of 2 cm at a temperature of 150°C. A tight fitting frictionless piston is kept over the water surface and a force of 10 kN is externally applied upon the piston. If 600 kJ of heat is supplied to water determine the dryness fraction of resulting steam and change in internal energy. Also find the work done.

Solution:

Heating of water in vessel as described above is a constant pressure heating. Pressure at which process occurs

$$= \frac{\text{Force}}{\text{area}} + \text{atmospheric pressure}$$

$$= \left(\frac{10}{\frac{\pi}{4}(0.2)^2} + 101.3 \right) \text{kPa}$$

$$= 419.61 \text{ kPa}$$

$$\text{Volume of water contained} = \frac{\pi}{4} \times (0.2)^2 \times (0.02)$$

$$\text{Volume} = 6.28 \times 10^{-4} \text{ m}^3$$

$$\text{Mass of water} = 6.28 \times 10^{-4} \times 10^3 = 0.628 \text{ kg}$$

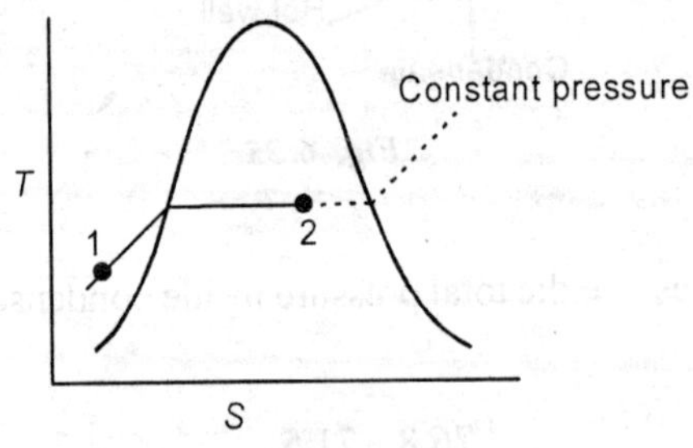

Fig. 6.26

Heat supplied shall cause sensible heating and latent heating.

Hence, Enthalpy change = Heat supplied

$$600 = \{(h_{f\text{ at }419.6\text{ kPa}} + x.h_{fg\text{ at }419.6\text{ kPa}}) - (4.18 \times 150)\} \times 0.628$$

$$600 = \{(612.1 + x.2128.7) - 627\} \times 0.628$$

Dryness fraction $\quad x = 0.456$

Dryness fraction of steam produced = 0.456 Ans.

Internal energy of water, initially

$$U_1 = mh_1 - p_1 V_1$$

$$= (0.628 \times 4.18 \times 150) - (419.61 \times 6.28 \times 10^{-4})$$

$$U_1 = 393.5 \text{ kJ}$$

Finally, internal energy of wet steam

$$U_2 = mh_2 - p_2V_2$$

Here $V_2 = m{\cdot}x{\cdot}v_{g \text{ at } 419.61 \text{ kPa}}$

$$= 0.456 \times 0.4435 \times 0.628$$

$$= 0.127 \text{ m}^3$$

Hence $U_2 = (0.628 \times 1582.8) - (419.61 \times 0.127)$

$$U_2 = 940.71 \text{ kJ}$$

Change in internal energy = $U_2 - U_1$.

Change in internal energy = **547.21 kJ Ans.**

Work done = $P \cdot (V_2 - V_1)$

$$= 419.61 \times (0.127 - 6.28 \times 10^{-4})$$

Work done = 53.03 kJ

Work done = **53.03 kJ** **Ans.**

17. In a separating and throttling calorimeter the total quantity of steam passed was 40 kg and 2.2 kg of water was collected from separator. Steam pressure before throttling was 1.47 MPa and temperature and pressure after throttling are 120°C and 107.88 kPa. Determine the dryness fraction of steam before entering to calorimeter. Specific heat of superheated steam may be considered as 2.09 kJ/kg.K.

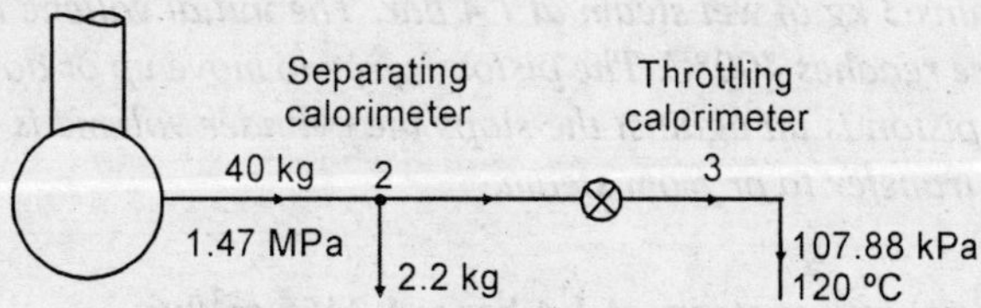

Fig. 6.27

Solution:

Consider throttling calorimeter alone,

Degree of superheat = 120 – 101.8 = 18.2°C

Enthalpy of superheated steam = (2673.95 + (18.2 × 2.09))

= 2711.988 kJ/kg

Enthalpy before throttling = Enthalpy after throttling

$$840.513 + x_2.1951.02 = 2711.988$$

or $x_2 = 0.9592$

For separating calorimeter alone, dryness fraction, $x_1 = \dfrac{40 - 2.2}{40}$

$$x_1 = 0.945$$

Overall dryness fraction = $(x_1 . x_2) = (0.945 \times 0.9592)$

= 0.9064

Dryness fraction : **0.9064** **Ans.**

18. A rigid vessel is divided into two parts A and B by means of frictionless, perfectly conducting piston. Initially, part A contains 0.4 m³ of air (ideal gas) at 10 bar pressure and part B contains 0.4 m³ of wet steam at 10 bar. Heat is now added to both parts until all the water in part B is evaporated. At this condition the pressure in part B is 15 bars. Determine the initial quality of steam in part B and the total amount of heat added to both parts. *[U.P.S.C. 1995]*

Solution:

Here heat addition to part B shall cause evaporation of water and subsequently the rise in pressure. Final, part B has dry steam at 15 bar. In order to have equilibrium the part A shall also have pressure of 15 bar. Thus, heat added

$$Q = V(P_2 - P_1)$$
$$= 0.4(15 - 10) \times 10^2$$
$$Q = 200 \text{ kJ}$$

Final enthalpy of dry steam at 15 bar, $h_2 = h_{g \text{ at } 15 \text{ bar}}$

$$h_2 = 2792.2 \text{ kJ/kg}$$

Let initial dryness fraction be x_1. Initial enthalpy,

$$h_1 = h_{f \text{ at } 10\text{bar}} + x_1 \cdot h_{fg \text{ at } 10 \text{ bar}}$$
$$h_1 = 762.83 + x_1 \cdot 2015.3$$

Heat balance yields,

$$h_1 + Q = h_2$$
$$(762.83 + x_1 \cdot 2015.3) + 200 = 2792.2$$
$$x_1 = 0.907$$

Heat added	= **200 kJ**	
Initial quality	= **0.907**	**Ans.**

19. A piston-cylinder contains 3 kg of wet steam at 1.4 bar. The initial volume is 2.25 m³. The steam is heated until its' temperature reaches 400°C. The piston is free to move up or down unless it reaches the stops at the top. When the piston is up against the stops the cylinder volume is 4.65 m³. Determine the amounts of work and heat transfer to or from steam. *[U.P.S.C. 1994]*

Solution:

From steam table, specific volume of steam at 1.4 bar = 1.2455 m³/kg

$$= v_{g \text{ at } 1.4 \text{ bar}}$$

$$\text{Specific volume of wet steam in cylinder, } v_1 = \frac{2.25}{3} = 0.75 \text{ m}^3\text{/kg}$$

$$\text{Dryness fraction of initial steam, } x_1 = \frac{0.75}{1.2455} = 0.602$$

$$\text{Initial enthalpy of wet steam, } h_1 = h_{f \text{ at } 1.4 \text{ bar}} + x_1 \cdot h_{fg \text{ at } 1.4 \text{ bar}}$$
$$= 457.99 + (0.602 \times 2232.3) \Rightarrow h_1 = 1801.83 \text{ kJ/kg}$$

$$\text{At 400°C specific volume of steam, } v_2 = \frac{4.65}{3} = 1.55 \text{ m}^3\text{/kg}$$

For specific volume of 1.55 m³/kg at 400°C the pressure can be seen from the steam table. From superheated steam tables the specific volume of 1.55 m³/kg lies between the pressure of 0.10 MPa (specific volume 3.103 m³/kg at 400°C) and 0.20 MPa (specific volume 1.5493 m³/kg at 400°C). Actual pressure can be obtained by interpolation;

$$P_2 = 0.10 + \left(\frac{0.20 - 0.10}{(1.5493 - 3.103)}\right) \times (1.55 - 3.103)$$

$$P_2 = 0.199 \text{ MPa} \approx 0.20 \text{ MPa}$$

Saturation pressure at 0.20 MPa = 120.23°C

Finally the degree of superheat = 400 – 120.23

= 279.77°C

Final enthalpy of steam at 0.20 MPa and 400°C, h_2 = 3276.6 kJ/kg

Heat added during process = $m\ (h_2 - h_1)$

= 3 × (3276.6 – 1801.83)

ΔQ = 4424.31 kJ

Internal energy of initial wet steam, $u_1 = u_{f\text{ at 1.4 bar}} + x_1 \cdot u_{fg\text{ at 1.4 bar}}$

u_1 = 457.84 + (0.607 × 2059.34)

u_1 = 1707.86 kJ/kg

Internal energy of final state,

$u_2 = u_{\text{at 0.2 MPa, 400°C}}$

u_2 = 2966.7 kJ/kg

Change in internal energy ⇒ $\Delta U = m(u_2 - u_1)$

= 3 × (2966.7 – 1707.86)

ΔU = 3776.52 kJ

From first law of thermodynamics,

Work done $\Delta W = \Delta Q - \Delta U$

= 4424.31 – 3776.52

Work done, ΔW = 647.79 kJ

Heat transfer = **4424.31 kJ**
Work transfer = **647.79 kJ**

Ans.

20. An insulated vessel is divided into two compartments connected by a valve. Initially, one compartment contains steam at 10 bar, 500°C, and the other is evacuated. The valve is opened and the steam is allowed to fill the entire volume, achieving a final pressure of 1 bar. Determine the final temperature, in °C, the percentage of the vessel volume initially occupied by steam and the amount of entropy produced, in kJ/kg. K.

[U.P.S.C. 1993]

Solution:

Here throttling process is occurring therefore enthalpy before and after expansion remains same. Let initial and final states be given by 1 and 2. Initial enthalpy, from steam table.

$h_{1\text{ at 10 bar}}$ and 500°C = 3478.5 kJ/kg

$s_{1\text{ at 10 bar and 500°C}}$ = 7.7622 kJ/kg.K

$v_{1\text{ at 10 bar and 500°C}}$ = 0.3541 m³/kg

Finally pressure becomes 1 bar so the final enthalpy at this pressure (of 1 bar) is also 3478.5 kJ/kg which lies between superheat temperature of 400°C and 500°C at 1 bar. Let temperature be T_2,

$h_{\text{at 1 bar, 400°C}}$ = 3278.2 kJ/kg

$h_{\text{at 1 bar, 500°C}}$ = 3488.1 kJ/kg

$$h_2 = 3478.5 = h_{\text{at 1 bar, 400°C}} + \frac{(h_{\text{at 1 bar,500°C}} - h_{\text{at 1 bar,400°C}})}{(500-400)} (T_2 - 400)$$

$$3478.5 = 3278.2 + \left\{\frac{3488.1 - 3278.2}{100}\right\} (T_2 - 400)$$

$$T_2 = 495.43°C,$$

Final temperature = **495.43°C**

Ans.

Entropy for final state,

$$s_2 = s_{\text{at 1 bar, 400°C}} + \frac{(s_{\text{at 1 bar,500°C}} - s_{\text{at 1 bar,400°C}})}{(500-400)}(495.43 - 400)$$

$$s_2 = 8.5435 + \left\{\frac{8.8342 - 8.5435}{100}\right\} \times (95.43)$$

$$s_2 = 8.8209 \text{ kJ/kg. K}$$

Change in entropy, $\Delta s = 8.8209 - 7.7622$

$= 1.0587$ kJ/kg . K

Change in entropy = **1.0587 kJ/kg . K**

Ans.

Final specific volume, $v_2 = v_{\text{at 1 bar, 400°C}}$

$$+ \frac{(v_{\text{at 1 bar,500°C}} - v_{\text{at 1 bar,400°C}})}{(500-400)} \times (95.43)$$

$$= 3.103 + \left\{\frac{3.565 - 3.103}{100}\right\} \times 95.43$$

$$v_2 = 3.544 \text{ m}^3\text{/kg}$$

Percentage volume occupied by steam $= \dfrac{0.3541}{3.544} \times 100 = 9.99\%$

Percentage of vessel volume initially occupied by steam = **9.99%**

Ans.

EXERCISE

6.1 Discuss generation of steam from ice at –5°C at 1 atm with the help of *T–S* and *P–V* diagrams.

6.2 What is meant by mollier diagram? Explain.

6.3 Write short notes on the following;
Sensible heating, Latent heating, Critical point, Triple point

6.4 Discuss different zones on *T–V* diagram for steam.

6.5 Derive the expression for enthalpy change during steam generation from feed water to superheated steam.

6.6 Discuss the throttling calorimeter for dryness fraction measurement.

6.7 Give a neat sketch of "separating and throttling calorimeter" for dryness fraction measurement.

6.8 Sketch the throttling and superheating processes on *h–s* and *T–S* diagrams.

6.9 Determine the final condition of steam if it is passed through a reducing valve which lowers the pressure from 2 MPa to 1 MPa. Assume initial state of steam to be 15% wet. [0.87]

6.10 Determine the final condition of steam, workdone, heat transferred and change in entropy if 0.5 kg of steam at 1 MPa and 0.8 dry is heated at constant pressure until its volume gets doubled.
[408.6°C, 77.5 kJ, 453.5 kJ, 0.895 kJ/K]

6.11 Determine the state of substance if 3346 kJ of heat is added to wet steam in a closed rigid vessel of 3m^3 volume containing 5 kg of wet steam at a pressure of 200 kPa till its pressure become 304 kPa. [Dry]

6.12 Complete the following table from steam table.

	Pressure (MPa)	*Temperature (°C)*	*Enthalpy (kJ/kg)*	*Quality (x)*	*Specific volume (m^3/kg)*	*Entropy (kJ/kg.K)*
(*a*)	1	–	–	–	–	6.5865
(*b*)	–	250.4	–	0	–	–
(*c*)	10	–	–	0.8	–	–
(*d*)	20	700	–	–	–	–
(*e*)	15	800	–	–	–	–

(*a*) 179.9°C, 762.8 kJ/kg, 1, 0.1944 m^3/kg.
(*b*) 4 MPa, 1087.31 kJ/kg, 1.252 m^3/kg, 2.7964 kJ/kg.K
(*c*) 311.06°C, 2461.33 kJ/kg, 0.01442 m^3/kg, 5.1632 kJ/kg.K
(*d*) 3809 kJ/kg, 1, 0.02113 m^3/kg, 6.7993 kJ/kg.
(*e*) 4092.4 kJ/kg, 1, 0.0321 m^3/kg, 7.204 kJ/kg.K.

6.13 Determine the pressure in a rigid vessel and volume of rigid vessel if it contains 500 kg of water at 65°C. [25 kPa, 0.51 m^3]

6.14 Estimate the change in volume of water and the total heat required for its' vaporization in a boiler producing saturated steam at 75 kPa. One kg feed water is supplied to boiler as saturated water. [2.22 m^3, 2.28 MJ]

6.15 Determine enthalpy, entropy and specific volume for following cases
(*i*) Steam at 4 MPa and 80% wet. (*ii*) Steam at 10 MPa and 550°C.
(*iii*) Steam at 8 MPa and 295°C.
Also estimate the above properties using Mollier diagram and quantify the percentage variation
[1430.13 kJ/kg, 3.45 kJ/kg.K, 0.011 m^3/kg]
[3500.9 kJ/kg, 6.76 kJ/kg.K, 0.036 m^3/kg]
[2758 kJ/kg, 5.74 kJ/kg.K, 0.024 m^3/kg]

6.16 Determine the temperature of steam at 20 MPa if its specific volume is 0.0155m^3/kg. [520°C]

6.17 Steam undergoes reversible adiabatic expansion in steam turbine from 500 kPa, 300°C to 50 kPa. Determine the work output per kg of steam turbine and quality of steam leaving steam turbine. [357.64 kJ/kg, 0.98]

6.18 Steam flowing through two pipelines at 0.5 MPa are mixed together so as to result in a mixture flowing at 2.2 kg/s and 60°C. One stream has quality of 0.8. Determine the temperature of second stream so as to result in the final mixture having dryness fraction of 0.994. [300°C]

6.19 A steam turbine operates with isentropic efficiency of 90%. Turbine handles 6 kg/s of steam at 0.980 MPa and 200°C and leaves at 0.294 MPa. Determine the power developed in hp and change of entropy from inlet to exit. [1660 hp, 0.050 kJ/kg.K]

6.20 A boiler is fed with water velocity of 2m/s, 1.96 MPa, 100°C. Steam is produced at 400°C temperature and comes out with velocity of 50 m/s. Determine the rate at which heat should be supplied per kg of steam for above operation of boiler. [2824.8 kJ/kg]

6.21 A steam nozzle is supplied steam at 1 MPa, 200°C and 100 m/s. Expansion upto 0.3 MPa occurs in the nozzle. Assuming isentropic efficiency of nozzle to be 0.9 determine final steam velocity.

6.22 Combined separating and throttling calorimeter is used to determine quality of steam. Following observations are made;
Steam inlet pressure = 1.4 MPa
Pressure after throttling = 0.1 MPa
Temperature after throttling = 120°C
Water collected in separator = 0.45 kg
Steam condensed after throttling = 6.75 kg
Take specific heat of superheated steam = 2.1 kJ/kg.K
Also find the limiting quality of steam to be measured by above throttling calorimeter alone assuming that separating calorimeter is not there. [0.90, 0.94]

7

Availability and General Thermodynamic Relations

7.1 INTRODUCTION

In the present civilization the use of energy resources has increased tremendously. Fast depleting fossil fuel reserves have inevitably gathered the attention of one and all to think and devise for optimum energy utilization. In order to optimally use energy, the efforts are required for identification and elimination of the sources of inefficiency during it's use, which obviously requires in depth study and analysis. A look into the laws of thermodynamics shows that the first law of thermodynamics bases upon the series of experiments done by James Joules, demonstrating the bidirectional numerical equivalence of converting work into heat while second law of thermodynamics exhibits a unidirectional equivalence between work and heat, i.e. for a given amount of heat the equivalent amount of work cannot be obtained whereas vice-a-versa may be there. Thus, the concept of quality of energy came into existence and work is considered as high grade of energy and heat as low grade of energy. Other forms of high grade energy are electrical energy, wind energy, tidal energy etc. and low grade energy may be heat from nuclear reactions, heat from combustion of fuel etc. Engineers have been using the first law of thermodynamics stating the energy conservation, therefore it could be concluded that energy can not be destroyed and exists with matters in all forms everywhere. It is now quite convincing to understand that the scarcity of energy resources and energy crisis is a paradox. Still in real life we find scarcity of energy, as in practice one is interested in the ability to feed, drive machines and occurrence of energy processes etc. Such discussions gave birth to the concept of 'available energy' and 'unavailable energy' or a concept of 'maximum work'.

This concept became very important in phenomenological thermodynamics, as it referred to the possibilities of performing work in real conditions. G. Gouy and A. Stodola pioneered in the studies pertaining to effect of ambient temperature upon the obtainable work and law of the loss of maximum work. The law of the loss of maximum work says that the work obtained is always less than the maximum obtainable work due to the irreversibility in thermal processes. Available energy concept came out of these propositions. Quality of energy, its convertibility into other forms and capability to perform work etc. are quantitatively defined using availability analysis. New term 'exergy' was introduced by Z. Rant in 1956 so as to differentiate it from energy. 'Exergy' analysis or 'availability' analysis has capability to identify and quantify the causes of thermodynamic imperfections of thermodynamic processes and thus indicate about the possibilities of improving the processes. It is preferred over energy analysis as energy analysis can not detect majority of thermodynamic imperfections. Such as the irreversible heat transfer, throttling and adiabatic combustion etc. do not have any energy loss but make the quality of energy inferior. Energy entering with fuel, electricity, flowing streams of matter and so on can be accounted for in products and by products. Energy cannot be destroyed. The idea that something can be destroyed is useful but should not be applied to 'energy', however it could be applied to another variable 'exergy'. Moreover, it is exergy and not energy that properly gauges the quality (utility), say one kJ of electricity generated by a power plant versus one kJ in plant cooling water stream. Electricity obviously has greater quality and the greater economic value.

These phenomenon can be evaluated by second law analysis easily. Exergy analysis could be integrated with principles of engineering economics to determine the potential for cost effective improvement of existing systems. Exergy and costing principles can also be used at initial design stage to develop systems that are optimized in annualized cost, sparing in use of fossil fuels and environmentally friendly.

Let us see a electricity generating power cycle as shown.

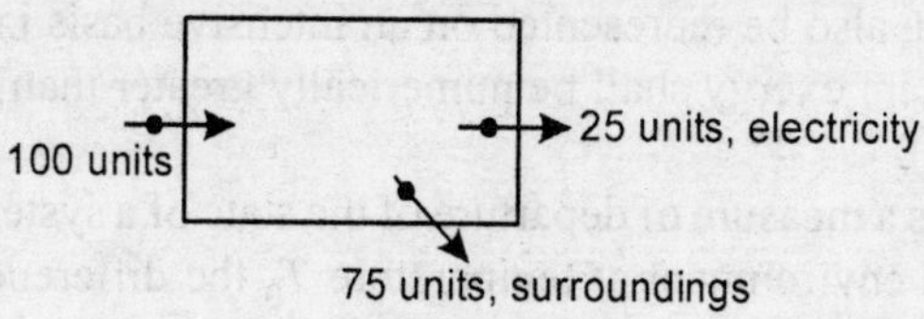

(*a*) Energy basis

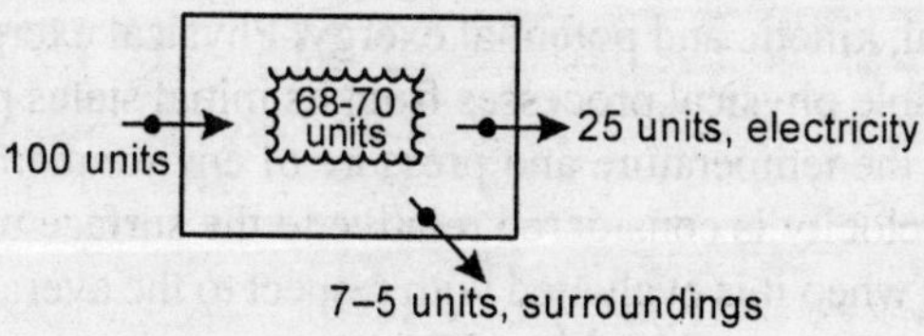

(*b*) Exergy basis

Fig. 7.1 Energy and exergy basis

Here Fig. 7.1(*a*) represents energy basis indicating that out of 100 energy units entering with fuel, 25 energy units are obtained as electricity and the remaining 75 units are discharged to surroundings. On exergy basis it may also be considered that 100 units of exergy enter with fuel are 25 units of exergy exit along with the electricity. For remaining 75 units it is seen that 68-70 units of exergy are destroyed within the plant due to irreversibilities and only 5-7 units are discharged to surroundings. Here it is worth noting from exergy basis that out of 75 units of energy considered to be discharged to surroundings in energy basis actually only 5-7 units are discharged to surroundings and the majority 68-70 units are lost due to irreversibilities. The loss of energy due to irreversibilities can be minimized by minimizing or eliminating the causes of irreversibility. Exergy analysis thus shows that significant performance improvement can come only by identifying and correcting the sources of inefficiency within system as the discharge is a minor area of concern.

7.2 AVAILABILITY OR EXERGY

From earlier discussions, it is obvious that energy can be conveniently categorised as low grade energy and high grade energy. Also, the second law of thermodynamics prohibits the complete conversion of low grade energy into high grade energy. The portion of low grade energy that can be converted is called 'available energy' or 'exergy' or 'availability' and the portion of energy not available for conversion is called 'unavailable energy' or 'anergy'. Mathematically;

$$\text{Anergy} = \text{Energy} - \text{Exergy}.$$

"Exergy can be quantified as the amount of work obtainable by bringing some matter to the state of thermodynamic equilibrium with common components of natural surroundings through reversible processes, thus involving interaction only with above mentioned components of nature."

As per Moran and Sciubba (1994), the "exergy refers to the maximum theoretical work that can be extracted from a combined system comprising of 'system' and 'environment' as the system passes from a given state to equilibrium with the environment—that is, system changes its' state to the dead state at which combined system possesses energy but no exergy."

Rickert defined "exergy as the shaft work or electrical energy necessary to produce a material in its specified state from materials common in the environment in a reversible way, heat being exchanged only with environment."

Exergy is an extensive property whose value is fixed by the state of system once the environment has been specified. Exergy can also be represented on an intensive basis i.e. per unit mass or per mole basis. For all states of the system exergy shall be numerically greater than or equal to zero.

$$\text{Exergy} \geq 0$$

Exergy as defined above is a measure of departure of the state of a system from that of environment. For state at emperature T and environment at temperature T_0 the difference $(T \sim T_0)$ shall decide the value of exergy i.e. greater the difference, the greater shall be exergy value. This exergy can be of basically two types i.e. chemical exergy and thermomechanical exergy. Thermomechanical exergy can be further classified as physical, kinetic and potential exergy. Physical exergy is the work obtainable by taking the substance by reversible physical processes from its initial states pressure 'p' and temperature 'T' to the state determined by the temperature and pressure of environment. Kinetic exergy is equal to the kinetic energy, when the velocity is considered relative to the surface of the earth. Potential exergy is equal to the potential energy when it is evaluated with respect to the average level of the surface of the earth in the locality of the process under consideration.

Chemical exergy refers to the work that can be obtained by taking a substance state at environmental pressure and temperature to the state of thermodynamic equilibrium with environment and bring system to restricted dead state.

Thermomechanical exergy refers to the maximum theoretical work obtainable as system passes from some given state to the restricted dead state.

Thermal exergy is defined as the sum of 'physical exergy' and 'chemical exergy'.

Rant defined, "exergy as that part of energy which could be fully converted into any other kind of energy". Exergy is function of state parameters of matter under consideration and of the state parameters of common components of environment as exergy results from the possibility of interaction between matter under consideration and common components of environment.

'Environment' here refers to the region or part of surroundings whose intensive properties do not change significantly with the occurrence of processes under consideration, while 'surroundings' comprise of everything that is not included in system. Environment is considered to be large and homogeneous in terms of pressure and temperature. Environment is regarded free of irreversibilities. All significant irreversibilities are present in the system and its' immediate surroundings. Irreversibilities present within system are called 'internal irreversibilities' while 'external irreversibilities' are those present in its' immediate surroundings.

'*Dead state*' refers to the state at which system and the environment are at mechanical, thermal and chemical equilibrium. Thus neither there can be any spontaneous change within the system or within the environment, nor any spontaneous interaction between the two. Dead state being a limiting state is also called '*restricted dead state*'. At dead state the system is at same temperature and pressure as that of its' surroundings and shall have no kinetic energy or potential energy relative to surroundings. A system shall thus have zero exergy (availability) at dead state and yield maximum possible work only when it follows a reversible process from its' state to the state of its' surroundings (dead state). Exergy or availability thus quantifies the maximum theoretical work available without violation of any laws of thermodynamics.

An engine operating with heat reservoir at T_1 and supplying Q_1 amount of heat and the environment temperature being T_0 shall give maximum amount of work when it operates between T_1 and T_0

$$\text{Maximum efficiency, } \eta_{max} = 1 - \frac{T_0}{T_1} = \eta_{rev}$$

$$\text{Maximum work} = Q_1 \cdot \eta_{max} = \text{Availability}$$

Let us consider a 'combined system' and find work done. Combined system comprises of control system and environment. Contents of control system do not mix with environment or have any reaction with environment. Maximum work is available when control system changes its state from initial state to dead state.

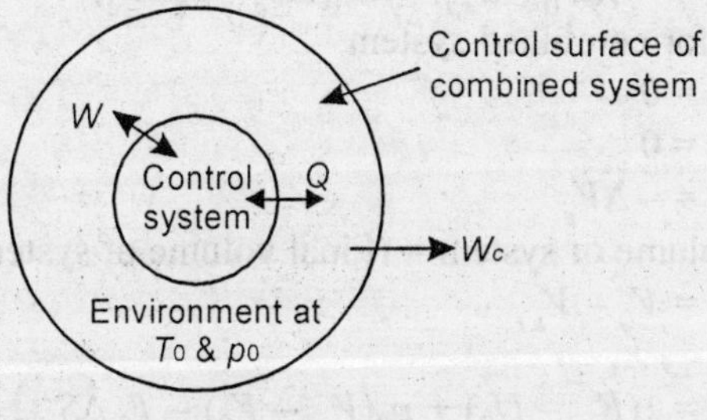

Fig. 7.2 Combined system

Let the control system have heat and work interaction Q and W with environment. Let us have only work interactions, W_c across the control surface of combined system. Let us use following nomenclature,

A = availability or exergy
E = energy, KE = kinetic energy,
PE = potential energy
p = pressure
Q = heat
S = entropy
T = temperature
U = internal energy
W = work

Subscript:
c = combined system
e = environment
s = control system
0 = dead state
i = initial state

Here for combined system

$\Delta V_c = 0$, where $V_c = V_s + V_e$, although V_s of system or V_e of environment may change but total volume of combined system shall remain constant.

Total work interaction of combined system can be given by total energy change of combined system.

$$W_c = -\Delta E_c$$

Energy change of combined system = Energy change in control system + Energy change in environment

$$\boxed{\Delta E_c = \Delta E_s + \Delta E_e}$$

Energy change in control system, ΔE_s=Energy of system at dead state – Energy of system at initial state

Energy of system at dead state,

as energy $$E = U + KE + PE$$

$$E_{s,\text{ at dead state}} = U_0, \text{ as } KE = PE = 0$$

Energy of system at initial state, $E_{s,\text{initially}} = E_{s,i}$

$$\Delta E_s = U_0 - E_{s,i}$$

Energy change in environment, shall be due to heat interaction and the work associated with its' volume change (*pdv* work). For example expansion inside a piston cylinder arrangement shall have piston also displacing the volume of environment (*pdV* work is boundary work). Change in extensive properties, internal energy, entropy, volume of environment can be given by first law of thermodynamics.

$$\Delta E_e = \Delta U_e = T_0 \Delta S_e - p_0 . \Delta V_e$$

Hence, work interaction of combined system,

$$W_c = -\{(U_0 - E_{s,i}) + (T_0 \Delta S_e - p_0 \Delta V_e)\}$$

Also, we have seen that for combined system

$$\Delta V_c = 0$$

$$\Delta V_s + \Delta V_e = 0$$

or, $$\Delta V_s = -\Delta V_e$$

Here, ΔV_s = change in volume of system = (Final volume of system at dead state – Initial volume)

$$\Delta V_s = V_0 - V_{s,i}$$

Substituting in W_c,

$$W_c = \{(E_{s,i} - U_0) + p_0(V_{s,i} - V_0) - T_0 \Delta S_e\}$$

Total entropy change of combined system shall be due to irreversibilities within the combined system;

$$\Delta S_c = \Delta S_s + \Delta S_e$$

$$\Delta S_c = (S_0 - S_{s,i}) + \Delta S_e$$

or, $$\Delta S_e = (-S_0 + S_{s,i}) + \Delta S_c$$

Substituting ΔS_e in work, W_c,

$$W_c = \{(E_{s,i} - U_0) + p_0(V_{s,i} - V_0) - T_0(-S_0 + S_{s,i}) - T_0 \Delta S_c)\}$$

For the combined system total entropy change shall be either zero for reversible process or more than zero for irreversible process;

Mathematically, $\Delta S_c \geq 0$

Hence,

$$W_c \leq \{(E_{s,i} - U_0) + p_0(V_{s,i} - V_0) - T_0(S_{s,i} - S_0)\}$$

For reversible processes there will be no entropy generation, i.e. $\Delta S_c = 0$ and work shall be maximum only when the combined system is internally reversible from all respects.

Thus, $$W_{c,\max} = \{(E_{s,i} - U_0) + p_0(V_{s,i} - V_0) - T_0(S_{s,i} - S_0)\}$$

In general terms for any initial state of system,

$$W_{c,\max} = \{(E - U_0) + p_0(V - V_0) - T_0(S - S_0)\}$$

Availability or exergy, $A = W_{c,\max} = \{(E - U_0) + p_0(V - V_0) - T_0(S - S_0)\}$

Availability or exergy cannot be less than zero as the maximum work interaction can not be less than zero.

Also it can be given as,

$$W_c = A - T_0 \cdot \Delta S_c$$

Above expression shows that some work done by combined system gets lost i.e. the irreversibilities causing entropy production keep work below its' maximum value.

Availability or exergy is not conserved like energy. Exergy gets destroyed by irreversibilities when the control system changes to dead state and no work is done by combined system as in case of spontaneous change.

Availability destruction is proportional to entropy generation due to irreversibilities in processes. Irreversibility can be given as the product of dead state temperature and entropy generation due to irreversible process.

$$I = T_0 \cdot \Delta S_c \quad \text{and} \quad W_c = A - I \quad \text{and} \quad I = A - W_c$$

Above discussion indicates that the *maximum work shall be obtained when a process takes place in reversible manner.* But in fact almost all the processes in real life occur in irreversible manner, so some portion of energy is always unavailable. As irreversible processes are continuously increasing therefore unavailable energy is also gradually increasing. This phenomenon is also called *principle of degradation of energy or law of degradation of energy.*

$$\text{Availability}, A = \{(E - U_0) - T_0(S - S_0) + p_0(V - V_0)\}, \text{ kJ}$$

Availability per unit mass,

$$\omega = \{(e - u_0) - T_0(s - s_0) + p_0(v - v_0)\}, \text{ kJ/kg}$$

Availability or exergy is thus a measure of departure of state of system from that of environment. Thus, it is an attribute of system and environment together. However, once the environment is specified, a numerical value can be assigned to availability in terms of property values for system only. Hence, *exergy can be regarded as property of the system.*

7.3 AVAILABILITY ASSOCIATED WITH HEAT AND WORK

Let us consider a reversible heat engine having heat transfer from environment to control system and vice-a-versa.

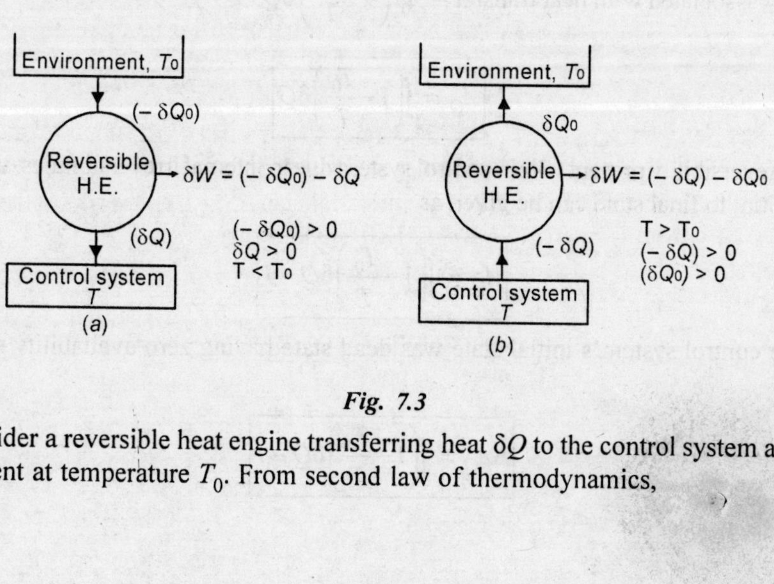

Fig. 7.3

Let us consider a reversible heat engine transferring heat δQ to the control system at temperature T from environment at temperature T_0. From second law of thermodynamics,

$$\frac{\delta Q}{T} = \left(\frac{-\delta Q_0}{T_0}\right)$$

work, $$\delta W = \left(\frac{T_0}{T} - 1\right).\delta Q$$

Let us now consider a reversible heat engine transferring heat δQ from control system to environment at T_0. From second law of thermodynamics,

$$\frac{\delta Q_0}{T_0} = \left(\frac{-\delta Q}{T}\right)$$

so work, $$\delta W = \left(1 - \frac{T_0}{T}\right)(-\delta Q)$$

$$= \left(\frac{T_0}{T} - 1\right).\delta Q$$

Availability associated with heat transfer : Let us consider a control system at dead state interacting with other system and there is heat interaction Q in control system. Let the final state of control system be given by 'f'. Due to heat interaction Q the control system may get heated up or cooled so that the final state is different from that of environment. Control system's temperature may increase from T_0 to T_f or decrease from T_f to T_0 but in every case availability shall increase. Availability of control system at final state gives maximum work that will be available from the combined system (control system + environment) as control system returns to dead state.

Work available from a reversible heat engine when control system gets heated or cooled by environment,

$$W_{max} = \int_f^0 \left(\frac{T_0}{T} - 1\right)\delta Q$$

Availability associated with heat transfer $= \int_0^f \left(1 - \frac{T_0}{T}\right)\delta Q$

$$\boxed{A_Q = \int\left(1 - \frac{T_0}{T}\right)\delta Q}$$

If there is irreversibility present within control system due to internal irreversibilities, then a availability change from initial to final state can be given as

$$\boxed{A_Q = \int\left(1 - \frac{T_0}{T}\right)\delta Q - I}$$

Since, here control system's initial state was dead state having zero availability so the change in availability

$$\boxed{\Delta A_Q = \int\left(1 - \frac{T_0}{T}\right)\delta Q - I}$$

Availability associated with work : Let us consider a control system initially at dead state. Control system has adiabatic compression occurring in it due to work interaction with some other system. $-W$ work is done on control system and it attains some final state, 'f'. Availability in this case shall be the maximum work available from the combined system of control system and environment as control system returns to the dead state.

If the work W is done by the control system as it returns from final state 'f' to dead state and the change in volume V_f to V_0 takes place by displacing the environment (pdV work), then availability associated with work,

$$\boxed{A_w = \Delta A_w = [W - p_0(V_f - V_0)]}$$

In case no boundary work is there, then $V_f = V_0$

$$\boxed{A_w = W = \Delta A_w}$$

Here it is also the availability change as system is returning to dead state. In case there is availability loss due to internal irreversibilities then change in availability,

$$\boxed{A_w = [W - p_0(V_f - V_0)] - I = \Delta A_w}$$

Similarly, availability associated with kinetic energy and potential energy can be given as,

$$\boxed{A_{KE} = \frac{1}{2}mV^2} \quad \text{; availability with K.E.}$$

and $\boxed{A_{PE} = mgz}$; availability with P.E.

Generalized availability equation : A general availability equation for a control system having heat and work interactions with other systems can be obtained using earlier formulations. Let us consider a control system interacting with other systems and also having irreversibilities causing availability destruction in it. For elemental change during a process the energy balance can be given as, $dE = \delta Q - \delta W$.

$$\text{Total entropy change } dS = \frac{\delta Q}{T} + \delta S_{\text{irrev}}$$

where T is temperature on control surface having δQ heat transfer and δS_{irrev} is entropy generated due to irreversibilities

Energy equation can be rewritten as,

$$dE + p_0 dV = \delta Q - \delta W + p_0 dV$$

Entropy equation can be rewritten as,

$$T_0\, dS = \frac{\delta Q \cdot T_0}{T} + T_0 . \delta S_{\text{irrev}}$$

Combining modified forms of energy and entropy equations by subtracting one from other,

$$dE + p_0 dV - T_0 \cdot dS = \delta Q - \delta W + p_0 dV - \frac{\delta Q \cdot T_0}{T} - T_0 \cdot \delta S_{\text{irrev}}$$

$$dE + p_0 dV - T_0 \cdot dS = \left(1 - \frac{T_0}{T}\right) \cdot \delta Q - (\delta W - p_0 \cdot dV) - T_0 \cdot \delta S_{\text{irrev}}$$

We have already seen in earlier article that the change in availability can be given as,

$$dA = dE + p_0.dV - T_0.dS$$

Hence,

$$dA = \left(1-\frac{T_0}{T}\right).\delta Q - (\delta W - p_0.dV) - T_0.\delta S_{irrev}$$

Here, $T_0 \cdot \delta S_{irrev} = I$

For any process in control system between states 1 and 2, availability change can be given as,

$$\Delta A_{1-2} = \underbrace{\int_1^2 \left(1-\frac{T_0}{T}\right) dQ}_{\text{Availability associated with heat transfer}} - \underbrace{(W_{1-2} - p_0\Delta V_{1-2})}_{\text{Availability associated with work interaction}} - \underbrace{I}_{\text{Irreversibility}}$$

Generally, for any control mass in control system the availability change can be given as,

$$\boxed{\Delta A_{1-2} = \int_1^2 \left(1-\frac{T_0}{T}\right) \cdot dQ - (W - p_0\Delta V) - I}$$

Above availability change can also be given on per unit time basis.

7.4 EFFECTIVENESS OR SECOND LAW EFFICIENCY

Performance of engineering systems are generally measured using efficiency as defined by first law of thermodynamics. Efficiency as defined by first law uses energy for its' quantification. *Second law efficiency* or *effectiveness* or *exergetic efficiency* is an analogous parameter defined using availability.

Energy balance for a system with steady state yields,

Energy in = Energy output + Energy loss

Availability equation shall yield,

Availability in = (Availability output + Availability loss + Availability destruction due to irreversibility)

Mathematically,

$$\text{(by first law), Efficiency } \eta = \frac{\text{Energy out in product (= Output)}}{\text{Energy in}}$$

$$\eta = \frac{\text{Energy input} - \text{Energy loss}}{\text{Energy input}}$$

$$\boxed{\eta = 1 - \frac{\text{Energy loss}}{\text{Energy input}}}$$

$$\text{(by second law), Effectiveness, } \varepsilon = \frac{\text{Availability output}}{\text{Availability in}}$$

$$\varepsilon = 1 - \left[\frac{\text{Availability loss} + \text{Availability destruction due to irreversibility}}{\text{Availability in}}\right]$$

Effectiveness can also be given as the ratio of thermal efficiency to the maximum possible thermal efficiency (reversible processes) under same conditions.

$$\varepsilon = \frac{\eta_{th}}{\eta_{th,rev}}$$

For work producing systems, effectiveness = $\left(\frac{\text{Useful work}}{\text{Reversible work or maximum work}}\right)$

For work absorbing systems, effectiveness = $\left(\frac{\text{Reversible work or maximum work}}{\text{Useful work}}\right)$

For refrigerators and heat pumps, effectiveness = $\left(\frac{\text{COP}}{\text{COP}_{\text{rev.}}}\right)$

In general terms, Second law efficiency or effectiveness = $\left(\frac{\text{Availability used}}{\text{Availability supplied}}\right)$

7.5 SECOND LAW ANALYSIS OF STEADY FLOW SYSTEMS

Let us consider a steady flow system as shown,

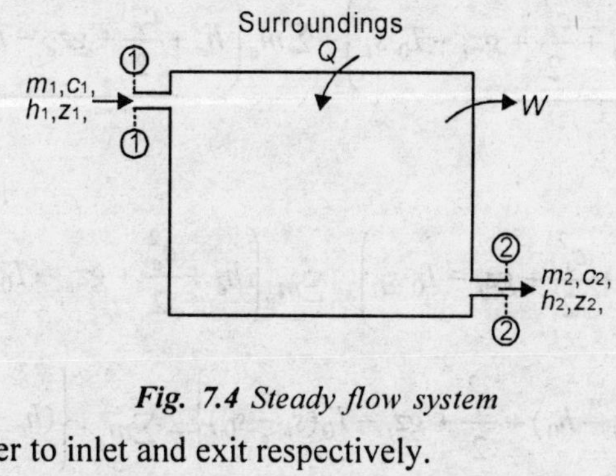

Fig. 7.4 Steady flow system

Section 1–1 and 2–2 refer to inlet and exit respectively.

Steady flow system interacts with surroundings at P_0 and T_0. Steady flow energy equation can be given as,

$$Q + m_1\left(h_1 + \frac{c_1^2}{2} + gz_1\right) = W + m_2\left(h_2 + \frac{c_2^2}{2} + gz_2\right)$$

Entropy generated,

$$S_{gen} = m_2 s_2 - m_1 s_1 + \frac{Q_{surr}}{T_0}$$

Heat transfer from control system, (–) = Heat gained by surrounding (+); $-Q = Q_{surr}$, from above two equations, substitution for Q yields,

$$-T_0 S_{gen} + m_1\left(h_1 + \frac{c_1^2}{2} + gz_1 - s_1 T_0\right) = W + m_2\left(h_2 + \frac{c_2^2}{2} + gz_2 - s_2 T_0\right)$$

$$\text{or, } W = m_1\left(h_1 + \frac{c_1^2}{2} + gz_1 - T_0 s_1\right) - m_2\left(h_2 + \frac{c_2^2}{2} + gz_2 - T_0 s_2\right) - T_0 \cdot S_{gen}$$

W can be quantified as above. This W shall be actual work available from system. Here entropy generation due to irreversibilities in processes reduce W and so for fully reversible processes $S_{gen} = 0$ and we get maximum available work;

$$W_{max} = m_1\left(h_1 + \frac{c_1^2}{2} + gz_1 - T_0 s_1\right) - m_2\left(h_2 + \frac{c_2^2}{2} + gz_2 - T_0 s_2\right)$$

In general terms, actual work available in this kind of systems where boundary work ($p.dV$) is absent can be given as under,

$$W = \Sigma m_i\,(h_i + c_i^2 + gz_i - T_0 \cdot s_i) - \Sigma m_e\left(h_e + \frac{c_e^2}{2} + gz_e - T_0 \cdot s_e\right) - T_0 \cdot S_{gen}$$

where subscript 'i' and 'e' refer to inlet and exit in system.

For no irreversibilities present or for reversible processes,

$$W_{max} = \Sigma m_i\left(h_i + \frac{c_i^2}{2} + gz_i - T_0 \cdot s_i\right) - \Sigma m_e\left(h_e + \frac{c_e^2}{2} + gz_e - T_0 \cdot s_e\right)$$

Expression 1

Change in availability,

$$\Delta A = \Sigma m_i\left(h_i + \frac{c_i^2}{2} + gz_i - T_0 \cdot s_i\right) - \Sigma m_e\left(h_e + \frac{c_e^2}{2} + gz_e - T_0 \cdot s_e\right)$$

$$\Delta A = \Sigma m_i\left\{(h_i - h_0) + \frac{c_i^2}{2} + gz_i - T_0(s_i - s_0)\right\} - \Sigma m_e \cdot \left\{(h_e - h_0) + \frac{c_e^2}{2} + gz_e - T_0(s_e - s_0)\right\}$$

on unit mass basis,

$$\Delta\omega = \{T_0 \cdot \Delta s - \Delta h - \Delta K.E. - \Delta P.E.\}, \text{ kJ/kg}$$

where $\Delta K.E.$ and $\Delta P.E.$ refer to kinetic energy and potential energy changes in system.

It indicates that the change in availability can be given by the difference of fluid stream availability at inlet and exit. "Fluid stream availability" can be defined in respect to dead state as, $c_0 = 0$, $z_0 = 0$

$$\psi = (h - h_0) + \frac{c^2}{2} + gz - T_0(s - s_0), \text{ kJ/kg}$$

$$\psi = \underline{(u - u_0) + p_0(v - v_0) - T_0(s - s_0)} + \frac{c^2}{2} + gz, \text{ kJ/kg.}$$

Here underlined terms are called physical exergy, $c^2/2$ is kinetic exergy and gz is potential exergy

$$\Delta A = W_{max} = \Sigma m_i\,\psi_i - \Sigma m_e \,.\, \psi_e$$

Change of availability can be obtained using stream availability as described above. "Stream availability" is quantification of availability at a point.

Irreversibility rate in steady flow process can be given as, $I = T_0 . S_{gen}$, (kW)

Exergy (Availability) and energy can be compared based upon their characteristics as given below.

Exergy (availability)	*Energy*
1. Exergy does not follow the law of conservation.	1. Energy follows the law of conservation.
2. It is function of states of the matter under consideration and the 'environment'.	2. It is function of the state of matter under consideration.
3. It is estimated with respect to the state of reference imposed by environment.	3. It may be calculated based upon the assumed state of reference.
4. Exergy always depends upon pressure.	4. In case of ideal gas energy does not depend upon pressure.
5. Exergy increases with temperature drop at low temperatures. For constant pressure processes exergy attains minimum value at the temperature of environment.	5. Energy increases with rise of temperature.
6. Exergy has positive value for ideal vacuum.	6. Energy is zero for an ideal vacuum.

7.6 GENERAL THERMODYNAMIC RELATIONS

Objective of this section is to develop mathematical relations for estimation of various thermodynamic properties such as *u*, *h*, *s* etc. for a compressible system. Thermodynamic properties such as pressure, volume and temperature (*P*, *V*, *T*) etc. can be directly measured experimentally while some other properties can not be measured directly and require thermodynamic relations for their determination. These thermodynamic relations are the basis for getting useful thermodynamic properties.

Important mathematical relations : To define state of a simple compressible system of known mass and composition one requires minimum two independent intensive properties. Thus, all intensive properties can be determined through functions of the two independent intensive properties such as,

$$p = p(T, v),\ u = u(T, v),\ h = h(T, v)...$$

Above are functions of two independent variables and can be given in general as, $z = z(x, y)$, where x, y are independent variables.

Exact differential : In earlier discussions we have seen that the differential of any property should be exact. Therefore, let us review calculus briefly.

Exact differential of any function *z* shall be as given below for *z* being continuous function of *x* and *y*.

$$dz = \left(\frac{\partial z}{\partial x}\right)_y dx + \left(\frac{\partial z}{\partial y}\right)_x dy$$

or,
$$dz = M \cdot dx + N \cdot dy$$

where, $M = \left(\frac{\partial z}{\partial x}\right)_y$, $N = \left(\frac{\partial z}{\partial y}\right)_x$ i.e. *M* is partial derivative of *z* with respect to *x* when variable *y* is held constant and *N* is partial derivative of *z* with respect to *y* when variable *x* is held constant.

Here, since *M* and *N* have continuous first partial derivative therefore, order of differentiation is immaterial for properties and second partial derivative can be given as,

$$\frac{\partial}{\partial y}\left[\left(\frac{\partial z}{\partial x}\right)_y\right]_x = \frac{\partial}{\partial x}\left[\left(\frac{\partial z}{\partial y}\right)_x\right]_y$$

or,
$$\left(\frac{\partial M}{\partial y}\right)_x = \left(\frac{\partial N}{\partial x}\right)_y$$

Thus, the test of exactness for any property shall be,

$$\boxed{\left(\frac{\partial M}{\partial y}\right)_x = \left(\frac{\partial N}{\partial x}\right)_y}$$

Reciprocity relation and cyclic relation : Let us consider three variables x, y, z such that any two of these are independent variables. Thus, we can write

$$x = x(y, z);\ y = y(x, z)$$

In differential form,
$$dx = \left(\frac{\partial x}{\partial y}\right)_z \cdot dy + \left(\frac{\partial x}{\partial z}\right)_y \cdot dz;\ dy = \left(\frac{\partial y}{\partial x}\right)_z dx + \left(\frac{\partial y}{\partial z}\right)_x dz$$

Combining above two relations we get

$$\left[1-\left(\frac{\partial x}{\partial y}\right)_z\left(\frac{\partial y}{\partial x}\right)_z\right] dx = \left[\left(\frac{\partial x}{\partial y}\right)_z\left(\frac{\partial y}{\partial x}\right)_x + \left(\frac{\partial x}{\partial z}\right)_y\right] dz$$

As x and z are independent variables so let us keep z constant and vary x, i.e. $dz = 0$ and $dx \neq 0$ which yields reciprocity relation as,

$$\left[1-\left(\frac{\partial x}{\partial y}\right)_z\left(\frac{\partial y}{\partial x}\right)_z\right] = 0$$

or,
$$\boxed{\left(\frac{\partial x}{\partial y}\right)_z\left(\frac{\partial y}{\partial x}\right)_z = 1}$$

or,
$$\boxed{\left(\frac{\partial x}{\partial y}\right)_z = \frac{1}{\left(\frac{\partial y}{\partial x}\right)_z}} \quad \textit{Reciprocity relation}$$

Similarly, let us keep x constant and vary z i.e. $dx = 0$, $dz \neq 0$ which shall be possible only when;

$$\left[\left(\frac{\partial x}{\partial y}\right)_z\left(\frac{\partial y}{\partial z}\right)_x + \left(\frac{\partial x}{\partial z}\right)_y\right] = 0$$

or
$$\boxed{\left(\frac{\partial x}{\partial y}\right)_z\left(\frac{\partial y}{\partial z}\right)_x\left(\frac{\partial z}{\partial x}\right)_y = -1} \quad \textit{Cyclic relation}$$

7.6.1 Gibbs and Helmholtz Functions

For a simple compressible system of fixed chemical composition thermodynamic properties can be given from combination of first law and second law of thermodynamics as,

$$du = T \cdot ds - pdv$$
$$dh = T \cdot ds + vdp$$

Gibbs function (g) and Helmholtz function (f) are properties defined as, Gibbs function,

$\boxed{g = h - T \cdot s}$, on unit mass basis i.e. specific Gibb's function

also, $\boxed{G = H - T \cdot S}$

Helmholtz function,

$\boxed{f = u - T \cdot s}$, on unit mass basis i.e. specific Helmholtz function

also, $\boxed{F = U - T \cdot S}$

In differential form Gibbs function can be given as below for an infinitesimal reversible process

$$dg = dh - T \cdot ds - s \cdot dT$$

$\boxed{dg = vdp - sdT}$ for a reversible isothermal process, $\int_1^2 dg = \int_1^2 vdp$

or, also $\boxed{dG = Vdp - SdT}$ for reversible isothermal process; $\int_1^2 dG = \int_1^2 Vdp$

For a "reversible isobaric and isothermal process", $dp = 0$, $dT = 0$ $dG = 0$

i.e. $G =$ constant

'Gibbs function' is also termed as 'Gibbs free energy'. For a reversible isobaric and isothermal process Gibbs free energy remains constant or Gibbs function of the process remains constant. Such reversible thermodynamic processes may occur in the processes involving change of phase, such as sublimation, fusion, vaporization etc., in which Gibbs free energy remains constant.

'Helmholtz function' is also called 'Helmholtz free energy'. For any infinitesimal reversible process Helmholtz function can be given in differential form as,

$$df = du - T \cdot ds - sdT$$

or, $\boxed{df = -pdv - sdT}$

or, $\boxed{dF = -pdV - SdT}$

For a reversible isothermal process

$$df = -pdv$$

or, $\int_1^2 df = -\int_1^2 p \cdot dv$

or, $\int_1^2 dF = -\int_1^2 pdV$

For a reversible isothermal and isochoric process, $dT = 0$, $dV = 0$

$$df = 0$$

or, $dF = 0$

or, $\boxed{F = \text{Constant}}$

Above includes that the Helmholtz free energy remains constant during a reversible isothermal and isochoric process. Such processes may occur during chemical reactions occurring isothermally and isochorically.

7.6.2 Maxwell Relations

Differential equations of thermodynamic properties, *u, h, f* and *g* can be given as function of *p, T, v, s* as below:

$$du = T \cdot ds - pdv$$
$$dh = T \cdot ds + vdp$$
$$df = -pdv - s \cdot dT$$
$$dg = vdp - s \cdot dT$$

Above equations can be used for defining the functions *u, h, f, g* based upon analogy with, $dz = M \cdot dx + N \cdot dy$, for $z = z(x, y)$ and $\left(\frac{\partial M}{\partial y}\right)_x = \left(\frac{\partial M}{\partial y}\right)_y$. From above four equations for properties to be exact differentials, we can write functions;

$$u = u(s, v)$$
$$h = h(s, p)$$
$$f = f(v, T)$$
$$g = g(p, T)$$

For differential of function '*u*' to be exact;

$$\left(\frac{\partial T}{\partial v}\right)_s = -\left(\frac{\partial p}{\partial s}\right)_v$$

For differential of function '*h*' to be exact; $\left(\frac{\partial T}{\partial p}\right)_s = \left(\frac{\partial v}{\partial s}\right)_p$

For differential of function '*f*' to be exact; $\left(\frac{\partial p}{\partial T}\right)_v = \left(\frac{\partial s}{\partial v}\right)_T$

For differential of function '*g*' to be exact; $\left(\frac{\partial v}{\partial T}\right)_p = -\left(\frac{\partial s}{\partial p}\right)_T$

Above four conditions for exact differentials of thermodynamic properties result into "Maxwell relations".

Thus, "Maxwell relations" are :

$$\left(\frac{\partial T}{\partial v}\right)_s = -\left(\frac{\partial p}{\partial s}\right)_v$$

$$\left(\frac{\partial T}{\partial p}\right)_s = \left(\frac{\partial v}{\partial s}\right)_p$$

$$\left(\frac{\partial p}{\partial T}\right)_v = \left(\frac{\partial s}{\partial v}\right)_T$$

$$\left(\frac{\partial v}{\partial T}\right)_p = -\left(\frac{\partial s}{\partial p}\right)_T$$

Maxwell relations have large significance as these relations help in estimating the changes in entropy, internal energy and enthalpy by knowing p, v and T. Some applications of these equations are discussed in subsequent articles.

7.6.3 Clapeyron Equation

Let us look upon phase change at fixed temperature and pressure and estimate changes in specific entropy, internal energy and enthalpy during phase change. Let us start with one of Maxwell relations;

$$\left(\frac{\partial p}{\partial T}\right)_v = \left(\frac{\partial s}{\partial v}\right)_T$$

From earlier discussions on pure substances we have seen that during phase transformation at some temperature the pressure is saturation pressure. Thus pressure is also independent of specific volume and can be determined by temperature alone. Hence, $p_{sat} = f(T_{sat})$

or
$$\left(\frac{\partial p}{\partial T}\right)_v = \left(\frac{dp}{dT}\right)_{sat}$$

Here $\left(\frac{dp}{dT}\right)_{sat}$ is the slope of saturation curve on pressure-temperature $(p - T)$ diagram at some point determined by fixed constant temperature during phase transformation and is independent of specific volume.

Substituting in the Maxwell relation.

$$\left(\frac{\partial s}{\partial v}\right)_T = \left(\frac{dp}{dT}\right)_{sat}$$

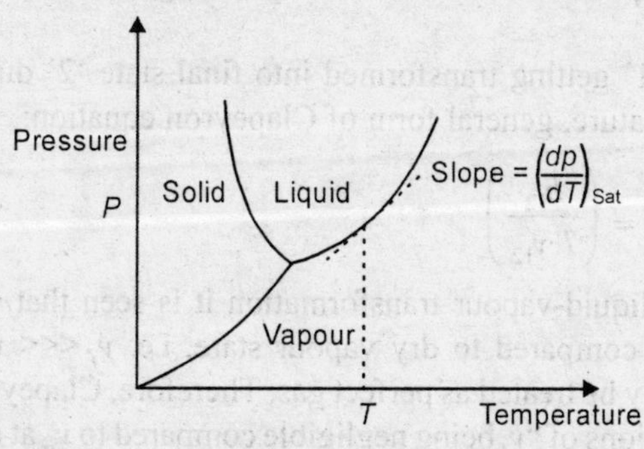

Fig. 7.5 Pressure-temperature diagram for pure substance

Thus, during vaporization i.e. phase transformation from liquid to vapour state, above relation can be given as,

$$\int_{\text{sat. liquid}}^{\text{dry vapour}} ds = \int_{\text{sat. liquid}}^{\text{dry vapour}} \left(\frac{dp}{dT}\right)_{sat} \cdot dv$$

Using notations for dry vapour and saturated liquid it can be given as,

$$(s_g - s_f) = \left(\frac{dp}{dT}\right)_{\text{sat}} \cdot (v_g - v_f)$$

or,
$$\left(\frac{dp}{dT}\right)_{\text{sat}} = \left(\frac{s_g - s_f}{v_g - v_f}\right)$$

$$\boxed{\left(\frac{dp}{dT}\right)_{\text{sat}} = \left(\frac{s_{fg}}{v_{fg}}\right)}$$

From differential form of specific enthalpy,

$$dh = T{\cdot}ds + v{\cdot}dp$$

for phase change occurring at constant pressure and temperature,

$$dh = T{\cdot}ds$$

for saturated liquid to dry vapour transformation,

$$(h_g - h_f) = T{\cdot}(s_g - s_f)$$
$$h_{fg} = T{\cdot}s_{fg}$$

Substituting $\frac{h_{fg}}{T}$ in place of entropy s_{fg} in $\left(\frac{dp}{dT}\right)_{\text{sat}}$

$$\boxed{\left(\frac{dp}{dT}\right)_{\text{sat}} = \left(\frac{h_{fg}}{T{\cdot}v_{fg}}\right)} \quad \textit{Clapeyron equation}$$

Above equation is called Clapeyron equation. It can be used for determination of change in enthalpy during phase change i.e. h_{fg} from the p, v and T values which can be easily measured. Thus, Clapeyron equation can also be used for "sublimation process" or " melting occurring at constant temperature and pressure" just by knowing slope of saturation curve on *p-T* diagram, temperature and change in specific volume.

Hence, for initial state '1' getting transformed into final state '2' due to phase transformation at constant pressure and temperature, general form of Clapeyron equation:

$$\left(\frac{dp}{dT}\right)_{\text{sat}} = \left(\frac{h_{12}}{T{\cdot}v_{12}}\right)$$

At low pressure during liquid-vapour transformation it is seen that specific volume of saturated liquid state is very small as compared to dry vapour state, i.e. $v_f <<< v_g$. Also at low pressure the substance in vapour phase may be treated as perfect gas. Therefore, Clapeyron equation can be modified in the light of two approximations of "v_f being negligible compared to v_g at low pressures" and "ideal gas equation of state during vapour phase at low pressure, $\left(v_g = \frac{RT}{p}\right)$".

Clapeyron equation thus becomes, Clausius-Clapeyron equation as given here,

$$\left(\frac{dp}{dT}\right)_{\text{sat}} = \frac{(h_g - h_f)}{(T{\cdot}v_g)}$$

or, $$\left(\frac{dp}{dT}\right)_{sat} = \frac{h_g - h_f}{T\cdot(RT/p)}$$

or, $$\left(\frac{dp}{dT}\right)_{sat} = \left(\frac{h_{fg}\cdot p}{RT^2}\right)$$

or, $$\boxed{\left(\frac{dp}{p}\right)_{sat} = h_{fg}\cdot\left(\frac{dT}{RT^2}\right)_{sat}}$$ *Clausius-Clapeyron equation*

Integrating between two states 1 and 2

$$\boxed{\ln\left(\frac{p_2}{p_1}\right)_{sat} = \frac{h_{fg}}{R}\left(\frac{1}{T_1}-\frac{1}{T_2}\right)_{sat}}$$

Clausius-Clapeyron equation is thus a modified form of Clapeyron equation based upon certain approximations and is valid for low pressure phase transformations of liquid-vapour or solid-vapour type.

7.6.4 General Relations for Change in Entropy, Enthalpy, Internal Energy and Specific Heats

Let us now derive expressions for changes in entropy, enthalpy, internal energy and specific heats as a function of thermodynamic properties, p, v and T. For defining a state any two of the properties amongst the p, v, and T may be regarded as independent properties. Let us take (T, p) and (T, v) as two sets of independent properties for defining other dependent properties.

Temperature and Pressure (T, p) as Independent Properties : By considering T and p as independent properties, dependent property say entropy can be given as,

$$s = s(T, p)$$

Writing differential form of entropy function,

$$ds = \left(\frac{\partial s}{\partial T}\right)_p \cdot dT + \left(\frac{\partial s}{\partial p}\right)_T \cdot dp$$

From Maxwell relations the partial derivative $\left(\frac{\partial s}{\partial p}\right)_T$ can be substituted by $-\left(\frac{\partial v}{\partial T}\right)_p$ as,

$$\boxed{ds = \left(\frac{\partial s}{\partial T}\right)_p \cdot dT - \left(\frac{\partial v}{\partial T}\right)_p \cdot dp}$$

Similarly, specific enthalpy can be given as function of T and p; $h = h(T, p)$

Writing differential form; $dh = \left(\frac{\partial h}{\partial T}\right)_p dT + \left(\frac{\partial h}{\partial p}\right)_T \cdot dp$

We have already seen that specific heat at constant pressure can be given as function of specific enthalpy and temperature at constant pressure.

$$C_p = \left(\frac{\partial h}{\partial T}\right)_p$$

Substituting C_p in dh,

$$dh = C_p \cdot dT + \left(\frac{\partial h}{\partial p}\right)_T \cdot dp$$

From definition of enthalpy, first and second law combined,

$$dh = T \cdot ds + vdp$$

Substituting dh and ds from above,

$$C_p \cdot dT + \left(\frac{\partial h}{\partial p}\right)_T \cdot dp = T \cdot \left(\frac{\partial s}{\partial T}\right)_p dT - T\left(\frac{\partial v}{\partial T}\right)_p \cdot dp + v \cdot dp$$

or,

$$\left\{\left(\frac{\partial h}{\partial T}\right)_T + T\left(\frac{\partial v}{\partial T}\right)_p - v\right\} dp = \left\{T \cdot \left(\frac{\partial s}{\partial T}\right)_p - C_p\right\} dT$$

Here T and p are considered to be independent variables so let us keep pressure constant and vary temperature i.e. $dp = 0$, $dT \neq 0$. It yields in modified form of above underlined equation as,

$$\left\{T \cdot \left(\frac{\partial s}{\partial T}\right)_p - C_p\right\} dT = 0$$

or, $$C_p = T \cdot \left(\frac{\partial s}{\partial T}\right)_p$$

or, $$\left(\frac{\partial s}{\partial T}\right)_p = \frac{C_p}{T}$$

Substituting $\left(\frac{\partial s}{\partial T}\right)_p$ in ds expression we get, $$ds = \frac{C_p dT}{T} - \left(\frac{\partial v}{\partial T}\right)_p \cdot dp$$

Similarly, temperature can be kept constant and pressure varied independently as,

$$dT = 0, \; dp \neq 0$$

Above underlined equation gets modified as,

$$\left\{\left(\frac{\partial h}{\partial p}\right)_T + T\left(\frac{\partial v}{\partial T}\right)_p - v\right\} dp = 0$$

or, $$\left(\frac{\partial h}{\partial p}\right)_T = v - T\left(\frac{\partial v}{\partial T}\right)_p$$

Substituting $\left(\frac{\partial h}{\partial p}\right)_T$ in expression for dh we get

or, $$dh = C_p \, dT + \left\{v - T\left(\frac{\partial v}{\partial T}\right)_p\right\} dp$$

$$h_2 - h_1 = \int_{T_1}^{T_2} C_p dT + \int_{p_1}^{p_2} \left\{ v - T\left(\frac{\partial v}{\partial T}\right)_p \right\} dp$$

Temperature and Specific Volume (T, v) as independent property: Considering T and v as independent properties the dependent properties can be given as,

$$u = u(T, v)$$

Writing differential form of specific internal energy.

$$du = \left(\frac{\partial u}{\partial T}\right)_v \cdot dT + \left(\frac{\partial u}{\partial v}\right)_T dv$$

From definition of specific heat at constant volume, $C_v = \left(\frac{\partial u}{\partial T}\right)_v$

or,
$$du = C_v \cdot dT + \left(\frac{\partial u}{\partial v}\right)_T dv$$

Writing specific entropy as function of T and v,

$$s = s(T, v)$$

$$ds = \left(\frac{\partial s}{\partial T}\right)_v dT + \left(\frac{\partial s}{\partial v}\right)_T dv$$

From Maxwell relations, $\left(\frac{\partial s}{\partial v}\right)_T = \left(\frac{\partial p}{\partial T}\right)_v$; substituting in above we get

$$ds = \left(\frac{\partial s}{\partial T}\right)_v dT + \left(\frac{\partial p}{\partial T}\right)_v dv$$

From I and II law combined, $du = T \cdot ds - pdv$

Substituting *du* and *ds* in above equation,

$$C_v\, dT + \left(\frac{\partial u}{\partial v}\right)_T dv = T\left(\frac{\partial s}{\partial T}\right)_v dT + T \cdot \left(\frac{\partial p}{\partial T}\right)_v dv - pdv$$

Rearranging terms,

$$\left\{\left(\frac{\partial u}{\partial v}\right)_T - T\left(\frac{\partial p}{\partial T}\right)_v + p\right\} dv = \left\{T \cdot \left(\frac{\partial s}{\partial T}\right)_v - C_v\right\} dT$$

As T and v are considered independent variables therefore let us keep T as constant and v as variable, i.e. $dT = 0$, $dv \neq 0$. It yields,

$$\left\{\left(\frac{\partial u}{\partial v}\right)_T - T\left(\frac{\partial p}{\partial T}\right)_v + p\right\} dv = 0$$

or,
$$\left(\frac{\partial u}{\partial v}\right)_T = T\left(\frac{\partial p}{\partial T}\right)_v - p$$

Similarly, let us keep v constant and T as variable i.e $dv = 0$ $dT \neq 0$. It yields,

$$\left\{T\left(\frac{\partial s}{\partial T}\right)_v - C_v\right\} dT = 0$$

or,

$$\left\{T\left(\frac{\partial s}{\partial T}\right)_v - C_v\right\} = 0$$

or,

$$\boxed{C_v = T\cdot\left(\frac{\partial s}{\partial T}\right)_v} \text{ or, } \boxed{\left(\frac{\partial s}{\partial T}\right)_v = \frac{C_v}{T}}$$

Let us now substitute $\left(\frac{\partial u}{\partial v}\right)_T$ in the differential function du which yields,

$$\boxed{du = C_v \cdot dT + \left\{T\cdot\left(\frac{\partial p}{\partial T}\right)_v - p\right\} dv}$$

For any state change from 1 to 2 we can get change in internal energy, as

$$\boxed{u_2 - u_1 = \int_{T_1}^{T_2} C_v dT + \int_{v_1}^{v_2} \left\{T\cdot\left(\frac{\partial p}{\partial T}\right)_v - p\right\} dv}$$

Also, let us substitute $\left(\frac{\partial s}{\partial T}\right)_v$ in the expression of entropy change 'ds'. It results in,

$$\boxed{ds = \frac{C_v dT}{T} + \left(\frac{\partial p}{\partial T}\right)_v dv}$$

Thus, number of expressions are available for getting the change in h, u and s, which are summarized as under.

$$\boxed{dh = C_p dT + \left\{v - T\left(\frac{\partial v}{\partial T}\right)_p\right\} dp}$$

$$\boxed{du = C_v\, dT + \left\{T\left(\frac{\partial p}{\partial T}\right)_v - p\right\} dv}$$

$$\boxed{ds = \frac{C_p .dT}{T} - \left(\frac{\partial v}{\partial T}\right)_p \cdot dp}$$

$$\boxed{ds = \frac{C_v dT}{T} + \left(\frac{\partial p}{\partial T}\right)_v \cdot dv}$$

From above two expressions of entropy change, the difference between C_p and C_v values can be obtained as,

$$(C_p - C_v)dT = T \cdot \left(\frac{\partial p}{\partial T}\right)_v dv + T\left(\frac{\partial v}{\partial T}\right)_p dp$$

Let us write from equation of state, the function $p = p(T, v)$

Differential, $$dp = \left(\frac{\partial p}{\partial T}\right)_v dT + \left(\frac{\partial p}{\partial v}\right)_T dv$$

Substituting dp in above equation of $(C_p - C_v)\, dT$ we get,

$$(C_p - C_v)\, dT = T\left(\frac{\partial p}{\partial T}\right)_v dv + T\left(\frac{\partial v}{\partial T}\right)_p \cdot \left(\frac{\partial p}{\partial T}\right)_v dT + T\left(\frac{\partial v}{\partial T}\right)_p \left(\frac{\partial p}{\partial v}\right)_T dv$$

or,

$$\left[(C_p - C_v) - T\left(\frac{\partial v}{\partial T}\right)_p \left(\frac{\partial p}{\partial T}\right)_v\right] dT = \left[T\left(\frac{\partial p}{\partial T}\right)_v + T\left(\frac{\partial v}{\partial T}\right)_p \left(\frac{\partial p}{\partial v}\right)_T\right] dv$$

Since T and v are independent so let us keep T as constant and v as variable i.e $dT = 0$ $dv \neq 0$.

$$\left[T\left(\frac{\partial p}{\partial T}\right)_v + T\left(\frac{\partial v}{\partial T}\right)_p \left(\frac{\partial p}{\partial v}\right)_T\right] dv = 0$$

or, $$\left(\frac{\partial p}{\partial T}\right)_v = -\left(\frac{\partial v}{\partial T}\right)_p \left(\frac{\partial p}{\partial v}\right)_T$$

Similarly keeping v as constant and T variable i.e., $dv = 0$, $dT \neq 0$

$$\left\{(C_p - C_v) - T\left(\frac{\partial v}{\partial T}\right)_p \left(\frac{\partial p}{\partial T}\right)_v\right\} dT = 0$$

or, $$\boxed{(C_p - C_v) = T\left(\frac{\partial v}{\partial T}\right)_p \left(\frac{\partial p}{\partial T}\right)_v}$$

Substituting for $\left(\frac{\partial p}{\partial T}\right)_v$ from above in $(C_p - C_v)$ we get,

$$\boxed{(C_p - C_v) = -T\left(\frac{\partial v}{\partial T}\right)_p^2 \cdot \left(\frac{\partial p}{\partial v}\right)_T}$$

In single phase region the specific volume can also be given as function of T & p and the differential of function v shall be,

$$v = v\,(T, p)$$

or,
$$dv = \left(\frac{\partial v}{\partial T}\right)_p dT + \left(\frac{\partial v}{\partial p}\right)_T dp$$

The above differential form of specific volume indicates that it depends upon partial derivatives of specific volume with respect to temperature and pressure. Partial derivatives of v with respect to temperature can be related to "volume expansivity" or "coefficient of volume expansion" as below,

$$\boxed{\text{Volume expansivity, } \beta = \frac{1}{v}\left(\frac{\partial v}{\partial T}\right)_p}$$

Partial derivative of specific volume with respect to pressure can be related to "isothermal compressibility", α as below.

$$\boxed{\text{Isothermal compressibility, } \alpha = \frac{-1}{v}\left(\frac{\partial v}{\partial p}\right)_T}$$

Inverse of isothermal compressibility is called "isothermal bulk modulus",

$$B_T = -v\left(\frac{\partial p}{\partial v}\right)_T$$

Thus, volume expansivity gives the change in volume that occurs when temperature changes while pressure remains constant. Isothermal compressibility gives change in volume when pressure changes while temperature remains constant. These volume expansivity and isothermal compressibility are thermodynamic properties.

Similarly, the change in specific volume with change in pressure isentropically is also called "isentropic compressibility" or "adiabatic compressibility", Mathematically $\alpha_s = \frac{-1}{v}\left(\frac{\partial v}{\partial p}\right)_s$. Reciprocal of isentropic compressibility is called "isentropic bulk modulus," $B_s = -v\left(\frac{\partial p}{\partial v}\right)_s$

Substituting β and α in $(C_p - C_v)$ expression;

$$C_p - C_v = -T(\beta^2 \cdot v^2)\left(\frac{-1}{\alpha \cdot v}\right)$$

or,

$$\boxed{C_p - C_v = \frac{v \cdot T \cdot \beta^2}{\alpha}} \quad \textit{Mayer relation}$$

Above difference in specific heat expression is called "Mayer relation" and helps in getting significant conclusion such as,

- The difference between specific heats is zero at absolute zero temperature i.e. specific heats at constant pressure and constant volume shall be same at absolute zero temperature (T = 0 K).
- Specific heat at constant pressure shall be generally more than specific heat at constant volume i.e., $C_p \geq Cv$. It may be attributed to the fact that 'α' the isothermal compressibility shall always

be +ve and volume expansivity 'β' being squared in $(C_p - C_v)$ expression shall also be +ve. Therefore $(C_p - C_v)$, shall be either zero or positive value depending upon magnitudes of v, T, β and α.

- For incompressible substances having $dv = 0$, the difference $(C_p - C_v)$ shall be nearly zero. Hence, specific heats at constant pressure and at constant volume are identical.

Let us obtain expression for ratio of specific heats. Earlier we have obtained C_p and C_v as below,

$$C_p = T \cdot \left(\frac{\partial s}{\partial T}\right)_p \quad \text{or,} \quad \frac{C_p}{T} = \left(\frac{\partial s}{\partial T}\right)_p$$

and
$$C_v = T \cdot \left(\frac{\partial s}{\partial T}\right)_v \quad \text{or} \quad \frac{C_v}{T} = \left(\frac{\partial s}{\partial T}\right)_v$$

By cyclic relation we can write for p, T and s properties;

$$\left(\frac{\partial s}{\partial T}\right)_p \cdot \left(\frac{\partial T}{\partial p}\right)_s \cdot \left(\frac{\partial p}{\partial s}\right)_T = -1$$

or,
$$\left(\frac{\partial s}{\partial T}\right)_p = \frac{-1}{\left(\frac{\partial T}{\partial p}\right)_s \cdot \left(\frac{\partial p}{\partial s}\right)_T}$$

Similarly for s, T and v properties we can write using cyclic relation;

$$\left(\frac{\partial s}{\partial T}\right)_v \cdot \left(\frac{\partial T}{\partial v}\right)_s \cdot \left(\frac{\partial v}{\partial s}\right)_T = -1$$

or,
$$\left(\frac{\partial s}{\partial T}\right)_v = \frac{-1}{\left(\frac{\partial T}{\partial v}\right)_s \cdot \left(\frac{\partial v}{\partial s}\right)_T}$$

Substituting in the relation for (C_p/T) and $\left(\frac{C_v}{T}\right)$.

$$\left(\frac{C_p}{T}\right) = \frac{-1}{\left(\frac{\partial T}{\partial p}\right)_s \cdot \left(\frac{\partial p}{\partial s}\right)_T}$$

and
$$\frac{C_v}{T} = \frac{-1}{\left(\frac{\partial T}{\partial v}\right)_s \cdot \left(\frac{\partial v}{\partial s}\right)_T}$$

Taking ratio of two specific heats,

$$\frac{C_p}{C_v} = \frac{\left(\frac{\partial T}{\partial v}\right)_s \cdot \left(\frac{\partial v}{\partial s}\right)_T}{\left(\frac{\partial T}{\partial p}\right)_s \cdot \left(\frac{\partial p}{\partial s}\right)_T}$$

or,
$$\left(\frac{C_p}{C_v}\right) = \left(\frac{\partial T}{\partial v}\right)_s \cdot \left(\frac{\partial v}{\partial s}\right)_T \cdot \left(\frac{\partial p}{\partial T}\right)_s \cdot \left(\frac{\partial s}{\partial p}\right)_T$$

$$\left(\frac{C_p}{C_v}\right) = \left\{\left(\frac{\partial v}{\partial s}\right)_T \cdot \left(\frac{\partial s}{\partial p}\right)_T\right\} \cdot \left\{\left(\frac{\partial p}{\partial T}\right)_s \cdot \left(\frac{\partial T}{\partial v}\right)_s\right\}$$

By chain rule of calculus we can write,

$$\left(\frac{\partial v}{\partial p}\right)_T = \left(\frac{\partial v}{\partial s}\right)_T \cdot \left(\frac{\partial s}{\partial p}\right)_T$$

and
$$\left(\frac{\partial p}{\partial v}\right)_s = \left(\frac{\partial p}{\partial T}\right)_s \cdot \left(\frac{\partial T}{\partial V}\right)_s$$

Upon substitution in specific heat ratio we get,

$$\left(\frac{C_p}{C_v}\right) = \left(\frac{\partial v}{\partial p}\right)_T \cdot \left(\frac{\partial p}{\partial v}\right)_s$$

or,
$$= \left\{\frac{-1}{v}\left(\frac{\partial v}{\partial p}\right)_T\right\} \cdot \left\{\frac{1}{\frac{-1}{v}\left(\frac{\partial v}{\partial p}\right)_s}\right\}$$

$$\frac{C_p}{C_v} = \frac{\alpha}{\alpha_s} = \frac{\text{Isothermal compressibility}}{\text{Isentropic compressibility}}$$

Thus, ratio of specific heats at constant pressure and constant volume can be given by the ratio of isothermal compressibility and isentropic compressibility.

7.6.5 Joule-Thomson Coefficient

Joule-Thomson coefficient is defined as the rate of change of temperature with pressure during an isenthalpic process or throttling process. Mathematically, Joule-Thomson coefficient (μ) can be given

as, $\mu = \left(\frac{\partial T}{\partial p}\right)_h$

It is defined in terms of thermodynamic properties and is itself a property. Joule-Thomson coefficient gives slope of constant enthalpy lines on temperature—pressure diagram. Thus, it is a parameter for characterizing the throttling process. Slope of isenthalpic line may be positive, zero or negative, i.e. $\mu > 0$, $\mu = 0$ and $\mu < 0$ respectively. Mathematically evaluating the consequence of μ we see,

– for $\mu > 0$, temperature decreases during the process.
– for $\mu = 0$, temperature remains-constant during the process.
– for $\mu < 0$, temperature increases during the process.

Joule-Thomson expansion can be shown as in Fig. 7.6. Here gas or liquid is passed through porous plug for causing isenthalpic process. Valve put near exit is used for regulating pressure after constant enthalpy process i.e. p_2.

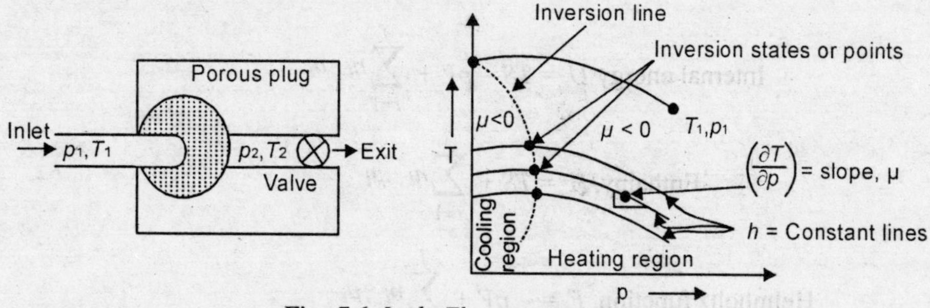

Fig. 7.6 Joule-Thomson expansion

If pressure p_2 is varied then the temperature variation occurs in the isenthalpic manner as shown in *T-p* diagram. This graphical representation of isenthalpic curve gives the Joule-Thomson coefficient by its slope at any point. Slope may be positive, negative or zero at different points on the curve. The points at which slope has zero value or Joule-Thomson coefficient is zero are called "inversion points" or "inversion states". Temperature at these inversion states is called "inversion temperature". Locii of these inversion states is called "inversion line". Thus, inversion line as shown divides *T-p* diagram into two distinct region i.e. one on the left of line and other on the right of line. For the states lying on left of the inversion line temperature shall decrease during throttling process while for the states on right of inversion line throttling shall cause heating of fluid being throttled. Temperature at the intersection of inversion line with zero pressure line is called "maximum inversion temperature".

7.6.6 Chemical Potential

In case of multicomponent systems such as non-reacting gas mixtures the partial molal properties are used for describing the behaviour of mixtures and solutions. Partial molal properties are intensive properties of the mixture and can be defined as,

$$X_i = \left(\frac{\partial X}{\partial n_i}\right)_{T,\,p,\,n_k}$$

where X is extensive property for multi component system in single phase.

$X = X(T, p, n)$ i.e. function of temperature, pressure and no. of moles of each component n_k refers to the all n values with varing k values and are kept constant except for n_i.

In multicomponent systems the partial molal Gibbs function for different constituents are called "chemical potential" for particular constituent. Chemical potential, μ can be defined for ith component as,

$$\mu_i = \left(\frac{\partial G}{\partial n_i}\right)_{T,\,p,\,n_k}$$

where G, n_i, n_k, T and P have usual meanings. Chemical potential being a partial molal property is intensive property.

Also, it can be given as,

$$\mathrm{G} = \sum_{i=1}^{j} (n_i \cdot \mu_i)$$

Thus, for non reacting gas mixture the expression for internal energy, enthalpy, Helmholtz function can be given using G defined as above,

$$\text{Internal energy } U = TS - pV + \sum_{i=1}^{j} n_i \ \mu_i$$

$$\text{Enthalpy, } H = TS + \sum_{i=1}^{j} n_i \cdot \mu_i$$

$$\text{Helmholtz function, } F = -pV + \sum_{i=1}^{j} n_i \cdot \mu_i$$

Writing differential of G considering it as function of $(T, p, n_1, n_2, \ldots n_j)$

$$G = G(T, p, n_1, n_2, n_3 \ldots n_j)$$

$$dG = \left(\frac{\partial G}{\partial p}\right)_{T,n} dp + \left(\frac{\partial G}{\partial T}\right)_{p,n} dT + \sum_{i=1}^{j}\left(\frac{\partial G}{\partial n_i}\right)_{T,p,n_k} dn_i$$

From definition of Gibbs function $dG = Vdp - SdT$, for T = constant,

$$V = \left(\frac{\partial G}{\partial p}\right)_{T,n}$$

$$\text{for pressure as constant, } -S = \left(\frac{\partial G}{\partial T}\right)_{p,n}$$

Therefore,

$$\boxed{dG = Vdp - SdT + \sum_{i=1}^{j}(\mu_i \cdot dn_i)}$$

Also from $G = \sum_{i=1}^{j}(n_i \cdot \mu_i)$ we can write differential as,

$$\boxed{dG = \sum_{i=1}^{j}(n_i \cdot d\mu_i) + \sum_{i=1}^{j}(\mu_i \cdot dn_i)}$$

From two differential of function G we get,

$$\boxed{Vdp - SdT = \sum_{i=1}^{j}(n_i \cdot d\mu_i)} \quad \textit{Gibbs-Duhem equation}$$

Above equation is also called Gibbs-Duhem equation.

7.6.7 Fugacity

From earlier discussions for a single component system one can write,

$$G = n \cdot \mu$$

or, $\mu = \dfrac{G}{n}$ ⇒ Chemical potential for pure substance = Gibbs function per mole.

or $$\bar{g} = \frac{G}{n} = \mu$$

For Gibbs function written on unit mole basis,

For constant temperature $$\bar{v} = \left(\frac{\partial \mu}{\partial p}\right)_T$$

If single component system is perfect gas then, $\bar{v} = \dfrac{\bar{R}T}{p}$

or, $$\left(\frac{\partial \mu}{\partial p}\right)_T = \frac{\bar{R}T}{p}$$

or $$\boxed{\mu_T = \bar{R}T \ln p + \text{constant}}$$

Here chemical potential may have any value depending upon the value of pressure. Above mathematical formulation is valid only for perfect gas behaviour being exhibited by the system. For a real gas above mathematical equation may be valid if pressure is replaced by some other property called 'fugacity'.

Fugacity was first used by Lewis.

Fugacity denoted by '$\mathcal{F}$' can be substituted for pressue in above equation,

$$\mu = \bar{R}T \ln \mathcal{F} + \text{Constant}$$

For constant pressure using $\bar{v} = \left(\dfrac{\partial u}{\partial p}\right)_T$ and above equation, we get

$$\bar{R}T\left(\frac{\partial \ln \mathcal{F}}{\partial p}\right)_T = \bar{v}$$

Thus, for a limiting case when ideal gas behaviour is approached the fugacity of a pure component shall equal the pressure in limit of zero pressure.

$$\lim_{p \to 0}\left(\frac{\mathcal{F}}{p}\right) = 1$$

For an ideal gas $$\mathcal{F} = p$$

For real gas, equation of state can be given using compressibility factor as,

$$p\bar{v} = Z\bar{R}T$$

or, $$\bar{v} = \frac{Z\bar{R}T}{p}$$

Substituting the fugacity function,

$$\frac{Z\bar{R}T}{p} = \bar{R}T\left(\frac{\partial \ln \mathcal{F}}{\partial p}\right)_T$$

$$\left(\frac{\partial \ln \mathcal{F}}{\partial p}\right)_T = \frac{Z}{p}$$

or, $$\left(\frac{\partial \ln \mathcal{F}}{\partial \ln p}\right)_T = Z. \text{ Here as } p \to 0 \text{ the } Z \to 1$$

Also we have seen

$$\mu_T = \bar{R}\,T \ln \mathcal{F} + \text{constant}$$

or, $$(d\mu)_T = \bar{R}\,T\,d\,(\ln \mathcal{F})_T$$

or, $$dg_T = \bar{R}\,Td\,(\ln \mathcal{F})_T$$

Integrating between very low pressure p^* and high pressure p.

$$\int_g^{g^*} dg_T = \int_{\mathcal{F}}^{\mathcal{F}^*} \bar{R}T\ d(\ln \mathcal{F})_T$$

or $$g = g^* + RT \ln\left(\frac{\mathcal{F}}{\mathcal{F}^*}\right)$$

Here for very low pressure, $\mathcal{F}^* = p^*$

or, $$g = g^* + RT \ln\left(\frac{\mathcal{F}}{p^*}\right)$$

When low pressure in 1 atm then the ratio $\left(\dfrac{\mathcal{F}}{\mathcal{F}^*}\right)$ is called "activity".

EXAMPLES

1. Steam at 1.6 MPa, 300°C enters a flow device with negligible velocity and leaves at 0.1 MPa, 150°C with a velocity of 150 m/s. During the flow heat interaction occurs only with the surroundings at 15°C and steam mass flow rate is 2.5 kg/s. Estimate the maximum possible output from device.

Solution:

Let us neglect the potential energy change during the flow.

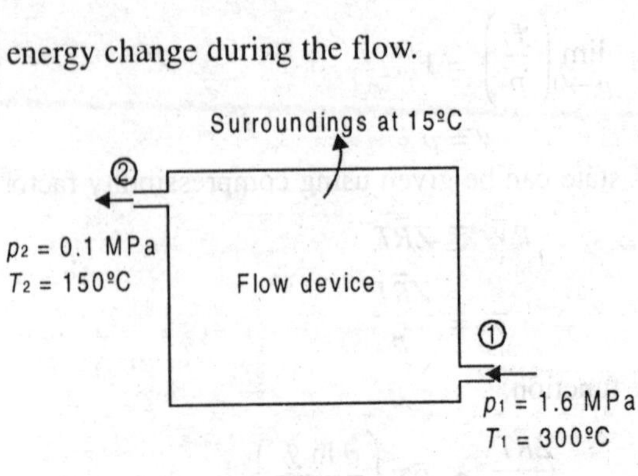

Fig. 7.7

Applying S.F.E.E., neglecting inlet velocity and change in potential energy,

$$W_{max} = (h_1 - T_0 s_1) - \left(h_2 + \frac{C_2^2}{2} - T_0 s_2 \right)$$

$$W_{max} = (h_1 - h_2) - T_0(s_1 - s_2) - \frac{C_2^2}{2}$$

From steam tables,

$$h_1 = h_{\text{at 1.6 MPa, 300°C}} = 3034.8 \text{ kJ/kg}$$

$$s_1 = s_{\text{at 1.6 MPa, 300°C}} = 6.8844 \text{ kJ/kg} \cdot \text{K}$$

$$h_2 = h_{\text{at 0.1 MPa, 150°C}} = 2776.4 \text{ kJ/kg}$$

$$s_2 = s_{\text{at 0.1 MPa, 150°C}} = 7.6134 \text{ kJ/kg} \cdot \text{K}$$

Given; $T_0 = 288\ K$

$$W_{max} = (3034.8 - 2776.4) - 288(6.8844 - 7.6134) - \frac{(150)^2}{2} \times 10^{-3}$$

$$= 457.1 \text{ kJ/kg}$$

Maximum possible work = 2.5 × 457.1 kJ/s = 1142.75 kW

Maximum possible work = **1142.75 kW**

Ans.

2. Two tanks A and B contain 1 kg of air at 1 bar, 50°C and 3 bar, 50°C when atmosphere is at 1 bar, 15°C. Identify the tank in which stored energy is more. Also find the availability of air in each tank.

Solution:

In these tanks the air stored is at same temperature of 50°C. Therefore, for air behaving as perfect gas the internal energy of air in tanks shall be same as it depends upon temperature alone. But the availability shall be different.

Both the tanks have same internal energy

Ans.

Availability of air in tank

$$A = \{E - U_0\} + p_0(V - V_0) - T_0(S - S_0)$$

$$= m\ \{(e - u_0) + p_0(v - v_0) - T_0(s - s_0)\}$$

$$= m\ \left\{ c_v(T - T_0) + p_0\left(\frac{RT}{p} - \frac{RT_0}{p_0} \right) - T_0\left(c_p \ln\frac{T}{T_0} - R\ln\frac{p}{p_0} \right) \right\}$$

$$A = m\ \left\{ c_v(T - T_0) + R\left(\frac{p_0}{p}T - T_0 \right) - T_0 c_p \ln\frac{T}{T_0} + T_0 R \ln\frac{p}{p_0} \right\}$$

For tank A, $m = 1$ kg, $c_v = 0.717$ kJ/kg · K, $T = 323$ K, $R = 0.287$ kJ/kg · K, $T_0 = 288$ K, $p_0 = 1$ bar, $c_p = 1.004$ kJ/kg · K

$$\text{Availability}_A = 1\ \{0.717\ (323 - 288) + 0.287\ (1 \times 323 - 288) - (288 \times 1.004) \ln\left(\frac{323}{288}\right)$$

$$+ 288 \times 0.287 \ln 1\}$$

$$= 1.98 \text{ kJ}$$

For tank B, T = 323 K, p = 3 bar

$$\text{Availability}_B = 1\left\{0.717\,(323-288) + 0.287\left(\frac{1}{3}\times 323 - 288\right) - \right.$$

$$\left.\left(288\times 1.004 \ln\frac{323}{288}\right) + \left(288\times 0.287 \ln\frac{3}{1}\right)\right\} = 30.98 \text{ kJ}$$

Availability of air in tank *B* is more than that of tank *A*.

Availability of air in tank *A* = **1.98 kJ**
Availability of air in tank *B* = **30.98 kJ** **Ans.**

3. 15 kg/s steam enters a perfectly insulated steam turbine at 10 bar, 300°C and leaves at 0.05 bar, 0.95 dry with velocity of 160 m/s. Considering atmospheric pressure to be 1 bar, 15°C. Determine (a) power output, (b) the maximum power for given end states, (c) the maximum power that could be obtained from exhaust steam. Turbine rejects heat to a pond having water at 15°C.

Solution:

From steam tables,

Enthalpy at inlet to turbine, $h_1 = 3051.2$ kJ/kg

$s_1 = 7.1229$ kJ/kg · K

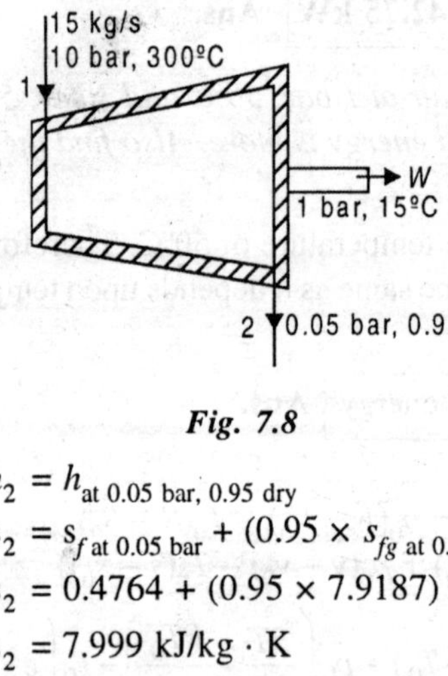

Fig. 7.8

Enthalpy at exit of turbine, $h_2 = h_{\text{at 0.05 bar, 0.95 dry}}$

$s_2 = s_{f\text{ at 0.05 bar}} + (0.95 \times s_{fg\text{ at 0.05 bar}})$

$s_2 = 0.4764 + (0.95 \times 7.9187)$

$s_2 = 7.999$ kJ/kg · K

Similarly, $h_2 = h_{f\text{ at 0.05 bar}} + (0.95 \times h_{fg\text{ at 0.05 bar}})$

$= 137.82 + (0.95 \times 2423.7)$

$h_2 = 2440.34$ kJ/kg

Neglecting the change in potential energy and velocity at inlet to turbine, the steady flow energy equation may be written as to give work output.

$$w = (h_1 - h_2) - \frac{V_2^2}{2}$$

$$w = (3051.2 - 2440.34) - \left(\frac{(160)^2}{2} \times 10^{-3}\right)$$

$$w = 598.06 \text{ kJ/kg}$$

$$\text{Power output} = \text{m.w} = 15 \times 598.06 = 8970.9 \text{ kW}$$

Power output = **8970.9 kW**	**Ans.**

Maximum work for given end states,

$$w_{max} = (h_1 - T_0 \cdot s_1) - \left(h_2 + \frac{V_2^2}{2} - T_0 \cdot s_2\right)$$

$$w_{max} = (3051.2 - 288 \times 7.1229) - \left(2440.34 + \frac{(160)^2 \times 10^{-3}}{2} - 288 \times 7.999\right)$$

$$w_{max} = 850.38 \text{ kJ/kg}$$

$$w_{max} = m \cdot w_{max} = 15 \times 850.38 = 12755.7 \text{ kW}$$

Maximum power output = **12755.7 kw**	**Ans.**

Maximum power that could be obtained from exhaust steam shall depend upon availability with exhaust steam and the dead state. Stream availability of exhaust steam,

$$A_{exhaust} = \left(h_2 + \frac{V_2^2}{2} - T_0 s_2\right) - (h_0 - T_0\, s_0)$$

$$= (h_2 - h_0) + \frac{V_2^2}{2} - T_0\,(s_2 - s_0)$$

Approximately the enthalpy of water at dead state of 1 bar, 15°C can be approximated to saturated liquid at 15°C.

$$h_0 = h_{f \text{ at } 15°C} = 62.99 \text{ kJ/kg}$$

$$s_0 = s_{f \text{ at } 15°C} = 0.2245 \text{ kJ/kg} \cdot \text{K}$$

Maximum work available from exhaust steam

$$= (2440.34 - 62.99) + \left(\frac{(160)^2}{2} \times 10^{-3}\right) - 288\,(7.999 - 0.2245)$$

$$A_{exhaust} = 151.1 \text{ kJ/kg}$$

Maximum power that could be obtained from exhaust steam

$$= m \times A_{exhaust}$$

$$= 15 \times 151.1 = 2266.5 \text{ kW}$$

Maximum power from exhaust steam = **2266.5 kW**	**Ans.**

4. 5 kg of steam, initially at elevation of 10 m and velocity of 25 m/s undergoes some process such that finally it is at elevation of 2m and velocity of 10 m/s. Determine the availability corresponding to the initial and final states. Also estimate the change in availability assuming environment temperature and pressure at 25°C and 100 kPa respectively. Thermodynamic properties u, v, s are as under.

Dead state of water

$$u_0 = 104.86 \text{ kJ/kg}$$
$$v_0 = 1.0029 \times 10^{-3} \text{ m}^3/\text{kg}$$
$$s_0 = 0.3673 \text{ kJ/kg} \cdot \text{K}$$

Initial state

$$u_1 = 2550 \text{ kJ/kg}$$
$$v_1 = 0.5089 \text{ m}^3/\text{kg}$$
$$s_1 = 6.93 \text{ kJ/kg} \cdot \text{K}$$

Final state

$$u_2 = 83.94 \text{ kJ/kg}$$
$$v_2 = 1.0018 \times 10^{-3} \text{ m}^3/\text{kg}$$
$$s_2 = 0.2966 \text{ kJ/kg} \cdot \text{K}$$

Solution:

Availability at any state can be given by,

$$A = m\left[(u - u_0) + P_0(v - v_0) - T_0(s - s_0) + \frac{V^2}{2} + gz\right]$$

Availability at initial state,

$$A_1 = 5\left[(2550 - 104.86)\times 10^3 + 100\times 10^3(0.5089 - 1.0029\times 10^{-3}) - 298(6.93 - 0.3673)\times 10^3 + \frac{(25)^2}{2} + (9.81\times 10)\right]$$

$$A_1 = 2704.84 \text{ kJ.}$$

$$\boxed{= \mathbf{2704.84\ kJ} \text{ Availability at initial state}}$$ **Ans.**

Availability at final state

$$A_2 = 5\left[(83.93 - 104.86)\times 10^3 + 100\times 10^3(1.0018\times 10^{-3} - 1.0029 \times 10^{-3}) - 298(0.2966 - 0.3673)\times 10^3 + \frac{(10)^2}{2} + (9.81\times 2)\right]$$

$$A_2 = 1.09 \text{ kJ}$$

$$\boxed{\text{Availability at final state} = \mathbf{1.09\ kJ}}$$ **Ans.**

Change in availability: $A_2 - A_1 = 1.09 - 2704.84$

$$= -2703.75 \text{ kJ}$$

$$\boxed{\text{Hence availability decreases by } \mathbf{2703.75\ kJ}}$$ **Ans.**

5. For a steady flow process as shown below, prove that irreversibility, $I = T_0 S_{gen}$, where T_0 and P_0 are temperature and pressure at dead state.

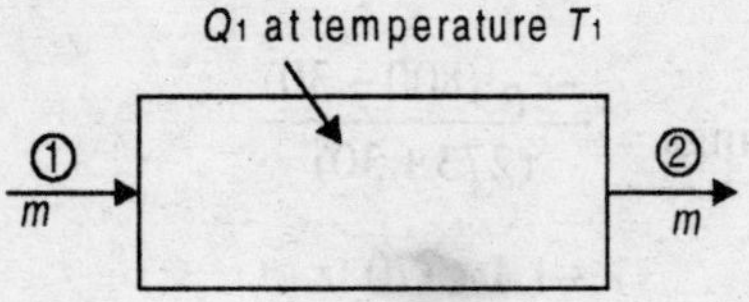

Fig. 7.9

Solution:

Let us assume changes in kinetic and potential energy to be negligible. Let us use subscript 1 for inlet and 2 for outlet.

From first law of thermodynamics;

$$Q_1 + m_1h_1 = m_2h_2,\ \text{here } m_1 = m_2 = m$$

or

$$Q_1 = m(h_2 - h_1)$$

From second law of thermodynamics,

$$S_{gen} + \frac{Q_1}{T_1} + ms_1 = ms_2;\ m(s_2 - s_1) = \frac{Q_1}{T_1} + S_{gen}$$

From availability considerations in control volume,

$$\left(1 - \frac{T_0}{T_1}\right) Q_1 + ma_1 = m \cdot a_2 + I$$

or, $$\left(1 - \frac{T_0}{T_1}\right) Q_1 + m\{(h_1 - h_0)\} - T_0(s_1 - s_0)\} = m\{(h_2 - h_0) - T_0(s_2 - s_0)\} + I$$

or,

Upon substituting from above equations,

$$m(s_2 - s_1) = \frac{Q_1}{T_1} + \frac{I}{T_0}$$

or it can be given that

$$\frac{I}{T_0} = S_{gen}$$

or, $$\boxed{I = T_0 \cdot S_{gen}}$$ Hence proved.

6. Exhaust gases leave an internal combustion engine at 800°C and 1 atmosphere, after having done 1050 kJ of work per kg of gas in engine. (c_p of gas = 1.1 kJ/kg · K). The temperature of surrounding is 30°C

(i) How much available energy per kg of gas is lost by throwing away the exhaust gases?

(ii) What is the ratio of the lost available exhaust gas energy to engine work?

Solution:

Loss of available energy = Irreversibility = $T_0 \cdot \Delta S_c$

Here, T_0 = 303 K = Temperature of surroundings

$$\Delta S_c = \Delta S_s + \Delta S_e$$

$$\text{Change in entropy of system} = \frac{1050}{(273+800)} = 0.9786 \text{ kJ/kg} \cdot \text{K}$$

$$\text{Change in entropy of surroundings} = \frac{-c_p \cdot (800-30)}{(273+30)}$$

$$= \frac{-1.1 \times 770}{303}$$

$$= -2.7954 \text{ kJ/kg} \cdot \text{K}$$

$$\text{Loss of available energy} = 303\,(-2.7954 + 0.9786)$$

$$= -550.49 \text{ kJ/kg}$$

$$\text{Loss of available energy} = 550.49 \text{ kJ/kg}$$

$$\text{Ratio of lost available exhaust gas energy to engine work} = \frac{550.49}{1050}$$

$$= \boxed{\frac{0.524}{1}} \quad \textbf{Ans.}$$

7. 10 kg of water undergoes transformation from initial saturated vapour at 150°C, velocity of 25 m/s and elevation of 10 m to saturated liquid at 20°C, velocity of 10 m/s and elevation of 3 m. Determine the availability for initial state, final state and change of availability considering environment to be at 0.1 MPa and 25°C and g = 9.8 m/s².

Solution:

Let us consider velocities and elevations to be given in reference to environment. Availability is given by

$$A = m\left[(u - u_0) + P_0(v - v_0) - T_0(s - s_0) + \frac{C^2}{2} + gz\right]$$

Dead state of water, $u_0 = 104.88$ kJ/kg

$v_0 = 1.003 \times 10^{-3}$ m³/kg

$s_0 = 0.3674$ kJ/kg · K

For initial state of saturated vapour at 150°C.

$u_1 = 2559.5$ kJ/kg, $v_1 = 0.3928$ m³/kg, $s_1 = 6.8379$ kJ/kg · K

For final state of saturated liquid at 20°C,

$u_2 = 83.95$ kJ/kg, $v_2 = 0.001002$ m³/kg, $s_2 = 0.2966$ kJ/kg · K

Substituting in the expression for availability

Initial state availability,

$$A_1 = 10 \times [(2559.5 - 104.88) + (0.1 \times 10^3 \times (0.3928 - 0.001003) - (298.15 \times (6.8379 - 0.3674)) + \left(\frac{(25)^2}{2} \times 10^{-3}\right) + (9.81 \times 10 \times 10^{-3})]$$

$$A_1 = 5650.28 \text{ kJ}$$

Final state availability

$$A_2 = 10[(83.95 - 104.88 + (0.1 \times 10^3 \times (0.001002 - 0.001003)) - (298.15 \times (0.2966 - 0.3674)) + \left(\frac{(10)^2}{2} \times 10^{-3}\right) + (9.81 \times 3 \times 10^{-3})\Big]$$

$$A_2 = 2.5835 \text{ kJ} \approx 2.58 \text{ kJ}$$

Change in availability, $\Delta A = A_2 - A_1$

$= 2.58 - 5650.28$

$= -5647.70$ kJ

Initial availability	= 5650.28 kJ
Final availability	= 2.58 kJ
Change in availability	= Decrease by **5647.70 kJ** **Ans.**

8. A steam turbine has steam flowing at steady rate of 5 kg/s entering at 5 MPa and 500°C and leaving at 0.2 MPa and 140°C. During flow through turbine a heat loss of 600 kJ/s occurs to the environment at 1 atm and 25°C. Determine

(i) the availability of steam at inlet to turbine,
(ii) the turbine output
(iii) the maximum possible turbine output, and
(iv) the irreversibility

Neglect the changes in kinetic energy and potential energy during flow.

Solution:

Let inlet and exit states of turbine be denoted as 1 and 2.

At inlet to turbine,

$$p_1 = 5 \text{ MPa}, T_1 = 500°\text{C}, h_1 = 3433.8 \text{ kJ/kg}, s_1 = 6.9759 \text{ kJ/kg} \cdot \text{K}$$

At exit from turbine.

$$p_2 = 0.2 \text{ MPa}, T_2 = 140°\text{C}, h_2 = 2748 \text{ kJ/kg}, s_2 = 7.228 \text{ kJ/kg} \cdot \text{K}$$

At dead state,

$$p_0 = 101.3 \text{ kPa}, T_0 = 25°\text{C}, h_0 = 104.96 \text{ kJ/kg}, s_0 = 0.3673 \text{ kJ/kg} \cdot \text{K}$$

Availability of steam at inlet, $A_1 = m[(h_1 - h_0) - T_0 (s_1 - s_0)]$

$$A_1 = 5 [(3433.8 - 104.96) - 298.15 (6.9759 - 0.3673)]$$

$$A_1 = 6792.43 \text{ kJ}$$

Availability of steam at inlet = **6792.43 kJ** **Ans.**

Applying first law of thermodynamics

$$Q + mh_1 = mh_2 + W.$$

$$W = m(h_1 - h_2) - Q$$

$$= 5(3433.8 - 2748) - 600$$

$$W = 2829 \text{ kJ/s}$$

Turbine output = **2829 kW** **Ans.**

Maximum possible turbine output will be available when irreversibility is zero.

$$W_{rev} = W_{max} = A_1 - A_2$$

$$= m [(h_1 - h_2) - T_0(s_1 - s_2)]$$

$$= 5[(3433.8 - 2748) - 298.15 (6.9759 - 7.228)]$$

$$W_{max} = 3053.18 \text{ kJ/s}$$

Maximum output = **3053.18 kW** **Ans.**

Irreversibility can be estimated by the difference between the maximum output and turbine output.

$$I = W_{max} - W = 224.18 \text{ kJ/s}$$

Irreversibility = **224.18 kW**	**Ans.**

9. Show that the sublimation line and vaporization lines have different slopes at triple point on the phase diagram of water.

Solution:

It is desired to show that slope of sublimation line shown by 0–1 is different than vaporization line 1 – 2.

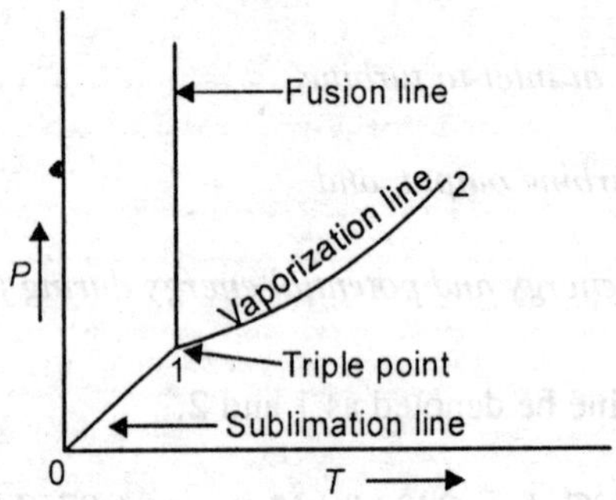

Fig 7.10

To show the slope let us find $\frac{dp}{dT}$ values at triple point 1. Here *i, f, g* subscripts refer to ice, water and steam states,

By Clapeyron equation.

$$\left(\frac{dp}{dT}\right)_{0-1} - \left(\frac{dp}{dT}\right)_{1-2} = \frac{s_{ig}}{v_{ig}} - \frac{s_{fg}}{v_{fg}}$$

For triple point state $s_{ig} = s_{if} + s_{fg}$

and $v_{ig} = v_{if} + v_{fg}$

Substituting in above slope difference expression

$$\left(\frac{dp}{dT}\right)_{0-1} - \left(\frac{dp}{dT}\right)_{1-2} = \left(\frac{s_{if} + s_{fg}}{v_{if} + v_{fg}}\right) - \frac{s_{fg}}{v_{fg}}$$

$$= \frac{s_{if} \cdot v_{fg} + s_{fg} \cdot v_{fg} - s_{fg} v_{if} - s_{fg} v_{fg}}{(v_{if} + v_{fg}) \cdot v_{fg}}$$

$$= \frac{s_{if} \cdot v_{fg} - s_{fg} \cdot v_{if}}{(v_{if} + v_{fg}) \cdot v_{fg}}$$

It is seen that $v_{if} <<< v_{fg}$ but the order of s_{if} being less than s_{if} is not very small as compared to $v_{if} <<< v_{fg}$. Neglecting smaller terms by order of magnitude

$$\left(\frac{dp}{dT}\right)_{0-1} - \left(\frac{dp}{dT}\right)_{1-2} = \frac{s_{if}}{v_{fg}}$$

Here, s_{if} and v_{fg} both are positive quantities so the ratio $\left(\frac{s_{if}}{v_{fg}}\right)$ is also positive and hence difference of slopes between sublimation line and vaporization line is positive. Thus, it shows that slope of sublimation line and vaporization line are different.

10. Obtain the expression for change in internal energy of gas obeying the Vander Waals equation of state.

Solution:

Van der Waals equation of state can be given as under,

$$\left(p+\frac{a}{v^2}\right)(v-b) = \text{RT} \Rightarrow p = \frac{RT}{v-b} - \frac{a}{v^2} \Rightarrow \frac{RT}{v-b} = p + \frac{a}{v^2}$$

Differentiating this equation of state, partially w.r.t. T at constant volume,

$$\left(\frac{\partial p}{\partial T}\right)_v = \frac{R}{v-b}$$

General expression for change in internal energy can be given as under,

$$du = C_v\, dT + \left\{T\left(\frac{\partial p}{\partial T}\right)_v - p\right\} dv$$

Substituting in the expression for change in internal energy

$$du = C_v \cdot dT + \left\{T \cdot \frac{R}{(v-b)} - p\right\} dv$$

Substituting for $\left(\frac{RT}{v-b}\right)$ is expression of du,

$$du = C_v \cdot dT + \left\{p + \frac{a}{v^2} - p\right\} dv$$

$$du = C_v \cdot dT + \frac{a}{v^2}\, dv$$

The change in internal energy between states 1 and 2,

$$\int_1^2 du = u_2 - u_1 = \int_1^2 C_v\, dT - a\left(\frac{1}{v_2} - \frac{1}{v_1}\right)$$

$$\boxed{u_2 - u_1 = \int_1^2 C_v.dT - a\left(\frac{1}{v_2} - \frac{1}{v_1}\right)} \quad \textbf{Ans.}$$

11. 500 kJ of heat is removed from a constant temperature heat reservoir maintained at 835 K. Heat is received by a system at constant temperature of 720 K. Temperature of the surroundings, the lowest available temperature is 280 K. Determine the net loss of available energy as a result of this irreversible heat transfer. *[U.P.S.C. 1992]*

Solution:

Here, $T_0 = 280$ K, i.e surrounding temperature.

$$\text{Availability for heat reservoir} = T_0 \cdot \Delta S_{\text{reservoir}}$$
$$= 280 \times \frac{500}{835}$$
$$= 167.67 \text{ kJ/kg} \cdot \text{K}$$
$$\text{Availability for system} = T_0 \cdot \Delta S_{\text{system}}$$
$$= 280 \times \frac{500}{720}$$
$$= 194.44 \text{ kJ/kg} \cdot \text{K}$$
$$\text{Net loss of available energy} = (167.67 - 194.44)$$
$$= -26.77 \text{ kJ/kg} \cdot \text{K}$$

Loss of available energy = **26.77 kJ/kg · K**

Ans.

12. Steam flows through an adiabatic steady flow turbine. The enthalpy at entrance is 4142 kJ/kg and at exit 2585 kJ/kg. The values of flow availability of steam at entrance and exit state are 1787 kJ/kg and 140 kJ/kg respectively, at state temperature T_0 is 300 K, determine per kg of steam, the actual work, the maximum possible work for the given change of state of steam and the change in entropy of steam. Neglect changes in kinetic and potential energy. *[U.P.S.C. 1993]*

Solution:

Here dead state is given as 300 K and the maximum possible work for given change of state of steam can be estimated by the difference of flow availability as given under:

$$W_{\max} = 1787 - 140 = 1647 \text{ kJ/kg}$$
$$\text{Actual work from turbine, } W_{\text{actual}} = 4142 - 2585$$
$$\text{Actual work} = 1557 \text{ kJ/kg}$$

Actual work = **1557 kJ/kg**
Maximum possible work = **1647 kJ/kg**

Ans.

13. What shall be second law efficiency of a heat engine having efficiency of 0.25 and working between reservoirs of 500°C and 20°C?

Solution:

$$\text{Reversible engine efficiency, } \eta_{\text{rev}} = 1 - \frac{T_{\min}}{T_{\max}} = 1 - \frac{293}{773} = 0.6209$$

$$\text{Second law efficiency} = \frac{\eta}{\eta_{\text{rev}}} = \frac{0.25}{0.6209} = 0.4026 \text{ or } 40.26\%$$

= **40.26%**

Ans.

14. An adiabatic cylinder of volume 10 m^3 is divided into two compartments A and B each of volume 6 m^3 and 4 m^3 respectively, by a thin sliding partition. Initially the compartment A is filled with air at 6 bar and 600 k, whilst there is a vacuum in the compartment B. Air expands and fills both the compartments. Calculate the loss in available energy. Assume atmosphere is at 1 bar and 300 K.

[U.P.S.C. 1997]

Solution:

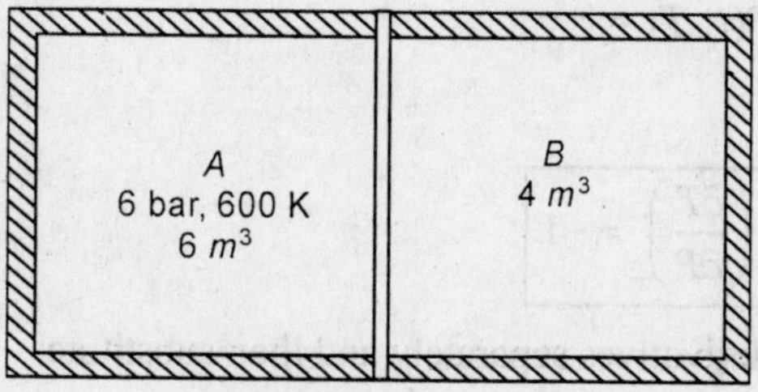

Fig 7.11

Here $T_0 = 300$ K, $P_0 = 1$ bar

$V_A = 6\ m^3$, $V_B = 4$ m^3

$P_1 = 6$ bar, $T_1 = 600$ K

Initially, $V_1 = V_A = 6$m^3, and finally, $V_2 = V_A + V_B = 10$m^3

Expansion occurs in adiabatic conditions.

Temperature after expansion can be obtained by considering adiabatic expansion.

$$\frac{T_2}{T_1} = \left(\frac{V_1}{V_2}\right)^{\gamma-1}$$

$$T_2 = 600\left(\frac{6}{10}\right)^{(1.4-1)} = 489.12 \text{ K}$$

$$\text{Mass of air, } m = \frac{P_1V_1}{RT_1} = \frac{6\times10^5\times6}{287\times600} = 20.91 \text{ kg}$$

Change in entropy of control system, $(S_2 - S_1) = mc_v \ln \dfrac{T_2}{T_1} + mR \ln \dfrac{V_2}{V_1}$

$$\Delta S_S = (S_2 - S_1) = 20.91 \left\{0.718\times\ln\left(\frac{489.12}{600}\right)+0.287\ln\left(\frac{10}{6}\right)\right\}$$

$$= -2.01 \times 10^{-3} \text{ kJ/K}$$

Here, there is no change in entropy of environment, $\Delta S_e = 0$

Total entropy change of combined system $= \Delta S_c = \Delta S_s + \Delta S_e$

$= -2.01 \times 10^{-3}$ kJ/K

Loss of available energy $=$ Irreversibility $= T_0 \cdot \Delta S_c$

$= 300 \times (-2.01 \times 10^{-3})$

$= -0.603$ kJ

Loss of available energy = **0.603 kJ**	**Ans.**

15. Prove that ideal gas equation satisfies the cyclic relation.

Solution:

Ideal gas equation, $Pv = RT$

Let us consider two variable (v, T) to be independent and P as dependent variable.

$$P = P(v, T) = \frac{RT}{v}$$

By cyclic relation,

$$\boxed{\left(\frac{\partial P}{\partial v}\right)_T \cdot \left(\frac{\partial v}{\partial T}\right)_p \cdot \left(\frac{\partial T}{\partial P}\right)_v = -1}$$

Let us find the three partial derivatives separately and then substitute.

$$\left(\frac{\partial P}{\partial v}\right)_T = \frac{-RT}{v^2}, \left(\frac{\partial v}{\partial p}\right)_p = \frac{R}{p}, \left(\frac{\partial T}{\partial P}\right)_v = \frac{v}{R}$$

Substituting

$$\left(\frac{-RT}{v^2}\right)\left(\frac{R}{P}\right)\left(\frac{v}{R}\right) = \frac{-RT}{Pv} = -1 \text{ Hence proved.}$$

16. A heat engine is working between 700° C and 30°C. The temperature of surroundings is 17°C. Engine receives heat at the rate of 2×10^4 kJ/min and the measured output of engine is 0.13 MW. Determine the availability, rate of irreversibility and second law efficiency of engine.

Solution:

$$\text{Availability or reversible work, } W_{rev} = \eta_{rev} \cdot Q_1 = \left(1 - \frac{303}{573}\right) \times 2 \times 10^4$$

$$= 1.38 \times 10^4 \text{ kJ/min}$$

$$\text{Rate of irreversibility, } I = W_{rev} - W_{useful}$$

$$= \left(\frac{1.38 \times 10^4}{60} - 0.13 \times 10^3\right) = 100 \text{ kJ/s}$$

$$\text{Second law efficiency} = \frac{W_{useful}}{W_{rev}}$$

$$= \frac{0.13 \times 10^3}{\left(\frac{1.38 \times 10^4}{60}\right)} = 0.5652 \text{ or } 56.52\%$$

Availability = **1.38×10^4 kJ/min,**
Rate of irreversibility = 100 kW, Second law efficiency = **56.52%** **Ans.**

17. A rigid tank contains air at 1.5 bar and 60°C. The pressure of air is raised to 2.5 bar by transfer of heat from a constant temperature reservoir at 400°C. The temperature of surroundings is 27°C. Determine per kg of air, the loss of available energy due to heat transfer. [U.P.S.C. 1998]

Solution:

Loss of available energy = Irreversibility = $T_0 \cdot \Delta S_c$

$T_0 = 300$ K, $\Delta S_c = \Delta S_s + \Delta S_e$

Change in entropy of system = ΔS_s

Change in entropy of environment/surroundings $= \Delta S_e$

Here heat addition process causing rise in pressure from 1.5 bar to 2.5 bar occurs isochorically. Let initial and final states be given by subscript 1 and 2.

$$\frac{P_1}{T_1} = \frac{P_2}{T_2}, \; T_1 = 333 \text{ K}, T_2 = ?, P_1 = 1.5 \text{ bar}, P_2 = 2.5 \text{ bar}$$

$$T_2 = \frac{2.5 \times 333}{1.5} = 555 \text{ K}$$

Heat addition to air in tank,

$$Q = m \cdot c_p \cdot \Delta T = 1 \times 1.005 \times (555 - 333)$$
$$Q = 223.11 \text{ kJ/kg}$$

$$\Delta S_s = \frac{Q}{T_1} = \frac{223.11}{333} = 0.67 \text{ kJ/kg} \cdot \text{K}$$

$$\Delta S_e = \frac{-Q}{T_{\text{reservoir}}} = \frac{-223.11}{673} = -0.332 \text{ kJ/kg} \cdot \text{K}$$

$$\Delta S_c = 0.67 - 0.332$$
$$\Delta S_c = 0.338 \text{ kJ/kg} \cdot \text{K}$$

Loss of available energy $= 300 \times (0.338)$

$= 101.4$ kJ/kg

Loss of available energy = **101.4 kJ/kg** **Ans.**

18. Using the Maxwell relation derive the following T · ds equation, $T \cdot ds = C_p \cdot dT - T \cdot \left(\frac{\partial v}{\partial T}\right)_p dp$

[U.P.S.C. 1998]

Solution:

Let $s = s(T, p)$

$$ds = \left(\frac{\partial s}{\partial T}\right)_p \cdot dT + \left(\frac{\partial s}{\partial p}\right)_T \cdot dp$$

or,

$$T \cdot ds = T \cdot \left(\frac{\partial s}{\partial T}\right)_p \cdot dT + T \cdot \left(\frac{\partial s}{\partial p}\right)_T \cdot dp$$

Using Maxwell's relation, $\left(\frac{\partial s}{\partial p}\right)_T = -\left(\frac{\partial v}{\partial T}\right)_p$

and $T \cdot \left(\frac{\partial s}{\partial T}\right)_p = C_p$

Substitution yields,

$$\boxed{T \cdot ds = C_p \cdot dT - T \cdot \left(\frac{\partial v}{\partial T}\right)_p \cdot \partial p}$$ Hence proved

19. *Determine the enthapy of vaporization of water at 200°C using Clapeyron equation. Compare it with tabulated value.*

Solution:

$$\text{Clapeyron equation says, } h_{fg} = T \cdot v_{fg} \cdot \left(\frac{dp}{dT}\right)_{sat}$$

From steam tables

$$v_{fg} = (v_g - v_f)_{at\ 200°C} = (0.12736 - 0.001157) = 0.126203 \text{ m}^3/\text{kg}$$

Let us approximate,

$$\left(\frac{dp}{dT}\right)_{sat,\ 200°C} = \left(\frac{\Delta p}{\Delta T}\right)_{sat,\ 200°C} = \frac{P_{sat\ at\ 205°C} - P_{sat\ at\ 195°C}}{(205-195)}$$

$$= \frac{(1.7230 - 1.3978)}{10} = 0.03252 \text{ MPa/°C}$$

Substituting in Clapeyron equation,

$$h_{fg} = (273 + 200) \times 0.126203 \times 0.03252 \times 10^3$$
$$= 1941.25 \text{ kJ/kg}$$

Calculated Enthalpy of vaporization = **1941.25 kJ/kg.**
Enthalpy of vaporization from steam table = **1940.7 kJ/kg.**

Ans.

20. *Determine h_{fg} of R –12 refrigerant at – 10°C using both Clapeyron equation and the Clapeyron-Clausius equation. Give the deviation in %. Take*

$$P_{sat\ at\ -5°C} = 260.96 \text{ kPa}$$
$$P_{sat\ at\ -15°C} = 182.60 \text{ kPa.}$$
$$v_{g\,at\ -10°C} = 0.07665 \text{ m}^3/\text{kg}$$
$$v_{f\,at\ -10°C} = 0.00070 \text{ m}^3/\text{kg}$$
$$R = 0.06876 \text{ kJ/kg} \cdot \text{K}$$
$$h_{fg\ at\ -10°C} = 156.3 \text{ kJ/kg from tables.}$$

Solution:

By Clapeyron equation

$$h_{fg} = T \cdot v_{fg} \left\{\frac{dP}{dT}\right\}_{sat}$$

$$= T \cdot (v_g - v_f) \left\{\frac{\Delta P}{\Delta T}\right\}$$

$$= (-5 + 273) \times (0.07665 - 0.0007) \times \left(\frac{P_{sat\ at\ -5°C} - P_{sat\ at\ -15°C}}{(-5-(-15))}\right)$$

$$= 268 \times 0.07595 \times \frac{(260.96 - 182.60)}{10}$$

$$h_{fg} = 159.49 \text{ kJ/kg}$$

By Clapeyron-Clausius equation,

$$\ln\left(\frac{P_2}{P_1}\right)_{sat} = \frac{h_{fg}}{R}\left[\frac{1}{T_1}-\frac{1}{T_2}\right]_{sat}$$

$$\ln\left(\frac{P_{sat\ at-5°C}}{P_{sat\ at-15°C}}\right) = \frac{h_{fg}}{R} = \left[\frac{1}{T_{sat\ at-15°C}}-\frac{1}{T_{sat\ at-5°C}}\right]$$

$$\ln\left(\frac{260.96}{182.60}\right) = \frac{h_{fg}}{0.06876}\times\left[\frac{1}{(-15+273)}-\frac{1}{(-5+273)}\right]$$

$\Rightarrow$ $h_{fg} = 169.76$ kJ/kg

% deviation from Clapeyron equation

$$= \left(\frac{169.76-159.49}{159.49}\right)\times 100$$

$$= 6.44\%$$

h_{fg} by Clapeyron equation = **159.49 kJ/kg**
h_{fg} by Clapeyron-Clausius equation = **169.76 kJ/kg**
% deviation in h_{fg} value by Clapeyron-Clausius equation from the value from Clapeyron equation = **6.44%**

Ans.

21. *Determine the volume expansivity and isothermal compressibility of steam at 300 kPa and 300°C.*

Solution:

$$\text{Volume expansivity} = \frac{1}{v}\left[\frac{\partial v}{\partial T}\right]_P$$

$$\text{Isothermal compressibility} = \frac{-1}{v}\left[\frac{\partial v}{\partial P}\right]_T$$

Let us write $\frac{\partial v}{\partial T} = \frac{\Delta v}{\Delta T}$ and $\frac{\partial v}{\partial P} = \frac{\Delta v}{\Delta P}$. The differences may be taken for small pressure and temperature changes.

Volume expansivity,

$$= \frac{1}{v}\left[\frac{\partial v}{\partial T}\right]_{300\ kpa}$$

$$= \frac{1}{v_{at\ 300\ kpa,\ 300°C}}\left[\frac{v_{350°C}-v_{250°C}}{(350-250)}\right]_{300\ kpa}$$

$$= \frac{1}{0.8753}\left[\frac{0.9534-0.7964}{100}\right]_{300\ kpa}$$

Volume expansivity = **1.7937 × 10^{-3} K^{-1}** **Ans.**

$$= \frac{-1}{v_{\text{at 300kpa, 300°C}}}\left[\frac{v_{350\text{kpa}} - v_{250\text{kpa}}}{(350-250)}\right]_{300°C}$$

$$= -\frac{1}{0.8753}\left[\frac{0.76505 - 1.09575}{100}\right]$$

$$= 3.778 \times 10^{-3}\ \text{KPa}^{-1}$$

Isothermal compressibility = 3.778×10^{-3} kPa^{-1} Ans.

22. An evacuated tank of 0.5 m³ is filled by atmospheric air at 1 bar till the pressure inside tank becomes equal to atmospheric temperature. Considering filling of tank to occur adiabatically determine final temperature inside the tank and also the irreversibility and change in entropy. Take atmospheric temperature as 25°C.

Solution:

Filling of the tank is a transient flow (unsteady flow) process. For the transient filling process, considering subscripts 'i' and 'f' for initial and final states,

$$h_i = u_f$$

$$c_p T_i = c_v T_f$$

$$T_f = \frac{c_p}{c_v} T_i$$

$$T_f = \frac{1.005}{0.718} \times 298.15$$

Inside final temperature, T_f = 417.33 · K Ans.

Change in entropy $\Delta S_{gen} = (S_f - S_i) + \Delta S_{surr}$

$$= c_p \ln \frac{T_f}{T_i} + 0$$

$$= 1.005 \times \ln \frac{417.33}{298.15}$$

Change in entropy ΔS_{gen} = 0.3379 kJ/kg · K Ans.

Irreversibility, $I = T_0 \cdot \Delta S_{gen}$

$$= 298.15 \times 0.3379$$

Irreversibility, I = 100. 74 kJ/kg Ans.

23. A closed vessel stores 75 kg of hot water at 400°C. A heat engine transfers heat from the hot water to environment which is maintained at 27°C. This operation of heat engine changes temperature of hot water from 400°C to 27°C over a finite time. Determine the maximum possible work output from engine. Take specific heat of water as 4.18 kJ/kg· K.

Solution:

Here the combined closed system consists of hot water and heat engine. Here there is no thermal reservoir in the system under consideration. For the maximum work output, irreversibility = 0

Therefore, $\frac{d}{dt}(E - T_0\ S) = W_{max}$

or $W_{max} = (E - T_0 S)_1 - (E - T_0 S)_2$

Here $E_1 = U_1 = m\ c_p\ T_1,\ E_2 = U_2 = m\ c_p\ T_2$

$T_1 = 400 + 273 = 673$ K, $T_2 = 27 + 273 = 300$ K $= T_0$

Therefore,

$$W_{max} = mc_p\,(T_1 - T_2) - T_0(S_1 - S_2)$$

$$= mc_p\,(T_1 - T_2) - T_0 \cdot m c_p \left(\ln\frac{T_1}{T_2}\right)$$

$$= 75 \times 4.18 \times \left\{(673 - 300) - 300 \times \ln\left(\frac{673}{300}\right)\right\}$$

$$= 40946.6 \text{ kJ}$$

Maximum work = **40946.6 kJ**

Ans.

24. In a steam turbine the steam enters at 50 bar, 600°C and 150 m/s and leaves as saturated vapour at 0.1 bar, 50 m/s. During expansion work of 1000 kJ/kg is delivered. Determine the inlet stream availability, exit stream availability and the irreversibility. Take dead state temperature as 15°C.

Solution:

$h_1 = h_{\text{at 50 bar, 600°C}} = 3666.5$ kJ/kg, $s_1 = s_{\text{at 50 bar, 600°C}} = 7.2589$ kJ/kg · K

$h_2 = h_{g \text{ at 0.1 bar}} = 2584.7$ kJ/kg, $s_2 = s_{g \text{ at 0.1 bar}} = 8.1502$ kJ/kg · K

$$\text{Inlet stream availability} = \left(h_1 + \frac{c_1^2}{2}\right) - T_0 s_1$$

$$= \left(3666.5 + \frac{(150)^2 \times 10^{-3}}{2}\right) - (298 \times 7.2589)$$

$$= 1514.59 \text{ kJ/kg}$$

Input stream availability is equal to the input absolute availability.

$$\text{Exit stream availability} = \left(h_2 + \frac{c_2^2}{2}\right) - T_0 s_2$$

$$= \left(2584.7 + \frac{(50)^2 \times 10^{-3}}{2}\right) - (298 \times 8.1502)$$

$$= 157.19 \text{ kJ/kg}$$

Exit stream availability is equal to the exit absolute availability.

$$W_{rev} = 1514.59 - 157.19 = 1357.4 \text{ kJ/kg}$$

$$\text{Irreversibility} = W_{rev} - W = 1357.4 - 1000 = 357.4 \text{ kJ/kg}$$

This irreversibility is in fact the availability loss.

Inlet stream availability = **1514.59 kJ/kg**	
Exit stream availability = **157.19 kJ/kg**	
Irreversibility = **357.4 kJ/kg**	**Ans.**

EXERCISE

7.1 Define 'available energy' and 'unavailable energy'.

7.2 What do you understand by second law efficiency? How does it differ from first law efficiency?

7.3 What is meant by a dead state? Discuss its' importance.

7.4 Define availability. Obtain an expression for availability of closed system.

7.5 Differentiate between useful work and maximum useful work in reference to the availability.

7.6 What do you understand by Gibbs function? How does it differ from the availability function?

7.7 Describe the Helmholtz function.

7.8 What are Maxwell relations? What is their significance?

7.9 Describe Clapeyron equation.

7.10 What do you understand by Joule-Thomson coefficient? Explain.

7.11 Describe chemical potential.

7.12 Write short notes on the following:
(*i*) Clapeyron-Clausius equation,
(*ii*) Volume expansivity
(*iii*) Fugacity,
(*iv*) Second law analysis of engineering systems.

7.13 Determine the loss of availability when 1 kg air at 260°C is reversibly and isothermally expanded from 0.145 m^3 initial volume to 0.58 m^3 final volume. [70.56 kJ/kg]

7.14 Determine the entropy generation and decrease in available energy when a heat source of 727°C transfers heat to a body at 127°C at the rate of 8.35 MJ/min. Consider the temperature of sink as 27°C. [12.54 kJ/K · min, 3762 kJ]

7.15 Determine the available energy of furnace having the gases getting cooled from 987°C to 207°C at constant temperature while the temperature of surroundings is 22°C. [–518.1 kJ/kg]

7.16 Determine the available amount of energy removed and the entropy increase of universe when 5 kg air at 1.38 bar, 500 K is cooled down to 300 K isobarically. The temperature of surroundings may be taken as 4°C. [–268.7 kJ. 3.316 kJ/K]

7.17 Determine the entropy change, unavailable energy and available energy for the process in which 2 kg air is heated isobarically so as to cause increase in its temperature from 21°C to 315°C. Take T_0 = 10°C. [1.393 kJ/K, 394.2 kJ, 196.6 kJ]

7.18 Steam enters in a steam turbine at 60 bar, 500°C and leaves at 0.1 bar, 0.89 dry with a flow rate of 3.2652 × 10^4 kg/hr. Determine the loss of available energy. [1286.2 kJ/s]

7.19 Determine the available portion of heat removed from 2.5 kg air being cooled from 2.1 bar, 205°C to 5°C at constant volume. The heat is rejected to surroundings at – 4°C. [– 97.2 kJ]

7.20 Prove that heat is an inexact differential. {Q (T, s)}.

7.21 Derive an expression for change in entropy of a gas obeying Vander Waals equation of state.

7.22 Determine the coefficient of thermal expansion and coefficient of isothermal compressibility for a gas obeying Vander Waals equation of state.

7.23 Determine the second law efficiency of a heat engine operating between 700°C and 30°C. The heat engine has efficiency of 0.40. [55.74%]

7.24 Determine the amount of heat that can be converted to the useful work if total heat at 1000 kJ is available at 500°C. The temperature of environment is 17°C. [624.84 kJ]

7.25 Determine the change in availability of air contained in an insulated vessel of 20×10^3 cm^3. The initial state of air is 1 bar. 40°C. The air is heated so as to arrive at temperature of 150°C. The temperature of surrounding environment may be considered as 20°C. [0.027 kJ]

7.26 Determine the enthalpy of vaporization of water at 50°C using the Clapeyron equation. [2396.44 kJ/kg]

7.27 Determine the % variation in the enthalpy of vaporization of steam at 500 kPa using Clapeyron equation as compared to value in steam table. [0.201%]

7.28 Air enters a compressor at 40°C, 500 kPa for being compressed upto 2000 kPa. Consider the compression to be at constant temperature without internal irreversibilities. Air flows into compressor at 6 kg/min. Neglecting the changes in kinetic energy and potential energy determine the availability transfers accompanying heat and work and irreversibility. Take T_0 = 25°C, P_0 = 1 bar. The control volume may be taken as under,

(*i*) Control volume comprises of compressor alone

(*ii*) Control volume comprises of compressor and its immediate surroundings so that heat transfer occurs at T_0.

[(*i*) – 0.597 kJ/s, –12.46 kJ/s, 0.597 kJ/s
(*ii*) 0 kJ/s, –12.46 kJ/s, 0.597 kJ/s]

7.29 Steam expands in a cylinder from 22 bar, 450°C to 4.5 bar, 250°C. The environment may be considered to be at 1 bar, 25°C. Determine

(*i*) the availability of steam at inlet, exit and change in availability.

(*ii*) the irreversibilities in the expansion process. [196 kJ, 132 kJ, –64 kJ, 5.5 kJ]

7.30 In a steam power cycle steam enters at 60 bar, 500°C into turbine and leaves at 0.04 bar. The isentropic efficiency of turbine is 85% and that of pump is 70%. Considering the environment to have T_0 = 25°C, P_0 =1 bar, determine the second law efficiency of constituent components i.e. boiler, turbine, condenser and pump. [67.6%, 99%, 84.4%, 65%]

8

Vapour Power Cycles

8.1 INTRODUCTION

Thermodynamic cycles can be primarily classified based on their utility such as for power generation, refrigeration etc. Based on this thermodynamic cycles can be

(*i*) Power cycles,

(*ii*) Refrigeration and heat pump cycles.

(*i*) ***Power cycles:*** Thermodynamic cycles which are used in devices producing power are called power cycles. Power production can be had by using working fluid either in vapour form or in gaseous form. When vapour is the working fluid then they are called vapour power cycles, whereas in case of working fluid being gas these are called gas power cycles. Thus, power cycles shall be of two types,

(*a*) Vapour power cycle,

(*b*) Gas power cycle.

Vapour power cycles can be further classified as,

1. Carnot vapour power cycle
2. Rankine cycle
3. Reheat cycle
4. Regenerative cycle.

Gas power cycles can be classified as,

1. Carnot gas power cycle
2. Otto cycle
3. Diesel cycle
4. Dual cycle
5. Stirling cycle
6. Ericsson cycle
7. Brayton cycle

Here in the present text Carnot, Rankine, Reheat and Regenerative cycles are discussed.

(*ii*) ***Refrigeration and heat pump cycles:*** Thermodynamic cycles used for refrigeration and heat pump are under this category. Similar to power cycles, here also these cycles can be classified as "air cycles" and "vapour cycles" based on type of working fluid used.

8.2 PERFORMANCE PARAMETERS

Some of commonly used performance parameters in cycle analysis are described here.

Thermal efficiency: Thermal efficiency is the parameter which gauges the extent to which the energy input to the working fluid is converted to net work output.

$$\text{Thermal efficiency} = \frac{\text{Net work in cycle}}{\text{Heat added in cycle}}$$

Heat rate: Heat rate refers to the amount of energy added by heat transfer to cycle to produce unit net work output. Usually energy added may be in kcal, unit of net work output in kW h and unit of heat rate may be in kcal/kW h. It is inverse of thermal efficiency.

Back work ratio: Back work ratio is defined as the ratio of pump work input (–ve work) to the work produced (+ve work) by turbine.

$$\text{Back work ratio} = \frac{W_{\text{pump}}}{W_{\text{turbine}}}$$

Generally, back work ratio is less than one and as a designer one may be interested in developing a cycle which has smallest possible back-work ratio. Small back-work ratio indicates smaller pump work (–ve work) and larger turbine work (+ve work).

Work ratio: It refers to the ratio of net work to the positive work.

$$\text{Mathematically, work ratio} = \frac{W_{\text{net}}}{W_{\text{turbine}}}$$

Specific steam consumption: It indicates the steam requirement per unit power output. It is generally given in kg/kW. h and has numerical value lying from 3 to 5 kg/kW. h

$$\text{Specific steam consumption} = \frac{3600}{W_{\text{net}}}, \text{ kg/kW.h}$$

8.3 CARNOT VAPOUR POWER CYCLE

Carnot cycle has already been defined earlier as an ideal cycle having highest thermodynamic efficiency. Let us use Carnot cycle for getting positive work with steam as working fluid. Arrangement proposed for using Carnot vapour power cycle is as follows.

1 – 2 = Reversible isothermal heat addition in the boiler
2 – 3 = Reversible adiabatic expansion in steam turbine
3 – 4 = Reversible isothermal heat rejection in the condenser
4 – 1 = Reversible adiabatic compression or pumping in feed water pump

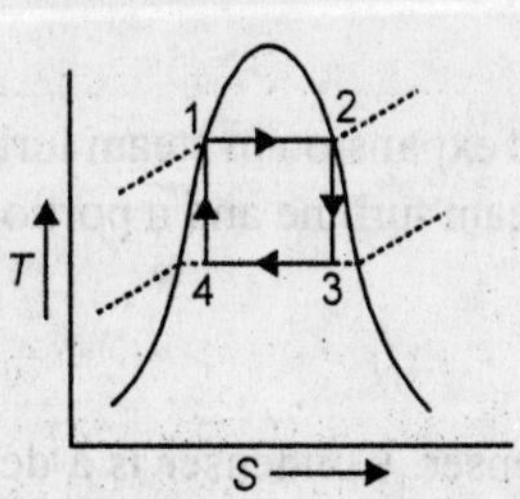

Fig 8.1 Carnot vapour power cycle

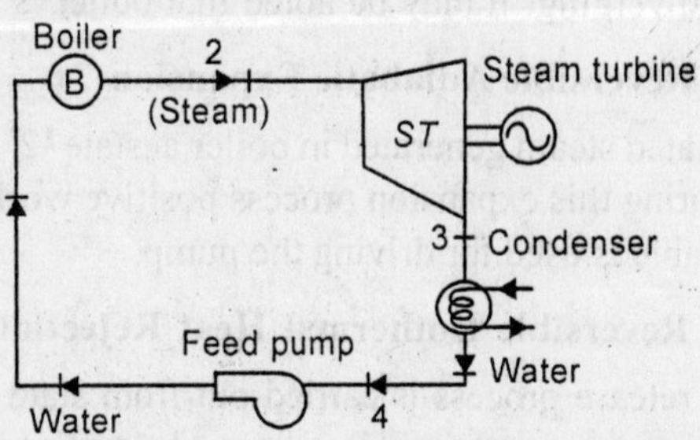

Fig 8.2 Arrangement for Carnot cycle

Assuming steady flow processes in the cycle and neglecting changes in kinetic and potential energies, thermodynamic analysis may be carried out.

$$\text{Thermal efficiency} = \frac{\text{Net work}}{\text{Heat added}}$$

Net work = Turbine work – Compression/Pumping work
For unit mass flow.

$$W = (h_2 - h_3) - (h_1 - h_4)$$

$$\text{Heat added, } Q_{add} = (h_2 - h_1)$$

$$\eta_{Carnot} = \frac{(h_2 - h_3) - (h_1 - h_4)}{(h_2 - h_1)}$$

$$= 1 - \frac{h_3 - h_4}{h_2 - h_1}$$

Here heat rejected, $Q_{rejected} = (h_3 - h_4)$

or $\eta_{Carnot} = 1 - \frac{Q_{rejected}}{Q_{add}}$

Also, heat added and rejected may be given as function of temperature and entropy as follows:

$$Q_{add} = T_1 \times (s_2 - s_1)$$

$$Q_{rejected} = T_3 \times (s_3 - s_4)$$

Also, $s_1 = s_4$ and $s_2 = s_3$

Therefore, substituting values: $\eta_{carnot} = 1 - \frac{T_3}{T_1}$

or $\eta_{carnot} = 1 - \frac{T_{minimum}}{T_{maximum}}$

Let us critically evaluate the processes in Carnot cycle and see why it is not practically possible.

1–2: Reversible Isothermal Heat Addition

Isothermal heat addition can be easily realised in boiler within wet region as isothermal and isobaric lines coincide in wet region. But the superheating of steam can't be undertaken in case of Carnot cycle as beyond saturated steam point isothermal heat addition can't be had inside boiler. This fact may also be understood from T–s diagram as beyond 2 the constant pressure line and constant temperature lines start diverging. It may be noted that boiler is a device which generates steam at constant pressure.

2–3: Reversible Adiabatic Expansion

Saturated steam generated in boiler at state '2' is sent for adiabatic expansion in steam turbine upto state 3. During this expansion process positive work is produced by steam turbine and a portion of this work available is used for driving the pump.

3–4: Reversible Isothermal Heat Rejection

Heat release process is carried out from state 3 to 4 in the condenser. Condenser is a device in which constant pressure heat rejection can be realized. Since expanded steam from steam turbine is available in wet region at state 3. Therefore, constant temperature heat rejection can be had as constant temperature and constant pressure lines coincide in wet region.

Heat rejection process is to be limited at state 4 which should be vertically below state 1. Practically it is very difficult to have such kind of control.

4–1: Reversible Adiabatic Compression (Pumping)

Carnot cycle has reversible adiabatic compression process occurring between 4 and 1, which could be considered for pumping of water into boiler.

In fact it is very difficult for a pump to handle wet mixture which undergoes simultaneous change in its phase as its pressure increases.

Above discussion indicates that Carnot vapour power cycle is merely theoretical cycle and cannot be used for a practical working arrangement. Also the maximum efficiency of Carnot cycle is limited by maximum and minimum temperatures in the cycle. Highest temperature attainable depends upon metallurgical limits of boiler material.

8.4 RANKINE CYCLE

Rankine cycle is a thermodynamic cycle derived from Carnot vapour power cycle for overcoming its limitations. In earlier discussion it has been explained that Carnot cycle cannot be used in practice due to certain limitations. Rankine cycle has the following thermodynamic processes.

1 – 2 = Isobaric heat addition (in boiler)

2 – 3 = Adiabatic expansion (in turbine)

3 – 4 = Isobaric heat release (in condenser)

4 – 1 = Adiabatic pumping (in pump)

$T-s$, $h-s$ and $P-V$ representations are as shown below.

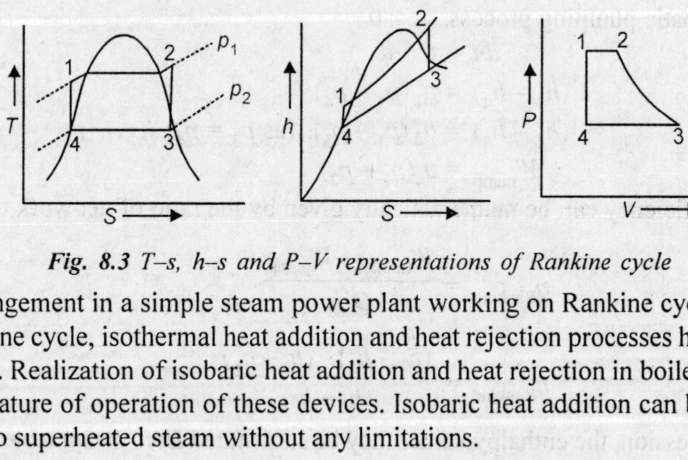

Fig. 8.3 T–s, h–s and P–V representations of Rankine cycle

Practical arrangement in a simple steam power plant working on Rankine cycle is shown ahead.

Thus in Rankine cycle, isothermal heat addition and heat rejection processes have been replaced by isobaric processes. Realization of isobaric heat addition and heat rejection in boiler and condenser is in conformity with nature of operation of these devices. Isobaric heat addition can be had in boiler from subcooled liquid to superheated steam without any limitations.

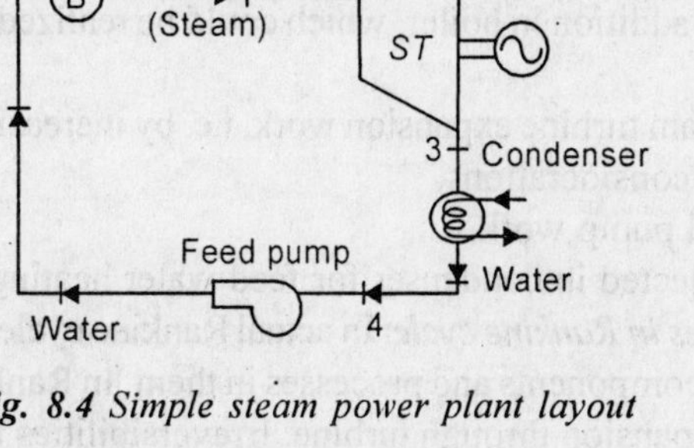

Fig. 8.4 Simple steam power plant layout

Let us understand the arrangement.

1 – 2: High pressure water supplied by feed pump is heated and transformed into steam with or without superheat as per requirement. This high pressure and temperature steam is sent for expansion in steam turbine. Heat added in boiler, for unit mass of steam.

$$Q_{add} = (h_2 - h_1)$$

2 – 3: Steam available from boiler is sent to steam turbine, where it's adiabatic expansion takes place and positive work is available. Expanded steam is generally found to lie in wet region. Expansion of steam is carried out to the extent of wet steam having dryness fraction above 85% so as to avoid condensation of steam on turbine blades and subsequently the droplet formation which may hit hard on blade with large force.

Turbine work, for unit mass, $W_{turbine} = (h_2 - h_3)$.

3 – 4: Heat rejection process occurs in condenser at constant pressure causing expanded steam to get condensed into saturated liquid at state 4.

Heat rejected in condenser for unit mass, $Q_{rejected} = (h_3 - h_4)$

4 – 1: Condensate available as saturated liquid at state 4 is sent to feed pump for being pumped back to boiler at state 1.

For unit mass, Pump work $W_{pump} = h_1 - h_4$.

Here pumping process is assumed to be adiabatic for the sake of analysis whereas it is not exactly adiabatic in the pump.

From first and second law combined together;

$$dh = T \cdot ds + v \cdot dp.$$

Here in this adiabatic pumping process. $ds = 0$

Therefore
$$dh = v \cdot dp.$$

or
$$(h_1 - h_4) = v_4\,(p_1 - p_4)$$

or
$$(h_1 - h_4) = v_4\,(p_1 - p_3). \; \{\text{as } p_3 = p_4\}$$

$$W_{pump} = v_4(p_1 - p_3)$$

Rankine cycle efficiency can be mathematically given by the ratio of net work to heat added.

$$\eta_{Rankine} = \frac{W_{turbine} - W_{pump}}{Q_{add}}$$

$$\eta_{Rankine} = \frac{(h_2 - h_3) - (h_1 - h_4)}{(h_2 - h_1)}$$

In the above expression, the enthalpy values may be substituted from steam table and mollier charts and by analysis for getting efficiency value.

Rankine cycle efficiency may be improved in the following ways:

(*a*) By reducing heat addition in boiler, which could be realized by preheating water entering into it.

(*b*) By increasing steam turbine expansion work, i.e. by increasing expansion ratio within limiting dryness fraction considerations.

(*c*) By reducing feed pump work.

(*d*) By using heat rejected in condenser for feed water heating. etc.

Irreversibilities and losses in Rankine cycle: In actual Rankine cycle there exist various irreversibilities and losses in its' constituent components and processes in them. In Rankine cycle the major irreversibility is encountered during the expansion through turbine. Irreversibilities in turbine significantly reduce the expansion work. Heat loss from turbine to surroundings, friction inside turbine and leakage losses contribute to irreversibilities. Due to this irreversible expansion there occurs an increase in entropy as compared to no entropy change during reversible adiabatic expansion process. This deviation of expansion from ideal to actual process can be accounted for by isentropic turbine efficiency. Ideal expansion in steam turbine is shown by 2–3 on *T–S* representation. Actual expansion process is shown by 2–3′.

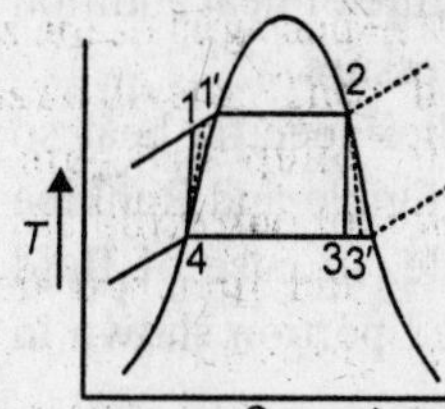

Fig. 8.5 Rankine cycle showing non-ideal expansion and pumping process

$$\text{Isentropic turbine efficiency, } \eta_{isen,\,t} = \frac{W_{t,actual}}{W_{t,ideal}} = \frac{W_{2-3'}}{W_{2-3}}$$

or,

$$\eta_{isen,\,t} = \left(\frac{h_2 - h_{3'}}{h_2 - h_3}\right)$$

Another important location for irreversibilities is the pump. During pumping some additional work is required to overcome frictional effects. Ideally pumping is assumed to take place with no heat transfer during pumping whereas actually it may not be so. Thus the pumping process as shown by ideal process 4–1 gets' modified to 4–1′ which is accompanied by increase in entropy across the pump. Isentropic efficiency of pump is a parameter to account for non-idealities of pump operation. Isentropic efficiency of pump is defined by;

$$\eta_{isen,\,p} = \frac{W_{p,ideal}}{W_{p,actual}} = \frac{W_{4-1}}{W_{4-1'}}$$

or,

$$\eta_{isen,\,p} = \left(\frac{h_1 - h_4}{h_{1'} - h_4}\right)$$

Thus, it indicates that actually pump work required shall be more than ideal pump work requirement.

Apart from the turbine and pump irreversibilities explained above there may be other sources of inefficiency too. These turbine and pump irreversibilities accounted for by isentropic efficiency of turbine and pump are called external irreversibility. Sources of internal irreversibilities are heat transfer from system to surroundings, frictional pressure loss in rest of components etc. There also occurs the steam pressure drop due to friction between pipe surface and working fluid.

8.5 DESIRED THERMODYNAMIC PROPERTIES OF WORKING FLUID

Working fluid being used in vapour power cycles must have following desirable properties. Generally water is used as working fluid in vapour power cycles as it is easily available in abundance and satisfies most of requirements. Other working fluids may be mercury, sulphur dioxide and hydrocarbons etc.

(*i*) Working fluid should be cheap and easily available.

(*ii*) Working fluid should be non-toxic, non-corrosive and chemically stable.

(*iii*) Fluid must have higher saturation temperature at moderate pressures as it shall yield high efficiency because most of heat will be added at high temperature. Thus, mean temperature of heat addition shall be high even at moderate pressure.

(*iv*) Working fluid should have smaller specific heat so that sensible heat supplied is negligible and Rankine cycle approaches to Carnot cycle. In case of fluid having small specific heat hatched portion shown in Fig. 8.6 will be absent.

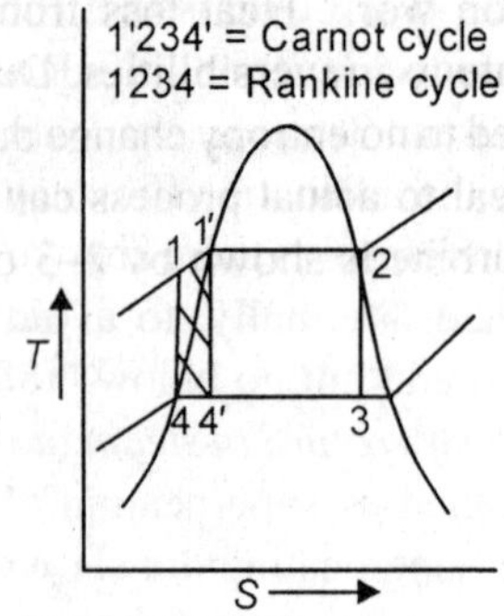

Fig. 8.6 Carnot cycle and Rankine cycle emphasizing for small specific heat of fluid

(*v*) Saturated vapour line should be steep enough so that state after expansion has high dryness fraction.

(*vi*) Working fluid density should be high so that the size of plant becomes smaller.

(*vii*) Working fluid should have its' critical temperature within metallurgical limits.

(*viii*) It should show significant decrease in volume upon condensation.

(*ix*) Working fluid should have its' freezing point much below atmospheric pressure so that there is no chance of freezing in condenser.

8.6 PARAMETRIC ANALYSIS FOR PERFORMANCE IMPROVEMENT IN RANKINE CYCLE

Let us carry out study of the influence of thermodynamic variables upon Rankine cycle performance.

(*i*) *Pressure of steam at inlet to turbine:* Steam pressure at inlet to turbine may be varied for same temperature of steam at inlet. Two different pressures of steam at inlet to turbine, also called throttle pressure are shown in Fig. 8.7. Comparative study shows that for back pressure and steam inlet temperatures being same the increase in steam inlet pressure from p'_1 to p_1 is accompanied by the reduction in net heat added as shown by hatched area $A_{2'3'37}$ and increase in net heat added by the amount shown by area $A_{1'1271'}$. Generally two areas $A_{2'3'37}$ and $A_{1'1271'}$ are nearly same which means that the increment in net heat added due to increasing throttle pressure p_1 is accompanied by decrease in net heat addition and the net heat added remains same as at lower throttle pressure $p_{1'}$. But increase in throttle pressure to p_1 also causes reduction in the heat rejected. At pressure $p_{1'}$, heat rejected is given by area $A_{43'6'54}$ while at pressure p_1 heat rejected is given by area A_{43654}

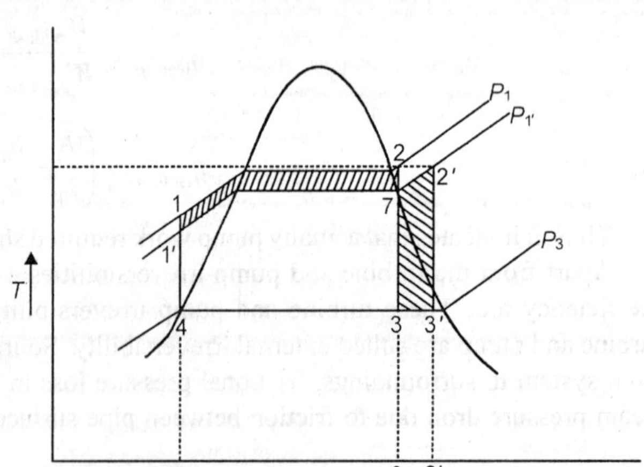

Fig. 8.7 Rankine cycle showing two different throttle pressures

$$A_{43654} < A_{43'6'54}$$

$$(\text{Heat rejected})_{\text{cycle } 1234} < (\text{Heat rejected})_{\text{cycle } 1'2'3'4}$$

Cycle efficiency is given by $\eta_{\text{cycle}} = 1 - \dfrac{\text{Heat rejected}}{\text{Heat added}}$

Hence, it is obvious that increasing steam pressure at inlet to steam turbine is accompanied by increase in cycle thermal efficiency. But this increase in pressure increases wetness of steam as shown by states 3 and 3′ i.e. $x_{3'} < x_3$, where x is dryness fraction. This increase in wetness of expanding steam decreases the adiabatic efficiency of turbine and also increases the chances of erosion of steam turbine blades. Therefore, as there are two contrary consequences of increasing throttle pressure so a compromise is to be had. Normally, to avoid erosion of turbine blades the minimum dryness fraction at turbine exhaust should not go below 0.88.

(*ii*) *Temperature of steam at inlet to turbine:* Increasing temperature of steam at inlet to turbine may also be called as superheating of steam at inlet to turbine. The two Rankine cycles having different degree of superheating are shown in Fig. 8.8.

For two different temperatures of steam at inlet to turbine i.e. T_2 and $T_{2'}$ while $T_2 < T_{2'}$ the comparison of two cycles 12341 and 12′3′41 shows the effect of increasing temperature at inlet to turbine.

Rankine cycle represented on *T–S* diagram show that increasing temperature from T_2 to $T_{2'}$ causes increase in net work by amount as shown by area $A_{22'3'32}$. This increase in steam temperature is also accompanied by increased heat addition as shown by area $A_{22'6'62}$. It is seen that this ratio of increase in net work to increase in heat addition in more than similar ratio for rest of cycle so the net effect is to improve the cycle thermal efficiency. It may also be stated that this increase in steam temperature from

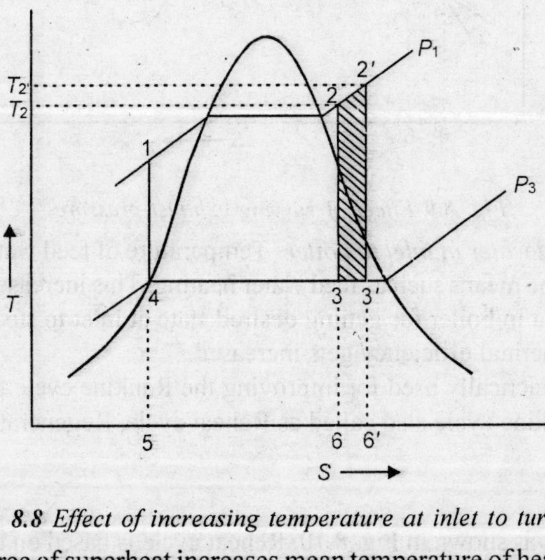

Fig. 8.8 Effect of increasing temperature at inlet to turbine

T_2 to $T_{2'}$ i.e. increase in degree of superheat increases mean temperature of heat addition, which increases thermal efficiency. With increased steam temperature the state of steam after expansion becomes more dry i.e. increase in temperature from 2 to 2' makes steam more dry after expansion i.e. from 3 to 3' having dryness fraction $x_{3'} > x_3$. This hotter steam supply to turbine is also advantageous from the specific work point of view. The work done per unit mass gets increased by superheating steam at inlet to turbine. Therefore, one is always interested in realizing highest possible temperature of steam provided it is within metallurgical temperature limits. Highest practical steam temperature at turbine inlet presently is 650°C.

(*iii*) *Pressure at the end of expansion:* Let us see the influence of pressure at the end of expansion from steam turbine. This pressure may also be called exhaust pressure or back pressure or condenser pressure. Rankine cycle with two different exhaust pressures is shown in Fig. 8.9, while the maximum pressure and temperature remains same.

With the lowering of back pressure from p_3 to $p_{3'}$ Rankine cycles get modified from 12341 to 1'23'4'1'. This reduction in back pressure causes increment in net work as shown by area $A_{1'1433'4'51'}$ and also the heat addition increases by the amount shown by area $A_{1'166'1''}$. It is seen that the two areas are such that the thermal efficiency of cycle increases by lowering back pressure as increase in heat addition is more than increase in heat rejection. This lowering back pressure is accompanied by increase in wetness of steam from 3 to 3' i.e. dryness fraction $x_3 > x_{3''}$. Practically there exists limitation of dryness fraction after expansion to avoid erosion of turbine blades so lowering exhaust pressure is limited by it, inspite of showing improvement in thermal efficiency.

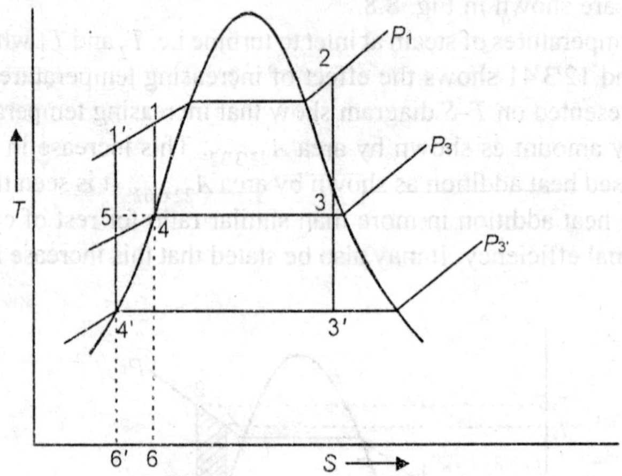

Fig. 8.9 Effect of varying exhaust pressure

(*iv*) *Temperature of feed water at inlet to boiler:* Temperature of feed water at inlet to boiler may be increased by employing some means such as feed water heating. This increase in feed water temperature reduces the heat requirement in boiler for getting desired state at inlet to steam turbine. Thus, with the reduced heat addition the thermal efficiency gets increased.

Different approaches practically used for improving the Rankine cycle performance have resulted into modified forms of Rankine cycle also called as Reheat cycle, Regenerative cycle etc.

8.7 REHEAT CYCLE

Schematic of reheat cycle is as shown in Fig. 8.10. Reheat cycle is based on the simple fact of realizing high efficiency with higher boiler pressure and yet avoid low quality of steam at turbine exhaust.

Here steam generated in boiler is supplied to high pressure steam turbine at state 2 and is expanded upto state 3. This steam is sent to boiler for being reheated so that its temperature gets increased, normally this temperature after reheating may be equal to temperature at inlet of high pressure steam turbine. Steam after reheating is supplied to subsequent turbine at state 4, say to low pressure steam turbine. Steam is now expanded upto the exhaust pressure say state '5'. Expanded steam is subsequently sent to condenser and condensate at state '6' is pumped back to the boiler employing feed pump at state '1'. Thus, it is possible to take advantage of high steam pressure at inlet to steam turbine as the problem of steam becoming excessively wet with increasing steam pressure could be regulated by reheating during the expansion. Expansion occurs in two stages one begining at high pressure and other occurring at low pressure with reheating in between. The principal advantage of reheat is to increase the quality of steam at turbine exhaust.

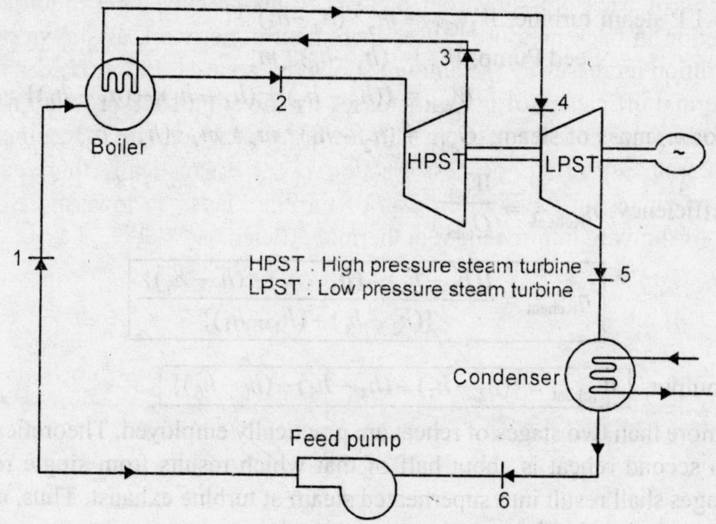

Fig 8.10 Reheat cycle

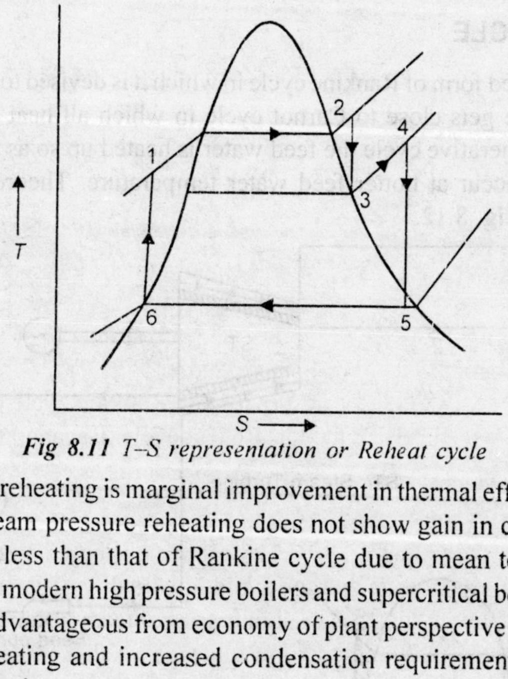

Fig 8.11 T–S representation or Reheat cycle

Secondary advantage of reheating is marginal improvement in thermal efficiency when steam pressure is above 100 bar. At low steam pressure reheating does not show gain in cycle thermal efficiency and even the efficiency may be less than that of Rankine cycle due to mean temperature of heat addition being lower. Generally, with modern high pressure boilers and supercritical boilers reheating is essentially employed. Reheating is disadvantageous from economy of plant perspective as the cost of plant increases due to arrangement for reheating and increased condensation requirements due to increased dryness fraction of steam after expansion.

Thermodynamic analysis of Reheat cycle as shown on T–S diagram may be carried out for estimation of different parameters as below,

$$\text{Total turbine work output} = W_{\text{HPST}} + W_{\text{LPST}}$$

$$\text{Net work, } W_{\text{net}} = (\text{Total turbine work output}) - (\text{Pump work})$$

$$W_{\text{net}} = W_{\text{HPST}} + W_{\text{LPST}} - W_p$$

where different works for m_s mass of steam are,

$$\text{HP steam turbine, } W_{\text{HPST}} = m_s \cdot (h_2 - h_3)$$

$$\text{LP steam turbine, } W_{LPST} = m_s \cdot (h_4 - h_5)$$

$$\text{Feed Pump, } W_p = (h_1 - h_6) \cdot m_s$$

$$W_{net} = \{(h_2 - h_3) + (h_4 - h_5) - (h_1 - h_6)\} \cdot m_s$$

$$\text{Heat supplied for } m_s \text{ mass of steam; } Q_{add} = (h_2 - h_1) \cdot m_s + m_s \cdot (h_4 - h_3)$$

$$\text{Cycle thermal efficiency, } \eta_{Reheat} = \frac{W_{net}}{Q_{add}}$$

$$\boxed{\eta_{Reheat} = \frac{\{(h_2 - h_3) + (h_4 - h_5) - (h_1 - h_6)\}}{\{(h_2 - h_1) + (h_4 - h_3)\}}}$$

Specific work output, $\boxed{W_{reheat} = \{(h_2 - h_3) + (h_4 - h_5) - (h_1 - h_6)\}}$

Generally not more than two stages of reheat are practically employed. Theoretically improvement in efficiency due to second reheat is about half of that which results from single reheat. Also more number of reheat stages shall result into superheated steam at turbine exhaust. Thus, mean temperature of heat rejection gets raised and efficiency drops.

8.8 REGENERATIVE CYCLE

Regenerative cycle is a modified form of Rankine cycle in which it is devised to increase mean temperature of heat addition so that cycle gets close to Carnot cycle in which all heat addition occurs at highest possible temperature. In regenerative cycle the feed water is heated up so as to reduce the heat addition in boiler and heat addition occur at hotter feed water temperature. Theoretically regenerative cycle arrangement is as shown in Fig. 8.12.

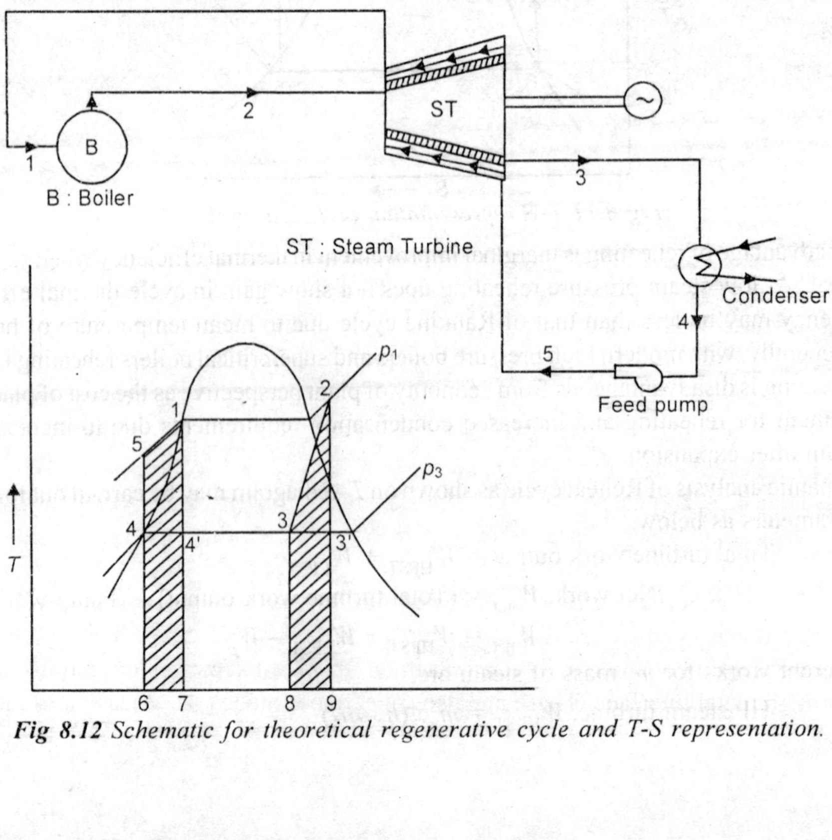

Fig 8.12 Schematic for theoretical regenerative cycle and T-S representation.

Theoretical arrangement shows that the steam enters the turbine at state 2 (temperature T_2) and expands to (temperature T_3) state 3. Condensate at state 5 enters the turbine casing which has annular space around turbine. Feed water enters turbine casing at state 5 and gets infinitesimally heated upto state 1 while flowing opposite to that of expanding steam. This hot feed water enters into boiler where steam generation occurs at desired state, say 2. Feed water heating in steam turbine casing is assumed to occur reversibly as the heating of feed water occurs by expanding steam with infinitesimal temperature difference and is called "regenerative heating". This cycle is called regenerative cycle due to regenerative heating employed in it. Regenerative heating refers to the arrangement in which working fluid at one state is used for heating itself and no external heat source is used for this purpose. Here feed water picks up heat from steam expanding in steam turbine, thus the expansion process in steam turbine shall get modified from 2-3′ ideally to 2-3. Heat picked up by feed water for getting heated up from state 5 to 1 is shown by hatched area A_{17651} on *T-S* diagram. Under ideal conditions for cent per cent heat exchange effectiveness the two areas i.e. A_{29832} indicating heat extraction from steam turbine and A_{17651} indicating heat recovered by feed water shall be same. Thus, *T-S* representation of regenerative cycle indicates that the cycle efficiency shall be more than that of Rankine cycle due to higher average temperature of heat addition.

But there exists serious limitation regarding realization of the arrangement described above. Limitations are due to impossibility of having a steam turbine which shall work as both expander for getting work output and heat exchanger for feed water heating. Also with the heat extraction from steam turbine the state of expanded steam at exhaust pressure shall be extremely wet, hence not desired. Due to theselimitations the regenerative cycle is realized employing the concept of bleeding out steam from turbine and using it for feed water heating in feed water heaters.

Feed Water Heaters: The feed water heater refers to the device in which heat exchange occurs between two fluids i.e. steam and feed water either in direct contact or indirect contact. Direct contact feed water heater is the one in which bled steam and feed water come in direct contact. These are also called open feed water heater.

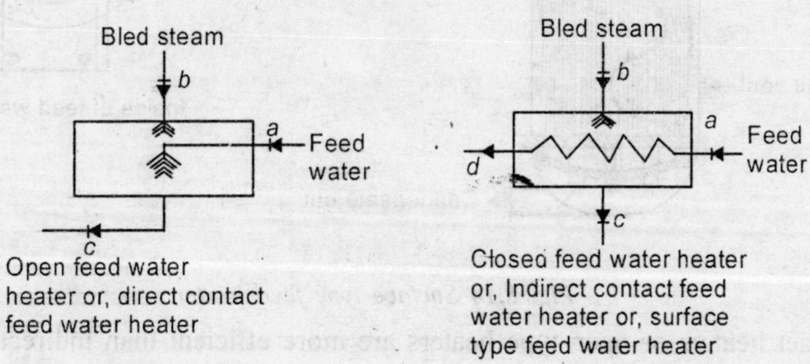

Fig. 8.13 Feed water heaters

In open feed water heater two fluids i.e. steam and feed water are at same pressure and adiabatic mixing is assumed to take place. Normally, it is considered that the mixture leaves open feed water heater as saturated liquid. Energy balance upon it shall be as follows,

$$m_a \cdot h_a + m_b \cdot h_b = (m_a + m_b) \cdot h_c$$

where subscripts *a*,*b* and *c* are for feed water, bled steam and mixture of the two as shown in Figure.

Indirect contact feed water heater as shown in Figure is the one in which two fluid streams i.e. bled steam and feed water do not come in direct contact, but the heat exchange between two streams occurs indirectly through metal interface. These are also called closed feed water heaters. In these feed water heaters since two fluids do not contact each other so they may be at different pressures. In these the arrangement comprises of steel, copper or brass tubes of solid drawn type placed in a shell. The heat transfer takes place through tube surface. Feed water flows inside the tube and is heated by extracted steam from outside. Steam enters in the shell and comes in contact with the tubes and then condenses over tubes. Steam condenses and trickles down and is collected in shell. Figure 8.14 gives schematic of surface type feed water heater. Performance of feed water heater is quantified using a parameter called "terminal temperature difference". Terminal temperature difference (T.T.D.) refers to the difference of temperature between temperature of feed water outlet and saturation temperature of steam entering the heater.

Terminal temperature difference = (Feed water outlet temperature – Saturation temperature of steam entering heater)

T.T.D. has its value lying around 5–8°C. T.T.D. shall be zero in desuperheater type heaters where superheated steam is used for feed water heating upto saturation temperature of steam.

In feed water heaters where steam pressure is quite high, the condensate from heater is expanded in an expander called 'flash tank' or 'drain expander' or 'drain cooler'. In this flash tank some portion of condensate gets converted into steam which is further used for heating feed water.

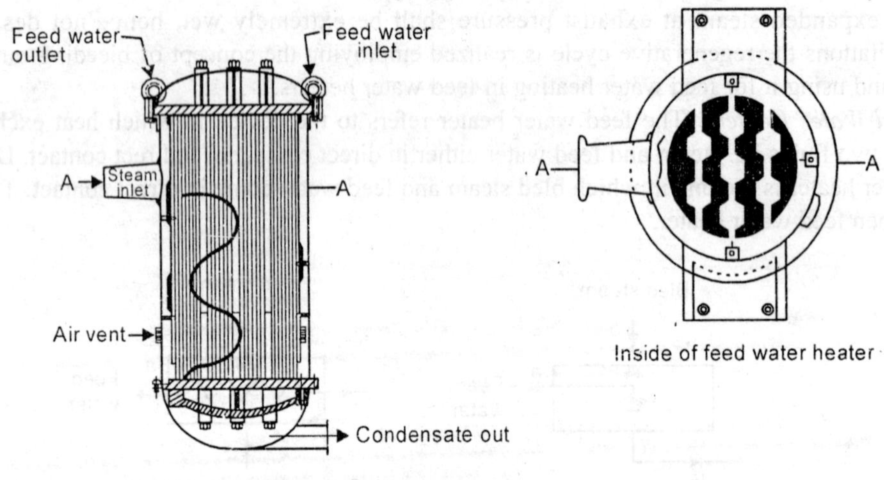

Fig. 8.14 Surface type feed heater

Direct contact heaters or open type heaters are more efficient than indirect contact type due to direct contact between two fluids. Feed water can achieve saturation temperature corresponding to the pressure of heating steam. In this case the terminal temperature difference is zero. During heating the non-condensable gases dissolved in water get released and are thrown out through vent passage. Deaerator is a type of open feed water heater. Schematic is shown in subsequent article on deaerator.

Deaerator: Deaerator is a type of open type feed water heater employed for the removal of dissolved oxygen and carbon dioxide from the feed water. The dissolved oxygen when not removed gets disintegrated into nascent oxygen at high temperature and pressure and forms iron oxide upon coming in contact with metal. This iron oxide formed causes pitting on the metal surface. At high temperature and pressure, dissolved CO_2 combines with metal and forms carbolic acid which causes mild pitting on metal surfaces. Therefore, it becomes necessary to remove dissolved gases from the feed water. Although feed water treatment plant is there but still the impurities may creep in along with the make up water added to compensate for loss of water from system due to leakages at valve, pipe flanges, steam valve spindles and boiler blow down etc. Normally added water is 3-5% of total boiler feed.

Deaeration of water is based on the principle of decreasing partial pressure of gas for removal of dissolved gases. Henery's law and Dalton's law of partial pressure may be considered in this reference for understanding the phenomenon. Henery's law states that, "the mass of gas dissolved in a definite mass of liquid at a given temperature is directly proportional to partial pressure of gas in contact with liquid". It is good for the gases having no chemical reaction with water.

Decrease in partial pressure of gas in water is achieved by increasing the vapour pressure by heating the water. Here feed water is heated by low pressure steam for heating it upto its saturation temperature. Feed water entering deaerator is broken into small particles so as to increase contact area for better heat exchange with high temperature steam. Constructional detail of deaerator is shown in Fig. 8.15 which has basically deaerator head and storage tank. Water enters deaerator head from top on to a distributor plate. The water trickles down from upper most tray to the bottom trays through tiny holes in these trays. Steam enters storage tank from one end and enters deaerator head after passing through water collected in tank. Steam gradually heats feed water flowing downward with its' portion getting condensed and remaining steam flowing along with liberated gases out of deaerator head. Steam and gas mixture vented out from the top of deaerator head may be used for preheating make up water or feed water entering deaerator, if economical. The deaerated water is collected in storage tank below the deaerator head. A bubbler line is also provided in deaerator tank for fast heating of deaerator system during start up of unit. Bubbler line is a perforated pipe laid at bottom of storage tank through which steam for heating the water is supplied.

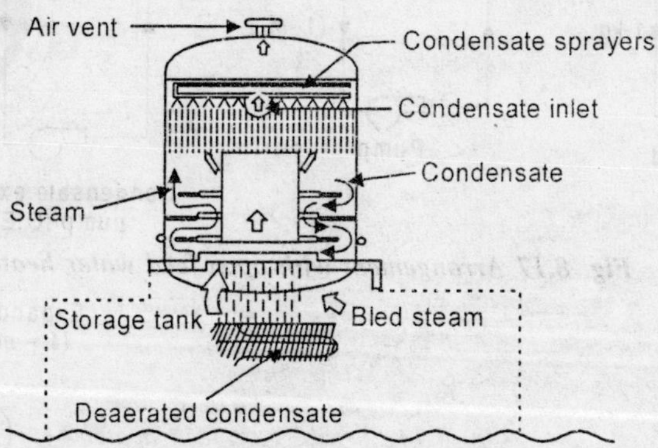

Fig. 8.15 Deaerator

Feed heater arrangements: In regenerative cycle feed water heaters of different types are employed. There are some generic arrangements frequently used in these cycles. The arrangements are discussed below with two bleed points from where m_1 and m_2 masses of steam are bled out at pressures p_1 and p_2 and expansion occurs upto pressure p_3. Total steam flowing is taken as 1 kg.

(*i*) *Surface type heaters method:* This employs surface type feed water heaters and the arrangement for them is as shown in Fig. 8.16.

Here two surface heaters are used for showing the arrangement. Condensate of the bled steam is drained out using drain pump and sent in the main line at high pressure. This arrangement is also called as drain pump method.

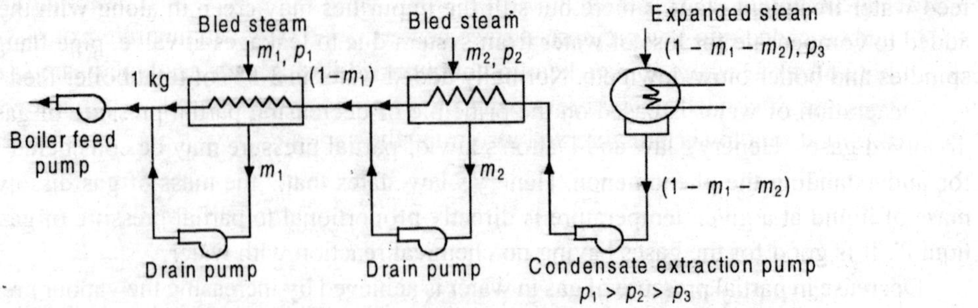

Fig. 8.16 Arrangement in drain pump method or surface type heaters method

(ii) *Open type heater method:* In this arrangement the open type feed water heaters are employed as shown. Here due to contact of two fluids, there occurs mixing of bled steam with water and is taken out using pump for being sent to next open feed water heater.

(iii) *Surface type heaters with hot well:* This arrangement employs a hot well with surface type heaters. Bled steam condensate leaving surface type heaters is sent to hot well from where it is picked up by pump and flown through heaters for getting heated up. Arrangement is shown in Fig. 8.18.

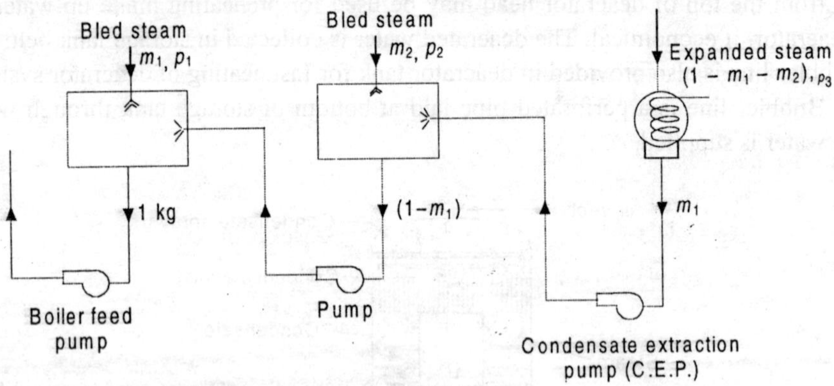

Fig. 8.17 Arrangement with open feed water heaters

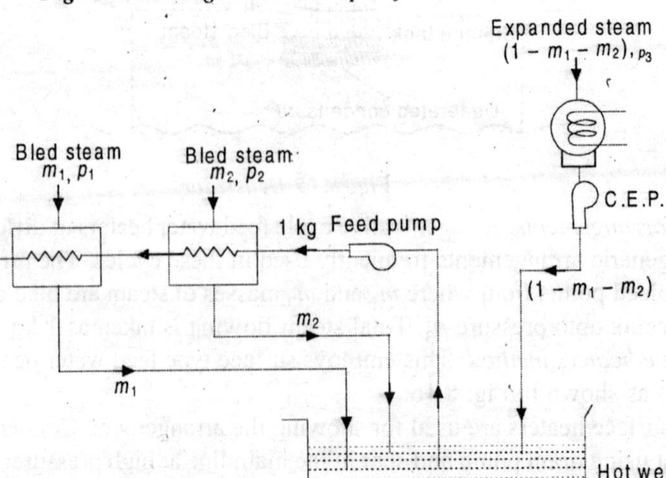

Fig 8.18 Arrangement with surface type heaters and hotwell

(*iv*) *Cascade method:* This arrangement is shown in Fig. 8.19. Here bled steam condensate is throttled and cascaded into low pressure surface heaters. Bled steam condensate from last heater is sent to hotwell from where it is picked up and pumped through surface type heaters. For lowering the pressure of condensate before mixing the traps may also be used. A trap allows the liquid to be throttled to a lower pressure and also traps the vapour.

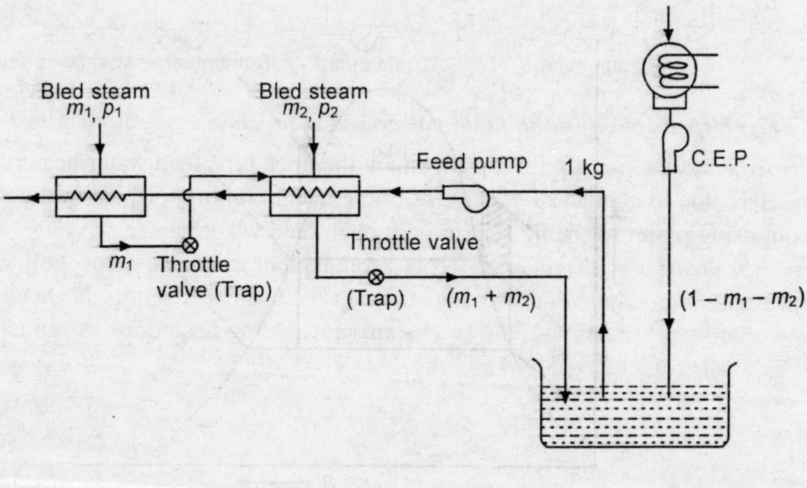

Fig. 8.19 Cascade method

Thermodynamics of regenerative cycle: Schematic of regenerative cycle with single feed water heater of open type is shown in Fig. 8.20. Arrangement shows that steam is bled out from turbine at state 6 and fed into feed heater.

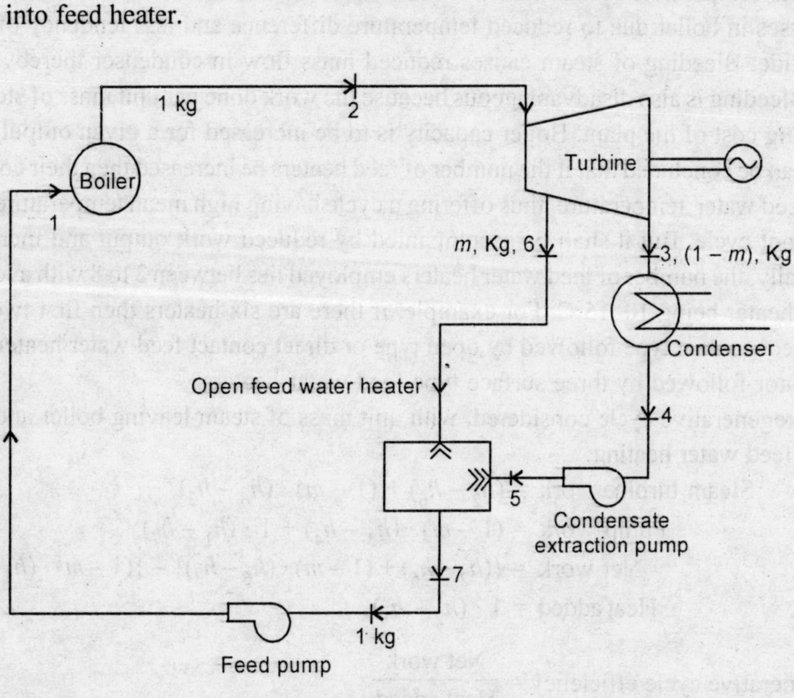

Fig 8.20 Schematic for regenerative cycle with one open feed water heater

Feed water leaving at state 7 as shown on *T-S* diagram is being pumped upto boiler pressure through feed pump. *T-S* diagram indicates that the amount of heat picked up by feed water is shown by hatched area $A_{1751'}$. In case of absence of bleeding and feed heater the feed water will enter into boiler at state 1′ as compared to state 1 when regenerative heating is employed. Thus, advantage of hotter feed water to boiler can be realized by bleeding expanding steam from turbine. Regeneration can be seen in the feed water heating as the bled steam gets mixed with feed water at state 5 thus resulting into hot feed water.

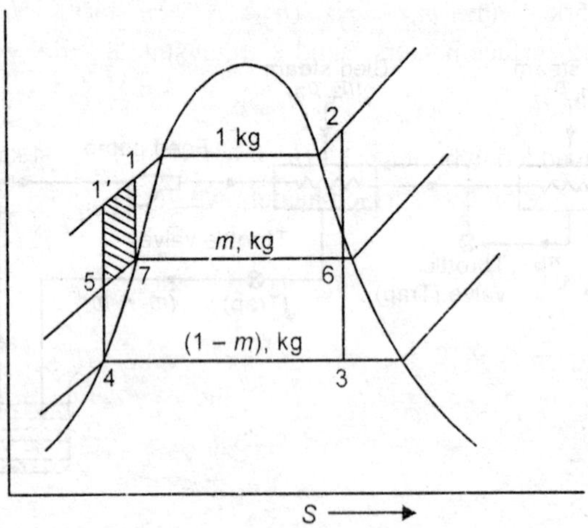

Fig 8.21 T-S representation for regenerative cycle with one open feed water heater

As described earlier the bleeding offers advantage in terms of increased cycle efficiency due to increased mean temperature of heat addition. Hotter feed water also offers advantage in terms of reduced thermal stresses in boiler due to reduced temperature difference and less tendency of condensation of sulphur dioxide. Bleeding of steam causes reduced mass flow in condenser thereby reducing size of condenser. Bleeding is also disadvantageous because the work done per unit mass of steam gets reduced, thus increasing cost of the plant. Boiler capacity is to be increased for a given output.

Here it can be concluded that if the number of feed heaters be increased then their could be substantial increase in feed water temperature, thus offering a cycle having high mean temperature of heat addition, close to Carnot cycle. But it shall be accompanied by reduced work output and increased cost of the plant. Generally, the number of feed water heaters employed lies between 3 to 8 with average temperature rise in each heater being 10–15°C. For example, if there are six heaters then first two may be surface type or indirect contact type followed by open type or direct contact feed water heater which shall also act as deaerator followed by three surface type feed water heaters.

For the regenerative cycle considered, with unit mass of steam leaving boiler and '*m*' kg of steam bled out for feed water heating:

$$\text{Steam turbine ork} = (h_2 - h_6) + (1 - m)\cdot(h_6 - h_3)$$

$$\text{Pump work} = (1 - m)\cdot(h_5 - h_4) + 1\cdot(h_1 - h_7)$$

$$\text{Net work} = \{(h_2 - h_6) + (1 - m)\cdot(h_6 - h_3)\} - \{(1 - m)\cdot(h_5 - h_4) + (h_1 - h_7)\}$$

$$\text{Heat added} = 1\cdot(h_2 - h_1)$$

$$\text{Hence, regenerative cycle efficiency} = \frac{\text{Net work}}{\text{Heat added}}$$

$$\eta_{\text{regenerative}} = \frac{\{(h_2 - h_6) + (1 - m)(h_6 - h_3)\} - \{(1 - m)(h_5 - h_4) + (h_1 - h_7)\}}{(h_2 - h_1)}$$

Example: Regenerative cycle with two surface type heaters, (Fig. 8.22)

Let us carry out thermodynamic analysis for 1 kg of steam generated in boiler at pressure p_1 and masses of bled steam be m_6 and m_7 at pressure p_6 and p_7.

Steam turbine work = $\{1 \cdot (h_2 - h_6) + (1 - m_6) \cdot (h_6 - h_7) + (1 - m_6 - m_7) \cdot (h_7 - h_3)\}$

Total pumping work = $\{(1 - m_6 - m_7) \cdot (h_5 - h_4) + m_7 \cdot (h_{10} - h_9) + m_6 \cdot (h_{14} - h_8)\}$

Bled steam leaving surface heaters 1 and 2 are at state 8 and 9 which are saturated liquid states as respective pressure, i.e., $h_8 = h_{f \text{ at } p_6}$, $h_9 = h_{f \text{ at } p_7}$

Heat added in boiler = $1 \cdot (h_2 - h_1)$

Applying heat balance on two surface heaters we get,

On surface heater 1,

$$m_6 \cdot h_6 + (1 - m_6) \cdot h_{12} = m_6 \cdot h_8 + (1 - m_5) \cdot h_{13}$$

On surface heater 2,

$$m_7 \cdot h_7 + (1 - m_6 - m_7) \cdot h_5 = m_7 \cdot h_9 + (1 - m_6 - m_7) \cdot h_{11}$$

At the point of mixing of output from surface heater and bled condensate the heat balance yields.

$$1 \cdot h_1 = (1 - m_6) \cdot h_{13} + m_6 \cdot h_{14}$$

$$h_1 = (1 - m_6) \cdot h_{13} + m_6 \cdot h_{14}$$

and

$$(1 - m_6 - m_7) \cdot h_{11} + m_7 \cdot h_{10} = (1 - m_6) \cdot h_{12}$$

Upon the pumps 1, 2 and 3;

On pump 1,

$$v_4(p_1 - p_3) = h_5 - h_4$$

and

$$h_5 = v_4\,(p_1 - p_3) + h_4$$

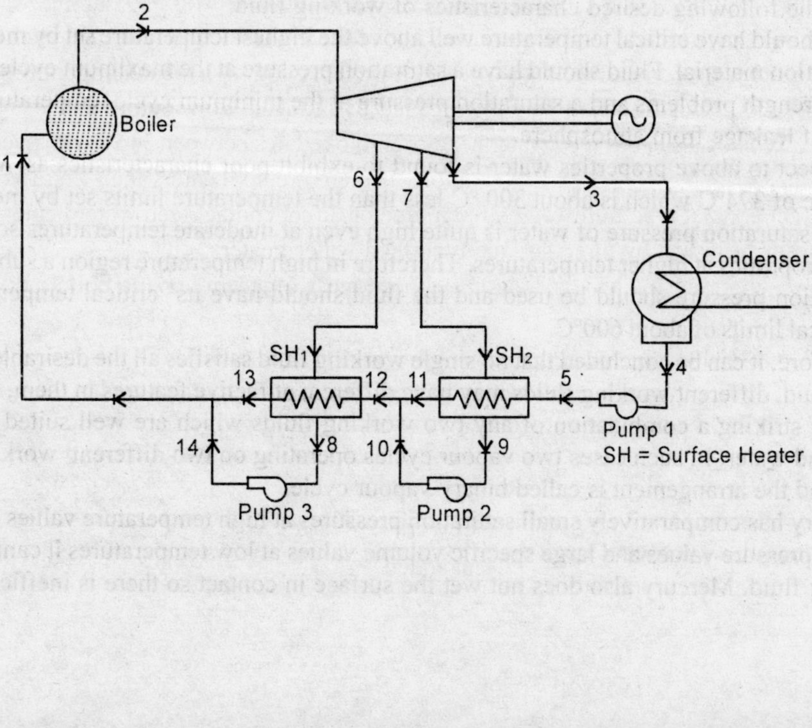

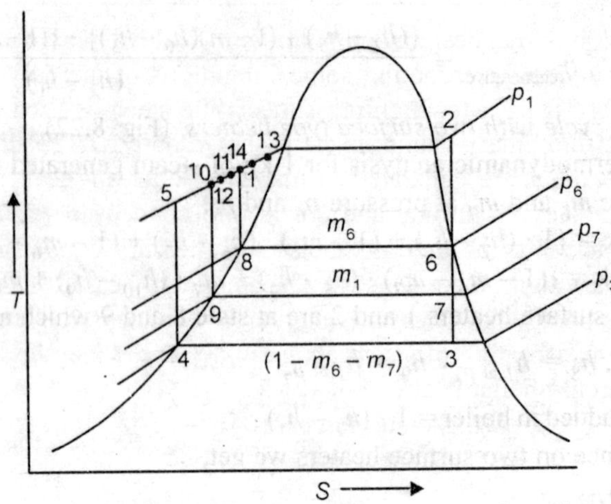

Fig. 8.22 Schematic and T-S representation for regenerative cycle with two surface type heaters

On pump 2,

$$v_9(p_1 - p_7) = h_{10} - h_9$$
$$h_{10} = v_9(p_1 - p_7) + h_9$$

On pump 3,

$$v_8(p_1 - p_6) = h_{14} - h_8$$
$$h_{14} = v_8(p_1 - p_6) + h_8$$

Above different enthalpy expressions can be used for getting enthalpy values at salient points of interest in order to get the net work and cycle efficiency of this arrangement.

8.9 BINARY VAPOUR CYCLE

Generally, water is used as working fluid in vapour power cycle as it is found to be better than any other fluid if looked from the point of view of desirable characteristics of working fluid. Water is poor in respect to the following desired characteristics of working fluid.

Fluid should have critical temperature well above the highest temperature set by metallurgical limits of construction material. Fluid should have a saturation pressure at the maximum cycle temperature that poses no strength problems and a saturation pressure at the minimum cycle temperature that posses no difficulty of leakage from atmosphere.

In respect to above properties water is found to exhibit poor characteristics as; Water has critical temperature of 374°C which is about 300 °C less than the temperature limits set by metallurgical properties. The saturation pressure of water is quite high even at moderate temperatures so it does not have desirable properties at higher temperatures. Therefore in high temperature region a substance which has low saturation pressure should be used and the fluid should have its' critical temperature well above metallurgical limits of about 600°C.

Therefore, it can be concluded that no single working fluid satisfies all the desirable requirements of working fluid, different working fluids may have different attractive features in them, but not all. So let us think of striking a combination of any two working fluids which are well suited together such as mercury and water. In such cases two vapour cycles operating on two different working fluids are put together and the arrangement is called binary vapour cycle.

Mercury has comparatively small saturation pressures at high temperature values but due to excessively low pressure values and large specific volume values at low temperatures it cannot be used alone as working fluid. Mercury also does not wet the surface in contact so there is inefficient heat transfer although 0.002% of solution of magnesium and potassium is added to give it wetting property of steel. Steam is used with mercury for overcoming some limitations of mercury. Thus in combination of mercury-steam, the mercury is used for high pressures while steam is used for low pressure region.

Layout for mercury-steam binary vapour cycle is shown on Fig. 8.23 along with depiction on *T-S* diagram. Here, mercury vapour are generated in mercury boiler and sent for expansion in mercury turbine and expanded fluid leaves turbine and enters into condenser. From condenser the mercury condensate is pumped back into the mercury boiler. In mercury condensate the water is used for extracting heat from mercury so as to condense it. The amount of heat liberated during condensation of mercury is too large to evaporate the water entering mercury condenser. Thus, mercury condenser also acts as steam boiler. For superheating of steam an auxilliary boiler may be employed or superheating may be realized in the mercury boiler itself.

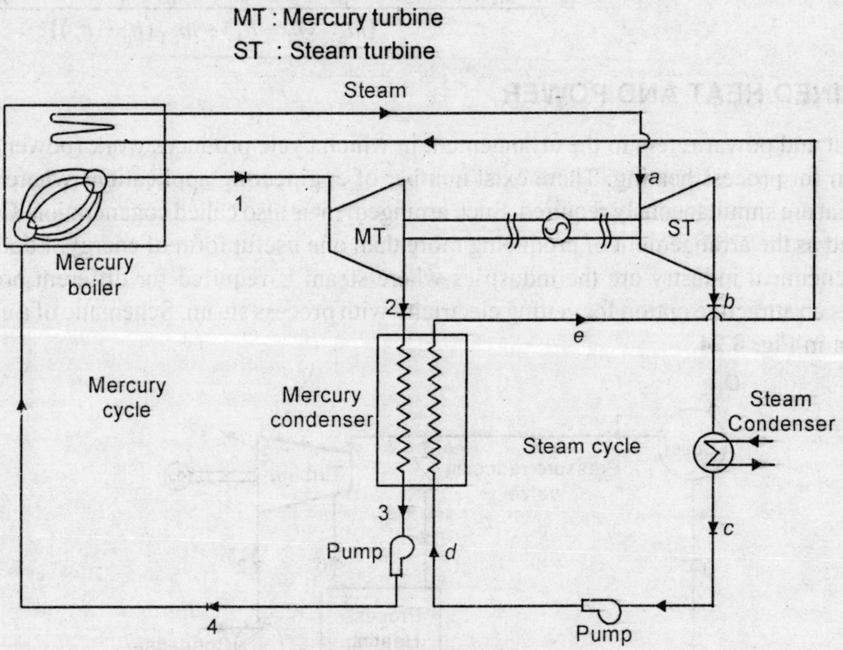

Mercury-Steam binary vapour cycle

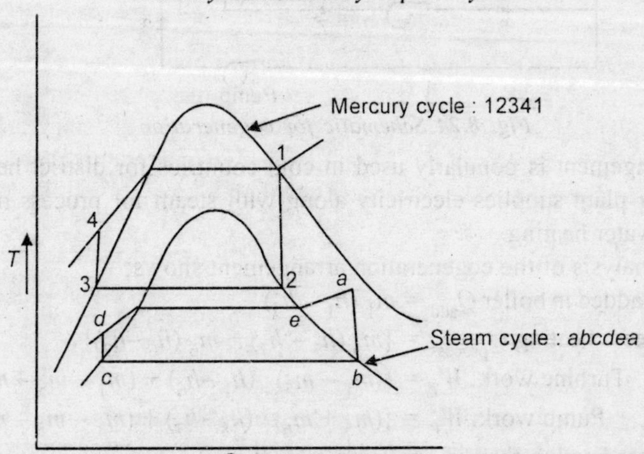

Thermodynamic cycle for mercury-steam binary vapour cycle

Fig. 8.23 Schematic of binary vapour cycle

Net work from cycle shall be, $W_{net} = W_{MT} + W_{ST} - \Sigma W_{pump}$

Work from mercury turbine, $W_{MT} = m_{MT} \cdot (h_1 - h_2)$

Work from steam turbine, $W_{ST} = m_{ST} \cdot (h_a - h_b)$

Pump work $= m_{MT} \cdot (h_4 - h_3) + m_{ST} \cdot (h_d - h_c)$

Heat added to the cycle, $Q_{add} = m_{MT} \cdot (h_1 - h_4) + m_{ST} \cdot (h_a - h_e)$

Binary Cycle efficiency, binary $= \dfrac{W_{net}}{Q_{add}}$

$$= \frac{\{m_{MT}(h_1 - h_2) + m_{ST}(h_a - h_b) - m_{MT}(h_4 - h_3) - m_{ST}(h_d - h_c)\}}{\{m_{MT}(h_1 - h_4) + m_{ST}(h_a - h_e)\}}$$

8.10 COMBINED HEAT AND POWER

Combined heat and power refers to the arrangement in which cycle produces work (power) along with heat utilization for process heating. There exist number of engineering applications where both power and process heat are simultaneously required. Such arrangement is also called cogeneration. Cogeneration may be defined as the arrangement of producing more than one useful form of energy. Food processing industry and chemical industry are the industries where steam is required for different processes and cogeneration is an attractive option for getting electricity with process steam. Schematic of a cogeneration plant is shown in Fig. 8.24.

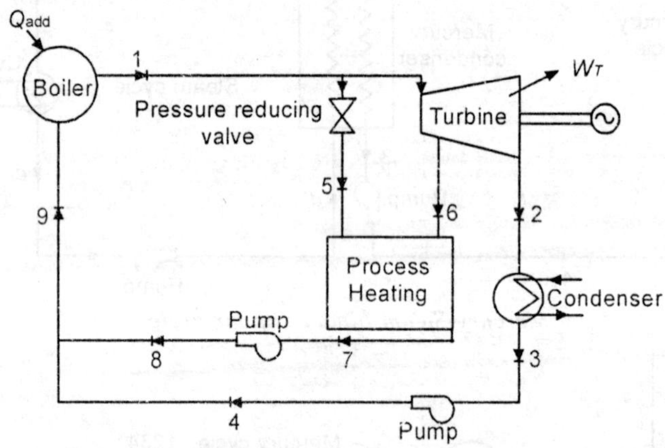

Fig. 8.24 Schematic for cogeneration

Cogeneration arrangement is popularly used in cold countries for district heating where, in this arrangement the power plant supplies electricity along with steam for process needs, such as space heating and domestic water heating.

Thermodynamic analysis of the cogeneration arrangement shows;

Heat added in boiler $Q_{add} = m_1 (h_1 - h_9)$

Heat used in process heating $Q_{process} = \{m_5 (h_5 - h_7) + m_6 (h_6 - h_7)\}$

Turbine work: $W_T = \{(m_1 - m_5) \cdot (h_1 - h_6) + (m_1 - m_5 - m_6) \cdot (h_6 - h_2)\}$

Pump work: $W_P = \{(m_5 + m_6) \cdot (h_8 - h_7) + (m_1 - m_5 - m_6) \cdot (h_4 - h_3)\}$

When there is no process heating then; $m_5 = 0$ and $m_6 = 0$

8.11 DIFFERENT STEAM TURBINE ARRANGEMENT

In certain applications simple steam turbines are unable to meet specific requirements. Back pressure turbine, pass out or extraction turbine and mixed pressure turbines are such special purpose turbines whose details are given ahead.

(*a*) *Back pressure turbine:* Back pressure turbine is the one in which steam is not expanded upto lowest pressure in steam turbine, instead steam leaves the turbine at higher pressure which is appropriate for the process steam/heating requirement. Thus, in back pressure turbine expansion is limited to high back pressure and steam leaving turbine goes for process heating. Generally, steam leaving turbine at high back pressure will be superheated. Since steam is to be used for process heating so the rate of heat transfer should be high. Superheated steam is not suitable for heating because of small rate of heat transfer therefore superheated steam should be desuperheated and brought to saturated steam state as saturated steam has high rate of heat transfer and also the control of temperature is convenient.

Thus, back pressure turbine has the provision of desuperheating as shown in Fig. 8.25. The steam leaving tubine enters into desuperheater where it is transformed into saturated steam. Saturated steam is subsequently sent for process heating where it gets condensed and condensate is sent back to the boiler through pump. A by pass valve is also provided so that if there is no power requirement then whole steam may be sent for process heating through desuperheater by closing turbine valve and opening by pass valve.

(*b*) *Pass out or extraction turbine:* Pass out turbine refers to the steam turbine having provision for extraction of steam during expansion. Such provision is required because in combined heat and power requirement the steam available from back pressure turbine may be more than required one or the power produced may be less than the required value. Pass out turbine has arrangement for continuous extraction of a part of steam at the desired pressure for process heating and left out steam goes into low pressure section of turbine through a pressure control valve. In the low pressure section of turbine control mechanism is provided so that the speed of turbine and pressure of steam extracted remains constant irrespective of the variations in power produced and process heating.

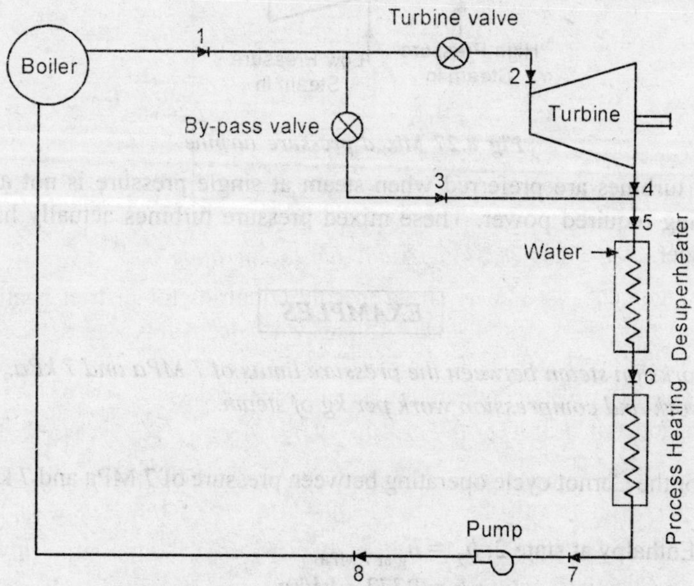

Fig. 8.25 Back pressure turbine

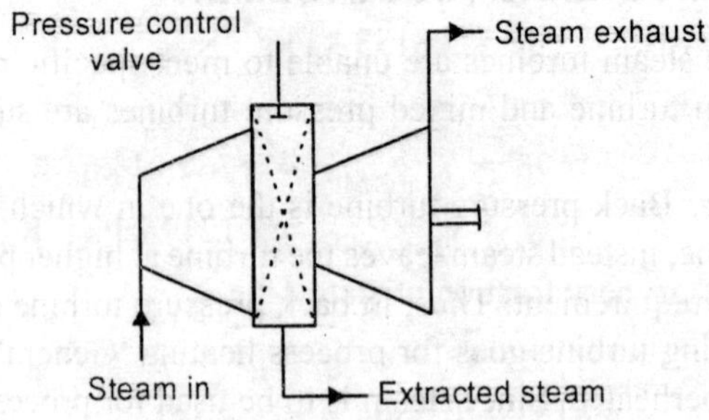

Fig. 8.26 Pass out turbine

The pass out turbines have to operate under widely varying load so its efficiency is quite poor. For facilitating the operation of pass out turbine from no extraction to full steam extraction conditions, nozzle control geverning or throttle control governing are used.

(c) *Mixed pressure turbine:* These are the turbines which have capability of admitting steam at more than one pressures and subjecting multiple pressure steam streams to expand. Generally, mixed pressure turbines utilize high pressure steam from a boiler and also low pressure steam from exhaust of a non-condensing engine or some auxiliary of the plant.

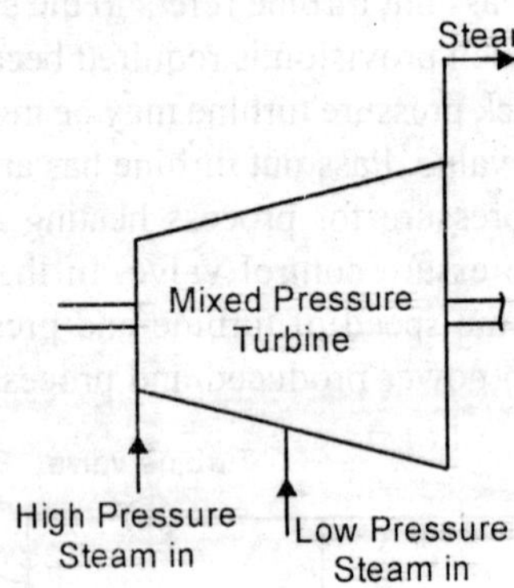

Fig 8.27 Mixed pressure turbine

Mixed pressure turbines are preferred when steam at single pressure is not available in desired quantity for producing required power. These mixed pressure turbines actually have more than one turbines in one cylinder.

EXAMPLES

1. A Carnot cycle works on steam between the pressure limits of 7 MPa and 7 kPa. Determine thermal efficiency, turbine work and compression work per kg of steam.

Solution:

T-s representation for the Carnot cycle operating between pressure of 7 MPa and 7 kPa is shown in Fig. 8.28

$$\text{Enthalpy at state 2, } h_2 = h_{g \text{ at } 7 \text{ MPa}}$$
$$h = 2772.1 \text{ kJ/kg}$$

Entropy at state 2, $s_2 = s_{g \text{ at 7 MPa}}$

$s_2 = 5.8133$ kJ/kg · K

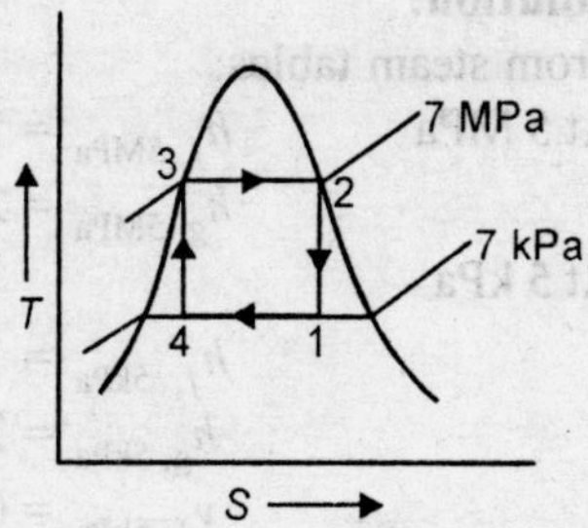

Fig. 8.28

Enthalpy and entropy at state 3,

$$h_3 = h_{f \text{ at 7 MPa}} = 1267 \text{ kJ/kg}$$

$$s_3 = s_{f \text{ at 7 MPa}} = 3.1211 \text{ kJ/kg} \cdot \text{K}$$

For process 2 –1, $s_1 = s_2$. Let dryness fraction at state 1 be x_1.

$$s_1 = s_2 = s_{f \text{ at 7 kPa}} + x_1 \cdot s_{fg \text{ at 7 kPa}}$$

$$5.8133 = 0.5564 + x_1 \cdot 7.7237$$

$$x_1 = 0.6806$$

Enthalpy of state 1, $h_1 = h_{f \text{ at 7 kPa}} + x_1 \cdot h_{fg \text{ at 7 kPa}}$

$$= 162.60 + (0.6806 \times 2409.54)$$

$$h_1 = 1802.53 \text{ kJ/kg}$$

Let dryness fraction at state 4 be x_4,

For process 4–3, $s_4 = s_3 = s_{f \text{ at 7 kPa}} + x_4 \cdot s_{fg \text{ at 7 kPa}}$

$$3.1211 = 0.5564 + x_4 \cdot 7.7237$$

$$x_4 = 0.3321$$

Enthalpy at state 4, $h_4 = h_{f \text{ at 7 kPa}} + x_4 \cdot h_{fg \text{ at 7 kPa}}$

$$= 162.60 + (0.3321 \times 2409.54)$$

$$h_4 = 962.81 \text{ kJ/kg}$$

$$\text{Thermal efficiency} = \frac{\text{Net work}}{\text{Heat added}}$$

Expansion work per kg $= h_2 - h_1 = (2772.1 - 1802.53) = 969.57$ kJ/kg

Compression work per kg $= h_3 - h_4 = (1267 - 962.81)$

$= 304.19$ kJ/kg (+ve)

Heat added per kg $= h_2 - h_3 = (2772.1 - 1267)$

$= 1505.1$ kJ/kg (–ve)

Net work per kg $= (h_2 - h_1) - (h_3 - h_4) = 969.57 - 304.19$

$= 665.38$ kJ/kg

$$\text{Thermal efficiency} = \frac{665.38}{1505.1} = 0.4421 \text{ or } 44.21\%$$

Thermal efficiency = **44.21%**
Turbine work = **969.57 kJ/kg (+ve)**
Compression work = **304.19 kJ/kg (–ve)** **Ans.**

2. *A steam power plant uses steam as working fluid and operates at a boiler pressure of 5 MPa, dry saturated and a condenser pressure of 5 kPa. Determine the cycle efficiency for (a) Carnot cycle (b) Rankine cycle. Also show the T-s representation for both the cycles.*

Solution:

From steam tables:

At 5 MPa $h_{f,\,5MPa} = 1154.23$ kJ/kg, $s_{f,\,5\,MPa} = 2.92$ kJ/kg · K

$h_{g,\,5MPa} = 2794.3$ kJ/kg, $s_{g,\,5\,MPa} = 5.97$ kJ/kg · K

At 5 kPa

$h_{f,\,5kPa} = 137.82$ kJ/kg, $s_{f,\,5kPa} = 0.4764$ kJ/kg · K

$h_{g,\,5kPa} = 2561.5$ kJ/kg, $s_{g,\,5kPa} = 8.3951$ kJ/kg · K

$v_{f,\,5kPa} = 0.001005$ m³/kg

As process 2-3 is isentropic, so $s_2 = s_3$

and

$$s_3 = s_{f,\,5kPa} + x_3 \cdot s_{fg,\,5kPa} = s_2 = s_{g,\,5MPa}$$

$$x_3 = 0.694$$

Hence enthalpy at 3,

$$h_3 = h_{f,\,5kPa} + x_3 \cdot h_{fg,\,5kPa}$$

$$h_3 = 1819.85 \text{ kJ/kg}$$

Enthalpy at 2, $h_2 = h_{g,\,5MPa} = 2794.3$ kJ/kg

Process 1-4 is isentropic, so $s_1 = s_4$

$$s_1 = 2.92 = 0.4764 + x_4 \cdot (8.3951 - 0.4764)$$

$$x_4 = 0.308$$

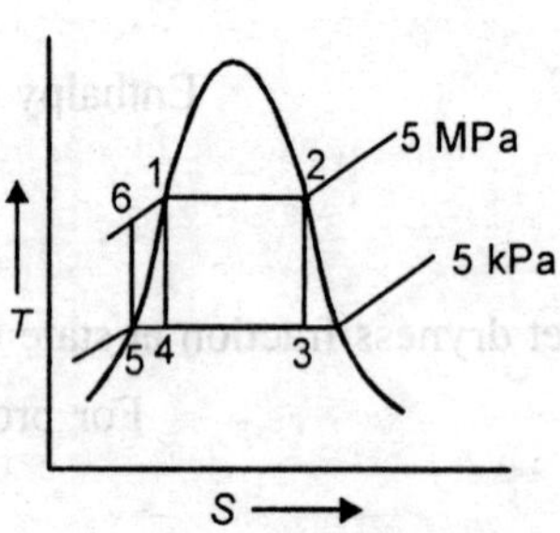

Fig. 8.29

Enthalpy at 4,

$$h_4 = 137.82 + (0.308 \times (2561.5 - 137.82))$$

$$h_4 = 884.3 \text{ kJ/kg}$$

Enthapy at 1,

$$h_1 = h_{f \text{ at } 5\text{ MPa}}$$

$$h_1 = 1154.23 \text{ kJ/kg}$$

Carnot cycle (1-2-3-4-1) efficiency:

$$\eta_{carnot} = \frac{\text{Net work}}{\text{Heat added}}$$

$$= \frac{(h_2 - h_3) - (h_1 - h_4)}{(h_2 - h_1)}$$

$$= \frac{\{(2794.3 - 1819.85) - (1154.23 - 884.3)\}}{(2794.3 - 1154.23)}$$

$$\eta_{carnot} = 0.4295$$

or

η_{carnot} = **42.95%** **Ans.**

In Rankine cycle, 1-2-3-5-6-1

Pump work, $h_6 - h_5 = v_{f,\,5}(p_6 - p_5)$

$= 0.001005\,(5000 - 5)$

$$h_6 - h_5 = 5.02$$
$$h_5 = h_{f \text{ at 5kPa}} = 137.82 \text{ kJ/kg}$$
$$\text{Hence } h_6 = 137.82 + 5.02 = 142.84 \text{ kJ/kg}$$
$$h_6 = 142.84 \text{ kJ/kg}$$
$$\text{Net work in Rankine cycle} = (h_2 - h_3) - (h_6 - h_5)$$
$$= 974.45 - 5.02$$
$$= 969.43 \text{ kJ/kg}$$
$$\text{Heat added} = h_2 - h_6$$
$$= 2794.3 - 142.84$$
$$= 2651.46 \text{ kJ/kg}$$
$$\text{Rankine cycle efficiency} = \frac{969.43}{2651.46}$$
$$\eta_{\text{Rankine}} = 0.3656$$

or

$$\boxed{\eta_{\text{Rankine}} = \mathbf{36.56\%}} \quad \textbf{Ans.}$$

3. A steam turbine plant operates on Rankine cycle with steam entering turbine at 40 bar, 350ºC and leaving at 0.05 bar. Steam leaving turbine condenses to saturated liquid inside condenser. Feed pump pumps saturated liquid into boiler. Determine the net work per kg of steam and the cycle efficiency assuming all processes to be ideal. Also show cycle on T-s diagram. Also determine pump work per kg of steam considering linear variation of specific volume.

Solution:

From steam table

$$h_2 = h_{\text{at 40 bar, 350ºC}} = 3092.5 \text{ kJ/kg}$$
$$s_2 = s_{\text{at 40 bar, 350ºC}} = 6.5821 \text{ kJ/kg}\cdot\text{K}$$
$$h_4 = h_{f \text{ at 0.05 bar}} = 137.82 \text{ kJ/kg}$$
$$s_4 = s_{f \text{ at 0.05 bar}} = 0.4764 \text{ kJ/kg}$$
$$v_4 = v_{f \text{ at 0.05 bar}} = 0.001005 \text{ m}^3\text{/kg}$$

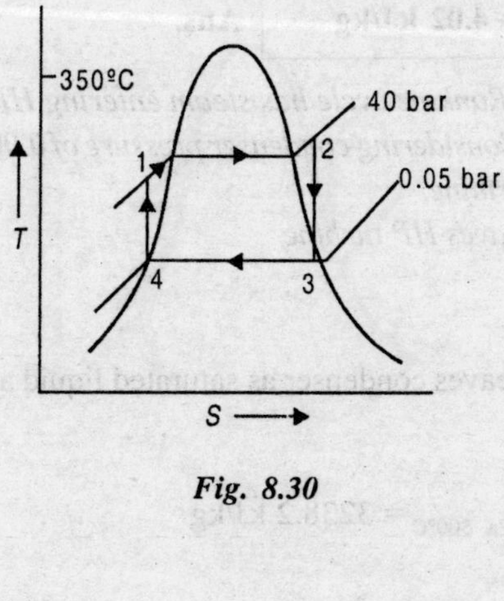

Fig. 8.30

Let dryness fraction at state 3 be x_3,

For ideal process, 2-3, $s_2 = s_3$

$$s_2 = s_3 = 6.5821 = s_{f\text{ at }0.05\text{ bar}} + x_3 \cdot s_{fg\text{ at }0.05\text{ bar}}$$

$$6.5821 = 0.4764 + x_3 \cdot 7.9187$$

$$x_3 = 0.7711$$

$$h_3 = h_{f\text{ at }0.05\text{ bar}} + x_3 \cdot h_{fg\text{ at }0.05\text{ bar}}$$

$$= 137.82 + (0.7711 \times 2423.7)$$

$$h_3 = 2006.74\text{ kJ/kg}$$

For pumping process

$$h_1 - h_4 = v_4 \cdot \Delta p = v_4 \times (p_1 - p_4)$$

$$h_1 = h_4 + v_4 \times (p_1 - p_4)$$

$$= 137.82 + (0.001005 \times (40 - 0.05) \times 10^2)$$

$$h_1 = 141.84\text{ kJ/kg}$$

Pump work per kg of steam $= (h_1 - h_4) = 4.02$ kJ/kg

Net work per kg of steam = (Expansion work – Pump work) per kg of steam

$$= (h_2 - h_3) - (h_1 - h_4)$$

$$= 1081.74\text{ kJ/kg}$$

$$\text{Cycle efficiency} = \frac{\text{Net work}}{\text{Heat added}}$$

$$= \frac{1081.74}{(h_2 - h_1)}$$

$$= \frac{1081.74}{(3092.5 - 141.84)}$$

$$= 0.3667\text{ or }36.67\%$$

Net work per kg of steam	= **1081.74 kJ/kg**
Cycle efficiency	= **36.67%**
Pump work per kg of steam	= **4.02 kJ/kg**

Ans.

4. A steam power plant running on Rankine cycle has steam entering HP turbine at 20 MPa, 500°C and leaving LP turbine at 90% dryness. Considering condenser pressure of 0.005 MPa and reheating occurring upto the temperature of 500°C determine,

(a) the pressure at wich steam leaves HP turbine

(b) the thermal efficiency

Solution:

Let us assume that the condensate leaves condenser as saturated liquid and the expansion in turbine and pumping processes are isentropic.

From steam tables,

$$h_2 = h_{\text{at }20\text{ MPa, }500°\text{C}} = 3238.2\text{ kJ/kg}$$

$$s_2 = 6.1401 \text{ kJ/kg} \cdot \text{K}$$

$$h_5 = h_{\text{at } 0.005 \text{ MPa, } 0.90 \text{ dry}}$$

$$h_5 = h_{f \text{ at } 0.005 \text{ MPa,}} + 0.9 \times h_{fg \text{ at } 0.005 \text{ MPa}}$$

$$= 137.82 + (0.9 \times 2423.7)$$

$$h_5 = 2319.15 \text{ kJ/kg}$$

$$s_5 = s_{f \text{ at } 0.005 \text{ MPa,}} + 0.9 \times s_{fg \text{ at } 0.005 \text{ MPa}}$$

$$= 0.4764 + (0.9 \times 7.9187)$$

$$s_5 = 7.6032 \text{ kJ/kg} \cdot \text{K}$$

$$h_6 = h_{f \text{ at } 0.005 \text{ MPa}} = 137.82 \text{ kJ/kg}$$

It is given that temperature at state 4 is 500°C and due to isentropic process $s_4 = s_5 = 7.6032$ kJ/kg ·K. The state 4 can be conveniently located on Mollier chart by the intersection of 500°C constant temperature line and entropy value of 7.6032 kJ/kg · K and the pressure and enthalpy obtained. But these shall be approximate.

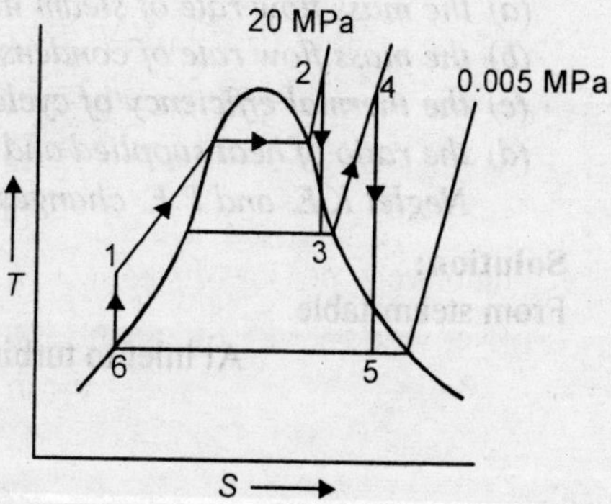

Fig. 8.31

The state 4 can also be located by interpolation using steam table. The entropy value of 7.6032 kJ/kg · K lies between the superheated steam states given under, $p = 1.20$ *MPa*, $s_{\text{at } 1.20 \text{ MPa and } 500°\text{C}} = 7.6027$ kJ/kg · K

$p = 1.40$ *MPa*, $s_{\text{at } 1.40 \text{ MPa and } 500°\text{C}} = 7.6027$ kJ/kg · K

By interpolation state 4 lies at pressure

$$= 1.20 + \frac{(1.40 - 1.20)}{(7.6027 - 7.6759)}(7.6032 - 7.6759)$$

$$= 1.399 \text{ MPa} \approx 1.40 \text{ MPa}$$

Thus, steam leaves HP turbine at 1.4 MPa

Enthalpy at state 4, $h_4 = 3474.1$ kJ/kg

For process 2-3, $s_2 = s_3 = 6.1401$ kJ/kg · K. The state 3 thus lies in wet region as $s_3 < s_{g \text{ at } 1.40 \text{ MPa.}}$ Let dryness fraction at state 3 be x_3.

$$s_3 = s_{f \text{ at } 1.4 \text{ MPa}} + x_3 \cdot s_{fg \text{ at } 1.4 \text{ MPa}}$$

$$6.1401 = 2.2842 + x_3 \cdot 4.1850$$

$$x_3 = 0.9214$$

$$h_3 = h_{f \text{ at } 1.4 \text{ MPa}} + x_3 \cdot h_{fg \text{ at } 1.4 \text{ MPa}}$$

$$= 830.3 + (0.9214 \times 1959.7) = 2635.97 \text{ kJ/kg}$$

Enthalpy at 1, $h_1 = h_6 + v_6(p_1 - p_6)$

$$= h_{f \text{ at } 0.005 \text{ MPa}} + v_{f \text{ at } 0.005 \text{ MPa}} (20 - 0.005) \times 10^3$$

$$= 137.82 + (0.001005 \times 19.995 \times 10^3)$$

$$h_1 = 157.91 \text{ kJ/kg}$$

Net work per kg of steam $= (h_2 - h_3) + (h_4 - h_5) - (h_1 - h_6)$

$$= 1737.09 \text{ kJ/kg}$$

Heat added per kg of steam $= (h_2 - h_1) = 3080.29$ kJ/kg

$$\text{Thermal efficiency} = \frac{\text{Net work}}{\text{Heat added}} = \frac{1737.09}{3080.29} = 0.5639 \text{ or } 56.39\%$$

Pressure of steam leaving HP turbine = **1.40 MPa**	**Ans.**
Thermal efficiency = **56.39%**	

5. In a steam turbine installation running on ideal Rankine cycle steam leaves the boiler at 10 MPa and 700°C and leaves turbine at 0.005 MPa. For the 50 MW output of the plant and cooling water entering and leaving condenser at 15°C and 30°C respectively determine

(a) the mass flow rate of steam in kg/s
(b) the mass flow rate of condenser cooling water in kg/s
(c) the thermal efficiency of cycle
(d) the ratio of heat supplied and rejected (in boiler and condenser respectively). Neglet K.E. and P.E. changes.

Solution:
From steam table

At inlet to turbine, $h_2 = h_{\text{at 10 MPa, 700°C}}$
$h_2 = 3870.5$ kJ/kg
$s_2 = 7.1687$ kJ/kg · K

For process 2-3, $s_2 = s_3$ and $s_3 < s_{f\text{ at 0.005 MPa}}$ so state 3 lies in wet region. Let dryness fraction at state 3 be x_3.

$$s_3 = 7.1687 = s_{f\text{ at 0.005 MPa}} + x_3 \cdot s_{fg\text{ at 0.005 MPa}}$$
$$7.1687 = 0.4764 + (x_3 \times 7.9187)$$
$$x_3 = 0.845$$
$$h_3 = h_{f\text{ at 0.005 MPa}} = x_3 \cdot h_{fg\text{ at 0.005 MPa}}$$
$$= 137.82 + (0.845 \times 2423.7)$$
$$h_3 = 2185.85 \text{ kJ/kg}$$
$$h_4 = h_{f\text{ at 0.005 MPa}} = 137.82 \text{ kJ/kg}$$

For pumping process, $(h_1 - h_4) = v_4 \times (p_1 - p_4)$

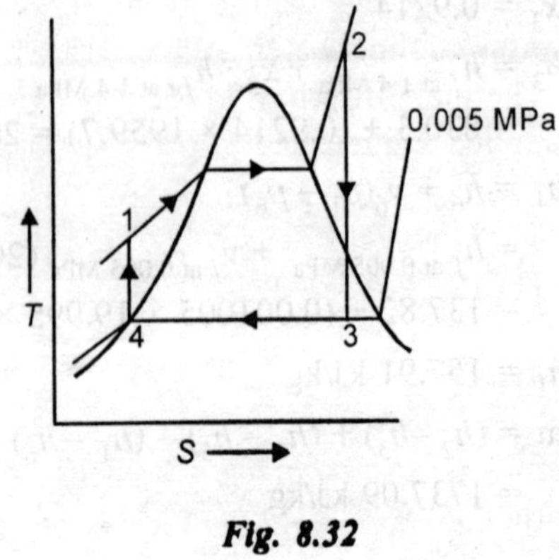

Fig. 8.32

$$v_4 = v_{f \text{ at } 0.005\text{ MPa}} = 0.001005 \text{ m}^3/\text{kg}$$

$$h_1 = 137.82 + (0.001005 \times (10 - 0.005)) \times 10^2$$

$$h_1 = 138.82 \text{ kJ/kg}$$

Net output per kg of steam, $w_{net} = (h_2 - h_3) - (h_1 - h_4)$

$$= (3870.5 - 2185.85) - (138.82 - 137.82)$$

$$w_{net} = 1683.65 \text{ kJ/kg}$$

Mass flow rate of steam, $m_s = \dfrac{50 \times 10^3}{1683.65} = 29.69$ kg/s

By heat balance on condenser, for mass flow rate of water being m_w kg/s.

$$(h_3 - h_4) \times m_s = m_w \cdot C_{p,\,w}\,(T_{w,\text{ out}} - T_{w,\text{ in}})$$

$$29.69 \times (2185.85 - 137.82) = m_w \times 4.18\,(15)$$

$$m_w = 969.79 \text{ kg/s}$$

The heat added per kg of steam $q_{add} = (h_2 - h_1) = 3731.68$ kJ/kg

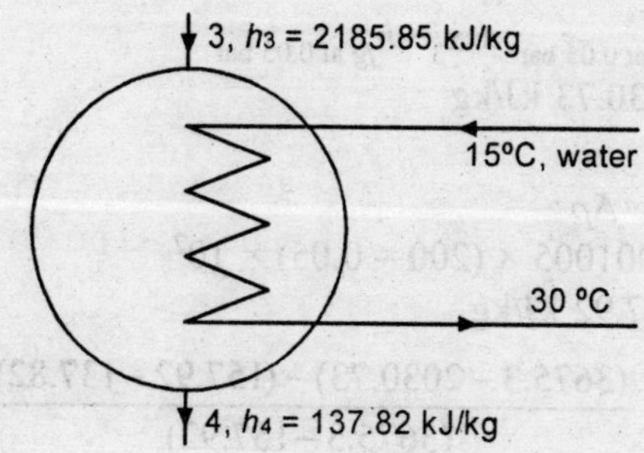

Fig. 8.33

Thermal efficiency $= \dfrac{w_{net}}{q_{add}} = \dfrac{1683.65}{3731.68} = 0.4512$ or 45.12%

Ratio of heat supplied and rejected $= \dfrac{(h_2 - h_1)}{(h_3 - h_4)} = 1.822$

Mass of flow rate of steam = **29.69 kg/s**
Mass flow rate of condenser cooling water = **969.79 kg/s**
Thermal efficiency = **45.12%**
Ratio of heat supplied and rejected = **1.822**

Ans.

6. A regenerative Rankine cycle has steam entering turbine at 200 bar, 650ºC and leaving at 0.05 bar. Considering feed water heaters to be of open type determine thermal efficiency for the following conditions;

(a) there is no feed water heater

(b) there is only one feed water heater working at 8 bar

(c) there are two feed water heaters working at 40 bar and 4 bar respectively.

Also give layout T-s representation for each of the case described above.

Solution:

Case (a) When there is no feed water heater

$$\text{Thermal efficiency of cycle} = \frac{(h_2 - h_3) - (h_1 - h_4)}{(h_2 - h_1)}$$

From steam tables,

$$h_2 = h_{\text{at 200 bar, 650°C}} = 3675.3 \text{ kJ/kg}$$
$$s_2 = s_{\text{at 200 bar, 650°C}} = 6.6582 \text{ kJ/kg} \cdot \text{K}$$
$$h_4 = h_{f \text{ at 0.05 bar}} = 137.82 \text{ kJ/kg}$$
$$v_4 = v_{f \text{ at 0.05 bar}} = 0.001005 \text{ m}^3\text{/kg.}$$
$$h_{f \text{ at 0.05 bar}} = 137.82 \text{ kJ/kg}, h_{fg \text{ at 0.05 bar}} = 2423.7 \text{ kJ/kg}$$
$$s_{f \text{ at 0.05 bar}} = 0.4764 \text{ kJ/kg} \cdot K, s_{fg \text{ at 0.05 bar}} = 7.9187 \text{ kJ/kg} \cdot \text{K}$$

For process 2 – 3, $s_2 = s_3$. Let dryness fraction at 3 be x_3.

$$s_3 = 6.6582 = s_{f \text{ at 0.05 bar}} + x_3 \cdot s_{fg \text{ at 0.05 bar}}$$
$$6.6582 = 0.4764 + x_3 \cdot 7.9187$$
$$x_3 = 0.781$$
$$h_3 = h_{f \text{ at 0.05 bar}} + x_3 \cdot h_{fg \text{ at 0.05 bar}}$$
$$= 2030.73 \text{ kJ/kg}$$

For pumping process 4-1,

$$h_1 - h_4 = v_4 \cdot \Delta p$$
$$h_1 - 137.82 = 0.001005 \times (200 - 0.05) \times 10^2$$
$$h_1 = 157.92 \text{ kJ/kg}$$

$$\text{Thermal efficiency of cycle} = \frac{(3675.3 - 2030.73) - (157.92 - 137.82)}{(3675.3 - 157.92)}$$

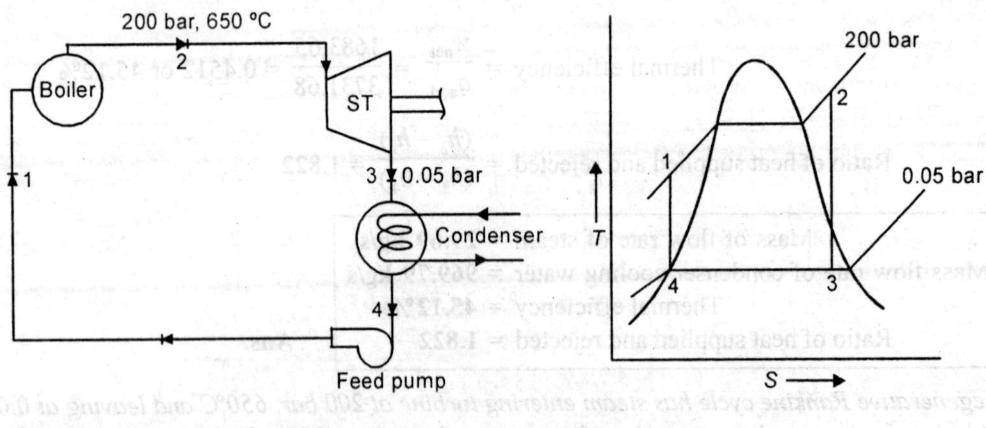

Fig. 8.34 Layout and T-s diagram, (Q 6.a)

$= 0.4618$ or 46.18%

Case (b) When there is only one feed water heater working at 8 bar

Here, let mass of steam bled for feed heating be m kg

For process 2-6, $s_2 = s_6 = 6.6582$ kJ/kg · K

Let dryness fraction at state 6 be x_6

$$s_6 = s_{f\text{ at 8 bar}} + x_6 \cdot s_{fg\text{ at 8 bar}}$$

From steam tables, $h_{f\text{ at 8 bar}} = 721.11$ kJ/kg

$$v_{f\text{ at 8 bar}} = 0.001115\ m^3/\text{kg},\ h_{fg\text{ at 8 bar}} = 2048\ \text{kJ/kg}$$

$$s_{f\text{ at 8 bar}} = 2.0462\ \text{kJ/kg}\cdot\text{K},\ s_{fg\text{ at bar}} = 4.6166\ \text{kJ/kg}\cdot\text{K}$$

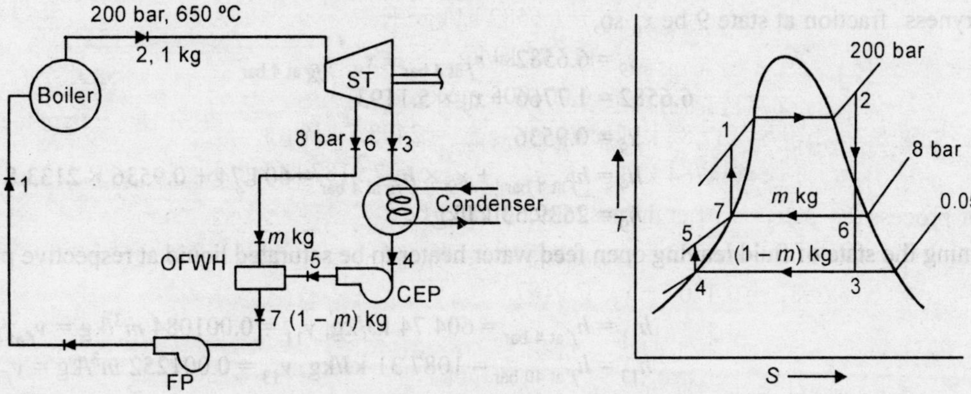

CEP = Condensate Extraction Pump
FP = Feed Pump
ST = Steam Turbine
OFWH = Open Feed Water Heater

Fig. 8.35 Layout and T-s diagram, (Q 6,b)

Substituting entropy values, $x_6 = 0.999$

$h_6 = h_{f\text{ at 8 bar}} + x_6 \cdot h_{fg\text{ at 8 bar}} = 721.11 + (0.999 \times 2048) = 2767.06$ kJ/kg

Assuming the state of fluid leaving open feed water heater to be saturated liquid at 8 bar. $h_7 = h_{f\text{ at 8 bar}} = 721.11$ kJ/kg.

For process 4-5, $h_5 = h_4 + v_4 \times (8 - 0.05) \times 10^2 = 137.82 + (0.001005 \times 7.95 \times 10^2) = 138.62$ kJ/kg

Applying energy balance at open feed water heater,

$$m \times h_6 + (1 - m) \times h_5 = 1 \times h_7$$

$$(m \times 2767.06) + ((1 - m) \times 138.62) = 721.11$$

$$m = 0.2216\ \text{kg}$$

For process 7-1, $h_1 = h_7 + v_7\,(200 - 8) \times 10^2$; here $h_7 = h_{f\text{ at 8 bar}},\ v_7 = v_{f\text{ at 8 bar}}$

$= h_7 + v_{f\text{ at 8 bar}}\,(200 - 8) \times 10^2 = 721.11 + (0.001115 \times 192 \times 10^2)$

$$h_1 = 742.518\ \text{kJ/kg}$$

$$\text{Thermal efficiency of cycle} = \frac{(h_2 - h_6) + (1 - m)\cdot(h_6 - h_3) - \{(1 - m)(h_5 - h_4) + (h_1 - h_7)\}}{(h_2 - h_1)}$$

$$= \frac{(3675.3 - 2767.06) + (1 - 0.2216) \times (2767.06 - 2030.73) - \{(1 - 0.2216) \times (138.62 - 137.82) + (742.518 - 721.11)\}}{(3675.3 - 742.518)}$$

Thermal efficiency of cycle = 0.4976 or 49.76%

Case (c) When there are two feed water heaters working at 40 bar and 4 bar

Here, let us assume the mass of steam at 40 bar, 4 bar to be m_1 kg, and m_2 kg respectively.

For process 2–10–9–3, $s_2 = s_{10} = s_9 = s_3 = 6.6582$ kJ/kg ·K

At state 10. $s_{10} > s_{g \text{ at 40 bar}}$ (6.0701 kJ/kg · K) so state 10 lies in superheated region at 40 bar pressure.

From steam table by interpolation, $T_{10} = 370.36°C$ so, $h_{10} = 3141.81$ kJ/kg

Let dryness fraction at state 9 be x_9 so,

$$s_9 = 6.6582 = s_{f \text{ at 4 bar}} + x_9 \cdot s_{fg \text{ at 4 bar}}$$

$$6.6582 = 1.7766 + x_9 \times 5.1193$$

$$x_9 = 0.9536$$

$$h_9 = h_{f \text{ at 4 bar}} + x_9 \times h_{fg \text{ at 4 bar}} = 604.74 + 0.9536 \times 2133.8$$

$$h_9 = 2639.53 \text{ kJ/kg}$$

Assuming the state of fluid leaving open feed water heater to be saturated liquid at respective pressures i.e.

$$h_{11} = h_{f \text{ at 4 bar}} = 604.74 \text{ kJ/kg}, v_{11} = 0.001084 \text{ m}^3\text{/kg} = v_{f \text{ at 4 bar}}$$

$$h_{13} = h_{f \text{ at 40 bar}} = 1087.31 \text{ kJ/kg}, v_{13} = 0.001252 \text{ m}^3\text{/kg} = v_{f \text{ at 40 bar}}$$

For process 4–8, i.e. in CEP.

$$h_8 = h_4 + v_4 \times (4 - 0.05) \times 10^2$$

$$= 137.82 + (0.001005 \times 3.95 \times 10^2)$$

$$h_8 = 138.22 \text{ kJ/kg}$$

For process 11-12, i.e. in FP_2,

$$h_{12} = h_{11} + v_{11}(40–4) \times 10^2$$

$$= 604.74 + (0.001084 \times 36 \times 10^2)$$

$$h_{12} = 608.64 \text{ kJ/kg}$$

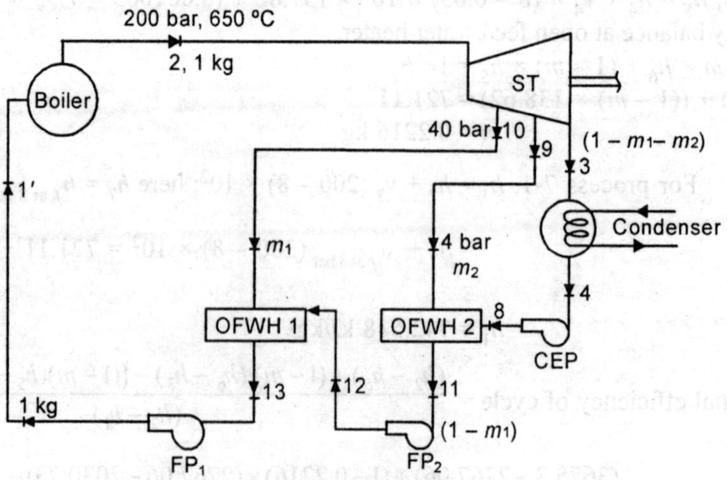

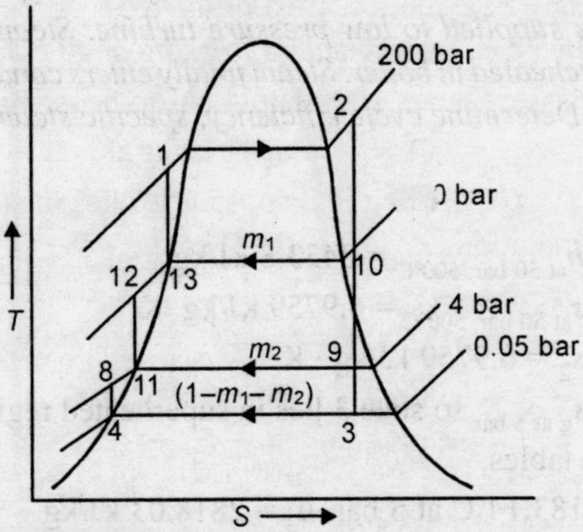

***Fig. 8.36** Layout and T-s diagram. (Q6.c)*

For process 13-1′ i.e. in FP_1, $h'_1 = h_{13} + v_{13}(200 - 40) \times 10^2$

$$= 1087.31 + (0.001252 \times 160 \times 10^2)$$

$$h'_1 = 1107.34 \text{ kJ/kg}$$

$$(m_1 \times 3141.81) + (1 - m_1) \times 608.64 = 1087.31$$

$$m_1 = 0.189 \text{ kg}$$

Applying energy balance an open feed water heater 1 (OFWH1)

$$(m_1 \times h_{10}) + (1 - m_1) \times h_{12} = 1 \times h_{13}$$

$$(m_1 \times 3141.81) + (1 - m_1) \times 608.64 = 1087.31$$

$$m_1 = 0.189 \text{ kg}$$

Applying energy balance an open feed water heater 2 (OFWH2)

$$m_2 \times h_9 + (1 - m_1 - m_2)\, h_8 = (1 - m_1) \times h_{11}$$

$$(m_2 \times 2639.53) + (1 - 0.189 - m_2) \times 138.22 = (1 - 0.189) \times 604.74, \; m_2 = 0.151 \text{ kg}$$

Thermal efficiency of cycle,

$$\eta = \frac{\{(h_2 - h_{10}) + (1 - m_1)(h_{10} - h_9) + (1 - m_1 - m_2)(h_9 - h_3)\} - \{W_{CEP} + W_{FP_1} + W_{FP_2}\}}{(h_2 - h_{1'})}$$

$$W_{CEP} = (1 - m_1 - m_2)\,(h_8 - h_4) = 0.264 \text{ kJ/kg steam from boiler}$$

$$W_{FP_1} = (h_{1'} - h_{13}) = 20.03 \text{ kJ/kg of steam from boiler}$$

$$W_{FP_2} = (1 - m_1)\,(h_{12} - h_{11}) = 3.16 \text{ kJ/kg of steam from boiler}$$

$$W_{CEP} + W_{FP_1} + W_{FP_2} = 23.454 \text{ kJ/kg of steam from boiler}$$

$$\eta = \frac{\{(3675.3 - 3141.81) + (1 - 0.189)(3141.81 - 2639.53) + (1 - 0.189 - 0.151)(2639.53 - 2030.73)\} - \{23.454\}}{(3675.3 - 1107.34)}$$

$$= 0.5137 \text{ or } 51.37\%$$

Cycle thermal efficiency η_a = **46.18%**	
η_b = **49.76%**	
η_c = **51.37%**	**Ans.**

Hence it is obvious that efficiency increases with increase in number of feed heaters.

7. A reheat cycle has steam generated at 50 bar, 500°C for being sent to high pressure turbine and expanded upto 5 bar before supplied to low pressure turbine. Steam enters at 5 bar, 400°C into low pressure turbine after being reheated in boiler. Steam finally enters condenser at 0.05 bar and subsequently feed water is sent to boiler. Determine cycle efficiency, specific steam consumption and work ratio.

Solution:

From steam table,

$$h_2 = h_{\text{at 50 bar, 500°C}} = 3433.8 \text{ kJ/kg}$$
$$s_2 = s_{\text{at 50 bar, 500°C}} = 6.9759 \text{ kJ/kg·K}$$
$$s_3 = s_2 = 6.9759 \text{ kJ/kg·K}$$

Since $s_3 > s_{g \text{ at 5 bar}}$ so state 3 lies in superheated region at 5 bar,

By interpolation from steam tables,

$$T_3 = 183.14°\text{C at 5 bar}, h_3 = 2818.03 \text{ kJ/kg}$$
$$h_4 = h_{\text{at 5 bar, 400°C}} = 3271.9 \text{ kJ/kg}$$
$$s_4 = s_{\text{at 5 bar, 400°C}} = 7.7938 \text{ kJ/kg·K}$$

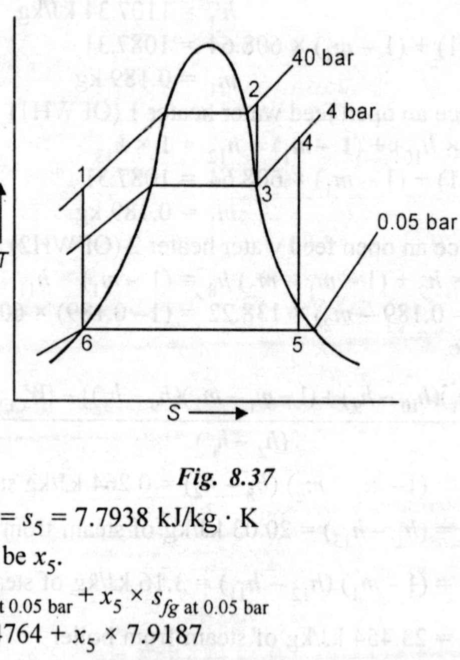

Fig. 8.37

For expansion process 4-5, $s_4 = s_5 = 7.7938$ kJ/kg · K

Let dryness fraction at state 5 be x_5.

$$s_5 = s_{f \text{ at 0.05 bar}} + x_5 \times s_{fg \text{ at 0.05 bar}}$$
$$7.7938 = 0.4764 + x_5 \times 7.9187$$
$$x_5 = 0.924$$
$$h_5 = h_{f \text{ at 0.05 bar}} + x_5 \times h_{fg \text{ at 0.05 bar}}$$
$$h_5 = 137.82 + 0.924 \times 2423.7 = 2377.32 \text{ kJ/kg}$$
$$h_6 = h_{f \text{ at 0.05 bar}} = 137.82 \text{ kJ/kg}$$
$$h_6 = v_{f \text{ at 0.05 bar}} = 0.001005 \text{ m}^3/\text{kg}$$

For process 6-1 in feed pump, $h_1 = h_6 + v_6 \times (50 - 0.05) \times 10^2$

$$h_1 = 137.82 + 0.001005 \times (49.95 \times 10^2)$$
$$h_1 = 142.84 \text{ kJ/kg}$$

$$\text{Cycle efficiency} = \frac{W_{net}}{Q_{add}}$$

$$W_T = (h_2 - h_3) + (h_4 - h_5)$$

$$= (3433.8 - 2818.03) + (3271.9 - 2377.32)$$

$$= 1510.35 \text{ kJ/kg}$$

$$W_{pump} = (h_1 - h_6) = 142.84 - 137.82 = 5.02 \text{ kJ/kg}$$

$$W_{net} = W_T - W_{pump} = 1505.33 \text{ kJ/kg}$$

$$Q_{add} = (h_2 - h_1) = 3433.8 - 142.84 = 3290.96 \text{ kJ/kg}$$

$$\text{Cycle efficiency} = \frac{1505.33}{3290.96} = 0.4574 \text{ or } 45.74\%$$

We know, 1 hp = 0.7457 kW

$$\text{Specific steam consumption} = \frac{0.7457 \times 3600}{1505.33} = 1.78 \text{ kg/hp} \cdot \text{hr}$$

$$\text{Work ratio} = \frac{\text{Net work}}{\text{Positive work}} = \frac{W_{net}}{W_T} = \frac{1505.33}{1510.35} = 0.9967$$

Cycle efficiency = 45.74%, Specific steam consumption = **1.78 kg/hp. hr.**
Work ratio = **0.9967.** **Ans.**

8. *In a steam power plant the high pressure turbine is fed with steam at 60 bar, 450°C and enters low pressure turbine at 3 bar with a portion of steam bled out for feed heating at this intermediate pressure. Steam finally leaves low pressure turbine at 0.05 bar for inlet to condenser. Closed feed heater raises the condensate temperature to 115°C. Bled steam leaving closed feed heater is passed through trap to mix with condensate leaving condenser. Consider actual alternator output to be 30 MW, boiler efficiency as 90% and alternator efficiency of 98%. Determine,*

(a) the mass of steam bled for feed heating,

(b) the capacity of boiler in kg/hr.

(c) the overall thermal efficiency of plant

Also give layout and T-s diagram.

Solution:

From steam tables,

At state 2, $h_2 = 3301.8$ kJ/kg, $s_2 = 6.7193$ kJ/kg · K

$h_5 = h_{f \text{ at } 0.05 \text{ bar}} = 137.82$ kJ/kg, $v_5 = v_{f \text{ at } 0.05 \text{ bar}} = 0.001005$ m^3/kg

Let mass of steam bled for feed heating be m kg/kg of steam generated in boiler. Let us also assume that condensate leaves closed feed water heater as saturated liquid i.e.

$$h_8 = h_{f \text{ at } 3 \text{ bar}}$$

$$h_8 = 561.47 \text{ kJ/kg}$$

For process 2–3–4, $s_2 = s_3 = s_4 = 6.7193$ kJ/kg · K

Let dryness fraction at state 3 and state 4 be x_3 and x_4 respectively.

$$s_3 = 6.7193 = s_{f \text{ at } 3 \text{ bar}} + x_3 . s_{fg \text{ at } 3 \text{ bar}}$$

$$= 1.6718 + x_3 \times 5.3201$$
$$x_3 = 0.949$$
$$s_4 = 6.7193 = s_{f\text{ at }0.05\text{ bar}} + x_4 \cdot s_{fg\text{ at }0.05\text{ bar}}$$
$$= 0.4764 + x_4 \times 7.9187$$
$$x_4 = 0.788$$

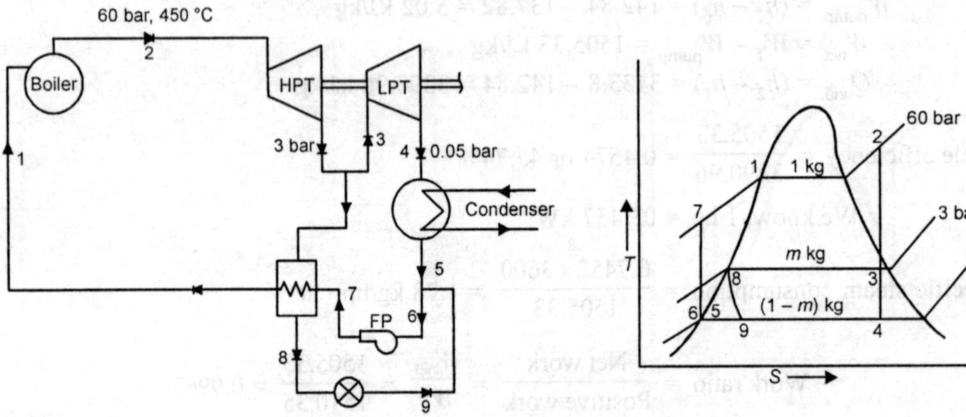

Fig. 8.38 Layout and T-s diagram

Thus, $h_3 = h_{f\text{ at }3\text{ bar}} + x_3 \cdot h_{fg\text{ at }3\text{ bar}} = 561.47 + (0.949 \times 2163.8) = 2614.92$ kJ/kg

$h_4 = h_{f\text{ at }0.05\text{ bar}} + x_4 \cdot h_{fg\text{ at }0.05\text{ bar}} = 137.82 + (0.788 \times 2423.7) = 2047.6$ kJ/kg

Assuming process across trap to be of throttling type so, $h_8 = h_9 = 561.47$ kJ/kg. Assuming $v_5 = v_6$,

Pumping work $(h_7 - h_6) = v_5 \cdot (60 - 0.05) \times 10^2$

$(h_7 - h_6) = 6.02$ kJ/kg

For mixing process between condenser and feed pump,

$$(1 - m) \cdot h_5 + m \cdot h_9 = 1 \cdot h_6$$
$$(1 - m) \times 137.82 + m \times 561.47 = h_6$$
$$h_6 = 137.82 + m \times 423.65$$

Therefore, $h_7 = h_6 + 6.02 = 143.84 + m \times 423.65$

Applying energy balance at closed feed water heater;

$$m \times h_3 + (1 - m)h_7 = m \cdot h_8 + (4.18 \times 115)$$
$$(m \times 2614.92) + (1 - m)(143.84 + m \times 423.65) = m \times 561.47 + 480.7, m = 0.144 \text{ kg}$$

Steam bled for feed heating = **0.144 kg/kg steam generated.**

Ans.

The net power output $W_{net} = (h_2 - h_3) + (1 - m)(h_3 - h_4) - (1 - m) \times (h_7 - h_6)$

$= 1167.28$ kJ/kg steam generated

Mass of steam required to be generated $= \dfrac{30 \times 10^3}{0.98 \times W_{net}}$

$$= \frac{30 \times 10^3}{0.98 \times 1167.28} = 26.23 \text{ kg/s}$$

$$= 94428 \text{ kg/hr}$$

Capacity of boiler required = **94428 kg/hr**	**Ans.**

$$\text{Overall thermal efficiency} = \frac{W_{net}}{Q_{add}}$$

$$Q_{add} = \frac{(h_2 - h_1)}{0.90} = \frac{(3301.8 - 4.18 \times 115)}{0.90}$$

$$= 3134.56 \text{ kJ/kg}$$

$$\text{Overall thermal efficiency} = \frac{1167.28}{3134.56} = 0.3724 \text{ or } 37.24\%$$

Overall thermal efficiency = **37.24%**	**Ans.**

9. *A steam power plant has expansion occurring stages in three stages with steam entering first stage at 30 bar, 400°C and leaving first stage at 6 bar for being sent to second stage with some steam being bled out for feed heating in closed feed water heater. Steam leaves second stage at 1 bar and enters third stage with some more steam being bled out for feed heating in closed feed water heater. Steam finally leaves third stage at 0.075 bar after complete expansion and ent condenser. Condensate temperature is 38°C at inlet to second heater, temperature of feed water after first heater and second heater is 150°C and 95°C respectively. Bled steam is condensed to saturated liquid with no undercooling in each of feed heater. Drain from first heater is passed through steam trap into second feed heater and combined drain from second heater is pumped by drain pump into feed line after second heater. Considering efficiency ratio of turbine as 0.8 and turbine output of 15 MW determine the capacity of drain pump. Neglect drain pump work.*

Solution:

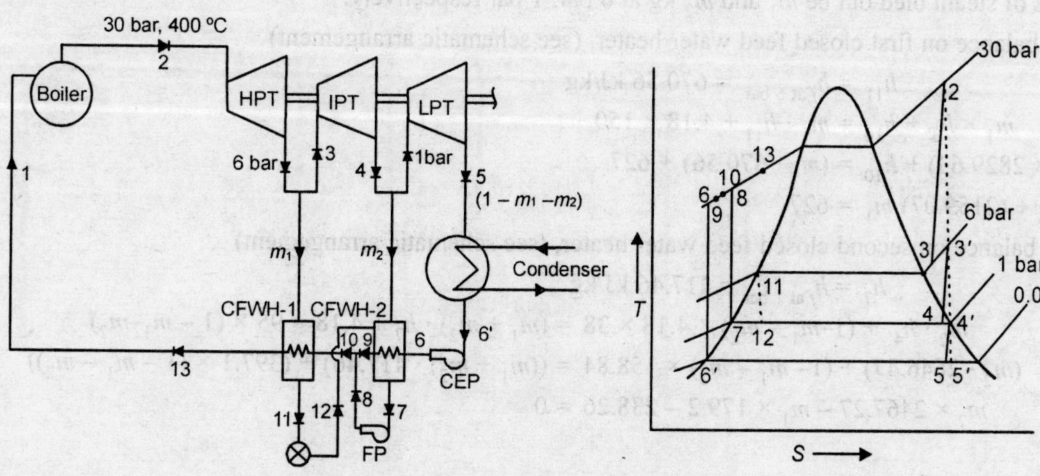

Fig. 8.39 Layout and T-s diagram

At inlet to first turbine stage, $h_2 = 3230.9$ kJ/kg, $s_2 = 6.9212$ kJ/kg ·K

For ideal expansion process $s_2 = s_3$

By interpolation, $T_3 = 190.97$°C from superheated steam tables at 6 bar

$$h_3 = 2829.63 \text{ kJ/kg}$$

actual state at exit of first stage $h'_3 = h_2 - 0.8 \times (h_2 - h_3)$

$$h'_3 = 2909.88 \text{ kJ/kg}$$

Actual state 3′ shall be at 232.78°C, 6 bar, so $s'_3 = 7.1075$ kJ/kg · K

For second stage $s'_3 = s_4$; By interpolation, $s_4 = 7.1075$

$$= s_{f\text{ at 1 bar}} + x_4 \cdot s_{fg\text{ at 1 bar}}$$

$$7.1075 = 1.3026 + x_4 \cdot 6.0568$$

$$x_4 = 0.958$$

$$h_4 = h_{f\text{ at 1 bar}} + x_4 \cdot h_{fg\text{, at 1 bar}}$$

$$= 417.46 + (0.958 \times 2258)$$

$$h_4 = 2580.62 \text{ kJ/kg}$$

Actual enthalpy at exit from second stage, $h_{4'} = h_{3'} - 0.8\,(h_{3'} - h_4)$

$$h_{4'} = 2646.47 \text{ kJ/kg}$$

Actual dryness fraction, $x_{4'} \Rightarrow h_{4'} = h_{f\text{ at 1 bar}} + x_{4'} \cdot h_{fg\text{ at 1 bar}}$

$x_{4'} = 0.987$, Actual entropy, $s_{4'} = 7.2806$ kJ/kg · K

For third stage, $s_{4'} = s_5 = 7.2806 = s_{f\text{ at 0.075 bar}} + x_5 \cdot s_{fg\text{ at 0.075 bar}}$

$$x_5 = 0.8735$$

$$h_5 = 2270.43 \text{ kJ/kg}$$

Actual enthalpy at exit from third stage, $h_{5'} = h_{4'} - 0.8(h_{4'} - h_5)$

$$h_{5'} = 2345.64 \text{ kJ/kg}$$

Let mass of steam bled out be m_1 and m_2 kg at 6 bar, 1 bar respectively.

By heat balance on first closed feed water heater, (see schematic arrangement)

$$h_{11} = h_{f\text{ at 6 bar}} = 670.56 \text{ kJ/kg}$$

$$m_1 \times h_{3'} + h_{10} = m_1 \cdot h_{11} + 4.18 \times 150$$

$$(m_1 \times 2829.63) + h_{10} = (m_1 \cdot 670.56) + 627$$

$$h_{10} + (2159.07)\, m_1 = 627$$

By heat balance on second closed feed water heater, (see schematic arrangement)

$$h_7 = h_{f\text{ at 1 bar}} = 417.46 \text{ kJ/kg}$$

$$m_2 \cdot h_{4'} + (1 - m_1 - m_2) \times 4.18 \times 38 = (m_1 + m_2) \cdot h_7 + 4.18 \times 95 \times (1 - m_1 - m_2)$$

$$(m_2 \cdot 2646.47) + (1 - m_1 - m_2) \times 158.84 = ((m_1 + m_2) \cdot 417.46) + (397.1 \times (1 - m_1 - m_2))$$

$$m_2 \times 2467.27 - m_1 \times 179.2 - 238.26 = 0$$

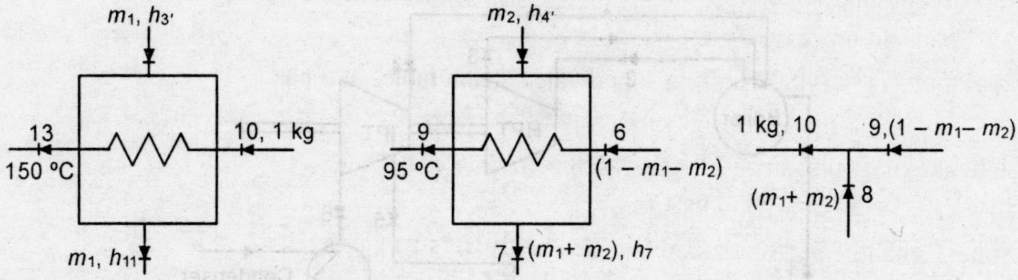

Heat balance at point of mixing.

$$h_{10} = (m_1 + m_2) \cdot h_8 + (1 - m_1 - m_2) \times 4.18 \times 95$$

Neglecting pump work, $h_7 = h_8$

$$h_{10} = m_2 \times 417.46 + (1 - m_1 - m_2) \times 397.1$$

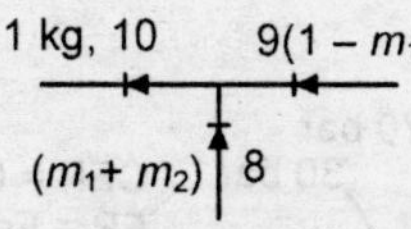

Substituting h_{10} and solving we get, $m_1 = 0.1293$ kg

$$m_2 = 0.1059 \text{ kg/kg of steam generated.}$$

Turbine output per kg of steam generated,

$$w_T = (h_2 - h_{3'}) + (1 - m_1)(h_{3'} - h_{4'}) + (1 - m_1 - m_2) \cdot (h_{4'} - h_{5'})$$

$$w_T = 780.446 \text{ kJ/kg of steam generated.}$$

$$\text{Rate of steam generation required} = \frac{15 \times 10^3}{780.446} = 19.22 \text{ kg/s}$$

or $= 69192$ kg/hr

Capacity of drain pump i.e. FP shown in layout $= (m_1 + m_2) \times 69192$

$= 16273.96$ kg/hr

Capacity of drain pump = **16273.96 kg/hr**	**Ans.**

10. *A steam power plant has steam entering at 70 bar, 450°C into HP turbine. Steam is extracted at 30 bar and reheated upto 400°C before being expanded in LP turbine upto 0.075 bar. Some portion of steam is bled out during expansion in LP turbine so as to yield saturated liquid at 140°C at the exit of open feed water heater. Considering HP and LP turbine efficiencies of 80% and 85% determine the cycle efficiency. Also give layout and T-s diagram.*

Solution:

At inlet to HP turbine, $h_2 = 3287.1$ kJ/kg, $s_2 = 6.6327$ kJ/kg.K

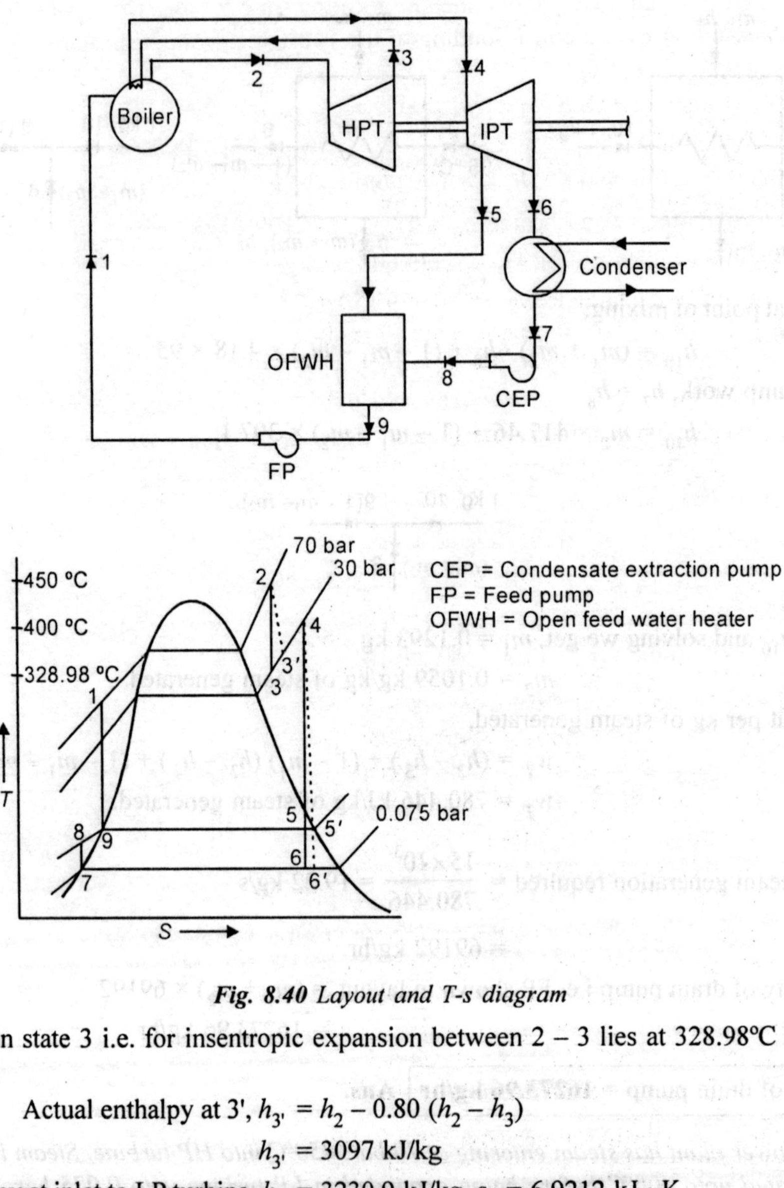

Fig. 8.40 Layout and T-s diagram

By interpolation state 3 i.e. for insentropic expansion between 2 – 3 lies at 328.98°C at 30 bar. h_3 = 3049.48 kJ/kg.

$$\text{Actual enthalpy at 3'},\ h_{3'} = h_2 - 0.80\ (h_2 - h_3)$$

$$h_{3'} = 3097 \text{ kJ/kg}$$

$$\text{Enthalpy at inlet to LP turbine},\ h_4 = 3230.9 \text{ kJ/kg},\ s_4 = 6.9212 \text{ kJ} \cdot \text{K}$$

For ideal expansion from 4-6, $s_4 = s_6$ Let dryness fraction at state 6 be x_6.

$$s_6 = 6.9212 = s_{f \text{ at } 0.075 \text{ bar}} + x_6 \cdot s_{fg \text{ at } 0.075 \text{ bar}}$$

$$x_6 = 0.827$$

$$h_6 = h_{f \text{ at } 0.075 \text{ bar}} + x_6 \cdot h_{fg \text{ at } 0.075 \text{ bar}} = 2158.55 \text{ kJ/kg}$$

For actual expansion process in LP turbine.

$$h_{6'} = h_4 - 0.85\ (h_4 - h_6)$$

$$h_{6'} = 2319.4 \text{ kJ/kg}$$

Ideally, enthalpy at bleed point can be obtained by locating state 5 using $s_5 = s_4$. The pressure at bleed point shall be saturation pressure corresponding to the 140°C i.e. from steam table p_5 = 3.61 bar. Let dryness fraction at state 5 be x_5.

$$s_{5'} = 6.9212 = s_{f\text{ at }140°C} + x_5 \cdot s_{fg\text{ at }140°C} \Rightarrow x_5 = 0.99$$

$$h_5 = h_{f\text{ at }140°C} + x_5 \cdot h_{fg\text{ at }140°C} \Rightarrow h_5 = 2712.38 \text{ kJ/kg}$$

Actual exthalpy $h_{5'} = h_4 - 0.85\,(h_4 - h_5) = 2790.16$ kJ/kg

Enthalpy at exit of open feed water heater, $h_9 = h_{f\text{ at }30\text{ bar}} = 1008.42$ kJ/kg

Specific volume at inlet of CEP, v_7 = 0.001008 m³/kg,

Enthalpy at inlet of CEP, h_7 = 168.79 kJ/kg

For pumping process 7-8

$$h_8 = h_7 + v_7\,(3.61 - 0.075) \times 10^2$$

$$h_8 = 169.15 \text{ kJ/kg}$$

Applying energy balance at open feed water heater. Let mass of bled steam be m kg per kg of steam generated.

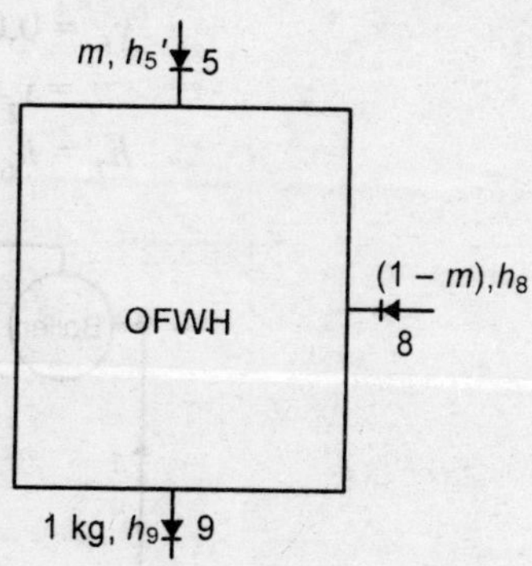

$$m \times h_{5'} + (1 - m) \cdot h_8 = h_9$$

$$(m \times 2790.16) + ((1 - m) \cdot 169.15) = 1008.42$$

$$m = 0.32 \text{ kg/kg of steam generated}$$

For process on feed pump, 9 – 1, $v_9 = v_{f\text{ at }140°C} = 0.00108$

$$h_1 = h_9 + v_9 \times (70 - 3.61) \times 10^2$$

$$h_1 = 1015.59 \text{ kJ/kg}$$

Net work per kg of steam generated,

$$W_{net} = (h_2 - h_{3'}) + (h_4 - h_5') + (1 - m) \cdot (h_{5'} - h_{6'}) - \{(1 - m)(h_8 - h_7) + (h_1 - h_9)\}$$

$$= 181.1 + 440.74 + 320.117 - \{0.2448 + 7.17\}$$

$$W_{net} = 934.54 \text{ kJ/kg steam generated}$$

Heat added per kg of steam generated,

$$q_{add} = (h_2 - h_1) + (h_4 - h_{3'})$$

$$q_{add} = 2262.51 + 133.9 = 2396.41 \text{ kJ/kg of steam generated}$$

Thermal efficiency, $\eta = \dfrac{W_{net}}{q_{add}} = \dfrac{934.54}{2396.41}$

$$\eta = 0.3899 \text{ or } 38.99\%$$

Thermal efficiency = **38.99%** **Ans.**

11. *A steam power plant works on regenerative cycle with steam entering first turbine stage at 150 bar, 500°C and getting expanded in three subsequent stages upto the condenser pressure of 0.05 bar. Some steam is bled out between first and second stage for feed heating in closed feed water heater at 10 bar with the saturated liquid condensate being pumped ahead into the boiler feed water line. Feed water leaves closed feed water heater at 150 bar, 150°C. Steam is also taken out between second and third stages at 1.5 bar for being fed into an open feed water heater working at that pressure. Saturated liquid at 1.5 bar leaves open feed water heater for being sent to closed feed water heater. Considering mass flow rate of 300 kg/s into the first stage of turbine determine cycle thermal efficiency and net power developed in kW. Also give lay out and T-s representation.*

Solution:

Enthalpy of steam entering ST_1, $h_2 = 3308.6$ kJ/kg, $s_2 = 6.3443$ kJ/kg · K

For isentropic expansion 2-3-4-5, $s_2 = s_3 = s_4 = s_5$

Let dryness fraction of states 3, 4 and 5 be x_3, x_4 and x_5

$$s_3 = 6.3443 = s_{f \text{ at } 10 \text{ bar}} + x_3 \cdot s_{fg \text{ at } 10 \text{ bar}}$$

$$\Rightarrow \quad x_3 = 0.945$$

$$h_3 = 2667.26 \text{ kJ/kg}$$

$$s_4 = 6.3443 = s_{f \text{ at } 1.5 \text{ bar}} + x_4 \cdot s_{fg \text{ at } 1.5 \text{ bar}}$$

$$\Rightarrow \quad x_4 = 0.848 \Rightarrow h_4 = 2355.18 \text{ kJ/kg}$$

$$s_5 = 6.3443 = s_{f \text{ at } 0.05 \text{ bar}} + x_5 \cdot s_{fg \text{ at } 0.05 \text{ abr}}$$

$$\Rightarrow \quad x_5 = 0.739 \cdot h_5 = 1928.93 \text{ kJ/kg}$$

$$h_6 = h_{f \text{ at } 0.05 \text{ bar}} = 137.82 \text{ kJ/kg}$$

$$v_6 = 0.001005 \; m^3/\text{kg}$$

$$= v_{f \text{ at } 0.05 \text{ bar}}$$

$$h_7 = h_6 + v_6\,(1.5 - 0.05) \times 10^2 = 137.96 \text{ kJ/kg}$$

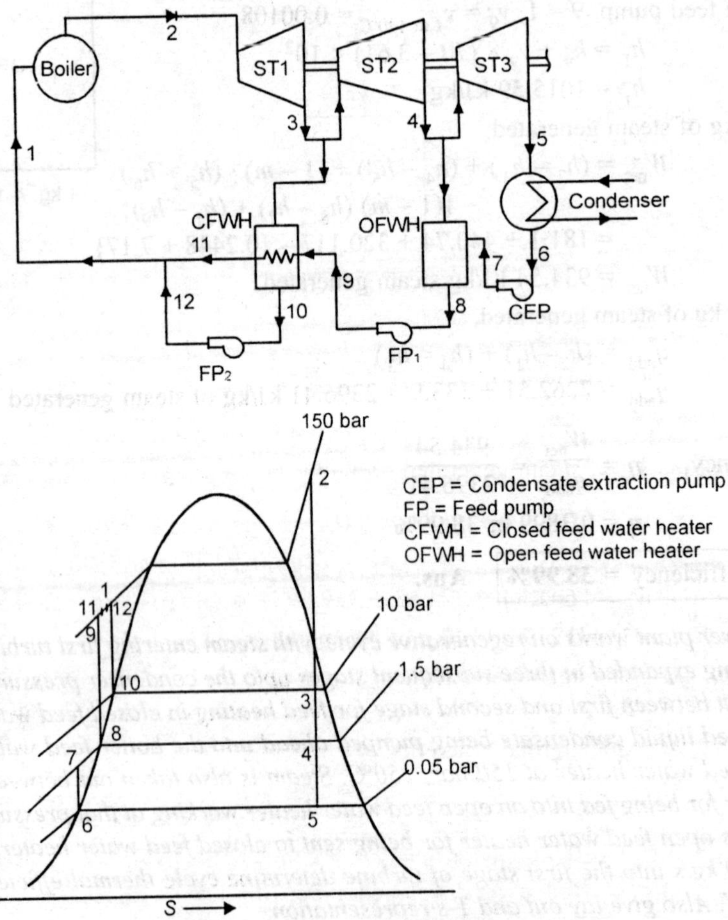

Fig. 8.41 Layout and T-s diagram

$$h_8 = h_{f\text{ at }1.5\text{ bar}} = 467.11 \text{ kJ/kg},$$

$$v_8 = 0.001053 \text{ m}^3/\text{kg} = v_{f\text{ at }1.5\text{ bar}}$$

$$h_9 = h_8 + v_8(150 - 1.5) \times 10^2 = 482.75 \text{ kJ/kg}$$

$$h_{10} = h_{f\text{ at }150\text{ bar}} = 1610.5 \text{ kJ/kg}$$

$$v_{10} = 0.001658 \text{ m}^3/\text{kg} = v_{f\text{ at }150\text{ bar}}$$

$$h_{12} = h_{10} + v_{10}(150 - 10) \times 10^2 = 1633.71 \text{ kJ/kg}$$

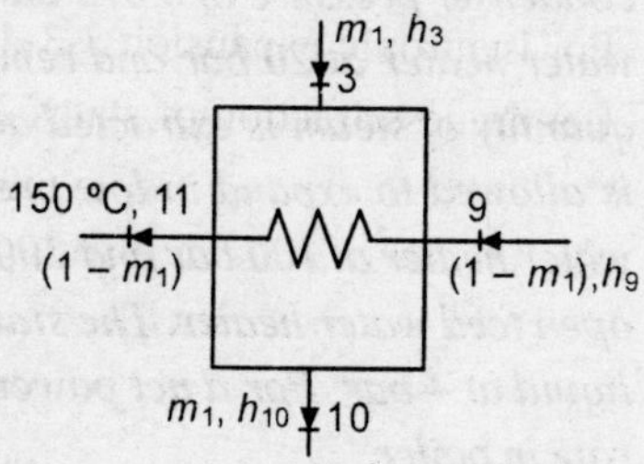

Let mass of steam bled out at 10 bar, 1.5 bar be m_1 and m_2 per kg of steam generated.

Heat balance on closed feed water heater yields,

$$m_1 \cdot h_3 + (1 - m_1)\, h_9 = m_1 \cdot h_{10} + (1 - m_1) \times 4.18 \times 150$$

$$(m_1 \times 2667.26) + (1 - m_1) \times 482.75 = (m_1 \times 1610.5) + (627 \cdot (1 - m_1))$$

$$m_1 = 0.12 \text{ kg/kg of steam generated.}$$

Heat balance on open feed water can be given as under

$$m_2 \cdot h_4 + (1 - m_1 - m_2) \cdot h_7 = (1 - m_1) \cdot h_8$$

$$(m_2 \times 2355.18) + (1 - m_1 - m_2) \times 137.96 = (1 - m_1) \times 467.11$$

$$(m_2 \times 2355.18) + (1 - 0.12 - m_2) \times 137.96 = (1 - 0.12) \times 467.11$$

$$m_2 = 0.13 \text{ kg/kg of steam}$$

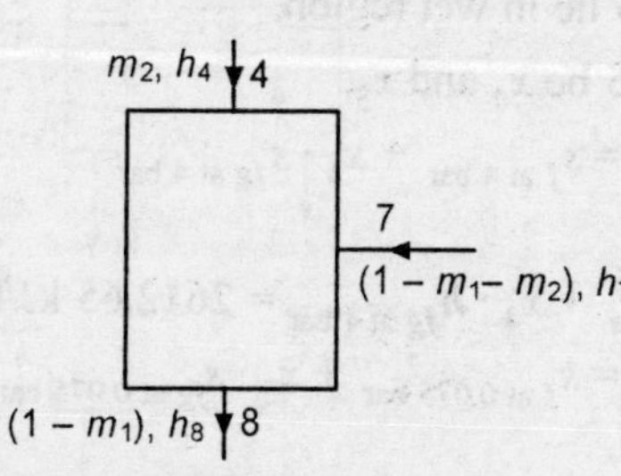

For mass flow rate of 300 kg/s $\Rightarrow m_1 = 36$ kg/s, $m_2 = 39$ kg/s

For mixing after closed feed water heater,

$$h_1 = (4.18 \times 150) \cdot (1 - m_1) + m_1 \times h_{12} = 747.81 \text{ kJ/kg}$$

Net work output per kg of steam generated $= w_{ST_1} + w_{ST_2} + w_{ST_3} - \{w_{CEP} + w_{FP} + w_{FP_2}\}$

$$w_{net} = (h_2 - h_3) + (1 - m_1)(h_3 - h_4) + (1 - m_1 - m_2)(h_4 - h_5) - \{(1 - m_1 - m_2) \cdot (h_7 - h_6) + (1 - m_1) \cdot (h_9 - h_8) + (m_1 \cdot (h_{12} - h_{10}))\}$$

$$w_{net} = 641.34 + 274.63 + 319.69 = \{0.105 + 13.76 + 2.7852\}$$

$$w_{net} = 1219.00 \text{ kJ/kg of steam generated.}$$

Heat added per kg of steam generated. $q_{add} = (h_2 - h_1) = 2560.79$ kJ/kg

$$\text{Cycle thermal efficiency, } \eta = \frac{w_{net}}{q_{add}} = 0.4760 \text{ or } 47.6\%$$

Net power developed in kW $= 1219 \times 300 = 365700$ kW

Cycle thermal efficiency = **47.6%**
Net power developed = **365700 kW**

Ans.

12. A steam power plant has expansion of steam leaving boiler at 100 bar, 500°C occurring in three stages i.e. HPT, IPT and LPT (high pressure, intermediate pressure and low pressure turbine) upto condenser pressure of 0.075 bar. At exit of HPT some steam is extracted for feed heating in closed feed water heater at 20 bar and remaining is sent to IPT for subsequent expansion upto 4 bar. Some more quantity of steam is extracted at 4 bar for feed heating in open feed water heater and remaining steam is allowed to expand in low pressure turbine upto condenser pressure. Feed water leaves closed feed water heater at 100 bar and 200°C. The condensate leaving as saturated liquid at 20 bar is trapped into open feed water heater. The state of liquid leaving open feed water heater may be considered saturated liquid at 4 bar. For a net power output of 100 MW determine thermal efficiency and steam generation rate in boiler.

(a) Modify the above arrangement by introducing reheating of steam entering IPT at 20 bar upto 400°C. Obtain thermal efficiency of modified cycle and compare it with non-reheat type arrangement.

Solution:

At inlet to HPT, $h_2 = 3373.7$ kJ/kg, $s_2 = 6.5966$ kJ/kg · K

For isentropic expansion between 2-3-4-5, $s_2 = s_3 = s_4 = s_5$

State 3 lies in superheated region as $s_3 > s_{g \text{ at } 20 \text{ bar}}$. By interpolation from superheated steam table, $T_3 =$ 261.6°C. Enthalpy at 3. $h_3 = 2930.57$ kJ/kg.

Since $s_4 < s_{g \text{ at } 4 \text{ bar}}$ so states 4 and 5 lie in wet region.

Let dryness fraction at state 4 and 5 be x_4 and x_5.

$$s_4 = 6.5966 = s_{f \text{ at } 4 \text{ bar}} + x_4 \cdot s_{fg \text{ at } 4 \text{ bar}}$$

$$x_4 = 0.941$$

$$h_4 = h_{f \text{ at } 4 \text{ bar}} + x_4 \cdot h_{fg \text{ at } 4 \text{ bar}} = 2612.65 \text{ kJ/kg}$$

for state 5,

$$s_5 = 6.5966 = s_{f \text{ at } 0.075 \text{ bar}} + x_5 \cdot s_{fg \text{ at } 0.075 \text{ bar}}$$

$$x_5 = 0.784$$

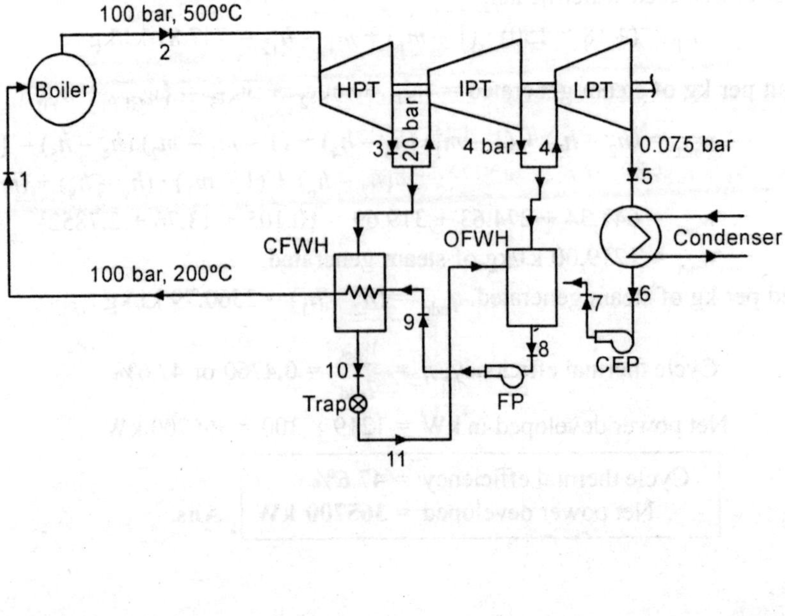

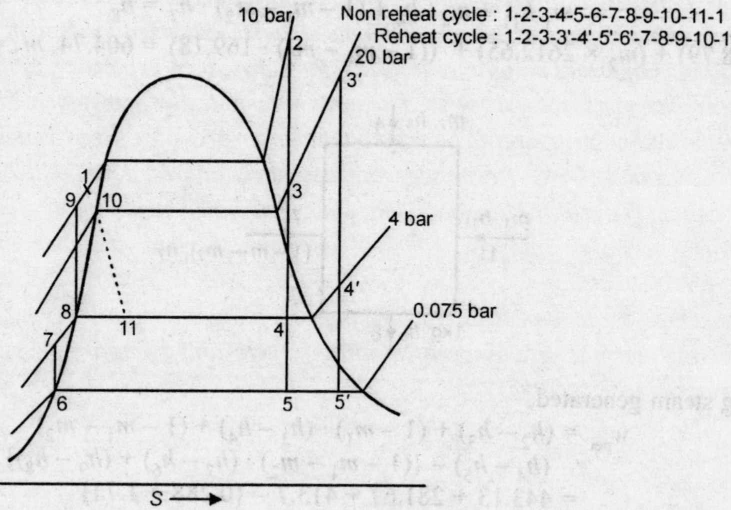

Fig. 8.42 Layout and T-s diagram

$$h_5 = h_{f \text{ at } 0.075 \text{ bar}} + x_5 \cdot h_{fg \text{ at } 0.075 \text{ bar}} = 2055.09 \text{ kJ/kg}$$

Let mass of steam bled at 20 bar, 4 bar be m_1 and m_2 per kg of steam generated.

$$h_{10} = h_{f \text{ at } 20 \text{ bar}} = 908.79 \text{ kJ/kg},\ h_8 = h_{f \text{ at } 4 \text{ bar}} = 604.74 \text{ kJ/kg}$$

$$\text{At trap } h_{10} = h_{11} = 908.79 \text{ kJ/kg}$$

At condensate extraction pump, (CEP), $h_7 - h_6 = v_6\,(4 - 0.075) \times 10^2$

$$v_6 = v_{f \text{ at } 0.075 \text{ bar}} = 0.001008 \text{ m}^3/\text{kg}$$

$$h_6 = h_{f \text{ at } 0.075 \text{ bar}} = 168.79 \text{ kJ/kg} \Rightarrow h_7 = 169.18 \text{ kJ/kg}$$

At feed pump, (FP), $h_9 - h_8 = v_8\,(20 - 4) \times 10^2$

$$h_8 = h_{f \text{ at } 4 \text{ bar}} = 604.74 \text{ kJ/kg}$$

$$v_8 = v_{f \text{ at } 4 \text{ bar}} = 0.001084 \text{ m}^3/\text{kg}$$

$$h_9 = 604.74 + (0.001084 \times 16 \times 10^2) = 606.47 \text{ kJ/kg}$$

Let us apply heat balance at closed feed water heater,

$$m_1 \cdot h_3 + h_9 = m_1 \cdot h_{10} + 4.18 \times 200$$

$$(m_1 \times 2930.57) + 606.47 = (m_1 \times 908.79) + 836$$

$$m_1 = 0.114 \text{ kg}$$

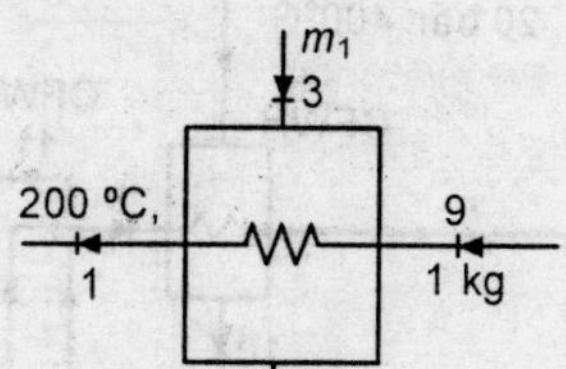

Applying heat balance at open feed water,

$$m_1\,h_{11} + m_2 \cdot h_4 + (1 - m_1 - m_2) \cdot h_7 = h_8$$

$$(m_1 \cdot 908.79) + (m_2 \times 2612.65) + ((1 - m_1 - m_2) \cdot 169.18) = 604.74,\ m_2 = 0.144 \text{ kg}$$

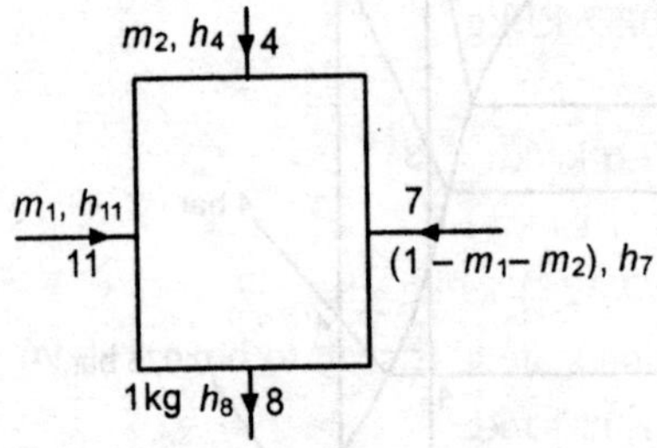

Net work per kg steam generated,

$$\begin{aligned} w_{net} &= (h_2 - h_3) + (1 - m_1) \cdot (h_3 - h_4) + (1 - m_1 - m_2) \\ &\quad (h_4 - h_5) - \{(1 - m_1 - m_2) \cdot (h_7 - h_6) + (h_9 - h_8)\} \\ &= 443.13 + 281.67 + 413.7 - \{0.288 + 1.73\} \end{aligned}$$

$$w_{net} = 1136.48 \text{ kJ/kg}$$

Heat added per kg steam generated,

$$q_{add} = (h_2 - h_1) = (3373.7 - 4.18 \times 200) = 2537.7 \text{ kJ/kg}$$

$$\text{Thermal efficiency} = \frac{w_{net}}{q_{add}} = 0.4478 \text{ or } 44.78\%$$

$$\text{Steam generation rate} = \frac{100 \times 10^3}{w_{net}} = 87.99 \text{ kg/s}$$

Thermal efficiency = **44.78%**
Steam generation rate = **87.99 kg/s**

Ans.

(a) For the reheating introduced at 20 bar up to 400°C: The modified cycle representation is shown on T-s diagram by 1-2-3-3'-4'-5'-6-7-8-9-10-11

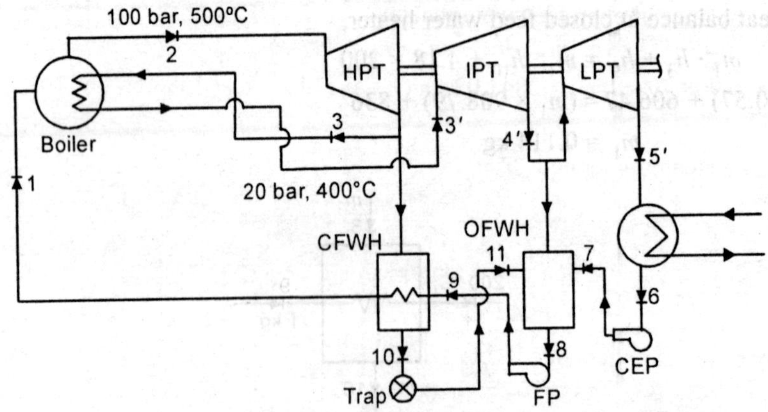

Fig. 8.43 Reheat cycle

At state 2,

$$h_2 = 3373.7 \text{ kJ/kg},$$
$$s_2 = 6.5966 \text{ kJ/kg} \cdot \text{K}$$

At state 3,

$$h_3 = 2930.57 \text{ kJ/kg}$$

At state 3',

$$h_{3'} = 3247.6 \text{ kJ/kg}$$
$$s_{3'} = 7.1271 \text{ kJ/kg} \cdot \text{K}$$

At state 4' and 5' $s_{3'} = s_{4'} = s_{5'} = 7.1271$ kJ/kg · K

From steam tables by interpolation state 4' is seen to be at 190.96ºC at 4 bar

$$h_{4'} = 2841.02 \text{ kJ/kg}$$

Let dryness fraction at state 5' be $x_{5'}$.

$$s_{5'} = 7.1271 = s_{f \text{ at } 0.07 \text{ bar}} + x_{5'} \cdot s_{fg \text{ at } 0.075 \text{ bar}}$$
$$\Rightarrow \quad x_{5'} = 0.853$$
$$h_{5'} = h_{f \text{ at } 0.075 \text{ bar}} + x_{5'} \cdot h_{fg} \text{ at } 0.075 \text{ bar}$$
$$h_{5'} = 2221.11 \text{ kJ/kg}$$

Let mass of bled steam at 20 bar and 4 bar be $m_{1'}$, $m_{2'}$ per kg of steam generated. Applying heat balance at closed feed water heater.

$$m_{1'} \cdot h_3 + h_9 = m_{1'} \cdot h_{10} + 4.18 \times 200$$
$$\Rightarrow \quad m_{1'} = 0.114 \text{ kg}$$

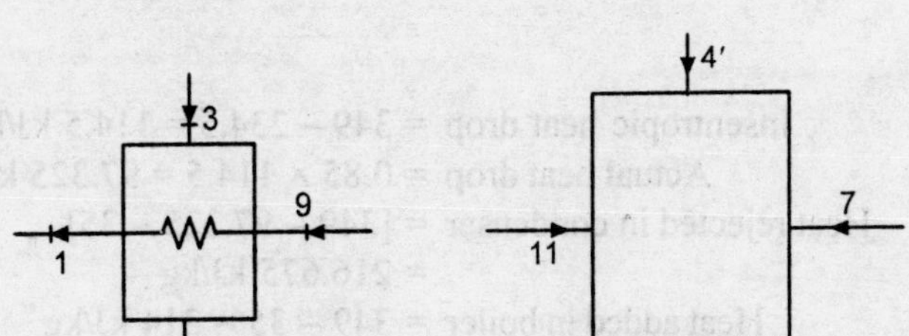

Applying heat balance at open feed water heater

$$m_{1'} \cdot h_{11} + m_{2'} \cdot h_{4'} + (1 - m_{1'} - m_{2'}) \cdot h_7 = h_8$$
$$(0.114 \times 908.79) + (m_{2'} \cdot 2841.02) + (1 - 0.114 - m_{2'}) \cdot 169.18 = 604.74$$
$$m_{2'} = 0.131 \text{ kg}$$

Net work per kg steam generated

$$w_{net} = (h_2 - h_3) + (1 - m_{1'}) \cdot (h_{3'} - h_{4'}) + (1 - m_{1'} - m_{2'})\,(h_{4'} - h_{5'}) - \{(1 - m_{1'} - m_{2'}) \cdot (h_7 - h_6) + (h_9 - h_8)\}$$
$$w_{net} = 443.13 + 360.22 + 468.03 - \{0.293 + 1.73\}$$
$$w_{net} = 1269.36 \text{ kJ/kg}$$

Heat added per kg steam generated, $q_{add} = (h_2 - h_1) + (1 - m_{1'})\,(h_{3'} - h_3)$

$$= 2537.7 + 280.88$$
$$q_{add} = 2818.58 \text{ kJ/kg}$$

Thermal efficiency $\eta = \dfrac{w_{net}}{q_{add}} = 0.4503$ or 45.03%

% Increase in thermal efficiency due to reheating $= \left(\dfrac{0.4503 - 0.4478}{0.4478}\right) \times 100$

$= 0.56\%$

Thermal efficiency of reheat cycle = **45.03%**
% Increase in efficiency due to reheating = **0.56%**

Ans.

13. In a binary vapour cycle working on mercury and steam, the mercury vapour is generated dry saturated at 8.45 bar and expanded upto 0.07 bar in mercury turbine. The condenser or mercury cycle is used for generating steam at 40 bar, 0.98 dry. The steam is superheated separately upto 450°C and then supplied into steam turbine for being expanded upto 0.075 bar. Two closed feed water heaters are used by bleeding out steam at 8 bar and 1 bar so as to provide feed water leaving at 150°C and 90°C respectively. Condensate leaves feed water heater as saturated liquid at respective pressures and is mixed with the hot feed water leaving the respective feed heater. The turbine running on mercury has capability of converting 85% of available heat into work. The enthalpies of mercury may be taken as,

enthalpy of dry saturated vapour at 8.45 bar = 349 kJ/kg
enthalpy after isentropic expansion to 0.07 bar = 234.5 kJ/kg
enthalpy of saturated liquid at 0.07 bar = 35 kJ/kg

Assume feed water to enter at 150°C into mercury condenser. Neglect pump work for getting efficiency. Determine the steam generation rate per kg of mercury and efficiency of cycle.

Solution:

For mercury cycle,

Insentropic heat drop = 349 – 234.5 = 114.5 kJ/kg Hg
Actual heat drop = 0.85 × 114.5 = 97.325 kJ/kg Hg
Heat rejected in condenser = [349 – 97.325 – 35]
= 216.675 kJ/kg
Heat added in boiler = 349 – 35 = 314 kJ/kg

For steam cycle,

Enthalpy of steam generated $= h_{\text{at 40 bar, 0.98 dry}} = 2767.13$ kJ/kg

Enthalpy of inlet to steam turbine $h_2 = h_{\text{at 40 bar, 450°C}} = 3330.3$ kJ/kg

Entropy of steam at inlet to steam turbine, $s_2 = 6.9363$ kJ/kg · K

Therefore, heat added in condenser of mercury cycle

$= h_{\text{at 40 bar, 0.98 dry}} - h_{\text{feed at 40 bar}}$

$= 2767.13 - 4.18 \times 150 = 2140.13$ kJ/kg

Therefore, mercury required per kg of steam $= \dfrac{2140.13}{\text{Heat rejected in condenser}}$

$= \dfrac{2140.13}{216.675} = 9.88$ kg per kg of steam

For isentropic expansion, $s_2 = s_3 = s_4 = s_5 = 6.9363$ kJ/kg · K

State 3 lies in superheated region, by interpolation the state can be given by, temperature 227.07°C at 8 bar, $h_3 = 2899.23$ kJ/kg

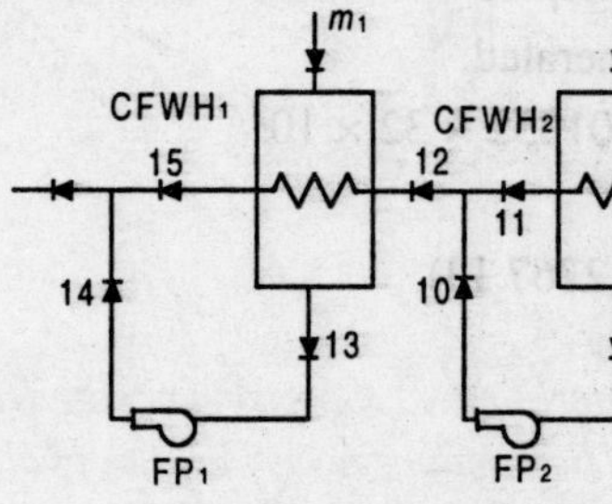

Regenerative arrangement

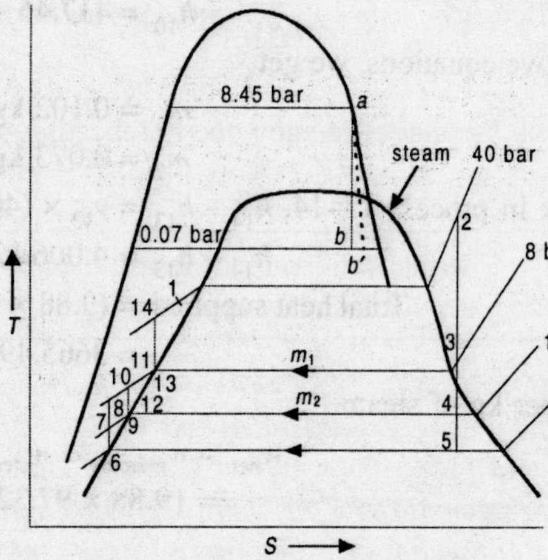

Fig. 8.44 T-s diagram showing expansion and bleeding

State 4 lies in wet region, say with dryness fraction x_4,

$$s_4 = 6.9363 = s_{f \text{ at 1 bar}} + x_4 \times s_{fg \text{ at 1 bar}} \Rightarrow x_4 = 0.93$$

$$h_4 = h_{f \text{ at 1 bar}} + x_4 \cdot h_{fg \text{ at 1 bar}} = 2517.4 \text{ kJ/kg}$$

Let state 5 lie in wet region with dryness fraction x_5,

$$s_5 = 6.9363 = s_{f \text{ at 0.075 bar}} + x_5 \cdot s_{fg \text{ at 0.075 bar}}$$

$$x_5 = 0.828$$

$$h_5 = 2160.958 \text{ kJ/kg}$$

Let mass of steam bled at 8 bar and 1 bar be m_1 and m_2 per kg of steam generated.

$$h_7 = h_6 + v_6(1 - 0.075) \times 10^2$$

$$= h_{f \text{ at 0.075 bar}} + v_{f \text{ at 0.075 bar}} (1 - 0.075) \times 10^2$$

$$= 168.79 + 0.001008 \times (1 - 0.075) \times 10^2$$

$$h_7 = 168.88 \text{ kJ/kg}$$

$$h_9 = h_{f \text{ at 1 bar}} = 417.46 \text{ kJ/kg}$$

$$h_{13} = h_{f \text{ at 8 bar}} = 721.11 \text{ kJ/kg}$$

Applying heat balance on CFWH$_1$, $T_1 = 150$ºC and also $T_{15} = 150$ºC

$$m_1 \times h_3 + (1 - m_1) \times h_{12} = m_1 \times h_{13} + (4.18 \times 150) \times (1 - m_1)$$

$$(m_1 \times 2899.23) + (1 - m_1) \times h_{12} = (m_1 \times 721.11) + 627 \cdot (1 - m_1)$$

Applying heat balance on CFWH$_2$, $T_{11} = 90$ºC

$$m_2 \times h_4 + (1 - m_1 - m_2) \times h_7 = m_2 \times h_9 + (1 - m_1 - m_2) \times 4.18 \times 90$$

$$(m_2 \times 2517.4) + (1 - m_1 - m_2) \times 168.88 = (m_2 \times 417.46) + 376.2\,(1 - m_1 - m_2)$$

Heat balance at mixing between CFWH$_1$ and CFWH$_2$,

$$(1 - m_1 - m_2) \times 4.18 \times 90 + m_2 \times h_{10} = (1 - m_1) \times h_{12}$$

$$376.2\,(1 - m_1 - m_2) + m_2 \times h_{10} = (1 - m_1) \times h_{12}$$

For pumping process 9–10, $h_{10} = h_9 + v_9(8 - 1) \times 10^2$

$$h_{10} = h_{f \text{ at bar}} + v_{f \text{ at 1 bar}} \, (7 \times 10^2)$$
$$h_{10} = 417.46 + 0.001043 \times 700 = 418.19 \text{ kJ/kg}$$

Solving above equations, we get

$$m_1 = 0.102 \text{ kg per kg steam generated}$$
$$m_2 = 0.073 \text{ kg per kg steam generated.}$$

Pump work in process 13–14, $h_{14} - h_{13} = v_{13} \times (40 - 8) \times 10^2 = 0.001252 \times 32 \times 10^2$

$$h_{14} - h_{13} = 4.006 \text{ kJ/kg}$$
$$\text{Total heat supplied} = (9.88 \times 314) + (3330.3 - 2767.13)$$
$$= 3665.49 \text{ kJ/kg of steam}$$

Net work per kg of steam,

$$w_{net} = w_{mercury} + w_{steam}$$
$$= \{9.88 \times 97.325\} + \{(h_2 - h_3) + (1 - m_1) \cdot (h_3 - h_4) + (1 - m_1 - m_2) \cdot (h_4 - h_5) - (1 - m_1 - m_2)(h_4 - h_6) - m_2 (h_{10} - h_9) - m_1 (h_{14} - h_{13})\}$$
$$= \{961.571\} + \{431.07 + 342.88 + 294.06 - 0.074 - 0.053 - 0.408\}$$
$$w_{net} = 2029.046 \text{ kJ/kg}$$

$$\text{Thermal efficiency of binary vapour cycle} = \frac{2029.046}{3665.49} = 0.5536 \text{ or } 55.36\%$$

Thermal efficiency = **55.36%** **Ans.**

14. A steam power plant has mixed pressure turbine of output 2500 hp with high pressure steam entering at 20 bar, 300°C and low pressure steam entering at 2 bar and dry saturated. The steam leaves turbine at 0.075 bar and efficiency ratio of both HP and LP stages are 0.8. The Willan's line for both are straight line and steam consumption at no load is 10% of full load. Determine the HP steam required for producing 1000 hp if low pressure steam is available at the rate of 1.5 kg/s.

Solution:

This is a mixed pressure turbine so the output of turbine shall be sum of the contributions by HP and LP steam streams.

For HP: At Inlet of HP steam, $\Rightarrow h_1 = 3023.5$ kJ/kg, $s_1 = 6.7664$ kJ/kg · K

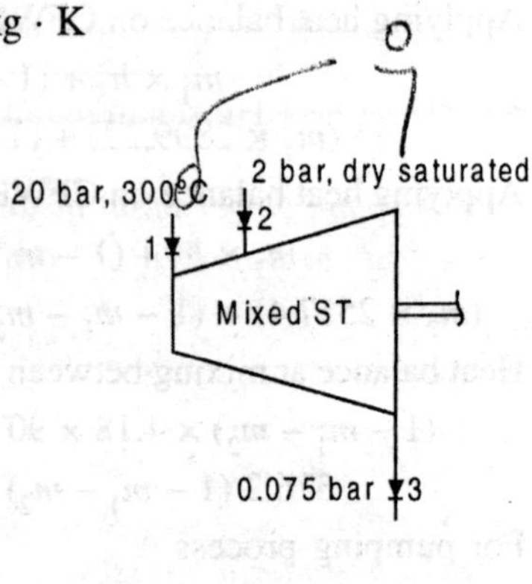

Ideally, $s_2 = s_1 = 6.7664$

$$s_2 = 6.7664 = s_{f \text{ at 0.075 bar}} + x_3 \times s_{fg \text{ at 0.075 bar}}$$
$$\Rightarrow \quad x_3 = 0.806$$
$$h_{3HP} = h_{f \text{ at 0.075 bar}} + x_3 \cdot h_{fg \text{ at 0.075 bar}} = 2108.03 \text{ kJ/kg}$$

Actual enthalpy drop in HP = $(h_1 - h_{3HP}) \times 0.8 = 732.38$ kJ/kg

For LP: At inlet of LP steam

$$h_2 = 2706.7 \text{ kJ/kg}, \; s_2 = 7.1271 \text{ kJ/kg} \cdot \text{K}$$
$$\text{Enthalpy at exit, } h_{3LP} = 2222.34 \text{ kJ/kg}$$
$$\text{Actual enthalpy drop in LP} = (h_1 - h_{3LP}) \times 0.8 = 387.49 \text{ kJ/kg}$$

$$\text{HP steam consumption at full load} = \frac{2500 \times 0.7457}{732.38} = 2.54 \text{ kg/s}$$

HP steam consumption at no load = 0.1 × 2.54 = 0.254 kg/s

$$\text{LP steam consumption at full load} = \frac{2500 \times 0.7457}{387.49} = 4.81 \text{ kg/s}$$

LP steam consumption at no load = 0.1 × 4.81 = 0.481 kg/s

The problem can be solved geometrically by drawing Willan's line as per scale on graph paper and finding out the HP stream requirement for getting 1000 hp if LP steam is available at 1.5 kg/s.

or,

Analytically the equation for Willan's line can be obtained for above full load and no load conditions for HP and LP separately.

Willan's line for HP: $y = mx + C$, here y = steam consumption, kg/s

x = load, hp

$$y_{HP} = m_{HP} \cdot x + C_{HP}$$
$$2.54 = m_{HP} \cdot 2500 + C_{HP}$$
$$0.254 = m_{HP} \cdot 0 + C_{HP} \Rightarrow C_{HP} = 0.254 \text{ and } m_{HP} = 9.144 \times 10^{-4}$$
$$\Rightarrow \quad y_{HP} = 9.144 \times 10^{-4} \cdot x_{HP} + 0.254$$

Willan's line for LP: $y_{LP} = m_{LP} \cdot x_{LP} + C_{LP}$

$$4.81 = m_{LP} \cdot 2500 + C_{LP}$$
$$0.481 = m_{LP} \cdot C_{LP} \Rightarrow C_{LP} = 0.481,\ m_{LP} = 1.732 \times 10^{-3}$$
$$y_{LP} = 1.732 \times 10^{-3} \cdot x_{LP} + 0.481$$

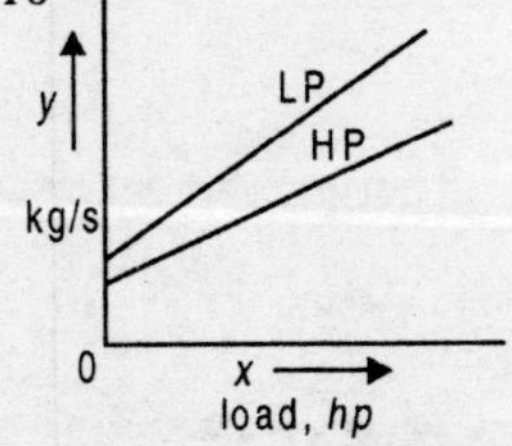

Fig. 8.45 Tentative representation of Willan's line

Total output (load) from mixed turbine, $x = x_{HP} + x_{LP}$

For load of 1000 hp to be met by mixed turbine, let us find out the load shared by LP for steam flow rate of 1.5 kg/s

$$1.5 = 1.732 \times 10^{-3} \cdot x_{LP} + 0.481$$
$$x_{LP} = 588.34 \text{ hp}$$

Since by 1.5 kg/s of LP steam only 588.34 hp output contribution is made so remaining (1000 – 588.34 = 411.66 hp), 411.66 hp should be contributed by HP steam. By Willan's line for HP turbine,

$$y_{HP} = (9.144 \times 10^{-4} \times 411.66) + 0.254 = 0.63 \text{ kg/s}$$

So, HP steam requirement = 0.63 kg/s

HP steam required = **0.63 kg/s**	**Ans.**

15. A steam power plant installation has steam leaving boiler at 40 bar, 300°C and expanding in HP turbine upto 2 bar. Half of steam leaving HP turbine is sent for process heating and remaining enters a separator where all the moisture is removed. Dry steam from separator is sent to low pressure LP turbine at 2 bar and gets expanded upto 0.075 bar.

The drain (moisture) of separator gets mixed with condensate from process heater and combined flow enters the hot well at 90°C. Trap is provided at exit of both process heater and separator. Condensate extraction pump extracts condensate from condenser and sends it to hot well at 40°C. Neglecting pump work and radiation losses etc. determine temperature of water leaving hotwell and heat transferred per kg in process heater. Also find out thermal efficiency of installation and give layout.

Solution:

Let us carry out analysis for 1 kg of steam generated in boiler.

Enthalpy at inlet to HPT, $h_2 = 2960.7$ kJ/kg, $s_2 = 6.3615$ kJ/kg · K

State at 3 i.e. exit from HPT can be identified by $s_2 = s_3 = 6.3615$ kJ/kg · K

Let dryness fraction be X_3, $s_3 = 6.3615 = s_{f \text{ at 2 bar}} + X_3 \cdot s_{fg \text{ at 2 bar}} \Rightarrow x_3 = 0.863$

$$h_3 = 2404.94 \text{ kJ/kg}$$

If one kg of steam is generated in boiler then at exit of HPT, 0.5 kg goes into process heater and 0.5 kg goes into separator

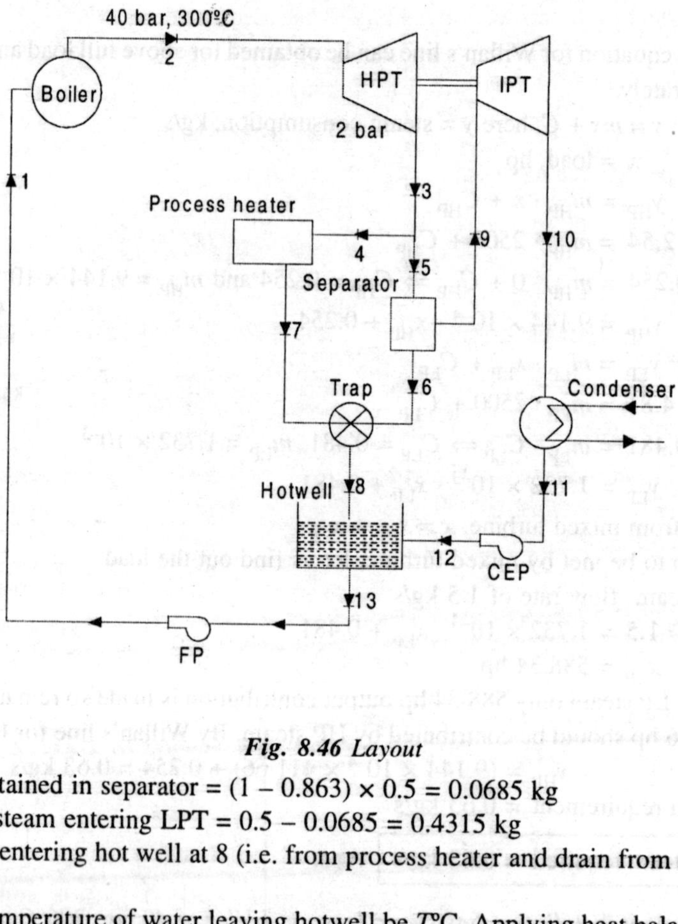

Fig. 8.46 Layout

Mass of moisture retained in separator = (1 – 0.863) × 0.5 = 0.0685 kg

Therefore, mass of steam entering LPT = 0.5 – 0.0685 = 0.4315 kg

Total mass of water entering hot well at 8 (i.e. from process heater and drain from separator) = 0.5685 kg

Let us assume the temperature of water leaving hotwell be T°C. Applying heat balance for mixing;

$(0.5685 \times 4.18 \times 90) + (0.4315 \times 4.18 \times 40) = (1 \times 4.18 \times T)$, $T = 68.425$°C

Temperature of water leaving hotwell = **68.425°C**	**Ans.**

Applying heat balanced on trap

$$0.5 \times h_7 + 0.0685 \times h_{f \text{ at 2 bar}} = (0.5685 \times 4.18 \times 90)$$

$$h_7 = 358.59 \text{ kJ/kg}$$

Therefore, heat transferred in process heater = $0.5 \times (h_3 - h_7)$

$= 1023.175$ kJ/kg steam generated

Heat transferred per kg steam generated = **1023.175 kJ/kg** steam generated

Ans.

For state 10 at exit of LPT, $s_{10} = s_3 = s_2 = 6.3615$ kJ/kg · K

Let dryness fraction be x_{10}

$$s_{10} = 6.3615 = s_{f \text{ at } 0.075 \text{ bar}} + x_{10} \cdot s_{fg \text{ at } 0.075 \text{ bar}}$$

$$\Rightarrow \quad x_{10} = 0.754$$

$$\Rightarrow \quad h_{10} = h_{f \text{ at } 0.075 \text{ bar}} + x_{10} \cdot h_{fg \text{ at } 0.075 \text{ bar}}$$

$$h_{10} = 1982.91$$

Net work output, neglecting pump work per kg of steam generated,

$$w_{net} = (h_2 - h_3) \times 1 + 0.4315 \times (h_3 - h_{10})$$

$$= 555.76 + 182.11$$

$$w_{net} = 737.87 \text{ kJ/kg steam generated}$$

Heat added in boiler per kg steam generated, $q_{add} = (h_2 - h_1)$

$$= (2960.7 - 4.18 \times 68.425)$$

$$q_{add} = 2674.68 \text{ kJ/kg}$$

$$\text{Thermal efficiency} = \frac{w_{net}}{q_{add}} = 0.2758 \text{ or } 27.58\%$$

Thermal efficiency = **27.58%**

Ans.

16. *In a steam power plant operating on Rankine cycle, the steam enters the turbine at 70 bar and 550°C with a velocity of 30 m/s. It discharges to the condenser at 0.20 bar with a velocity 90 m/s. If the steam flow rate is 35 kg/s, find the thermal efficiency and net power produced.* *[U.P.S.C. 1992]*

Solution:

From steam tables, $h_1 = 3530.9$ kJ/kg, $s_1 = 6.9486$ kJ/kg · K

Assuming isentropic expansion in nozzle, $s_1 = s_2 = 6.9486$

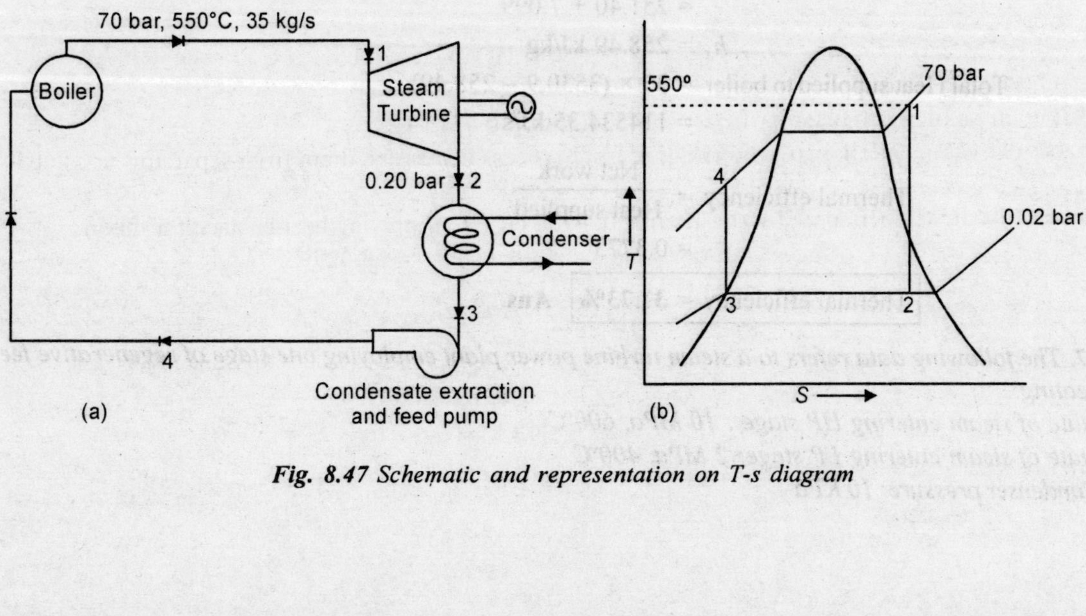

Fig. 8.47 Schematic and representation on T-s diagram

Let dryness fraction at state 2 be x_2 then;

$$s_2 = s_{f \text{ at } 0.2 \text{ bar}} + x_2 \times s_{fg \text{ at } 0.2 \text{ bar}}$$

$$6.9486 = 0.8320 + x_2 \cdot (7.0766)$$

Dryness fraction at state 2, $x_2 = 0.864$

Hence,

$$h_2 = h_{f \text{ at } 0.2 \text{ bar}} + x_2 \cdot h_{fg \text{ at } 0.2 \text{ bar}}$$

$$= 251.40 + (0.864 \times 2358.3)$$

$$h_2 = 2288.97 \text{ kJ/kg}$$

Considering pump work to be of isentropic type, $\Delta h_{34} = v_3 \times \Delta p_{34}$

From steam stable, $v_3 = v_{f \text{ at } 0.2 \text{ bar}} = 0.001017 \; m^3/\text{kg}$

or

$$\Delta h_{34} = 0.001017 \times \frac{(70 - 0.20) \times 10^5}{10^3}$$

$$\text{Pump work, } W_P = \Delta h_{34} = 7.099 \text{ kJ/kg}$$

$$\text{Turbine work, } W_T = \Delta h_{12} = (h_1 - h_2) = (3530.9 - 2288.97)$$

$$W_T = 1241.93 \text{ kJ/kg}$$

$$\text{Net work} = W_T - W_P$$

$$= 1241.93 - 7.099$$

$$W_{net} = 1234.831 \text{ kJ/kg}$$

$$\text{Power produced} = \text{mass flow rate} \times W_{\text{net}}$$

$$= 35 \times 1234.831$$

$$= 43219.085 \text{ kJ/s}$$

Net Power = 43.22 MW **Ans.**

$$\text{Heat supplied in boiler} = (h_1 - h_4), \text{ kJ/kg}$$

$$\text{Enthalpy at state 4, } h_4 = h_3 + \Delta h_{34}$$

$$= h_{f \text{ at } 0.2 \text{ bar}} + \Delta h_{34}$$

$$= 251.40 + 7.099$$

$$h_4 = 258.49 \text{ kJ/kg}$$

$$\text{Total Heat supplied to boiler} = 35 \times (3530.9 - 258.49)$$

$$= 114534.35 \text{ kJ/s}$$

$$\text{Thermal efficiency} = \frac{\text{Net work}}{\text{Heat supplied}}$$

$$= 0.3773$$

Thermal efficiency = 37.73% **Ans.**

17. The following data refers to a steam turbine power plant employing one stage of regenerative feed heating:

State of steam entering HP stage : 10 MPa, 600°C

State of steam entering LP stage: 2 MPa, 400°C

Condenser pressure: 10 KPa

The correct amount of steam is bled for feed heating at exit from the HP stage. Calculate the mass of steam bled per kg of steam passing through the HP stage and the amount of heat supplied in the boiler per second for an output of 10 MW. Neglect pump work, *[U.P.S.C. 1993]*

Solution:

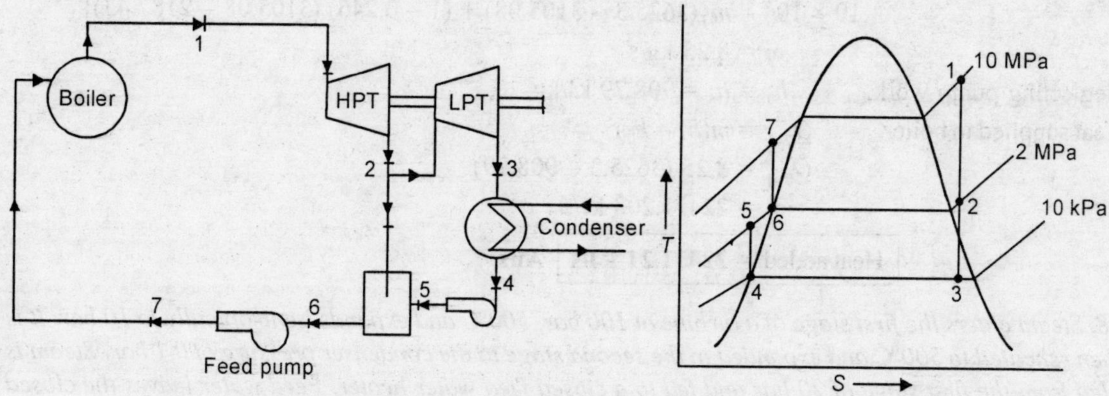

Fig. 8.48 Possible arrangement and T-s representation

From steam tables:

$h_1 = 3625.3$ kJ/kg, $s_1 = 6.9029$ kJ/kg · K

Due to isentropic expansion, $s_1 = s_2 = s_3 = 6.9029$ kJ/kg · K

At state 2, i.e. at pressure of 2 MPa and entropy 6.9029 kJ/kg · K.

By interpolating state for s_2 between 2 MPa, 300°C and 2 MPa, 350°C from steam tables,

$$h_2 = 3105.08 \text{ kJ/kg}$$

For state 3, i.e. at pressure of 0.01 MPa entropy, s_3 lies in wet region as $s_3 < s_g$ at 0.01 MPa. Let dryness fraction be x_3 at this state

$$s_3 = s_{f \text{ at } 0.01 \text{ MPa}} + x_3 \cdot s_{fg \text{ at } 0.01 \text{ MPa}}$$

$$6.9029 = 0.6493 + x_3 \times 7.5009$$

$$x_3 = 0.834$$

Enthalpy at state 3, $h_3 = h_{f \text{ at } 0.01 \text{ MPa}} + x_3 \cdot h_{fg \text{ at } 0.01 \text{ MPa}}$

$$= 191.83 + (0.834 \times 2392.8)$$

$$h_3 = 2187.43 \text{ kJ/kg}$$

Let the mass of steam bled be m_b per kg of steam from exit of HP for regenerative feed heating.

Considering state at exit from feed heater being saturated liquid the enthalpy at exit of feed heater will be, $h_{f \text{ at } 2 \text{ MPa}}$.

$$h_6 = h_{f \text{ at } 2 \text{ MPa}} = 908.79 \text{ kJ/kg}$$

For adiabatic mixing in feed heater, for one kg of steam leaving boiler, the heat balance yields,

$$(1 - m_b) \cdot h_5 + m_b \cdot h_2 = h_6$$

While neglecting pump work,

$$h_5 = h_4 = h_{f \text{ at } 0.01 \text{ MPa}} = 191.83 \text{ kJ/kg}$$

Substituting in heat balance on the feed heater,

$$(1 - m_b) \cdot 191.83 + m_b \cdot 3105.08 = 908.79$$

$$m_b = 0.246 \text{ kg per kg of steam entering HP turbine}$$

Steam bled per kg of steam passing through HP stage = **0.246 kg** **Ans.**

Let mass of steam leaving boiler be m kg/s.

$$\text{Output} = 10 \times 10^3 = m(h_1 - h_2) + m(1 - m_b)(h_2 - h_3)$$

$$10 \times 10^3 = m\{(3625.3 - 3105.08) + (1 - 0.246)(3105.08 - 2187.43)\}$$

$$m = 8.25 \text{ kg/s}$$

Neglecting pump work, $h_7 = h_6 = 908.79$ kJ/kg

Heat supplied to boiler, $Q_{7-1} = m(h_1 - h_7)$

$$Q_{7-1} = 8.25\,(3625.3 - 908.79)$$

$$= 22411.208 \text{ kJ/s}$$

Heat added = **22411.21 kJ/s** **Ans.**

18. Steam enters the first stage of a turbine at 100 bar, 500°C and expands isentropically to 10 bar. It is then reheated to 500°C and expanded in the second stage to the condenser pressure of 0.1 bar. Steam is bled from the first stage at 20 bar and fed to a closed feed water heater. Feed water leaves the closed heater at 100 bar, 200°C (enthalpy = 856.8 kJ/kg), while the condensate is supplied to the open heater into which steam is bled at 4 bar pressure. Saturated liquid at 4 bar exits from the open heater and enters the closed heater. The net output of the turbine is 50 MW. Assuming the turbine and pump processes to be isentropic, determine the mass of steam bled at each feed water heater per kg of steam entering the first stage, the mass of steam entering the first stage per second, and the thermal efficiency.

[U.P.S.C. 1995]

Solution:

From steam table, at inlet to first stage of turbine,

$h_1 = h_{\text{at 100 bar, 500°C}} = 3373.7$ kJ/kg

$s_1 = s_{\text{at 100 bar, 500°C}} = 6.5966$ kJ/kg · K

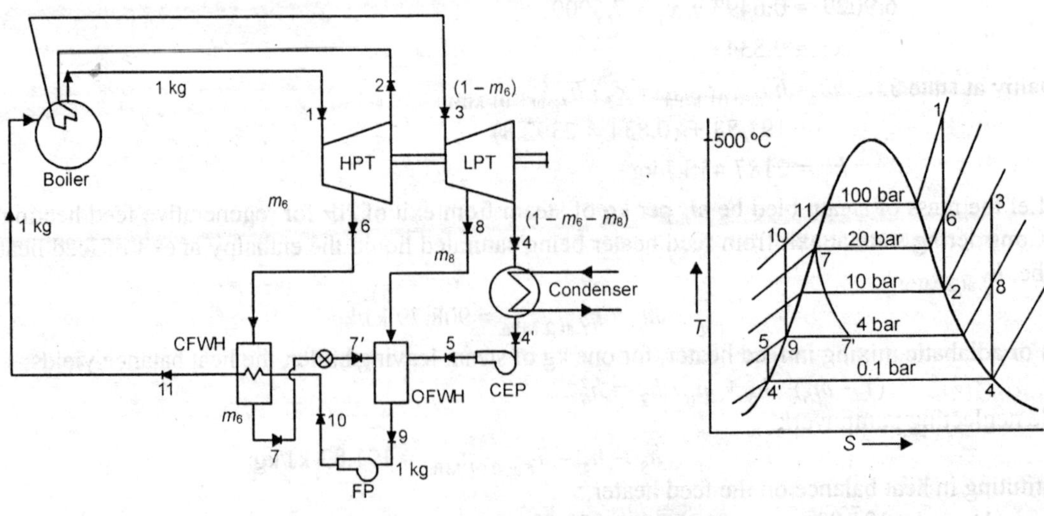

Fig. 8.49 Arrangement and T-s representation

Due to isentropic expansion, $s_1 = s_6 = s_2$ and $s_3 = s_8 = s_4$

State at 6 i.e. bleed state from HP turbine, Temperature by interpolation from steam table = 261.6°C

At inlet to second stage of turbine, $h_6 = 2930.572$

$$h_3 = h_{\text{at 10 bar, 500°C}} = 3478.5 \text{ kJ/kg}$$

$$s_3 = s_{\text{at 10 bar, 500°C}} = 7.7622 \text{ kJ/kg} \cdot \text{K}$$

At exit from first stage of turbine i.e. at 10 bar and entropy of 6.5966 kJ/kg · K Temperature by interpolation from steam table at 10 bar and entropy of 6.5966 kJ/kg · K

$$T_2 = 181.8°\text{C}$$

$$h_2 = 2782.8 \text{ kJ/kg}$$

State at 8, i.e. bleed state from second stage of expansion, i.e. at 4 bar and entropy of 7.7622 kJ/kg · K Temperature by interpolation from steam table, $T_8 = 358.98°\text{C} \simeq 359°\text{C}$

$$h_8 = 3188.7 \text{ kJ/kg}$$

State at 4 i.e. at condenser pressure of 0.1 bar and entropy of 7.7622 kJ/kg. K the state lies in wet region. So let the dryness fraction be x_4.

$$s_4 = s_{f\text{ at 0.1 bar}} + x_4 \cdot s_{fg\text{ at 0.1 bar}}$$

$$7.7622 = 0.6493 + x_4 \cdot 7.5009$$

$$x_4 = 0.95$$

$$h_4 = h_{f\text{ at 0.1 bar}} + x_4 \cdot h_{fg\text{ at 0.1 bar}}$$

$$= 191.83 + (0.95 \times 2392.8)$$

$$h_4 = 2464.99 \text{ kJ/kg}$$

Given, $h_{11} = 856.8 \text{ kJ/kg}, h_9 = h_{f\text{ at 4 bar}}$

$$h_9 = 604.74 \text{ kJ/kg}$$

Considering pump work, the net output can be given as,

$$W_{\text{net}} = W_{\text{HPT}} + W_{\text{LPT}} - (W_{\text{CEP}} + W_{\text{FP}})$$

where $W_{\text{HPT}} = \{(h_1 - h_6) + (1 - m_6)(h_6 - h_2)\}$ per kg of steam from boiler.

$W_{\text{LPT}} = \{(1 - m_6) + (h_3 - h_8)(1 - m_6 - m_8)(h_8 - h_4)\}$ per kg of steam from boiler.

For closed feed water heater, energy balance yields;

$$m_6 \cdot h_6 + h_{10} = m_6 \cdot h_7 + h_{11}$$

Assuming condensate leaving closed feed water heater to be saturated liquid,

$$h_7 = h_{f\text{ at 20 bar}} = 908.79 \text{ kJ/kg}$$

Due to throttline, $h_7 = h_{7'} = 908.79 \text{ kJ/kg}$

For open feed water heater, energy balance yields,

$$m_6 \cdot h_{7'} + m_8 \cdot h_8 + (1 - m_6 - m_8) \cdot h_5 = h_9$$

For condensate extraction pump, $h_5 - h_{4'} = v_{4'} \cdot \Delta p$

$$h_5 - h_{f\text{ at 0.1 bar}} = v_{f\text{ at 0.1 bar}} \cdot (4 - 0.1) \times 10^2$$

$$h_5 - 191.83 = (0.001010) \times (3.9 \times 10^2)$$

$$h_5 = 192.224 \text{ kJ/kg}$$

For feed pump, $h_{10} - h_9 = v_9 \cdot \Delta p$

$$h_{10} - 604.74 = v_{f\text{ at 4 bar}} \times (100 - 4) \times 10^2$$

$$h_{10} - 604.74 = 0.001084 \times 96 \times 10^2$$

$$h_{10} = 615.15 \text{ kJ/kg}$$

Substituting in energy balance upon closed feed water heater,

$$m_6 \times 2701.2 + 615.15 = m_6 \times 908.79 + 856.8$$

$$m_6 = 0.135 \text{ kg per kg of steam from boiler.}$$

Substituting in energy balance upon feed water heater,

$$m_6 \cdot h_{7'} + m_8 \cdot h_8 + (1 \cdot m_6 - m_8) \cdot h_5 = h_9$$

$$(0.135 \times 908.79) + (m_8 \times 3188.7) + (1 - 0.135 - m_8) \times 192.224 = 604.74$$

$$m_8 = 0.105 \text{ kg per kg of steam from boiler}$$

Let mass of steam entering first stage of turbine be m kg, then

$$W_{HPT} = m\{(h_1 - h_6) + (1 - m_6)(h_6 - h_2)\}$$

$$= m\{(3373.7 - 2930.572) + (1 - 0.135)\}(2930.572 - 2782.8)$$

$$W_{HPT} = m\{570.95\}, \text{ kJ}$$

Also,

$$W_{LPT} = \{(1 - m_6)(h_3 - h_8) + (1 - m_6 - m_8) \cdot (h_8 - h_4)\}, \text{ per kg of steam from boiler}$$

$$W_{LPT} = \{(1 - 0.135)(3478.5 - 3188.7) + (1 - 0.135 - 0.105) \cdot (3188.7 - 2464.99)\}$$

$$W_{LPT} = m\{800.69\} \text{ kJ}$$

Pump works (negative work)

$$W_{CEP} = m \cdot (1 - m_6 - m_8)(h_5 - h_{4'})$$

$$= m \cdot (1 - 0.135 - 0.105) \cdot (192.224 - 191.83)$$

$$W_{CEP} = \{0.299\ m\}$$

$$W_{FP} = m\{h_{10} - h_9\}$$

$$= m\{615.45 - 604.74\}$$

$$W_{FP} = \{10.71\ m\}$$

Net output

$$W_{net} = W_{HPT} + W_{LPT} - W_{CEP} - W_{FP}$$

$$50 \times 10^3 = \{570.95\ m + 800.69\ m - 0.299\ m - 10.71\ m\}$$

$$m = 36.75 \text{ kg/s}$$

Heat supplied in boiler, $Q_{add} = m(h_1 - h_{11})$

$$= 36.75\ (3373.7 - 856.8)$$

$$= 92496.075 \text{ kJ/s}$$

$$\text{Thermal efficiency} = \frac{W_{net}}{Q_{add}}$$

$$= \frac{50 \times 10^3}{92496.075}$$

$$= 0.54056 \text{ or } 54.06\%$$

Mass of steam bled at 20 bar = **0.135 kg per kg** of steam entering first stage
Mass of steam bled at 4 bar = **0.105 kg per kg** of steam entering first stage
Mass of steam entering first stage = **36.75 kg/s**
Thermal efficiency = **54.06%** **Ans.**

EXERCISE

8.1 Explain Carnot vapour power cycle and its relevance.

8.2 Give limitations of Carnot vapour power cycle and explain how Rankine cycle helps in overcoming them.

8.3 Discuss the limitations of maximum and minimum temperatures in a steam power cycle.

8.4 What should be the desired properties of a working fluid in a Rankine cycle?

8.5 Compare the efficiency and specific steam consumption of ideal Carnot cycle with Rankine cycle for given pressure limits.

8.6 Discuss the effect of pressure of steam at inlet to turbine, temperature at inlet to turbine and pressure at exit from turbine upon Rankine cycle performance.

8.7 Describe reheat cycle and compare it with simple Rankine cycle.

8.8 What do you understand by ideal regenerative cycle? Why is it not possible in practice? Also give actual regenerative cycle.

8.9 Compare performance of regenerative cycle with simple Rankine cycle.

8.10 What do you understand by binary vapour power cycles?

8.11 Discuss advantages of binary vapour cycles over single vapour cycle.

8.12 What is mixed pressure turbine?

8.13 Describe pass out turbines and back pressure turbines.

8.14 Discuss briefly the factors limiting thermal efficiency of a steam power plant.

8.15 Describe combined heat and power cycles.

8.16 A steam engine working on Rankine cycle operates between 1.96 MPa, 250ºC and 13.7 kPa. If engine consumes steam at the rate of 0.086 kg per second, determine Rankine cycle efficiency, neglecting pump work. Also, find Rankine cycle efficiency considering pump work. [58.8%]

8.17 In a steam power plant working on boiler pressure of 80 bar and condenser pressure of 0.075 bar determine cycle efficiency considering it to work on Carnot cycle. [44.9%]

8.18 A reheat cycle operates between 80 bar and 0.075 bar. Temperature of steam entering turbine is 600ºC and first stage of expansion is carried out till the steam becomes dry saturated. Subsequently steam is reheated upto the initial temperature at inlet. Neglecting pump work determine efficiency and specific steam consumption in kg/hp·hr. [42.5%, 141 kg/hp-hr]

8.19 A steam turbine installation of 60 MW capacity has steam entering turbine at 7 MPa, 500ºC with steam bleeding at 2 MPa and 0.2 MPa for feed heating. Remaining steam at 2 MPa is reheated upto 480ºC. Steam finally leaves turbine at 36 mm Hg (absolute). Give layout, mass fraction of steam bled out per kg of steam generated, cycle efficiency and mass flow rate of steam entering turbine. [0.159, 0.457, 46.58 kg/s]

8.20 In a steam turbine steam enters at 1.4 MPa, 320ºC and exhausts at 175 kPa. Determine the steam flow rate considering turbine internal efficiency of 70% and load requirement of 800 kW. [2.6 kg/s]

8.21 During trial of turbine in steam power plant the steam is found to enter turbine at 2.1 MPa, 260ºC with flow rate of 226 kg/s. Mechanical efficiency is 95% and turbine output is 100 MW. The exhaust pressure is 140 kPa and condensate temperature is 70ºC.

Unfortunately the old boiler fails and is to be replaced by another boiler generating steam at 4.9 MPa, 320ºC. This unit may drive an additional turbine with an internal efficiency of 90% which exhausts at 2.1 MPa. A part of exhaust is reheated to above trial conditions, including flow rate and reminder heats feed water to 205ºC. Determine (*a*) total plant output, (*b*) new total steam flow rate, kg/s, (*c*) old turbine internal efficiency, (*d*) the overall thermal efficiency. [14.9 MW, 293 kg/s, 94.3%, 22.9%]

8.22 A steam turbine has steam entering at 40 bar, 450ºC and leaving at 0.06 bar. Turbine develops 4000 hp at 3000 rpm with expansion occurring in two stages and reheating in between at 2 bar and both stages give equal output. The stages have efficiency ratio of 0.8. Determine thermal efficiency of cycle. [32.2%]

8.23 A regenerative cycle has steam supplied at 28 bar, 300°C in first stage. A steam fraction is bled out at 3 bar for feed heating. The feed heater drains are pumped into feed line downstream of heater at the same temperature as the bled steam. The steam expansion may be considered isentropic throughout. Condenser works at 0.15 bar. Determine mass of steam bled per kg of steam generated and thermal efficiency. Also give layout and *T-s* diagram. [0.143 kg/kg steam generated, 31.2%]

8.24 A steam power plant has steam generated at 140 bar, 400°C and throttled to 70 bar before supplied to turbine. Steam expands isentropically to the pressure of 15 bar and at this pressure some quantity of steam is bled out for feed heating which leaves as condensate at 40°C. Remaining steam is reheated at constant pressure by mixing it with the throttled steam taken from boiler in such amount the resultant mixture has temperature of 250°C. This mixture then expands isentropically upto condenser pressure of 0.075 bar. Determine horse power developed by turbine for steam generation rate of 50 ton/hr. Also give layout and *T-s* diagram. [14400 hp]

8.25 Steam is required for process heating at 3 bar, dry saturated along with power output of 1 MW. A back pressure turbine with 70% internal efficiency is used for this purpose. For the steam supplied at the rate of 2.78 kg/s determine the pressure and temperature of steam to be generated from boiler. [37.3 bar, 344°C]

8.26 A steam turbine has steam being generated at 30 bar, 400°C for being expanded upto 0.04 bar with bleeding of some steam at 3 bar for feed heating in closed feed water heater. Feed water leaves feed heater at 130°C and the condensate leaving feed heater as saturated liquid is sent to drain cooler where it gets cooled to 27°C and finally discharged in hotwell. Condensate from condenser is also discharged in hotwell from where feed water is picked by a pump and passed through drain cooler and subsequently closed feed water heater. Assuming no losses and negligible pump work determine the thermal efficiency of cycle and mass of steam bled out per kg of steam generated. Draw layout and *T-s* diagram. [38.82%, 0.167 kg/kg steam]

8.27 In a steam turbine plant steam enters turbine at 20 bar, 250°C and leaves at 0.05 bar. During expansion some amount of steam is bled out at 5 bar, 1.5 bar and 0.3 bar for feed heating in closed feed water heaters. Condensate leaves each feed heater as saturated liquid and is passed through traps into next feed heater at lower pressure. Combined drain at 0.3 bar is cooled in drain cooler upto condenser temperature and then put into hot well which also collects condensate coming from condenser, Determine,

(*a*) the masses of steam bled per kg of steam generated
(*b*) the thermal efficiency of cycle
(*c*) the net work output per kg of steam generated
(*d*) the layout and *T-s* diagram. Neglect pump work.

[0.088 kg, 0.0828 kg, 0.046 kg, 35.63%, 807 kJ/kg]

8.28 A binary vapour power cycle works on mercury and steam such that dry saturated mercury vapour at 4.5 bar is supplied to mercury turbine and leaves at 0.04 bar. Steam is generated as dry saturated at 15 bar and supplied to steam turbine for being expanded upto condenser pressure of 0.04 bar. Determine thermal efficiency of cycle. For mercury take,

$h_{f\text{ at 4.5 bar}} = 62.93$ kJ/kg, $h_{g\text{ at 4.5 bar}} = 355.98$ kJ/kg,

$v_{f\text{ at 0.04 bar}} = 0.0000765$ m³/kg

$h_{f\text{ at 0.04 bar}} = 29.98$ kJ/kg, $h_{g\text{ at 0.04 bar}} = 329.85$ kJ/kg

$s_{g\text{ at 4.5 bar}} = 0.5397$ kJ/kg · K, $s_{g\text{ at 0.04 bar}} = 0.6925$ kJ/kg · K,

$s_{f\text{ at 0.04 bar}} = 0.0808$ kJ/kg · K

9

Gas Power Cycles

9.1 INTRODUCTION

In the preceding chapter brief description is made for thermodynamic cycles comprising of both power and refrigeration cycles. In the present chapter the thermodynamic cycles operating on gas as working fluid and producing power are studied. Gas power cycles are commonly used for power production as in gas turbine plants, internal combustion engines and other applications. Here performance prediction based on thermodynamic analysis is presented for ideal cycles of different types used for gas turbine installations and internal combustion engines.

9.2 AIR-STANDARD CYCLE

Air standard cycle refers to thermodynamic cycle being studied with certain assumptions, so as to use the principles of thermodynamics conveniently. It is the most simplified form of thermodynamic cycle under consideration. General assumptions made for a cycle to be air-standard cycle are as follows:

(*i*) Air is the working fluid and behaves as a perfect gas.

(*ii*) Working fluid does not get changed in its mass and composition.

(*iii*) Thermodynamic processes constituting cycle are reversible.

(*iv*) There is no heat loss from system to surrounding and vice-versa.

(*v*) During heat addition process, heat is assumed to be supplied from a high temperature source.

(*vi*) During heat rejection process, heat is assumed to be rejected to a low temperature sink.

(*vii*) Specific heats of working fluid do not change throughout the cycle.

After incorporating above assumptions in thermodynamic cycle analysis, the analysis done is called air standard cycle analysis. Results obtained from such analysis disagree quite largely from the actual performances due to over simplification, but to begin with, such analysis is necessary. Gradually the assumptions made may be removed and analysis becomes realistic.

Air standard or ideal efficiency can be given as follows:

$$= \frac{\text{Ideal work done i.e. (heat supplied – heat rejected)}}{\text{Heat supplied}}$$

$$\eta_{\text{air}} = 1 - \frac{\text{heat rejected}}{\text{heat supplied}}$$

Mean effective pressure (m.e.p.) is the parameter which indicates the average pressure throughout the cycle. Mathematically, it is given as follows:

$$\text{m.e.p.} = \frac{\text{work done per cycle}}{\text{stroke volume}}$$

Graphically it is given by mean height of $P - V$ diagram.

Air standard cycle analysis for Carnot gas power cycle has already been discussed in chapter 4.

Otto Cycle : This is a modified form of Carnot cycle in order to make it a realistic cycle. Otto cycle has two constant volume and two adiabatic processes as shown below.

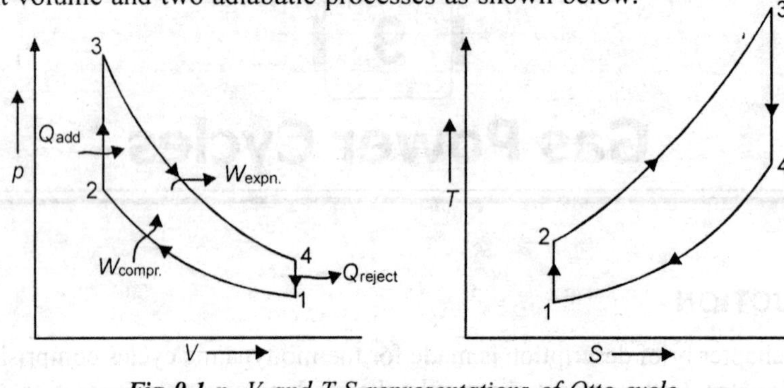

Fig 9.1 p -V and T-S representations of Otto cycle

Thermodynamic processes constituting Otto cycle are

1 – 2 = Adiabatic compression process, (–ve work, W_{compr})

2 – 3 = Constant volume heat addition process (+ve heat, Q_{add})

3 – 4 = Adiabatic expansion process, (+ve work, W_{expan})

4 – 1 = Constant volume heat rejection process (–ve heat, Q_{reject})

In order to have an engine based on Otto cycle let us find out the relevance of above processes. Spark ignition type internal combustion engines are based on this cycle.

Process 1 – 2, adiabatic compression process can be realized by piston moving from volume V_1 to V_2 and therefore compressing air.

Process 2 – 3, heat addition process can be undertaken in constant volume manner with piston at volume V_2 and heat added to woking fluid.

Heat addition is practically realized by combustion of fuel and air. As a result of heat addition the compressed air attains state 3 and it is allowed to expand from 3–4 adiabatically. After expansion air is brought back to original state 1 by extracting heat from it at volume V_1.

Internal combustion engine based on Otto cycle is explained ahead. Let us find air-standard thermal efficiency of Otto cycle.

Compression ratio for the cycle shown can be given by the ratio of volumes of air before and after compression. It is generally denoted by *r*. For unit mass of air and properties at states given with subscript 1, 2, 3, 4, we can write,

$$r = \frac{V_1}{V_2} = \frac{V_4}{V_3}$$

Heat added during 2–3, constant volume process

$$q_{add} = c_v \times (T_3 - T_2)$$

Heat rejected during 4–1, constant volume process

$$q_{rejected} = c_v \times (T_4 - T_1)$$

Air standard efficiency of Otto cycle

$$\eta_{otto} = \frac{\text{Net work}}{\text{Heat added}}$$

For a cycle,

Net work = Heat added – Heat rejected

$$= c_v \{(T_3 - T_2) - (T_4 - T_1)\}$$

Substituting in the expression for efficiency;

$$\eta_{otto} = \frac{c_v\{(T_3 - T_2) - (T_4 - T_1)\}}{c_v(T_3 - T_2)}$$

or $$\eta_{otto} = 1 - \frac{(T_4 - T_1)}{(T_3 - T_2)}$$

For perfect gas, by gas laws,

$$\frac{T_2}{T_1} = \left(\frac{V_1}{V_2}\right)^{\gamma-1} = r^{\gamma-1}$$

and $$\frac{T_3}{T_4} = \left(\frac{V_4}{V_3}\right)^{\gamma-1} = r^{\gamma-1}$$

from above $$\frac{T_2}{T_1} = \frac{T_3}{T_4} \Rightarrow \frac{T_2}{T_3} = \frac{T_1}{T_4}$$

or $$1 - \frac{T_2}{T_3} = 1 - \frac{T_1}{T_4}$$

or $$\frac{T_3 - T_2}{T_4 - T_1} = \frac{T_3}{T_4} = r^{\gamma-1}$$

Substituting in the expression for η_{otto}

$$\eta_{otto} = 1 - \frac{1}{r^{\gamma-1}}$$

Air standard efficiency of Otto cycle thus depends upon compression ratio (r) alone.

For $r = 1$, $\eta_{otto} = 0$

As compression ratio increases, the Otto cycle efficiency increases as shown in Fig 9.2.

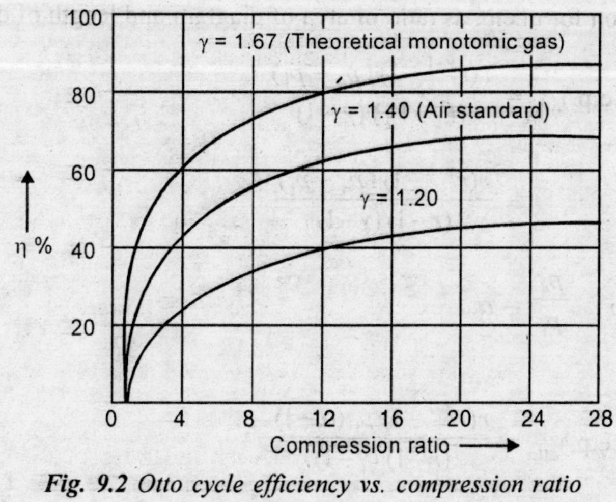

Fig. 9.2 Otto cycle efficiency vs. compression ratio

Mean effective pressure for Otto cycle can be estimated by knowing the area of diagram (work) and length of diagram (volume).

Mean effective pressure (m.e.p.)

$$= \frac{\text{Net Work (W)}}{\text{Volume change }(\Delta V)}$$

Net work can be obtained from the area enclosed by cycle on P–V diagram.

Thus, Net work = Area of diagram

$$= \left\{\frac{p_3 V_3 - p_4 V_4}{(\gamma - 1)}\right\} - \left\{\frac{p_2 V_2 - p_1 V_1}{(\gamma - 1)}\right\}$$

or
$$= \frac{1}{(\gamma - 1)}\left\{p_4 V_4\left(\frac{p_3 V_3}{p_4 V_4} - 1\right) - p_1 V_1\left(\frac{p_2 V_2}{p_1 V_1} - 1\right)\right\}$$

If $V_2 = V_3$ 1, i.e. unity clearance volume,

Then $V_1 = r$ and $V_4 = r$

Therefore, Net work

$$= \frac{1}{(\gamma - 1)}\left\{p_4 r\left(\frac{p_3}{p_4 \cdot r} - 1\right) - p_1 \cdot r\left(\frac{p_2}{p_1 \cdot r} - 1\right)\right\}$$

Also, for adiabatic expansion and compression process;

$$\frac{p_2}{p_1} = \frac{p_3}{p_4} = (r)^\gamma$$

Substituting in expression for Net work,

$$\text{Net work} = \frac{1}{(\gamma - 1)}\ \{p_4 \cdot r \cdot (r^{\gamma - 1} - 1) - p_1 \cdot r\,(r^{\gamma - 1} - 1)\}$$

or
$$= \frac{r}{(\gamma - 1)}\ (r^{\gamma - 1} - 1) \cdot (p_4 - p_1)$$

Writing expression for m.e.p. as ratio of area of diagram and length of diagram ($V_1 - V_2$).

$$(\text{m.e.p.})_{\text{otto}} = \frac{r(r^{\gamma-1} - 1)\cdot(p_4 - p_1)}{(V_1 - V_2)\cdot(\gamma - 1)}$$

or
$$= \frac{r(r^{\gamma-1} - 1)\cdot(p_4 - p_1)}{(r - 1)\,(\gamma - 1)}$$

Let us write ratio $\dfrac{p_4}{p_1} = \alpha$,

Therefore

$$(\text{m.e.p.})_{\text{otto}} = \frac{r(r^{\gamma-1} - 1)\cdot p_1\cdot(\alpha - 1)}{(r - 1)\cdot(\gamma - 1)}$$

Otto cycle has limitations such as isochoric heat addition and rejection is difficult in piston cylinder arrangement. Also, adiabatic compression and expansions are difficult to be realized in practice.

Diesel cycle : Diesel cycle is modified form of Otto cycle and heat addition process is replaced from constant volume type to constant pressure type. In a piston cylinder arrangement heat addition with piston at one position allows very little time for heat supply in Otto cycle. By having heat addition at constant pressure the sufficient time is available for heat supply in Diesel cycle. Compression ignition engines work based on diesel cycles.

Thermodynamic processes constituting Diesel cycle are as given below.

1 – 2 = Adiabatic compression, (–ve work, W_{compr})

2 – 3 = Heat addition at constant pressure (+ve heat, Q_{add})

3 – 4 = Adiabatic expansion, (+ve work, W_{expn})

4 – 1 = Heat rejection at constant volume (–ve heat, $Q_{rejected}$)

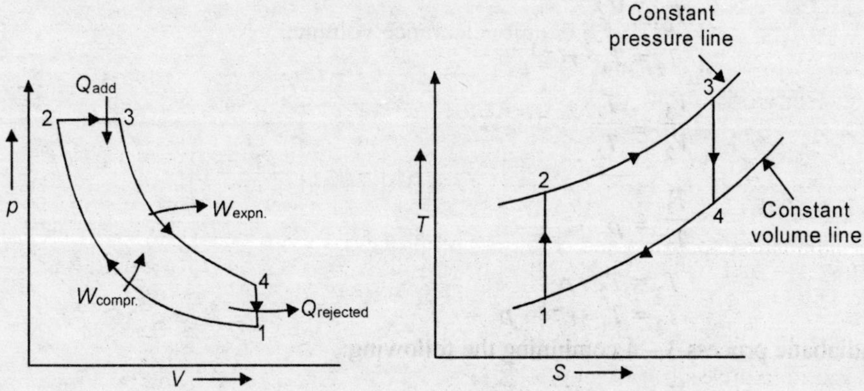

***Fig 9.3** p –V and T–S representation for Diesel cycle*

$P – V$ and $T – S$ representations for the cycle are shown in Fig. 9.3.

Thermodynamic analysis of the cycle for unit mass of air shows;

$$\text{Heat added} = c_p (T_3 - T_2)$$

$$\text{Heat rejected} = c_v (T_4 - T_1)$$

Let us assume; Compression ratio, $r = \dfrac{V_1}{V_2}$

$$\text{Cut off ratio, } \rho = \frac{V_3}{V_2}$$

$$\text{Expansion ratio} = \frac{V_4}{V_3}$$

Air standard efficiency for Diesel cycle may be given as,

$$\eta_{\text{diesel}} = \frac{\text{heat added} - \text{heat rejected}}{\text{heat added}}$$

$$= \frac{c_p(T_3 - T_2) - c_v(T_4 - T_1)}{c_p(T_3 - T_2)}$$

$$\eta_{\text{diesel}} = 1 - \frac{1}{\gamma}\left\{\frac{(T_4 - T_1)}{(T_3 - T_2)}\right\}$$

Using perfect gas equation and governing equation for thermodynamic process 1 –2;

$$\frac{p_1 V_1}{T_1} = \frac{p_2 V_2}{T_2}$$

and $$p_1 V_1^{\gamma} = p_2 V_2^{\gamma}$$

Combining above two, we get $\frac{T_2}{T_1} = \left(\frac{V_1}{V_2}\right)^{\gamma-1}$

$$\frac{T_2}{T_1} = (r)^{\gamma-1}$$

$\Rightarrow$ $$T_2 = T_1 \cdot r^{\gamma-1}$$

also as $$\frac{V_3}{V_2} = \frac{T_3}{T_2}$$

or $$\frac{T_3}{T_2} = \rho$$

or $$T_3 = T_2 \cdot \rho$$

or $$T_3 = T_1 \cdot r^{\gamma-1} \cdot \rho$$

Also for adiabatic process 3 – 4 combining the following:

$$\frac{p_3 V_3}{T_3} = \frac{p_4 V_4}{T_4} \cdot \text{and } p_3 V_3^{\gamma} = p_4 V_4^{\gamma}$$

We get, $$\frac{T_3}{T_4} = \left(\frac{V_4}{V_3}\right)^{\gamma-1}$$

or $$\frac{T_3}{T_4} = \left(\frac{V_4}{V_2} \times \frac{V_2}{V_3}\right)^{\gamma-1}$$

$$= \left(\frac{V_1}{V_2} \times \frac{V_2}{V_3}\right)^{\gamma-1}$$

$$\frac{T_3}{T_4} = \left(\frac{r}{\rho}\right)^{\gamma-1}$$

$$T_4 = T_1 \cdot \rho \cdot r^{\gamma-1} \times \frac{\rho^{\gamma-1}}{r^{\gamma-1}}$$

$$T_4 = T_1 \cdot \rho^{\gamma}$$

Substituting T_2, T_3 and T_4 as function of T_1, r, ρ, γ in the expression for air standard efficiency of diesel cycle

$$\eta_{\text{diesel}} = 1 - \frac{1}{\gamma}\left\{\frac{\rho^{\gamma}-1}{r^{\gamma-1}(\rho-1)}\right\}$$

on

$$\eta_{\text{diesel}} = 1 - \frac{1}{\gamma \cdot r^{\gamma-1}}\left\{\frac{\rho^{\gamma}-1}{(\rho-1)}\right\}$$

Thus, diesel cycle efficiency depends upon r, ρ and γ.

For air as working fluid the variation of diesel cycle efficiency with compression ratio for varying cut-off ratios is given in Figure 9.4.

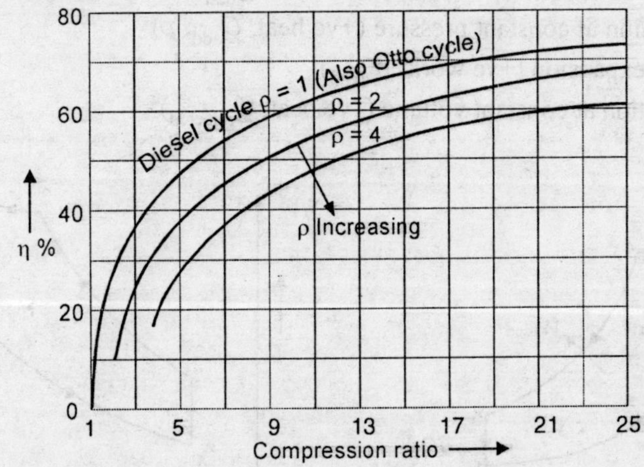

Fig. 9.4 Diesel cycle efficiency vs. compression ratio

Mean effective pressure for diesel cycle can be obtained by ratio of area enclosed on $P - V$ diagram (work) and length of $P - V$ diagram (volume). Assuming clearance volume to be unity i.e. $V_2 = 1$

Net Work = Area of diagram

$$= \frac{p_3 V_3 - p_4 V_4}{\gamma - 1} + p_2 (V_3 - V_2) - \frac{p_2 V_2 - p_1 V_1}{\gamma - 1}$$

Substituting in terms of cut off ratio and compression ratio:

$$= \frac{p_2 \rho - p_4 \cdot r - (p_2 - p_1 r)}{\gamma - 1} + p_2 (\rho - 1)$$

or

$$\text{Net work} = \frac{p_2}{\gamma - 1}\ \{\gamma(\rho-1) - r^{1-\gamma}\cdot(r^{\gamma}-1)\}$$

$$\text{Mean effective pressure} = \frac{\text{Area of } P - V \text{ diagram}}{\text{Length of } P - V \text{ diagram}}$$

$$= \left(\frac{p_2}{\gamma - 1}\right)\ \{\gamma\cdot(\rho-1) - r^{1-\gamma}(\rho^{\gamma}-1)\} \times \frac{1}{(r-1)}$$

or

$$(\text{m.e.p.})_{\text{diesel}} = \frac{p_1 \cdot r^{\gamma}\left[\gamma(\rho-1) - r^{1-\gamma}\cdot(\rho^{\gamma}-1)\right]}{(\gamma-1)(r-1)}$$

Dual cycle : It is also called 'mixed cycle' or 'limited pressure cycle.' Dual cycle came up as a result of certain merits and demerits associated with Otto cycle and Diesel cycle due to heat addition occurring at constant pressure respectively.

Dual cycle is the combination of Otto cycle and Diesel cycle in which heat addition takes place partly at constant volume and partly at constant pressure.

Thermodynamic processes involved in Dual cycle are given as under.

1 – 2 = Adiabatic compression (–ve work, W_{compr})

2 – 3 = Heat addition at constant volume (+ve heat, Q_{add}, v)

3 – 4 = Heat addition at constant pressure (+ve heat, Q_{add}, p)

4 – 5 = Adiabatic expansion (+ve work, W_{expn})

5 – 1 = Heat rejection at constant volume (–ve heat, Q_{rejected})

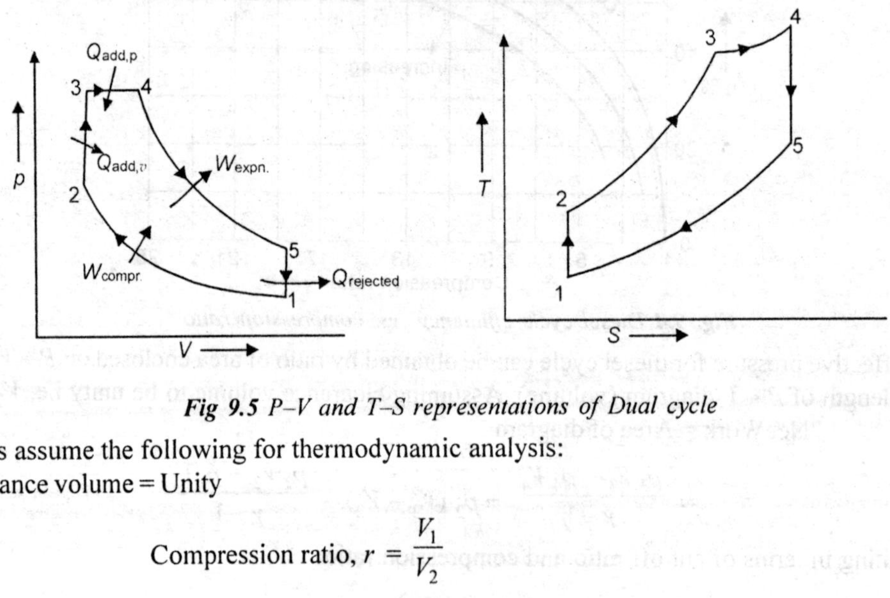

Fig 9.5 P–V and T–S representations of Dual cycle

Let us assume the following for thermodynamic analysis:

Clearance volume = Unity

$$\text{Compression ratio, } r = \frac{V_1}{V_2}$$

$$\text{Cut-off ratio, } \rho = \frac{V_4}{V_3}$$

$$\text{Pressure ratio during heat addition, } \alpha = \frac{p_3}{p_2}$$

For unit mass of air as working fluid throughout the cycle.

Total Heat added = Heat added at constant volume (2 –3) + Heat added at constant pressure (3 – 4)

$$Q_{\text{add}} = c_v\,(T_3 - T_2) + c_p\,(T_4 - T_3)$$

$$\text{Heat rejected, } Q_{\text{rejected}} = c_v\,(T_5 - T_1)$$

Air standard efficiency for Dual cycle can be given as;

$$\eta_{\text{dual}} = \frac{\text{Heat added} - \text{Heat rejected}}{\text{Heat added}}$$

$$= \frac{c_v(T_3 - T_2) + c_p(T_4 - T_3) - c_v(T_5 - T_1)}{c_v(T_3 - T_2) + c_p(T_4 - T_3)}$$

$$= 1 - \frac{c_v(T_5 - T_1)}{c_v(T_3 - T_2) + c_p(T_4 - T_3)}$$

or
$$= 1 - \frac{(T_5 - T_1)}{(T_3 - T_2) + \gamma(T_4 - T_3)}$$

From gas laws applied to process 2–3,

$$\frac{p_3}{T_3} = \frac{p_2}{T_2}$$

or
$$T_2 = T_3 \times \frac{p_2}{p_3}$$

$$T_2 = \frac{T_3}{\alpha}$$

For process 3–4,

$$\frac{V_4}{T_4} = \frac{V_3}{T_3}$$

or
$$T_4 = T_3 \times \frac{V_4}{V_3}$$

$$T_4 = \rho \cdot T_3$$

For adiabatic process 4–5

$$\frac{T_4}{T_5} = \left(\frac{V_5}{V_4}\right)^{\gamma-1}$$

or
$$T_5 = \frac{T_4}{\left(\frac{V_5}{V_4}\right)^{\gamma-1}}$$

$$T_5 = \frac{T_4 \cdot \rho^{\gamma-1}}{r^{\gamma-1}}$$

Substituting T_4

or
$$T_5 = \frac{T_3 \cdot \rho^{\gamma}}{r^{\gamma-1}}$$

For adiabatic process 1–2

$$\frac{T_2}{T_1} = \left(\frac{V_1}{V_2}\right)^{\gamma-1}$$

or
$$T_1 = \frac{T_2}{r^{\gamma-1}}$$

Substituting for T_2

$$T_1 = \frac{T_3}{\alpha \cdot r^{\gamma-1}}$$

Substituting for T_1, T_2, T_4, T_5 in expression for efficiency.

$$\eta_{\text{dual}} = 1 - \frac{1}{r^{\gamma-1}}\left[\frac{\alpha \cdot \rho^{\gamma} - 1}{(\alpha-1) + \alpha \cdot \gamma(\rho-1)}\right]$$

For unity cut off ratio i.e. absence of 3 – 4 process, cycle becomes equal to Otto cycle.

i.e. for
$$\rho = 1,\ \eta_{\text{dual}} = 1 - \frac{1}{r^{\gamma-1}} = \eta_{\text{otto}}$$

For the pressure ratio α being unity, cycle gets modified to Diesel cycle.

i.e. for
$$\alpha = 1,\ \eta_{\text{dual}} = 1 - \frac{1}{\gamma \cdot r^{\gamma-1}}\left(\frac{\rho^{\gamma} - 1}{(\rho-1)}\right) = \eta_{\text{diesel}}$$

Mean effective pressure of dual cycle can be obtained as follows:
Net work = Area enclosed in $P - V$ diagram

$$= p_3 (V_4 - V_3) + \frac{p_4 V_4 - p_5 V_5}{\gamma - 1} - \frac{p_2 V_2 - p_1 V_1}{\gamma - 1}$$

or
$$= p_3(\rho - 1) + \frac{p_3 \rho - p_5 \cdot r - p_2 + p_1 \cdot r}{(\gamma - 1)}$$

Making suitable substitutions for p_1, p_2 and p_5, it gets modified as

$$\text{Net work} = \frac{p_3\left[\alpha \cdot \gamma(\rho-1) + (\alpha-1) - r^{1-\gamma} \cdot (\alpha \cdot \rho^{\gamma} - 1)\right]}{\alpha(\gamma-1)}$$

Substituting for mean effective pressure

$$(\text{m.e.p.})_{\text{dual}} = \frac{\text{Area of diagram}}{\text{Length of diagram}}$$

$$= \frac{1}{(r-1)} \times \left\{\frac{p_3\left[\alpha \cdot \gamma(\rho-1) + (\alpha-1) - r^{1-\gamma} \cdot (\alpha \cdot \rho^{\gamma} - 1)\right]}{\alpha(\gamma-1)}\right\}$$

Substituting p_3 as function of p_1

or

$$(\text{m.e.p.})_{\text{dual}} = \frac{p_1 \cdot r^{\gamma}\left[\alpha \cdot \gamma(\rho-1) + (\alpha-1) - r^{1-\gamma} \cdot (\alpha \cdot \rho^{\gamma} - 1)\right]}{(\gamma-1)(r-1)}$$

9.3 BRAYTON CYCLE

Brayton cycle, popularly used for gas turbine power plants comprises of adiabatic compression process, constant pressure heat addition, adiabatic expansion process and constant pressure heat release process. A schematic diagram for air-standard Brayton cycle is shown in Fig. 9.6. Simple gas turbine power plant working on Brayton cycle is also shown here.

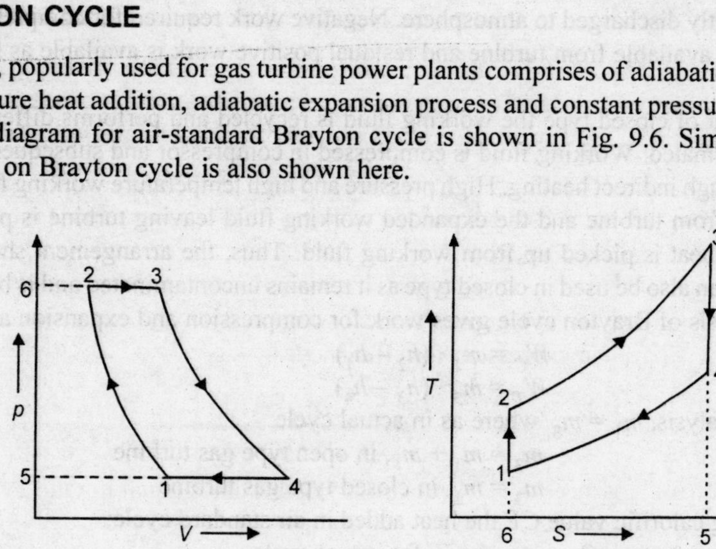

Fig. 9.6 Brayton cycle on P–V and T–S diagram

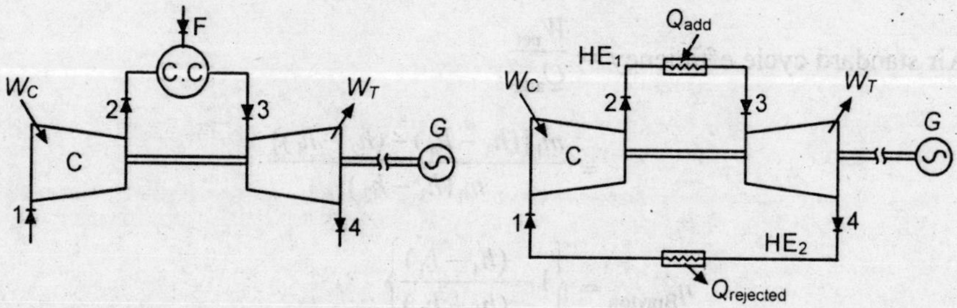

C : Compressor
CC : Combustion chamber
(a) Open type

F : Fuel
G : Generator
HC : Heat exchanger
(b) Closed type

Fig. 9.7 Simple gas turbine plant

Thermodynamic cycle shows following processes:

1-2 : Adiabatic compression, involving (–ve) work, W_C in compressor.
2-3 : Constant pressure heat addition, involving heat Q_{add} in combustion chamber or heat exchanger.
3-4 : Adiabatic expansion, involving (+ve) work, W_T in turbine.
4-1 : Constant pressure heat rejection, involving heat, $Q_{rejected}$ in atmosphere or heat exchanger.

In the gas turbine plant layout shown process 1–2 (adiabatic compression) is seen to occur in compressor, heat addition process 2–3 occurs in combustion chamber having open type arrangement and in heat exchanger in closed type arrangement. Process 3–4 of adiabatic expansion occurs in turbine. In open type arrangement exhaust from turbine is discharged to atmosphere while in closed type, heat rejection occurs in heat exchanger. In gas turbine plant of open type air entering compressor gets compressed and subsequently brought up to elevated temperature in combustion chamber where fuel is added to high pressure air and combustion occurs. High pressure and high temperature combustion products are sent for expansion in turbine where its' expansion yields positive work. Expanded combustion products are subsequently discharged to atmosphere. Negative work required for compression is drawn from the positive work available from turbine and residual positive work is available as shaft work for driving generator.

In gas turbine plant of closed type the working fluid is recycled and performs different processes without getting contaminated. Working fluid is compressed in compressor and subsequently heated up in heat exchanger through indirect heating. High pressure and high temperature working fluid is sent for getting positive work from turbine and the expanded working fluid leaving turbine is passed through heat exchanger where heat is picked up from working fluid. Thus, the arrangement shows that even costly working fluids can also be used in closed type as it remains uncontaminated and is being recycled.

Air standard analysis of Brayton cycle gives work for compression and expansion as;

$$W_C = m_1 \cdot (h_2 - h_1)$$
$$W_T = m_3 \cdot (h_3 - h_4)$$

for air standard analysis, $m_1 = m_3$, where as in actual cycle

$$m_3 = m_1 + m_f,\ \text{in open type gas turbine}$$
$$m_3 = m_1,\ \text{in closed type gas turbine}$$

For the fuel having calorific value CV the heat added in air standard cycle;
$Q_{add} = m_1(h_3 - h_2)$, whereas $Q_{add} = m_f \times CV$ for actual cycle.

$$\text{Net work} = W_T - W_C$$
$$W_{net} = \{m_3 (h_3 - h_4) - m_1(h_2 - h_1)\}$$

$$\text{Air standard cycle efficiency} = \frac{W_{net}}{Q_{add}}$$

$$= \frac{m_1\{(h_3 - h_4) - (h_2 - h_1)\}}{m_1 (h_3 - h_2)}$$

$$\eta_{Brayton} = \left\{1 - \frac{(h_4 - h_1)}{(h_3 - h_2)}\right\}$$

$$\eta_{Brayton} = \left\{1 - \frac{(T_4 - T_1)}{(T_3 - T_2)}\right\} \text{ as, } (h_4 - h_1) = c_p\,(T_4 - T_1)$$

$$\text{and } (h_3 - h_2) = c_p (T_3 - T_2)$$

For processes 1–2 and 3–4, which are of isentropic type,

$$\frac{T_2}{T_1} = \left(\frac{p_2}{p_1}\right)^{\frac{(\gamma-1)}{\gamma}} \text{ and } \frac{T_3}{T_4} = \left(\frac{p_2}{p_1}\right)^{\frac{(\gamma-1)}{\gamma}}$$

Let the pressure ratio be, $\dfrac{p_2}{p_1} = r$,

$$\frac{T_2}{T_1} = (r)^{\frac{(\gamma-1)}{\gamma}} \text{ and } \frac{T_3}{T_4} = (r)^{\frac{(\gamma-1)}{\gamma}}$$

so, $$\frac{T_2}{T_1} = \frac{T_3}{T_4}$$

$$\frac{T_4}{T_1} = \frac{T_3}{T_2}$$

$$\frac{T_4 - T_1}{T_1} = \frac{T_3 - T_2}{T_2}$$

$$\frac{(T_4 - T_1)}{(T_3 - T_2)} = \frac{T_1}{T_2} = \left(\frac{p_1}{p_2}\right)^{\left(\frac{\gamma-1}{\gamma}\right)}$$

Making substitution for $\left\{\frac{(T_4 - T_1)}{(T_3 - T_2)}\right\}$ in cycle efficiency:

$$\eta_{\text{Brayton}} = \left(1 - \frac{T_1}{T_2}\right) = 1 - \left(\frac{p_1}{p_2}\right)^{\left(\frac{\gamma-1}{\gamma}\right)} = 1 - \frac{1}{r^{\frac{\gamma-1}{\gamma}}}$$

Air standard Brayton cycle efficiency: $\boxed{\eta_{\text{Brayton}} = 1 - \frac{1}{r^{\frac{\gamma-1}{\gamma}}}}$

Thus, it is obvious from the expression of efficiency that it depends only on pressure ratio (r) and nature of gas (γ). For pressure ratio of unity, efficiency shall be zero. For a particular gas the cycle efficiency increases with increasing pressure ratio. Here the variation of efficiency with pressure ratio is shown for air ($\gamma = 1.4$) and monoatomic gas as argon ($\gamma = 1.66$).

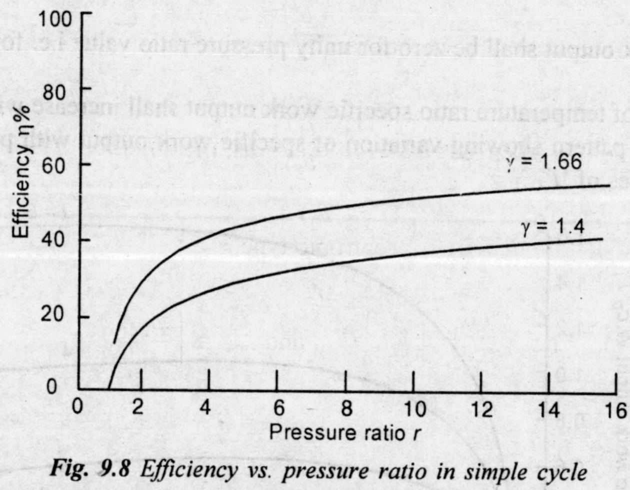

Fig. 9.8 Efficiency vs. pressure ratio in simple cycle

Specific work output for the plant can be given as below:

$$w = \{(h_3 - h_4) - (h_2 - h_1)\}, \text{ as } m_1 = 1$$

$$w = c_p \{(T_3 - T_4) - (T_2 - T_1)\}$$

or,
$$\frac{w}{c_p T_1} = \left\{\left(\frac{T_3}{T_1} - \frac{T_4}{T_1}\right) - \left(\frac{T_2}{T_1} - 1\right)\right\}$$

$$= \left\{ \left(\frac{T_3}{T_1} \right) \left(1 - \frac{T_4 \cdot T_1}{T_3 \cdot T_1} \right) - \left(\frac{T_2}{T_1} - 1 \right) \right\}$$

In the cycle states 3 and 1 refer to the states having highest temperature and pressure and lowest temperature and pressure states. T_3 is also referred as turbine inlet temperature while T_1 is ambient temperature in case of open type arrangement. Let the ratio $\frac{T_3}{T_1}$ be given by some constant 't' i.e. $\frac{T_3}{T_1} = t$.

Therefore, upon making substitutions,

$$\frac{w}{c_p T_1} = \left\{ t \cdot \left(1 - \frac{T_4}{T_3} \right) - \left(\frac{T_2}{T_1} - 1 \right) \right\}$$

as,

$$\frac{T_4}{T_3} = \frac{1}{r^{\frac{\gamma-1}{\gamma}}} \text{ and } \frac{T_2}{T_1} = r^{\frac{\gamma-1}{\gamma}}$$

$$\boxed{\frac{w}{c_p T_1} = \left\{ t \cdot \left(1 - \frac{1}{r^{\frac{\gamma-1}{\gamma}}} \right) - \left(r^{\frac{\gamma-1}{\gamma}} - 1 \right) \right\}}$$

Here $\left(\frac{w}{c_p T_1} \right)$ is a non-dimensional form of specific work output.

In general for a given arrangement T_1 being ambient temperature remains fixed (mostly atmospheric temperature). So specific work output may be said to depend upon 't' (temperature ratio) and pressure ratio 'r'. Specific work output shall be zero for unity pressure ratio value i.e. for $r = 1$, $\frac{w}{c_p T_1} = 0$.

For certain value of temperature ratio specific work output shall increase with increasing pressure ratio value. Graphical pattern showing variation of specific work output with pressure ratio is shown below for certain values of 't'.

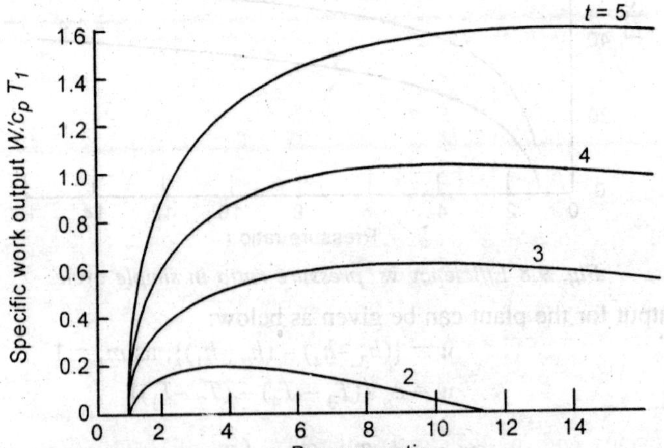

Fig. 9.9 Specific work output vs. pressure ratio for simple cycle

The temperature ratio 't' is generally used as design parameter. Higher is the 't' value higher shall be specific work output. Maximum value of 't' depends upon the maximum temperature which the highly stressed parts of turbine can withstand for the required working life (also called metallurgical limit).

In the modern gas turbine value of 't' ranges from 3.5 to 6. Further higher values of turbine inlet temperature and thus 't' value can be realized with improved gas turbine blade cooling.

Specific work output vs. pressure ratio variation shows that there exist some optimum pressure ratio value at which specific work output is maximum at given 't' value. Let us differentiate expression for specific work output with respect to $r^{\left(\frac{\gamma-1}{\gamma}\right)}$.

$$\text{Let } r^{\left(\frac{\gamma-1}{\gamma}\right)} = a,\ t = \text{constant}$$

$$w = c_p\, T_1 \left\{ t\cdot\left(1 - \frac{1}{r^{\frac{\gamma-1}{\gamma}}}\right) - \left(r^{\frac{\gamma-1}{\gamma}} - 1\right) \right\}$$

$$w = c_p\, T_1 \left\{ t\cdot\left(1 - \frac{1}{a}\right) - (a-1) \right\}$$

$$\frac{dw}{da} = c_p\, T_1 \left\{ t\cdot\left(\frac{1}{a^2}\right) - 1 \right\}$$

$$\text{Equating } \frac{dw}{da} = 0, \text{ so } \frac{t}{a^2} - 1 = 0$$

$$t = a^2$$

$$a = \sqrt{t}, \text{ or } \boxed{r = t^{\gamma/2(\gamma-1)}}$$

Thus,
$$r^{\frac{\gamma-1}{\gamma}} = \sqrt{t} \text{ or, } r^{\frac{\gamma-1}{\gamma}} \cdot r^{\frac{\gamma-1}{\gamma}} = t$$

Substituting for $r^{\frac{\gamma-1}{\gamma}}$ and t;

$$\frac{T_2}{T_1} \times \frac{T_3}{T_4} = \frac{T_3}{T_1}$$

or,
$$\boxed{T_2 = T_4}$$

Thus, the specific work output is seen to be maximum for a given 't' value when the pressure ratio is such that temperature at compressor outlet and turbine outlet are equal.

For all values of pressure ratio 'r' lying between 1 and $t^{\gamma/2(\gamma-1)}$ temperature T_4 is greater than T_2. In order to reduce heat transfer from external source for improving efficiency a heat exchanger can be incorporated. This concept of using heat exchanger gives birth to modified form of simple gas turbine cycle with heat exchange. This modified form of cycle is also called regenerative gas turbine cycle. Simple gas turbine cycles with modifications of reheat and intercooling are also used for improved performance.

9.4 REGENERATIVE GAS TURBINE CYCLE

In earlier discussions it is seen that for the maximization of specific work output the gas turbine exhaust temperature should be equal to compressor exhaust temperature. The turbine exhaust temperature is normally much above the ambient temperature. Thus, their exist potential for tapping the heat energy getting lost to surroundings with exhaust gases. Here it is devised to use this potential by means of a heat exchanger called regenerator, which shall preheat the air leaving compressor before entering the combustion chamber, thereby reducing the amount of fuel to be burnt inside combustion chamber (combustor).

Regenerative air standard gas turbine cycle shown ahead in Fig. 9.10 has a regenerator (counter flow heat exchanger) through which the hot turbine exhaust gas and comparatively cooler air coming from compressor flow in opposite directions. Under ideal conditions, no frictional pressure drop occurs in either fluid stream while turbine exhaust gas gets cooled from 4 to 4′ while compressed air is heated from 2 to 2′. Assuming regenerator effectiveness as 100% the temperature rise from 2–2′ and drop from 4 to 4′ is shown on *T-S* diagram.

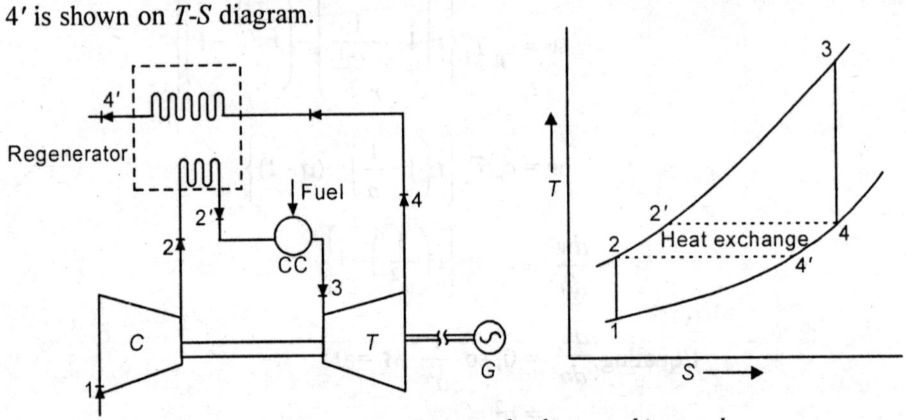

Fig. 9.10 Regenerative air standard gas turbine cycle.

Regenerator effectiveness, $\varepsilon = \dfrac{h_{2'} - h_2}{h_4 - h_2}$, where '*h*' refers to specific enthalpy values.

Thus, thermodynamically the amount of heat now added shall be

$$Q_{\text{add, regen}} = m(h_3 - h_{2'})$$

where as without regenerator the heat added; $Q_{\text{add}} = m(h_3 - h_2)$

Here it is obvious that, $Q_{\text{add, regen}} < Q_{\text{add}}$.

This shows an obvious improvement in cycle thermal efficiency as every thing else remains same.

Net work produced per unit mass flow is not altered by the use of regenerator.

$$\text{Air standard cycle thermal efficiency, } \eta_{\text{regen}} = \frac{(h_3 - h_4) - (h_2 - h_1)}{(h_3 - h_{2'})}$$

$$\eta_{\text{regen}} = \frac{c_p \cdot (T_3 - T_4) - c_p(T_2 - T_1)}{c_p(T_3 - T_{2'})}$$

Thus, regenerative gas turbine cycle efficiency approaches to Carnot cycle efficiency at $r = 1$ which is because in this limiting situation the Carnot cycle requirement of total heat addition and rejection at maximum and minimum cycle temperature is satisfied. Variation of efficiency with cycle pressure ratio for regenerative cycle is shown in Fig. 9.11 for fixed '*t*' values.

Cycle efficiency is seen to decrease with increasing pressure ratio till pressure ratio becomes equal to $r = t^{\left(\frac{\gamma-1}{2\gamma}\right)}$. In earlier discussions it has been shown that at this pressure ratio value the specific work output becomes maximum. Same specific work output variation is true for this case too. For further higher values of pressure ratio the heat exchanger shall cool the air leaving compressor and lower the cycle efficiency than simple cycle efficiency.

This is because of which the constant 't' lines are not extended beyond the point where they meet efficiency curve for simple cycle.

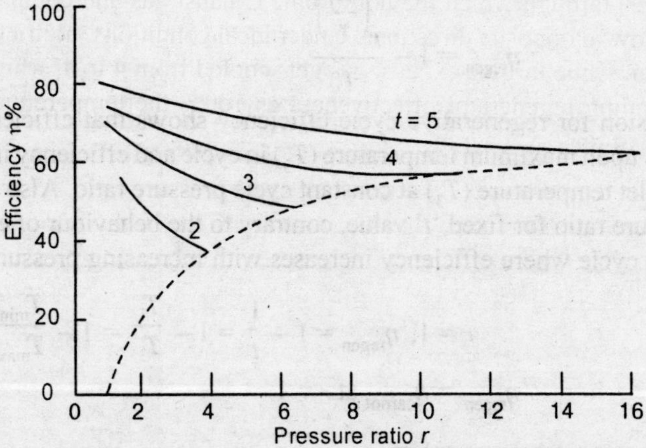

Fig. 9.11 Efficiency vs. pressure ratio for regenerative gas turbine cycle.

Thus, it can be concluded that for getting improvement in efficiency by incorporating regenerator the pressure ratio 'r' appreciably less than the optimum one for maximum specific work output should be used. Also maximum turbine inlet temperature should be used.

For ideal heat exchange in regenerator with 100% effectiveness,

Upon substitution it yields $T_2' = T_4$

$$\eta_{\text{regen}} = \frac{(T_3 - T_4) - (T_2 - T_1)}{(T_3 - T_4)}$$

$$\eta_{\text{regen}} = 1 - \frac{(T_2 - T_1)}{(T_3 - T_4)}$$

From thermodynamic processes

$$\frac{T_2}{T_1} = \frac{T_3}{T_4} = r^{\left(\frac{\gamma-1}{\gamma}\right)}$$

or,
$$\frac{T_2 - T_1}{T_3 - T_4} = \frac{T_1}{T_4}$$

or,
$$\frac{T_2 - T_1}{T_3 - T_4} = \frac{T_1}{T_3} \times \frac{T_3}{T_4}$$

Also we know

$$\frac{T_3}{T_1} = t;$$

or,
$$\frac{T_2 - T_1}{T_3 - T_4} = \frac{r^{\left(\frac{\gamma-1}{\gamma}\right)}}{t}$$

Substituting in expression for efficiency,

$$\eta_{regen} = 1 - \frac{r^{\left(\frac{\gamma-1}{\gamma}\right)}}{t}$$

Thus, the expression for regenerative cycle efficiency shows that efficiency of regenerative gas turbine cycle depends upon maximum temperature (T_3) in cycle and efficiency increases with increasing 't' value or turbine inlet temperature (T_3) at constant cycle pressure ratio. Also the efficiency decreases with increasing pressure ratio for fixed 't' value, contrary to the behaviour of efficiency shown in case of simple gas turbine cycle where efficiency increases with increasing pressure ratio.

At
$$r = 1, \ \eta_{regen} = 1 - \frac{1}{t} = 1 - \frac{T_1}{T_3} = 1 - \frac{T_{min}}{T_{max}}$$

i.e.
$$\eta_{regen} = \eta_{carnot} \text{ at } r = 1$$

9.5 REHEAT GAS TURBINE CYCLE

Reheat gas turbine cycle arrangement is shown in Fig. 9.12. In order to maximize the work available from the simple gas turbine cycle one of the option is to increase enthalpy of fluid entering gas turbine and extend its expansion upto the lowest possible enthalpy value.

C : Compressor
CC : Combustion chamber
G : Generator
f : Fuel
HPT : High pressure turbine
LPT : Low pressure turbine
RCC : Reheat combustion chamber

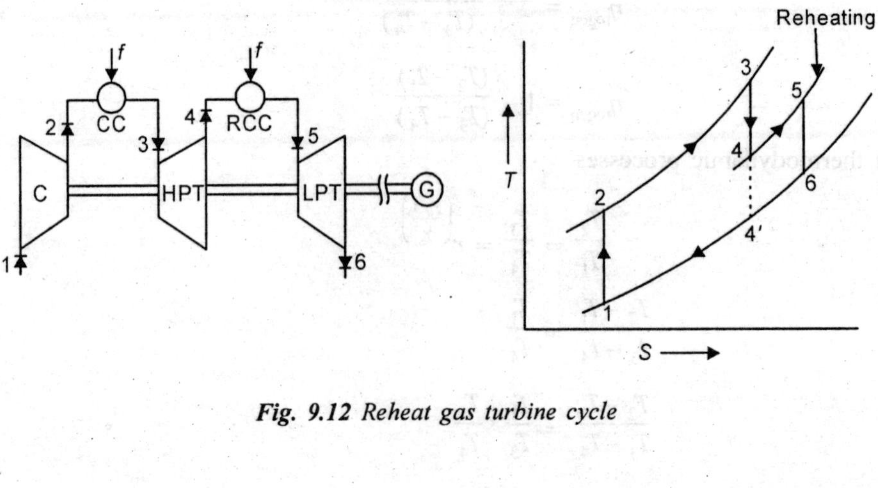

Fig. 9.12 Reheat gas turbine cycle

This can also be said in terms of pressure and temperature values i.e. inject fluid at high pressure and temperature into gas turbine and expand upto lowest possible pressure value. Upper limit at inlet to turbine is limited by metallurgical limits while lower pressure is limited to near atmospheric pressure in case of open cycle. For further increasing the net work output the positive work may be increased by using multistage expansion with reheating in between. In multistage expansion the expansion is divided into parts and after part expansion working fluid may be reheated for getting larger positive work in left out expansion. For reheating another combustion chamber may be used.

Here in the arrangement shown ambient air enters compressor and compressed air at high pressure leaves at 2. Compressed air is injected into combustion chamber for increasing its temperature upto desired turbine inlet temperature at state 3. High pressure and high temperature fluid enters high pressure turbine (HPT) for first phase of expansion and expanded gases leaving at 4 are sent to reheat combustion chamber (reheater) for being further heated. Thus, reheating is a kind of energising the working fluid. Assuming perfect reheating (in which temperature after reheat is same as temperature attained in first combustion chamber), the fluid leaves at state 5 and enters low pressure turbine (LPT) for remaining expansion upto desired pressure value. Generally, temperature after reheating at state 5 is less than temperature at state 3. In the absence of reheating the expansion process within similar pressure limits goes upto state 4′. Thus, reheating offers an obvious advantage of work output increase since constant pressure lines on *T-S* diagram diverge slightly with increasing entropy, the total work of the two stage turbine is greater than that of single expansion from state 3 to state 4′. i.e., $(T_3 - T_4) + (T_5 - T_6) > (T_3 - T_{4'})$.

Here it may be noted that the heat addition also increases because of additional heat supplied for reheating. Therefore, despite the increase in net work due to reheating the cycle thermal efficiency would not necessarily increase. Let us now carry out air standard cycle analysis.

Net work output in reheat cycle, $W_{\text{net, reheat}} = W_{\text{HPT}} + W_{\text{LPT}} - W_{\text{C}}$

$$W_{\text{HPT}} = m(h_3 - h_4),\ W_{\text{LPT}} = (h_5 - h_6),\ W_{\text{C}} = m(h_2 - h_1)$$

$$W_{\text{net, reheat}} = m\{(h_3 - h_4) + (h_5 - h_6) - (h_2 - h_1)\}$$

$$W_{\text{net, reheat}} = m\,c_p\,\{(T_3 - T_4) + (T_5 - T_6) - (T_2 - T_1)\}$$

assuming $T_3 = T_5$ i.e. perfect reheating

$$W_{\text{net, reheat}} = m\,c_p\,\{(2T_3 - T_4) - T_6 - (T_2 - T_1)\}$$

$$W_{\text{net, reheat}} = m\,c_p\,T_1\left\{\left(\frac{2T_3}{T_1} - \frac{T_4}{T_1}\right) - \left(\frac{T_6}{T_1}\right) - \left(\frac{T_2}{T_1} - 1\right)\right\}$$

Here $\dfrac{T_3}{T_1} = t$; pressure ratio, $r = \left(\dfrac{T_2}{T_1}\right)^{\left(\frac{\gamma}{\gamma-1}\right)}$

Let pressure ratio for HP turbine be r_1 and for LP turbine be r_2 then;

$$r = r_1 \times r_2$$

$$r_1 = \left(\frac{T_3}{T_4}\right)^{\left(\frac{\gamma}{\gamma-1}\right)},\ r_2 = \left(\frac{T_5}{T_6}\right)^{\left(\frac{\gamma}{\gamma-1}\right)} = \frac{r}{r_1};$$

or,

$$\left(\frac{T_5}{T_6}\right) = \left(\frac{r}{r_1}\right)^{\left(\frac{\gamma-1}{\gamma}\right)} = \left(\frac{T_3}{T_6}\right)$$

$$\left(\frac{W_{\text{net, reheat}}}{mc_p\,T_1}\right) = \left\{\left(2t - \frac{T_4}{T_3}\cdot\frac{T_3}{T_1}\right) - \left(\frac{T_6}{T_3}\times\frac{T_3}{T_1}\right) - \left(r^{\frac{\gamma-1}{\gamma}} - 1\right)\right\}$$

Let $\left(\frac{W_{\text{net, reheat}}}{mc_p\,T_1}\right) = w$ then

$$w = \left\{\left(2t - \frac{t}{(r_1)^{\left(\frac{\gamma-1}{\gamma}\right)}}\right) - \left(\left(\frac{r_1}{r}\right)^{\left(\frac{\gamma-1}{\gamma}\right)}\times t - r^{\left(\frac{\gamma-1}{\gamma}\right)} - 1\right)\right\}$$

$$\frac{dw}{dr_1} = 0 - \left(\left(\frac{1-\gamma}{\gamma}\right)\times\frac{t}{r_1^{\left(\frac{2\gamma-1}{\gamma}\right)}}\right) - \left(\left(\frac{\gamma-1}{\gamma}\right)\frac{t}{r_1^{1/\gamma}}\times\frac{1}{r^{\left(\frac{\gamma-1}{\gamma}\right)}}\right) - 0$$

Equating $\frac{dw}{dr_1}$ to zero, $\frac{1}{r_1^{2\gamma-1/\gamma}} = \frac{1}{r^{\left(\frac{\gamma-1}{\gamma}\right)}\cdot r_1^{1/\gamma}} \Rightarrow r^{\left(\frac{\gamma-1}{\gamma}\right)} = r_1^{2\left(\frac{\gamma-1}{\gamma}\right)}$

or, $r_1 = \sqrt{r} = r_2$

Thus, expansion pressure ratio for high pressure turbine and low pressure turbine should be same for maximum work output from the cycle. With such optimum division the expression for work output and efficiency can be given as,

$$w = \left\{\left(2t - \frac{t}{(r)^{\left(\frac{\gamma-1}{2\gamma}\right)}}\right) - \left(\frac{t}{(r)^{\frac{\gamma-1}{2\gamma}}}\right) - r^{\left(\frac{\gamma-1}{\gamma}\right)} - 1\right\}$$

Let, $r^{\left(\frac{\gamma-1}{\gamma}\right)} = c$ then,

$$\left(\frac{W_{\text{net, reheat}}}{mc_p\,T_1}\right) = \left\{\left(2t - \frac{2t}{\sqrt{c}}\right) - (c-1)\right\}$$

$$\boxed{\frac{W_{\text{net, reheat}}}{mc_pT_1} = \left\{2t - \frac{2t}{\sqrt{c}} - c + 1\right\}}$$

Efficiency for reheat cycle

$$\eta_{\text{reheat}} = \frac{W_{\text{net, reheat}}}{m(h_3 - h_2) + m(h_5 - h_4)}$$

$$= \frac{W_{\text{net, reheat}}}{mc_p\{(T_3 - T_2) + (T_5 - T_4)\}}$$

$$\eta_{\text{reheat}} = \frac{W_{\text{net, reheat}}}{mc_pT_1\left\{\left(\frac{T_3}{T_1} - \frac{T_2}{T_1}\right) + \left(\frac{T_5}{T_1} - \frac{T_4}{T_1}\right)\right\}}$$

Let us simplify denominator first,

$$= m\,c_p\,T_1\left\{\left(\frac{T_3}{T_1} - \frac{T_2}{T_1}\right) + \left(\frac{T_5}{T_1} - \frac{T_4}{T_1}\right)\right\}$$

We know $T_5 = T_3$ and $\frac{T_3}{T_1} = t$

$$= m\,c_p\,T_1\left\{\left(t - \frac{T_2}{T_1}\right) + \left(t - \frac{T_4}{T_1}\right)\right\}$$

Also $\frac{T_2}{T_1} = r^{\left(\frac{\gamma-1}{\gamma}\right)}, \frac{T_3}{T_4} = r^{\left(\frac{\gamma-1}{2\gamma}\right)}$

Substituting,

$$= m\,c_p\,T_1\left\{\left(t - r^{\left(\frac{\gamma-1}{\gamma}\right)}\right) + \left(t - \frac{T_4}{T_3} \times \frac{T_3}{T_1}\right)\right\}$$

$$= m\,c_p\,T_1\left\{2t - r^{\frac{\gamma-1}{\gamma}} - \frac{t}{r^{\left(\frac{\gamma-1}{2\gamma}\right)}}\right\}$$

$$\text{Denominator} = m\,c_p\,T_1\left\{2t - c - \frac{t}{\sqrt{c}}\right\}$$

Substituting in expression for efficiency for optimum division of pressure ratio;

$$\eta_{\text{reheat}} = \frac{W_{\text{net, reheat}}}{mc_pT_1\left\{2t - c - \frac{t}{\sqrt{c}}\right\}}$$

$$\boxed{\eta_{\text{reheat}} = \frac{\left\{2t - \frac{2t}{\sqrt{c}} - c + 1\right\}}{\left\{2t - c - \frac{t}{\sqrt{c}}\right\}}}$$

A plot showing variation of efficiency with pressure ratio '*r*' is shown in Fig. 9.13 along with simple cycle efficiency variation. It indicates that reheating offers increase in specific work output at the

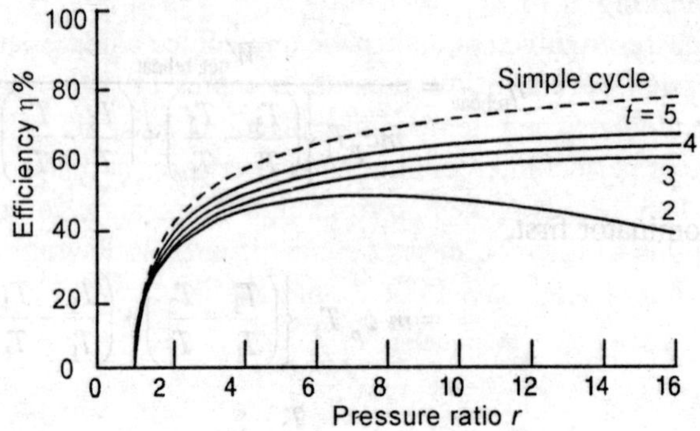

Fig. 9.13 Reheat cycle efficiency vs. cycle pressure ratio

cost of cycle efficiency. This reduction in efficiency may be attributed to the addition of a less efficient cycle 4564′ to a simple cycle 4565′ is a less efficient cycle since it operates over a smaller temperature range. Variation of specific work output with pressure ratio is shown in Fig. 9.14. It shows how specific work output shows increase with increasing pressure ratio upto optimum pressure ratio. It may also be noted that in reheat cycle the temperature of exhaust gases at exit of gas turbine gets increased as compared to simple cycle within similar limits. Therefore, reheat cycle offers potential for use of regenerator for harnessing the hotter exhaust from gas turbine.

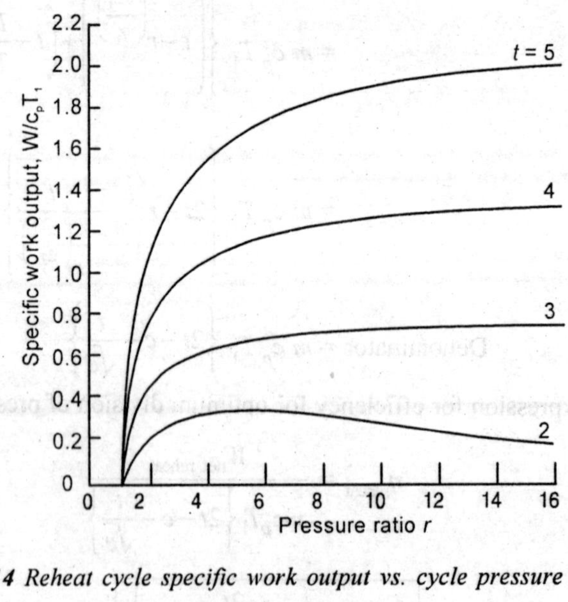

Fig. 9.14 Reheat cycle specific work output vs. cycle pressure ratio

9.6 GAS TURBINE CYCLE WITH INTERCOOLING

Net work output from gas turbine cycle can also be increased by reducing negative work i.e. compressor work. Multistaging of compression process with intercooling in between is one of the approach for reducing compression work. It is based on the fact that for a fixed compression ratio higher is the inlet temperature higher shall be compression work requirement and vice-a-versa. Schematic for intercooled gas turbine cycle is given in Fig. 9.15.

Thermodynamic processes involved in multistage intercooled compression are shown in Figs. 9.16, 9.17. First stage compression occurs in low pressure compressor (LPC) and compressed air leaving LPC at '2' is sent to intercooler where temperature of compressed air is lowered down to state 3 at constant pressure. In case of perfect intercooling the temperatures at 3 and 1 are same. Intercooler is a kind of heat exchanger where heat is picked up from high temperature compressed air. The amount of compression work saved due to intercooling is obvious from *p-V* diagram and shown by area 2342′. Area 2324′ gives the amount of work saved due to intercooling between compression.

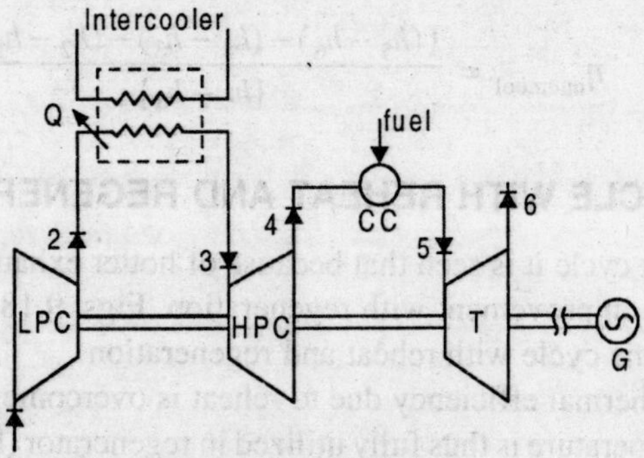

CC: Combustion chamber
LPC: Low pressure compressor
HPC: High pressure compressor
T: Turbine
G: Generator

***Fig. 9.15** Gas turbine cycle with intercooling*

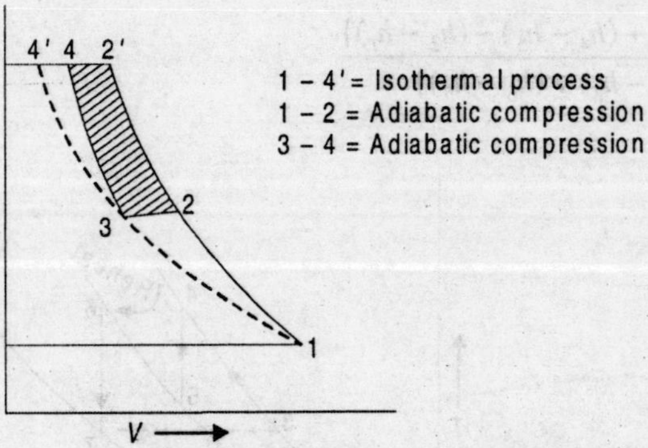

***Fig. 9.16** Intercooled compression*

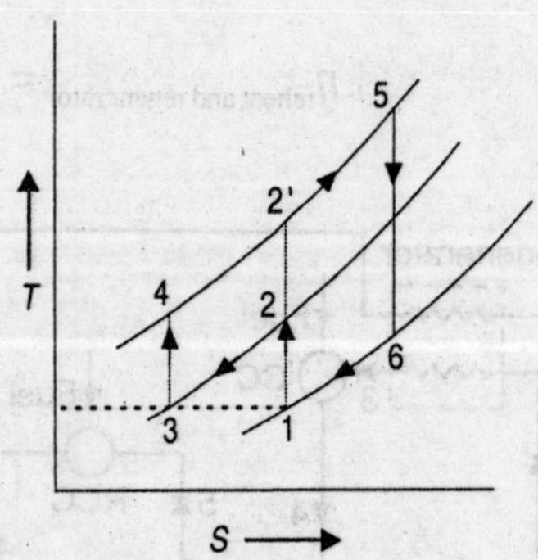

***Fig. 9.17** T-S diagram for gas turbine cycle with intercooling*

Some large compressors have several stages of compression with intercooling between stages. Use of multistage compression with intercooling in a gas turbine power plant increases the net work produced because of reduction in compressor work. Intercooled compression results in reduced temperature at the end of final compression. *T-S* diagram for gas turbine cycle with intercooling shows that in the absence of intercooling within same pressure limits the state at the end of compression would be 2′ while with perfect intercooling this state is at 4 i.e., $T_{2'} > T_4$. The reduced temperature at compressor exit leads to additional heat requirement in combustion chamber i.e. more amount of fuel is to be burnt for attaining certain turbine inlet temperature as compared to simple cycle without intercooling.

Thus, intercooled cycle thermal efficiency may not increase with intercooling because of simultaneous increase in heat addition requirement. The lower temperature at compressor exit enhances the potential for regeneration so when intercooling is used in conjunction with regeneration an appreciable increase in thermal efficiency can result.

Net work output in gas turbine cycle with intercooling;

$$W_{\text{net, intercool}} = m\{(h_5 - h_6) - (h_4 - h_3) - (h_2 - h_1)\}$$

$$W_{\text{net, intercool}} = mc_p\{(T_5 - T_6) - (T_4 - T_3) - (T_2 - T_1)\}$$

Cycle thermal efficiency;

$$\eta_{\text{intercool}} = \frac{\{(h_5 - h_6) - (h_4 - h_3) - (h_2 - h_1)\}}{\{h_5 - h_4\}}$$

9.7 GAS TURBINE CYCLE WITH REHEAT AND REGENERATION

During discussion of reheat cycle it is seen that because of hotter exhaust from gas turbine, there exists possibility for performance improvement with regeneration. Figs. 9.18, 9.19 give schematic and cycle representation for gas turbine cycle with reheat and regeneration.

Here the reduction in thermal efficiency due to reheat is overcome by adding regeneration. Higher gas turbine exhaust gas temperature is thus fully utilized in regenerator (heat exchanger). Additional heat supplied for reheating in order to increase work output is supplemented to some extent by regenerator. It is seen that in this kind of combustion thermal efficiency is increased.

Net work output,

$$W_{\text{net}} = m\{(h_6 - h_7) + (h_4 - h_5) - (h_2 - h_1)\}$$

Heat added,

$$Q_{\text{add}} = m\{(h_4 - h_3) + (h_6 - h_5)\}$$

Thermal efficiency;

$$\eta_{\text{reheat and regenerator}} = \frac{\{(h_6 - h_7) + (h_4 - h_5) - (h_2 - h_1)\}}{\{(h_4 - h_3) + (h_6 - h_5)\}}$$

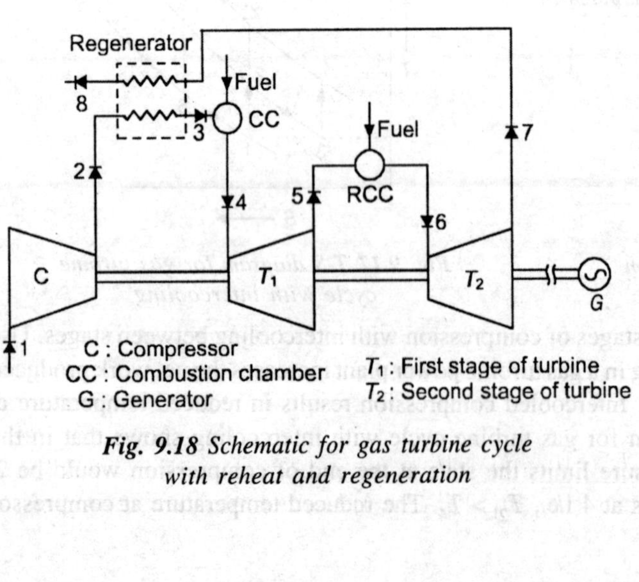

Fig. 9.18 Schematic for gas turbine cycle with reheat and regeneration

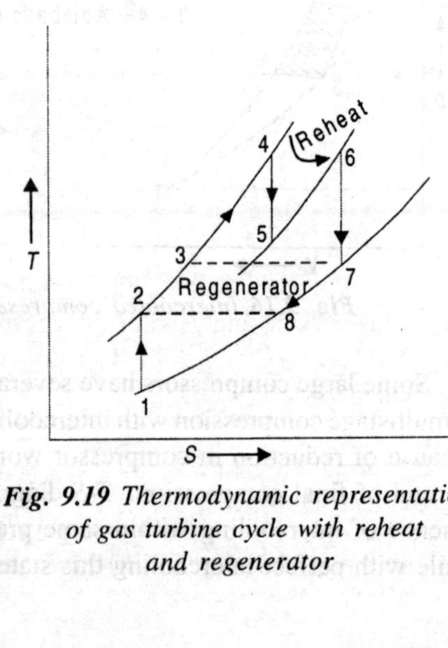

Fig. 9.19 Thermodynamic representation of gas turbine cycle with reheat and regenerator

9.8 GAS TURBINE CYCLE WITH REHEAT AND INTERCOOLING

Schematic arrangement for gas turbine cycle with reheating during expansion and intercooling during compression is given in Fig. 9.20 along with *T-S* diagram. Here reheating and intercooling offer appreciable increase in specific work output but the thermal efficiency deteriorates as reheating and intercooling both have consequence of additional heat requirement. Although due to higher temperature the gas turbine exhaust offers potential for regeneration.

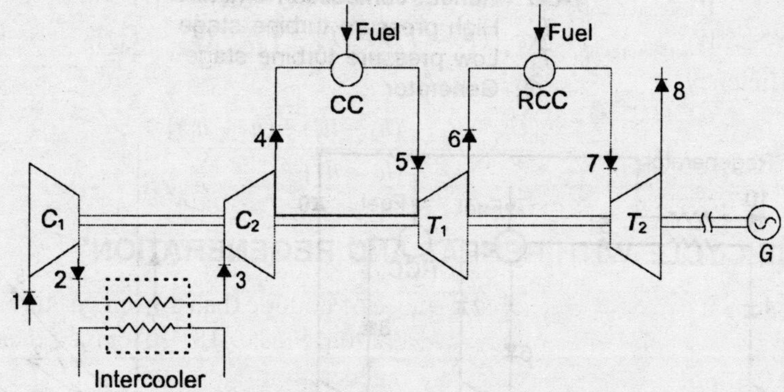

C_1 : Low pressure compressor stage
C_2 : High pressure compressor stage
CC : Combustion chamber
G : Generator
RCC : Reheat combustion chamber
T_1 : High pressure turbine stage
T_2 : Low pressure turbine stage

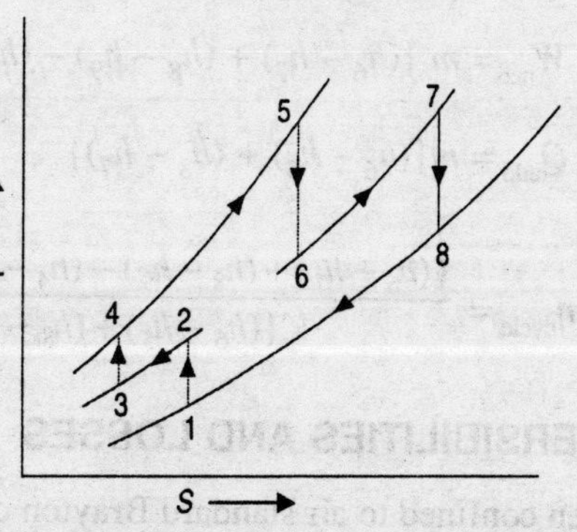

Fig. 9.20 Schematic for gas turbine cycle with reheat and intercooling and T-S diagram representation

Air standard cycle analysis for this arrangement gives,

$$W_{net} = m\ \{(h_6 - h_5)\} + (h_7 - h_8) - (h_4 - h_3) - (h_2 - h_1)\}$$

Heat added

$$Q_{add} = m\{(h_5 - h_4) + (h_7 - h_6)\}$$

Cycle thermal efficiency,

$$\eta_{cycle} = \frac{\{(h_6 - h_5) + (h_7 - h_8) - (h_4 - h_3) - (h_2 - h_1)\}}{\{(h_5 - h_4) + (h_7 - h_6)\}}$$

9.9 GAS TURBINE CYCLE WITH REGENERATION, REHEAT AND INTERCOOLING

Regenerative gas turbine employing reheating during expansion and intercooling during compression is considered here as shown in Fig. 9.21. This combination offers considerable increase in net work output and thermal efficiency.

C_1: Low pressure compressor stage
C_2: High pressure compressor stage
CC: Combustion chamber
RCC: Reheat combustion chamber
T_1 : High pressure turbine stage
T_2 : Low pressure turbine stage
G: Generator

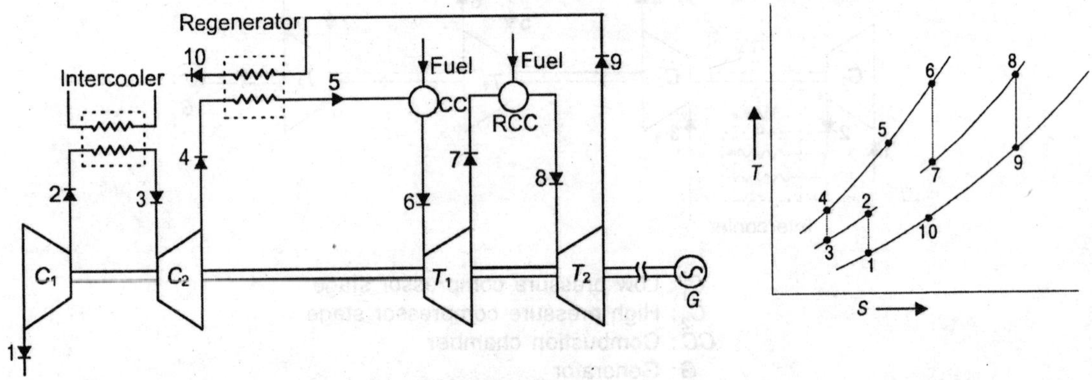

Fig. 9.21 Schematic for gas turbine cycle with regeneration, reheat and intercooling and T-S diagram

Based upon air standard cycle considerations thermodynamic analysis gives the Net work output from cycle,

$$W_{net} = m \{(h_6 - h_7) + (h_8 - h_9) - (h_4 - h_3) - (h_2 - h_1)\}$$

Heat added

$$Q_{add} = m\{(h_6 - h_5) + (h_8 - h_7)\}$$

Cycle thermal efficiency,

$$\eta_{cycle} = \frac{\{(h_6 - h_7) + (h_8 - h_9) - (h_4 - h_3) - (h_2 - h_1)\}}{\{(h_6 - h_5) + (h_8 - h_7)\}}$$

9.10 GAS TURBINE IRREVERSIBILITIES AND LOSSES

Till now the discussions have been confined to air standard Brayton cycle. But the realistic gas turbine cycle has deviations from air standard cycle due to,

(*i*) frictional effects within compressor and turbine which causes increase in specific entropy of working fluid across these components.

(*ii*) friction which shall cause drop in pressure of working fluid across the constant pressure processes.

Apart from above irreversibilities of the gas turbine power plant the irreversibilities of combustion chamber are quite significant.

Salient state points of realistic gas turbine Brayton cycle with above irreversibilities and losses are shown below:

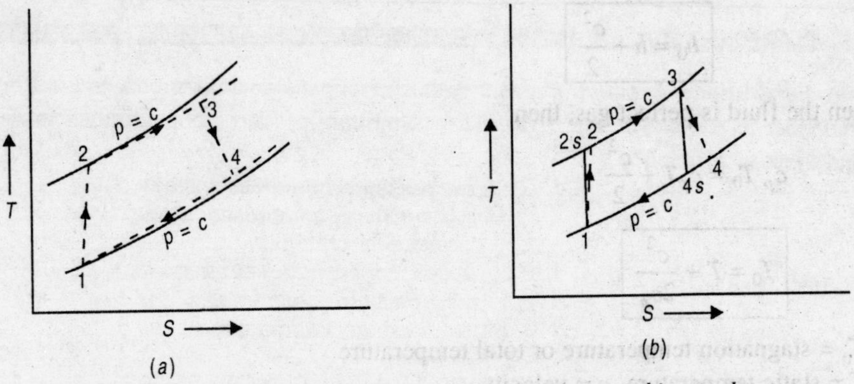

Fig. 9.22 Effect of irreversibilities and losses on gas turbine cycle.

Isentropic efficiency of turbine and compressor can be mathematically given as below:

Isentropic efficiency of turbine, $\eta_{\text{isen}, t} = \left\{\dfrac{h_3 - h_4}{h_3 - h_{4s}}\right\}$

$$\text{i.e. } \eta_{\text{isen}, t} = \frac{\text{Actual expansion work}}{\text{Ideal expansion work}}$$

Isentropic efficiency of compressor, $\eta_{\text{isen}, c} = \left\{\dfrac{h_{2s} - h_1}{h_2 - h_1}\right\}$

$$\text{i.e. } \eta_{\text{isen}, c} = \frac{\text{Ideal compressor work}}{\text{Actual compressor work}}$$

Other factors causing the real cycle to be different from ideal cycle are as below:

(*i*) Fluid velocities in turbomachines are very high and there exists substantial change in kinetic energy between inlet and outlet of each component. In the analysis carried out earlier the changes in kinetic energy have been neglected whereas for exact analysis it cannot be.

(*ii*) In case of regenerator the compressed air cannot be heated to the temperature of gas leaving turbine as the terminal temperature difference shall always exist.

(*iii*) Compression process shall involve work more than theoretically estimated value in order to overcome bearing and windage friction losses.

Different factors described above can be accounted for by stagnation properties, compressor and turbine isentropic efficiency and polytropic efficiency.

Stagnation properties: Stagnation properties are properties corresponding to stagnation states. Stagnation state of a fluid refers to the state at which fluid is adiabatically brought to the state of rest without work transfer. By using stagnation properties for themodynamic analysis the effect of kinetic energy variation is taken care and analysis becomes realistic. Let us write down the steady flow energy equation considering no heat and work interactions between states 1 and state 0. Let state '2' be the stagnation state. Generally stagnation states are denoted by using subscript '0'.

At stagnation state velocity of fluid shall be zero.

$$h_1 + \frac{c_1^2}{2} = h_0 + 0, \text{ or, } h_0 = h_1 + \frac{c_1^2}{2}$$

or, $$h_0 = h + \frac{c^2}{2}$$

When the fluid is perfect gas, then

$$c_p\, T_0 = c_p T + \frac{c^2}{2}$$

$$T_0 = T + \frac{c^2}{2c_p}$$

where T_0 = stagnation temperature or total temperature
T = static temperature, c = velocity

$\frac{c^2}{2c_p}$ = dynamic temperature

When a gas stream is slowed down and the temperature rises then there is simultaneous rise in pressure. The stagnation pressure p_0 is defined similar to T_0 with certain modification that the gas is brought to rest isentropically Stagnation pressure can be mathematically given as,

$$\frac{p_0}{p} = \left(\frac{T_0}{T}\right)^{\gamma/(\gamma-1)}$$

Let us use stagnation properties for an isentropic compression process between states 1 and 2. Stagnation pressure ratio,

$$\frac{p_{02}}{p_{01}} = \frac{p_{02}}{p_2} \times \frac{p_1}{p_{01}} \times \frac{p_2}{p_1}$$

$$= \left(\frac{T_{02}}{T_2} \times \frac{T_1}{T_{01}} \times \frac{T_2}{T_1}\right)^{\gamma/(\gamma-1)}$$

$$\frac{p_{02}}{p_{01}} = \left(\frac{T_{02}}{T_{01}}\right)^{\left(\frac{\gamma}{\gamma-1}\right)}$$

Thus, it shows p_0 and T_0 can be used similar to static values. Stagnation pressures and temperatures can be used with static values to determine combined thermodynamic and mechanical state of stream. Representation of compression process on *T-S* diagram using static and stagnation properties gives better conceptual information.

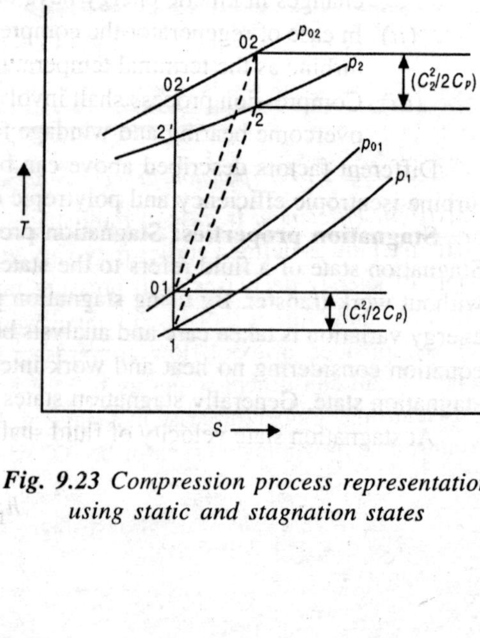

Fig. 9.23 Compression process representation using static and stagnation states

9.11 COMPRESSOR AND TURBINE EFFICIENCY

Isentropic efficiency of compressor and turbine as defined earlier can also be given in terms of stagnation properties.

$$\text{Isentropic compressor efficiency, } \eta_{\text{isen, c}} = \frac{W_{\text{ideal}}}{W_{\text{actual}}} = \frac{\Delta h_{0'}}{\Delta h_0}$$

In terms of temperature values, assuming mean specific heat value over a range of temperature, the efficiency can be given as,

$$\eta_{\text{isen, c}} = \frac{T_{02'} - T_{01}}{T_{02} - T_{01}}$$

$$\text{Isentropic turbine efficiency, } \eta_{\text{isen, t}} = \frac{W_{\text{actual}}}{W_{\text{ideal}}}$$

$$\eta_{\text{isen, t}} = \frac{T_{03} - T_{04}}{T_{03} - T_{04'}}$$

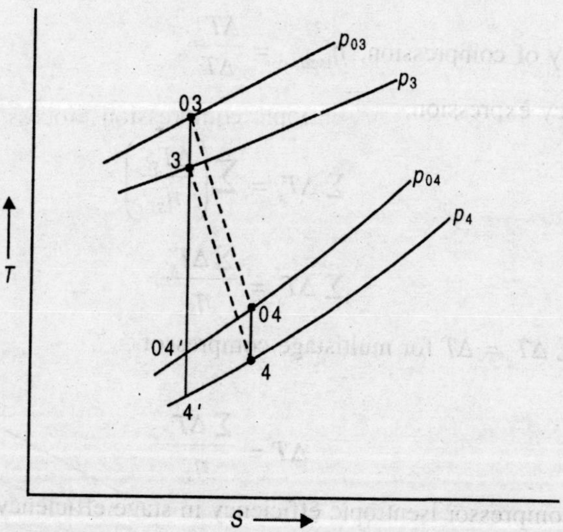

***Fig. 9.24** Expansion process representation using static and stagnation states*

Polytropic efficiency: In case of multistage compression and expansion the isentropic efficiency of complete machine as defined earlier may vary with pressure ratio. Actually it is found that compressor isentropic efficiency decreases with pressure ratio while turbine isentropic efficiency increases with pressure ratio, therefore use of fixed typical values of these efficiencies is not proper.

Let us consider an axial flow compressor comprising of number of stages having equal stage efficiency for all constituent stages as η_s.

Let isentropic temperature rise in a stage be $\Delta T_{s'}$ while actual temperature rise is ΔT_s. Let isentropic temperature rise in whole compressor be $\Delta T'$ while actual temperature rise be ΔT.

$$\text{For a stage, } \eta_s = \frac{\Delta T_{s'}}{\Delta T_s}$$

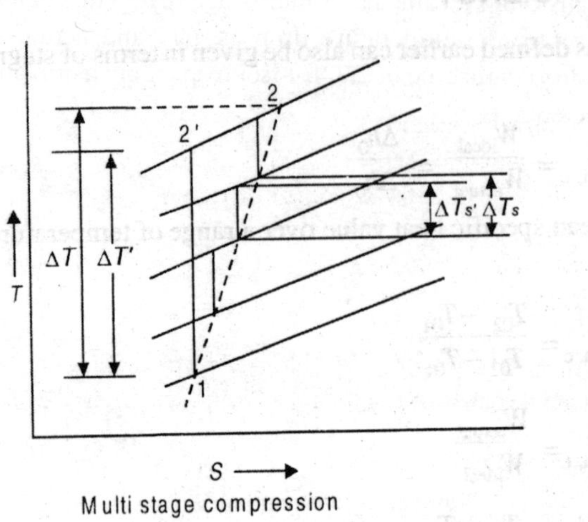

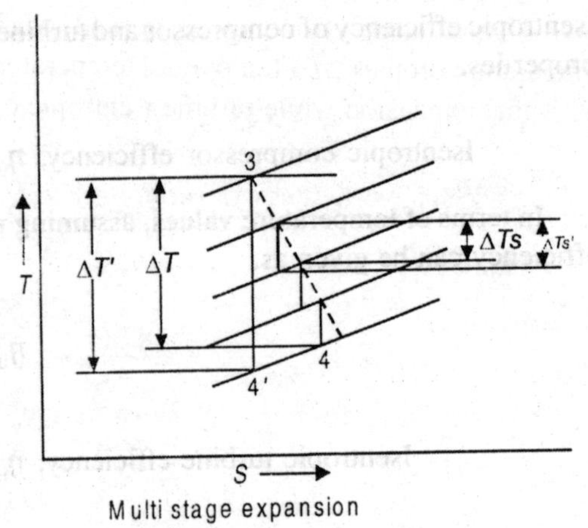

Fig. 9.25

Isentropic efficiency of compression, $\eta_{\text{isen, }c} = \dfrac{\Delta T'}{\Delta T}$

From stage efficiency expression,

$$\Sigma\,\Delta T_s = \Sigma\left(\frac{\Delta T_{s'}}{\eta_s}\right)$$

or

$$\Sigma\,\Delta T_s = \frac{\Sigma\,\Delta T_{s'}}{\eta_s}$$

It is obvious that, $\Sigma\,\Delta T_s = \Delta T$ for multistage compressor

so,

$$\Delta T = \frac{\Sigma\,\Delta T_{s'}}{\eta_s}$$

Substituting from compressor isentropic efficiency in stage efficiency expression

$$\frac{\Delta T'}{\eta_{\text{isen, c}}} = \frac{\Sigma\,\Delta T_{s'}}{\eta_s}$$

or

$$\boxed{\eta_{\text{isen, c}} = \eta_s\left(\frac{\Delta T'}{\Sigma\,\Delta T_{s'}}\right)}$$

From the nature of constant pressure lines on *T-S* diagram it could be concluded that constant pressure lines are of diverging type.

Therefore, $\Delta T' < \Sigma\,\Delta T_s$

Hence, $\boxed{\eta_{\text{isen, c}} < \eta_s}$

With the increasing pressure ratio this difference between $\Sigma\,\Delta T_s{}'$ and $\Delta T'$ goes on increasing and so with increasing pressure ratio the $\eta_{\text{isen, c}}$ goes on reducing despite constant stage efficiency. This phenomenon may be attributed to the fact that the increase in temperature due to friction in a stage results in more work requirement on next stage. This effect is also termed as the 'preheat' effect.

For multistage expansion in turbine: Let expansion in turbine be of multistage tpye with each stage having equal stage efficiency of η_s and stage isentropic temperature drop as $\Delta T_s'$ and actual stage temperature drop as ΔT_s. Let overall temperature drop in turbine be $\Delta T'$ for isentropic expansion and ΔT for actual expansion while turbine isentropic efficiency is $\eta_{\text{isen, t.}}$

$$\text{Stage efficiency of turbine, } \eta_s = \frac{\Delta T_s}{\Delta T_s'}$$

Cumulatively

$$\Sigma\, \Delta T_s' = \frac{\Sigma \Delta T_s}{\eta_s}$$

It is obvious from Fig. 9.25 showing multistage expansion that,

$$\Sigma\, \Delta T_s = \Delta T$$

So,

$$\Sigma\, \Delta T_s' = \frac{\Delta T}{\eta_s}$$

or,

$$\eta_s = \frac{\Delta T}{\Sigma \Delta T_s'}$$

$$\text{Turbine isentropic efficiency, } \eta_{\text{isen, t.}} = \frac{\Delta T}{\Delta T'}$$

Combining two expressions for turbine isentropic efficiency and stage efficiency, it yields,

$$\eta_{\text{isen, t}} = \eta_s \cdot \frac{\Sigma \Delta T_s'}{\Delta T'}$$

It may also be seen that due to diverging nature of constant pressure lines on *T-S* diagram, $\Delta T' < \Sigma\, \Delta T_s'$.

Therefore, $\boxed{\eta_{\text{isen, t}} > \eta_s}$

With increasing pressure ratio this difference between $\Delta T'$ and $\Sigma\, \Delta T_s'$ goes on increasing and thus with increasing pressure ratio $\eta_{\text{isen, t}}$ goes on increasing despite constant stage efficiency.

This may be attributed to the fact that frictional reheating in one stage is partially recovered as work in next stage. Graphical pattern showing behaviour of $\eta_{\text{isen, c}}$ and $\eta_{\text{isen, t}}$ shown in Fig. 9.26.

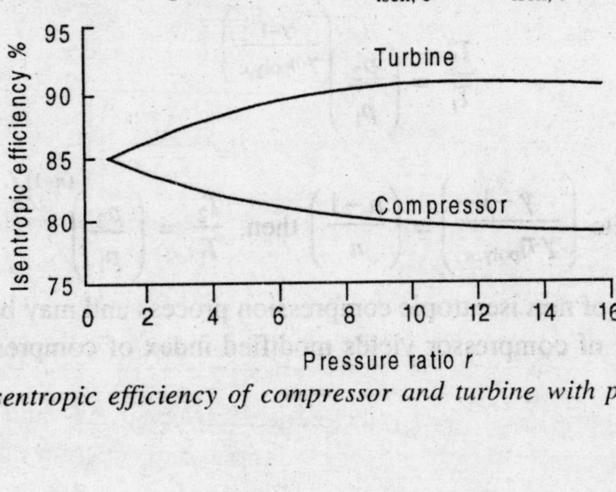

Fig. 9.26 Isentropic efficiency of compressor and turbine with pressure ratio

Thus, it is seen that isentropic efficiency of turbine and compressor are not suitable parameters to account for non-idealities in turbines and compressors. Hence the polytropic efficiency, also called small-stage efficiency is used for turbines and compressors. "Polytropic efficiency is defined as the isentropic efficiency of an elemental stage in the compression or expansion process such that it remains constant throughout the whole process." Polytropic efficiency for compressor and turbine may be denoted as $\eta_{poly,\,c}$ and $\eta_{poly,\,t}$. In an elemental stage let the isentropic temperature rise be dT' while actual temperature rise is dT.

For compressor, $$\eta_{poly,\,c} = \frac{dT'}{dT}$$

Isentropic compression processes shall follow; $\dfrac{T}{p^{\gamma-1/\gamma}}$ = constant

Taking log of both sides, $\log T = \left(\dfrac{\gamma-1}{\gamma}\right)\log p + \log(\text{constant})$

Differentiating partially,

$$\frac{dT'}{T} = \left(\frac{\gamma-1}{\gamma}\right)\frac{dp}{p}$$

Substituting dT' from polytropic efficiency expression,

$$\eta_{poly,\,c}\frac{dT}{T} = \left(\frac{\gamma-1}{\gamma}\right)\frac{dp}{p}$$

Integrating between states 1 and 2;

$$\eta_{poly,\,c}\ln\left(\frac{T_2}{T_1}\right) = \ln\left(\frac{p_2}{p_1}\right)^{(\gamma-1/\gamma)}$$

$$\eta_{poly,\,c} = \frac{\ln\left(\dfrac{p_2}{p_1}\right)^{(\gamma-1/\gamma)}}{\ln\left(\dfrac{T_2}{T_1}\right)}$$

or $$\frac{T_2}{T_1} = \left(\frac{p_2}{p_1}\right)^{\left(\frac{\gamma-1}{\gamma\cdot\eta_{poly,c}}\right)}$$

Let us write $\left(\dfrac{\gamma-1}{\gamma\cdot\eta_{poly,c}}\right) = \left(\dfrac{n-1}{n}\right)$ then, $\dfrac{T_2}{T_1} = \left(\dfrac{p_2}{p_1}\right)^{(n-1)/n}$

Hence 'n' is new index of non isentropic compression process and may be called polytropic index. Thus, polytropic efficiency of compressor yields modified index of compression to account for non-isentropic compression.

$$\frac{n-1}{n} = \left(\frac{\gamma - 1}{\gamma \cdot \eta_{\text{poly, c}}}\right)$$

Similarly for polytropic efficiency of expansion in turbine. In elemental stage of turbine the isentropic temperature rise dT' and actual temperature rise dT occurs.

$$\eta_{\text{poly, t}} = \left(\frac{dT}{dT'}\right)$$

Expansion in elemental stage is governed by $\dfrac{T}{p^{\frac{\gamma-1}{\gamma}}}$ = Constant

Taking log and differentiating partially,

$$\frac{dT'}{T} = \left(\frac{\gamma - 1}{\gamma}\right) \frac{dp}{p}$$

Substituting from polytropic efficiency of turbine,

$$\frac{dT}{\eta_{\text{poly,t}} \cdot T} = \left(\frac{\gamma - 1}{\gamma}\right) \frac{dp}{p}$$

$$\frac{dT}{T} = \left(\frac{(\gamma - 1) \cdot \eta_{\text{poly, t}}}{\gamma}\right) \cdot \frac{dp}{p}$$

Integrating between states 3 and 4

$$\frac{T_3}{T_4} = \left(\frac{p_3}{p_4}\right)^{\left(\frac{\eta_{\text{poly,t}} \cdot (\gamma - 1)}{\gamma}\right)}$$

Here also the polytropic index for non-isentropic expansion process may be given as,

$$\frac{T_3}{T_4} = \left(\frac{p_3}{p_4}\right)^{(n-1)/n}$$

$$\boxed{\frac{n-1}{n} = \left(\frac{\eta_{\text{poly,t}} \cdot (\gamma - 1)}{\gamma}\right)}$$

Thus, the polytropic efficiency may be assumed constant over a range of pressure ratio as it takes care of variation of isentropic efficiency with pressure ratio.

9.12 ERICSSON CYCLE

From earlier discussions, we have seen that the thermal efficiency of gas turbine power plant may be increased by regeneration, reheat and intercooling. But there exists limit to the number of stages practically. If a large number of multistages of compression with intercooling and large number of multistages of expansion with reheating and regeneration are employed then the cycle is called Ericsson cycle. Ericsson cycle is shown below in Fig. 9.27. Here 1-2 is isothermal compression while 2-3 is constant pressure heat addition process. 3-4 is isothermal expansion and 4-1 is constant pressure heat rejection process. 3-4 is isothermal expansion and 4-1 is constant pressure heat rejection process.Heat addition and heat rejection processes may be considered to occur inheat exchanger having 100% effectiveness.

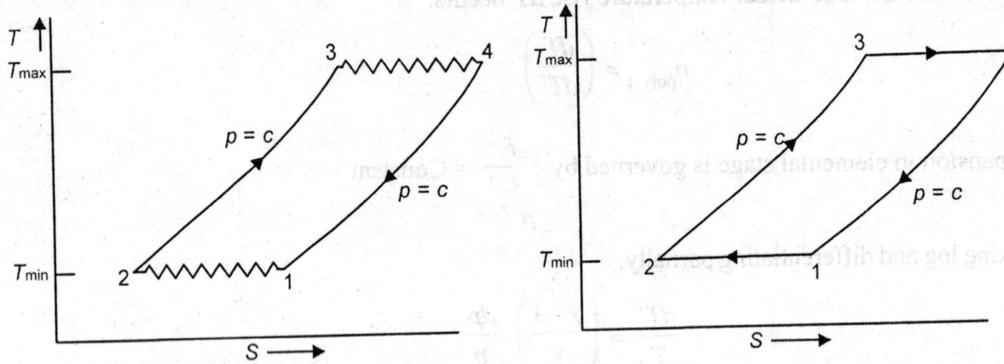

Fig. 9.27 Ericsson cycle

The Ericsson engine involves;

(*i*) a turbine through which gas expands isothermally doing work and absorbing heat from an energy reservoir at T_{max}. (process 3-4).

(*ii*) a compressor that compresses gas isothermally while heat is rejected from the gas at T_{min} (process 1-2) to the energy reservoir at T_{min}.

(*iii*) a counter flow heat exchanger which is used as regenerator and the gas coming from compressor is heated in process 2-3 while the gas leaving turbine is cooled in process 4-1. In this type of regenerator the desired heat exchange is possible only when an infinitesimal temperature difference exists between two gas streams at any section and regenerator should operate reversibly. Although such regenerator is not practically possible.

Thus, all processes in Ericsson cycle are reversible processes and cycle approaches to Carnot cycle.

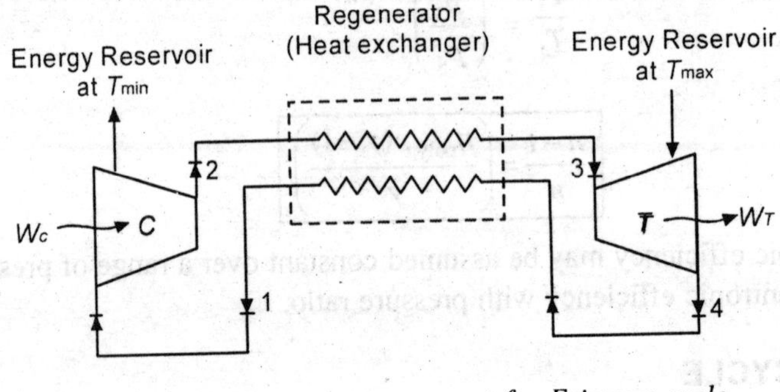

Fig. 9.28 Schematic arrangement for Ericsson cycle

$$\text{Heat added, } Q_{add} = RT_{max} \cdot \ln \frac{V_4}{V_3} = RT_{max} \cdot \frac{p_3}{p_4}$$

Heat rejected, $$Q_{\text{rejected}} = -RT_{\min}\cdot\ln\frac{V_2}{V_1} = RT_{\min}\cdot\ln\frac{p_2}{p_1}$$

Here,

$$p_2 = p_3 \ \&\ p_1 = p_4, \text{ so } \frac{p_3}{p_4} = \frac{p_2}{p_1}$$

Also, $$W_{\text{net}} = Q_{\text{add}} - Q_{\text{rejected}}$$

Ericsson cycle efficiency, $$\eta_{\text{ericsson}} = \frac{Q_{\text{add}} - Q_{\text{rejected}}}{Q_{\text{add}}}$$

Substitution yields,

$$\eta_{\text{ericsson}} = \frac{T_{\max} - T_{\min}}{T_{\max}}$$

$$\eta_{\text{ericsson}} = 1 - \frac{T_{\min}}{T_{\max}} = \eta_{\text{Carnot}}$$

9.13 STIRLING CYCLE

Stirling cycle consists of four reversible processes as given below:

(*i*) Isothermal compression from state 1 to 2 at temperature $T_{\min}$.

(*ii*) Constant volume heat addition from state 2 to 3.

(*iii*) Isothermal expansion from state 3 to 4 at temperature $T_{\max}$.

(*iv*) Constant volume heat rejection from state 4 to 1.

Here also similar to Ericsson cycle a regenerator with 100% effectiveness is employed. Regenerator facilitates for the heat rejected during process 4–1 to be used as heat input in process 2-3. Efficiency of Stirling cycle is similar to that of Carnot cycle.

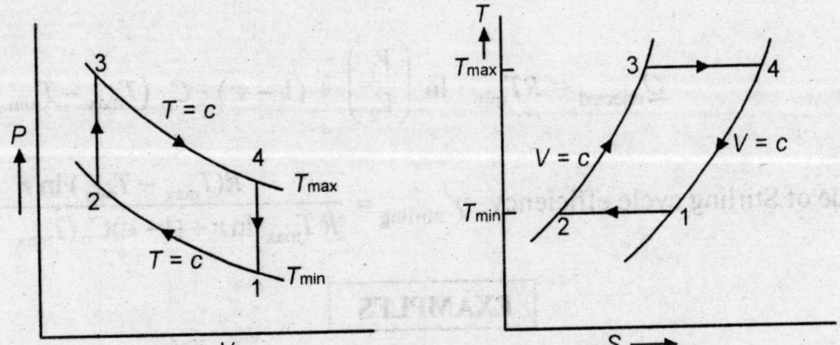

Fig. 9.29 Stirling cycle

For a regenerator having 100% effectiveness the heat added in process 2-3 and heat rejected in process 4-1 will be same. If we consider regenerator as part of system then heat interactions with surroundings shall be taking place at temperatures $T_{\min}$ and $T_{\max}$ only. Heat added in cycle will be during process 3–4 at temperature $T_{\max}$ and heat rejected in cycle will be during process 1-2 at temperature $T_{\min}$.

Heat added, $$Q_{add} = RT_{max} \cdot \ln\left(\frac{V_4}{V_3}\right)$$

Heat rejected, $$Q_{rejected} = RT_{min} \cdot \ln\left(\frac{V_1}{V_2}\right)$$

Here, $$V_4 = V_1 \text{ and } V_3 = V_2 \text{ so, } \frac{V_4}{V_3} = \frac{V_1}{V_2} = r$$

Stirling cycle thermal efficiency, $$\eta_{stirling} = \frac{W_{net}}{Q_{add}} = \frac{Q_{add} - Q_{rejected}}{Q_{add}}$$

$$= 1 - \frac{RT_{min} \cdot \ln\left(\frac{V_1}{V_2}\right)}{RT_{max} \cdot \ln\left(\frac{V_4}{V_3}\right)}$$

$$\eta_{stirling} = 1 - \frac{T_{min}}{T_{max}} = \eta_{Carnot}$$

In case of regenerator operating with effectiveness less than 100% the heat added and rejected in processes 2-3 and 4-1 will not be same and some heat would be lost. Let us consider regenerator effectiveness as '$\in$'.

Heat added, $$Q_{add} = RT_{max} \cdot \ln\left(\frac{V_4}{V_3}\right) + (1 - \in) \cdot C_v \, (T_{max} - T_{min})$$

Heat rejected, $$Q_{rejected} = RT_{min} \cdot \ln\left(\frac{V_1}{V_2}\right) + (1 - \in) \cdot C_v \, (T_{max} - T_{min})$$

Modified value of Stirling cycle efficiency, $$\eta'_{stirling} = \frac{R(T_{max} - T_{min}) \cdot \ln r}{R\,T_{max} \cdot \ln r + (1 - \in) C_v (T_{max} - T_{min})}$$

EXAMPLES

1. A four stroke SI engine has the compression ratio of 6 and swept volume of 0.15 m^3. Pressure and temperature at the beginning of compression are 98 kPa and 60°C. Determine the pressure, volume and temperatures at all salient points if heat supplied to it is 150 kJ/kg. Also find out entropy change, work done, efficiency and mean effective pressure of cycle assuming c_p = 1 kJ/kg · K, c_v = 0.71 kJ/kg · K. Also plot the cycle on T-S diagram.

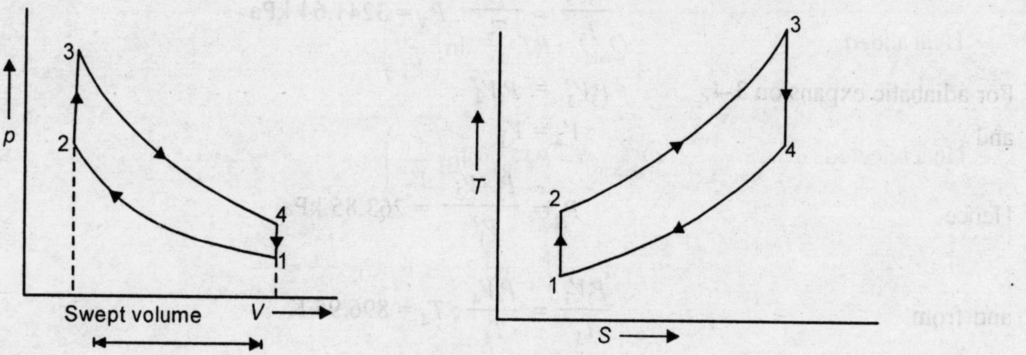

Fig. 9.30

Solution:

SI engines operate on Otto cycle. Consider working fluid to be perfect gas.

Here $\gamma = \dfrac{c_p}{c_v} = 1.4$

$$c_p - c_v = R = 0.29 \text{ kJ/kg} \cdot \text{K}.$$

Given:

$$P_1 = 98 \text{ kPa}$$

$$T_1 = 60 + 273.15 = 333.15 \text{ K}$$

$$Q_{2\text{-}3} = 150 \text{ kJ/kg}$$

$$\text{Compression ratio, } r = \frac{V_1}{V_2} = \frac{0.15 + V_2}{V_2} = 6$$

Therefore,

$$V_2 = 0.03 \text{ m}^3$$

$$\text{Total cylinder volume} = 0.18 \text{ m}^3 = V_1$$

From perfect gas law, $PV = \text{mRT}$

$$m = \frac{P_1 V_1}{RT_1} = 0.183 \text{ kg.}$$

From state 1 to 2 by PV^γ = constant

$$P_1 V_1^\gamma = P_2 V_2^\gamma \text{ or } P_2 = P_1 \times \left(\frac{V_1}{V_2}\right)^\gamma = 1204.03 \text{ kPa}$$

Also $\dfrac{P_1 V_1}{T_1} = \dfrac{P_2 V_2}{T_2}$ yields, $T_2 = 682.18$ K

From heat addition process 2 – 3

$$Q_{2\text{-}3} = mc_v (T_3 - T_2)$$

$$150 = 0.183 \times 0.71 (T_3 - 682.18)$$

$$T_3 = 1836.65 \text{ K}$$

Also from

$$\frac{P_3V_3}{T_3} = \frac{P_2V_2}{T_2}, \; P_3 = 3241.64 \text{ kPa}$$

For adiabatic expansion 3-4, $P_3V_3^{\gamma} = P_4V_4^{\gamma}$

and $V_4 = V_1$

Hence $$P_4 = \frac{P_3 \cdot V_3^{\gamma}}{V_1^{\gamma}} = 263.85 \text{ kPa}$$

and from $$\frac{P_3V_3}{T_3} = \frac{P_4V_4}{T_4}; \; T_4 = 896.95 \text{ K}$$

Entropy change from 2-3 and 4-1 are same, and can be given as,

$$S_3 - S_2 = S_4 - S_1 = mc_v \ln\frac{T_4}{T_1} = 0.183 \times 0.71 \times \ln\left(\frac{896.95}{333.15}\right)$$

$$\text{Entropy change } \Delta S_{3\text{-}2} = \Delta S_{4\text{-}1} = 0.1287 \text{ kJ/K.}$$

$$\text{Heat rejected, } Q_{4\text{-}1} = mc_v \times (T_4 - T_1)$$

$$= 0.183 \times 0.71\,(896.95 - 333.15)$$

$$Q_{4\text{-}1} = 73.25 \text{ kJ}$$

$$\text{Net Work} = (Q_{2\text{-}3}) - (Q_{4\text{-}1}) = 150 - 73.25$$

$$\text{Net Work} = 76.75 \text{ kJ}$$

$$\text{Efficiency} = \frac{\text{Net work}}{\text{Heat added}}$$

$$= \frac{76.75}{150} = 0.5117 \text{ or } \boxed{\eta = 51.17\%}$$

$$\text{Mean effective pressure} = \frac{\text{Work}}{\text{Volume change}} = \frac{76.75}{0.15} = 511.67 \text{ kPa}$$

m.e.p. = 511.67 kPa **Ans.**

2. In a Diesel engine during the compression process, pressure is seen to be 138 kPa at $\frac{1}{8}$ th of stroke and 1.38 MPa at $\frac{7}{8}$ th of stroke. The cut-off occurs at $\frac{1}{15}$ th of stroke. Calculate air standard efficiency and compression ratio assuming indicated thermal efficiency to be half of ideal efficiency, mechanical efficiency as 0.8, calorific value of fuel = 41800 kJ/kg and $\gamma = 1.4$. Also find fuel consumption bhp/hr.

Solution:

As given

$$V_A = V_2 + \frac{7}{8}(V_1 - V_2)$$

$$V_B = V_2 + \frac{1}{8}(V_1 - V_2)$$

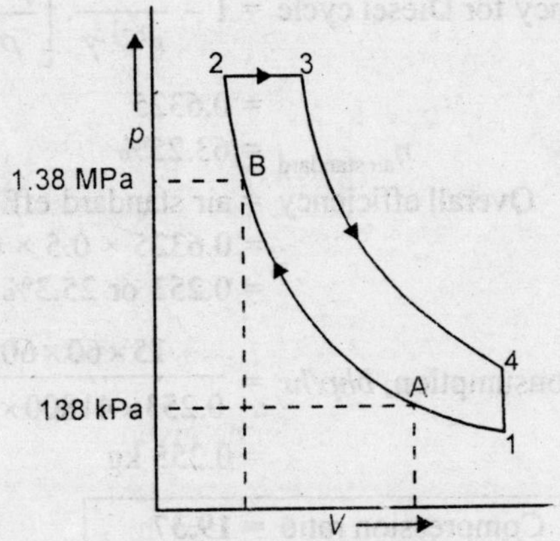

Fig. 9.31

and also $$p_A \cdot V_A^{\gamma} = p_B \cdot V_B^{\gamma}$$

or $$\left(\frac{V_A}{V_B}\right) = \left(\frac{P_B}{P_A}\right)^{1/\gamma}$$

$$= \left(\frac{1380}{138}\right)^{1/1.4}$$

$$\frac{V_A}{V_B} = 5.18$$

Also substituting for V_A & V_B

$$\frac{V_2 + \frac{7}{8}(V_1 - V_2)}{V_2 + \frac{1}{8}(V_1 - V_2)} = 5.18$$

It gives $$r = 19.37$$

$$\frac{V_1}{V_2} = 19.37 \text{ Compression ratio} = 19.37$$

As given; Cut off occurs at $\left\{\frac{V_1 - V_2}{15}\right\}$ volume

or $$V_3 = \left\{\frac{V_1 - V_2}{15}\right\} + V_2$$

Cut off ratio, $$\rho = \frac{V_3}{V_2} = 2.22$$

$$\text{Air standard efficiency for Diesel cycle} = 1 - \frac{1}{r^{\gamma-1}\cdot\gamma}\left[\frac{\rho^{\gamma}-1}{\rho-1}\right]$$

$$= 0.6325$$

$$\eta_{\text{air standard}} = 63.25\%$$

$$\text{Overall efficiency} = \text{air standard efficiency} \times 0.5 \times 0.8$$

$$= 0.6325 \times 0.5 \times 0.8$$

$$= 0.253 \text{ or } 25.3\%$$

$$\text{Fuel consumption, } bhp/hr = \frac{75\times 60\times 60}{0.253\times 41800\times 10^2}$$

$$= 0.255 \text{ kg}$$

Compression ratio	= **19.37**
Air standard efficiency	= **63.25%**
Fuel consumption, *bhp/hr*	= **0.255 kg**

Ans.

3. In an IC engine using air as working fluid, total 1700 kJ/kg of heat is added during combustion and maximum pressure in cylinder does not exceed 5 MPa. Compare the efficiency of following two cycles used by engine:

(a) cycle in which combustion takes place isochorically.

(b) cycle in which half of heat is added at constant volume and half at constant pressure.

Temperature and pressure at the beginning of compression are 100°C and 103 kPa. Compression and expansion processes are adiabatic. Specific heat at constant pressure and volume are 1.003 kJ/kg · K and 0.71 kJ/kg · K.

Solution:

$$1\text{-}2\text{-}3\text{-}4 = \text{cycle } (a)$$

$$1\text{-}2'\text{-}3'\text{-}4'\text{-}5 = \text{cycle } (b)$$

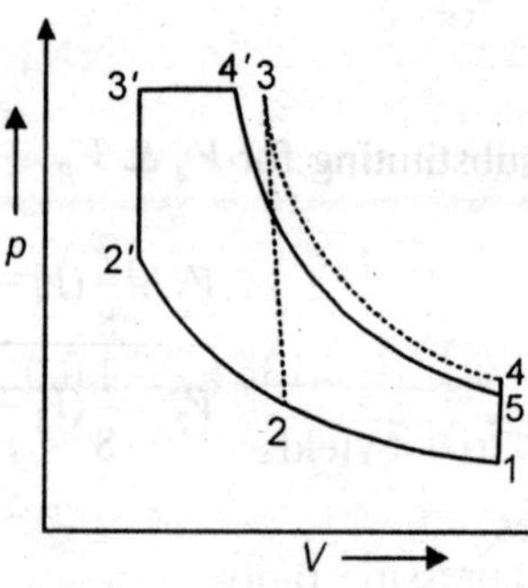

Fig. 9.32

Here $\dfrac{c_p}{c_v} = \gamma = 1.4$

and $R = 0.293 \text{ kJ/kg} \cdot \text{K}$

Let us consider 1 kg of air for perfect gas,

$$PV = mRT$$

or

$$V_1 = \frac{mRT_1}{p_1}$$

$$= \frac{1\times 0.293\times 373.15}{103}$$

$$V_1 = 1.06 \text{ m}^3$$

at state 3, $p_3V_3 = mRT_3$ or $T_3 = \dfrac{5000\times V_3}{1\times 0.293} = 17064.8\ V_2$

for cycle (*a*) and also for cycle (*b*)

$$T'_3 = 17064.8\ V'_2$$

(a) For Otto cycle

$$Q_{23} = c_v(T_3 - T_2);\ \text{given } Q_{23} = 1700 \text{ kJ/kg}$$

$$1700 = 0.71(T_3 - T_2)$$

or
$$T_3 - T_2 = 2328.77, \text{ or } T_2 = T_3 - 2328.77$$

From gas law

$$\frac{p_2 V_2}{T_2} = \frac{p_3 V_3}{T_3}$$

$$\frac{p_2 \cdot V_2}{(T_3 - 2328.77)} = \frac{5000 \times V_2}{T_3}, \ \{\text{as } V_2 = V_3\}$$

or
$$\frac{p_2}{(T_3 - 2328.77)} = \frac{5000}{T_3}$$

Substituting T_3 as function of V_2

$$\frac{p_2}{(17064.8V_2 - 2328.77)} = \frac{5000}{(17064.8V_2)}$$

or
$$p_2 = \frac{5000(17064.8V_2 - 2328.77)}{(17064.8V_2)}$$

Also
$$p_1 V_1^{\gamma} = p_2 V_2^{\gamma}$$

or
$$103 \times (1.06)^{1.4} = \frac{5000(17064.8V_2 - 2328.77)}{(17064.8V_2)} \cdot (V_2^{1.4})$$

upon solving it yields

$$381.4V_2 = 17064.8\ V_2^{2.4} - 2328.77\ V_2^{1.4}$$

$$V_2 = 44.7\, V_2^{2.4} - 6.1\, V_2^{1.4}$$

$$44.7\, V_2^{1.4} - 6.1\, V_2^{0.4} - 1 = 0$$

or
$$V_2^{1.4} - 0.136\, V_2^{0.4} - 0.022 = 0$$

By hit and trial it yields $V_2 = 0.18 \text{ m}^3$

Thus, compression ratio
$$r = \frac{V_1}{V_2} = \frac{1.06}{0.18} = 5.89$$

Otto cycle efficiency,
$$\eta_{otto} = 1 - \frac{1}{r^{\gamma - 1}} = 0.5083$$

or
$$\eta_{otto} = 50.83\%$$

(b) For mixed or Dual Cycle

Given:

$$c_p (T_{4'} - T_{3'}) = c_v (T_{3'} - T_{2'}) = \frac{1700}{2} = 850$$

or $$T_{3'} - T_{2'} = 1197.2$$

or $$T_{2'} = T_{3'} - 1197.2$$

Also $$\frac{P_{2'}V_{2'}}{T_{2'}} = \frac{P_{3'}V_{3'}}{T_{3'}}$$

$$\Rightarrow \quad \frac{P_{2'}V_{2'}}{(T_{3'} - 1197.2)} = \frac{5000 \cdot V_{2'}}{T_{3'}}$$

or $$\frac{P_{2'}}{(T_{3'} - 1197.2)} = \frac{5000}{T_{3'}}$$

Also we had seen earlier that $T_{3'} = 17064.8\, V_{2'}$

$$\frac{P_{2'}}{(17064.8V_{2'} - 1197.2)} = \frac{5000}{(17064.8\, V_{2'})}$$

or $$P_{2'} = \frac{5000(17064.8\, V_{2'} - 1197.2)}{(17064.8\, V_{2'})}$$

For adiabatic process 1-2′

$$P_1V_1^{\gamma} = P_{2'}V_{2'}^{\gamma}$$

or $$103 \times (1.06)^{1.4} = (V_{2'})^{1.4} \cdot \left(5000 - \frac{359.78}{V_{2'}}\right)$$

$$111.75\, V_{2'} = 5000\, V_{2'}^{2.4} - 350.78\, V_{2'}^{1.4}$$

$$1 = 44.74\, V_{2'}^{1.4} - 3.14\, V_{2'}^{0.4}$$

or $$V_{2'}^{1.4} - 0.07\, V_{2'}^{0.4} - 0.022 = 0$$

By hit and trial

$$V_{2'} = 0.122 \text{ m}^3$$

Therefore upon substituting $V_{2'}$

$$P_{2'} = 2124.75 \text{ kPa}$$
$$T_{3'} = 2082 \text{ K}$$
$$T_{2'} = 884.8 \text{ K}$$

From constant pressure heat addition

$$c_p(T_{4'} - T_{3'}) = 850$$
$$1.003\,(T_{4'} - 2082) = 850$$
$$T_{4'} = 2929.5 \text{ K}$$

Also $$\frac{P_{4'}\, V_{4'}}{T_{4'}} = \frac{P_{3'}\, V_{3'}}{T_{3'}}$$ {here $p_{3'} = p_{4'}$ and $V_{2'} = V_{3'}$}

or $$V_{4'} = \frac{0.122 \times 2929.5}{2082}$$

$$V_{4'} = 0.172 \text{ m}^3$$

Using adiabatic formulations $v_{4'} = 0.172\ m^3$

$$\left(\frac{V_5}{V_{4'}}\right)^{\gamma-1} = \left(\frac{T_{4'}}{T_5}\right) \quad \{\text{here } V_5 = V_1\}$$

or
$$T_5 = 2929.5 \times \left(\frac{0.172}{1.06}\right)^{0.4}$$

$$T_5 = 1415.4 \text{ K}$$

Heat rejected in process 5 – 1, $Q_{5-1} = c_v\,(T_5 - T_1)$

$$= 0.71\ (1415.4 - 373.15)$$

$$Q_{5-1} = 739.99 \text{ kJ}$$

$$\text{Efficiency of mixed cycle} = \frac{1700 - 739.99}{1700} = 0.5647 \text{ or } 56.47\%$$

$\boxed{\eta_{\text{otto cycle}} = 50.83\%;\ \eta_{\text{mixed cycle}} = \mathbf{56.47\%}}$ **Ans.**

***4.** In an air standard Brayton cycle the minimum and maximum temperature are 300 K and 1200 K, respectively. The pressure ratio is that which maximizes the net work developed by the cycle per unit mass of air flow. Calculate the compressor and turbine work, each in kJ/kg air, and thermal efficiency of the cycle.* *[U.P.S.C. 1994]*

Solution:

Maximum temperature, $T_3 = 1200$ K

Minimum temperature, $T_1 = 300$ K

Optimum pressure ratio for maximum work output,

$$r_p = \left(\frac{T_{max}}{T_{min}}\right)^{\frac{\gamma}{2(\gamma-1)}}$$

$$r_p = \left(\frac{T_3}{T_1}\right)^{\frac{\gamma}{2(\gamma-1)}}$$

$$r_p = 11.3$$

$$\frac{p_2}{p_1} = r_p = 11.3, \text{ For process 1–2, } \frac{T_2}{T_1} = \left(\frac{P_2}{T_1}\right)^{\frac{\gamma}{\gamma-1}}$$

$$T_2 = 300 \cdot (11.3)^{\frac{1.4-1}{1.4}}$$

$$T_2 = 600.21 \text{ K}$$

For process 3–4,

$$\frac{T_3}{T_4} = \left(\frac{P_3}{P_4}\right)^{\frac{\gamma-1}{\gamma}} = \left(\frac{P_2}{P_1}\right)^{\frac{\gamma-1}{\gamma}}, \Rightarrow T_4 = 599.79 \text{ K}$$

Heat supplied, $Q_{23} = c_p (T_3 - T_2) = 602.79$ kJ/kg

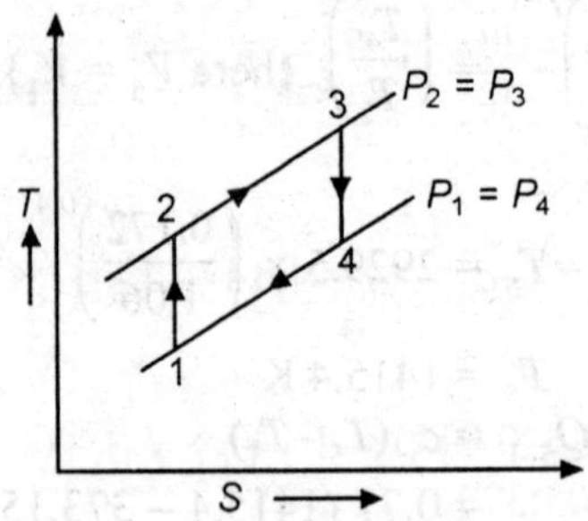

Fig. 9.33

Compressor work, $W_C = c_p \cdot (T_2 - T_1) = 1.005 \times (600.21 - 300) = 301.71$ kJ/kg

Turbine work $W_T = c_p \cdot (T_3 - T_4) = 1.005\,(1200 - 599.79) = 603.21$ kJ/kg

$$\text{Thermal efficiency} = \frac{\text{Net work}}{\text{Heat supplied}} = \frac{W_T - W_c}{Q_{23}} = \frac{603.21 - 301.71}{602.79} = 0.5002$$

Compressor work = **301.71 kJ/kg**, Turbine work = **603.21 kJ/kg**, Thermal efficiency = **50.02%**	**Ans.**

5. A gas turbine unit receives air at 1 bar, 300 K and compresses it adiabatically to 6.2 bar. The compressor efficiency is 88%. The fuel has a heating value of 44186 kJ/kg and the fuel-air ratio is 0.017 kg fuel/kg of air. The turbine internal efficiency is 90%. Calculate the work of turbine and compressor per kg of air compressed and thermal efficiency. For products of combustion $c_p = 1.147$ kJ/kg K, $\gamma = 1.33$. *[U.P.S.C. 1992]*

Solution:

Gas turbine cycle is shown by 1–2–3–4 on T-S diagram,

Given:

$P_1 = 1$ bar, $P_2 = P_3 = 6.2$ bar, F/A ratio = 0.017

$T_1 = 300$ K, $\eta_{compr.} = 88\%$, $\eta_{turb} = 90\%$

Heating value of fuel = 44186 kJ/kg

For process 1–2 being isentropic,

$$\frac{T_2}{T_1} = \left(\frac{P_2}{P_1}\right)^{\frac{\gamma-1}{\gamma}}$$

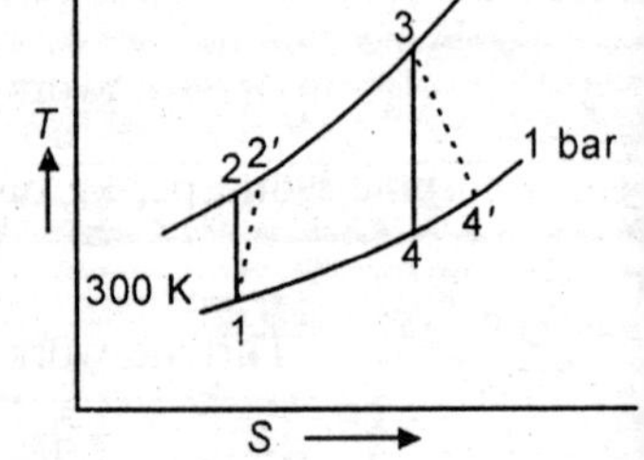

Fig. 9.34

$$T_2 = 505.26 \text{ K}$$

Considering compressor efficiency, $\eta_{compr} = \dfrac{T_2 - T_1}{T_{2'} - T_1}$, $0.88 = \dfrac{(505.26 - 300)}{(T_{2'} - 300)}$

Actual temperature after compression, $T'_2 = 533.25$ K

During process 2–3 due to combustion of unit mass of air compressed the energy balance shall be as under,

Heat added = $m_f \times$ Heating value

$$= ((m_a + m_f) \cdot c_{p,\,\text{comb}} \cdot T_3) - (m_a \cdot c_{p,\,\text{air}} \cdot T_{2'})$$

or $$\left(\frac{m_f}{m_a}\right) \times 44186 = \left(\left(1+\frac{m_f}{m_a}\right) \cdot c_{p,\,\text{comb}} \cdot T_3\right) - (c_{p,\,\text{air}} \times 533.25)$$

Here, $\frac{m_f}{m_a} = 0.017$, $c_{p,\,\text{comb}} = 1.147$ kJ/kg · K, $c_{p,\,\text{air}} = 1.005$ kJ/kg · K

Upon substitution

$$(0.017 \times 44186) = ((1 + 0.017) \times 1.147 \times T_3) - (1.005 \times 533.25)$$

$$T_3 = 1103.37 \text{ K}$$

For expansion 3–4 being

$$\frac{T_4}{T_3} = \left(\frac{P_4}{P_3}\right)^{\frac{n-1}{n}}$$

$$T_4 = 1103.37 \times \left(\frac{1}{6.2}\right)^{\frac{0.33}{1.33}}$$

$$T_4 = 701.64 \text{ K}$$

Actual temperature at turbine inlet considering internal efficiency of turbine,

$$\eta_{\text{turb}} = \frac{T_3 - T_{4'}}{T_3 - T_4}\,;\ 0.90 = \frac{(1103.37 - T_{4'})}{(1103.37 - 701.64)}$$

$$T_{4'} = 741.81 \text{ K}$$

Compressor work, per kg of air compressed $= c_{p,\,\text{air}} \cdot (T_{2'} - T_1)$

$$W_C = 1.005 \times (533.25 - 300)$$

$$W_C = 234.42 \text{ kJ/kg or air}$$

Compressor work = **234.42 kJ/kg of air** **Ans.**

Turbine work, per kg of air compressed $= c_{p,\,\text{comb}} \cdot (T_3 - T_{4'})$

$$= 1.147 \times (1103.37 - 741.81)$$

Turbine work = **414.71 kJ/kg of air** **Ans.**

$$W_T = 414.71, \text{ kJ/kg of air}$$

$$\text{Net work} = W_T - W_C = (414.71 - 234.42)$$

$$W_{\text{net}} = 180.29, \text{ kJ/kg of air}$$

Heat supplied = 0.017 × 44186 = 751.162 kJ/kg of air.

$$\text{Thermal efficiency} = \frac{W_{\text{net}}}{\text{Heat supplied}} = \frac{180.29}{751.162} = 24\%$$

Thermal efficiency = **24%** **Ans.**

6. In a Brayton cycle gas turbine power plant the minimum and maximum temperature of the cycle are 300 K and 1200 K. The compression is carried out in two stages of equal pressure ratio with intercooling of the working fluid to the minimum temperature of the cycle after the first-stage of compression. The entire expansion is carried out in one stage only. The isentropic efficiency of both compressors is 0.85 and that of the turbine is 0.9. Determine the overall pressure ratio that would give the maximum net work per kg working fluid. Take γ = 1.4. *[U.P.S.C. 1995]*

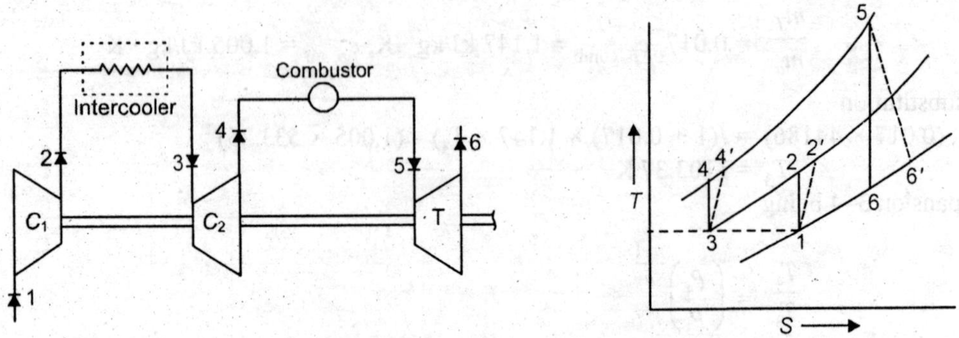

Fig. 9.35 Arrangement and T-S representation

Given: $T_5 = 1200\text{ K},\ T_1 = 300\text{ K},\ \dfrac{P_2}{P_1} = \dfrac{P_4}{P_3},\ T_3 = 300\text{ K},\ \eta_{\text{isen, c}} = 0.85$

$$\eta_{\text{isen, T}} = 0.9,\ \text{As } \frac{P_2}{P_1} = \frac{P_4}{P_3},\ \text{so } \frac{P_4}{P_3} = \frac{P_2}{P_1} = \sqrt{\frac{P_4}{P_1}} = \sqrt{r_p}$$

For process, 1–2,

$$\frac{T_2}{T_1} = \left(\frac{P_2}{P_1}\right)^{\frac{\gamma-1}{\gamma}}$$

$$\frac{T_2}{T_1} = (r_p)^{\frac{\gamma-1}{2\gamma}}$$

For process 3–4,

$$\frac{T_4}{T_3} = \left(\frac{P_4}{P_3}\right)^{\frac{\gamma-1}{\gamma}} = (r_p)^{\frac{\gamma-1}{2\gamma}}$$

$$\frac{T_4}{T_1} = (r_p)^{\frac{\gamma-1}{2\gamma}}$$

For process 5–6,

$$\frac{T_5}{T_6} = \left(\frac{P_4}{P_1}\right)^{\frac{\gamma-1}{\gamma}}$$

$$\frac{T_5}{T_6} = (r_p)^{\frac{\gamma-1}{\gamma}} \text{ or, } T_6 = \frac{T_5}{(r_p)^{\frac{\gamma-1}{\gamma}}}$$

Compressor work, $W_C = W_{C1} + W_{C2}$

$$= c_p (T_{2'} - T_1) + c_p(T_{4'} - T_3)$$
$$= c_p (T_{2'} - T_1) + c_p(T_{4'} - T_1)$$

Using Isentropic efficiency of compressor,

$$\eta_{\text{isen, c}} = \frac{T_2 - T_1}{T_{2'} - T_1} = \frac{T_4 - T_3}{T_{4'} - T_3} = \frac{T_4 - T_1}{T_{4'} - T_1}$$

Thus,

$$W_C = c_p \cdot \frac{(T_2 - T_1)}{\eta_{\text{isen, c}}} + c_p \frac{(T_4 - T_1)}{\eta_{\text{isen, c}}}$$

$$= \frac{c_p}{\eta_{\text{isen, c}}}\left\{\left(r_p^{\frac{(\gamma-1)}{2\gamma}} \cdot T_1 - T_1\right) + \left(r_p^{\frac{(\gamma-1)}{2\gamma}} \cdot T_1 - T_1\right)\right\}$$

$$W_C = \frac{2c_p \cdot T_1\left(r_p^{\frac{(\gamma-1)}{2\gamma}} - 1\right)}{\eta_{\text{isen, c}}}$$

Turbine work, $W_T = c_p \cdot (T_5 - T_{6'})$

Using isentropic efficiency of turbine, $\eta_{\text{isen, T}} = \dfrac{T_5 - T_{6'}}{T_5 - T_6}$

$$W_T = c_p \cdot (T_5 - T_6)\ \eta_{\text{isen, T}} = c_p \cdot T_5 \cdot \eta_{\text{isen, T}}\left(1 - \frac{1}{r_p^{\frac{\gamma-1}{\gamma}}}\right)$$

Net work $W_{\text{net}} = W_T - T_C$

$$= c_p \cdot T_5 \cdot \eta_{\text{isen, T}} \cdot \left(1 - \frac{1}{r_p^{\frac{\gamma-1}{\gamma}}}\right) - \frac{2c_p \cdot T_1\left(r_p^{\frac{\gamma-1}{2\gamma}} - 1\right)}{\eta_{\text{isen, c}}}$$

Differentiating new work with respect to r_p.

$$\frac{d\,W_{\text{net}}}{d\,r_p} = c_p \cdot T_5 \cdot \eta_{\text{isen, T}} \cdot \left(\frac{\gamma - 1}{\gamma}\right) \cdot r_p^{\frac{1-2\gamma}{\gamma}} - \frac{2c_p \cdot T_1}{\eta_{\text{isen, c}}} \cdot \left(\frac{\gamma - 1}{2\gamma}\right) \cdot r_p^{\frac{-1-\gamma}{2\gamma}}$$

Putting, $\dfrac{d\,W_{\text{net}}}{d\,r_p} = 0$

$$c_p \cdot T_5 \cdot \eta_{\text{isen, T}}\ r_p^{\frac{1-2\gamma}{\gamma}} = \frac{c_p \cdot T_1}{\eta_{\text{isen, c}}} \cdot r_p^{\frac{(-1-\gamma)}{2\gamma}}$$

$$r_p^{\frac{3(1-\gamma)}{2\gamma}} = \frac{T_1}{T_5 \cdot \eta_{\text{isen, c}}\ \eta_{\text{isen, T}}} \Rightarrow r_p = \left(\frac{T_1}{T_5 \cdot \eta_{\text{isen, c}} \cdot \eta_{\text{isen, T}}}\right)^{\frac{2\gamma}{3(1-\gamma)}}$$

$$r_{p,\,opt} = \left(\frac{T_1}{T_5 \cdot \eta_{\text{isen, c}}\ \eta_{\text{isen, T}}}\right)^{\frac{2\gamma}{3(1-\gamma)}}$$

Substituting known values,

$$r_{p,\,opt} = \left(\frac{300}{1200 \times 0.85 \times 0.90}\right)^{\frac{2\times 1.4}{3(1-1.4)}} = 13.6$$

Overall optimum pressure ratio = **13.6** **Ans.**

7. An air compressor has eight stages of equal pressure 1.35. The flow rate through the compressor and its' overall efficiency are 50 kg/s and 82% respectively. If the air enters the compressor at a pressure of 1 bar and temperature of 313 K, determine;

(i) State of air at the exit of compressor

(ii) Polytropic or small stage efficiency

(iii) Efficiency of each stage

(iv) Power required to drive the compressor assuming overall efficiency as 90%.

[U.P.S.C. 1992]

Solution:

(*i*) Theoretically state of air at exit can be determined by the given stage pressure ratio of 1.35. Let pressure at inlet to first stage be P_1 and subsequent intermediate pressure be P_2, P_3, P_4, P_5, P_6, P_7, P_8, and exit pressure being P_9.

Therefore,

$$\frac{P_2}{P_1} = \frac{P_3}{P_2} = \frac{P_4}{P_3} = \frac{P_5}{P_4} = \frac{P_6}{P_5} = \frac{P_7}{P_6} = \frac{P_8}{P_7} = \frac{P_9}{P_8} = 1.35$$

or

$$\frac{P_9}{P_1} = (1.35)^8 = 11.03$$

Theoretically, the temperature at exit of compressor can be predicted considering isentropic compression of air ($\gamma = 1.4$)

$$\frac{T_9}{T_1} = \left(\frac{P_9}{P_1}\right)^{\frac{\gamma-1}{\gamma}} = (11.03)^{\frac{0.4}{1.4}}$$

$$T_9 = 621.9 \text{ K}$$

Considering overall efficiency of compression 82% the actual temperature at compressor exit can be obtained

$$\frac{T_9 - T_1}{T_{9,\text{actual}} - T_1} = 0.82$$

$T_{9,\text{ actual}} = 689.71$ K

Let the actual index of compression be 'n', then

$$\left(\frac{T_{9,\text{ actual}}}{T_1}\right) = \left(\frac{P_9}{P_1}\right)^{\frac{n-1}{n}}$$

$$\left(\frac{689.71}{313}\right) = (11.03)^{\frac{n-1}{n}}$$

$$n = 1.49$$

State of air at exit of compressor, Pressure = **11.03 bar** Temperature = **689.71 K**	**Ans.**

(*ii*) Let polytropic efficiency be $\eta_{\text{polytropic}}$ for compressor, then;

$$\frac{n-1}{n} = \left(\frac{\gamma-1}{\gamma}\right) \times \frac{1}{\eta_{\text{polytropic}}}$$

$$\frac{(1.49-1)}{1.49} = \left(\frac{1.4-1}{1.4}\right) \times \frac{1}{\eta_{\text{polytropic}}}$$

$$\eta_{\text{polytropic}} = 0.8688 \text{ or } 86.88\%$$

Polytropic efficiency = **86.88%**	**Ans.**

(*iii*) Stage efficiency can be estimated for any stage. Say first stage.

Ideal state at exit of compressor stage $\Rightarrow \dfrac{T_2}{T_1} = \left(\dfrac{P_2}{P_1}\right)^{\frac{\gamma-1}{\gamma}}$

$$T_2 = 341.05 \text{ K}$$

Actual temperature at exit of first stage can be estimated using polytropic index 1.49.

$$\frac{T_{2,\text{ actual}}}{T_1} = \left(\frac{P_2}{P_1}\right)^{\frac{n-1}{n}}$$

$$T_{2,\text{ actual}} = 345.47 \text{ K}$$

Stage efficiency for first stage, $\eta_{s,\,1} = \dfrac{(T_2 - T_1)}{(T_{2,\text{ actual-1}})}$

$$= \frac{(341.05-313)}{(345.47-313)} = 0.8638 \text{ or } 86.38\%$$

Actual temperature at exit of second stage,

$$\frac{T_{3,\text{ actual}}}{T_{2,\text{ actual}}} = \left(\frac{P_3}{P_2}\right)^{\frac{n-1}{n}}$$

$$T_{3,\,actual} = 381.30\text{ K}$$

Ideal temperature at exit of second stage

$$\frac{T_3}{T_{2,\,actual}} = \left(\frac{P_3}{P_2}\right)^{\frac{\gamma-1}{\gamma}}$$

$$T_3 = 376.43\text{ K}$$

Stage efficiency for second stage, $\eta_{s,2} = \dfrac{T_3 - T_{2,\,actual}}{T_{3,\,actual} - T_{2,\,actual}} = \dfrac{376.43 - 345.47}{381.3 - 345.47}$

$$\eta_{s,2} = 0.8641 \text{ or } 86.41\%$$

Actual temperature at exit of third stage,

$$\frac{T_{4,\,actual}}{T_{3,\,actual}} = \left(\frac{P_4}{P_3}\right)^{\frac{n-1}{n}}$$

$$T_{4,\,actual} = 420.85\text{ K}$$

Ideal temperature at exit of third stage,

$$\frac{T_4}{T_{3,\,actual}} = \left(\frac{P_4}{P_3}\right)^{\frac{\gamma-1}{\gamma}},\; T_4 = 415.47\text{ K}$$

Stage efficiency for third stage, $\eta_{s,\,3} = \dfrac{T_4 - T_{3,\,actual}}{T_{4,\,actual} - T_{3,\,actual}} = \dfrac{(-381.3)}{(420.85 - 381.3)}$

$$= 0.86396 \text{ or } 86.40\%$$

Stage efficiency = 86.4% **Ans.**

(*iv*) From steady flow energy equation,

$$W_C = \int_1^9 dw = \int_1^9 dh, \text{ and } dh = du + pdv + vdp$$

$$dh = dq + vdp$$

for $dq = 0$ in adiabatic process

$$dh = vdp$$

$$W_C = \int_1^9 vdp$$

Here for polytropic compression

$$Pv^{1.49} = \text{constant i.e. } n = 1.49$$

$$W_C = \left(\frac{n}{n-1}\right) mRT_1 \left[\left(\frac{P_9}{T_1}\right)^{\frac{n-1}{n}} - 1\right]$$

$$= \left(\frac{1.49}{0.49}\right) \times 50 \times 0.287 \times 313 \left[\left(\frac{11.03}{1.0}\right)^{\frac{1.49-1}{1.49}} - 1\right]$$

$$W_C = 16419.87 \text{ kJ/s}$$

Due to overall efficiency being 90% the actual compressor work = $W_C \times 0.90$

$$W_{C,\text{ actual}} = 14777.89 \text{ kJ/s}$$

Power required to drive compressor = **14777.89 kJ/s** **Ans.**

8. A thermodynamic cycle has following processes occurring sequentially;

- *Adiabatic compression from 1 to 2*
- *Constant volume heat addition from 2 to 3 (explosion)*
- *Adiabatic expansion from 3 to 4*
- *Isothermal heat rejection from 4 to 1*

Obtain expression for air standard cycle efficiency considering compression ratio of 'r' and expansion ratio of 'e'.

Solution:

Let us analyze for 1 kg of air,

Given,

$$\frac{V_1}{V_2} = r,\ \frac{V_4}{V_3} = e;\ V_2 = V_3,\ T_1 = T_4$$

$$\frac{V_1}{V_4} = \frac{r}{e}, \text{ For process 4–1, } P_1V_1 = P_4V_4 = RT_1$$

$$\text{Heat added} = c_v(T_3 - T_2)$$

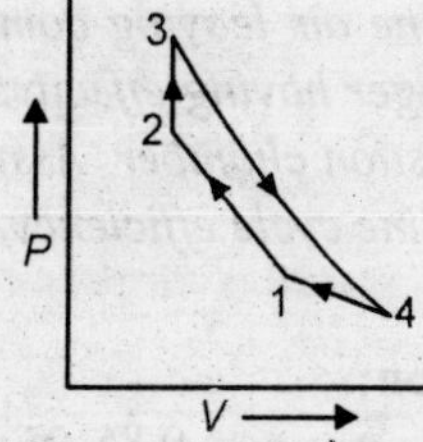

Fig. 9.36

Heat rejected in isothermal process, 4–1 = $P_4V_4 \ln\left(\frac{V_4}{V_1}\right)$

$$\text{Net work} = \text{Heat added} - \text{Heat rejected}$$

$$W_{\text{net}} = c_v(T_3 - T_2) - P_4V_4 \ln\left(\frac{e}{r}\right)$$

$$\text{Efficiency, } \eta = \frac{\text{Net work}}{\text{Heat added}} = \frac{c_v(T_3 - T_2) - P_4V_4 \ln\left(\frac{e}{r}\right)}{c_v(T_3 - T_2)}$$

$$\eta = 1 - \frac{RT_1 \ln\left(\frac{e}{r}\right)}{c_v(T_3 - T_2)}$$

As, $c_v = \frac{R}{(\gamma - 1)}$, substituting in expression for efficiency,

$$\eta = 1 - \frac{(\gamma-1)\cdot T_1 \ln\left(\frac{e}{r}\right)}{(T_3 - T_2)}$$

$$\eta = 1 - \frac{(\gamma-1)\ln\left(\frac{e}{r}\right)}{\left(\frac{T_3}{T_1} - \frac{T_2}{T_1}\right)} = 1 - \frac{(\gamma-1)\ln\left(\frac{e}{r}\right)}{\left(\frac{T_3}{T_4} - \frac{T_2}{T_4}\right)}$$

For process 3–4, $\frac{T_3}{T_4} = \left(\frac{V_4}{V_3}\right)^{\gamma-1} = e^{\gamma-1}$

For process 1–2, $\frac{T_2}{T_1} = \left(\frac{V_1}{V_2}\right)^{\gamma-1} \Rightarrow \frac{T_2}{T_4} = \left(\frac{V_1}{V_2}\right)^{\gamma-1} = r^{\gamma-1}$

Substituting in expression of η

$$\boxed{\eta = 1 - \frac{(\gamma-1)\ln\left(\frac{e}{r}\right)}{(e^{\gamma-1} - r^{\gamma-1})}}$$ **Ans.**

***9.** In a gas turbine installation air is supplied at 1 bar, 27°C into compressor having compression ratio of 8. The air leaving combustion chamber is heated upto 1100 K and expanded upto 1 bar. A heat exchanger having effectiveness of 0.8 is fitted at exit of turbine for heating the air before its inlet into combustion chamber. Assuming polytropic efficiency of the compressor and turbine as 0.85 and 0.90 determine cycle efficiency, work ratio and specific work output of plant. Take c_p = 1.0032 kJ/kg ·K for air.*

Solution:

$\gamma = 1.4$, $\eta_{poly,\,c} = 0.85$, $\eta_{poly,\,T} = 0.90$

Using polytropic efficiency the index of compression and expansion can be obtained as under,

Let compression index be n_c

$$\left(\frac{n_c - 1}{n_c}\right) = \left(\frac{\gamma - 1}{\gamma \cdot \eta_{poly,\,c}}\right)$$

$$\eta_c = 1.506$$

Let expansion index be n_T.

$$\left(\frac{n_T - 1}{n_T}\right) = \left(\frac{\eta_{poly,T}\cdot(\gamma-1)}{\gamma}\right)$$

$\Rightarrow$ $$n_T = 1.346$$

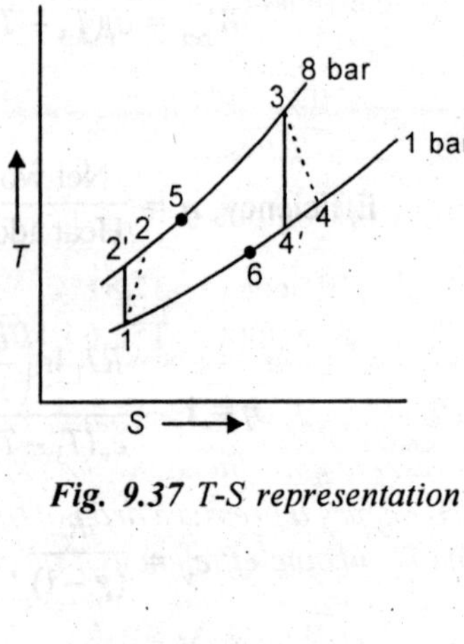

Fig. 9.37 T-S representation

For process 1–2

$$\frac{T_2}{T_1} = \left(\frac{P_2}{P_1}\right)^{\frac{n_c-1}{n_c}}$$

$$\frac{T_2}{300} = (8)^{\frac{1.506-1}{1.506}}$$

$$T_2 = 603.32 \text{ K}$$

$$T_3 = 1100 \text{ K}$$

$$\frac{T_4}{T_3} = \left(\frac{P_4}{P_3}\right)^{\frac{n_T-1}{n_T}}$$

$$\frac{T_4}{1100} = \left(\frac{1}{8}\right)^{\frac{1.346-1}{1.346}}$$

$$T_4 = 644.53 \text{ K}$$

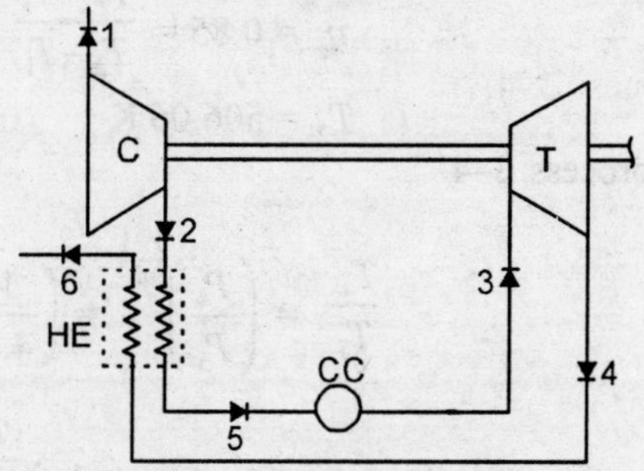

Fig. 9.38 Layout of regenerative cycle

Using heat exchanger effectiveness,

$$\varepsilon = 0.8 = \frac{T_5 - T_2}{T_4 - T_2}$$

$$0.8 = \frac{T_5 - 603.32}{(644.53 - 603.32)}$$

$$T_5 = 636.28 \text{ K}$$

Heat added in combustion chambers, $q_{\text{add}} = c_p\,(T_3 - T_5)$

$$q_{\text{add}} = 1.0032\,(1100 - 636.28)$$

$$q_{\text{add}} = 465.204 \text{ kJ/kg}$$

Compressor work $W_C = c_p\,(T_2 - T_1) = 1.0032\,(603.32 - 300) = 304.29$ kJ/kg

Turbine work, $W_T = c_p\,(T_3 - T_4) = 1.0032\,(1100 - 644.53) = 456.93$ kJ/kg

$$\text{Cycle efficiency} = \frac{W_T - W_C}{q_{\text{add}}} = \frac{456.93 - 304.29}{465.204} = 0.3281 \text{ or } 32.81\%$$

$$\text{Work ratio, } \frac{W_T - W_C}{W_T} = 0.334$$

Specific work output $= W_T - W_C = 152.64$ kJ/kg

Cycle efficiency = **32.81%**, Work ratio = **0.334**, **Ans.**
Specific work output = **152.64 kJ/kg**

***10.** A gas turbine plant has air being supplied at 1 bar, 27°C to compressor for getting compressed upto 5 bar with isentropic efficiency of 85%. Compressed air is heated upto 1000 K in combustion chamber where also occurs a pressure drop of 0.2 bar. Subsequently expansion occurs to 1 bar in turbine. Determine isentropic efficiency of turbine, if thermal efficiency of plant is 20%.*

Neglect the air property variation throughout cycle. Take γ = 1.4

Solution:

$T_1 = 300$ K, $P_1 = 1$ bar, $P_2 = 5$ bar, $\eta_c = 0.85$, $T_3 = 1000$ K

For process 1–2′

$$\frac{T_{2'}}{T_1} = \left(\frac{P_{2'}}{P_1}\right)^{\frac{\gamma-1}{\gamma}} \Rightarrow T_{2'} = 300\,(500)^{\frac{(1.4-1)}{1.4}}$$

$$T_{2'} = 475.15 \text{ K}$$

$$\eta_c = 0.85 = \frac{T_{2'} - T_1}{T_2 - T_1}$$

$$T_2 = 506.06 \text{ K}$$

For process 3–4′

$$\frac{T_{4'}}{T_3} = \left(\frac{P_4}{P_3}\right)^{\frac{\gamma-1}{\gamma}} = \left(\frac{1}{4.8}\right)^{\frac{1.4-1}{1.4}}$$

$$T_{4'} = 1000 \times \left(\frac{1}{4.8}\right)^{\frac{1.4-1}{1.4}}$$

$$T_{4'} = 638.79 \text{ K}$$

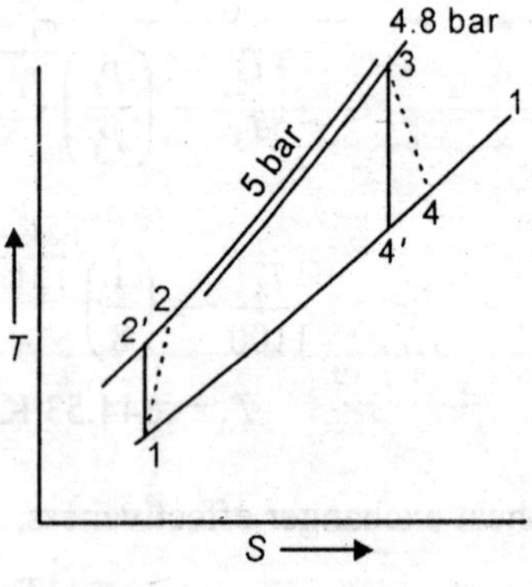

Fig. 9.39 T-S representation

Compressor work per kg $W_C = c_p\,(T_2 - T_1) = 1.0032 \times (506.06 - 300) = 206.72$ kJ/kg

Turbine work per kg, $W_T = c_p(T_3 - T_4) = 1.0032 \times (1000 - T_4)$, kJ/kg

Net output, $W_{net} = W_T - W_C = \{206.72 - (1.0032\,(1000 - T_4))\}$, kJ/kg

Heat added, $q_{add} = c_p\,(T_3 - T_2) = 1.0032\,(1000 - 506.06) = 495.52$ kJ/kg

$$\text{Thermal efficiency,} \quad \eta = \frac{W_{net}}{q_{add}} \Rightarrow 0.20 = \frac{206.71 - (1.0032 \times (1000 - T_4))}{495.52}$$

$$T_4 = 892.73 \text{ K}$$

$$\text{Therefore, isentropic efficiency of turbine, } \eta_T = \frac{T_3 - T_4}{T_3 - T_{4'}} = \frac{(1000 - 892.73)}{(1000 - 638.79)}$$

$$\eta_T = 0.2969 \text{ or } 29.69\%$$

Turbine isentropic efficiency = **29.69%**	**Ans.**

***11.** A gas turbine plant has air supplied at 1 bar, 27°C for being compressed through pressure ratio of 10. Compression of air is achieved in two stages with perfect intercooling in between at optimum pressure. The maximum temperature in cycle is 1000 K and compressed air at this temperature is sent for expansion in two stages of gas turbine. First stage expansion occurs upto 3 bar and is subsequently reheated upto 995 K before being sent to second stage. Fuel used for heating in combustion chamber has calorific value of 42,000 kJ/kg. Considering c_p = 1.0032 kJ/kg. K throughout cycle determine, net output, thermal efficiency and air fuel ratio when air flows into compressor at 30 kg/s. Take isentropic efficiency of compression and expansion to be 85% and 90% respectively.*

Solution:

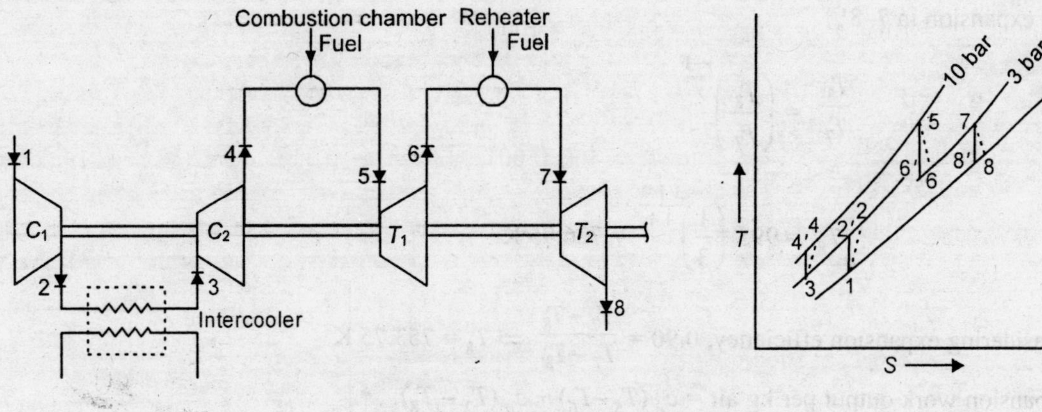

Fig. 9.40 Layout and T-S diagram

For perfect intercooling the pressure ratio of each compression stage = $\sqrt{10}$ = 3.16

For process 1–2′, $\frac{T_{2'}}{T_1} = \left(\frac{P_2}{P_1}\right)^{\frac{\gamma-1}{\gamma}} \Rightarrow T_{2'} = (3.16)^{\frac{1.4-1}{1.4}} \times 300 = 416.76$ K

Considering isentropic efficiency of compression,

$$\eta_c = 0.85 = \frac{T_{2'} - T_1}{T_2 - T_1} = \frac{416.76 - 300}{T_2 - 300}$$

For perfect intercooling, $T_3 = T_1 = 300$ K, $T_2 = 437.36$ K

For process 3–4,

$$\frac{T_{4'}}{T_3} = \left(\frac{P_4}{P_3}\right)^{\frac{1.4-1}{1.4}} \Rightarrow T_{4'} = 300 \times (3.16)^{\frac{1.4-1}{1.4}} = 416.76 \text{ K}$$

Again due to compression efficiency, $\eta_c = 0.85 = \frac{T_{4'} - T_3}{T_4 - T_3}$

$\Rightarrow$ $T_4 = 437.36$ K

Total compressor work, $W_C = 2 \times c_p \times (437.36 - 300) = 275.59$ kJ/kg

$T_5 = 1000$ K, For Expansion process 5–6′

$$\frac{T_{6'}}{T_5} = \left(\frac{P_6}{P_5}\right)^{\frac{\gamma-1}{\gamma}}$$

$$T_{6'} = 1000 \times \left(\frac{3}{10}\right)^{\frac{1.4-1}{1.4}} = 708.93 \text{ K}$$

Considering expansion efficiency, $0.90 = \frac{T_5 - T_6}{T_5 - T_{6'}}$

$\Rightarrow \quad T_6 = 738.4$ K

$T_7 = 995$ K

For expansion in 7–8′,

$$\frac{T_{8'}}{T_7} = \left(\frac{P_8}{P_7}\right)^{\frac{\gamma-1}{\gamma}}$$

$$\Rightarrow \quad T_{8'} = 995\left(\frac{1}{3}\right)^{\frac{1.4-1}{1.4}} = 726.95 \text{ K}$$

Considering expansion efficiency, $0.90 = \dfrac{T_7 - T_8}{T_7 - T_{8'}} \Rightarrow T_8 = 753.75$ K

Expansion work output per kg air $= c_p\,(T_5 - T_6) + c_p\,(T_7 - T_8)$

$W_T = 1.0032\ \{(1000 - 738.04) + (995 - 753.75)\}$

$W_T = 514.85$ kJ/kg

Heat added per kg air $= c_p(T_5 - T_4) + c_p(T_7 - T_6)$

$q_{add} = 1.0032\ \{(1000 - 437.36) + (995 - 738.04)\}$

$q_{add} = 822.22$ kJ/kg

Fuel required per kg of air, $m_f = \dfrac{822.22}{42000} = 0.01958$

Air-fuel ratio $= \dfrac{1}{0.01958} = 51.07$

Net output $= W_T - W_C = 239.26$ kJ/kg

Output for air flowing at 30 kg/s, $= 239.26 \times 30 = 7177.8$ kW

Thermal efficiency $= \dfrac{W_T - W_C}{q_{add}} = \dfrac{239.26}{822.22} = 0.2909$ or 29.09%

Thermal efficiency = 20.09%, Net output = **7177.8** kW, A/F ratio = **51.07**	**Ans.**

***12.** A regenerative-reheat cycle has air entering at 1 bar, 300 K into compressor having intercooling in between the two stages of compression. Air leaving first stage of compression is cooled upto 290 K at 4 bar pressure in intercooler and subsequently compressed upto 8 bar. Compressed air leaving second stage compressor is passed through a regenerator having effectiveness of 0.80. Subsequent combustion chamber yields 1300 K at inlet to turbine having expansion upto 4 bar and then reheated upto 1300 K before being expanded upto 1 bar. Exhaust from turbine is passed through regenerator before discharged out of cycle. For the fuel having heating value of 42000 kJ/kg determine fuel-air ratio in each combustion chamber, total turbine work and thermal efficiency. Consider compression to be isentropic and air as working fluid throughout the cycle.*

Solution:

$T_1 = 300$ K, $P_1 = 1$ bar, $P_2 = P_3 = 4$ bar, $T_3 = 290$ K, $T_6 = 1300$ K,
$P_6 = P_4 = 8$ bar, $T_8 = 1300$ K, $P_8 = 4$ bar

For process 1–2,

$$\frac{T_2}{T_1} = \left(\frac{P_2}{P_1}\right)^{\frac{\gamma-1}{\gamma}} \Rightarrow T_2 = 445.79 \text{ K}$$

For process 3–4,

$$\frac{T_4}{T_3} = \left(\frac{P_4}{P_3}\right)^{\frac{\gamma-1}{\gamma}} \Rightarrow T_4 = 353.51 \text{ K}$$

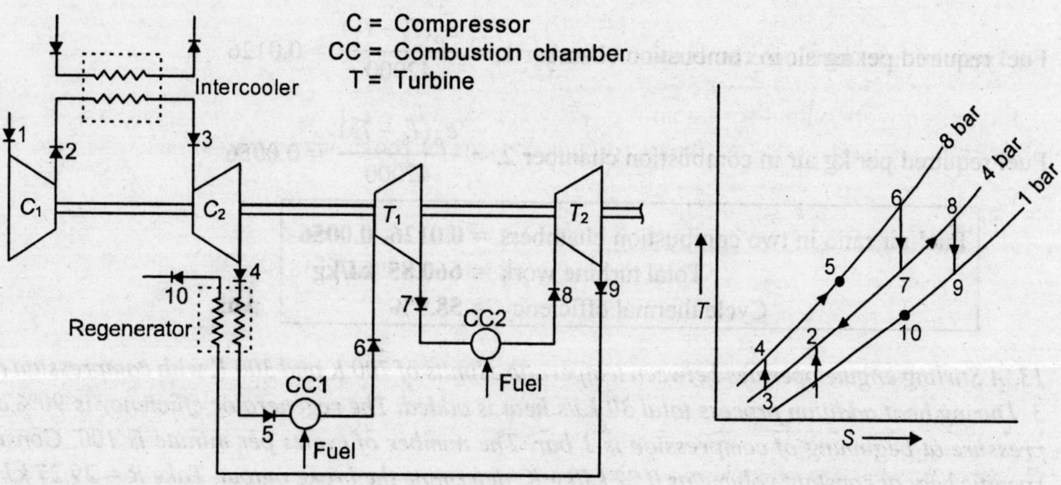

Fig. 9.41 Layout and T-s diagram

For process 6–7,

$$\frac{T_7}{T_6} = \left(\frac{P_7}{P_6}\right)^{\frac{\gamma-1}{\gamma}} = \left(\frac{4}{8}\right)^{\frac{1.4-1}{1.4}} \Rightarrow T_7 = 1066.43 \text{ K}$$

For process 8–9,

$$\frac{T_9}{T_8} = \left(\frac{P_9}{P_8}\right)^{\frac{\gamma-1}{\gamma}} \Rightarrow T_9 = 1300 \times \left(\frac{1}{4}\right)^{\frac{1.4-1}{1.4}} = 874.83 \text{ K}$$

In regenerator, effectiveness $= 0.80 = \frac{T_5 - T_4}{T_9 - T_4} = \frac{T_5 - 353.51}{(874.83 - 353.51)}$

$\Rightarrow \quad T_5 = 770.56$ K

Compressor work per kg air, $W_c = c_p\,(T_2 - T_1) + c_p\,(T_4 - T_3)$

$W_c = 209.97$ kJ/kg

Turbine work per kg air, $W_T = c_p\,(T_6 - T_7) + c_p\,(T_8 - T_9)$

$W_T = 660.85$ kJ/kg

Heat added per kg air, $q_{add} = c_p (T_6 - T_5) + c_p (T_8 - T_7)$

$q_{add} = 765.45$ kJ/kg

$$\text{Total fuel required per kg of air} = \frac{765.45}{42000} = 0.0182$$

Net work $W_{net} = W_T - W_C = 450.88$ kJ/kg

$$\text{Cycle thermal efficiency,} \quad \eta = \frac{W_{net}}{q_{add}} = \frac{450.88}{765.45} = 0.5890 \text{ or } 58.9\%$$

$$\text{Fuel required per kg air in combustion chamber 1,} = \frac{c_p(T_8 - T_7)}{42000} = 0.0126$$

$$\text{Fuel required per kg air in combustion chamber 2,} = \frac{c_p(T_6 - T_5)}{42000} = 0.0056$$

Fuel-air ratio in two combustion chambers =	**0.0126, 0.0056**
Total turbine work =	**660.85 kJ/kg**
Cycle thermal efficiency =	**58.9%**

Ans.

13. A Stirling engine operates between temperature limits of 700 K and 300 K with compression ratio of 3. During heat addition process total 30 kJ/s heat is added. The regenerator efficiency is 90% and the pressure at beginning of compression is 1 bar. The number of cycles per minute is 100. Considering specific heat at constant volume as 0.72 kJ/kg·K, determine the brake output. Take R = 29.27 kJ/kg · K. Also determine stroke volume.

Solution:

Work done per kg of air

$= R(T_2 - T_1). \ln r$

$= 29.27 (700 - 300) \ln 3$

$W = 12862.55$ kJ/kg

Heat added per kg of air

$= RT_2 \ln r + (1 - \varepsilon) c_v \cdot (T_2 - T_1)$

$= (29.27 \times 700 \times \ln 3) + \{(1 - 0.9) \times 0.72 \times (700 - 300)\}$

$q = 22538.27$ kJ/kg

$$\text{For 30 kJ/s heat supplied, the mass of air/s} = \frac{30}{22538.27}$$

$= 1.33 \times 10^{-3}$ kg/s

$$\text{Mass of air per cycle} = \frac{1.33 \times 10^{-3}}{100} = 1.33 \times 10^{-4} \text{ kg/cycle}$$

Brake output $= 12862.55 \times 1.33 \times 10^{-3} = 17.11$ kW

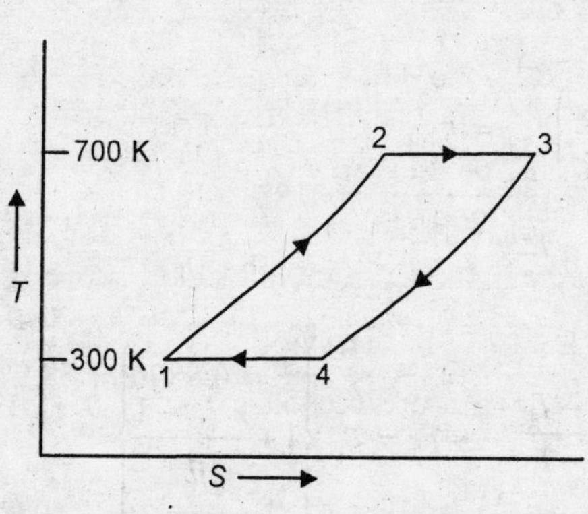

Fig. 9.42 Stirling cycle

Stroke volume, $V = \frac{mRT}{P} = \frac{(1.33\times10^{-4})\times29.27\times300}{1\times10^{2}} = 0.0116 \text{ m}^3$

Brake output = **17.11 kW**
Stroke volume = **0.0116 m³**

Ans.

14. *A gas turbine installation operates with fixed maximum and minimum temperatures T_3 and T_1 respectively. Show that the optimum specific work output shall be obtained at same overall pressure ratio for each of following arrangement. Also give expression for this pressure ratio for air as working fluid throughout.*

(i) there is single stage compression followed by two stages of expansion in turbine. The expansion ratio in two stages is equal and reheating is done upto the maximum temperature at inlet of second stage of expansion.

(ii) there occurs compression in two stages of equal compression ratio with intercooling upto the minimum cycle temperature at inlet to second stage of compression followed by single stage expansion in turbine.

Take isentropic efficiency of compressor and turbine stages as η_c and η_T respectively.

Solution:

Let the overall pressure ratio be *r*.

For arrangement (*i*)

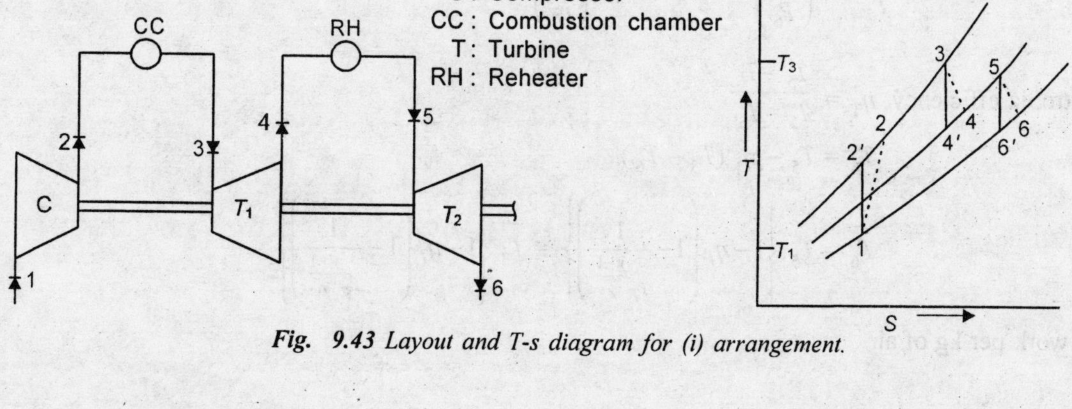

Fig. 9.43 Layout and T-s diagram for (i) arrangement.

In compressor,

$$\frac{T_{2'}}{T_1} = \left(\frac{P_2}{P_1}\right)^{\frac{\gamma-1}{\gamma}} = r^{\frac{\gamma-1}{\gamma}}$$

$$T_{2'} = T_1 \cdot r^{\frac{\gamma-1}{\gamma}}$$

$$\eta_c = \frac{T_{2'} - T_1}{T_2 - T_1} \Rightarrow T_2 = T_1 \left\{1 + \frac{r^{\frac{\gamma-1}{\gamma}} - 1}{\eta_c}\right\}$$

Let expansion ratio in turbine stages be $r_T = \sqrt{r}$

For first stage expansion

$$\frac{T_{4'}}{T_3} = \left(\frac{P_4}{P_3}\right)^{\frac{\gamma-1}{\gamma}}$$

$$T_{4'} = T_3 \left(\frac{1}{r_T}\right)^{\frac{\gamma-1}{\gamma}}$$

Isentropic efficiency of turbine,

$$\eta_T = \frac{T_3 - T_4}{T_3 - T_{4'}} \Rightarrow T_4 = T_3 - \eta_T (T_3 - T_{4'})$$

$$\Rightarrow \quad T_4 = T_3 \left\{1 - \eta_T \left(1 - \frac{1}{r_T^{\frac{\gamma-1}{\gamma}}}\right)\right\}$$

For second stage expansion, $T_5 = T_3$

$$\frac{T_5}{T_{6'}} = \left(\frac{P_5}{P_6}\right)^{\frac{\gamma-1}{\gamma}}, \quad T_{6'} = T_3 \cdot \left(\frac{1}{r_T}\right)^{\frac{\gamma-1}{\gamma}}$$

Isentropic efficiency, $\eta_T = \dfrac{T_5 - T_6}{T_5 - T_{6'}}$

$$T_6 = T_5 - \eta_T (T_5 - T_{6'})$$

$$T_6 = T_5 \left\{1 - \eta_T \left(1 - \frac{1}{r_T^{\frac{\gamma-1}{\gamma}}}\right)\right\} = T_3 \left\{1 - \eta_T \left(1 - \frac{1}{r_T^{\frac{\gamma-1}{\gamma}}}\right)\right\}$$

Net work per kg of air

$$W_{\text{net}} = \{c_p(T_3 - T_4) + c_p(T_5 - T_6)\} - \{c_p(T_2 - T_1)\}$$

$$W_{\text{net}} = c_p \left[T_3 - T_3 \left\{ 1 - \eta_T \left(1 - \frac{1}{r_T^{\frac{\gamma-1}{\gamma}}} \right) \right\} + T_3 - T_3 \right.$$

$$\left. \left\{ 1 - \eta_T \left(1 - \frac{1}{r_T^{\frac{\gamma-1}{\gamma}}} \right) \right\} - T_1 \left\{ 1 + \frac{r^{\frac{\gamma-1}{\gamma}} - 1}{\eta_c} \right\} + T_1 \right]$$

$$W_{\text{net}} = c_p \left[2\, T_3 \cdot \eta_T \left(1 - \frac{1}{r^{\frac{\gamma-1}{2\gamma}}} \right) - T_1 \cdot \frac{r^{\frac{\gamma-1}{\gamma}} - 1}{\eta_c} \right]$$

Differentiating with respect to r.

$$\frac{d\,W_{\text{net}}}{dr} = -\,T_3 \cdot \eta_T \left(\frac{1-\gamma}{2\gamma} \right) r^{\frac{1-3\gamma}{2\gamma}} - \frac{T_1}{\eta_c} \left(\frac{\gamma-1}{\gamma} \right) r^{\frac{-1}{\gamma}}$$

Substituting it equal to zero. $\dfrac{d\,W_{\text{net}}}{dr} = 0$

$$\Rightarrow \qquad \boxed{r = \left(\frac{\eta_c \cdot \eta_T \cdot T_3}{T_1} \right)^{\frac{2}{3}\left(\frac{\gamma}{\gamma-1}\right)}}$$

Optimum pressure ratio
For arrangement (*ii*)

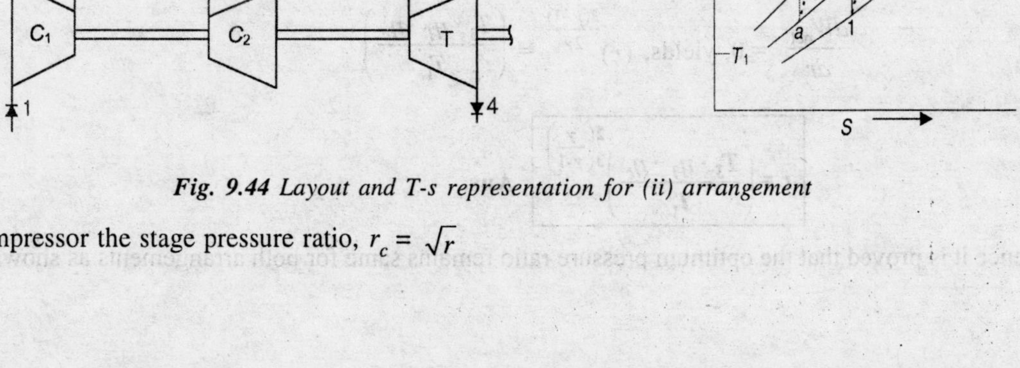

Fig. 9.44 Layout and T-s representation for (ii) arrangement

In compressor the stage pressure ratio, $r_c = \sqrt{r}$

For first stage compressor, $\frac{T_{2'}}{T_1} = \left(\frac{P_2}{P_1}\right)^{\frac{\gamma-1}{\gamma}}$, $T_{2'} = T_1 \cdot (r_c)^{\frac{\gamma-1}{\gamma}}$

Isentropic compression efficiency, $\eta_c = \frac{T_{2'} - T_1}{T_2 - T_1}$, $\Rightarrow T_2 = T_1 \left\{1 + \frac{r_c^{\frac{\gamma-1}{\gamma}} - 1}{\eta_c}\right\}$

For second stage compressor, $T_a = T_1$

$$\frac{T_{b'}}{T_a} = \left(\frac{P_b}{P_a}\right)^{\frac{\gamma-1}{\gamma}} \quad T_{b'} = T_1 \, (r_c)^{\frac{\gamma-1}{\gamma}}$$

Isentropic efficiency of compression, $\eta_c = \frac{T_{b'} - T_a}{T_b - T_a} \Rightarrow T_b = T_1 \left\{1 + \frac{r_c^{\frac{\gamma-1}{\gamma}} - 1}{\eta_c}\right\}$

For expansion in turbine,

$$\frac{T_3}{T_{4'}} = \left(\frac{P_3}{P_4}\right)^{\frac{\gamma-1}{\gamma}} \Rightarrow T_{4'} = \frac{T_3}{(r)^{\frac{\gamma-1}{\gamma}}}$$

Isentropic expansion efficiency, $\eta_T = \frac{T_3 - T_4}{T_3 - T_{4'}}$

$$\Rightarrow \qquad T_4 = T_3 \left\{1 - \eta_T \left(1 - \frac{1}{r^{\frac{\gamma-1}{\gamma}}}\right)\right\}$$

Net work per kg of air,

$$W_{\text{net}} = \{c_p \, (T_2 - T_1) + c_p(T_b - T_a)\} - \{c_p(T_3 - T_4)\}$$

$$= c_p \left\{T_1 \left(1 + \frac{r_c^{\frac{\gamma-1}{\gamma}} - 1}{\eta_c}\right) - T_1 + T_1 \left(1 + \frac{r_c^{\frac{\gamma-1}{\gamma}} - 1}{\eta_c}\right) - T_1\right\} - c_p \left\{T_3 - T_3 \left(1 - \eta_T \left(1 - \frac{1}{r^{\frac{\gamma-1}{\gamma}}}\right)\right)\right\}$$

Differentiating W_{net} with respect to 'r' and equating it to zero.

$$\frac{dW_{\text{net}}}{dr} = 0, \text{ yields, } (r)^{\frac{3(\gamma-1)}{2\gamma}} = \left(\frac{T_3 \cdot \eta_T \cdot \eta_c}{T_1}\right)$$

$$\Rightarrow \qquad \boxed{r = \left(\frac{T_3 \cdot \eta_T \cdot \eta_c}{T_1}\right)^{\frac{2}{3}\left(\frac{\gamma}{\gamma-1}\right)}} \quad \textbf{Ans.}$$

Hence it is proved that the optimum pressure ratio remains same for both arrangements as shown here.

15. *In a gas/steam combined cycle power plant exhaust from gas turbine leaves at 420 K and generates steam in heat recovery steam generator (HRSG) at 6 MPa and 400ºC. Ambient air at 1 bar, 17ºC enters the compressor operating at pressure ratio of 10. Turbine inlet temperature of gas turbine is 1400 K. The steam turbine operates with condenser pressure of 15 kPa. Consider air as working fluid throughout gas turbine cycle and combined cycle output to be 37.3 MW. Determine the overall efficiency, mass flow rate of steam per kg of air. Take* $c_{p,\,air} = 1.0032$ *kJ/kg · K*

Solution:

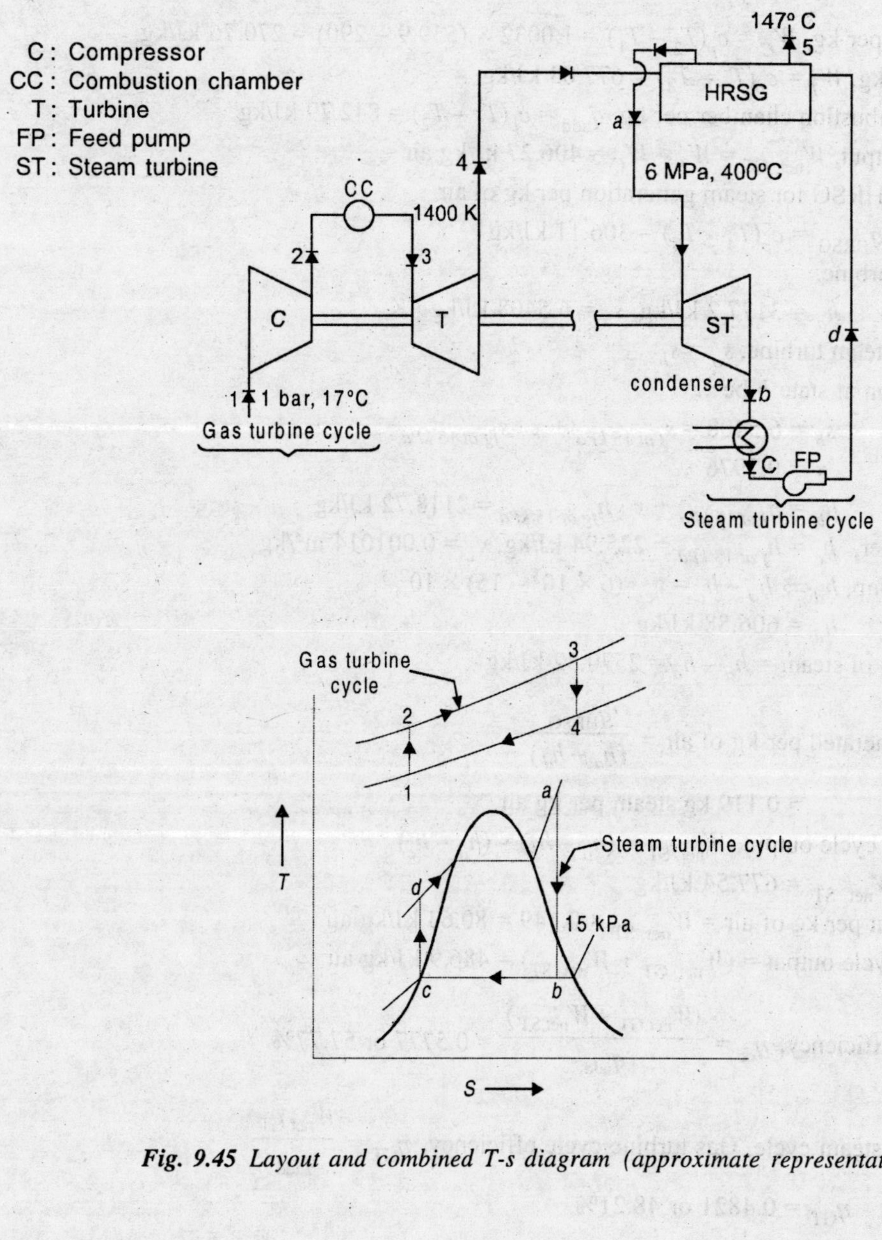

Fig. 9.45 Layout and combined T-s diagram (approximate representation)

In gas turbine cycle, $\frac{T_2}{T_1} = \left(\frac{P_2}{P_1}\right)^{\frac{\gamma-1}{\gamma}} \Rightarrow T_2 = 290 \times (10)^{\frac{1.4-1}{1.4}} = 559.9$ K

$T_3 = 1400$ K,

$$\frac{T_4}{T_3} = \left(\frac{P_4}{P_3}\right)^{\frac{\gamma-1}{\gamma}} \Rightarrow T_4 = 1400\left(\frac{1}{10}\right)^{\frac{1.4-1}{1.4}} = 725.13 \text{ K}$$

Compressor work per kg, $W_C = c_p(T_2 - T_1) = 1.0032 \times (559.9 - 290) = 270.76$ kJ/kg

Turbine work per kg, $W_T = c_p(T_3 - T_4) = 677.03$ kJ/kg

Heat added in combustion chamber per kg, $q_{add} = c_p(T_3 - T_2) = 842.79$ kJ/kg

Net gas turbine output, $W_{net,\,GT} = W_T - W_C = 406.27$ kJ/kg air

Heat recovered in HRSG for steam generation per kg of air

$$q_{HRSG} = c_p(T_4 - T_5) = 306.11 \text{ kJ/kg}$$

At inlet to steam turbine,

$$h_a = 3177.2 \text{ kJ/kg},\ s_a = 6.5408 \text{ kJ/kg} \cdot \text{K}$$

For expansion in steam turbine, $s_a = s_b$

Let dryness fraction at state b be x.

$$s_b = 6.5408 = s_{f\,at\,15\,kPa} + x \cdot s_{fg\,at\,15\,kPa}$$

$$x = 0.7976$$

$$h_b = h_{f\,at\,15\,kPa} + x \cdot h_{fg\,at\,15\,kPa} = 2118.72 \text{ kJ/kg}$$

At exit of condenser, $h_c = h_{f\,at\,15\,kPa} = 225.94$ kJ/kg, $v_c = 0.001014$ m^3/kg

At exit of feed pump, $h_d \Rightarrow h_d - h_c = v_c \cdot (6 \times 10^3 - 15) \times 10^2$

$$h_d = 606.88 \text{ kJ/kg}$$

Heat added per kg of steam $= h_a - h_d = 2570.32$ kJ/kg

$$\text{Mass of steam generated per kg of air} = \frac{q_{HRSG}}{(h_a - h_d)}$$

$= 0.119$ kg steam per kg air.

Net steam turbine cycle output, $W_{net,\,ST} = (h_a - h_b) - (h_d - h_c)$

$$W_{net,\,ST} = 677.54 \text{ kJ/kg}$$

Steam cycle output per kg of air $= W_{net,\,ST} \times 0.149 = 80.63$ kJ/kg air

Total combined cycle output $= (W_{net,\,GT} + W_{net,\,ST}) = 486.9$ kJ/kg air

$$\text{Combined cycle efficiency, } \eta_{cc} = \frac{(W_{net,GT} + W_{net,ST})}{q_{add}} = 0.5777 \text{ or } 57.77\%$$

$$\text{In the absence of steam cycle, Gas turbine cycle efficiency, } \eta_{GT} = \frac{W_{net,GT}}{q_{add}}$$

$$\eta_{GT} = 0.4821 \text{ or } 48.21\%$$

Thus, efficiency is seen to increase in combined cycle upto 57.77% as compared to gas turbine offering 48.21% efficiency.

Overall efficiency = **57.77%**
Steam per kg of air = **0.119 kg** steam/kg air **Ans.**

16. In an I.C. engine operating on the dual cycle (limited pressure cycle), the temperature of the working fluid (air) at the beginning of compression is 27°C. The ratio of the maximum and minimum pressures of the cycle is 70 and the compression ratio is 15. The amount of heat added at constant volume and at constant pressure are equal. Compute the air standard thermal efficiency of the cycle. State three main reasons why the actual thermal efficiency is different from the theoretical value.

[U.P.S.C. 1993]

Solution:

$$T_1 = 27°C = 300 \text{ K}$$

$$\frac{P_4}{P_1} = \frac{P_3}{P_1} = 70$$

Compression ratio, $\frac{V_1}{V_2} = 15 = \frac{V_1}{V_3}$

Heat added at constant volume = Heat added at constant pressure

$$Q_{23} = Q_{34}$$

$$m \cdot c_v(T_3 - T_2) = m \cdot c_p \cdot (T_4 - T_3)$$

$$(T_3 - T_2) = \gamma \cdot (T_4 - T_3)$$

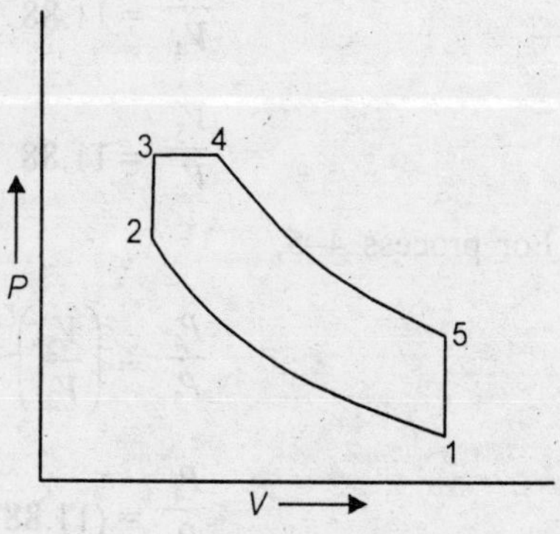

Fig. 9.46

For process 1–2;

$$\frac{T_2}{T_1} = \left(\frac{P_2}{P_1}\right)^{\frac{\gamma-1}{\gamma}}$$

$$\frac{T_2}{T_1} = \left(\frac{V_1}{V_2}\right)^{\gamma-1}$$

$$\frac{T_2}{300} = (15)^{0.4}$$

$$T_2 = 886.25$$

and

$$\left(\frac{P_2}{P_1}\right) = \left(\frac{V_1}{V_2}\right)^{\gamma} = (15)^{1.4}$$

$$P_2 = P_1 \cdot (15)^{1.4}$$

$$P_2 = 44.3\ P_1$$

For process 2–3

$$\frac{P_2}{P_3} = \frac{T_2}{T_3}$$

$$\frac{44.3P_1}{70P_1} = \frac{886.25}{T_3}$$

or, $$T_3 = 1400.39$$

Using equal heat additions for processes 2–3 and 3–4,

$$(T_3 - T_2) = \gamma\,(T_4 - T_3)$$

$$(1400.39 - 886.25) = 1.4(T_4 - 1400.39)$$

$$T_4 = 1767.63$$

For process 3–4,

$$\frac{V_3}{V_4} = \frac{T_3}{T_4}$$

$$\frac{V_3}{V_1} \times \frac{V_1}{V_4} = \frac{T_3}{T_4}$$

$$\frac{1}{15} \times \left(\frac{V_1}{V_4}\right) = \left(\frac{1400.39}{1767.63}\right)$$

$$\frac{V_1}{V_4} = 11.88$$

$$\frac{V_5}{V_4} = 11.88$$

For process 4–5,

$$\frac{P_4}{P_5} = \left(\frac{V_5}{V_4}\right)^{\gamma} \text{ or, } \frac{T_4}{T_5} = \left(\frac{V_5}{V_4}\right)^{\gamma-1}$$

$$\frac{P_4}{P_5} = (11.88)^{1.4} \text{ or, } \frac{T_4}{T_5} = (11.88)^{0.4}$$

$$T_5 = \frac{1767.63}{(11.88)^{0.4}} = 656.85$$

$$\text{Air standard thermal efficiency} = 1 - \frac{\text{Heat rejected}}{\text{Heat added}}$$

$$= 1 - \frac{m\,c_v(T_5 - T_1)}{m\,c_p(T_4 - T_3) + m\,c_v(T_3 - T_2)}$$

$$= 1 - \frac{(T_5 - T_1)}{\gamma \cdot (T_4 - T_3) + (T_3 - T_2)}$$

$$= 1 - \frac{(656.85 - 300)}{1.4 \times (1767.63 - 1400.39) + (1400.39 - 886.25)}$$

Air standard thermal efficiency = 0.6529

Air standard thermal efficiency = **65.29%**	**Ans.**

Actual thermal efficiency may be different from theoretical efficiency due to following reasons;

(*a*) Air standard cycle analysis considers air as the working fluid while in actual cycle it is not air throughout the cycle. Actual working fluid which are combustion products do not behave as perfect gas.

(*b*) Heat addition does not occur isochorically in actual process. Also combustion is accompanied by inefficiency such as incomplete combustion, dissociation of combustion products, etc.

(*c*) Specific heat variation occurs in actual processes where as in air standard cycle analysis specific heat variation is neglected. Also during adiabatic process theoretically no heat loss occurs while actually these processes are accompanied by heat losses.

EXERCISE

9.1 What do you mean by air standard cycles? Discuss its' significance.

9.2 Derive the expression for Carnot cycle efficiency.

9.3 Discuss limitations of Carnot cycle and explain Otto cycle, Diesel cycle and Dual cycle in light of these limitations.

9.4 Derive expression for efficiency of Otto cycle, Dual cycle and Diesel cycle.

9.5 What is meant by mean effective pressure? Discuss its' relevance.

9.6 Obtain expressions for mean effective pressure of Otto cycle, Diesel cycle and Dual cycle.

9.7 Discuss gas turbine cycles. Obtain efficiency of Brayton cycle.

9.8 What are possible modifications in gas turbine cycle? Explain.

9.9 Explain different irreversibilities and losses in gas turbine cycle.

9.10 Define polytropic efficiency and its significance mathematically for compressors.

9.11 Obtain optimum pressure ratio condition for minimum compressor work requirement in two stage perfect intercooled compression.

9.12 Compare the influence of reheating, regeneration and intercooling on performance of gas turbine cycle.

9.13 Discuss Ericsson cycle and obtain its efficiency.

9.14 Obtain efficiency of Stirling cycle.

9.15 Write short notes on compressor and turbine efficiencies.

9.16 An Otto cycle operates between maximum and minimum pressures of 600 *kPa* and 100 *kPa*. The minimum and maximum temperatures in the cycle are 27ºC and 1600 K. Determine thermal efficiency of cycle and also show it on T-s and P-V diagram. [48%]

9.17 In an air standard Carnot cycle operating between temperatures of 57ºC and 1327ºC, determine the heat added if cycle efficiency is 79.4% and the minimum pressure of cycle is 100 kPa. [300 kJ/kg]

9.18 Determine air standard efficiency for a diesel engine having L/D ratio of 1.6 and bore as 25 cm. Take the clearance volume and pressure at end of suction as 1.2 litres and 1 atm. respectively. Also obtain mean effective pressure. Volume after combustion is 1.8 times clearance volume. [63.5, 6 bar]

9.19 An engine operates on Dual cycle with a compression ratio of 15. At the end of suction the air is available at 1 atm and 27ºC. Total heat added is 430 kJ/kg. Heat supply is in ratio of (0.536:1) for heat supply at constant volume and constant pressure. Determine cycle efficiency and mean effective pressure. [65%, 0.342 MPa]

9.20 A regenerative cycle has heat exchanger effectiveness of 75% and pressure ratio of 4 with compression occurring in two stages of equal pressure ratio with intercooling back to initial temperature of 15ºC. Maximum temperature in cycle is 650º C and expansion occurs with efficiency of 0.88. Compression process has isentropic efficiency of 0.85 in each stage. Considering air as working fluid throughout the cycle determine efficiency of cycle. [34.6%]

9.21 A gas turbine installation has intercooling, regeneration and reheating with air entering compressor at 100 kPa, 290 K and compressed to 0.41 MPa. Subsequently it is cooled until temperature drops to 13°C in intercooler and finally compressed to 0.75 MPa. Regenerator effectiveness is 0.70 and turbine inlet temperature (TIT) is 1350 K and expands to 0.41 MPa where it is reheated upto 1350 K. The exhaust pressure is 0.1 MPa. Determine overall efficiency considering expansion and compression processes to be of isentropic type. Take air as working fluid throughout. [54.3%]

9.22 In an actual gas turbine the compressor requires 300 kJ/kg of work to quadruple the inlet pressure. For inlet air temperature of 100°C determine, the compressor exit air temperature and compressor efficiency. [671.6 K, 60.6%]

9.23 In a gas turbine installation the exhaust from gas turbine is sent for process heating at 200 kPa. Turbine produces power just sufficient to drive the compressor. Turbine inlet temperature is 815°C and air is supplied to compressor at 1 bar, 17°C. Determine compressor pressure ratio for air as working fluid throughout. [3]

9.24 A Brayton cycle producing 75 kW is designed for maximum work. The compressor inlet conditions are 100 kPa and 27°C. Compression ratio is 5.5. For air standard Brayton cycle determine the turbine inlet temperature, cycle efficiency and air flow rate. [794 K, 38.6%, 0.631 kg/s]

9.25 A gas turbine cycle has reheating and heat exchanger employed. The air is supplied at 1 bar, 15°C into compressor where it is compressed upto 4 bar with isentropic efficiency of 82%. Turbine inlet temperature is 700°C and expansion occurs in two stages with equal pressure ratio upto 1 bar. The reheating between two stages occurs upto 700°C. Isentropic efficiency of expansion is 85%. Exhaust from low pressure turbine is passed into a heat exchanger having 0.75 effectiveness. Heat exchanger heats the discharge from compressor before being supplied to combustion chamber. Considering air as woking fluid throughout determine cycle efficiency. [32.1%]

9.26 In a gas turbine plant air is supplied at 1 bar, 290 K to compressor having isentropic efficiency of 0.84 and pressure ratio of 5. The air flow rate to compressor is 111600 kg/hr. Compressed air is passed through a heat exchanger and it leaves exchanger at 360°C, from where it is supplied to combustion chamber. Fuel having calorific value of 41800 kJ/kg is supplied at the rate of 890 kg/hr in combustion chamber. The combustion products expand from 4.7 bar to 1.95 bar, 410°C in turbine. Take $c_{p\,air}$ = 1.0032 kJ/kg K, γ_{air} = 1.4, $c_{p\,comb,\,products}$ = 1.1286 kJ/kg K, $\gamma_{comb,\,products}$ = 1.34. Determine *hp* developed. [3000 hp]

9.27 Show that the necessary condition for positive work output from an open cycle gas turbine can be given by,

$$\eta_c \times \eta_T \times T_{max} > T_{min} \cdot r^{\frac{\gamma-1}{\gamma}}$$

The simple gas turbine cycle has compressor with isentropic efficiency of η_c and pressure ratio of '*r*'. Turbine has isentropic efficiency of η_T and expansion ratio similar to that of compressor. The maximum and minimum temperature are T_{max} and T_{min} and air is working fluid throughout with γ as ratio of specific heat.

10

Introduction to Internal Combustion Engines

10.1 INTRODUCTION

History of internal combustion engines dates back to year 1680 when Christian Huygens developed an engine using gun powder for explosion inside a cylindrical tube. Subsequently, in year 1860 a non compression engine utilising coal gas was developed by 'Lenoir' in France. Engine was called Lenoir engine and it was operationally similar to steam engine. In 1866 'Otto-Langen free piston engine' was developed in Germany and it had thermal efficiency more than 'Lenoir engine.' In 1876 four stroke engine based on Otto cycle was developed by Nikolous Otto in Germany which revolutionised the developments of internal combustion engines and are even used till date. The compression ignition engine technology and engines based on it evolved in year 1892. Such engines are credited to Rudolf Diesel another German engineer and are so named as Diesel engine. Two stroke engines came up in 1881 in Scotland and were first developed by Dugald Clerke. 'Brayton engine' based on Brayton cycle and 'Atkinson engine' based on Atkinson cycle were developed in year 1873 and 1885 respectively. Apart from conventional two stroke and four stroke engines described earlier the 'Wankel engine' which was rotary engine and the 'Stirling engine' which was external combustion engine came up in years 1957 and 1938 respectively. In 1923 an engine called as free piston engine was developed by Pateras Pescara in France. Free piston engine is actually a combination of reciprocating piston-cylinder engine and turbine. "Engine refers to a device which transforms one form of energy into the other form". "Heat engine is a modified form of engine used for transforming chemical energy of fuel into thermal energy and subsequently for producing work". Based on the mechanism used for adding thermal energy they can be classified into the following:

(*a*) External combustion engine

(*b*) Internal combustion engine.

External combustion engine have combustion occuring outside engine and adding heat to the working fluid used in the engine. Thus, in external combustion engines heat released during combustion is indirectly utilized by the working fluid in external combustion engine. Internal combustion engines have combustion occuring in engine itself and heat released during combustion is directly utilized for getting shaft work. Internal combustion engines have numerous advantages over external combustion engines such as lower weight to power output ratio, simplicity, smaller initial cost, and higher efficiency etc. Due to indirect heat transfer the external combustion engines permit for the use of any cheaper fuel such as coal, wood, oil etc. for combustion.

Internal combustion engines are exhaustively used in automobiles gas turbine etc. and external combustion engines are used in steam turbine, steam engine, nuclear power plant etc. Internal combustion engines are efficient than external combustion engines and also are smaller than external combustion engine of similar capacity.

10.2 CLASSIFICATION OF IC ENGINES

Internal combustion engines can be classified on the following basis.

(*a*) *Based on number of strokes* : Number of strokes involved in a cycle of IC engine can be two strokes or four strokes. Such engine can be;
  (*i*) Two stroke engines
  (*ii*) Four stroke engines

(*b*) *Based on thermodynamic cycle* : Depending upon thermodynamic cycle used in the internal combustion engines these can be classified as:
  (*i*) Engines based on Otto cycle ('Spark-Ignition engine')
  (*ii*) Engines based on Diesel or Dual cycle ('Compression-Ignition engines')

(*c*) *Based on mechanism of ignition:* Internal combustion engines have combustion as the basic process. Combustion process may be initiated using externally assisted ignition (spark ignition) or it may get initiated on its' own due to excessive compression (compression ignition). Such engines are called:
  (*i*) Spark ignition engines
  (*ii*) Compression ignition engines.

  The spark ignition engines may have "magneto ignition system" or "battery ignition system" for creating necessary electric potential for producing spark.

(*d*) *Based on type of fuel used:* IC engines may be classified depending upon the type of fuel being used. These can be:
  (*i*) Petrol engines (petrol being used as fuel)
  (*ii*) Gas engines (gaseous fuel being used)
  (*iii*) Diesel engines (diesel being used as fuel)
  (*iv*) Multi-fuel engines (more than one fuel being used)

(*e*) *Based on fuel admission:* IC engines can be of different types depending upon arrangement used for fuel admission:
  (*i*) Carburettor type engines (use carburettor fuel metering)
  (*ii*) Injection type engines (use fuel injector and injection system)

(*f*) *Based on type of cooling:* IC engines have inherent requirement of continuous cooling of engine. Based on type of cooling these can be classified as:
  (*i*) Air cooled engines (Generally used in small sized engines)
  (*ii*) Water cooled engines (Generally used in large sized engines)

(*g*) *Based on type of motion:* IC engines may have reciprocating motion of piston or it may also have rotary motion. Such engines can be:
  (*i*) Reciprocating engines
  (*ii*) Rotary engines

  Reciprocating engines may have different cylinder arrangements such as:
  (*i*) Opposed cylinder engines
  (*ii*) Inclined cylinder engines
  (*iii*) V-shaped cylinder arrangement.

  Rotary engines may be further classified as single rotor engines or multirotor engines i.e.
  (*i*) Single rotor engine
  (*ii*) Multi rotor engine

10.3 IC ENGINE TERMINOLOGY

Some of the basic components and generally used terms in internal combustion engines are given as under.

1. Cylinder: It is a cylindrical block having cylindrical space inside for piston to make reciprocating motion. Upper portion of cylinder which covers it from the top is called cylinder head. This is manufactured by casting process and materials used are cast iron or alloy steel.

2. Piston and Piston rings: Piston is a cylindrical part which reciprocates inside the cylinder and is used for doing work and getting work. Piston has piston rings tightly fitted in groove around piston and provides a tight seal so as to prevent leakage across piston and cylinder wall during piston's reciprocating motion. Pistons are manufactured by casting or forging process. Pistons are made of cast iron, aluminum alloy. Piston rings are made of silicon, cast iron, steel alloy by casting process.

3. Combustion space: It is the space available between the cylinder head and top of piston when piston is at farthest position from crankshaft (*TDC*).

4. Intake manifold: It is the passage/duct connecting intake system to the inlet valve upon cylinder. Through intake manifold the air/air-fuel mixture goes into cylinder.

5. Exhaust manifold: It is the passage/duct connecting exhaust system to the exhaust valve upon cylinder. Through exhaust manifold burnt gases go out of cylinder.

6. Valves: Engine has both intake and exhaust type of valves which are operated by valve operating mechanism comprising of cam, camshaft, follower, valve rod, rocker arm, valve spring etc. Valves are generally of spring loaded type and made out of special alloy steels by forging process.

7. Spark plug: It is the external ignitor used for initiating combustion process. Spark plug is activated by electrical energy fed by electrical system with engine. It delivers spark with suitable energy to initiate combustion at appropriate time for suitable duration.

8. Bearing: Bearings are required to support crank shaft. Bearings are made of white metal leaded bronze.

9. Connecting rod: It is the member connecting piston and crankshaft. It has generally I section and is made of steel by forging process.

10. Crank: It is the rigid member connecting the crankshaft and connecting rod. Crank is mounted on crankshaft. Crank transfers motion from connecting rod to crankshaft as it is linked to connecting rod through crank pin.

11. Crankshaft: It is the shaft at which useful positive work is available from the piston-cylinder arrangement. Reciprocating motion of piston gets converted into rotary motion of crankshaft. Crankshaft are manufactured by forging process from alloy steel.

12. Crankcase: Crankcase actually acts like a sump housing crank, crankshaft, connecting rod and is attached to cylinder. These are made of aluminium alloy, steel, cast iron etc. by casting process.

13. Gudgeon pin: It is the pin joining small end of the connecting rod and piston. This is made of steel by forging process.

14. Cams and Camshafts: Cams are mounted upon camshaft for opening and closing the valves at righ timings and for correct duration. Camshaft gets motion from crankshaft through timing gears.

15. Carburettor: Carburettor is device to prepare the air fuel mixture in right proportion and supply at right time.

16. Bore: It is nominal inner diameter of the cylinder.

17. Piston area: It is the area of a circle of diameter equal to bore.

18. Stroke: It is the nominal distance travelled by the piston between two extreme positions in the cylinder.

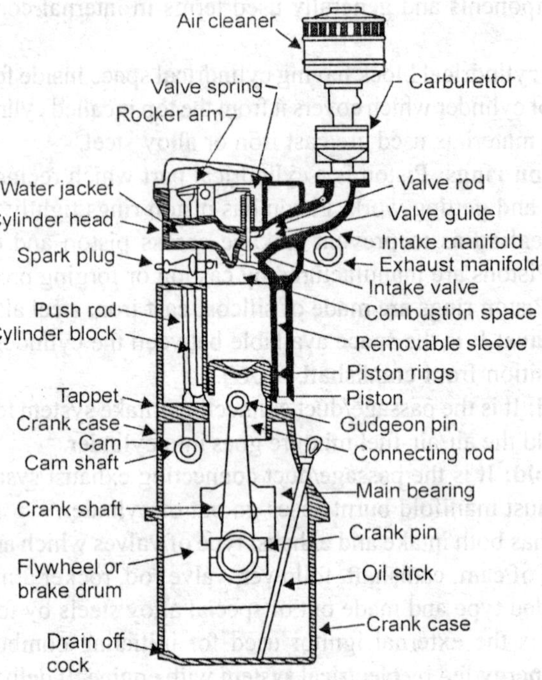

Fig. 10.1 Internal combustion engine

19. Dead centre: It refers to the extreme end positions inside the cylinder at which piston reverses it's motion. Thus, there are two dead centres in cylinder, called as 'top dead centre' or 'inner dead centre' and 'bottom dead centre' or 'outer dead centre'.

Top dead centre (TDC) is the farthest position of piston from crankshaft. It is also called inner dead centre (IDC).

Bottom dead centre (BDC) refers to the closed position of piston from crankshaft. It is also called outer dead center (ODC).

20. Swept volume : It is the volume swept by piston while travelling from one dead centre to the other. It may also be called stroke volume or displacement volume.

Mathematically, Swept volume = Piston area × Stroke.

21. Clearance volume: It is the volume space above the piston inside cylinder, when piston is at top dead centre. It is provided for cushioning considerations and depends, largely upon compression ratio.

22. Compression ratio: It is the ratio of the total cylinder volume when piston is at BDC to the clearance volume.

$$\text{Compression ratio} = \frac{\text{Swept volume} + \text{Clearance volume}}{\text{Clearance volume}}$$

10.4 4-STROKE SI ENGINE

Spark ignition (SI) engines employ external ignition system for initiating the combustion process. Spark plug is the most commonly used ignition method used in spark ignition engines. Let us first understand working of 4-stroke SI engine. Fig. 10.3 shows the different strokes and processes associated with them.

State	*Valve position*		*Action*	*Work*	*Heat*
	Inlet	*Exhaust*			
1	Just opened	Just closed			
1-2	Open	Closed	Air-fuel intake	Air to piston	None
2	Just closed	Closed			
2-3	Closed	Closed	Compression	Piston to air	None
3	Closed	Closed	Ignition		
3-4	Closed	Closed	Combustion		Input to gas
4	Closed	Closed	Max. *T*, *P*		
4-5	Closed	Closed	Expansion	Gas to piston	None
5	Closed	Just opened			
5-6	Closed	Open	Exhaust blowdown	None	Gas to atmosphere
6	Closed	Open	Pressure drop to atmosphere	None	Gas to atmosphere
6-1	Closed	Open	Exhaust	Piston to gas	Gas to atmosphere

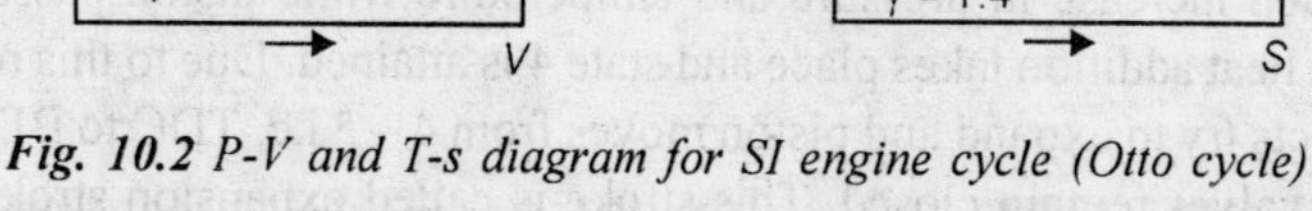

Fig. 10.2 P-V and T-s diagram for SI engine cycle (Otto cycle)

Let us start with piston at TDC, state 1. As piston moves for TDC to BDC, the inlet vlave gets opened and fresh air-fuel mixture prepared in carburettor enters the cylinder. This supply of air-fuel mixture into cylinder is called suction process or suction stroke during which inlet valve is open while exhaust valve remains closed. After the piston reaches BDC, it reverses its motion and moves towards TDC. During this piston travel both inlet and exhaust valves remain closed. Thus, the air-fuel mixture inside cylinder gets compressed till piston reaches TDC. This is the second stroke and called compression stroke or compression process (2 – 3).

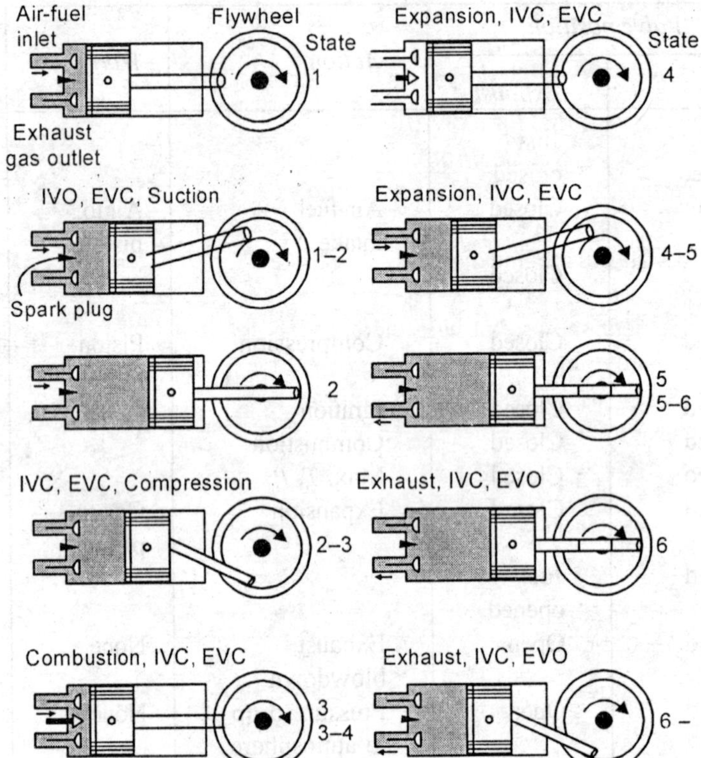

IVC : Inlet valve closed, IVO: Inlet valve open, EVC: Exhaust valve closed, EVO: Exhaust valve open.

***Fig. 10.3** Working of 4-stroke S.I. engine*

Now highly compressed air-fuel mixture is available inside the cylinder and ready for combustion. With piston at TDC, the spark plug is activated and it releases spark for igniting air-fuel mixture. Spark plug used in the SI engines gets activated by suitable mechanism in the engine and provides suitable amount of energy in the form of spark for initiating combustion process. This burning of mixture is accompanied by sudden increase in pressure and temperature while piston passes the TDC position. Here constant volume heat addition takes place and state 4 is attained. Due to this release of fuel energy, the combustion products try to expand and piston moves from 4 – 5 i.e. TDC to BDC. During this travel the inlet and exhaust valves remain closed. This stroke is called expansion stroke or power stroke or expansion process. This is the stroke accompanied by positive work available at shaft. Now while piston is at BDC the exhaust valve gets opened and combustion products are exhausted out during 5–6. Cylinder is further emptied and made ready for being recharged while piston travels from BDC to TDC, 6 – 1 pushing out burnt gases. This is called exhaust stroke.

Here we have seen that out of suction, compression, expansion and exhaust strokes only expansion stroke is accompanied by the production of positive work, rest three strokes are work absorbing strokes. Work requirement for the three strokes is met from the work available during expansion stroke. For storing the excess energy and releasing it when required, a flywheel is mounted over the crankshsaft. Cycle gets completed in two revolutions of crankshaft. Examples for four stroke engines are petrol engines used in cars etc.

10.5 2-STROKE SI ENGINE

2-stroke SI engine is a modified form of 4-stroke engine where all the four processes required for completion of one cycle of SI engine get completed in two strokes. Thus, obviously in each stroke two processes get completed. Fig. 10.4 shows the line diagram of 2-stroke SI engine.

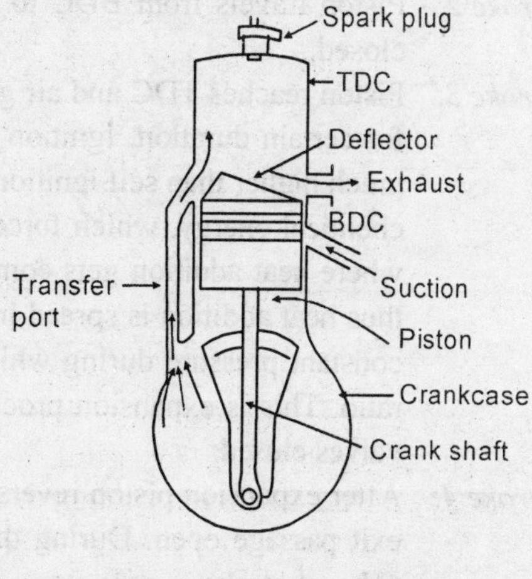

Fig. 10.4 2-stroke SI engine

General arrangement shows that here there are no valves as in case of 4-stroke engines, instead it has exhaust and suction ports. Piston has a projection on its top, which acts like deflector. Mixture of air-fuel goes into crank case first and then gets transferred to top of piston at appropriate time. Let us start piston movement from TDC to BDC. When piston reverses its motion from BDC to TDC then the suction port gets uncovered and fresh mixture enters and goes into crank case. With piston moving from TDC to BDC and during covered position of suction port the mixture gets transferred to the top of piston through transfer port. Upon reversal during piston travelling from BDC to TDC, the air fuel mixture on top of piston gets compressed and subsequently gets ignited by spark from spark plug. The combustion of fuel-air mixture results in release of excessive energy which forces piston to move from TDC to BDC. Simultaneously as piston uncovers exhaust port the burnt gases go out through exhaust port. Again when piston reaches BDC it reverses its motion and during travel from BDC to TDC the suction takes place as explained above at the bottom of piston while compression of fuel air mixture takes place on top of piston. Thus, suction and compression, both processes get completed during travel of piston from BDC to TDC. Expansion and exhaust processes occur during travel of piston from TDC to BDC along with transfer of fresh fuel air mixture from crankcase to top of piston.

Here all four processes occur during two strokes and one revolution of crank shaft. Theremodynamic cycle followed by 2-stroke SI engine is also Otto cycle. Scooter engines are generally two stroke engines. 2-stroke SI engines are used for smaller applications.

10.6 4-STROKE CI ENGINE

Compression ignition (CI) engines operate generally on "Diesel"/"Dual" cycle. In these engines the combustion is realized due to excessive compression and are so called compression ignition engines. Here air alone is sucked inside the cylinder during suction stoke and compressed. Degree of compression is much more than that of spark ignition (SI) engines. After compression of air the fuel is injected into the high pressure and high temperature compressed air. Due to high temperature of air the combustion of fuel gets set on its' own. Self ignition of fuel takes place due to temperature of air-fuel mixture being higher than self ignition temperature of fuel. Thus in CI engines, larger amount of compression causes high temperature, therefore unassisted combustion.

Schematic of 4-stroke CI engine is quite similar to that of 4-stroke SI engine with the only major difference that spark plug is replaced by fuel injector for injecting fuel at high pressure into compressed air. 4-Stroke CI engine works with following four processes getting completed in separate strokes. General arrangement in CI engine is similar to that of SI engine with spark plug replaced by fuel injector.

Stroke 1: Piston travels from TDC to BDC and air is sucked.

Stroke 2: Piston travels from BDC to TDC, while air is compressed with inlet and exit passages closed.

Stroke 3: Piston reaches TDC and air gets compressed. Fuel injector injects fuel into compressed air for certain duration. Ignition of fuel also takes place simultaneously as air temperature is much higher than self ignition temperature of fuel. Burning of fuel results in release of fuel chemical energy, which forces piston to travel from TDC to BDC. Contrary to SI engine where heat addition gets completed near instantaneously in CI engines, fuel injection and thus heat addition is spread in certain stroke travel of piston i.e. heat addition take place at constant pressure during which piston travels certain stroke length as decided by cut-off ratio. This is expansion process and piston comes down to BDC with both inlet and exit valves closed.

Stroke 4: After expansion piston reverses its motion upon reaching BDC and travels upto TDC with exit passage open. During this piston travel burnt gases are expelled out of cylinder i.e. exhaust stroke.

Completion of above four stroke requires two revolutions of crankshaft.

10.7 2-STROKE CI ENGINE

Fig. 10.5 shows the general arrangement in 2-stroke CI engine. Here the structure of CI engine is very much similar to that of SI engine with the major difference that spark plug is replaced by fuel injector, structure is made more sturdy to withstand high compression ratio.

Working of 2-stroke compression ignition engine is shown in (*a*), (*b*), (*c*), (*d*) explaining the suction, compression, expansion and exhaust processes. During piston travel from BDC to TDC air enters crankcase. When piston reaches TDC and reverses its motion to BDC air in crankcase gets partly compressed and is transferred from crankcase to top of piston through transfer port. Upon reversal of piston motion from BDC to TDC the compression of air occurs by the top side of piston while on the back of piston air again enters into crankcase. Upon piston reaching TDC fuel is injected into compressed air which is at high temperature and pressure. AS fuel is injected into compressed air the fuel ignition gets set on its own due to temperature being more than self ignition temperature of fuel, i.e. compression ignition. Fuel injection is continued for some duration along with it's ignition which causes release of excessive fuel energy. This energy release causes piston to go back from TDC to BDC, i.e. the expansion process as shown in (*c*). As piston reaches BDC it simultaneously forces air in crank case to get transferred to cylinder space and forces burnt gases out of cylinder i.e. exhaust process.

Here also cycle gets completed in single revolution of crankshaft i.e two processes occurring simultaneously in each stroke.

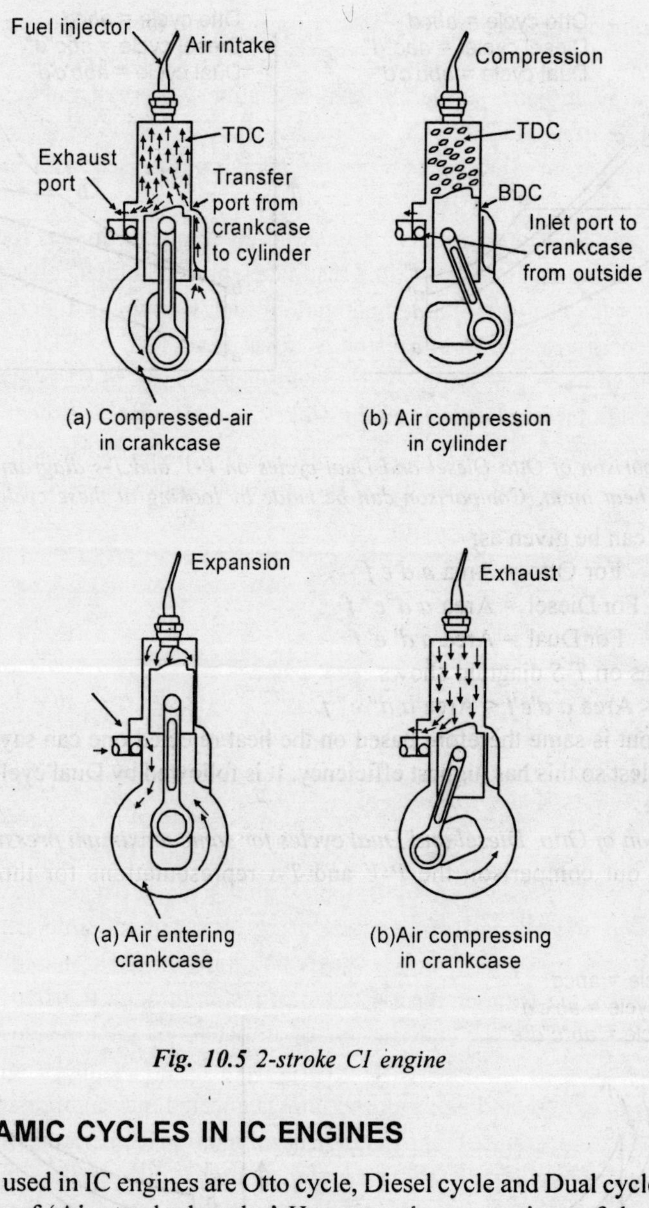

Fig. 10.5 2-stroke CI engine

10.8 THERMODYNAMIC CYCLES IN IC ENGINES

Thermodynamic cycles used in IC engines are Otto cycle, Diesel cycle and Dual cycle and are described in Chapter 9 under title of 'Air standard cycles' However, the comparison of these cycles with one another for different reference conditions is presented here.

(a) Comparison of Otto, Diesel and Dual cycles for same compression ratio and heat input:

Fig. 10.6 shows the three cycles on *P-V* and *T-S* diagram for same compression ratio and heat input.

For same heat input,

$$\underset{\text{(Otto)}}{\text{Area } bcef} = \underset{\text{(Diesel)}}{\text{Area } bc''e''f} = \underset{\text{(Dual)}}{\text{Area } bb'c'e'f}$$

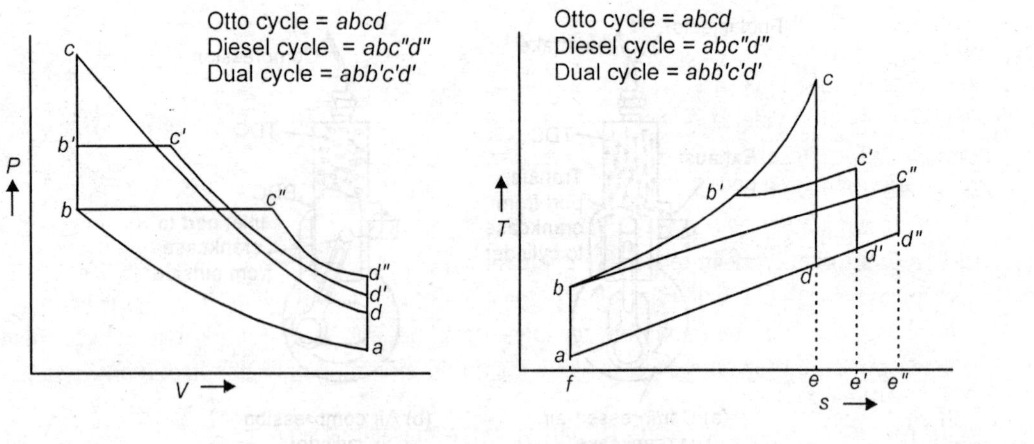

Fig. 10.6 Comparison of Otto Diesel and Dual cycles on P-V and T-s diagram for same compression ratio and heat input. Comparison can be made by looking at these cycle representations.

Heat rejected can be given as;

For Otto = Area $a\,d\,e\,f$

For Diesel = Area $a\,d''e''\,f$

For Dual = Area $a\,d'\,e'\,f$

A look at areas on *T-S* diagram shows

Area $a\,d\,e\,f$ < Area $a\,d'e'f$ < Area $a\,d''\,e''\,f$.

Since heat input is same therefore based on the heat rejected one can say that as heat rejected for Otto cycle is smallest so this has highest efficiency. It is followed by Dual cycle and smallest efficiency is for Diesel cycle.

(*b*) *Comparison of Otto, Diesel and Dual cycles for same maximum pressure and same heat input:*

For carrying out comparison the *P-V* and *T-s* representations for three cycles are shown in Fig. 10.7.

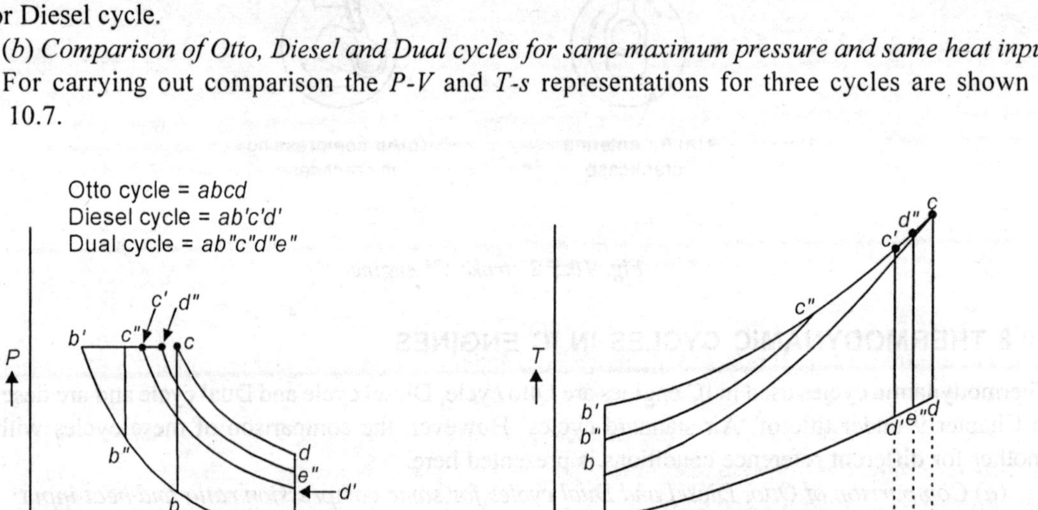

Fig. 10.7 Comparison of Otto, Diesel and Dual cycles on P-V and T-s diagram for same maximum pressure and same heat input.

Here heat input is same so,

Area $bcfg$ = Area $b'c'f'g$ = Area $b''c''d''f''g$.
(Otto) (Diesel) (Dual)

Heat rejected shall be given by;

Area $adfg$ for Otto cycle

Area $ad'f'g$ for Diesel cycle

Area $ae''f''g$ for Dual cycle

Area $adfg$ > Area $ae''f''g$ > Area $ad'f'g \Rightarrow W_{Diesel} > W_{Dual} > W_{Otto}$

Here W refers to net work in respective cycle.

Hence, it can be seen that heat rejected is highest in Otto cycle followed by Dual cycle and Diesel cycle. Thus, efficiency shall be maximum for Diesel cycle due to largest work available followed by Dual cycle and minimum for Otto cycle at same maximum pressure and same heat input.

10.9 INDICATOR DIAGRAM AND POWER MEASUREMENT

Internal combustion engines have combustion taking place inside and power is available at crankshaft. The shaft work available is less than the total energy released inside the cylinder due to frictional and other losses.

For performance evaluation of internal combustion engine one is interested in following different powers.

(a) **Indicated power**: It refers to the power available inside the cylinder i.e. the power provided to piston.

Mathematically:

Indicated power = (Energy in fuel) – (Energy loss in exhaust, coolant, radiation etc.)

It is measured from the indicator diagram which is obtained using indicator mechanism.

(b) **Brake power:** It refers to the power available at crankshaft i.e. it is the useful shaft work.

Mathematically:

Brake power = (Indicated power) – (Energy loss in friction, pumping and unaccounted losses etc.)

Brake power is usually measured by absorption or transmission type dynamometers. It can be given as:

$$\text{Brake power} = \frac{2\pi NT}{60}\ W$$

where {N is speed of rotation of shaft in rpm, T is torque, N.m.}

(c) **Friction power:** It refers to the power lost due to friction and other reasons. It is quantified by the difference between indicated power and brake power.

Friction power = Indicated power – Brake power.

Indicator diagram: Indicator diagram is the graphical description of pressure and volume variations occurring inside cylinder. An indicator diagram for a four-stroke internal combustion engine is shown in Fig. 10.8 along with the four different processes.

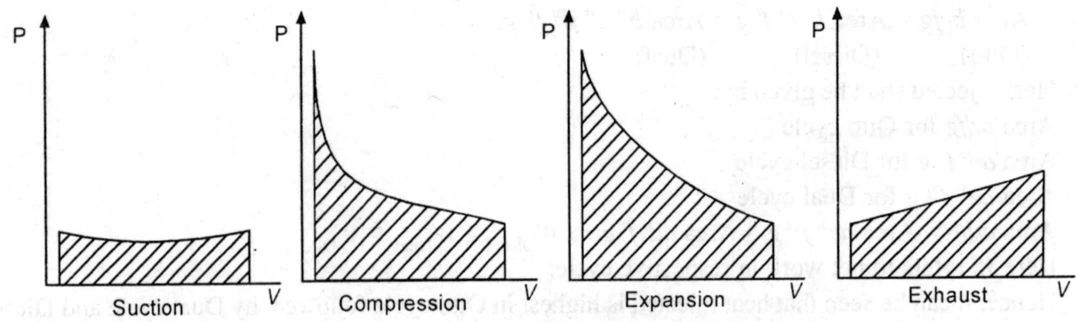

(a) Four processes in 4-stroke IC engine (Otto cycle)

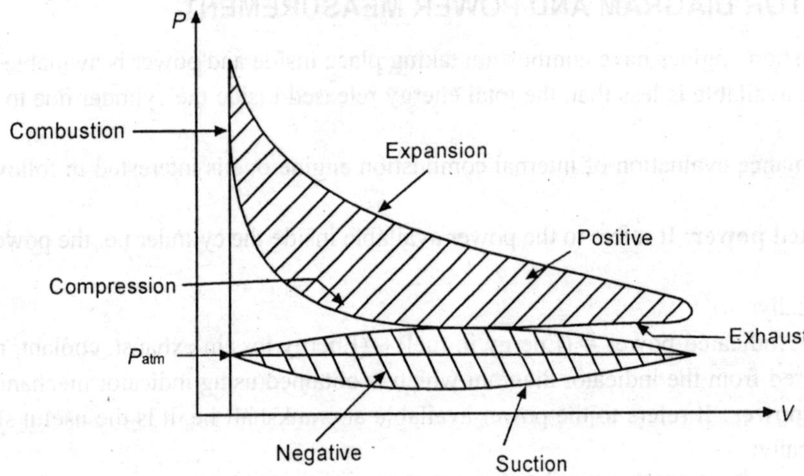

(b) Indicator diagram of 4-stroke IC engine (Otto cycle).

Fig 10.8 Indicator diagram and processes in IC engine.

For getting the indicator diagram, "indicator diagram mechanism" is put over the IC engine and real variation of pressure and volume states obtained.

Indicator mechanism shown in Fig. 10.9, has number of links 'g' and chord for transferring linear displacement of piston rod into the rotation of drum 'f'. A stylus 'd' is operated over the drum through a parallel motion link 'e'. Displacement of 'e' is caused along y-axis by an indicator piston rod 'h', which is connected to indicator piston 'b' put in indicator cylinder 'a' as shown. Indicator piston is resisted by an indicator spring 'c'. In between indicator cylinder and engine cylinder a cock is provided to isolate it after indicator diagram is obtained. Indicator mechanism has thus drum 'f' rotating about its axis and stylus 'd' operated by pressure variation inside cylinder along y-axis. Thus, displacement along volume axis is made available by rotation of drum and along pressure axis it is made available by vertical reciprocatory movement of stylus depending upon pressure inside engine cylinder.

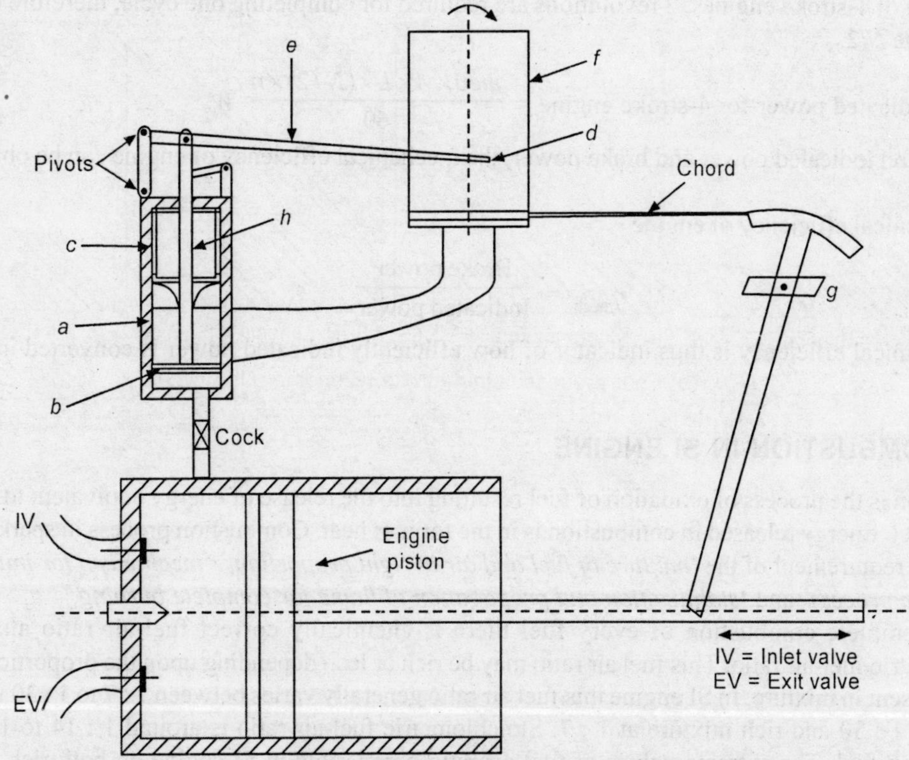

Fig. 10.9 Indicator diagram mechanism

For getting indicated power we need mean effective pressure (average pressure) and volume displaced. From the indicator diagram obtained the mean effective pressure can be obtained, knowing area of diagram, length of diagram and indicator spring constant. Area of diagram can be estimated using planimeter and length of diagram measured.

Mean effective pressure, $mep = \frac{A_d}{L_d} \times C$

where A_d is area enclosed in diagram

L_d is length of diagram

C is indicator spring constant given as N/cm^2 × cm travel.

Using mean effective pressure the indicated power can be obtained by product of pressure and volume displaced.

Indicated power for 2-stroke engine, $= \frac{mep \times A \times L \times N \times n}{60}\ W$

Where mep is mean effective pressure in Pa

A is cross-section area of cylinder.

$$A = \frac{\pi}{4} D^2, \text{ where } D \text{ is bore.}$$

L is length of stroke.

N is rpm

n is no. of cylinders.

In case of 4-stroke engines, 2 revolutions are required for completing one cycle, therefore effective rpm shall be $N/2$.

$$\text{Indicated power for 4-stroke engine} = \frac{mep \times A \times L \times (N/2) \times n}{60} \ W$$

Based on indicated power and brake power, the mechanical efficiency of engine can be obtained as below.

Mechanical efficiency of engine

$$\eta_{mech} = \frac{\text{Brake power}}{\text{Indicated power}}$$

Mechanical efficiency is thus indicator of how efficiently indicated power is converted into brake power.

10.10 COMBUSTION IN SI ENGINE

Combustion is the process of oxidation of fuel resulting into the release of energy equivalent to calorific value of fuel. Energy released in combustion is in the form of heat. Combustion process in spark ignition engine has requirement of the '*mixture of fuel and air in right proportion,*' '*mechanism for initiation of combustion process*' and '*stabilization and propagation of flame for complete burning.*'

For complete combustion of every fuel there is chemically correct fuel-air ratio also called stoichiometric fuel-air ratio. This fuel air ratio may be rich or lean depending upon the proportion of fuel and air present in mixture. In SI engine this fuel air ratio generally varies between 1 : 7 to 1 : 30 with lean mixture at 1 : 30 and rich mixture at 1 : 7. Stoichiometric fuel-air ratio is around 1 : 14 to 1 : 15 for hydrocarbon fuel. The extreme values of fuel-air ratio permissible in SI engine on both rich and lean ends put limits as '*lower ignition limit*' and '*upper ignition limit*'. Varying fuel-air ratio is required in SI engine due to varying loads on engine between no load to full load on engine. The ratio of actual fuel-air ratio to stoichiometic fuel-air ratio is given by 'equivalence ratio' or 'relative fuel-air ratio'. Appropriate fuel-air ratio is maintained in SI engines through 'carburettor' (the fuel metering system).

Combustion in SI engine may be described to be occurring in following significant phase:

(*i*) After compression of fuel-air mixture in cylinder the high temperature spark is delivered by spark plug in the compressed fuel-air mixture. Temperature at the tip of spark plug electrode may go even more than 10,000°C at the time of release of spark. Sparkles released have sufficiently high temperature to initiate the combustion of fuel. For complete combustion of fuel mere initiation of combustion does not serve the purpose instead a sustainable combustion process is required. After setting up of combustion a sustainable flame front or flame nuclei is needed so that it proceeds across the combustion space to ensure complete combustion. Thus, this phase in which spark is first released followed by setting up of sustainable flame front is called "preparation phase" and my consume around 10° of crank angle rotation. Crank angle rotation consumed in "preparation phase" depends upon the speed of engine, constructional feature of cylinder, piston, location of spark plug, strength of spark, characteristics of fuel, fuel-air ratio etc. Designer always wishes to complete combustion in smallest possible time as, theoretically Otto cycle requires combustion to occur instantaneously. Therefore, this first phase termed as preparation phase should be as small as possible. Fig. 10.10 shows different phases of combustion.

Preparation phase is shown to occur from '*a*' to '*b*' with small or negligible pressure rise as initially rate of burning is very small.

(*ii*) After sustainable combustion flame is set up then the flame nuclei get scattered due to excessive turbulence in combustion space causing pressure to rise from '*b*' to '*c*'. This phase of combustion

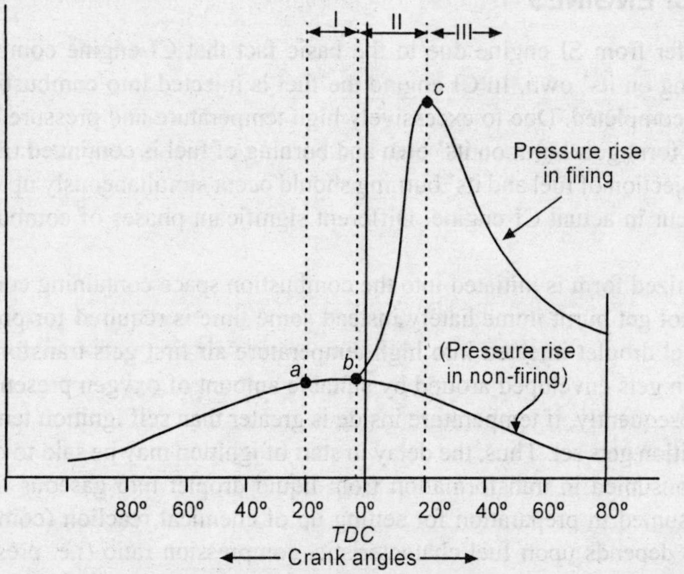

Fig. 10.10 Phases of combustion in SI engine

depends upon the turbulence inside cylinder, strength of compbustion nuclei, fuel-air ratio, strength of spark, cylinder geometry, fuel properties etc. This phase of combustion is called as "flame propagation phase" and is accompanied by the excessive pressure rise. Flame propagation phase should also be as small as possible.

(*iii*) After the maximum amount of fuel-air mixture is burnt, the residual gets burnt after the piston has moved across the TDC. This last phase is termed as "after burning phase" and occurs during the expansion stroke.

Hence, it can be summarised that the complete combustion in SI engine occurs in three distinct zones i.e. preparation phase, flame propagation phase and after burning phase. In order to complete combustion in smallest possible time the flame propagation phase and preparation phase should be shortened. It is seen that out of total distance travelled in combustion space in first phase i.e. preparation phase about 10% of combustion space length is covered in about 20-30% of total time for combustion. Flame propagation phase is spread in about 80% of combustion space length and is covered in 60-70% of total time of combustion. After burning occurs in less than 10% of combustion space in less than 10% of total combustion time as evident from Fig. 10.11. Combustion may also sometimes occur abnormally. "Abnormal combustion" is said to occur when combustion begins inside the cylinder on its' own before the stipulated time for it. This abnormal combustion may be due to preignition (i.e. ignition of fuel even before spark plug ignites it) and results in uncontrolled pressure rise. Abnormal combustion is also termed as detonation or knocking and can be felt by jerky operation of engine, excessive noise, reduced power output etc.

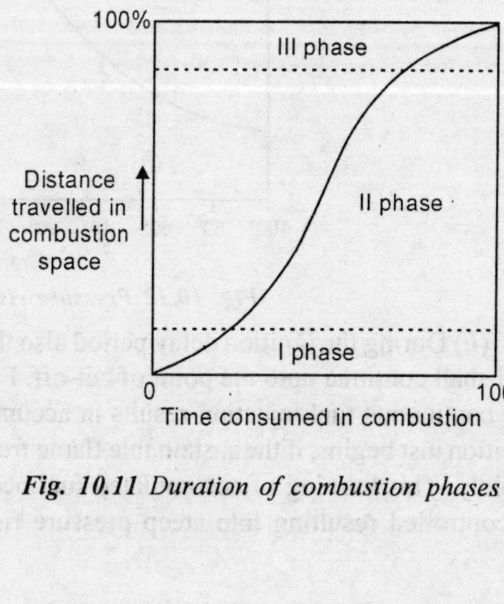

Fig. 10.11 Duration of combustion phases

10.11 COMBUSTION IN CI ENGINES

Combustion in CI engines differ from SI engine due to the basic fact that CI engine combustion is unassisted combustion occurring on its' own. In CI engine the fuel is injected into combustion space after the compression of air is completed. Due to excessively high temperature and pressure of air the fuel when injected in atomised form gets burnt on its' own and burning of fuel is continued till the fuel is injected. Theoretically this injection of fuel and its' burning should occur simultaneously upto the cut-off point, but this does not occur in actual CI engine. Different significant phases of combustion are explained as under.

(*i*) Injection of fuel in atomized form is initiated into the combustion space containing compressed air. Fuel upon injection does not get burnt immediately instead some time is required for preparation before start of combustion. Fuel droplet injected into high temperature air first gets transformed into vapour (gaseous form) and then gets enveloped around by suitable amount of oxygen present so as to form combustible mixture. Subsequently, if temperature inside is greater than self ignition temperature at respective pressure then ignition gets set. Thus, the delay in start of ignition may be said to occur due to 'physical delay' i.e. time consumed in transformation from liquid droplet into gaseous form, and 'chemical delay' i.e. time consumed in preparation for setting up of chemical reaction (combustion). The duration of ignition delay depends upon fuel characteristic, compression ratio (i.e. pressure and temperature after compression), fuel injection, ambient air temperature, speed of engine, and geometry of combustion chamber etc. Ignition delay is inevitable stage and in order to accommodate it, the fuel injection is advanced by about 20° before TDC. Ignition delay is shown by $a - b$ in Fig. 10.12, showing pressure rise during combustion. Fuel injection begins at 'a' and ignition begins at 'b'. Theoretically, this ignition delay should be as small as possible.

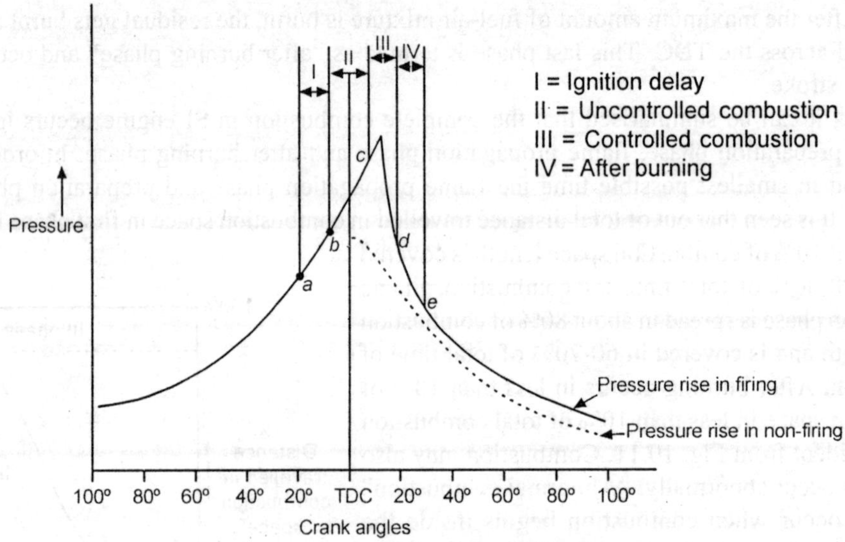

Fig. 10.12 Pressure rise during combustion in CI engines

(*ii*) During the ignition delay period also the injection of fuel is continued as it has begun at point 'a' and shall continue upto the point of cut-off. For the duration in which preparation for ignition is made, the continuous fuel injection results in accumulation of fuel in combustion space. The moment when ignition just begins, if the sustainable flame front is established then this accumulated fuel also gets burnt rapidly. This burning of accumulated fuel occurs in such a manner that combustion process becomes uncontrolled resulting into steep pressure rise as shown from 'b' to 'c'. The uncontrolled burning continues till the collected fuel gets burnt. During this 'uncontrolled combustion' phase if the pressure rise is very abrupt then combustion is termed as 'abnormal combustion' and may even lead to damage of engine parts in extreme conditions.

Thus, it is obvious that 'uncontrolled combustion' depends upon the 'ignition delay' period as during ignition delay itself the accumulation of unburnt fuel occurs and its' burning results in steep pressure rise. Hence in order to have minimum uncontrolled combustion the ignition delay should be as small as possible. During this uncontrolled combustion phase about one-third of total fuel heat is released.

(*iii*) After the 'uncontrolled combustion' is over then the rate of burning matches with rate of fuel injection and the combustion is termed as 'controlled combustion'. Controlled combustion is shown between '*c*' to '*d*' and during this phase maximum of heat gets evolved in controlled manner. In controlled combustion phase rate of combustion can be directly regulated by the rate of fuel injection i.e. through fuel injector. Controlled combustion phase has smooth pressure variation and maximum temperature is attained during this period. It is seen that about two-third of total fuel heat is released during this phase.

(*iv*) After controlled combustion, the residual if any gets burnt and the combustion is termed as 'after burning'. This after burning may be there due to fuel particles residing in remote position in combustion space where flame front could not reach. 'After burning' is spread over 60 – 70° of crank angle rotation and occurs even during expansion stroke.

Thus, it is seen that the complete combustion in CI engines may be comprising of four distinct phases i.e. 'ignition delay' followed by 'uncontrolled combustion,' 'controlled combustion' and 'after burning'. Combustion generally becomes abnormal combustion in CI engines when the ignition delay is too large resulting into large uncontrolled combustion and zig-zag pressure rise. Abnormal combustion in CI engines may also be termed as 'knocking' in engines and can be felt by excessive vibrations, excessive noise, excessive heat release, pitting of cylinder head and piston head etc. In order to control the knocking some additives are put in CI engine fuel so as to reduce its' self ignition temperature and accelerate ignition process. Also, the combustion chambers are properly designed so as to have reduced physical and chemical delay.

10.12 IC ENGINE FUELS

Fuels used in SI engines and CI engines are different as the nature of combustion process is different in the two engines. Normally hydrocarbon fuels are used for both the applications and should possess desirable properties such as high calorific value, suitable combustion characteristics, ease of handling, environment friendly etc. In SI engines the fuel used is generally 'gasoline' also called as petrol and is mixture of different hydrocarbons. Gasoline is available from refining process and its' exact composition depends upon the degree of refining and source etc. Gasoline should have capability of mixing rapidly with air, suitable volatility, resistance to abnormal combustion etc.

SI engine fuels are characterised in respect to its' resistance to abnormal combustion by its' rating in terms of "Octane number". Octane number of fuel is determined by comparing the combustion performance of actual fuel with that of reference fuel. Octane number is defined as the percentage by volume of iso-octane in a mixture of iso-octane (C_8H_{18}) and *n*-heptane (C_7H_{16}). Iso octane is low boiling point fuel having very high resistance to abnormal combustion and is arbitrarily assigned octane number of 100. '*n*-heptane' is seen to have very poor resistance to abnormal combustion and is so assigned octane number of 0. These 'iso-octane' and '*n*-heptane' are also called as 'primary reference fuels'. In order to determine octane number of any fuel its combustion characteristics are matched with those of a variable mixture of 'iso-octane' and '*n*-heptane'. The composition of iso-octane fraction and *n*-heptane fraction which yields the similar combustion characteristics is used to know the octane number. Say, for example in a particular test the mixture of 80% by volume of iso-octane and 20% by volume of

n-heptane offers the combustion similar to that of test fuel. Then the test fuel shall have octane number of 80. Octane number generally lies between 0 and 100 but can be extended beyond 100 by using certain additives. 'Terra ethyl lead' is the most popular additive used in SI engines which further increases resistance to abnormal combustion i.e. even more than that of iso-octane.

In CI engines the Diesel fuel is generally used which is also available during pertroleum refining but is impure as compared to SI engine fuel. CI engine fuels are also characterized by a rating given in terms of 'Cetane number' which also indicates the resistance of fuel to knocking. Cetane number for any fuel is given by percentage by volume of cetane ($C_{16}H_{34}$) in a mixture of cetane and α-methyl napthalene ($C_{10}H_7CH_3$) which offers the combustion characteristics similar to that of test fuel.

Cetane is arbitrarily assigned cetane number of 100 as it offers maximum knock resistance and α-methyl napthalene is assigned cetane number of 0 as it has mimimum knock resistance. Cetane rating is also obtained by the test similar to that for octane rating.

10.13 MORSE TEST

In IC engines there are three powers namely 'indicated power' which is developed inside cylinder, 'brake power' which is available at crank shaft and 'friction power' which is lost in overcoming friction and other losses. Brake power can be measured by using dynamometer at the crank shaft of engine. Friction power can be experimentally determined by number of methods namely 'Willan's line method', 'Motoring test' and by the numerical difference between indicated power and brake power.

Morse test is experimental method for estimating the friction power of the multi cylinder engine. Here engine is first run upto certain speed of rotation. Subsequently, one by one each cylinder of the engine is isolated from contributing to shaft work either by switching off electric supply to spark plug or by stopping fuel supply. When one cylinder is not producing power i.e. due to its isolation then also piston reciprocates in this cylinder similar to other cylinders thus the frictional losses remain constant for the shaft running at same speed. When second cylinder is cut then also frictional losses remain same but the brake power and indicated power get reduced accordingly. This process of cutting off each cylinder one by one is carried out till last cylinder and the brake power is measured by dynamometer attached to crank shaft which is maintained to run at same speed of rotation when cylinders are cut off one by one.

Mathematically, if there are n number of cylinders in a multicylinder engine. Then initially when all cylinders are working.

Indicated power from n cylinders, $(ip)_n = (bp)_n + (fp)_n$; here ip, bp and fp refer to indicated power, brake power and friction power respectively.

When one cylinder is cut then,

$$(ip)_{n-1} = (bp)_{n-1} + (fp)_n$$

When second cylinder is cut then,

$$(ip)_{n-2} = (bp)_{n-2} + (fp)_n$$

From above the power produced from first cylinder being cut off can be obtained as:

$$(ip)_{nth\ cylinder} = (bp)_n - (fp)_{n-1}$$

Total indicated power from engine: $(ip)_n = \sum (ip)_{nth\ cylinder}$

Thus, after knowing indicated power of engine the friction power can be determined as under, because brake power is also known through dynamometer.

$$(fp)_n = (ip)_n - (bp)_n$$

Friction power estimated using Morse test gives fairly correct estimate. Although, there may be slight deviation because when each cylinder is cut then there is substantial increase in engine vibrations and the friction losses may change due to change in back pressure etc.

10.14 COMPARATIVE STUDY OF IC ENGINES

Comparative study of IC engines with other prime movers is described here.

(a) Difference between SI and CI engine

SI engine	*CI engine*
1. It works on Otto cycle	2. It works on Diesel/Dual cycle
2. Compression ratio is from 5 to 15.	2. Compression ratio is from 12 to 30.
3. Petrol (Gasoline) is used as fuel.	3. Diesel is used as fuel.
4. Fuel should have high self ighition temperature.	4. Fuel should have low self ignition temperature.
5. A mixture of air and fuel is sucked inside the cylinder during suction process.	5. Air alone is sucked during suction process.
6. Carburettor is required for preparing air-fuel mixture.	6. Carburettor is not required.
7. Fuel goes into cylinder alongwith air during suction.	7. Fuel is injected after compression. For injection of fuel a fuel pump and injector is required
8. Spark plug is required to initiate combustion.	8. Ignition of fuel takes place on its' own due to high temperature of air inside cylinder.
9. Combustion takes place isochorically.	9. Combustion takes place isobarically.
10. Due to low compression ratio the engine structure is light.	10. To withstand high Compression ratio the engine structure is sturdy.
11. Engine speed is higher.	11. Engine speed is lower.
12. Efficiency is low.	12. Efficiency is high.
13. Generally used in 2-wheeler automobiles and smaller and lighter engines.	13. Generally used in 2 wheeler motorbikes, 4 wheelers and bigger engines.

(b) Comparison of 2-stroke engine with 4-stroke engine

2-stroke engine	*4-stroke engine*
1. One cycle is completed in one revolution of crankshaft.	1. One cycle is completed in two revolution of crankshaft.
2. Power stroke is available in each revolution of crankshaft.	2. Power stroke is available in alternate revolution of crankshaft.
3. Thermal efficiency is low.	3. Thermal efficiency is high.
4. Volumetric efficiency is less due to less time available for suction.	4. Volumetric efficiency is high due to large time available for suction.
5. Lighter flywheel is required due to more uniform torque available.	5. Heavier flywheel is required due to less uniform torque available.
6. Engine is compact and light in weight.	6. Engine is heavy.
7. Cooling requirements are less.	7. Cooling requirements are high.
8. Valves are not required.	8. Complex valve operating mechanism is required.
9. Cost is low.	9. Cost is high.
10. Used in light vehicles and small engines.	10. Used in heavy duty vehicles and bigger engines.

(c) Comparison of IC engines with gas turbine

IC engines	*Gas turbine*
1. IC engine have good part load efficiency. These have internal combustion.	1. Gas turbines have poor part load efficiency. These may have external combustion or internal combustion both as per requirement.
2. IC engines are well suited for smaller applications such as automobiles. IC engines can be used for small capacity power plants.	2. Gas turbines are not well suited for smaller applications, specially automobiles. These are well suited for aviation applications and land applications in power plants etc. as prime movers. It has small specific weight and small size capability and are so suited for aeroplanes.
3. Balancing is poor due to reciprocating parts. Mechanical efficiency is also poor.	3. Balancing is good due to absence of reciprocating parts. Mechanical efficiency is good due to rotary parts and no sliding members
4. Maintenance requirement is high.	4. Poor maintenance requirement.
5. Type of fuels which can be used is limited in number.	5. Large number of different types of fuels can be used.
6. Operation is noisy and large vibrations are present.	6. Operation is silent and smooth.
7. No modification is possible for improving efficiency and output in existing IC engines.	7. Modifications such as reheating, regeneration and intercooling etc. are possible in order to improve efficiency and output.

(d) Comparison of gas turbine with steam turbines

Gas turbine	*Steam turbines*
1. These are generally internal combustion type, but may be of external combustion type too.	1. These are of external combustion type.
2. High quality fuel is being used.	2. Any type of poor or good quality fuel can be utilized.
3. Thermal efficiency is better.	3. Thermal efficiency is poor than gas turbines.
4. It does not require boiler, feed water supply, condensing plant etc. Number of components involved is small.	4. It requires boiler, feed water supply, feed pump, condensing plant etc. instead of compressor, combustion chamber as in gas turbines. Number of components involved is large.
5. These can be started quickly.	5. These take time to start because of time required in generating steam.
6. Maintenance cost is small.	6. Maintenance cost is high.
7. These have smaller specific weight and size i.e. small weight per kW output and so these can be used in aviation applications.	7. These have large specific weight and size and are exclusively suited for land applications.
8. Initial cost and running cost are small.	8. Initial cost and running cost are large in steam turbines.

(e) Comparison between Diesel engine, Gas turbine and Steam turbine based plants: (Parametric form)

Parameter	*Diesel engine*	*Steam turbine plant*	*Gas turbine*
1. Power generation capacity	10 kW to 10 MW	1 MW to 1000 MW	100 kW to 100 MW
2. Efficiency	0.35 to 0.40	0.30 to 0.40	0.20 to 0.40
3. Fuel	High grade fuel, liquid fuel of diesel grade	Medium to low grade fuel, may be gaseous, liquid or solid fuel.	High to medium grade, may be liquid or gas fuel.
4. Size and weight	Medium	Large	Small
5. Capital cost	\$US 110-180 US\$ per kW	\$US 140-1200 per kW	\$US 80-300 per kW
6. Lubrication requirement	Frequent oil and filter changes	Negligible	Oil topping up.
7. Maintenance period	Weekly and monthly	Daily	Monthly and annual
8. Specific installation needs	Very heavy foundations, vibration isolation, enclosures for making its operation quiet etc.	Coal handling facility, Ash handling facility, enough supplies of water, exhaust emission control etc.	Intake and exhaust mufflers etc.
9. Plant life	10-12 years	25 to 35 years	15 to 30 years

EXAMPLES

1. A four cylinder diesel engine of 4-stroke type has stroke to bore ratio as 1.2 and the cylinder diameter is 12 cm. Estimate indicated power of the engine using the indicator diagram arrangement. Indicator card shows the diagram having area of 30 cm² and length as half of stroke. Indicator spring constant is 20 × 10³ kN/m² and engine is running at 2000 rpm. Also find out mechanical efficiency of engine if 10% of power is lost in friction and other losses.

Solution:

From stroke to bore ratio i.e. $\frac{L}{D}$ = 1.2 and cylinder diameter = bore, i.e. D = 12 cm, so L = 14.4 cm

Area of indicator diagram = 30 cm² = 30×10^{-4} m²

$$\text{Length of indicator diagram} = \frac{1}{2} \times \text{stroke}$$

$$= 7.2 \text{ cm or } 7.2 \times 10^{-2} \text{ m}$$

$$\text{Mean effective pressure} = \frac{30\times10^{-4}\times20\times10^{3}\times10^{3}}{7.2\times10^{-2}}$$

$$\text{m.e.p.} = 8333.333 \times 10^{2} \text{ N/m}^2$$

$$\text{Cross-section area of piston} = \frac{\pi}{4}D^2 = 0.01131 \text{ m}^2$$

For one cylinder indicated power

$$= \frac{8333.333 \times 10^2 \times 0.01131 \times 14.4 \times 10^{-2} \times 2000}{2 \times 60}$$

$$= 22619.9 \text{ W}$$

For four cylinders total indicated power = 90479.6 W.

Frictional power loss = 0.10 × 90479.6 = 9047.96 W.

Brake power available = Indicated power – Frictional power

$$= (90479.6 - 9047.96)$$

$$= 81431.64 \text{ W}$$

Therefore, mechanical efficiency of engine $= \frac{\text{brake power}}{\text{indicated power}}$

$$= \frac{81431.64}{90479.6}$$

$$= 0.90$$

Indicated power = **90479.6 W**
Mechanical efficiency = **90%**

Ans.

2. Determine the power required to drive a double acting reciprocating pump having indicator diagram with area 40 cm² and length 8 cm. Bore and stroke of the pump are 15 cm and 20 cm. The pump motor runs at 100 rpm. Indicator spring constant is 1.5 × 10⁸ Pa per m.

Solution:

Reciprocating Pump is work absorbing machine having its mechanism similar to the piston-cylinder arrangement in IC engine.

$$\text{Mean effective pressure} = \frac{40 \times 10^{-4} \times 1.5 \times 10^8}{8 \times 10^{-2}}$$

$$\text{m.e.p.} = 7.5 \times 10^6 \text{ Pa}$$

Indicated power = (Piston area × Stroke × mep × Speed × No. of cylinder)

(It is double acting so let us assume total power to be double of that in single acting)

$$= \left(\frac{\pi}{4} \times (0.15)^2 \times 0.20 \times 7.5 \times 10^6 \times \frac{100}{60} \times 1 \times 2\right)$$

$$= 88357.29 \text{ W}$$

Power required to drive = **88.36 kW.**

Ans.

3. An engine with 90% mechanical efficiency has rating of 38 kW brake power. Estimate its indicated power and frictional power loss. Also determine the mechanical efficiency at quarter load assuming frictional power to remain same.

Solution:

$$\text{Indicated power} = \frac{\text{Brake power}}{\text{Mechanical efficiency}} = \frac{38}{0.9}$$

$$= 42.22 \text{ kW}$$

$$\text{Frictional power loss} = \text{Indicated power} - \text{Brake power}$$

$$= 42.22 - 38$$

$$= 4.22 \text{ kW}$$

$$\text{Brake power at quarter load} = 0.25 \times 38$$

$$= 9.5 \text{ kW}$$

$$\text{Mechanical efficiency} = \frac{\text{Brake power}}{\text{Indicated power}} = \frac{9.5}{9.5 + 4.22}$$

$$= 0.6924 \text{ or } 69.24\%$$

Indicated power = **42.22 kW**	
Frictional power loss = **4.22 kW**	
Mechanical efficiency = **69.24%**	**Ans.**

***4.** An eight cylinder diesel engine of two stroke type has specific fuel consumption of 0.25 kg/kWh. The brake mean effective pressure of each cylinder is 1.5 MPa and engine run at 100 rpm. The bore and stroke of cylinder are 85 cm and 220 cm respectively. Considering the calorific value of diesel as 43 MJ/kg determine the brake power of engine, fuel consumption in kg/hr and brake thermal efficiency of engine.*

Solution:

$$\text{Brake power of engine} = P_{b\,\text{mep}} \times L \times A \times N$$

$$= \frac{1.5 \times 10^3 \times (2.2) \times \left(\frac{\pi}{4} \times (0.85)^2\right) \times 100}{60}$$

$$= 3120.97 \text{ kW or } 31.21 \text{ MW}$$

Brake power = **31.21 MW** **Ans.**

The fuel consumption in kg/hr = 0.25 × 3120.97 = 780.24 kg/hr.

In order to find out brake thermal efficiency the heat input from fuel per second is required.

$$\text{Heat from fuel, kJ/s} = \frac{780.24 \times 43 \times 10^3}{3600} = 9319.53 \text{ kJ/s}$$

$$\text{Energy to brake power} = 3120.97 \text{ kW}$$

$$\text{Brake thermal efficiency} = \frac{3120.97}{9319.53} = 0.33488 \text{ or } 33.49\%$$

Fuel consumption = 780.24 kg/hr, Brake thermal efficiency = **33.49%** **Ans.**

***5.** A four cylinder, four stroke diesel engine has brake mean effective pressure of 6 bar at full load speed of 600 rpm and specific fuel consumption of 0.25 kg/kWh. The cylinder has bore of 20 cm and stroke length of 30 cm. The air fuel ratio is measured as 26 from the exhaust gas analysis. The ambient conditions are 1 bar, 27°C. Assuming the calorific value of fuel as 43 MJ/kg determine the brake thermal efficiency and the volumetric efficiency. Also find out brake power.*

Solution:

$$\text{Brake thermal efficiency} = \frac{3600}{(\text{brake specific fuel consumption} \times \text{Calorific value})}$$

$$= \frac{3600}{(0.25 \times 43 \times 10^3)} = 0.3349 \text{ or } 33.49\%$$

Brake thermal efficiency = **33.49%**	**Ans.**

$$\text{Brake power} = \frac{4 \times 6 \times 10^2 \times 0.30 \times \left(\frac{\pi}{4} \times (0.20)^2\right) \times 600}{60} = 226.19 \text{ kW}$$

Brake power = **226.19 kW**	**Ans.**

$$\text{Brake specific fuel consumption} = \frac{m_f}{\text{brake power}}$$

$$0.25 = \frac{m_f}{226.19} \Rightarrow m_f = 56.55 \text{ kg/hr}$$

Air consumption from given air fuel ratio = 26 × 56.55 = 1470.3 kg/hr

or

$$m_a = 24.505 \text{ kg/min}$$

Using perfect gas equation,

$$P \cdot V_a = m_a RT$$

$$1 \times 10^2 \times V_a = 24.505 + 0.287 \times 300$$

$$V_a = 21.098 \text{ m}^3/\text{min}$$

$$\text{Swept volume, } V_s = \frac{\pi}{4} D^2 L = \frac{\pi}{4} \times (0.30)^2 \times 0.4 = 0.02827 \text{ m}^3$$

$$\text{Volumetric efficiency, } \eta_{\text{vol.}} = \frac{V_a}{V \times \frac{N}{2} \times \text{number or cylinders}}$$

$$= \frac{21.09 \times 2}{0.02827 \times 600 \times 4}$$

$$= 0.6217 \text{ or } 62.17\%$$

Volumetric efficiency = **62.17%**	**Ans.**

***6.** A two stroke two cylinder engine runs with speed of 3000 rpm and fuel consumption of 5 litres/hr. The fuel has specific gravity of 0.7 and air-fuel ratio is 19. The piston speed is 500 m/min and indicated mean effective pressure is 6 bar. The ambient conditions are 1.013 bar, 15°C. The volumetric efficiency is 0.7 and mechanical efficiency is 0.8. Determine brake power output considering R for gas = 0.287 kJ/kg · K*

(Take piston speed, m/min = 2 LN where L is stroke (m) and N is rpm)

Solution:

Let the bore of cylinder be '*D*' *m*.

Using piston speed, $500 = 2 \times L \times 3000$

$$\Rightarrow \quad L = 0.0833 \text{ m}$$

$$\text{Volumetric efficiency, } \eta_{vol.} = \frac{\text{Air sucked (m}^3\text{/min)}}{\text{Swept volume} \times N \times \text{number of cylinders}}$$

$$0.7 = \frac{\text{Air sucked}}{\left(\frac{\pi}{4} D^2 \cdot L\right) \times 3000 \times 2}$$

$$\Rightarrow \quad \text{Air sucked} = 274.78\, D^2 \text{ m}^3\text{/min}$$

$$\text{Air requirement, kg/min} = A/F \text{ ratio} \times \text{Fuel consumption per minute}$$

$$= 19 \times \frac{5 \times 0.7 \times 10^3 \times 10^{-3}}{60}$$

$$m_a = 1.11 \text{ kg/min}$$

Using perfect gas equation, $PV_a = m_a RT$

$$V_a = \frac{1.11 \times 0.287 \times 288}{1.013 \times 10^2}$$

$$V_a = 0.906 \text{ m}^3\text{/min}$$

$$\text{Ideally, Air sucked} = V_a$$

$$274.78\, D^2 = 0.906$$

$$\Rightarrow \quad D = 0.0574 \text{ m or } 57.4 \text{ mm}$$

$$\text{Indicated power} = P_{imep} \times L \times A \times N \times \text{no. of cylinders}$$

$$= 6 \times 10^2 \times 0.0833 \times \frac{\pi}{4} \times \frac{(0.0574)^2 \times 3000 \times 2}{60}$$

$$= 12.93 \text{ kW}$$

$$\text{Brake power} = \text{Indicated power} \times \text{Mechanical efficiency}$$

$$= 12.93 \times 0.8 = 10.34 \text{ kW}$$

Brake power = **10.34 kW**	**Ans.**

7. During trial of four stroke single cylinder engine the load on dynamometer is found 20 kg at radius of 50 cm. The speed of rotation is 3000 rpm. The bore and stroke are 20 cm and 30 respectively. Fuel is supplied at the rate of 0.15 kg/min. The calorific value of fuel may be taken as 43 MJ/kg. After some time the fuel supply is cut and the engine is rotated with motor which required 5 kW to maintain the same speed of rotation of engine. Determine the brake power, indicated power, mechanical efficiency, brake thermal efficiency, indicated thermal efficiency, brake mean effective pressure, indicated mean effective pressure.

Solution:

After switching off fuel supply the capacity of motor required to run engine will be the friction power required at this speed of engine.

$$\text{Friction power} = 5 \text{ kW}$$

$$\text{Brake power} = 2\pi NT = \frac{2\times\pi\times3000\times20\times9.81\times0.5\times10^{-3}}{60}$$

$$= 30.82 \text{ kW}$$

$$\text{Indicated power} = \text{Brake power} + \text{Friction power} = 35.82 \text{ kW}$$

$$\text{Mechanical efficiency} = \frac{\text{Brake power}}{\text{Indicated power}} = \frac{30.82}{35.82} = 0.8604 \text{ or } 86.04\%$$

$$\text{Brake specific fuel consumption} = \frac{\text{Specific fuel consumption}}{\text{Brake power}}$$

$$= \frac{0.15\times60}{30.82} = 0.292 \text{ kg/kW hr}$$

$$\text{Brake thermal efficiency} = \frac{3600}{\text{Brake specific fuel consumption}\times\text{calorific value}}$$

$$= \frac{3600}{0.292\times43\times10^3} = 0.2867 \text{ or } 28.67\%$$

$$\text{Also, Mechanical efficiency} = \frac{\text{Brake thermal efficiency}}{\text{Indicated thermal efficiency}}$$

$$\Rightarrow \quad \text{Indicated thermal efficiency} = \frac{0.2867}{0.8604} = 0.3332 \text{ or } 33.32\%$$

$$\text{Indicated power} = P_{\text{imep}} \times L \times A \times N$$

$$35.82 = P_{\text{imep}} \times 0.30 \times \frac{\pi}{4} \times (0.20)^2 \times \frac{3000}{60}$$

$$\Rightarrow \quad P_{\text{imep}} = 76.01 \text{ kPa}$$

$$\text{Also, Mechanical efficiency} = \frac{P_{bmep}}{P_{imep}}$$

$$\text{Brake mean effective pressure} = P_{imep} \times \text{Mechanical efficiency}$$

$$P_{imep} = 76.01 \times 0.8604 = 65.39 \text{ kPa}$$

Brake power = **30.82 kW**, Indicated power = **35.82 kW**
Mechanical efficiency = **86.04%.** Brake thermal efficiency = **28.67%**
Indicated thermal efficiency = **33.32%,** Brake mean effective pressure = **65.39 kPa,**
Indicated mean effective pressure = **76.01 kPa** **Ans.**

8. *A four stroke four cylinder diesel engine running at 300 rpm produces 250 kW of brake power. The cylinder dimensions are 30 cm bore and 25 cm stroke. Fuel consumption rate is 1 kg/min while air fuel ratio is 10. The average indicated mean effective pressure is 0.8 MPa. Determine indicated power, mechanical efficiency, brake thermal efficiency and volumetric efficiency of engine. The calorific value of fuel is 43 MJ/kg. The ambient conditions are 1.013 bar, 27°C.*

Solution:

Given, $D = 0.3$ m, $L = 0.25$ m, $N = 300$ rpm, $m_f = 1$ kg/min, $F/A = 20$,
$P_{imep} = 0.8$ MPa, Brake power = 250 kW.

$$\text{Indicated power kW} = P_{imep} \times L \times A \times \frac{N \times 4}{60} = 0.8 \times 0.25 \times \frac{\pi}{4} \times (0.3)^2$$

$$\times \frac{300}{60} \times 4 \times 10^3 = 282.74 \text{ kW}$$

$$\text{Mechanical efficiency} = \frac{\text{Brake power}}{\text{Indicated power}} = \frac{250}{282.74} = 0.8842 \text{ or } 88.42\%$$

$$\text{Brake specific fuel consumption} = \frac{1 \times 60}{250} = 0.24 \text{ kg/kW hr.}$$

$$\text{Brake thermal efficiency} = \frac{3600}{bsfc \times CV} = \frac{3600}{0.24 \times 43000} = 0.3488 \text{ or } 34.88\%$$

$$\text{Swept volume, } V_s \frac{\pi}{4} D^2 L = \frac{\pi}{4} \times (0.30)^2 \times 0.25 = 0.01767 \text{ m}^3$$

Mass of air corresponding to above swept volume, using Perfect gas equation

$$PV_S = m_a RT$$

$$m_a = \frac{1.013 \times 10^2 \times 0.017}{0.287 \times 300}$$

$$= 0.02 \text{ kg}$$

$$\text{Volumetric efficiency}, \quad \eta_{vol} = \frac{\text{Mass of air taken/minute}}{\text{Mass corresponding to swept volume per minute}}$$

$$\text{Mass of air taken per minute} = 1 \times 10 = \text{kg/min}$$

$$\text{Mass corresponding to swept volume per minute} = 0.02 \times 4 \times \frac{300}{2} = 12 \text{ kg/min.}$$

$$\text{Volumetric efficiency} = \frac{10}{12} = 0.8333 \text{ or } 83.33\%$$

Indicated power = 282.74 kW, Mechanical efficiency = 88.42%

Brake thermal efficiency = **34.88%,** Volumetric efficiency = **83.33%**	**Ans.**

9. *During an experiment on four stroke single cylinder engine the indicator diagram obtained has average height of 1 cm while indicator constant is 25 kN/m² per mm. The engine run at 300 rpm and the swept volume is 1.5 × 10⁴ cm³. The effective brake load upon dynamometer is 60 kg while the effective brake drum radius is 50 cm. The fuel consumption is 0.12 kg/min and the calorific value of fuel oil is 42 MJ/kg. The engine is cooled by circulating water around it at the rate of 6 kg/min.*

The cooling water enters at 35° C and leaves at 70°C. Exhaust gases leaving have energy of 30 kJ/s with them. Take specific heat of water as 4.18 kJ/kg K. Determine indicated power output, brake power output and mechanical efficiency. Also draw the overall energy balance in kJ/s.

Solution:

Indicated mean effective pressure = 10 × 25 = 250 kPa

$$\text{Indicated power} = P_{imep} \times L \times A \times \frac{N}{2} \quad \left\{\text{here } N = \frac{N}{2} \text{ as it is four stroke}\right\}$$

$$= 250 \times 1.5 \times 10^4 \times \frac{300}{2\times 60} \times 10^{-6} = 9.375 \text{ kW}$$

$$\text{Brake power} = 2\pi\ NT = 2 \times \pi \times \frac{300}{2\times 60} \times (60 \times 9.81 \times 0.50) \times 10^{-3} = 4.62 \text{ kW}$$

$$\text{Mechanical efficiency} = \frac{\text{Brake power}}{\text{Indicated power}} = \frac{4.62}{9.375}$$

$$= 0.4928 \text{ or } 49.28\%$$

Indicated power	= **9.375 kW,**
Brake power	= **4.62 kW,**
Mechanical efficiency	= **49.28%**

Ans.

$$\text{Energy liberated from fuel} = \frac{42\times 10^3 \times 0.12}{60} = 84 \text{ kJ/s}$$

Energy available as brake power = 4.62 kW

$$\text{Energy to coolant} = \frac{6}{60} \times 4.18 \times (70 - 35) = 14.63 \text{ kW}$$

Energy carried by exhaust gases = 30 kJ/s

Unaccounted energy loss = 84 – 4.62 – 14.63 – 30 = 34.75 kW

Energy Balance on per second basis

Energy input, kJ/s	*Energy consumed, kJ/s*	*in %*
Heat from fuel = 84 kJ/s, 100%	Energy consumed as brake power = 4.62 kW	5.5%
	Energy carried by coolant = 14.63 kW	17.42%
	Energy carried by exhaust gases = 30 kW	35.71%
	Unaccounted losses = 34.75 kW	41.37%

10. During 15 minutes trial of an internal combustion engine of 2-stroke single cylinder type the total 4 kg fuel is consumed while the engine is run at 1500 rpm. Engine is cooled employing water being circulated at 15 kg/min with its inlet and exit temperatures as 27°C and 50°C. The total air consumed is 150 kg and the exhaust temperature is 400°C. The atmospheric temperature is 27°C. The mean specific heat of exhaust gases may be taken as 1.25 kJ/kg K. The mechanical efficiency is 0.9. Determine, the brake power, brake specific fuel consumption and indicated thermal efficiency. Also draw energy balance on per minute basis. Brake torque is 300 Nm and the fuel calorific value is 42 MJ/kg.

Solution:

$$\text{Brake power} = 2\pi NT = \frac{2\times\pi\times 1500\times 300}{60} = 47.124 \text{ kW}$$

Brake power = 47.124 kW **Ans.**

$$\text{Brake specific fuel consumption} = \frac{4\times 60}{15\times 47.124} = 0.339 \text{ kg/kW hr.}$$

Brake specific fuel consumption = 0.339 kg/kW hr. **Ans.**

$$\text{Indicated power} = \frac{\text{Brake power}}{\text{Mechanical efficiency}} = \frac{47.124}{0.9} = 52.36 \text{ kW}$$

$$\text{Indicated thermal efficiency} = \frac{52.36}{\left(\dfrac{4\times 42\times 10^3}{15\times 60}\right)} = 0.2805 \text{ or } 28.05\%$$

Indicated thermal efficiency = 28.05% **Ans.**

$$\text{Heat available from fuel, kJ/min} = \frac{4}{15}\times 42\times 10^3 = 11200 \text{ kJ/min}$$

$$\text{Energy consumed as brake power} = 47.124 \times 60 = 2827.44 \text{ kJ/min}$$

$$\text{Energy carried by cooling water} = 15 \times 4.18 \times (50 - 27) = 1442.1 \text{ kJ/min}$$

$$\text{Energy carried by exhaust gases} = \left(\frac{150+4}{15}\right)\times 1.25\times (400-27)$$

$$= 4786.83 \text{ kJ/min}$$

$$\text{Unaccounted energy loss} = 11200 - (2827.44 + 1442.1 + 4786.83)$$

$$= 2143.63 \text{ kJ/min.}$$ Heat balance on per minute basis

Heat input, kJ/min	*%*	*Heat consumed, kJ/min,*	*%*
Heat from fuel = 11200	100%	1. Brake power = 2827.44	25.24%
		2. Cooling water = 1442.1	12.88%
		3. Exhaust gases = 4786.83	42.74%
		4. Unaccounted loss = 2143.63 (by difference)	19.14%
Total = 11200 kJ/min	100%	Total = 11200 kJ/min	100%

11. *During Morse Test experiment on a six cylinder petrol engine the brake power output was found 50 kW when all cylinders run at full load. When one by one each cylinder is cut and load is reduced to bring engine back to original speed, the measured brake power outputs are as under. Determine the indicated power of engine and mechanical efficiency of engine.*

No. of cylinders cut	*Brake power, kW*
1	40.1
2	39.5
3	39.1
4	39.6
5	39.8
6	40

Solution:

Indicated power of Ist cylinder = 50 – 40.1 = 9.9 kW
Indicated power of IInd cylinder = 50 – 39.5 = 10.5 kW
Indicated power of IIIrd cylinder = 50 – 39.1 = 10.9 kW
Indicated power of IVth cylinder = 50 – 39.6 = 10.4 kW
Indicated power of Vth cylinder = 50 – 39.8 = 10.2 kW
Indicated power of VIth cylinder = 50 – 40 = 10 kW
Total indicated power = 61.9 kW

$$\text{Mechanical efficiency} = \frac{\text{Brake power}}{\text{Indicated power}} = \frac{50}{61.9} = 0.8077 \text{ or } \mathbf{80.77\%}$$

Indicated power = **61.9 kW**
Mechanical efficiency = **80.77%** **Ans.**

12. During trial of a four cylinder four stroke petrol engine running at full load it has speed of 1500 rpm and brake load of 250 N when all cylinders are working. After some time each cylinder is cut one by one and then again brought back to same speed of engine. The brake readings are measured as 175 N, 180 N, 182 N and 170 N. The brake drum radius is 50 cm. The fuel consumption rate is 0.189 kg/min with the fuel whose calorific value is 43 MJ/kg and A/F ratio of 12. Exhaust gas temperature is found to be 600°C. The cooling water flows at 18 kg/min and enters at 27°C and leaves at 50°C. The atmospheric air temperature is 27°C. Take specific heat of exhaust gas as 1.02 kJ/kg K. Determine the brake power output of engine, its indicated power and mechanical efficiency. Also draw a heat balance on per minute basis.

Solution:

$$\text{Brake power output of engine} = \frac{2\pi NT}{60} = \frac{2\pi \times 1500 \times 250 \times 0.50 \times 10^{-3}}{60} = 19.63 \text{ kW}$$

$$\text{Brake power when cylinder 1 is cut} = \frac{2 \times \pi \times 1500 \times 175 \times 0.5 \times 10^{-3}}{60} = 13.74 \text{ kW}$$

Indicated power of first cylinder = 19.63 – 13.74 = 5.89 kW

$$\text{Brake power when cylinder 2 is cut} = \frac{2 \times \pi \times 1500 \times 180 \times 0.5 \times 10^{-3}}{60} = 14.14 \text{ kW}$$

Indicated power of second cylinder = 19.63 – 14.14 = 5.49 kW

$$\text{Brake power when cylinder 3 is cut} = \frac{2 \times \pi \times 1500 \times 182 \times 0.5 \times 10^{-3}}{60} = 14.29 \text{ kW}$$

Indicated power of third cylinder = 19.63 – 14.29 = 5.34 kW

$$\text{Brake power when cylinder 4 is cut} = \frac{2 \times \pi \times 1500 \times 170 \times 0.5 \times 10^{-3}}{60} = 13.35 \text{ kW}$$

Indicated power of fourth cylinder = 19.63 – 13.35 = 6.28 kW
Total indicated power output = 5.89 + 5.49 + 5.34 + 6.28 = 23 kW

$$\text{Engine's mechanical efficiency} = \frac{\text{Brake power}}{\text{Indicated power}} = \frac{19.63}{23} = 0.8535 \text{ or } 85.35\%$$

B.P. = 19.63 kW, I.P. = 23 kW, η_{mech} = **83.35%**	**Ans.**

Heat liberated by fuel, kJ/min $= 0.189 \times 43 \times 10^3 = 8127$ kJ/min

Heat carried by exhaust gases $= (12 + 1) \times 0.189 \times 1.02 \times (600 - 27) = 1436.02$ kJ/min

Heat carried by cooling water $= 18 \times 4.18 \times (50 - 27) = 1730.52$ kJ/min

Energy to brake power $= 19.63 \times 60 = 1177.8$ kJ/min

Unaccounted losses $= 8127 - (1436.02 + 1730.52 + 1177.8) = 3782.66$ kJ/min

Heat balance sheet on per minute basis

Energy in, kJ/min	%	*Energy out, kJ/min,*	%
Heat released by fuel = 8127	100%	(a) Brake power = 1177.8 (b) Exhaust gases = 1436.02 (c) Cooling water = 1730.52 (d) Unaccounted = 3782.66 (by difference)	14.50 17.67 21.29 40.54
Total = 8127, kJ/min	100%	Total = 8127 kJ/min	100%

13. *During the trial of a single acting oil engine, cylinder diameter is 20 cm, stroke 28 cm, working on two stroke cycle and firing every cycle, the following observations were made:*

Duration of trial :1 hour
Total fuel used :4.22 kg
Calorific value :44670 kJ/kg
Proportion of hydrogen in fuel : 15%
Total number of revolutions : 21000
Mean effective pressure : 2.74 bar
Net brake load applied to a drum of 100 cm diameter : 600 N
Total mass of cooling water circulated : 495 kg
Total mass of cooling water : inlet 13°C, outlet 38°C
Air used : 135 kg
Temperature of air in test room : 20°C
Temperature of exhaust gases : 370°C
Assume $C_{p,\, gases} = 1.005$ *kJ/kg K,*
$C_{p,\, steam}$ *at atmospheric pressure = 2.093 kJ/kg K*
Calculate thermal efficiency and draw up the heat balance. *[U.P.S.C. 1997]*

Solution:

Brake power $= 2\pi NT$

$$N = \frac{21{,}000}{60} \text{ rpm} = 350 \text{ rpm}$$

$$T = 600 \times 0.5 = 300 \text{ N.m}$$

$$\text{Brake power} = \frac{2\pi \times 350 \times 300}{60 \times 1000} = 10.996 \text{ kW}$$

$$\text{Indicated power} = \frac{\text{mep} \times L.A.N}{60 \times 1000} \text{ kW}$$

$$= \frac{2.74 \times 10^5 \times 0.28 \times \pi \times (0.2)^2 \times 300}{60 \times 1000 \times 4} = 14.06 \text{ kW}$$

(A) $$\text{Heat added} = \frac{4.22}{3600} \times 44670 = 52.36 \text{ kJ/s or } 3141.79 \text{ kJ/min}$$

$$\text{Thermal efficiency, } \eta = \frac{\text{Indicated power}}{\text{Heat added}}$$

$$= \frac{14.06}{\frac{4.22}{3600} \times 44670}$$

$$= 0.2685 \text{ or } 26.85\%$$

(B) Heat equivalent of brake power = 10.996 × 60 = 659.76 kJ/min

(C) Heat lost to cooling water = $m_w \times C_{pw} \times \Delta T$

$$= \frac{495}{60} \times 4.18 \times (38 - 13)$$

$$= 862.13 \text{ kJ/min}$$

Heat carried by exhaust gases = Heat carried by steam in exhaust gases + Heat carried by fuel gases (dry gases) in exhaust gases

Mass of exhaust gases = mass of air + mass of fuel

$$= \frac{(135 + 4.22)}{60}, \text{ kg/min}$$

$$= 2.32 \text{ kg/min}$$

$$\text{Mass of steam in exhaust gases} = 9 \times \left(0.15 \times \frac{4.22}{60}\right)$$

$$= 0.095 \text{ kg/min}$$

Mass of dry exhaust gases = 2.32 – 0.095

= 2.225 kg/min

(D) Heat carried by steam in exhaust = 0.095 × {4.18 (100 –20) + 2256.9 + 2.093 × (370 – 100)}

= 299.86 kJ/min

(E) Heat carried by fuel gases (dry gases) in exhaust gases

= 2.225 × 1.005 (370 – 20)

= 782.64 kJ/min.

(F) Unaccounted loss = A – B – C – D – E

= (3141.79) – (659.76) – (862.13) – (299.86) – (782.64)

= 537.4 kJ/min

Heat balance sheet (on per minute basis) **Ans.**

Input		Expenditure	
Heat, kJ/min	%, Percentage	Heat, kJ/min	%, Percentage
(A) Heat added by fuel = 3141.79	100	(B) Heat equivalent to brake power = 659.76	21%
		(C) Heat lost to cooling water = 862.13	27.44%
		(D) Heat carried by steam in exhaust gases = 299.86	9.54%
		(E) Heat carried by dry fuel gases in exhaust gases = 782.64	24.92%
		(F) Unaccounted losses = 537.4	17.10%

EXERCISE

10.1 Give classification of internal combustion engines.

10.2 Give a sketch of simple one cylinder engine and label important parts.

10.3 What is meant by spark ignition and compression ignition engines?

10.4 Explain the working of 4-stroke SI engine.

10.5 Describe the working of 4-stroke CI engine.

10.6 Explain working of 2-stroke CI engine.

10.7 Explain working of 2-stroke SI engine.

10.8 What is meant by indicator diagram? Describe the arrangement for getting it?

10.9 Explain the power measurement using indicator diagram arrangement.

10.10 Explain the combustion in SI engines.

10.11 Explain the combustion in CI engines.

10.12 Describe some IC engine fuels.

10.13 What is More test? How is it carried out?

10.14 Compare SI engines with CI engines.

10.15 Compare 2 stroke SI engine with 4-stroke SI engines.

10.16 Write short notes on the following:
Brake power, Indicated power, Brake specific fuel consumption, Indicated specific fuel consumption, Brake mean effective pressure, Indicated mean effective pressure, Mechanical efficiency. Brake thermal efficiency, Indicated thermal efficiency.

10.17 A single cylinder four stroke CI engine runs at 200 rpm and the cylinder has bore and stroke of 10 cm and 12.5 cm respectively. The indicator diagram taken has length of 10 cm and area of 20 cm^2. The indicator constant is 30 kPa per mm. Considering mechanical efficiency of 0.8 determine indicated mean effective pressure and indicated power. [600 kPa, 5.89 kW]

10.18 Determine the dimensions of cylinder, brake thermal efficiency and indicated mean effective pressure for the single cylinder four stroke diesel engine producing 100 kW of brake power. Engine runs at speed of 400 rpm, brake mean effective pressure of 850 kPa, brake specific fuel consumption of 0.335 kg/kW hr. The mechanical efficiency is 0.8 and stroke to bore ratio is 1.25.
The calorific value of fuel may be taken as 43.5 MJ/kg. [D = 33 cm, L = 41.25 cm, 24.7%, 10.62 bar]

10.19 Determine thermal efficiency and mean effective pressure of thermodynamic cycle used by a 4-stroke petrol engine. Details of cycle are as follows.
Compression ratio = 7
Initial state = 100 kPa and 90ºC
Swept volume = 0.1 m^3
Heat added to cycle at constant volume = 100 kJ/cycle.
Consider air as working fluid. [54%, 530 kPa]

10.20 A 4-stroke four cylinder petrol engine has bore and stroke as 0.15 m and 0.17 m respectively. Indicator diagram arrangement when put on a cylinder yields indicator diagram having area of 25 cm^2 and length of diagram as 6 cm. The speed of engine is 2500 rpm. Determine the indicated power and frictional power if mechanical efficiency is 88%. Take indicator spring constant as 20×10^3 kN/m^3. [208.6 kW, 25 kW]

10.21 Determine the cylinder diameter and stroke length for a 8 cylinder, single acting, 2-stroke diesel engine for which details of engine and indicator diagram are as under;
Indicated power = 16266.7 kW
Speed = 150 rpm
L:D ratio = 1.5
Area of indicator diagram = 550 mm^2
Length of diagram = 60 mm
Indicator spring constant = 14.7×10^7 Pa per m [80 cm, 120 cm]

10.22 Determine the capacity of motor required for running a single-cylinder, double acting reciprocating water pump at 50 rpm. Indicator diagram arrangement gives an indicator diagram with area of 37.5 cm^2 and length 7.5 cm. Indicator spring constant is 14.7×10^7 Pa per m. Pump cylinder diameter and stroke are 15 cm and 20 cm Mechanical efficiency of arrangement is 90%. [33.66 kW]

10.23 During a test on single cylinder four stroke diesel engine having bore of 15 cm, stroke of 18 cm gives indicator diagram of 10 cm length and area of 9 cm^2. Engine has speed of 480 rpm, fuel consumption of 0.05 kg/min and cooling water flow rate of 250 kg/hr. Cooling water enters at 27°C and leaves at 67°C. The calorific value of fuel is 44000 kJ/kg. Brake torque is 200 Nm. Determine the indicator constant, brake specific fuel consumption and brake thermal efficiency. Take mechanical efficiency as 87.8%. [0.298 kg.kW hr, 27.4%]

10.24 A two cylinder four stroke engine runs at 240 rpm developing a torque of 5.16 kN.m. The bore and stroke of cylinder are 30 cm and 58.5 cm respectively.
Engine runs with gaseous fuel having calorific value of 16.8 MJ/m^3. The gas and air mixture is supplied in proporation of 1 : 7 by volume.
The volumetric efficiency is 0.85. Determine
(*i*) the brake power
(*ii*) the mean piston speed in m/s
(*iii*) the brake mean effective pressure
(*iv*) the brake thermal efficiency. [129.68 kg, 4.68 m/s, 4.89 bar, 27.37%]

10.25 A four cylinder petrol engine has output of 52 kW at 200 rpm. During the Morse test being performed each cylinder is cut one by one. The brake load readings are 35.47 kg, 36.08 kg, 34.25 kg and 34.66 kg when one by one all cylinders are cut and the engine is brought back to original running speed. The brake drum radius may be taken as 50 cm. The specific fuel consumption at this speed is 364 gm/kW hr. The calorific value of fuel may be taken as 42 MJ/kg. Determine the mechanical efficiency, brake thermal efficiency, and indicated thermal efficiency of engine. [81.61%, 23.54%, 28.87%]

10.26 In the testing of two stroke engine the following observations are made.
Bore and stroke = 20 cm, 25 cm
Brake drum radius = 60 cm
Speed = 360 rpm
Brake load = 500 N
Fuel consumption = 0.06 kg/min
Calorific value of fuel = 42000 kJ/kg
Cooling water flow rate = 350 kg/hr
Cooling water temperature rise = 30°C
Area of indicator diagram = 2.8 cm^2
Length of indicator diagram = 10 cm
Spring constant = 100 kPa/min
Determine indicated power, brake power thermal efficiency. [13.19 kW, 11.31 kW, 26.93%]

10.27 A four stroke engine has single cylinder with bore of 18 cm and stroke of 34 cm. Engine has indicated mean effective pressure of 555 kPa at 390 rpm. At this the brake load is 510 N at brake radius of 56.5 cm. Cooling water is flown at the rate of 270 kg/hr with its inlet and exit temperatures being 18ºC and 58ºC. The fuel has calorific value of 43000 kJ/kg and the fuel is supplied at 3.6 kg/ hr. The exhaust gases carry 38% of fuel energy. Determine mechanical efficiency and brake thermal efficiency. Also prepare a heat balance sheet on per second basis. [75.54%, 27.36%]

10.28 Trial on a four stroke single cylinder engine results in following observations:

Bore = 16.5 cm, Stroke = 30.5 cm
Speed = 300 rpm
Brake load = 360 N
Brake diameter = 0.9 m
Gas consumption = 4.95 m^3/hr
Calorific value of gaseous fuel = 18850 kJ/m^3
Cooling water supply = 204 kg/hr
Cooling water temperature rise = 30ºC
Indicated mean effective pressure = 455 kPa

Determine

(*i*) the mechanical efficiency.
(*ii*) the indicated thermal efficiency
(*iii*) the brake thermal efficiency

Draw the energy balance on per second basis. [68.86%, 28.51%, 19.63%]

11

Introduction to Refrigeration and Airconditioning

11.1 INTRODUCTION

Refrigeration refers to the use of mechanical devices or heat activated devices for producing and maintaining the temperature in a region less than that of surroundings. American Society of Refrigerating Engineers defines refrigeration as "the science of providing and maintaining temperature below that of the surrounding atmosphere". Refrigerating a region in reference to the human comfort is called air conditioning. Thus, air conditioning refers to maintaining a space for the human comfort i.e. temperature, humidity and ventilation are the parameters to be controlled. When refrigeration provides temperature below – 150ºC then this is known as cryogenics. Equipment used for removing the heat continuously for maintaining low temperature in a space is called 'refrigerator' and cycles on which it operates are called 'refrigeration cycles'. The working fluids used for carrying away heat are called 'refrigerants' which are used in both refrigeration and air conditioning equipments. Equipment used for air conditioning are called airconditioners. The air conditioning of space may also be the requirement of various goods, equipments and processes etc. When air conditioning is done for human comfort then it is called comfort air conditioning while when it is done for industrial purposes then it is called 'industrial air conditioning'. Comfort air conditioning can be further categorized as summer air conditioning and winter air conditioning. Summer air conditioning is used in summer season and generally requires cooling and dehumidification. Winter air conditioning used in winter generally requires heating and humidification.

Comfort air conditioning is extensively used in residential buildings, hospitals, offices, working spaces, vehicles, trains, aeroplanes etc.

Industrial air conditioning is used for spot cooling/heating, environmental laboratories, printing industry, textile industry, precision parts manufacturing, photographic product handling, computer rooms, control rooms of power plants etc.

Refrigeration is extensively used for increasing the storage life of perishable items specially food products, vegetables, fruits, milk beverages, chilling of water, ice formation etc. along with industrial applications in chemical manufacturing, petroleum refinery, petrochemical plants, paper and plup industry etc.

History of refrigeration system dates back to 1834 when Jacob Perkins had got first patent on mechanical type refrigeration system using 'ether' as refrigerant. First successful commercial refrigeration system was produced in 1857 by James Harrison and D.E. Siebe using ethyl ether as refrigerant.

Refrigeration can be achieved by using refrigeration cycle which may be open cycle type or closed cycle type. In open cycle the refrigerant passes through the system once for refrigerating the space and is thrown away as it gets contaminated. In closed cycle the refrigerant flows inside tubes while refrigerating the space and does not get contaminated. Thus, in closed cycle a definite mass of refrigerant can be used for quite long time while it is not so in case of open cycle where for every run fresh refrigerant is to be picked up.

Different methods for refrigeration are described below:

(*i*) *Refrigeration by evaporation:* Refrigeration effect can be realized by evaporation of liquid. Evaporation of liquid requires latent heat of evaporation which is provided by the surroundings of region where liquid is being evaporated. This causes the cooling of the region supplying latent heat for evaporation. Examples of this type of refrigeration are:

- Cooling of water kept in earthen pot. Earthen pot has porosity thereby allowing evaporation of water and cooling down the inside remaining water.
- Cooling of water kept in desert bag (made of thick canvas i.e tight woven fabric). Here some quantity of water always keeps on leaking out across the cloth surface. This water which appears on the surface of bag gets evaporated after absorbing heat from atmospheric air and water inside the bag, thus cooling the water kept in bag.
- Sweating from the human skin is also example of such refrigeration which keeps the human being cool.
- Evaporative condensers also use this concept for cooling.

(*ii*) *Refrigeration by ice:* Ice is being popularly used for refrigeration. In this kind of refrigeration ice is kept in between the walls of an insulated cabinet. Apart from above there can be some other arrangements too such as ice may be kept in a tray upon the insulated box having flow of air across ice blocks to the refrigerated space.

Temperature inside refrigerated space is controlled by flow rate of air and the temperature upto 5°C can be attained like this.

(*iii*) *Refrigeration by expansion of air:* Generally, the expansion of air is accompanied by the cooling air due to reduction in its temperature. Subsequently cool air is passed into the region where refrigeration is to be done. Generally the adiabatic expansion of air is used in air refrigeration system described ahead.

(*iv*) *Refrigeration by throttling process:* Throttling of certain gases show the reduction in temperature of gas after throttling. Effect of throttling is evident by Joule-Thomson coefficient which shows whether the throttling is accompanied by cooling or heating of gases. Throttling is constant enthalpy process for which Joule-Thomson coefficient (μ) is given as:

$$\mu = \left(\frac{\partial p}{\partial T}\right)_h \text{; for } \mu > 0 \text{ the cooling is seen.}$$

(*v*) *Refrigeration by dry ice:* Dry ice is oftenly used in packaging of frozen foods for maintaining them at low temperature. Carbon dioxide in solid phase is termed as dry ice. It has typical property of getting transformed directly from solid phase to vapour phase at its sublimation temperature of –78°C at atmospheric pressure. Dry ice blocks in suitable sizes are stacked around the frozen food for keeping them in frozen state.

(*vi*) *Refrigeration using liquid gases:* Liquid gases such as liquid nitrogen, liquid carbon dioxide are used for refrigeration in refrigerated cargo vehicles. Liquid gases are expanded thereby causing its transformation from liquid to gas accompanied by lowering of temperature of refrigerated space.

Liquid gases are sprayed inside the cooling area and it gets transformed into vapour which is exhausted out from vehicle. Nitrogen and carbon dioxide being cheaply available are generally used as liquid refrigerant. It facilitates attainable temperature upto –30°C.

(*vii*) *Vapour Refrigeration system:* Vapour refrigeration system based on vapour compression refrigeration system and vapour absorption refrigeration system are discussed ahead.

11.2 PERFORMANCE PARAMETER

Refrigerator's performance is given by coefficient of performance (COP), COP for a refrigerator shown in block diagram, Fig. 11.1 can be given as under. COP is the ratio of refrigeration effect and net work done upon refrigerator. COP is dimensionless quantity so the two quantities should have same units.

$$\text{C.O.P. of refrigeration} = \frac{\text{Desired effect}}{\text{Net work done}}$$

Here the desired effect is the continuous removal of heat at the rate of Q_2 for maintaining body at lower temperature T_2. Net work done for getting this refrigeration effect is W.

$$\text{C.O.P.} = \frac{Q_2}{W} = \frac{Q_2}{(Q_1 - Q_2)}$$

COP of refrigerator may have any magnitude i.e. less than unity or greater than unity.

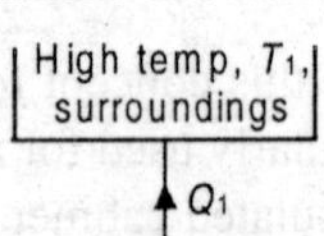

Fig. 11.1 Refrigerator

Heat pump is the device which runs exactly similar to the refrigerator but its purpose is to maintain the body at temperature higher than that of surroundings. Therefore, heat pump supplies heat at constant rate to the body to be maintained at higher temperature. Performance of heat pump is also given by the parameter defined exactly like COP but instead of calling it COP, it is called as energy performance ratio. Different name of EPR is assigned to differentiate it from COP of refrigerator.

Thus, COP of heat pump = Energy performance ratio (EPR)

$$= \frac{Q_1}{W}$$

$$\text{EPR} = \frac{Q_1}{Q_1 - Q_2}$$

or, $\boxed{\text{E.P.R.} = \text{COP} + 1}$

EPR shall always have its magnitude greater than unity.

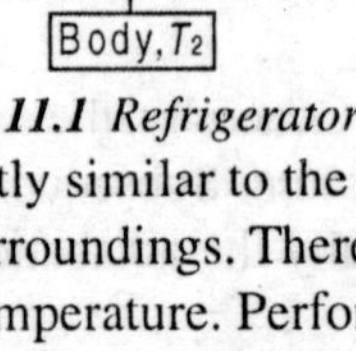

Fig. 11.2 Heat pump

11.3 UNIT OF REFRIGERATION

Refrigeration effect is the amount of heat extracted by refrigerator from the refrigerated space. The higher capacity of refrigerator means higher shall be the refrigeration effect. This refrigeration is defined by the unit of refrigeration called 'Ton' of refrigeration. 'Ton' as unit of refrigeration has been defined based on formation of ice as described ahead.

"One 'Ton' of refrigeration can be defined by the amount of heat being removed from one ton of water at 0ºC to form one ton of ice at 0ºC within 24 hours." Thus, a Ton of refrigeration shall quantify the latent heat required to be removed for solidification of water at 0ºC.

Latent heat of solidification from water to ice at 0ºC = 334.5 kJ/kg

1 Ton of refrigeration = mass of water × latent heat at 0ºC from water to ice (in SI units)

$$= \frac{(1000 \times 334.5)}{24} \text{ kJ/hr}$$

$$= \frac{334.5 \times 10^3}{24 \times 3600} = 3.87 \text{ kJ/sec}$$

Generally, 1 Ton is taken as 3.5 kJ/sec. This deviation from 3.87 to 3.5 kJ/sec is there because Ton was originally defined in fps units and the approximations during conversion yielded numerical value of 3.5 kJ/sec which is now universally accepted.

$$\text{1 Ton of refrigeration (in MKS units)} = \frac{1000 \times 80}{24 \times 60} = 55.4 \text{ kcal/min}$$

$$= 50 \text{ kcal/min}$$

In MKS units this is commonly taken as 50 kcal/min.

Therefore,

$$1 \text{ Ton} = 3.5 \text{ kJ/sec}$$

and

$$1 \text{ Ton} = 50 \text{ kcal/min}$$

11.4 CARNOT REFRIGERATION CYCLE

Carnot cycle being the most efficient and ideal cycle can also be used for getting the refrigeration effect upon its reversal. Reversed carnot cycle is shown in Fig. 11.3

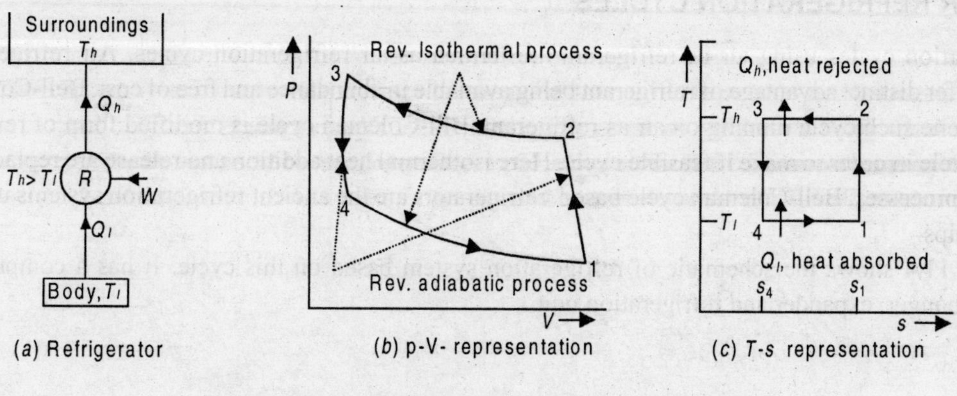

(a) Refrigerator (b) p-V- representation (c) T-s representation

Fig. 11.3 Reversed Carnot cycle for refrigeration

Here, the refrigerated space/body is to be maintained at low temperature T_l for which heat Q_l should be removed at constant rate and rejected to surroundings at high temperature, T_h. Amount of heat rejected to surroundings is Q_h while the net work done upon refrigerator is W.

Different processes in reversed Carnot cycle are,

1–2 Reversible adiabatic compression

2–3 Reversible isothermal heat rejection, Q_h at temperature, T_h

3–4 Reversible adiabatic expansion

4–1 Reversible isothermal heat absorption, Q_l at temperature, T_l

Refrigeration effect is seen during process 4–1.

For cyclic process; Net work $W = Q_h - Q_l$

$$W = T_h (S_1 - S_4) - T_l (S_1 - S_4)$$

$$W = (S_1 - S_4)(T_h - T_l)$$

For reversed Carnot cycle

$$\text{COP} = \frac{Q_l}{W} = \frac{T_l(S_1 - S_4)}{(T_h - T_l)(S_1 - S_4)} = \left\{\frac{T_l}{T_h - T_l}\right\}$$

The coefficient of performance of refrigerator depends upon the two temperature values i.e. low temperature T_l and high temperature T_h. For COP value to be high the low temperature T_l should be high while higher temperature T_h should be small. With increasing low temperature value the COP goes on increasing, while with decreasing high temperature T_h value the COP goes on increasing. Practically, the lower temperature requirement is decided by the user while higher temperature is generally fixed by the atmospheric temperature value. Thus, it could be said that for certain low temperature to be maintained, COP of refrigerator shall be more during cold days as compared to hot days.

$$\text{COP}_{\text{cold days}} > \text{COP}_{\text{hot days}} \text{ because } T_{h,\text{ cold days}} < T_{h,\text{ hot days}}$$

Similar to the carnot power cycles, reversed carnot cycle for refrigeration is not feasible in practice. Therefore, number of modified cycles are practically used for refrigeration as discussed ahead. But the reversed carnot cycle provides the basis for comparison and provides the benchmark for achieving maximum COP.

11.5 AIR REFRIGERATION CYCLES

Refrigeration cycles using air as refrigerant are termed as air refrigeration cycles. Air refrigeration cycles offer distinct advantage of refrigerant being available in abundance and free of cost. Bell-Coleman cycle is one such cycle running on air as refrigerant. Bill-Coleman cycle is modified form of reversed carnot cycle in order to make it feasible cycle. Here isothermal heat addition and release are replaced by isobaric processes, Bell-Coleman cycle based refrigerators are the ancient refrigeration systems used in cargo ships.

Fig. 11.4 shows the schematic of refrigeration system based on this cycle. It has a compressor, heat exchanger, expander and refrigeration unit.

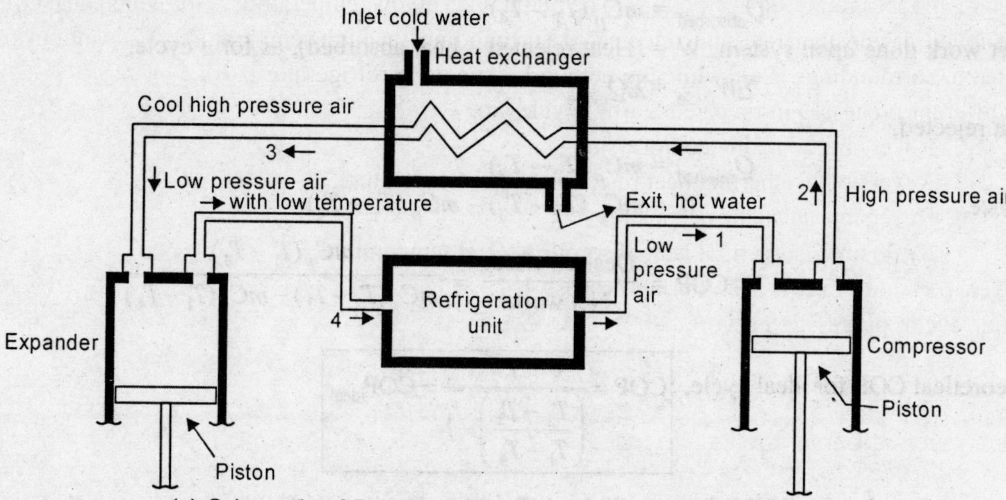

(a) Schematic of Bell Coleman cycle based refrigeration system

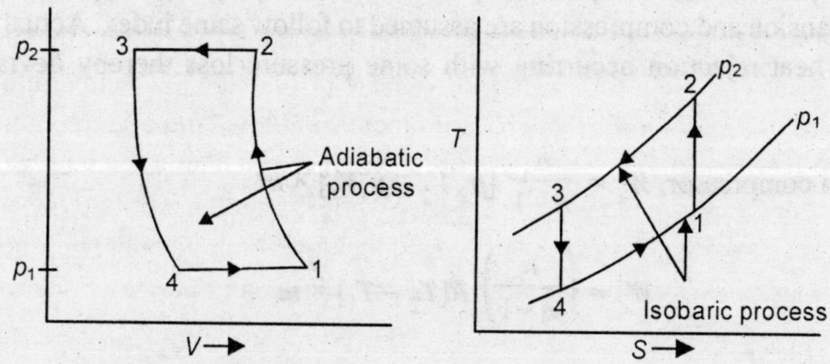

(b) *p-V* diagram (c) *T-s* diagram

Fig. 11.4 Bell Coleman cycle

Bell-Coleman cycle is actually the reversed Brayton cycle.

Different processes in Bell-Coleman cycle are as under.

1-2: Adiabatic compression of air causing rise in its pressure and temperature inside compressor. Ideally, this compression process is considered isentropic whereas actually it may not be so.

2-3: Isobaric heat rejection causing cooling of high pressure and high temperature air coming from compressor making it low temperature and high pressure air inside heat exchanger. Water may be used as cooling fluid inside heat exchanger.

3-4: Adiabatic expansion inside expander causing further cooling of high pressure low temperature air entering expander. This very cool air enters the refrigeration unit subsequently. Ideally expansion can be considered to be of isentropic type whereas actually if may not be isentropic.

4-1: Low temperature cool air enters refrigeration space and picks up heat from space thereby showing refrigeration effect. Subsequently low temperature air gets heated up and leaves the space after showing desired refrigeration to enter in to compressor. Thus, process 4–1 is the actual process which is giving the desired refrigeration effect.

Thermodynamically, the refrigeration effect, net work supplied and COP estimated are as under.

Heat picked up from refrigerated space:

$$Q_{\text{absorbed}} = mC_p\,(T_1 - T_4)$$

Net work done upon system, W = (Heat rejected – heat absorbed), as for a cycle.

$$\Sigma W_{\text{cycle}} = \Sigma Q_{\text{cycle}}$$

heat rejected,

$$Q_{\text{rejected}} = mC_p\,(T_2 - T_3)$$

Hence,

$$W = mC_p\,(T_2 - T_3) - mC_p\,(T_1 - T_4)$$

$$\text{COP} = \frac{\text{Desired effect}}{\text{Net work}} = \frac{mC_p(T_1 - T_4)}{mC_p(T_2 - T_3) - mC_p(T_1 - T_4)}$$

Theoretical COP for ideal cycle,

$$\boxed{\text{COP} = \frac{1}{\left(\dfrac{T_2 - T_3}{T_1 - T_4}\right) - 1} = \text{COP}_{\text{ideal}}}$$

Above expression for COP is correct till the different processes are of isobaric or isentropic type. In actual practice the expansion and compression processes are of polytropic type and given by pV^n = constant i.e. both expansion and compression are assumed to follow same index. Actual cycle shall have heat absorption and heat rejection occurring with some pressure loss thereby deviating from ideal isobaric process.

Work required in compressor,

$$W_c = \frac{n}{n-1}\,[p_2 V_2 - p_1 V_1] \times m$$

$$W_c = \left(\frac{n}{n-1}\right) R[T_2 - T_1] \times m$$

Work available from expander,

$$W_e = \frac{n}{n-1}\,[p_2 V_3 - p_1 V_4] \times m$$

$$W_c = \left(\frac{n}{n-1}\right) R[T_3 - T_4] \times m$$

Net work,

$$W_{\text{actual}} = W_c - W_e$$

$$W_{\text{actual}} = \left(\frac{n}{n-1}\right) \cdot m \cdot R\,[(T_2 - T_1) - (T_3 - T_4)]$$

as

$$C_p = \frac{\gamma R}{(\gamma - 1)}, \text{ so}$$

$$W_{\text{actual}} = \left(\frac{n}{n-1}\right) \cdot m \cdot C_p \cdot \left(\frac{\gamma - 1}{\gamma}\right) [(T_2 - T_1) - (T_3 - T_4)]$$

Therefore,

$$\text{COP}_{\text{actual}} = \frac{mC_p(T_1 - T_4)}{mC_p\left(\dfrac{n}{n-1}\right)\left(\dfrac{\gamma - 1}{\gamma}\right)[(T_2 - T_1) - (T_3 - T_4)]}$$

$$COP_{actual} = \frac{1}{\left(\frac{n}{n-1}\right)\left(\frac{\gamma-1}{\gamma}\right)\left[\left(\frac{T_2-T_3}{T_1-T_4}\right)-1\right]}$$

For isentropic process, $\gamma = n$ and $COP_{ideal} = COP_{actual}$

Using isentropic considerations, $COP_{ideal} = \frac{1}{\left[\left(\frac{T_2-T_3}{T_1-T_4}\right)-1\right]}$

$$COP_{ideal} = \frac{1}{\left[\left(\frac{T_2}{T_1}\right)\left(\frac{1-\frac{T_3}{T_2}}{1-\frac{T_4}{T_1}}\right)-1\right]}$$

For isentropic process between states 1 and 2, 3 and 4.

$$\left(\frac{p_2}{p_1}\right)^{\frac{\gamma-1}{\gamma}} = \frac{T_2}{T_1} \text{ and } \left(\frac{p_2}{p_1}\right)^{\frac{\gamma-1}{\gamma}} = \left(\frac{T_3}{T_4}\right)$$

Upon equating we get,

$$\frac{T_2}{T_1} = \frac{T_3}{T_4}$$

or,

$$\frac{T_4}{T_1} = \frac{T_3}{T_2} \text{ or } \frac{T_1-T_4}{T_2-T_3} = \frac{T_1}{T_2}$$

Therefore, upon substitution,

$$COP_{ideal} = \frac{1}{\left(\frac{T_2}{T_1}-1\right)}$$

$$\boxed{COP_{ideal} = \left(\frac{T_1}{T_2-T_1}\right)}$$

$$\boxed{COP_{actual} = \frac{T_1}{\left(\frac{n}{n-1}\right)\left(\frac{\gamma-1}{\gamma}\right)(T_2-T_1)}}$$

Comparing the COP of ideal Bell Coleman cycle with reversed Carnot cycle COP we find that COP of Bell Coleman cycle will be less than that of COP of reversed Carnot cycle.

$$COP_{Bell\ Coleman} < COP_{Carnot}$$

Carnot cycle operating between the temperature limits of Bell Coleman cycle shall have COP given

by, $COP_{carnot} = \frac{T_1}{T_3 - T_1}$

Since $T_2 > T_3$ so, $COP_{Bell\ Coleman} < COP_{Carnot}$.

Constant pressure heat transfer results in operating range which is greater than minimum range given by temperature of those parts of the surroundings with which heat is exchanged.

Air refrigeration systems offer advantages as these run on air as refrigerant which is cheap, non-inflammable and non toxic. Also, air refrigeration systems and light weight per ton of refrigeration compared to other refrigeration systems and is used in aircraft applications. Leakage of air refrigerant may not be dangerous problem.

Air refrigeration systems are disadvantageous due to lower COP compared to other refrigeration systems. There is significant possibility of freezing of moisture accompanied by atmospheric air due to low temperature in refrigeration space. Air may also pick up moisture while passing through refrigeration space. Quantity of air refrigerant required is more compared to other refrigerants for similar refrigeration capacity. This is because of fact that air only carries heat in the form of sensible heat and so its large quantity is required compared to other refrigerants.

11.6 VAPOUR COMPRESSION CYCLES

Vapour compression cycle based refrigeration systems are extensively used in refrigeration systems. These cycles are used for most of small domestic and large industrial applications.

The vapour compression cycle has the refrigerant being circulated in closed circuit through compressor, condenser, throttle valve or expansion valve and evaporator as shown in Fig. 11.5. Refrigerant (in gas/vapour) is compressed isentropically in compressor from state 1 to 2. High pressure and high temperature refrigerant enters the condenser at state 2 where its condensation occurs and refrigerant is available in liquid form at state 3. Refrigerant in the form of saturated liquid at high pressure is passed through expansion valve where isenthalpic expansion occurs. Refrigerant leaving expansion valve at state 4 is in the form of low pressure wet mixture of liquid and vapour. Low pressure liquid-vapour mixture is passed through evaporator section in which refrigerant picks up heat from surroundings thereby showing refrigeration effect. As a result of this heat absorption the liquid-vapour mixture refrigerant gets transformed into dry gaseous refrigerant in case of *dry compression* as shown in Fig. 11.5 (*b*). Compression of dry and saturated or superheated refrigerant is called dry compression.

Compression of dry refrigerant yields superheated state of refrigerant as shown by state 2. (Fig. 11.5 (*b*))

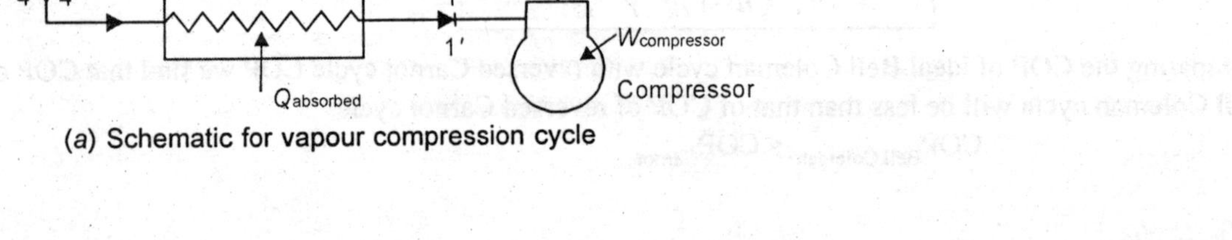

(*a*) Schematic for vapour compression cycle

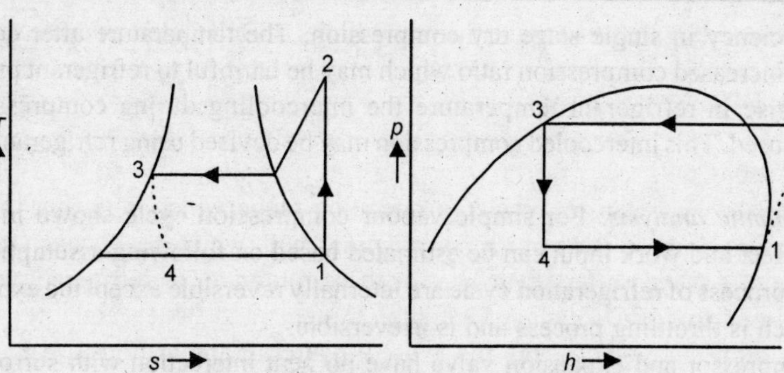

(b) *T-s* and *p-h* diagram for dry compression

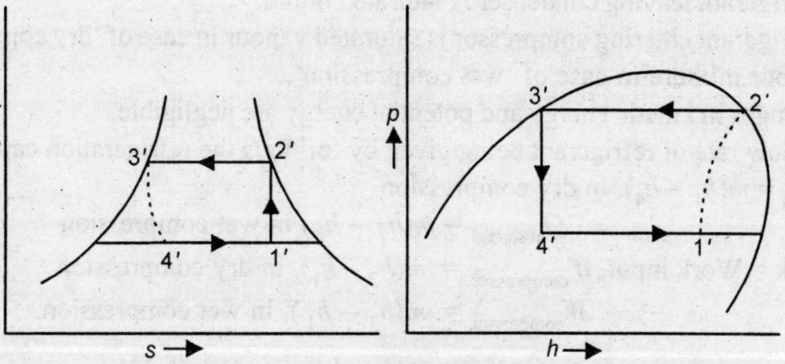

(c) *T-s* and *p-h* diagram for wet compression

Fig. 11.5 Vapour Compression Cycle

Wet compression: It is also possible that the refrigerant is in wet state i.e. liquid-vapour mixture at inlet of compressor, state 1'. Compression of wet mixture gets transformed into dry refrigerant (gaseous form) as shown by state 2'. Dry refrigerant at high pressure and high temperature is passed through condenser where refrigerant gets condensed and condensate is available in saturated liquid form at high pressure. Subsequently refrigerant is throttled from high pressure to low pressure inside expansion valve from state 3' to 4'. Low pressure refrigerant in wet state is passed through evaporator from state 4' to 1' where it picks up heat from surroundings and some of its liquid fraction gets transformed into vapour but it does not become dry (gas) refrigerant at inlet to compressor. This wet refrigerant is compressed inside compressor.

In case of dry compression the compressor efficiency is found to be more than that of wet compression due to higher volumetric efficiency with dry refrigerant and also chances of damage by liquid refrigerant are absent. During dry compression the temperature of compressed refrigerant becomes more than condensation temperature. Due to this high temperature after compression the compressor becomes very hot therefore requiring cooling. Coolant used in condenser may also be used for cooling compressor as its temperature is lower than compressor temperature. This cooling of compressor reduces compression work.

T-s and *p-h* diagram for vapour compression cycle shows that the refrigeration capacity can be increased by subcooling the condensate before it enters the expansion valve and also by increasing the degree of expansion in expansion valve.

Here in second option the condensation temperature is fixed and evaporator temperature is lowered which increases expansion ratio and compression ratio. This increase in compression ratio shows poor volumetric efficiency in single stage dry compression. The temperature after compression also gets increased with increased compression ratio which may be harmful to refrigerant properties. To regulate the excessive rise in refrigerant temperature the intercooling during compression with multistage compression is used. This intercooled compression may be devised using refrigerant itself as intercooling medium.

Thermodynamic analysis: For simple vapour compression cycle shown in Fig. 11.6 the COP, refrigeration effect and work input can be estimated based on following assumptions:

(*i*) All process of refrigeration cycle are internally reversible except the expansion through valve which is throttling process and is irreversible.

(*ii*) Compressor and expansion valve have no heat interaction with surroundings during their operation i.e. they operate adiabatically.

(*iii*) Refrigerant leaving condenser is saturated liquid.

(*iv*) Refrigerant entering compressor is saturated vapour in case of 'dry compression' and liquid-vapour mixture in case of 'wet compression'.

(*v*) Changes in kinetic energy and potential energy are negligible.

For mass flow rate of refrigerant being given by 'm' kg/s the refrigeration capacity or refrigeration effect. $Q_{\text{absorbed}} = m(h_1 - h_4)$, in dry compression

$$Q_{\text{absorbed}} = m(h_{1'} - h_{4'}) \text{ in wet compression}$$

$$\text{Net work} = \text{Work input, } W_{\text{compressor}} = m(h_2 - h_1), \text{ in dry compression}$$

$$W_{\text{compressor}} = m(h_{2'} - h_{1'}), \text{ in wet compression.}$$

Therefore,

$$\text{COP} = \frac{m(h_1 - h_4)}{m(h_2 - h_1)} = \left\{\frac{(h_1 - h_4)}{(h_2 - h_1)}\right\}, \text{ in dry compression}$$

$$\text{COP} = \frac{m(h_{1'} - h_{4'})}{m(h_{2'} - h_{1'})} = \left\{\frac{(h_{1'} - h_{4'})}{(h_{2'} - h_{1'})}\right\}, \text{ in wet compression.}$$

Some calculations in vapour compression cycle: Let us take three different cases of vapour compression cycle similar to the dry and wet compression.

Case 1: Compression yields dry and saturated vapour at the end of compression.

Vapour compression cycle in which the compression process produces vapour dry and saturated, is shown on *T-s* diagram in Fig. 11.6.

Different processes are,

1–2 : Isentropic compression

2–3 : Heat rejection in condenser

3–4 : Isenthalpic expansion in expansion valve (throttling)

4–1 : Heat absorption in evaporator or refrigerated space.

Work done in the cycle can be given by the area enclosed 1–2–3–5–1 on T-s diagram. The work required for compression process.

The refrigeration effect in evaporator can be given by area enclosed 1–4–9–10–1 on *T–s* diagram.

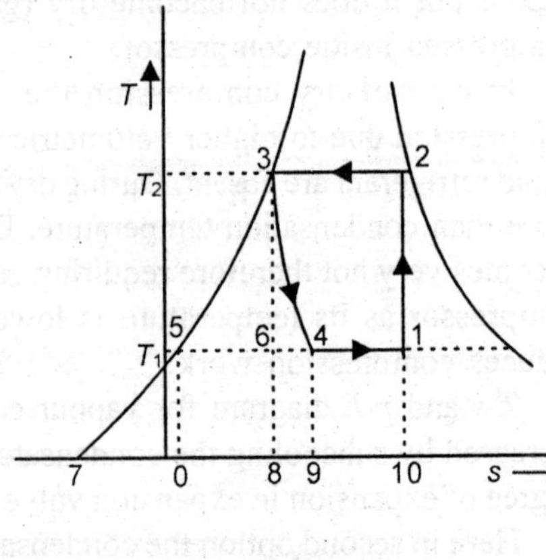

Fig. 11.6

Compressor work, W = Area 1 – 2 – 3 – 5 – 1

= (Area 2 – 3 – 5 – 7 – 10 – 2) – (Area 1 – 5 – 7 – 10 – 1)

= (Enthalpy at 2) – (Enthalpy at 1)

$$W = h_2 - h_1$$

Here, $h_2 = h_{g \text{ at } T_2}$

Refrigeration effect = Area 1 – 4 – 9 – 10 – 1 = (Area 1 – 5 – 7 – 10 – 1) – (Area 4 – 5 – 7 – 9 – 4)

= (Enthalpy at 1) – (Enthalpy at 4)

Refrigeration effect $= h_1 - h_4$

Enthalpy at 3 = Enthalpy at 4

For throttling process 3 – 4, $h_3 = h_4$

(Area 3 – 7 – 8 – 3) = (Area 4 – 5 – 7 – 9 – 4)

or,

(Area 3 – 5 – 6 – 3) = (Area 6 – 5 – 7 – 8 – 6)

= (Area 6 – 5 – 7 – 8 – 6) – (Area 4 – 6 – 8 – 9 – 4)

or,

(Area 3 – 5 – 6 – 3) = (Area 4 – 6 – 8 – 9 – 4)

(Area 3 – 5 – 7 – 8 – 3) – (Area 5 – 7 – 0 – 5) – (Area 6 – 5 – 0 – 8 – 6) = (Area 4 – 6 – 8 – 9 – 4)

(Enthalpy at 3) – (Enthalpy at 5) – $(T_1 \times \Delta s_{56}) = (T_1 \times \Delta s_{46})$

$\Rightarrow$ (Area 3 – 5 – 6 – 3) = $\{(h_{f \text{ at } T_2} - h_{f \text{ at } T_1}) - T_1 \times (s_{f \text{ at } T_2} - s_{f \text{ at } T_1})\}$

Here enthalpy at 3 and enthalpy at 5 are enthalpies for saturated liquid states at temperature T_2 and T_1 respectively.

Enthalpy at 3 = $h_{f \text{ at } T_2} = h_3$

Enthalpy at 5 = $h_{f \text{ at } T_1} = h_s$

State 6 lies vertically below 3 so, $s_3 = s_6$ and,

Entropy at 5 = $s_{f \text{ at } T_1} = s_5$

Entropy at 3 = $s_{f \text{ at } T_2} = s_3 = s_6$

$$\Delta s_{56} = s_6 - s_5 = (s_3 - s_5) = (s_{f \text{ at } T_2} - s_{f \text{ at } T_1})$$

So, for known h_3, h_5, Δs_{56} the entropy change Δs_{46} can be obtained as under;

$$\Delta s_{46} = \frac{1}{T_1}(h_{f \text{ at } T_2} - h_{f \text{ at } T_1}) - \Delta s_{56}$$

Therefore, for getting work,

W = Area 1 – 2 – 3 – 5 –1

= (Area 3 – 5 – 6 – 3) + Area (2 – 3 – 6 – 1 – 2)

$= \{(h_{f \text{ at } T_2} - h_{f \text{ at } T_1}) - T_1 \times (s_{f \text{ at } T_2} - s_{f \text{ at } T_1})\} + \{(T_2 - T_1) \times \Delta s_{23}\}$

$$W = \{(h_{f \text{ at } T_2} - h_{f \text{ at } T_1}) - T_1 \times (s_{f \text{ at } T_2} - s_{f \text{ at } T_1})\} + \{(T_2 - T_1) \times (s_2 - s_3)\}$$

$$W = \{(h_{f \text{ at } T_2} - h_{f \text{ at } T_1}) - T_1 \times (s_{f \text{ at } T_2} - s_{f \text{ at } T_1})\} + \{(T_2 - T_1) \times (s_{g \text{ at } T_2} - s_{f \text{ at } T_2})\}$$

The refrigeration effect, = (Area 1 – 4 – 9 – 10 – 1)

$= T_1 \times \Delta s_{14} = T_1 \times (\Delta s_{16} - \Delta s_{46})$

here, $\Delta s_{16} = s_1 - s_6 = s_{g \text{ at } T_2} - s_{f \text{ at } T_2}$

Substituting Δs_{16} and Δs_{46} values in expression for refrigeration effect.

$$\text{Refrigeration effect} = T_1 \left\{(s_{g \text{ at } T_2} - s_{f \text{ at } T_2}) - \left(\frac{1}{T_1}(h_{f \text{ at } T_2} - h_{f \text{ at } T_1}) - \Delta s_{56}\right)\right\}$$

$$= T_1 \left\{(s_{g \text{ at } T_2} - s_{f \text{ at } T_2}) - \frac{1}{T_1}(h_{f \text{ at } T_2} - s_{f \text{ at } T_1}) + (s_{f \text{ at } T_2} - s_{f \text{ at } T_1})\right\}$$

$$= T_1 \left\{(s_{g \text{ at } T_2} - s_{f \text{ at } T_1}) - \frac{1}{T_1}(h_{f \text{ at } T_2} - h_{f \text{ at } T_1})\right\}$$

$$\text{Here, COP of cycle} = \frac{\text{Refrigeration effect}}{\text{Work done}}$$

$$\text{COP} = \frac{T_1(s_{g \text{ at } T_2} - s_{f \text{ at } T_1}) - (h_{f \text{ at } T_2} - h_{f \text{ at } T_1})(1/T_1)}{\left\{(h_{f \text{ at } T_2} - h_{f \text{ at } T_1}) - T_1(s_{f \text{ at } T_2} - s_{f \text{ at } T_1}) + (T_2 - T_1)\cdot(s_{g \text{ at } T_2} - s_{f \text{ at } T_2})\right\}}$$

Case 2 : Compression yields wet vapour at the end of compression

Fig. 11.7 gives *T-s* representation of the vapour compression cycle having compression process yielding wet vapour after compression i.e. at state 2. Let wetness after compression be x_2.

Work done = Area 1 – 2 – 3 – 5 – 1
= (Area 3 – 5 – 6 – 3) + (Area 2 – 3 – 6 – 1 –2)

From earlier analysis,

Area 3 – 5 – 6 – 3 = Area (4 – 6 – 8 – 9 – 4)

$$\text{Area } 3-5-6-3 = \{(h_{f \text{ at } T_2} - h_{f \text{ at } T_1}) - T_1(s_{f \text{ at } T_2} - s_{f \text{ at } T_1})\}$$

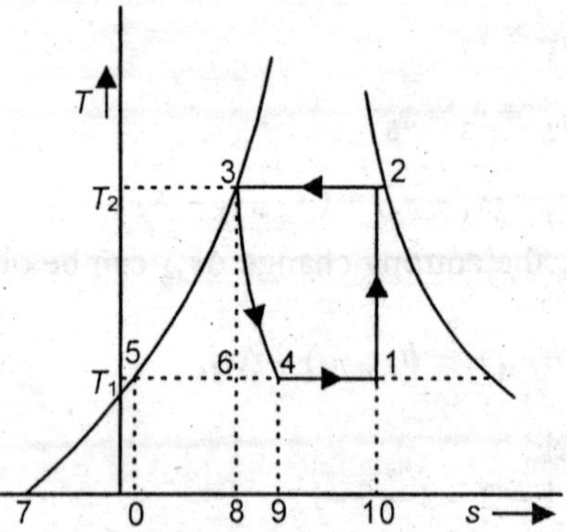

Fig. 11. 7

Work = {Area 3 – 5 – 6 – 3} + {Area 2 – 3 – 6 – 1 – 2}

$$\text{Work} = \{(h_{f \text{ at } T_2} - h_{f \text{ at } T_1}) - T_1(s_{f \text{ at } T_2} - s_{f \text{ at } T_1})\} + \left\{(T_2 - T_1) \times \frac{x_2 \cdot h_{fg \text{ at } T_2}}{T_2}\right\}$$

Refrigeration effect = Area 1 – 4 – 9 – 10 – 1

$$= \{\text{Area } 1-6-8-10-1\} - \{\text{Area } 4-6-8-9-4\}$$

$$= \{\text{Area } 1-6-8-10-1\} - \{\text{Area } 3-5-6-3\}$$

$$= \{T_1 \cdot (x_2 \cdot s_{fg \text{ at } T_2})\} - \{(h_{f \text{ at } T_2} - h_{f \text{ at } T_1}) - T_1\,(s_{f \text{ at } T_2} - s_{f \text{ at } T_1})\}$$

$$\text{Refrigeration effect} = \{T_1.((x_2 . s_{fg \text{ at } T_2} + s_{f \text{ at } T_2}) - s_{f \text{ at } T_1})\} - \{h_{f \text{ at } T_2} - h_{f \text{ at } T_1}\}$$

COP may be obtained by taking the ratio of refrigeration effect and work.

Case 3: *Compression yields superheated vapour after compression*

Vapour compression cycle having the compression process yielding superheated vapour at the end of compression is shown on *T-s* diagram in Fig. 11.8.

Cycle is shown as 1–2–3–4–1.

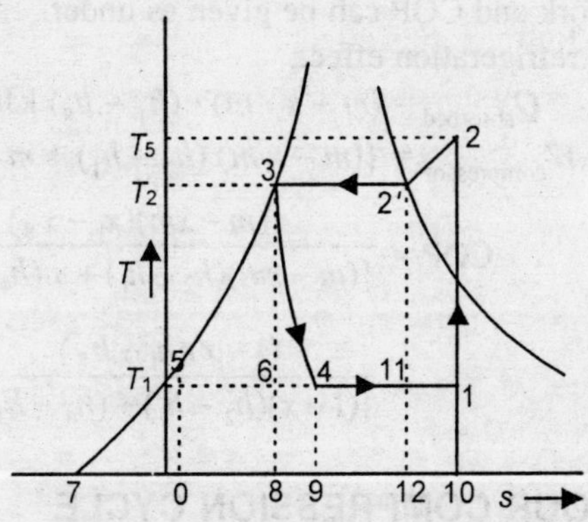

Fig. 11.8

In this cycle additional work shown by area 2 – 2′ – 11 – 1 – 2 is done and also the additional refrigeration effect shown by area 1 – 11 – 12 – 10 – 1 is available.

Additional work W' = Area 2 – 2′ – 11 – 1 – 2

= (Area 2 – 2′ – 12 – 10 – 2) – (Area 1 – 11 – 12 – 10 – 1)

$$W' = C_p\,(T_5 - T_2) - \left(T_1 C_p \cdot \ln\left(\frac{T_5}{T_2}\right)\right)$$

Additional refrigeration effect = (Area 1 – 11 – 12 – 10 – 1)

$$= T_1,\, C_p \ln\left(\frac{T_5}{T_1}\right)$$

Therefore, COP of system could be obtained by ratio of total refrigeration effect and total work done.

$$\text{COP} = \frac{\text{Refrigeration effect } (= \text{Area } 1-4-9-10-1)}{\text{Work done } (= \text{Area } 2-2'-3-5-1-2)}$$

This ratio of mass fraction of gas (vapour) to liquid i.e. $\left(\frac{m_9}{m_7}\right)$ can be estimated by energy balance on flash chamber. State 3 can be determined by the energy balance on mixing box.

Thermodynamic analysis

Thermodynamic analysis of multistage vapour compression cycle can also be carried out for the dry compression considered here based on the assumptions as under

(*i*) All processes of refrigeration cycle are internally reversible except the processes in expansion valve, flash chamber and mixing box.

(*ii*) The processes in expansion valve, flash chamber, mixing box and compressors operate adiabatically.

(*iii*) Refrigerant leaving condenser is saturated liquid and the refrigerant entering low pressure compressor is saturated vapour.

(*iv*) There are no changes in kinetic and potential energies in different processes.

For mass flow rate of m kg/s and dryness fraction of refrigerant at inlet of flash chamber being x, the refrigeration effect, net work and COP can be given as under.

Refrigeration capacity or refrigeration effect,

$$Q_{\text{absorbed}} = (m - x \cdot m) \cdot (h_1 - h_8) \text{ kJ/s}$$

Net work or Work input, $W_{\text{compressor}} = \{(m - x{\cdot}m)(h_2 - h_1) + m(h_4 - h_3)\}$kJ/s

$$\text{COP} = \frac{(m - xm)(x_1 - x_8)}{\{(m - xm)(h_2 - h_1) + m(h_4 - h_3)\}}$$

$$= \frac{(1 - x)(h_1 - h_8)}{\{(1 - x)(h_2 - h_1) + (h_4 - h_3)\}}$$

11.7 MULTISTAGE VAPOUR COMPRESSION CYCLE

Intercooled compression using flash chamber is shown in Fig. 11.9. Here compression for total compression ratio occurs in two stages i.e. LP compressor and HP compressor.

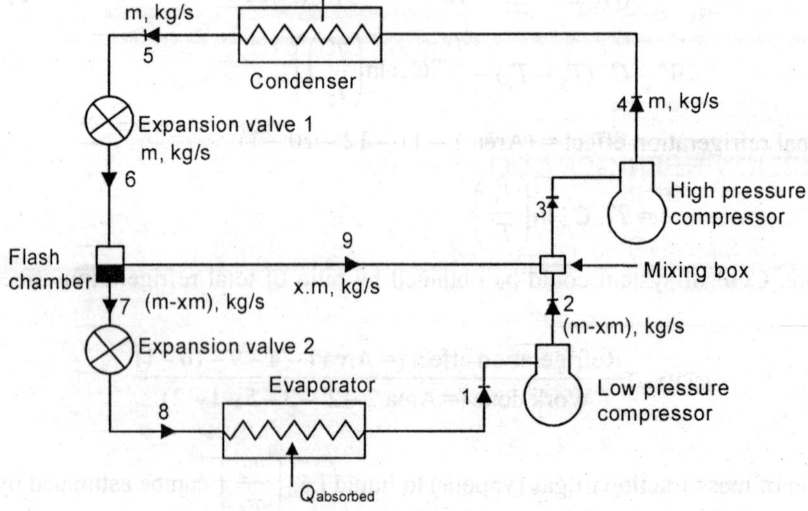

(*a*) Two stage intercooled vapour compression cycle with flash chamber

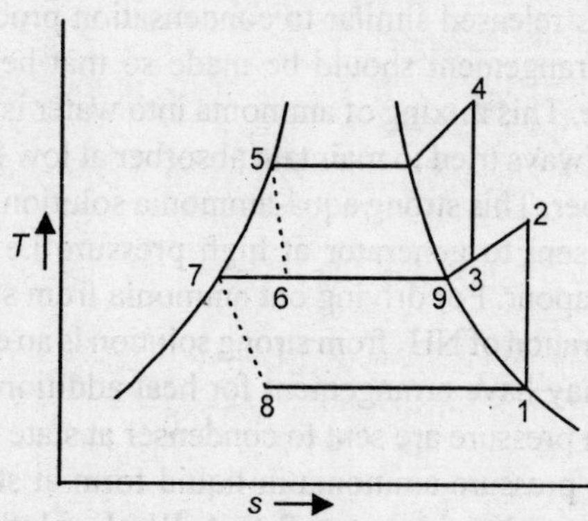

1–2, 3–4 Reversible adiabatic compression
5–6, 7–8 Irreversible throttling process
6–7, 6–9 In flash chamber
4–5 Isobaric condensation
9–3–2 Mixing of refrigerant from LP compressor with cool vapour from flash chamber.

(*b*) T-s representation of two stage intercooled vapour compression cycle

Fig. 11.9 Multistage vapour compression cycle

Refrigerant is partly compressed in LP stage from 1 to 2. Compressed refrigerant leaving compressor is cooled by mixing it with the cool vapour leaving flash chamber at stage 9 and finally yielding low pressure refrigerant at low temperature, stage 3 (Temperature at 3 < Temperature at 2). This reduced temperature of refrigerant from 2 to 3 results in smaller compressor work requirement. Subsequently, refrigerant is compressed in high pressure compressor upto state 4. High pressure refrigerant is passed through condenser where it gets condensed into saturated liquid form of refrigerant and passed through expansion valve (expansion valve 1) from stage 5 to state 6. Expansion is carried out till the refrigerant's pressure gets reduced to interstage pressure and sent to flash chamber. Liquid vapour mixture entering flash chamber is separated such that liquid fraction of refrigerant passes into next expansion valve (expansion valve 2) at state 7 and the gaseous (vapour) fraction of refrigerant leaves flash chamber at state 9 for intercooling refrigerant in mixing box between state 2 and 3. Fraction of liquid refrigerant entering expansion valve is expanded from state 7 to 8. As a result of expansion the refrigerant's pressure gets reduced and low pressure liquid refrigerant is passed through evaporator between state 8 to 1 and its phase transformation occurs from liquid to vapour. Inside evaporator this phase transformation from liquid to vapour (gas) occurs by absorbing latent heat of vaporization from the refrigerated space thereby showing cooling affect.

11.8 ABSORPTION REFRIGERATION CYCLE

In vapour compression cycle it is seen that the power input is made for compressing refrigerant from low pressure to high pressure. For constant refrigerating capacity the COP of refrigeration system can be improved by reducing the work input to cycle. Here in order to reduce work input to cycle the compressor is being replaced by pump if the refrigerant is in liquid form instead of being in gas (vapour) form. Since the average specific volume of liquid solution is much less than that of refrigerant in vapour form so considerably less amount of work is required for increasing its pressure in pump.

The absorption refrigeration cycle is based on this premise of employing pump for raising pressure of refrigerant in order to take advantage of less work requirement. For realizing it entirely different arrangement is employed as shown in Fig. 11.10. Such arrangement is called absorption refrigeration system. Here aqua-ammonia absorption refrigeration system has been discussed. Ammonia works as refrigerant and water acts as carrier for refrigerant in this cycle. Ammonia vapour leaving evaporator enter the absorber where ammonia vapours are released through perforated pipe into water for getting absorbed in water. This dissolving of NH_3 in water result in strong aqua-ammonia solution. Mixing of NH_3 in water is an exothermic process where heat is released similar to condensation process of pure substance. Due to heat release in absorber some arrangement should be made so that heat could be extracted out for maintaining its constant temperature. This mixing of ammonia into water is more when temperature of water mixing is low. Therefore, it is always tried to maintain absorber at low temperature and for this cooling water may be circulated in absorber. This strong aqua-ammonia solution prepared in absorber is pumped employing a pump for being sent to generator at high pressure i.e. condenser pressure. Generator is used for separating out NH_3 vapour. For driving out ammonia from strong-aqua-ammonia solution heat is added to generator. The separation of NH_3 from strong solution is an endothermic process similar to evaporation process. Generator may have arrangement for heat addition employing burners. Ammonia vapours leaving generator at high pressure are sent to condenser at state 2 where the condensation of ammonia takes place yielding high pressure ammonia in liquid form at state 3. High pressure ammonia liquid is passed through expansion valve from state 3 to 4. Weak solution of aqua-ammonia left inside generator is sent to absorber at state *d*. Thus, it is seen that the combination of absorber, pump and generator helps in getting high pressure refrigerant (ammonia) without using compressor.

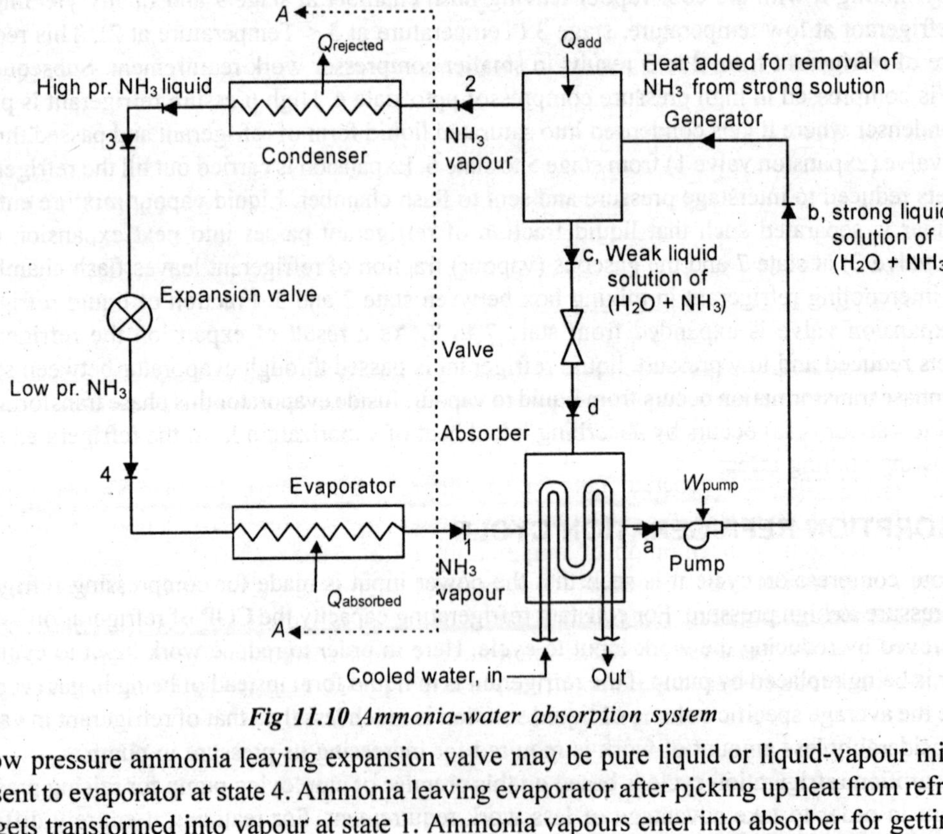

Fig 11.10 Ammonia-water absorption system

Low pressure ammonia leaving expansion valve may be pure liquid or liquid-vapour mixture for being sent to evaporator at state 4. Ammonia leaving evaporator after picking up heat from refrigerating space gets transformed into vapour at state 1. Ammonia vapours enter into absorber for getting mixed with weak aqua-ammonia solution to result in strong aqua-ammonia solution at state *a*. It is obvious from the arrangement that work input to absorption cycle gets considerably reduced but the cost incurred in replacing compressor by absorber, pump and generator and heat source etc. is quite large and nullifies the advantage of small work input. Here on absorption system the arrangement of components between stages 2, 3, 4 and 1 is similar to that of vapour compression system i.e. condenser followed by expansion valve and evaporator as shown on left of section A–A on schematic of ammonia water absorption system (Fig. 11.10).

Performance of aqua-ammonia absorption depends largely upon the characteristics of binary mixtures i.e. water and ammonia in present case. The generic phase equilibrium diagram for ammonia-water mixture is shown in Fig. 11.11.

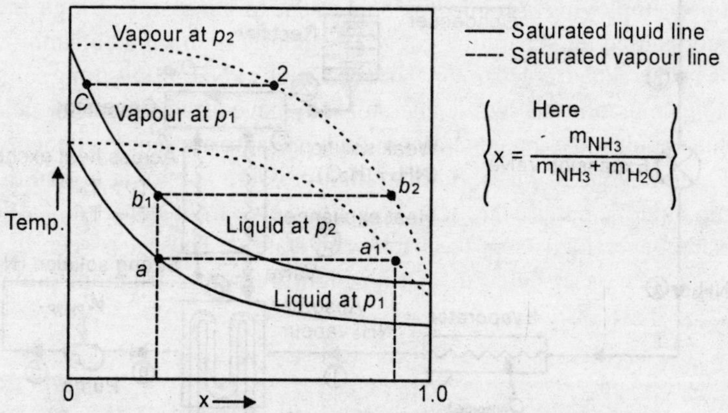

Fig. 11.11 Ammonia-water mixture equilibrium diagram.

Phase equilibrium diagram has temperature on y axis and mass fraction of ammonia, x on x-axis. The saturated liquid states and saturated vapour states are shown on saturated liquid line and saturated vapour lines given by firm lines and dotted lines respectively. Saturation curves are shown for two pressures p_1 and p_2 between which vapour absorption cycle operates. Here when $x = 0$ then mixture shall be complete water and the temperature shall be saturation temperature of water at respective pressure. When $x = 1$ then mixture shall be complete ammonia and temperature is saturation temperature of ammonia at respective pressure. Phase equilibrium diagram shows that the saturated liquid and saturated vapour can exist together in equilibrium as evident from states a and a_1 respectively at any temperature. It also shows that at any temperature the concentration of ammonia is more in vapour than liquid i.e. $((x_{a_1} > x_a))$. For the absorption cycle shown in Fig. 11.11 the saturated liquid at state a leaving absorber at pressure p_1 enters the pump for being pumped upto pressure p_2. For causing evaporation of NH_3 the mixture is heated in generator upto state b. Further heat addition in generator causes evaporation of liquid to vapour yielding state b_2. Transformation from b_1 to b_2 results in formation of ammonia rich vapour thereby reducing concentration of ammonia in liquid. If the pressure and temperature are not changed then evaporation gets stopped as lower concentration liquid is not at its boiling point. With further heat addition the evaporation of mixture continues and the mixture's temperature is raised showing new liquid and vapour states as C and 2. Liquid at C is weak aqua-ammonia mixture and is passed through valve back to absorber. NH_3 vapour (gas) leaves at state 2 from generator to the condenser for being condensed.

11.9 MODIFIED ABSORPTION REFRIGERATION CYCLE

In the absorption cycle discussed earlier it is seen that the strong ammonia solution preparation requires low temperature in absorber while vaporization of NH_3 from strong aqua-ammonia solution needs heating up. Thus, weak solution leaving generator shall be at higher temperature while strong solution leaving absorber shall be at lower temperature. It may be noted that it will be desirable to send strong solution into generator at high temperature so that less external heat addition is required. Also, it is desirable to send weak solution into absorber at lower possible temperature so that less external cooling is needed for effective absorption. Using above two facts the absorption cycle can be modified by employing a heat exchanger between generator and absorber as shown in Fig. 11.12.

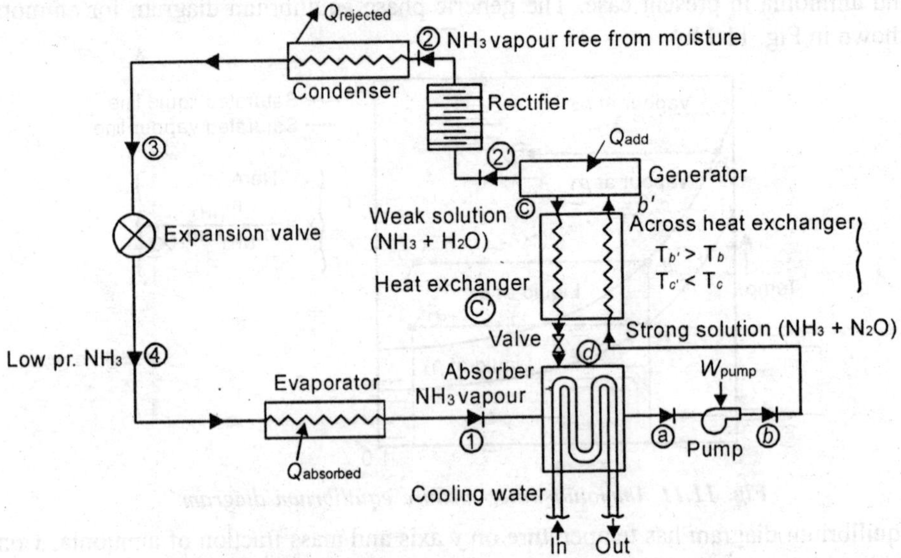

Fig. 11.12 Modified ammonia water absorption system

Ammonia vapour leaving generator are outcome of evaporation from aqua-ammonia mixture due to its heating and there is possibility of water vapour accompanying the ammonia vapour leaving generator. This moisture in refrigerant may undergo freezing (probably at expansion valve or evaporator) and choke the refrigerant flow circuit which is not desirable. Therefore, for preventing moisture accompanying refrigerant from entering into circuit the device called rectifier is put after generator where traces of water are separated out and retained. Rectifier is shown in schematic of modified absorption cycle, Fig. 11.12. Thus, modified absorption cycle employing heat exchanger and rectifier permits to take advantage of heat of weak solution and make refrigerant free of moisture content.

Such type of absorption cycle can also be run using 'lithium-bromide' as absorbent and 'water' as refrigerant on the principle similar to 'water-ammonia system' discussed here. This lithium bromide and water system may not offer very low temperature refrigeration and so this cycle may be cascaded with another cycle using refrigerant having good low temperature characteristic, such as ammonia. This combination is termed as cascade refrigeration system.

11.10 HEAT PUMP SYSTEMS

Heat pump refers to the device which is used for maintaining a region or body at temperature more than that of surroundings. Heat pumps have their working similar to refrigerator with only difference of objective i.e. heat pump maintains temperature more than surroundings contrary to refrigerator which maintains temperature less than surroundings temperature. Heat pump systems can be based upon vapour compression cycle, absorption cycle etc. and are extensively used for space heating applications in domestic and industrial premises.

Reversed Carnot cycle is the ideal thermodynamic cycle for heat pumps similar to the refrigerators. Heat pump system based on vapour compression cycle is described here in Fig. 11.13.

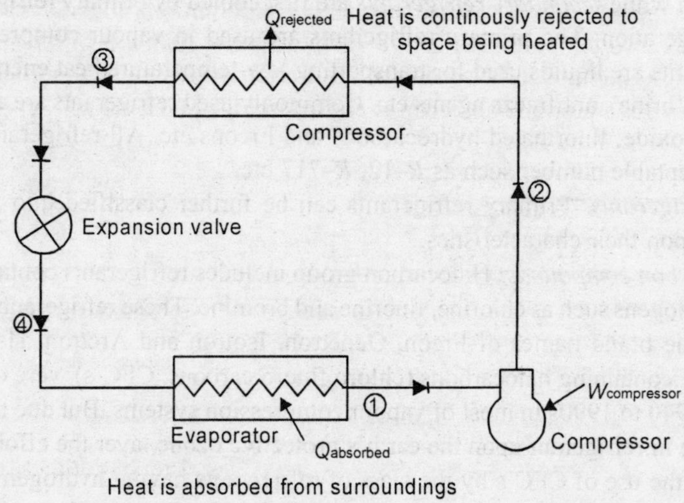

Fig. 11.13 Heat pump system based on vapour compression cycle

Here compressor compresses the refrigerant from low pressure state 1 to high pressure state 2. High pressure gaseous refrigerant is sent to condenser where the condensation of refrigerant occurs rejecting heat equal to the latent heat of condensation to the space being heated up. Subsequently refrigerant in saturated liquid form passes into expansion valve from state 3 to 4. Low pressure liquid refrigerant leaving expansion valve subsequently enters the evaporator where its evaporation occurs upto state 1. Evaporation of refrigerant occurs due to heat being provided by surroundings, water in rivers/ponds, wells or solar collectors etc. Hence the performance of heat pump can be given by:

$$\text{COP of heat pump} = \frac{\text{Desired effect (heating of space)}}{\text{Net work}}$$

$$= \frac{m(h_2 - h_3)}{m(h_2 - h_1)}$$

$$\text{COP of heat pump} = \left\{\frac{(h_2 - h_3)}{(h_2 - h_1)}\right\}$$

Here, heat pump works between the low temperature surroundings or other heat sources giving heat of $Q_{absorbed}$ and the region/body acting as high temperature body where heat $Q_{rejected}$ is rejected to region/body. These heat pumps are not very attractive options when used for only direct heating but they become very attractive when used with some other application.

11.11 REFRIGERANTS

Refrigerant is the working fluid used in refrigeration/air conditioning equipments having capability of carrying heat/rejecting heat in the form of sensible heat or latent heat. Refrigerants carrying heat/rejecting heat in the form of latent heat are more effective and for this refrigerant should possess suitable properties to get transformed from liquid to gas and vice-versa. During selection of refrigerant its chemical, physical and other general properties are being looked into along with refrigeration cycle requirements and application. Refrigerants can be classified into two main categories as 'Primary refrigerants' and 'Secondary refrigerants'. The *primary refrigerants* and those refrigerants which are directly involved in refrigeration system while *secondary refrigerants* are first cooled by primary refrigerants and then used for imparting refrigeration. The primary refrigerants are used in vapour compression systems while secondary refrigerants are liquids used for transporting low-temperature heat energy from one place to another as done by 'brine', antifreeze agents etc. Commonly used refrigerants are air, ammonia carbon-dioxide, sulphur-dioxide, fluorinated hydrocarbons and Freons etc. All refrigerants are assigned with internationally acceptable number such as *R*–12, *R*–717 etc.

(*i*) *Primary refrigerants:* Primary refrigerants can be further classified into following categories depending upon their characteristics.

(*a*) *Halocarbon compounds:* Halocarbon group includes refrigerants containing one or more of three halogens such as chlorine, fluorine and bromine. These refrigerants are traded in market under the brand names of Freon, Genetron, Isotron and Arctron. Halocarbons, specially chlorine containing halocarbons (chloro fluoro carbons, CFC's) were commonly used from about 1940 to 1990s in most of vapour compression systems. But due to disastrous effect of chlorine in refrigerant upon the earth's protective ozone layer the efforts are being made to replace the use of CFC's by the class of refrigerants having hydrogen in place of chlorine. Such new ecofriendly refrigerants are called hydrofluoro carbons (HFC's) for example R–134a ($CF_3\,CH_2F$) can replace R-12 (CCl_2F_2).

Table 11.1: *Halocarbon refrigerants*

Name of refrigerant	*Chemical name*	*Chemical formula*
R–11	Trichloro mono fluoro-methane	CCl_3F
R–12	Dichloro difluoro-methane	CCl_2F_2
R–13	Monochloro trifluoro-methane	$CClF_3$
R–21	Dichloro monofluoro-methane	$CHCl_2F$
R–22	Monochloro difluoro-methane	$CHClF_2$
R–30	Methylene-chloride	CH_2Cl_2
R–40	Methly chloride	CH_3Cl
R–100	Ethyl chloride	CH_3CH_2Cl
R–113	Trichloro tetrafluoro-ethane	$C_2F_3Cl_3$
R–114	Dichloro tetrafluoro-ethane	$CClF_2CClF_2$
R–152	Difluoro-ethane	$CH_3CH_3F_2$

The numbering of halocarbon group refrigerants is based on;

– first digit on right gives number of fluorine atoms.
– second digit from right is one more than the number of hydrogen atoms.
– third digit from right is one less than the number of carbon atoms. Third digit is not given when it is zero.

Some halocarbon refrigerants are as given in Table 11.1.

(*b*) *Inorganic compounds:* Refrigerants used in olden days were inorganic compounds. Some of such refrgerants are still used in different applications such as ice plants, steam refrigeration, aeroplanes and ship refrigeration etc. due to their inherent thermodynamic and physical properties. These inorganic compounds are listed in Table 11.2.

Table 11.2 *Inorganic compound refrigerants*

Name of refrigerant	*Chemical name*	*Chemical formula*
R–717	Ammonia	NH_3
R–718	Water	H_2O
R–729	Air	
R–744	Carbondioxide	CO_2
R–764	Sulfurdioxide	SO_2

(*c*) *Hydrocarbons*: Some of hydrocarbons as given in Table 11.3, are also used as refrigerants particularly in petroleum and petrochemical industry.

Table 11.3 *Hydrocarbon refrigerants*

Name of refrigerant	*Chemical name*	*Chemical formula*
R–50	Methane	CH_4
R–170	Ethane	C_2H_6
R–290	Propane	C_3H_8

(*d*) *Azeotropes:* Azeotropes are those mixture of different refrigerants which do not separate into their components by distillation or with the change in pressure and temperature. Azeotrope refrigerants behave differently from their constituents such as azeotrope evaporates and condenses as single substance having properties that are different from those of either constituents.

Examples of azeotrope are; R–502 is azeotrope which is mixture of 48.8%, *R*–22 and 51.2% of *R*–115. *R*–500 is azeotrope which is mixture of 73.8% *R*–12 and 26.2% of *R*–152.

(*e*) *Unsaturated organic compounds:* These are refrigerants from hydrocarbon group and have ethylene or propylene as the base. Table 11.4 gives-such refrigerants.

Table 11.4 *Unsaturated organic compounds*

Name of refrigerant	*Chemical name*	*Chemical formula*
R–1120	Trichloro ethylene	$C_2H_4Cl_3$
R–1130	Dichloro ethylene	$C_2H_4Cl_2$
R–1150	Ethylene	C_2H_4
R–1270	Propylene	C_3H_6

(*ii*) *Secondary refrigerants:* Secondary refrigerants are indirect refrigerants that carry heat from the substance being cooled to the evaporator. Secondary refrigerants undergo change in their temperature by absorbing heat and rejecting it at evaporator without any phase change. Examples of secondary refrigerants are water, brine and antifreezes etc. Antifreezing agents may be solution of water and ethylene glycol, propylene glycol or calcium chloride, with freezing temperature well below 0°C.

Secondary refrigerants offer advantages of ease of handling and control, ease of maintaining temperature of large buildings by controlling flow of secondary refrigerants such as brine etc. Secondary refrigerants also eliminate requirement of long primary refrigerant lines.

11.12 DESIRED PROPERTIES OF REFRIGERANTS

A refrigerant being used in refrigeration should have following thermodynamic properties, physical properties and working properties as:

(*i*) Boiling temperature of refrigerant should be quite low at atmospheric conditions for effective refrigerant. For refrigerants having higher boiling temperatures at atmospheric conditions the compressor is run at higher vacuum.

(*ii*) For an ideal refrigerant the freezing temperature of refrigerant should be quit low so as to prevent its freezing at evaporator temperature. Freezing point temperature should be less than evaporator temperature. For example, refrigerant R–22 has freezing point of – 160°C and normally most of refrigerants have freezing point below – 30°C.

(*iii*) Critical temperature of the ideal refrigerant should be higher than the condenser temperature for the ease of cndensation.

(*iv*) Refrigerant should have large latent heat at evaporator temperature as this shall increase the refrigerating capacity per kg of refrigerant.

(*v*) Refrigerant should have small specific volume at inlet to compressor as this reduces compressor size for same refrigeration capacity.

(*vi*) Specific heat of refrigerant in liquid from should be small and it should be large for refrigerant in vapour form, since these increase the refrigerating capacity per kg of refrigerant.

(*vii*) Thermal conductivity of refrigerant should be high.

(*viii*) Viscosity of refrigerants should be small for the ease of better heat transfer and small pumping work requirement.

(*ix*) Refrigerant should be chemically inert and non toxic.

(*x*) Refrigerant should be non-flammable, non explosive and do not have any harmful effect upon coming in contact with material stored in refrigeration space.

(*xi*) Refrigerant may have pleasant distinct odour so as to know about its leakage.

(*xii*) Refrigerant should be readily available at lesser price.

Choice of refrigerant: Selection of suitable refrigerant depends upon the number of parameters as there can not be single refrigerant well suited for all kinds of refrigeration systems. Thermodynamic properties, physical properties and other properties are considered in reference to the refrigeration system during choice of suitable refrigerant. Apart from these properties, due consideration should be given to the working pressure and temperature range and pressure ratio, space restrictions, corrosiveness and inflammability, oil miscibility etc. before selecting refrigerant with suitable compressor and other equipments of refrigeration system.

11.13 PSYCHROMETRY

Psychrometry or Psychrometrics refers to the study of system involving dry air and water. Properties of mixture of air and water are called psychrometric properties. Psychrometry becomes very important from air conditioning point of view as in air conditioning the human comfort conditions defined in terms of temperature, humidity and air circulation/ventilation are being considered. Different psychrometric properties, charts and processes generally used are defined ahead.

(*i*) *Dry air:* Atmospheric air having 79% nitrogen and 21% oxygen by volume is considered dry air. Its, molecular weight is taken as 29.

(*ii*) *Moist air:* Moist air is the mixture of dry air and water vapour in which dry air is treated as if it were pure component. Quantity of water vapour present in the mixture depends upon the temperature of air and it may vary from zero in dry air to the maximum quantity when mixture is saturated of water vapour (called saturation capacity of air). Moist air is assumed to behave as ideal gas for the purpose of analysis. Mixture pressure is the sum of partial pressures of dry air and water vapour. When the partial pressure of water vapour corresponds to the saturation pressure of water at mixture temperature then mixture is said to be saturated. Saturated air is the mixture of dry air and saturated water vapour. When the temperature of mixture of air and vapour is above the saturation temperature of water vapour then the vapour is called superheated vapour.

(*iii*) *Humidity ratio or specific humidity:* This is given by the ratio of the mass of water vapour to the mass of dry air. It can also be defined as mass of water vapour present in per kg of dry air.

It is given in terms of grams per kg of dry air. Mathematically, humidity ratio or specific humidity,

$$\omega = \frac{m_v}{m_a}$$

where m_v and m_a are mass of water vapour and dry air respectively.

Humidity ratio can also be given in terms of partial pressures and molecular weights.

$$m_v = \frac{m_v \cdot p_v \cdot V}{\bar{R}T}$$

$$m_a = \frac{m_a \cdot p_a \cdot V}{\bar{R}T}$$

where p_v and p_a are partial pressures of water vapour and dry air respectively, M_v and M_a are molecular weights of water vapour and dry air respectively. For total pressure of mixture being p, the Dalton's law of partial pressure says,

$$p = p_a + p_v$$

or,
$$p_a = p - p_v$$

Hence, upon substitution,

$$\omega = \frac{M_v \cdot p_v}{M_a \cdot p_a} = \frac{M_v \cdot p_v}{M_a \cdot (p - p_v)}$$

or,
$$\boxed{\omega = 0.622\left(\frac{p_v}{p - p_v}\right)}$$

Humidity ratio is measured using an instrument called 'hygrometer' in which moist air sample is processed till the moisture present in it gets absorbed and moisture absorbed can be quantified.

(*iv*) *Absolute humidity:* Absolute humidity refers to the weight of water vapour present in unit volume of air.

(*v*) *Relative humidity:* Relative humidity gives an account of moisture content in an actual mixture as compared to the mixture in saturated state at same temperature and pressure. It can be given by the ratio of actual mass of water vapour in given volume to the mass of water vapour if the air is saturated at the same temperature and pressure.

Mathematically, relative humidity (ϕ) can be given as:

$$\phi = \frac{\text{(Actual mass of water vapour in given volume mixture)}}{\left(\begin{array}{l}\text{Maximum mass of water vapour in same volume of mixture being staurated at} \\ \text{same temperature and pressure}\end{array}\right)}$$

$$\phi = \frac{m_v}{m_{v,\,sat}},$$ where m_v and $m_{v,\,sat}$ are the actual and maximum mass of water vapour in same volume of mixture as defined earlier.

$$\phi = \frac{\left(\dfrac{M_v \cdot p_v \cdot V}{\bar{R}T}\right)}{\left(\dfrac{M_v \cdot p_{v,\,sat} \cdot V}{\bar{R}T}\right)}$$

or,

$$\phi = \frac{p_v}{p_{v,\,sat}},$$ where p_v and $p_{v,\,sat}$ are the partial pressures of 'actual mass of water vapour' and 'maximum mass of water vapour in same volume of mixture as defined earlier'.

(*vi*) *Enthalpy of moist air or Mixture enthalpy:* The enthalpy of moist air can be estimated by addition of the enthalpy contribution of each constituent component i.e water vapour and dry air at the state of mixture. Thus, enthalpy of mixture (moist air) of dry air and water can be given as,

$$H = H_a + H_v = m_a h_a + m_v h_v, \text{ or, } \frac{H}{m_a} = h_a + \left(\frac{m_v}{m_a}\right) h_v$$

(Enthalpy of moist air per unit mass of dry air) = (Enthalpy of unit mass of dry air)
+ (Enthalpy of water vapour associated with unit mass of dry air)

Here H and h refer to total enthalpy and specific enthalpy values respectively. Subscripts a and v refer to dry air and water vapour respectively.

(*vii*) *Dry bulb temperature*: Dry bulb temperature refers to the temperature of air measured with ordinary thermometer having its bulb open.

(*viii*) *Wet bulb temperature:* Wet bulb temperature refers to the temperature measurement with the thermometer having its bulb covered with wet cloth and exposed to air stream whose temperature is being measured. For getting wet bulb temperature thermometer bulb is covered with water wet cloth/wick moistened with water and then temperature is measured. Difference between dry bulb temperature and wet bulb temperature is called wet bulb depression. Wet bulb depression is zero in case of saturated air as the dry bulb temperature and wet bulb temperatures are equal.

(*ix*) *Dew point temperature*: It is the temperature at which condensation of water vapour present in moist air just begins. It could be understood from different examples that if the moist air temperature is reduced then water vapour gets condensed. Examples for such condensation is formation of dew on grass, condensation of water vapour on exterior surface of steel tumbler having cold water inside, condensation of water vapour on glass window-panes during winter season etc. Dew point temperature shall be equal to the saturation temperature corresponding to partial pressure of water vapour in moist air. Difference between dew point temperature and dry bulb temperature is called dew point depression.

(*x*) *Humid specific volume:* It refers to the volume of mixture per kg of dry air in mixture and is given by m^3/kg of dry air.

(*xi*) *Psychrometric charts:* Psychrometric chart gives the graphical representation of different important properties of moist air. These charts are readily available for different mixture pressure. Fig. 11.14 shows psychrometric chart. Abscissa (*x*-axis) of chart gives the dry bulb temperature (°C) and the ordinate (*y*-axis) has humidity ratio (ω) in kg or gram of water vapour per kg of dry air.

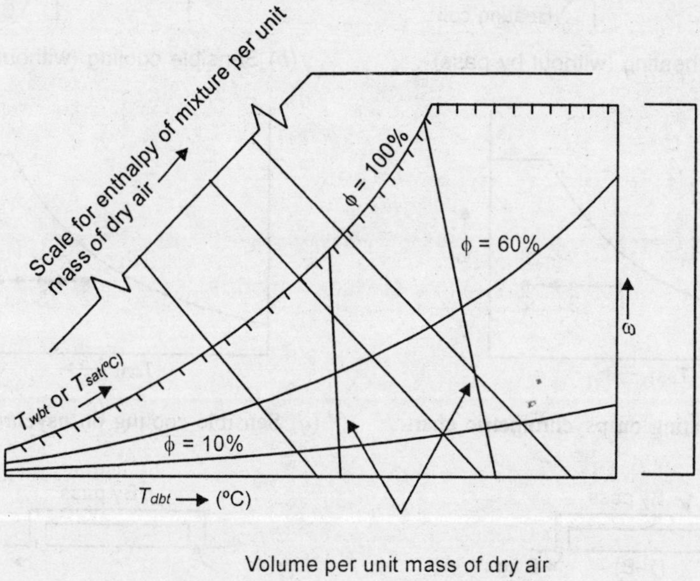

Fig. 11.14 Psychrometric chart

Since humidity ratio can be directly related to partial pressure of water vapour so vapour pressure (p_v) can also be shown on ordinate. Constant relative humidity (ϕ) curves are also shown on psychrometric chart for different ϕ values such as $\phi = 10\%$, ..., 60%, ..., 100%. Mixture enthalpy per unit mass of dry air (kJ/kg of dry air) is also available on psychrometric chart. Dew point temperature for the moist air can be known by following the line of constant ω (or constant p_v) upto saturation line, $\phi = 100\%$ as the dew point refers to the state where mixture becomes saturated when cooled at constant vapour pressure. Wet bulb temperature is also available on psychrometric chart where constant wet bulb temperature is also available on psychrometric chart where constant wet bulb temperature lines run from upper left to the lower right of chart. These constant wet bulb temperature lines approximate to the lines of constant mixture enthalpy per unit mass of dry air. Pscychrometric chart also has lines representing volume per unit mass of dry air (m^3/kg). These specific volume lines can be approximated as state giving volume of dry air or water vapour per unit mass of dry air since each component of mixture i.e. air and water vapour occupy the same volume.

(*xii*) *Psychrometric processes*: Different psychrometric processes commonly encountered in air conditioning are as follows.

(*a*) *Sensible heating or sensible cooling:* Sensible heating or cooling refers to the heating or cooling without phase change i.e. heating or cooling of air without increase or decrease of moisture content. For sensible heating the air is passed over heating coils (electrical resistance type or steam type) while for sensible cooling the air is passed over cooling coils (such as evaporator coil of refrigeration cycle), as shown in Fig. 11.15.

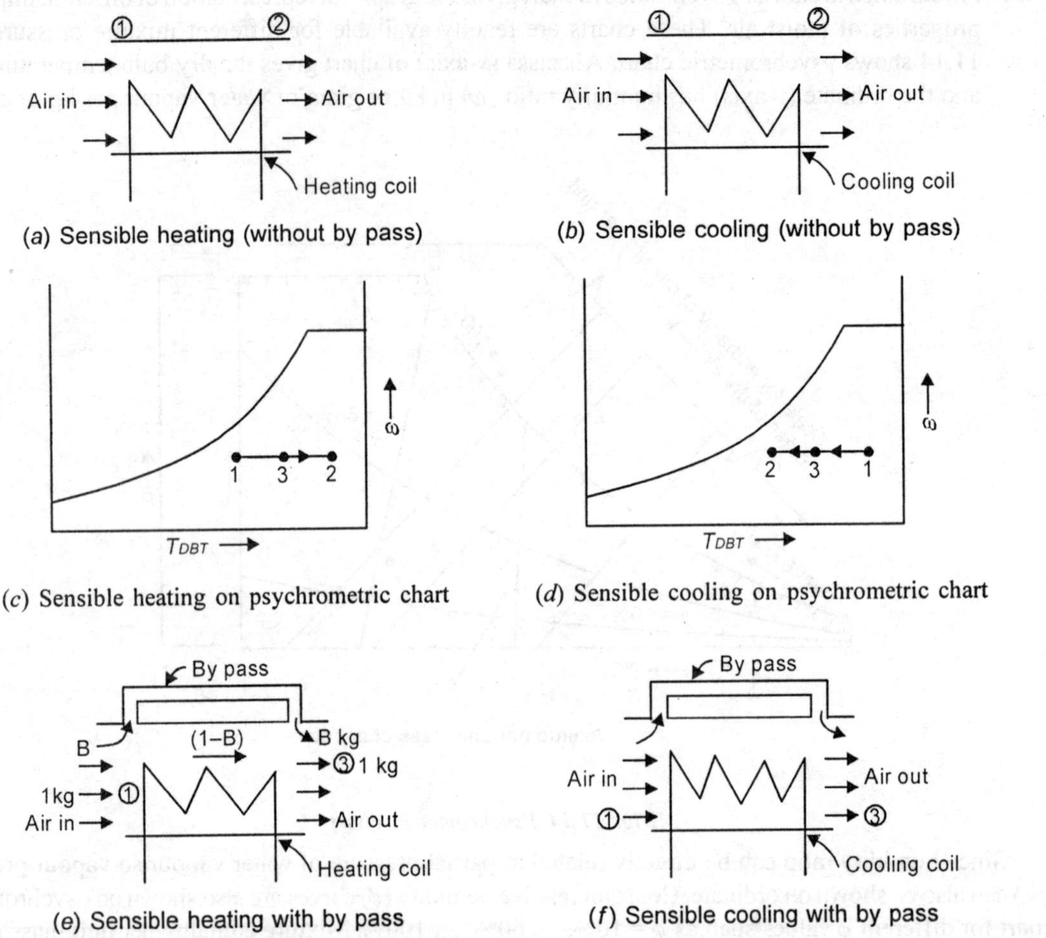

Fig. 11.15 Sensible heating and sensible cooling

This heating and cooling may be for complete amount of air being passed over the coil as shown in Fig. 11.15 *a* and *b*, or the portion of air flowing can be by passed through by-pass passage thus, only a fraction is passed over coil. Later arrangement is called sensible heating or sensible cooling with by pass as shown in Fig. 11.15 *e* and *f*. Sensible heating and sensible cooling without by pass is shown on psychrometric charts Fig. 11.15 *c* and *d*, respectively between states 1 and 2. Thus, for inlet air temperature being T_1, the heating/cooling causes change in temperature upto T_2. When some amount of air is by passed then the temperature of air coming out shall be different from T_2. In case of sensible heating with by pass, temperature of air coming out is less than temperature of air wihout by pass i.e. $T_3 < T_2$. Similarly in case of sensible cooling with by pass temperature of air leaving is more than temperature of air without by pass i.e. $T_3 > T_2$.

Heat added can be quantified by the amount of heat gained by dry air and water vapour between dry bulb temperature T_1 and T_2.

$$Q_{\text{sensible heating}} = C_{pa}\,(T_2 - T_1) + \omega \cdot C_{pv}\,(T_2 > T_1)$$

$$= (Cp_a + \omega\, Cp_v)\,(T_2 - T_1)$$

$$Q_{\text{sensible heating}} = C_{pm} \cdot (T_2 - T_1),\ \text{Hence } C_{pm} \text{ is mean specific heat.}$$

In case of sensible heating with by pass, if the amount of air by passed is B kg per kg of air then heat balance can be given as below. Here $(1 - B)$ kg of air passes over the coil and later on mixes with by passed air at exit.

$$B \cdot (C_{pa} + \omega\, C_{pv})\, T_1 + (1 - B)\,(C_{pa} + \omega\, C_{pv})\, T_2 = 1 \times (C_{pa} + \omega\, C_{pv}) \cdot T_3$$

or,

$$\text{By pass factor, } B = \frac{T_2 - T_3}{T_2 - T_1} = \frac{\text{length}_{2\text{-}3} \text{ on psychrometric chart}}{\text{length}_{2\text{-}1} \text{ on psychrometric chart}}$$

This amount of air by passed per kg of air flowing is also called as By-pass factor of the coil. Factor $(1 - B)$ is called contact factor.

(*b*) *Cooling with dehumidification*: This process refers to the reducing air temperature along with removal of water vapour from air. Removal of moisture content can be done by reducing the temperature of air below dew point temperature so as to allow condensation of water vapour. Arrangement for cooling with humidification is shown in Fig. 11.16 where moist air entering is first passed over cooling coil having cool refrigerant inside. Water vapour present in moist air gets condensed and saturated moist air leaves dehumidification section at state 2.

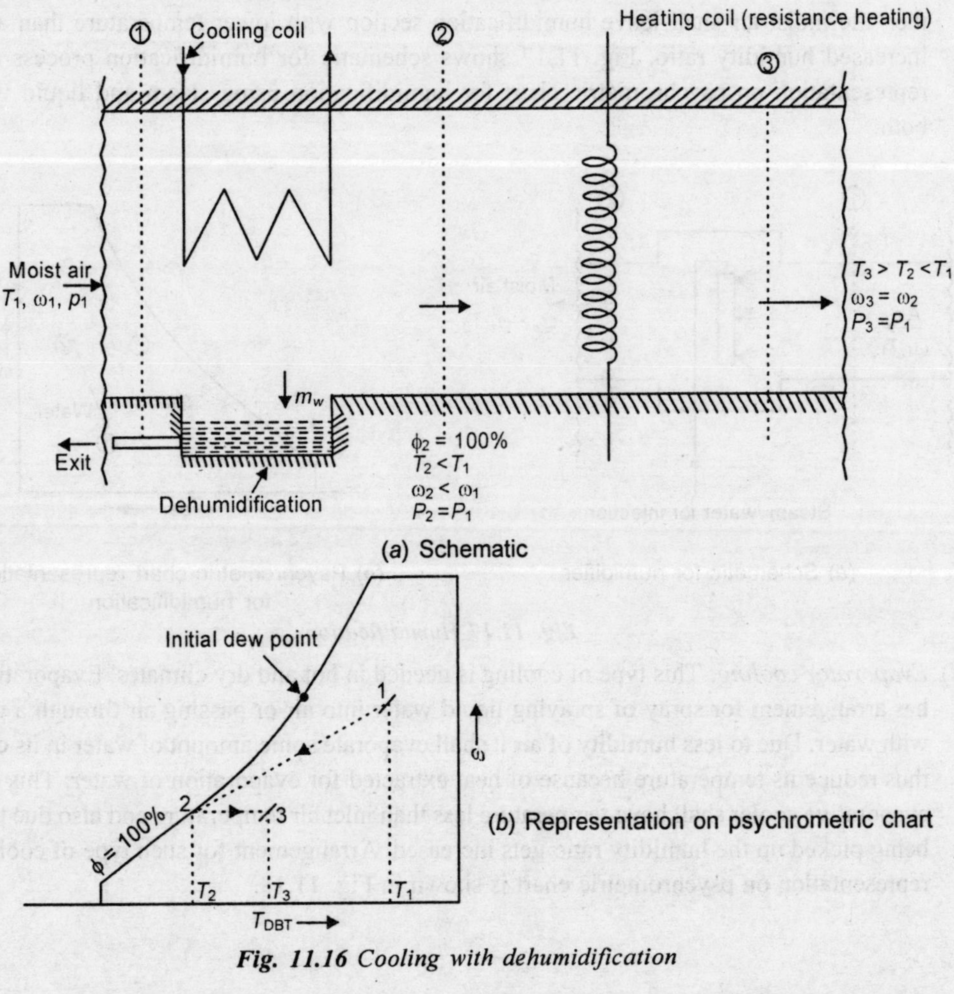

Fig. 11.16 *Cooling with dehumidification*

Condensate also gets cooled upto temperature T_2 before leaving dehumidification section. Moist air at state (2) may be excessively cooled so that it can be subsequently heated by passing it over heating coil. Thus, at state (3) the available air shall be at least temperature and in dehumidified form. States (1), (2) and (3) are also shown on psychrometric chart.

Mass balance shows,

Dry air at section (1) and (2) $m_{a1} = m_{a2} = m_a$

Vapour content at section (1) and (2); $m_{v1} = m_w + m_{v2}$, or $m_w = m_{v1} - m_{v2}$,

It can also be given: $m_{v1} = \omega_1 \cdot m_a$; $m_{v2} = \omega_2 \cdot m_a$

Mass of water condensed per unit mass of dry air; $\omega_1 - \omega_2 = \dfrac{m_w}{m_a}$

Between section (2) and (3), $\omega_2 = \omega_3$.

(*c*) *Humidification*: This is the process of adding moisture to the air. Humidification may be required during air conditioning when air being circulated in occupied space may have little moisture in it. Available means for increasing humidity are to inject steam or spray liquid water into air. Normally the steam being injected shall be at temperature more than that of air and humidification shall be accompanied by increase in humidity ratio and dry-bulb temperature. When liquid water is sprayed then the moist air shall leave humidification section with lower temperature than at inlet and increased humidity ratio. Fig. 11.17 shows schematic for humidification process along with representation on psychrometric chart for humidification using steam and liquid water spray both.

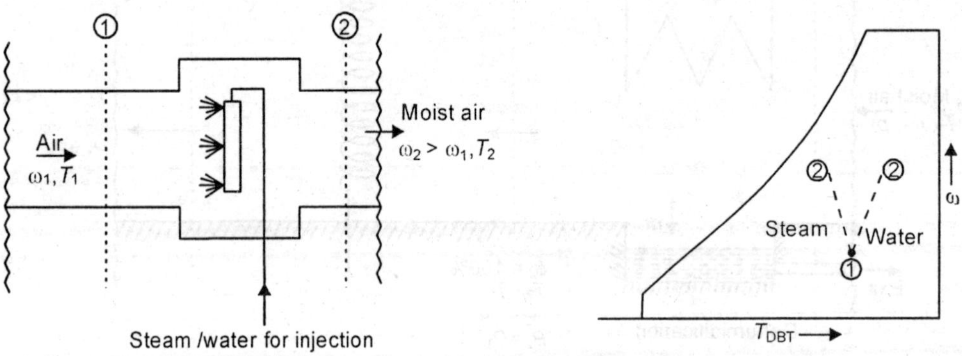

(*a*) Schematic for humidifier (*b*) Psychrometric chart representation for humidification

Fig. 11.17 Humidification

(*d*) *Evaporator cooling:* This type of cooling is needed in hot and dry climates. Evaporative cooling has arrangement for spray of spraying liquid water into air or passing air through a pad soaked with water. Due to less humidity of air it shall evaporate some amount of water in its contact and thus reduce its temperature because of heat extracted for evaporation of water. This air leaving evaporative cooler shall have temperature less than inlet air temperature and also due to moisture being picked up the humidity ratio gets increased. Arrangement for such type of cooling and its representation on psychrometric chart is shown in Fig. 11.18.

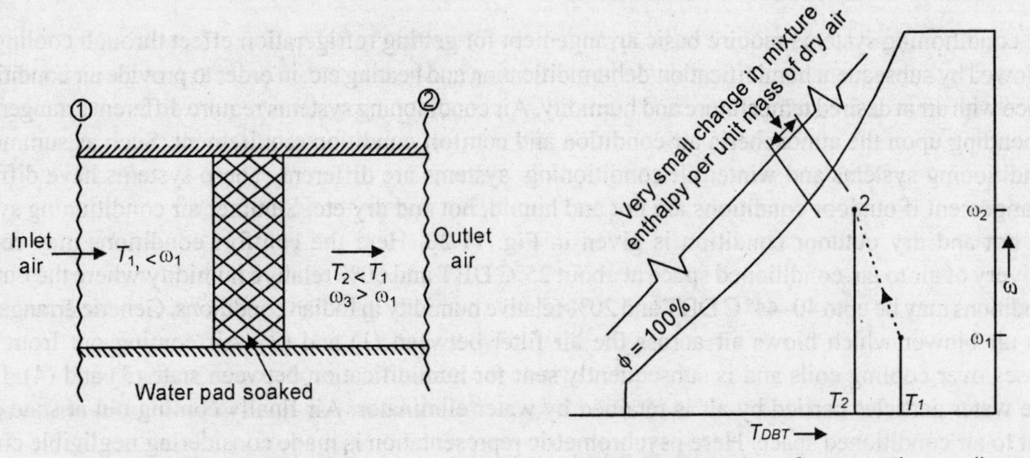

(*a*) Evaporative cooler (*b*) Representation of evaporative cooling on psychrometric chart

Fig. 11.18 *Evaporative cooling*

(*e*) *Adiabatic mixing of moist air stream:* Adiabatic mixing of moist air streams is the one in which there is no heat interaction during mixing.

Obviously there is no work interaction during mixing. This process of mixing of moist air streams is used very frequently in air conditioning systems. Fig. 11.19. shows adiabatic mixing of two streams entering at (1) and (2) and leaving at (3).

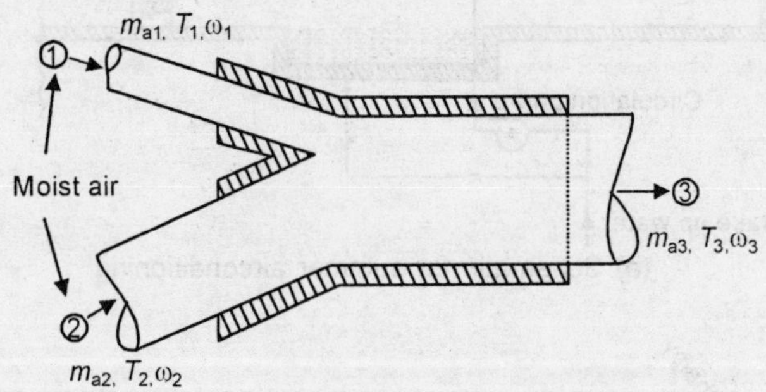

Fig. 11.19 *Adiabatic mixing*

If dry air entering be m_{a1} & m_{a2} and water vapour flow rate be m_{v1} & m_{v2} with streams (1) and (2) respectively at steady state then mass balances and energy balances can be given as under:

$$\text{For dry air, } m_{a1} + m_{a2} = m_{a3}$$

$$\text{For water vapour, } m_{v1} + m_{v2} = m_{v3}$$

Since $m_{v1} = \omega_1 \cdot m_{a1}$, $m_{v2} = \omega_2 \cdot m_{a2}$, $m_{v3} = \omega_3 \cdot m_{a3}$, then substitution yields;

$$\omega_1 \cdot m_{a1} + \omega_2 \cdot m_{a2} = \omega_3 \cdot m_{a3}$$

or,

$$\omega_1 \cdot m_{a1} + \omega_2 \cdot m_{a2} = \omega_3 (m_{a1} + m_{a2})$$

Energy balance yields; following as $Q = 0$, $W = 0$.

$$m_{a1} \cdot (h_{a1} + \omega_1 h_{g1}) + m_{a2} \cdot (h_{a2} + \omega_2 h_{g2}) = m_{a3} \cdot (h_{a3} + \omega_3 h_{g3})$$

11.14 AIR CONDITIONING SYSTEMS

Air conditioning systems require basic arrangement for getting refrigeration effect through cooling coil followed by subsequent humidification/dehumidification and heating etc. in order to provide air conditioned space with air at desired temperature and humidity. Air conditioning systems require different arrangements depending upon the atmospheric air condition and comfort condition requirement. Such as summer air conditioning systems and winter air conditioning systems are different. These systems have different arrangement if outdoor conditions are hot and humid, hot and dry etc. Summer air conditioning system for hot and dry outdoor condition is given in Fig. 11.20. Here the comfort conditions may require delivery of air to air-conditioned space at about 25°C DBT and 60% relative humidity where the outdoor conditions may be upto 40–44° C DBT and 20% relative humidity in Indian conditions. Generic arrangement has air blower which blows air across the air filter between (1) and (2). Air coming out from filter passes over cooling coils and is subsequently sent for humidification between state (3) and (4). Large size water particles carried by air is retained by water eliminator. Air finally coming out at state (5) is sent to air conditioned space. Here psychrometric representation is made considering negligible change in humidity in water eliminator.

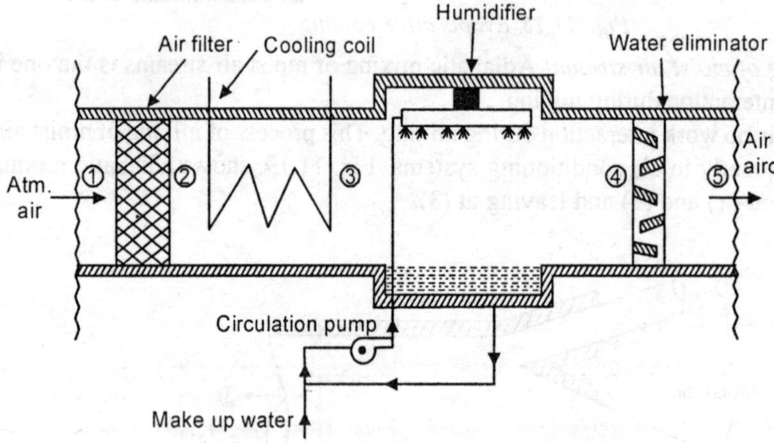

(*a*) Schematic for summer airconditioning

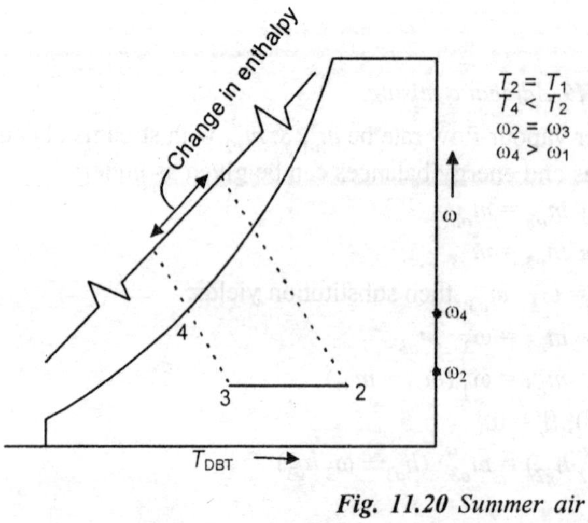

(*b*) Representation on psychrometric chart

Fig. 11.20 Summer air conditioning

Winter air conditioning has slightly different arrangement as compared to summer air conditioning. Fig. 11.21 shows the schematic for winter air-conditioning along with representation on psychrometric chart.

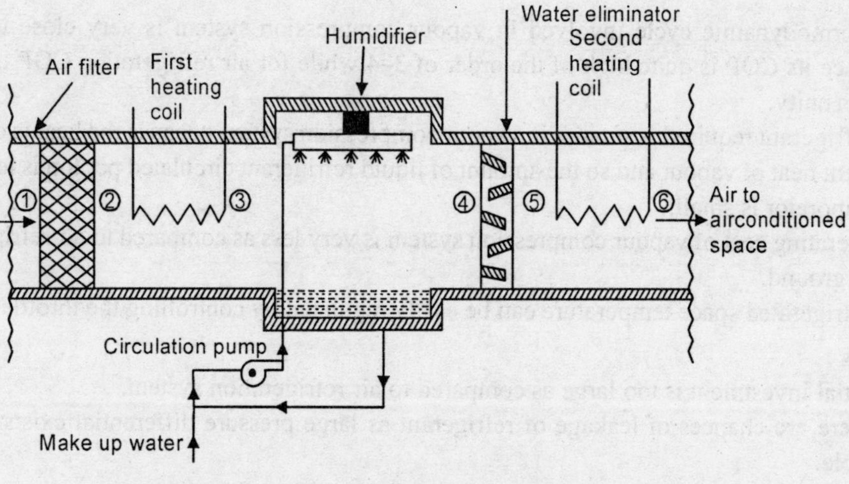

(*a*) Schematic for winter airconditioning

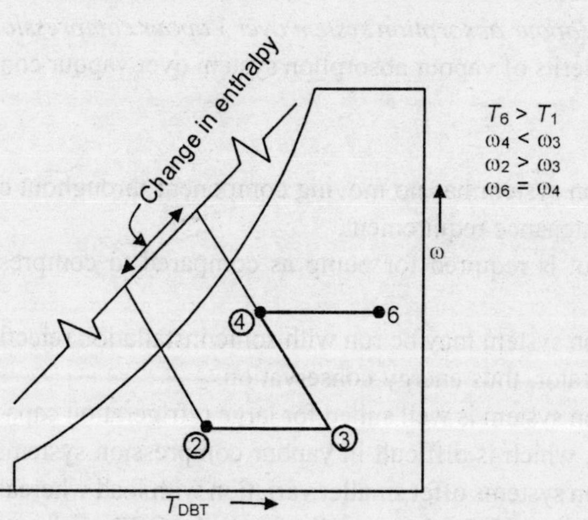

(*b*) Representation on psychrometric chart

Fig. 11.21 Winter air conditioning

Let us assume air from atmosphere to be available at 10ºC and 80% relative humidity while comfort conditions are 25ºC DBT and 60% relative humidity.

Here atmospheric air is blown into air filter and it comes out at (2). Air is passed across heating coil for being heated. Hot air at (3) passes through humidifier section where its humidity increases as shown on psychrometric chart by state (4). It is further passed through water eliminator to retain large sized water particles. Air may again be passed through heating coil section so as to compensate for temperature reduction in humidifier section and attain desired temperature.

11.15 COMPARISON OF VARIOUS REFRIGERATION METHODS

Merits and demerits of Vapour compression system over air-refrigeration system.

Vapour compression system when compared with air refrigeration system possess following merits and demerits.

Merits

- Thermodynamic cycle involved in vapour compression system is very close to Carnot cycle hence its COP is quite high of the order of 3–4 while for air refrigeration COP is generally less than unity.
- Refrigerant required is smaller in vapour compression system because the heat is carried away by latent heat of vapour and so the amount of liquid refrigerant circulated per ton is less. Also size of evaporator is small.
- Operating cost of vapour compression system is very less as compared to air refrigeration system on ground.
- Refrigerated space temperature can be easily regulated by controlling the throttle valve.

Demerits

- Initial investment is too large as compared to air refrigeration system.
- There are chances of leakage of refrigerant as large pressure differential exists throughout the cycle.
- Air leakage into refrigerant offers reduced performance of vapour compression system and so it should be prevented.

Merits and demerits of Vapour absorption system over Vapour compression system.

The merits and demerits of vapour absorption system over vapour compression system are listed as under:

Merits

- Vapour absorption system has no moving component throughout cycle, so it works quietly and has reduced maintenance requirement.
- Lesser work input is required for pump as compared to compression in vapour compression system.
- Vapour absorption system may be run with some installation rejecting waste heat, which may be used to run generator, thus energy conservation.
- Vapour absorption system is well suited for large refrigeration capacity applications such as even above 1000 tons, which is difficult in vapour compression systems.
- Vapour absorption systems offer smaller variation with load whereas COP of vapour compression system varies largely with load. At smaller loads the COP of absorption system is more than that of vapour absorption system.

Demerits

- Size of absorption system is large and is bulky to handle.
- If heating source (such as electric heating) is costly then the vapour absorption system becomes costly than vapour compression system.
- At very high loads the COP of vapour compression system is more than that of vapour absorption system.

EXAMPLES

1. *A refrigerator operating on reversed Carnot extracts 500 kJ/min heat from a refrigerated space being maintained at –16ºC and rejects heat to the atmosphere at 27ºC. Determine the work input required to run the refrigerator.*

Solution:

For refrigerator working on reversed Carnot cycle.

$$\frac{Q_1}{(-16+273)} = \frac{Q_2}{(27+273)}$$

$$\frac{Q_1}{Q_2} = \frac{257}{300}$$

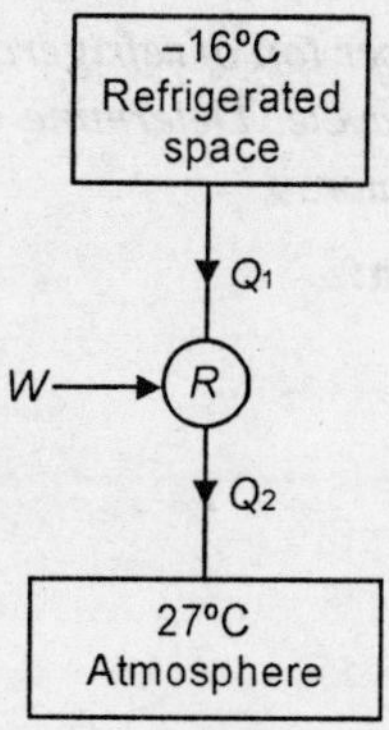

Fig 11.22

Also it is given that $Q_1 = 500$ kJ/min

$Q_2 = 583.66$ kJ/min

Work input required, $W = Q_2 - Q_1$

$= 83.66$ kJ/min

Work input = **83.66 kJ/min**

Ans.

2. *A refrigerator operating between –7ºC and 27ºC has the capacity of 800 tons. Determine the mass rate of ice formation considering water being supplied at atmospheric temperature of 27ºC. Also find the minimum hp required. Consider the refrigerator to work upon reversed Carnot cycle. Latent heat for ice formation from water may be taken as 335 kJ/kg.*

Solution:

Refrigeration capacity = 800 × 3.5 = 2800 kJ/s = Heat extraction rate

Let the ice formation rate be '*m*' kg/s.

Heat to be removed from per kg of water for its transformation into ice.

$= 4.18 \times (27 - 0) + 335$ kJ/kg

$= 447.86$ kJ/kg

$$\text{Ice formation rate} = \frac{\text{Refrigeration capacity}}{\text{Heat removed for getting per kg of ice}}$$

$$= \frac{2800}{447.86} = 6.25 \text{ kg/s}$$

Mass rate of ice formation = **6.25 kg/s**

Ans.

$$\text{COP of refrigerator,} = \frac{T_1}{T_2 - T_1} = \frac{\text{Refrigeration capacity}}{\text{Work done}}$$

Here $T_1 = 273 - 5 = 268$ K, $T_2 = 27 + 273 = 300$ K

$$\text{COP} = \frac{268}{300-268} = \frac{2800}{W}$$

$$W = 334.33 \text{ kJ/s}$$

$$= \frac{334.33}{0.7457} \text{ hp} \qquad \{\text{as } 1 \text{ hp} = 0.7457 \text{ kW}\}$$

$$= 448.34 \text{ hp}$$

$$\boxed{\text{HP required} = \mathbf{448.34\ hp}} \quad \textbf{Ans.}$$

3. 3 hp per ton of refrigeration is required to maintain temperature of –27°C in refrigerator working on Carnot cycle. Determine COP of cycle and temperature of surroundings to which heat is rejected by refrigerator.

Solution:

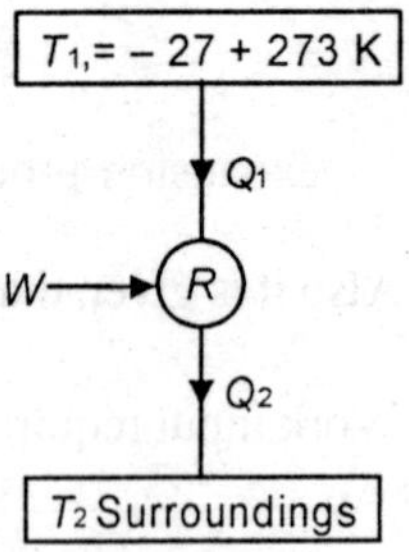

Fig. 11.23

$$\text{COP} = \frac{T_1}{T_2 - T_1} = \frac{246}{T_2 - 246}$$

Also,

$$\text{COP} = \frac{\text{Refrigeration effect}}{\text{Work input}}$$

$$= \frac{1 \times 3.5}{3 \times 0.7457}$$

$$\text{COP} = 1.56$$

$$\boxed{\text{COP} = \mathbf{1.56}} \quad \textbf{Ans.}$$

Equating COP,

$$1.56 = \frac{246}{T_2 \times 246}$$

$$T_2 = 403.69 \text{ K}$$

$$\boxed{\text{Temperature of surroundings} = \mathbf{403.69\ K}} \quad \textbf{Ans.}$$

4. An air refrigeration cycle has pressure ratio of 8, temperature of air at beginning of compression is –30°C and air is cooled upto 27°C after compression. Compression and expansion processes have isentropic efficiency of 85% and 90% respectively. Determine the refrigeration capacity and coefficient of performance. Consider C_p = 1.005 kJ/kg ·K, γ = 1.4 and air flow rate of 1 kg/s.

Solution:

Given:

$T_1 = -30$°C, $T_3 = 27$°C or 300 K

During process 1–2′,

$$\frac{p_2}{p_1} = \left(\frac{T_{2'}}{T_1}\right)^{\frac{\gamma}{\gamma-1}}$$

$$T_1 = -30 + 273 = 240 \text{ K}$$

$$\frac{T_{2'}}{T_1} = \left(\frac{P_2}{P_1}\right)^{\frac{\gamma-1}{\gamma}}$$

$$T_{2'} = 240 \times (8)^{\frac{(1.4-1)}{1.4}}$$

Theoretical temperature after compression $T_{2'}$ = 434.75 K

For compression process,

$$\eta_{\text{isen,compr.}} = 0.85 = \frac{T_{2'} - T_1}{T_2 - T_1}$$

$$(T_2 - 240) = \frac{(434.75 - 240)}{0.85}$$

$$T_2 = 469.12 \text{ K}$$

For expansion process, 3–4',
$$\frac{T_{4'}}{T_3} = \left(\frac{P_1}{P_2}\right)^{\frac{\gamma-1}{\gamma}}$$

$$T_{4'} = 300 \times \left(\frac{1}{8}\right)^{\frac{1.4-1}{1.4}}$$

$$T_{4'} = 165.61 \text{ K}$$

$$\eta_{\text{isen, expansion}} = 0.90 = \frac{T_3 - T_4}{T_3 - T_{4'}} = \frac{300 - T_4}{300 - 165.61}$$

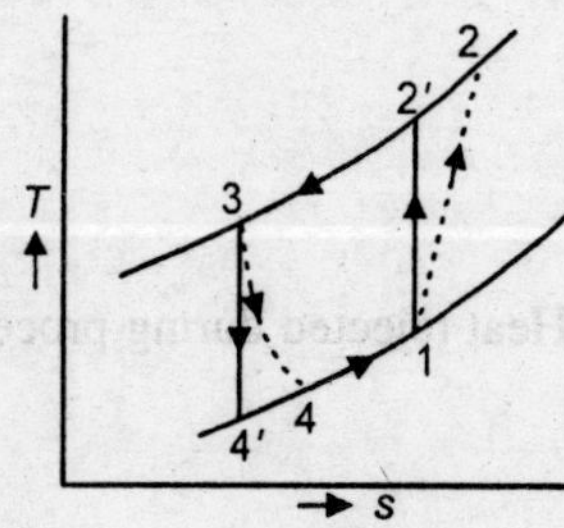

Fig. 11.24

or, $T_4 = 179.05$ K

Work during compression, $W_c = 1 \times C_p \times (T_2 - T_1)$

$= 1 \times 1.005\ (469.12 - 240)$

$= 230.27$ kJ/s.

Work during expansion $W_T = 1 \times C_p \times (T_3 - T_4)$

$W_T = 1 \times 1.005 \times (300 - 179.05)$

$= 121.55$ kJ/s

Refrigeration effect is realized during process, 4–1. So refrigeration effect shall be,

$Q_{\text{ref}} = 1 \times C_p \times (T_1 - T_4)$

$= 1 \times 1.005\ (240 - 179.05)$

$Q_{\text{ref}} = 61.25$ kJ/s or 17.5 ton

Net work required $= W_C - W_T = 230.27 - 121.55 = 108.72$ kJ/s.

$$\text{COP} = \frac{Q_{\text{ref}}}{W_C - W_T} = \frac{61.25}{108.72} = 0.56$$

Refrigeration capacity = **17.5** ton or **61.25 kJ/s**
COP = **0.56**

Ans.

5. A refrigeration system has air leaving refrigerated space at 7°C and 1 bar and subsequently air is compressed isentropically to 5 bar. After being compressed air is cooled upto 27°C at this pressure and then expanded isentropically upto 1 bar and discharged into refrigerated space. Determine COP of system.

Solution:
Given

$$T_1 = 7°C \text{ or } 280 \text{ K}$$
$$T_3 = 27°C \text{ or } 300 \text{ K}$$
$$P_1 = 1 \text{ bar}, P_2 = 5 \text{ bar}$$

For isentropic compression process:

$$\left(\frac{P_2}{P_1}\right)^{\frac{\gamma-1}{\gamma}} = \frac{T_2}{T_1}$$

$$T_2 = 280 \times (5)^{\frac{(1.4-1)}{1.4}}$$
$$T_2 = 443.47 \text{ K}$$

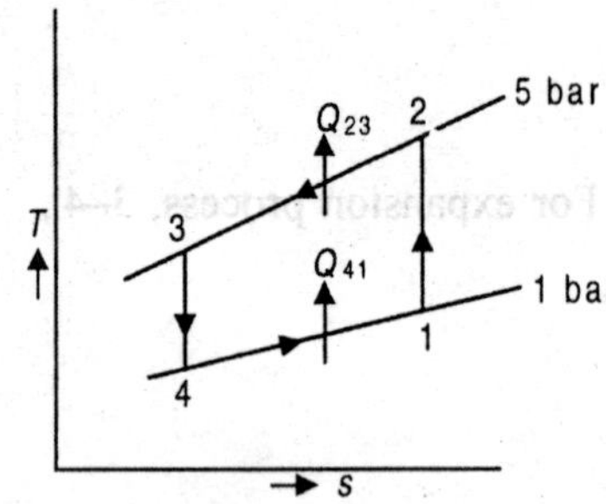

Fig. 11.25

For isentropic expansion process:

$$\left(\frac{P_3}{P_4}\right)^{\frac{\gamma-1}{\gamma}} = \left(\frac{T_3}{T_4}\right) = \left(\frac{P_2}{P_1}\right)^{\frac{\gamma-1}{\gamma}}$$

$$\left(\frac{T_3}{T_4}\right) = (5)^{\frac{(1.4-1)}{1.4}}$$

$$T_4 = 189.42 \text{ K}$$

Heat rejected during process 2 – 3, $Q_{23} = m \times C_p \times (T_2 - T_3)$

$$= m \times 1.005 \times (443.47 - 300)$$
$$Q_{23} = 144.19 \text{ m}$$

Refrigeration process, heat picked during process 4 –1, $Q_{41} = m \times C_p \times (T_1 - T_4)$

$$= m \times 1.005 \times (280 - 189.42)$$
$$Q_{41} = 91.03 \text{ m}$$
$$\text{Net work} = Q_{23} - Q_{41}$$
$$= (144.19 - 91.03) \text{ m}$$
$$W = 53.16 \text{ m kJ}$$

$$\text{COP} = \frac{\text{Refrigeration effect}}{\text{Net work}} = \frac{91.03\text{m}}{53.16\text{m}} = 1.71$$

COP = **1.71** **Ans.**

6. In a refrigerator working on Bell Coleman cycle the air enters compressor at 1 bar and –10°C and gets compressed upto 5.5 bar. Compressed air is cooled to 27°C at same pressure before being sent to expander for expansion upto 1 bar and then passes through refrigerated space. Determine refrigeration capacity, hp required to run compressor and COP of system if air flow rate is 0.8 kg/s.

Solution:
Given:
$T_1 = -10°C$ or 263 K, $P_1 = 1$ bar, $T_3 = 27°C$ or 300 K, $P_2 = 5.5$ bar, $m = 0.8$ kg/s.

For process 1–2,

$$\left(\frac{P_2}{P_1}\right)^{\frac{\gamma-1}{\gamma}} = \frac{T_2}{T_1}$$

$$T_2 = 263 \times (5.5)^{\frac{(1.4-1)}{1.4}}$$

$$T_2 = 428.04 \text{ K}$$

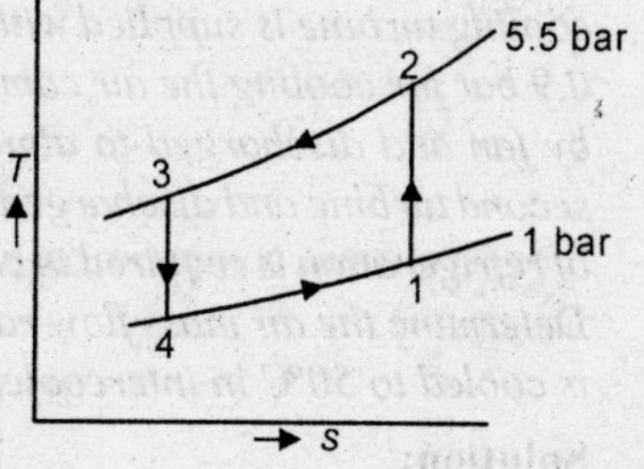

Fig. 11.26

For process 3–4,

$$\left(\frac{P_3}{P_4}\right)^{\frac{\gamma-1}{\gamma}} = \frac{T_3}{T_4}$$

$$T_4 = T_3 \times \left(\frac{P_4}{P_3}\right)^{\frac{\gamma-1}{\gamma}} = T_3 \times \left(\frac{P_1}{P_2}\right)^{\frac{\gamma-1}{\gamma}}$$

$$T_4 = 184.33 \text{ K}$$

Refrigeration capacity $= mC_p\,(T_1 - T_4)$
$= 0.8 \times 1.005\,(263 - 184.33)$
$= 63.25$ kJ/s or 18.07 ton

Refrigeration capacity = **18.07 ton** **Ans.**

$$\text{Work required to run the compressor} = \frac{m \times n}{n-1}\,(p_2v_2 - p_1v_1)$$

$$= \frac{m \times n}{n-1}\,\text{R}(T_2 - T_1)$$

$$= \frac{0.8 \times 1.4}{(0.4)} \times 0.287\,(428.04 - 263)$$

$$= 132.63 \text{ kJ/s}$$

or

$$\text{HP required to run compressor} = \frac{132.63}{0.7457} = 177.86 \text{ hp.}$$

HP required to run compressor = **177.86 hp** **Ans.**

Net work input $= mC_p\,\{(T_2 - T_3) - (T_1 - T_4)\}$
$= 0.8 \times 1.005\,\{(428.04 - 300) - (263 - 184.33)\}$
$W = 39.69$ kJ/s

$$\text{COP} = \frac{\text{Refrigeration capacity}}{\text{Work}} = \frac{63.25}{39.69} = 1.59$$

COP = **1.59** **Ans.**

7. Air refrigeration system for an aircraft has two cooling turbines, one intercooler and a fan. First cooling turbine is supplied with ram air at 1.2 bar, 15°C and delivers it after expansion to intercooler at 0.9 bar for cooling the air coming from compressor at 4 bar. The cooling air from intercooler is sucked by fan and discharged to atmosphere and the cooled air from intercooler is expanded upto 1 bar in second turbine and discharged into passenger cabin. The air is exhausted from cabin at 25°C and 10 ton of refrigeration is required in cabin. Index of compression and expansion are. 1.45 and 1.30 respectively. Determine the air mass flow rate in cabin, ram air mass flow rate, and COP of system if compressed air is cooled to 50°C in intercooler and temperature rise in intercooler for ram air is limited to 30°C.

Solution:

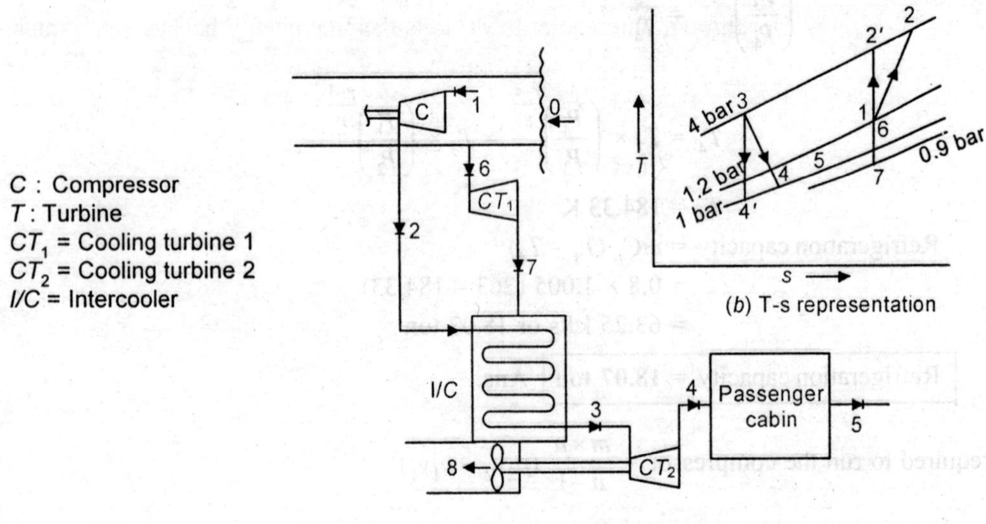

Fig. 11.27 Schematic of air refrigeration cycle

Given:

$P_1 = P_6 = 1.2$ bar, $T_1 = T_6 = 15°C = 288$ K, $P_7 = 0.9$ bar

$P_3 = P_2 = 4$ bar, $P_4 = 1$ bar, $T_5 = 25°C$, $T_3 = 50°C$ or 323 K, $T_8 = 30°C$

For process 1–2, $n = 1.45$;

$$\frac{T_2}{T_1} = \left(\frac{P_2}{P_1}\right)^{\frac{n-1}{n}}$$

$$T_2 = 288\left(\frac{4}{1.2}\right)^{\frac{1.45-1}{1.45}}$$

$$T_2 = 418.47 \text{ K}$$

For process 3–4, $n' = 1.3$,

$$\frac{T_4}{T_3} = \left(\frac{P_4}{P_3}\right)^{\left(\frac{n'-1}{n'}\right)}$$

$$T_4 = 323\left(\frac{1}{4}\right)^{\frac{1.3-1}{1.3}}$$

$$T_4 = 234.57 \text{ K}$$

Refrigeration effect in passenger cabin with m kg/s mass flow rate of air.

$$10 \times 3.5 = m \times C_P\,(T_5 - T_4)$$

$$10 \times 3.5 = m \times 1.005\,(298 - 234.57)$$

$$m = 0.055 \text{ kg/s}$$

Air mass flow rate in cabin = **0.55 kg/s**	**Ans.**

Let the mass flow rate of air through intercooler be m_1 kg/s then the energy balance upon intercooler yields,

$$m_1 \times C_P \times (T_8 - T_7) = m \times C_P\,(T_2 - T_3)$$

During process 6 – 7, $\frac{T_7}{T_6} = \left(\frac{P_7}{P_6}\right)^{\frac{1.3-1}{1.3}}$

$$T_7 = 288\left(\frac{0.9}{1.2}\right)^{\frac{1.3-1}{1.3}} = 269.5 \text{ K}$$

Substituting T_2, T_3, T_7, T_8 and m in energy balance on intercooler,

$$m_1 = 1.005 \times (303 - 269.5) = 0.55 \times 1.005 \times (418.47 - 323)$$

$$m_1 = 1.57 \text{ kg/s}$$

$$\text{Total ram air mass flow rate} = m + m_l = 0.55 + 1.57$$

$$= 2.12 \text{ kg/s}$$

Ram air mass flow rate = **2.12 kg/s**	**Ans.**

$$\text{Work input to compressor} = m \times C_P\,(T_2 - T_1)$$

$$= 0.55 \times 1.005\,(418.47 - 288)$$

$$= 72.12 \text{ kJ/s}$$

$$\text{COP} = \frac{\text{Refrigeration effect}}{\text{Work input}}$$

$$= \frac{10 \times 3.5}{72.12}$$

$$\text{COP} = 0.485$$

COP = **0.485**	**Ans.**

8. *An air cooling system for aeroplane has refrigeration capacity of 15 tons. Atmospheric air is available at 0.9 bar, 3°C and due to ramming action pressure rises upto 1 bar. The ram air leaves main compressor at 4 bar and subsequently 66% of heat of compressed air is removed in air cooled heat exchanger and then passed through an evaporator for further cooling. Temperature of air gets lowered by 10°C in evaporator and this cool air is then expanded in cooling turbine before being supplied to passenger cabin maintained at 1.03 bar. Air leaves cabin at 25°C. Considering isentropic efficiency of compressor and turbine as 0.9 and 0.8 determine, (a) HP required to run the system, (b) COP of system.*

Solution:

Given:

$P_0 = 0.9$ bar, $T_0 = 3°C$ or 276 K, $P_2 = P_3 = P_4 = 4$ bar, $P_5 = 1.03$ bar.

$T_3 - T_4 = 10$, $T_6 = 25°C$ or 298 K

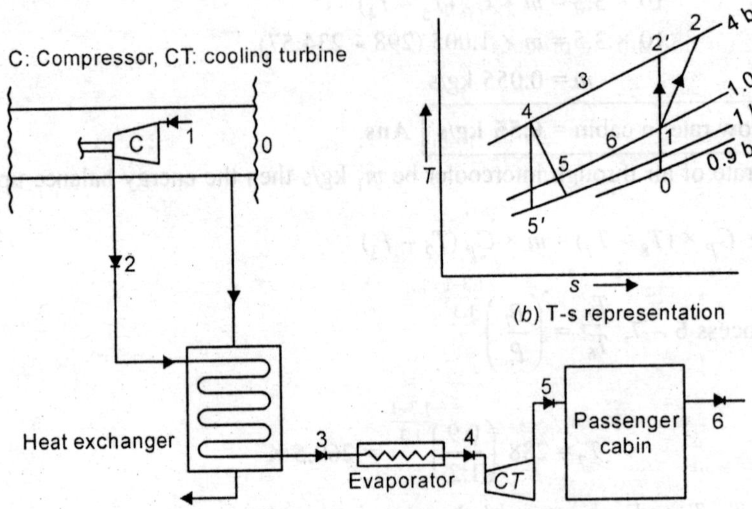

Schematic or air refrigeration system

Fig. 11.28

Considering index of compression and expansion as 1.4.

During ramming action, process 0–1,

$$\frac{T_1}{T_0} = \left(\frac{P_1}{P_0}\right)^{\frac{\gamma-1}{\gamma}}$$

$$T_1 = 276 \times \left(\frac{1}{0.9}\right)^{\frac{1.4-1}{1.4}}$$

$$T_1 = 284.43 \text{ K}$$

During compression process 1–2′

$$\frac{T_{2'}}{T_1} = \left(\frac{P_2}{P_1}\right)^{\frac{\gamma-1}{\gamma}}$$

$$T_{2'} = 284.43 \left(\frac{4}{1}\right)^{\frac{1.4-1}{1.4}}$$

$$T_{2'} = 422.66 \text{ K}$$

$$\eta_{isen,\,c} = 0.9 = \frac{T_{2'} - T_1}{T_2 - T_1} \quad \text{or, } T_2 = 438.02 \text{ K}$$

In heat exchanger 66% of heat loss shall result in temperature at exit from heat exchanger to be, $T_3 = 0.34 \times 438.02 = 148.93$ K. Subsequently for 10°C temperature drop in evaporator,

$$T_4 = 148.93 - 10$$
$$T_4 = 138.93 \text{ K}$$

Expansion in cooling turbine during process 4–5′,

$$\frac{T_{5'}}{T_4} = \left(\frac{P_5}{P_4}\right)^{\frac{\gamma-1}{\gamma}}$$

$$T_{5'} = 138.93 \left(\frac{1.03}{4}\right)^{\frac{1.4-1}{1.4}}$$

$$T_{5'} = 94.29 \text{ K}$$

$$\eta_{isen,\,T} = \frac{T_4 - T_5}{T_4 - T_{5'}}$$

$$0.8 = \frac{138.93 - T_5}{138.93 - 94.29}$$

$$T_5 = 103.22 \text{ K}$$

Let the mass flow rate of air through cabin be m kg/s. Using refrigeration capacity heat balance yields.

$$15 \times 3.5 = m \times C_P (T_6 - T_5)$$

$$m = \frac{15 \times 3.5}{1.005 \times (298 - 103.22)}$$

$$m = 0.268 \text{ kg/s}$$

$$\text{Work input to compressor} = 0.268 \times 1.005\,(T_2 - T_1)$$
$$= 41.37 \text{ kJ/s or } 55.48 \text{ hp}$$

$$\text{COP} = \frac{\text{Refrigeration effect}}{\text{Work input}}$$

$$= \frac{15 \times 3.5}{41.37}$$

$$\text{COP} = 1.27$$

COP = **1.27**
HP required = **55.48 hp**

Ans.

9. *A refrigerator working on simple vapour compression cycle operates between the temperature of 25°C and – 15°C with NH_3, refrigerant. Ammonia is found to be dry after compression and no under cooling of liquid refrigerant occurs in cycle. Using following data obtain COP of system.*

Properties of NH_3

Temperature	*Enthalpy, kJ/kg*		*Entropy, kJ/kg · K*	
	h_f	h_g	s_f	s_g
– 15ºC	– 54.51	1303.74	– 0.2132	5.0536
25ºC	99.94	1317.95	0.3386	4.4809

Solution:

$$T_1 = -15°C \text{ or } 258 \text{ K}, \; T_2 = 25°C \text{ or } 298 \text{ K}$$

$$h_2 = h_{g \text{ at } 25°C} = 1317.95 \text{ kJ/kg}$$

$$s_2 = s_{g \text{ at } 25°C} = 4.4809 \text{ kJ/kg.K}$$

$$h_3 = h_{f \text{ at } 25°C} = 99.94 \text{ kJ/kg}$$

$$s_3 = s_{f \text{ at } 25°C} = 0.3386 \text{ kJ/kg.K}$$

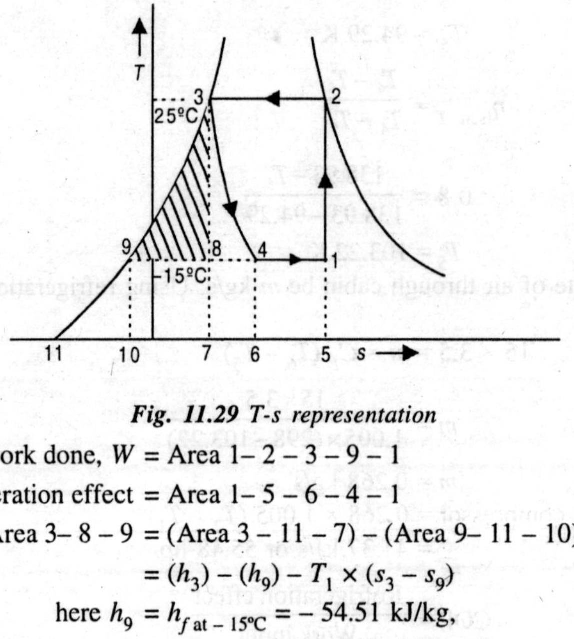

Fig. 11.29 T-s representation

Here work done, W = Area 1– 2 – 3 – 9 – 1

Refrigeration effect = Area 1– 5 – 6 – 4 – 1

$$\text{Area } 3-8-9 = (\text{Area } 3-11-7) - (\text{Area } 9-11-10) - (\text{Area } 9-8-7-10)$$

$$= (h_3) - (h_9) - T_1 \times (s_3 - s_9)$$

$$\text{here } h_9 = h_{f \text{ at } -15°C} = -54.51 \text{ kJ/kg},$$

$$s_9 = h_{f \text{ at } -15°C} = -0.2132 \text{ kJ/kg. K}$$

Substituting

$$\text{Area } 3-8-9 = (99.94 - (-54.51) - 258\,(0.3386 + 0.2132)$$

$$= 12.09 \text{ kJ/kg}$$

During throttling process between 3 and 4.

$$h_3 = h_4$$

$$(\text{Area} = 3-11-7-3) = (\text{Area } 4-9-11-6-4)$$

$$(\text{Area } 3-8-9) + (\text{Area } 8-9-11-7-8)$$

$$= (\text{Area } 4-6-7-8-4) + (\text{Area } 8-9-11-7-8)$$

or $$(\text{Area } 3-8-9) = (\text{Area } 4-6-7-8-4)$$

$$(\text{Area } 4-6-7-8-4) = 12.09 \text{ kJ/kg}$$

$$\text{Also, } (\text{Area } 4-6-7-8-4) = T_1 \times (s_4 - s_8)$$

$$12.09 = 258 \times (s_4 - s_8)$$

$$s_4 - s_8 = 0.0469 \text{ kJ/kg} \cdot \text{K}$$

Also, $s_3 = s_8 = 0.3386$ kJ/kg · K,

therefore, $s_4 = 0.3855$ kJ/kg · K

Also $s_1 = s_2 = 4.4809$ kJ/kg.

$$\text{Refrigeration effect} = \text{Area } (1-5-6-4-1) = T_1 \times (s_1 - s_4)$$

$$= 258 \times (4.4809 - 0.3855)$$

$$= 1056.61 \text{ kJ/kg}$$

$$\text{Work done} = \text{Area } 1-2-3-9-1$$

$$= (\text{Area } 3-8-9) + ((T_2 - T_1) \times (s_1 - s_8))$$

$$= 12.09 + ((298 - 258) \times (4.4809 - 0.3386))$$

$$= 177.782 \text{ kJ/kg}$$

$$\text{COP} = \frac{\text{Refrigeration effect}}{\text{Work done}} = \frac{1056.61}{177.782} = 5.94$$

COP = **5.94** **Ans.**

10. *A vapour compression cycle works on Freon-12 refrigerant with condensation temperature of 40°C and evaporator temperature of – 20°C. Refrigeration effect of 2.86 ton is desired from the cycle. The compressor runs with 1200 rpm and has clearance volume of 2%. Considering compression index of 1.13 and following data determine.*

(a) the COP

(b) the piston displacement in the reciprocating compressor used for compression.

Properties of Freon - 12

Temperature °C	*Saturation Pressure,* bar	*Specific Volume,* $v_g m^3/kg$	*Enthalpy, kJ/kg* h_f	h_g	*Entropy, kJ/kg·K* s_f	s_g	*specific heat kJ/kg·K* C_{pf}	C_{pg}
– 20	1.51	0.1088	17.8	178.61	0.0730	0.7082	–	0.605
40	9.61	–	74.53	203.05	0.2716	0.682	0.976	0.747

Solution:

Given $T_1 = -20°C = 253$ K. During expansion (throttling) between 3 and 4,

$h_3 = h_4$, $h_3 = h_{f\text{ at }40°C} = 74.53 \text{ kJ/kg} = h_4$

Process 1–2 is adiabatic compression so

$$s_1 = s_2$$

and $s_1 = s_{g\text{ at }-20°C} = 0.7082$ kJ/kg·K

$$0.7082 = s_{g\text{ at }40°C} + C_{pg\text{ at }40°C} \cdot \ln\left(\frac{T_2}{313}\right)$$

$$0.7082 = 0.682 + 0.747 \ln\left(\frac{T_2}{313}\right)$$

Temperature after compression $T_2 = 324.17$ K

Enthalpy after compression, $h_2 = h_{g\text{ at }40°C} + C_{pg\text{ at }40°C} \cdot (324.17 - 313)$

$= 203.05 + (0.747 \times (324.17 - 313))$

$h_2 = 211.39$ kJ/kg

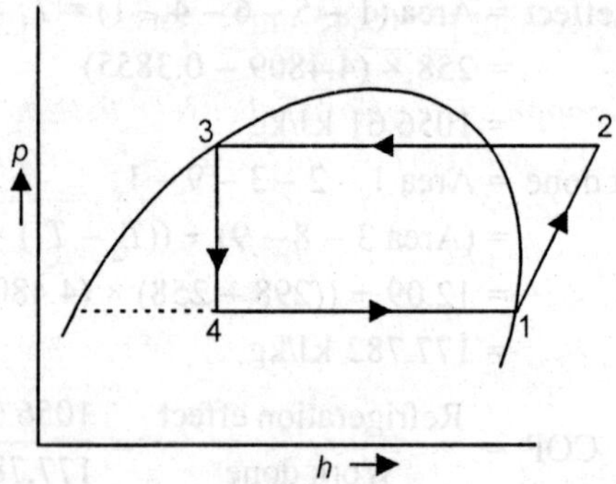

Fig. 11.30

Compression work required, per kg $= h_2 - h_1$

$= 211.39 - h_{g\text{ at }-20°C}$

$= 211.39 - 178.61$

$W_c = 32.78$ kJ/kg

Refrigeration effect during cycle, per kg $= h_1 - h_4$

$= 178.61 - 74.53$

$= 104.08$ kJ/kg

Mass flow rate of refrigerant, $m = \dfrac{2.86 \times 3.5}{104.08}$

$m = 0.096$ kg/s.

$$\text{COP} = \frac{104.08}{32.78} = 3.175$$

Volumetric efficiency of reciprocating compressor, given $C = 0.02$

$$\eta_{vol} = 1 + C - C\left(\frac{P_2}{P_1}\right)^{\frac{1}{n}}$$

$$= 1 + 0.02 - 0.02\left(\frac{9.61}{1.51}\right)^{\frac{1}{1.13}}$$

$$\eta_{vol} = 0.9171$$

Let piston displacement by V, m^3.

$$\text{Mass flow rate, } 0.096 = \frac{V \times 0.9171 \times 1200}{60 \times v_{g\,\text{at}-20^\circ\text{C}}}$$

$$0.096 = \frac{V \times 0.9171 \times 1200}{60 \times 0.1088}$$

$$V = 569.45 \text{ cm}^3$$

COP = **3.175**
Piston displacement = **569.45 cm³** **Ans.**

***11.** A vapour compression refrigeration cycle operates between the condensation temperature of 20°C and evaporator temperature of –10°C with carbon dioxide as refrigerant. Temperature after isentropic compression is 40°C and condensate leaves at 10°C before being passed through expansion valve. Determine COP of system and mass flow rate of CO_2 required to get refrigeration effect of 2 kW. Properties of CO_2 are given as under:*

Properties of CO_2

Temperature °C	*Saturation Pressure, bar*	*Specific Volume, v_g vm^3/kg*	*Enthalpy, kJ/kg*		*Entropy, kJ/kg·K*		*Specific heat kJ/kg·K*	
			h_f	h_g	s_f	s_g	C_{pf}	C_{pg}
20	57.27	–	144.11	299.62	0.523	1.0527	2.889	2.135
– 10	26.49	0.014	60.78	322.28	0.2381	1.2324	–	–

Solution:

Processes of vapour compression cycle are shown on *T-s* diagram

1 – 2 : Isentropic compression process

2 – 3 – 4: Condensation process

4 – 5: Isenthalpic expansion process

5 – 1: Refrigeration process in evaporator

$$h_1 = h_{g\,\text{at}\,-\,10^\circ\text{C}} = 322.28 \text{ kJ/kg}$$

$$h_2 = h_{g\,\text{at}\,20^\circ\text{C}} + C_{pg\,\text{at}\,20^\circ\text{C}} \times (40 - 20)$$

$$= 299.62 + (2.135 \times 20)$$

$$h_2 = 342.32 \text{ kJ/kg}$$

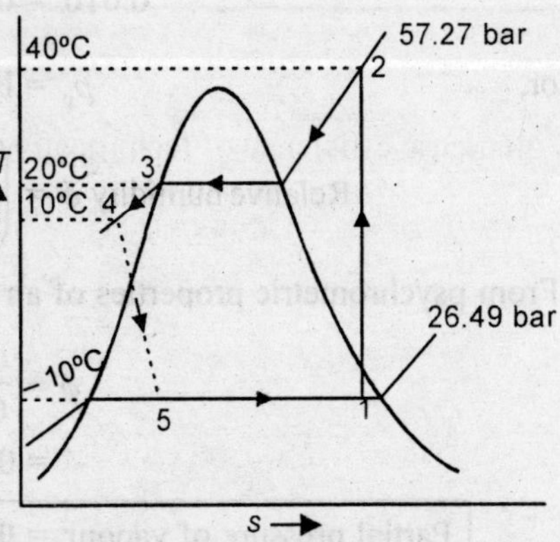

Fig. 11.31 T-s representation

$$\text{Entropy at state 2, } s_2 = s_{g\,\text{at}\,+\,20^\circ\text{C}} + C_{\text{pg at }20^\circ\text{C}} \times \ln\left(\frac{273+40}{273+20}\right)$$

$$= 1.0527 + 2.135 \times \ln\left(\frac{313}{293}\right)$$

$$s_2 = 1.1937 \text{ kJ/kg·K}$$

Entropy during isentropic process, $s_1 = s_2$

$$1.1937 = s_{f\,\text{at}\,-\,10^\circ\text{C}} + x_1 \times s_{fg\,\text{at}\,-\,10^\circ\text{C}}$$

$$1.1937 = 0.2381 + x_1 \times (1.2324 - 0.2381)$$

$$x_1 = 0.961$$

Enthalpy at state 1, $h_1 = h_{f\,at\,-10°C} + x_1 \times h_{fg\,at\,-10°C}$

$= 60.78 + (0.961 \times (322.28 - 60.78))$

$h_1 = 312.08$ kJ/kg

$h_3 = h_{f\,at\,20°C} = 144.11$ kJ/kg

Since undercooling occurs upto 10°C, so, $h_4 = h_3 - C_{pf} \cdot \Delta T$

$= 144.11 - (2.889 \times (20 - 10))$

$h_4 = 115.22$ kJ/kg

Also, $h_4 = h_5 = 115.22$ kJ/kg

Refrigeration effect per kg of refrigerant $= (h_1 - h_5) = (312.08 - 115.22)$

$= 196.86$ kJ/kg

Let refrigerant flow rate be m kg/s.

Refrigeration effect $\Rightarrow 2 = m \times 196.86$

$m = 0.1016$ kg/s

Compressor work, $w_c = h_2 - h_1 = 342.32 - 312.08$

$w_c = 30.24$ kJ/kg

$$\text{COP} = \frac{\text{Refrigeration effect per kg}}{\text{Compressor work per kg}} = \frac{196.86}{30.24} = 6.51$$

COP = **6.51**, Mass flow rate = **0.1016 kg/s** **Ans.**

12. Determine partial pressure of vapour and relative humidity in the atmospheric air having specific humidity of 16 gm/kg of air and 25°C DBT.

Solution:

Here pressure of atmospheric air may be taken as 1.013 bar

$$\text{Specific humidity, } \omega = 0.622 \left(\frac{p_v}{p - p_v}\right)$$

$$0.016 = 0.622 \left(\frac{p_v}{1.013 - p_v}\right)$$

or, p_v = Partial pressure of vapour = 0.0254 bar

$$\text{Relative humidity } \phi = \left(\frac{p_v}{p_{v,\,sat}}\right)$$

From psychrometric properties of air table $p_{v,\,sat\,at\,25°C} = 0.03098$ bar

$$\phi = \frac{0.0254}{0.03098}$$

$= 0.8198$ or 81.98%

Partial pressure of vapour = **0.0254**
Relative humidity = **81.98%** **Ans.**

***13.** For the atmospheric air at room temperature of 30°C and relative humidity of 60% determine partial pressure of air, humidity ratio, dew point temperature, density and enthalpy of air.*

Solution:

At 30°C from steam table, saturation pressure, $p_{v,\,sat} = 0.0425$ bar

$$\text{Partial pressure of vapour} = \text{Relative humidity} \times p_{v,\,sat}$$
$$= 0.6 \times 0.0425 = 0.0255 \text{ bar}$$

$$\text{Partial pressure of air} = \text{Total pressure of mixture} - \text{Partial pressure of vapour}$$
$$= 1.013 - 0.0255$$
$$= 0.9875 \text{ bar}$$

Partial pressure of air = **0.9875 bar** **Ans.**

$$\text{Humidity ratio, } \omega = 0.622 \frac{p_v}{p - p_v}$$
$$= 0.622 \left(\frac{0.0255}{1.013 - 0.0255}\right)$$
$$\omega = 0.01606 \text{ kg/kg of dry air.}$$

Humidity ratio = **0.01606 kg/kg of dry air** **Ans.**

Dew point temperature may be seen from the steam table. The saturation temperature corresponding to the partial pressure of vapour is 0.0255 bar. Dew point temperature can be approximated as 21.4°C by interpolation.

Dew point temperatrure = **21.4°C** **Ans.**

$$\text{Density of mixture} = \text{Density of air } (\rho_a) + \text{Density of vapour } (\rho_v)$$
$$= \rho_a + \rho_v$$
$$= \rho_a + \omega \cdot \rho_a$$
$$= \rho_a (1 + \omega)$$
$$= \frac{P_a}{RT_a} (1 + 0.01606)$$
$$= \frac{1.013 \times 10^2 \times 1.01606}{0.287 \times 303}$$
$$= 1.1835 \text{ kg/m}^3$$

Density = **1.1835 kg/m³** **Ans.**

$$\text{Enthalpy of mixture, } h = C_p \cdot T + \omega\,(h_{g\text{ at }30^\circ\text{C}} + 1.860\,(30 - 21.4))$$
$$= (1.005 \times 30) + (0.01606 \times (2540.1 + 1.86 \times 8.6))$$
$$= 71.2 \text{ kJ/kg of dry air}$$

Enthalpy of mixture = **71.2 kJ/kg** of dry air **Ans.**

***14.** Determine the mass of water added and heat transferred for conditioning atmospheric air at 15°C and 80% relative humidity to temperature of 25°C and relative humidity of 50%. Final volume of conditioned air is 0.8 m³/s.*

Solution:

Initial state at 15°C and 80% relative humidity is shown by point 1 and final state at 25°C and 50% relative humidity is shown by point 2 on psychrometric chart.

$$\omega_1 = 0.0086 \text{ kg/kg of air and } h_1 = 37 \text{ kJ/kg}$$
$$\omega_2 = 0.01 \text{ kg/kg of air and } h_2 = 50 \text{ kJ/kg}$$
$$v_2 = 0.854 \text{ m}^3\text{/kg}$$

Mass of water added between states 1 and 2 $= \omega_2 - \omega_1$
$= 0.01 - 0.0086$
$= 0.0014$ kg/kg of air

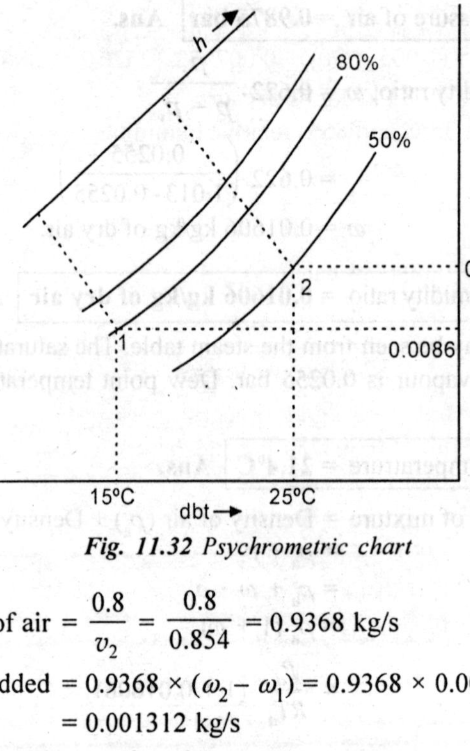

Fig. 11.32 Psychrometric chart

$$\text{Mass flow rate of air} = \frac{0.8}{v_2} = \frac{0.8}{0.854} = 0.9368 \text{ kg/s}$$

$$\text{Total mass of water added} = 0.9368 \times (\omega_2 - \omega_1) = 0.9368 \times 0.0014$$
$$= 0.001312 \text{ kg/s}$$

$$\text{Heat transferred} = m_a(h_2 - h_1)$$
$$= 0.9368 \times (50 - 37) = 12.18 \text{ kJ/s}$$

Mass of water added = 0.001312 kg/s, Heat transferred = **12.18 kW**

Ans.

15. Two streams of moist air, one having flow rate of 3 kg/s at 30°C and 30% relative humidity, other having flow rate of 2 kg/s at 35°C and 85% relative humidity get mixed adiabatically. Determine specific humidity and partial pressure of water vapour after mixing. Take $C_{p,\ stream} = 1.86$ kJ/kg · K

Solution:

Let temperature after mixing be T°C. For getting final enthalpy after adiabatic mixing the enthalpy of two streams are required.

For moist air stream at 30°C and 30% relative humidity.

$$\phi_1 = \frac{p_{v1}}{p_{v\text{sat at }30°C}}$$

$$p_{v\text{ sat at }30°C} = 0.04246 \text{ bar, so, } p_{v1} = 0.3 \times 0.04246 = 0.01274 \text{ bar}$$

Corresponding to vapour pressure of 0.01274 bar the dew point temperature shall be 10.5ºC.

$$\text{Specific humidity, } \omega_1 = \frac{0.622 \times p_{v1}}{(p - p_{v1})} = \frac{0.622 \times 0.01274}{(1.013 - 0.01274)}$$

$$\omega_1 = 0.007922 \text{ kg/kg of air.}$$

At dew point temperature of 10.5ºC, enthalpy, $h_{g\text{ at }10.5°C} = 2520.7$ kJ/kg.

$$h_1 = C_{p,\text{ air}} \times T_1 + \omega_1 \{h_{g\text{ at }10.5°C} - C_{p,\text{ stream}} \times (T_1 - T_{dp1})\}$$

$$= (1.005 \times 30) + 0.007922 \{2520.7 - 1.86 \times (30 - 10.5)\}$$

$$h_1 = 49.83 \text{ kJ/kg of dry air}$$

For second moist air stream at 35ºC and 85% relative humidity.

$$\phi_2 = 0.85 = \frac{p_{v2}}{p_{v\text{sat at }35°C}} = \frac{p_{v2}}{0.005628}$$

or,
$$p_{v2} = 0.004784 \text{ bar}$$

$$\text{Specific humidity } \omega_2 = \frac{0.622 \times p_{v2}}{(p - p_{v2})} = \frac{0.622 \times 0.004784}{(1.013 - 0.004784)}$$

$$\omega_2 = 0.002951 \text{ kg/kg of air}$$

Corresponding to vapour pressure of 0.004784 bar the dew point temperature is 32ºC.
So, enthalpy of second stream,

$$h_2 = C_{p,\text{ air}} \cdot T_2 + \omega_2 \{h_{g\text{ at }32°C} + C_{p,\text{ stream}} \times (T_2 - T_{dp2})\}$$

$$= 1.005 \times 35 + 0.002951 \{2559.9 + 1.86 \times (35 - 32)\}$$

$$h_2 = 42.75 \text{ kJ/kg of dry air}$$

Enthalpy of mixture after adiabatic mixing

$$= \frac{1}{(m_1 + m_2)} \times \left\{ \frac{h_1 \times m_1}{(1 + \omega_1)} + \frac{h_2 \times m_2}{(1 + \omega_2)} \right\}$$

$$= \frac{1}{5} \times \left\{ \left(\frac{49.83 \times 3}{1.007922} \right) + \left(\frac{42.75 \times 2}{1.002951} \right) \right\}$$

$$= 46.71 \text{ kJ/kg of moist air}$$

$$\text{Mass of vapour per kg of moist air} = \frac{1}{5} \left\{ \frac{\omega_1 \times m_1}{(1 + \omega_1)} + \frac{\omega_2 \times m_2}{(1 + \omega_2)} \right\}$$

$$= \frac{1}{5} \times \left\{ \left(\frac{0.007922 \times 3}{1.007922} \right) + \left(\frac{0.002951 \times 2}{1.002951} \right) \right\}$$

$$= 0.005893 \text{ kg per kg of moist air}$$

$$\text{Specific humidity of mixture} = \frac{0.005893}{1 - 0.005893}$$

$$= 0.00593 \text{ kg/kg of dry air.}$$

$$\omega = \frac{0.622 \times p_v}{p - p_v}$$

$$0.00593 = \frac{0.622 \times p_v}{1.013 - p_v}$$

Partial pressure of water vapour = 0.00957 bar

Specific humidity of mixture = **0.00593 kg/kg** dry air Partial pressure of water vapour in mixture = **0.00957 bar**	**Ans.**

16. Calculate the amount of heat added to moist air entering heating coil at 15°C dbt and 80% relative humidity at a rate of 3m³/s. Air leaves the heating coil at 30°C and atmospheric pressure is 1.013 bar.

Solution:

The type of heating involved is sensible heating. Locating state 1 on psychrometric chart corresponding to 15°C dbt and 80% relative humidity the other property values shall be,

$h_1 = 36.4$ kJ/kg, $\omega_1 = 0.0086$ kg/kg of air, $v_1 = 0.825$ m³/kg

Final state 2 has, $h_2 = 52$ kJ/kg

$$\text{Mass of air} = \frac{3}{0.825} = 3.64 \text{ kg/s}$$

$$\text{Amount of heat added} = 3.64 \times (52 - 36.4)$$

$$= 56.78 \text{ kJ/s.}$$

Heat added = **56.78 kJ/s**	**Ans.**

EXERCISE

11.1 Define refrigeration and compare it with air conditioning.

11.2 Explain reversed carnot cycle and its limitations.

11.3 Define unit of refrigeration.

11.4 Explain working of a Carnot refrigerator and why it cannot be used in practice.

11.5 Describe air-refrigeration cycles.

11.6 Explain simple air refrigeration cycle being used in aeroplanes.

11.7 What do you understand by vapour compression cycle? Explain the processes on *T-s* and *p-h* diagrams.

11.8 Compare Carnot refrigeration cycle with vapour compression cycle.

11.9 What is meant by wet compression and dry compression in vapour compression cycles? Explain.

11.10 What is there in multistage vapour compression cycle? List its advantages over simple vapour compression cycle.

11.11 Describe the vapour absorption cycle for ammonia-water system and its working.

11.12 Explain the modified absorption-refrigeration cycle and its working.

11.13 Describe heat pump system based on vapour compression cycle.

11.14 Classify refrigerants and list desired properties of refrigerants.

11.15 Explain sensible heating and sensible cooling on psychrometric chart.

11.16 Write short notes on relative humidity, specific humidity, dry bulb temperature, dew point temperature, wet bulb temperature.

11.17 Discuss the working of summer air-conditioning system.

11.18 Describe working of winter air- conditioning system.

11.19 Show that the moisture content 'ω' of the atmospheric air is given as,

$$\omega = \frac{0.622 \times p_v}{p_a}$$

where p_v and p_a are partial pressure of vapour and air respectively.

11.20 Determine power required to run the compressor and heat rejected by compressor per second from a refrigerator working on reversed Carnot cycle between temperatures of 266 K and 300 K. Refrigerator produces 0.139 kg/s ice at –5°C from water at 287 K. Also determine COP of refrigerator.

[9.76 hp, 63.19 kJ/s, 7.82]

11.21 A refrigerator working between –5°C and 20°C has capacity of 600 tons. Determine the minimum hp required for machine. [266 hp]

11.22 An air refrigeration cycle has pressure ratio of 4 and the air enters compressor at 15°C and expander at 30°C. Consider compression and expansion index to be 1.25 and 1.35 respectively. Determine COP of refrigeration cycle. [1.7]

11.23 Show that the ratio of maximum COP of Bell-Coleman cycle and COP of Carnot cycle is given by:

$$\frac{\text{COP}_{\text{Bell–Colemen}}}{\text{COP}_{\text{carnot}}} = \left(\frac{T_1 - T_2}{T_2}\right)^2$$

where T_1 and T_2 are the temperature of surroundings and refrigerated space. Assume that the rise in temperature of air passing through heat exchanger (cooler) is same in both cases and is constant.

11.24 A Bell-coleman refrigeration cycle of 75.8 kJ/s refrigeration capacity runs between pressures of 1 bar and 5 bar. Refrigerated space is maintained at 0°C and ambient temperature is 300 K. For isentropic efficiency of compression and expansion being 0.85 and 0.90 respectively. Determine COP of system.

[0.815]

11.25 An air refrigeration system working on reversed Brayton cycle has refrigeration capacity of 10 ton and pressure ratio of 3. Pressure in refrigerated space is 1.4 bar, temperature of air at inlet to compressor is 253 K and air is cooled to a temperature of 323 K in cooler. Considering C_p = 1.07 kJ/kg·K determine COP of system, air circulation per minute, theoretical piston displacement and power required.

[2.43, 114 kg/min, 59.13 m³/min, 13 kW]

11.26 An air refrigeration cycle works between 1 bar and 12 bar to produce refrigeration capacity of 25 tons when air leaves cooler at 25°C and air leaves refrigerated space at 0°C. Considering compression and expansion processes to have index of 1.35 determine.

(*i*) the COP of system

(*ii*) the HP required

(*iii*) the mass of air circulated per minute

(*iv*) the piston displacements of compressor and expanders.

[1.94, 116 hp, 133 kg/min, 35.5 m³/min, 20.35 m³/min]

11.27 Air cooled aeroplane system has pressure of air getting increased from atmospheric pressure of 0.9 bar to 1 bar due to ram action. Compressed air leaves compressor at 3.5 bar and temperature of compressed air leaving heat exchanger is 355 K and subsequently leaving the evaporator is 348 K. High pressure cool air leaving evaporator enters the cooling turbine and subsequently to passenger cabin at 1.03 bar. Considering isentropic efficiencies of compression and expansion to be 0.8 and 0.75 determine hp required for desired cooling and COP of system when stale air leaves passenger cabin at 298 K. Ambient air temperature is 296 K. [596 hp, 0.16]

11.28 The passenger cabin of an aircraft is maintained at 298 K when ambient air is at 0.35 bar and –15°C. Compressor in aircraft has pressure ratio of 3 and aircraft is moving at 1000 kmph. Compressed air leaving compressor is passed through a heat exchanger and cooled there upto ambient air temperature. Subsequently this cool air passes through cooling turbine so as to leave it at 1.06 bar. Considering cooling load to be 200 ton determine.

(*i*) the stagnation temperature and pressure

(*ii*) the COP of system
(*iii*) the mass flow rate of air
Take $C_{pair} = 1$ kJ/kg·K, $\gamma = 1.4$ [296.5 K, 0.57 bar, 0.31, 2 kg/s]

11.29 Calculate COP of vapour compression system running on ammonia (NH_3) as refrigerant between 25°C and – 10°C such that the refrigerant is dry saturated at the end of compression and there is no undercooling. Properties of NH_3 are;

Temperature, °C	*Enthalpy, kJ/kg* h_f	h_{fg}	*Entropy, kJ/kg·K* s_f
25	100.32	1228.92	0.3469
–10	–33.52	1345.96	–0.1379

11.30 A vapour compression cycle running on NH_3 had refrigeration capacity of 1990.5 ton while operating between temperature limits of 298 K and 258 K. The vapour is dry saturated at the end of compression and actual COP of cycle is only 60% of theoretical COP, then determine power required to drive compressor.
Properties of NH_3

Temperature, °C	*Enthalpy, kJ/kg* h_f	h_g	*Entropy, kJ/kg·K* s_f	s_g
25°C	99.94	1318	0.0347	4.4809
–15°C	–54.51	1303.7	–0.2132	5.0536

[43.8 hp]

11.31 Determine the COP of vapour compression system running with CO_2 and operating between temperature of 25°C and – 5°C such that carbon dioxide entering compressor is 0.6.
Properties of CO_2

Temperature, °C	*Enthalpy, kJ/kg* h_f	h_g	*Entropy, kJ/kg·K* s_f	s_g
25	81.09	202.31	0.2508	0.6287
– 5	– 7.52	236.59	–0.0418	0.8402

[3.3]

11.32 A water cooler working on Freon– 12 as refrigerant operates between 299 K and 275 K such that the vapour leaves evaporator dry saturated. Considering volumetric efficiency of compressor as 80%, mechanical efficiency of compressor as 85%, 20% of useful cooling getting lost and cooler cooling water from 299 K to 279 K at the rate of 100 kg/hr determine.
(*i*) the hp required in compressor
(*ii*) the volumetric capacity of compressor
Properties of Freon-12

Temperature, °C	*Sp. volume* v_g, m³/kg	*Enthalpy,* kJ/kg h_f	h_g	*Entropy,* kg/kg·K s_f	s_g	*Specific heat* kg/kg·K C_{pf}	C_{pg}
26	0.027	443.12	583.95	4.2665	4.737	60.9948	0.6729
2	0.053	419.88	573.54	4.1867	4.745	60.9405	0.618

[0.442 hp, 0.085 m³/min.]

12
Introduction to Heat Transfer

12.1 INTRODUCTION

Science of heat transfer deals with rate of heat transfer as a system undergoes from one state to another. During the heat transfer heat is transferred across the system boundary due to temperature difference prevailing between system and surroundings. Heat is not contained by the system instead it's energy in transit. Heat transfer is thus, the science that describes how the heat is transferred and how much is transferred in certain thermodynamic conditions. Temperature difference is the necessary condition for heat transfer to take place. Temperature difference offers the driving potential for heat transfer analogous to the fluid flow for which pressure difference offers driving potential/ driving force. Rate of heat transfer thus depends upon the temperature difference and the temperature gradient i.e. rate of change of temperature in that direction. Heat transfer has direction as well as magnitude.

Comparison between thermodynamics and heat transfer shows that thermodynamics deals with state changes occurring between equilibrium states while heat transfer deals with systems lacking in equilibrium. Heat transfer studies are therefore not based on laws of thermodynamics alone. Laws of thermodynamics only prepare background for heat transfer studies. There are number of examples from which heat transfer could be understood. For example, let us take boiling water at 100ºC in an open pan and keep it in the room. After some time water's temperature becomes equal to room temperature. Why this drop in temperature of water from 100ºC to room temperature and how? Answer for such questions lies in heat transfer. Human body also has continuous heat transfer as body continuously rejects heat to surroundings. In fact the human comfort depends largely upon rate of heat rejection and we regulate it by wearing clothes which shield body. In almost all engineering applications and appliances having energy interactions the study of heat transfer is very important e.g. refrigerator, heater, boilers, air conditioners, iron, heat exchangers, furnaces, solar collectors, computers, TV, radiators etc.

Heat transfer analysis of any system may involve either the determination of rate of heat transfer at given conditions or determination of size of system required for realizing desired heat transfer at given conditions. Heat transfer analysis can be performed either experimentally or analytically i.e. by calculations. For carrying out heat transfer analysis mathematically it is required to study physical phenomenon, identification of parameters & variables affecting phenomenon, determination of the relevant physical laws and principles which can be used for mathematical modelling with reasonable assumptions and approximations, and finally solving the mathematical equations to get meaningful results.

Heat transfer takes place in three different modes i.e. conduction, convection and radiation. These modes of heat transfer differ in respect to the mechanism by which heat is transferred across system boundary due to temperature difference. At macroscopic level three different modes of heat transfer can be understood as explained below.

Let us consider heat to be transferred from one end '*A*' to other end '*B*' of a body. Heat transfer may be realized by transferring it sequentially through molecules from one end to other end i.e. molecule(s) at end '*A*' receive heat and pass it on to the molecule(s) adjacent to it and this sequential transfer is continued till heat reaches to end '*B*'. This mode of heat transfer in which transfer of heat occurs from high energy particles (molecules) of the substance to the adjacent low energy particles (molecules) due to interactions between particles, is called 'conduction'.

Another mode of heat transfer could be one having transfer of heat from one place to another place due to displacement (movement) of molecules. This mode of heat transfer is called 'convection'. Thus, convection can be defined as the mode of heat transfer having physical movement of fluid molecules from one place to another in the direction of temperature gradient.

Heat transfer may also take place from one place to another without requiring any carrier medium in between i.e. heat may be transferred in the form of electromagnetic waves or photons from one place to another. Such type of heat transfer mode is called 'radiation'. Sun rays falling on earth are the obvious example of radiation as mode of heat transfer. Radiation mode of heat transfer does not require intervening medium for carrying photons.

Thus, heat transfer is the process of energy transfer due to temperature difference and may occur in three different modes of conduction, convection and radiation.

$$\text{Mathematically, heat transfer} = \frac{\text{Thermal potential difference}}{\text{Thermal resistance}}$$

Heat transfer problems could be classified as 'steady state' or 'unsteady state'. Steady state and unsteady state heat transfer can be characterized by the space time variation in heat exchange, i.e. $Q = f(x, y, z, t)$, where x, y, z are space coordinates and t is time coordinate.

In case of steady heat transfer the temperature or heat flux at each point of system remains constant with time and it is function of only space coordinates. Term steady means no change with time and so in case of steady heat transfer there shall be no change in internal energy.

$$\text{Mathematically, } \frac{dQ}{dt} = 0 \text{ and } Q = f(x, y, z).$$

Steady state heat transfer can be seen in case of heat transfer through walls of house if the inside and outside temperatures remains constant with time, heat exchange in heat exchangers etc.

Contrary to steady state in case of unsteady state heat transfer the temperature at each point in system changes with time. Unsteady state heat transfer is also called transient heat transfer. Here the rate of heat transfer changes with time and internal energy of system changes.

$$\text{Mathematically, } \frac{dQ}{dt} \neq 0,\ Q = f(x, y, z, t)$$

'Unsteady state heat transfer' may be seen in case of heat transfer across the walls of house if the inside and outside temperatures do not remain constant, cooling of vegetables kept in refrigerator etc. Here temperature or heat flux changes with temperature and position both. There could be some special cases when the temperature varies with time but not with position (x, y, z coordinates) and in such cases the heat transfer systems are called 'lumped systems'.

Generally, the heat transfer problems are solved with the assumptions for steady state due to ease of analysis even if the problems is transient in nature.

Heat transfer problems could also be classified as 'one dimensional heat transfer', 'two dimensional heat transfer' and 'three dimensional heat transfer'. Although the heat transfer in system occurs in all the directions but for the ease of analysis problem is modified as one dimensional or two dimensional depending upon relative magnitudes of heat transfer rates in different directions.

'One dimensional heat transfer' is the one in which the temperature in medium varies in one direction only and the heat is transferred in one direction while heat transfer in other directions is absent. Heat transfer will be 'two dimensional when the temperature varies mainly in two directions while variation in third direction may be negligible. Three dimensional heat transfer has temperature variation in three directions. Mathematically these can be given as described ahead for both steady and unsteady state;

	Steady state	*Unsteady state*
One dimensional heat transfer	$Q = f(x)$, or $Q_{,} = f(y)$ or, $Q = f(z)$	$Q = f(x, t)$, or $Q = f(y, t)$ or, $Q = f(z, t)$
Two dimensional heat transfer	$Q = f(x, y)$, or $Q_{,} = f(y, z)$, or, $Q = f(x, z)$	$Q = f(x, y, t)$, or $Q = f(y, z, t)$ or, $Q = f(x, z, t)$
Three dimensional heat transfer	$Q = f(x, y, z)$	$Q = f(x, y, z, t)$

12.2 CONDUCTION

Heat conduction is the mode of heat transfer in which heat transfer occurs across the medium (solid or fluid) from a region of high temperature to region of low temperature. Conduction has the energy transfer occurring through direct contact amongst molecules without their significant displacement. Thus, heat transfer by conduction is the transfer of energy from high energy particles to adjacent low energy particles and this energy transfer occurs sequentially from high temperature end to low temperature end. From kinetic theory all the molecules in the matter possess energy depending upon their velocity and relative position. This energy possessed by molecules is responsible for temperature of the matter. When temperature at some location is high as compared to other location than it means that there exist molecules with high and low energy respectively. Let us take case of heating of conducting rod at one end. With heating at one end the energy level of molecules increases at this end. These high energy molecules transfer the excess energy possesed by them to the neighbouring molecules through repeated collisions and intimate contact. This process of energy transfer through molecules continues from heating end to other end of rod. Here the conduction is seen to be realized through redistribution of molecular activity level in the medium. In solids conduction occurs due to combination of vibration of molecules and energy transport by free electrons while in liquids and gases it occurs due to collision and diffusion of molecules during random motion in matter.

Basic law of conduction is the Fourier's law of heat conduction and was proposed in 1822 by French scientist J. Fourier. "Fourier's law of heat conduction states that the rate of heat flow by conduction in solids is proportional to the temperature gradient and heat transfer area." Here, temperature gradient ($\Delta T/\Delta x$) refers to the rate of temperature change (ΔT) with respect to the distance (Δx) in the direction of heat flow and heat transfer is vector quantity.

Mathematically,

Rate of heat transfer by conduction $\propto$ (Area). (Temperature gradient)

$$\text{Heat transfer} = Q_{\text{conduction}} = k.A. \frac{(T_1 - T_2)}{\Delta x} = -k.A. \frac{\Delta T}{\Delta x}$$

Here $Q_{\text{conduction}}$ is in W, T_1 & T_2 are temperatures at two ends of large plane wall in °C, A is surface area in m^2 and proportionality constant k is thermal conductivity in W/m.°C. Thermal conductivity is the physical property of the material and indicates the ability of material to conduct heat. It depends largely upon the state of material and changes with structure, humidity, pressure and temperature conditions. Thermal conductivity of some substances is given is Table 12.1. For a limiting case having $\Delta x \to 0$ the heat conduction may be given in differential form as,

$$Q_{\text{conduction}} = -k \cdot A \cdot \frac{dT}{dx}$$

Here –ve sign is put arbitrarily to make $Q_{conduction}$ positive and signifies the transfer of heat from high temperature end to low temperature end as the temperature gradient shall be negative. Fig. 12.1 shows the one dimensional heat conduction across a plane wall.

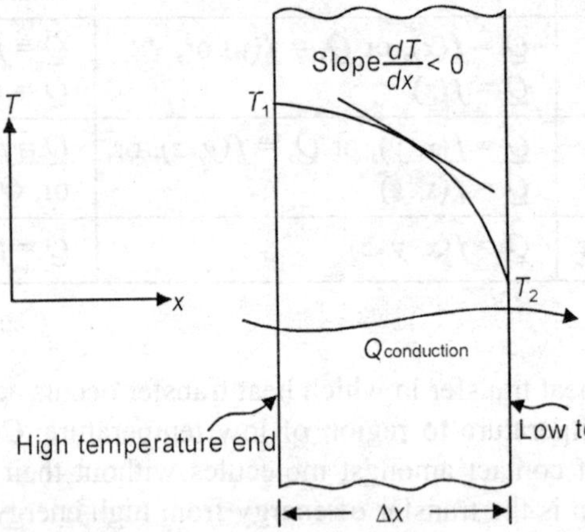

Fig. 12.1 Conduction heat transfer along x-direction

Thermal conductivity is defined as the rate of conduction heat transfer through a unit thickness of material per unit area per unit temperature difference. Generally, the experimentally obtained value of thermal conductivity is

$$k = \frac{Q_{conduction} \cdot \Delta x}{A \cdot \Delta T}$$

If $A = 1, \Delta T = 1$ & $\Delta x = 1$ then, $k = Q_{conduction}$

Thermal conductivity is higher for the materials which are good heat conductor and smaller for the materials which are insulator or poor conductor. Thermal conductivity of substance in solid phase is highest and smallest in gaseous phase due to typical molecular arrangement and mechanism of heat transfer in conduction. Thermal conductivity of gases is seen to increase with increase in temperature and decrease in molar mass. This may be attributed to the fact that higher temperature means higher kinetic energy of molecules which causes increase in randomness in motion and rate of collisions thereby increasing the rate of energy and momentum exchange.

Thermal conductivity of liquids decreases with increase in temperature and increase in molar mass. In case of solids thermal conductivity is decided by factors namely 'lattice vibrations' and 'energy transport by free electrons'. Thermal conductivity of pure metals is high due to the increased energy transport by free electrons. Different variations in thermal conductivity are seen with temperature variation due to varying molecular structure and arrangements. Fig. 12.2 shows the variation of thermal conductivity of solids, liquids & gases.

Since, there is too much of variation of thermal conductivity with temperature so thermal conductivity is considered at mean temperature in order to keep it constant throughout.

Table 12.1 *Thermal conductivity of some material*

Material	*Thermal conductivity, W/m·°C*
Air	0.026
Aluminium	237
Brick	0.72
Copper	401
Constantan (55% Cu & 45% Ni)	23
Diamond	2300
Glass	0.78
Glass fiber	0.043
Gold	317
Helium	0.152
Human skin	0.37
Iron	80.2
Mercury	8.54
Nickel	91
Wood	0.17

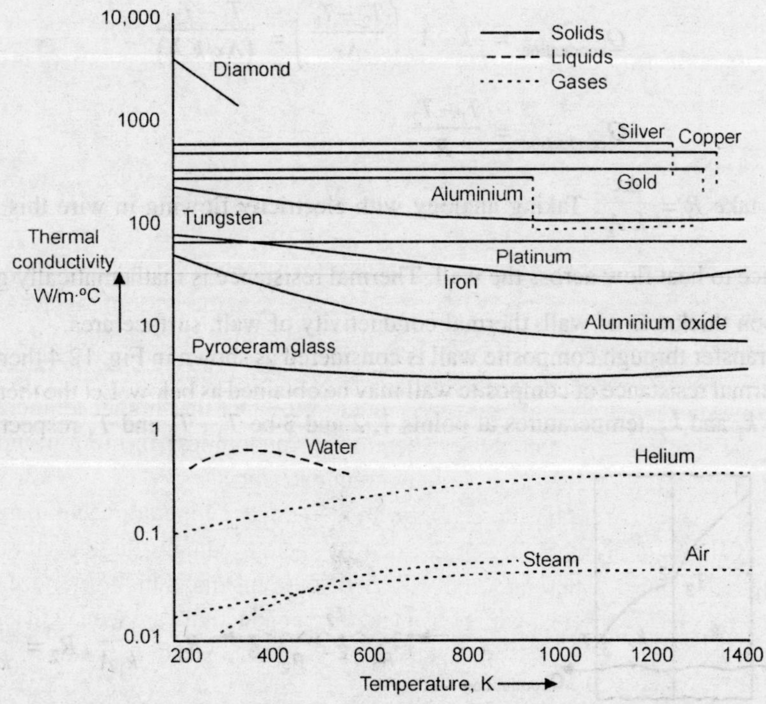

***Fig. 12.2** Variation of thermal conductivity of different solids, liquids and gases with temperature*

Heat transfer in wall as shown in Fig. 12.1 can be represented analogous to flow of current in electrical systems. By Ohm's law the current flowing through a wire is proportional to the potential difference across the wire. Mathematically, Current, $(I) \propto$ Potential difference, (V)

$$I \propto V$$

or, $$V = I \cdot R$$

or, $$I = \frac{V}{R}$$

Here current (I) is in amperes, potential difference (V) is in volts, proportionality constant (R) is resistance of wire in ohm.

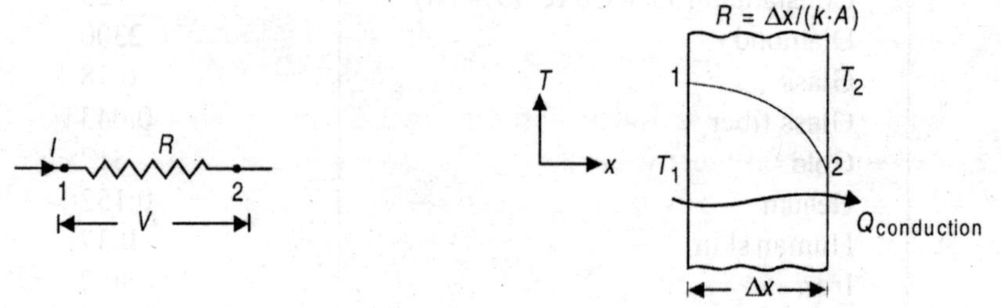

(*a*) Current flowing through wire (*b*) Heat transfer through wall

Fig. 12.3 Electrical system analogy to heat transfer across wall

In case of heat transfer across wall,

$$Q_{\text{conduction}} = -k \cdot A \cdot \left(\frac{T_2 - T_1}{\Delta x}\right) = \frac{T_1 - T_2}{(\Delta x / k.A)}$$

or, $$Q_{\text{conduction}} = \frac{T_1 - T_2}{R}$$

Here let us take $R = \dfrac{\Delta x}{k \cdot A}$. Taking analogy with electricity flowing in wire this R may be called thermal resistance to heat flow across the wall. Thermal resistance is mathematically given as $R = \dfrac{\Delta x}{k \cdot A}$ and depends upon thickness of wall, thermal conductivity of wall, surface area.

If the heat transfer through composite wall is considered as shown in Fig. 12.4 then using resistance concept total thermal resistance of composite wall may be obtained as below. Let the thermal conductivity of two walls be k_1 and k_2, temperatures at points 1, 2 and 3 be T_1, T_2 and T_3 respectively.

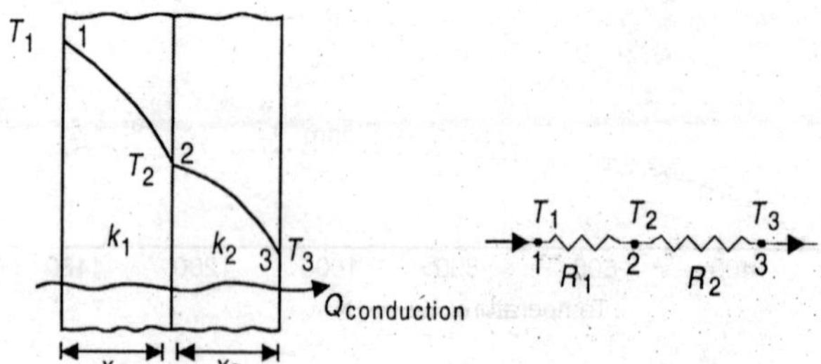

$$R_1 = \frac{x_1}{k_1 A}, \; R_2 = \frac{x_2}{k_2 A}$$

Fig. 12.4 Heat transfer through composite wall (thermal resistance in series)

In case of composite wall the total thermal resistance can be obtained from electrical resistance analogy as the two resistances R_1 and R_2 are in series.

Total thermal resistance, $R = R_1 + R_2 = \frac{x_1}{k_1 A} + \frac{x_2}{k_2 A}$

Heat transfer in conduction through composite wall;

$$Q_{\text{conduction}} = \frac{T_1 - T_3}{R}$$

Also, $$Q_{\text{conduction}} = \frac{T_1 - T_2}{R_1} = \frac{T_2 - T_3}{R_2} = \frac{T_1 - T_3}{R}$$

Let us take another composite wall having two different materials as shown in figure 12.5. Let the surface areas of two different walls be A_1 & A_2 and temperatures at two ends be T_1 & T_2.

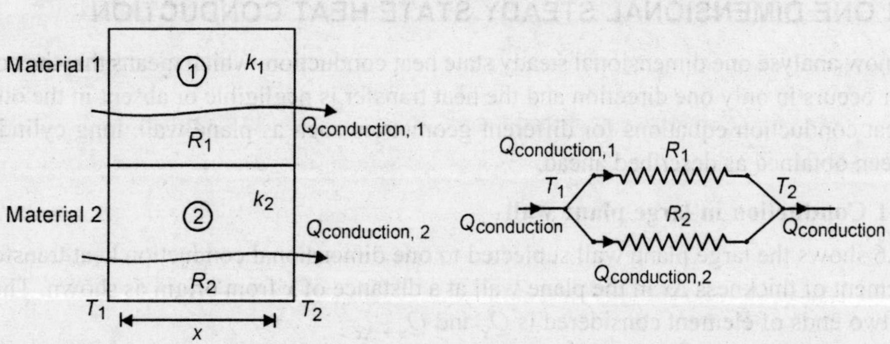

Fig. 12.5 Heat transfer through composite wall (thermal resistance in parallel)

Thermal resistance of two walls made of materials having thermal conductivity of k_1 & k_2 can be given by R_1 & R_2 as under.

$$R_1 = \frac{x}{k_1 A_1},\ R_2 = \frac{x}{k_2 A_2}$$

Total thermal resistance (R) can be obtained from electrical circuit analogy for two resistances in parallel.

$$\frac{1}{R} = \frac{1}{R_1} + \frac{1}{R_2}$$

$$\Rightarrow \qquad R = \frac{R_1 R_2}{R_1 + R_2}$$

Heat transfer in conduction through different walls.

$$Q_{\text{conduction, 1}} = \frac{T_1 - T_2}{R_1}$$

$$Q_{\text{conduction, 2}} = \frac{T_1 - T_2}{R_2}$$

Total heat transfer in conduction through composite wall.

$$Q_{\text{conduction}} = \frac{T_1 - T_2}{R}$$

Also,
$$Q_{\text{conduction}} = Q_{\text{conduction, 1}} + Q_{\text{conduction, 2}}$$

$$= (T_1 - T_2)\left(\frac{1}{R_1} + \frac{1}{R_2}\right)$$

$$Q_{\text{conduction}} = \frac{(T_1 - T_2)}{R}$$

Similarly any heat transfer problem can be suitably modeled on the basis of thermal resistance concept using electrical circuit analogy.

12.2.1 ONE DIMENSIONAL STEADY STATE HEAT CONDUCTION

Let us now analyse one dimensional steady state heat conduction, which means that the conduction heat transfer occurs in only one direction and the heat transfer is negligible or absent in the other directions.

Heat conduction equations for different geometries such as plane wall, long cylinder and sphere have been obtained as described ahead.

12.2.1.1 Conduction in large plane wall

Fig 12.6 shows the large plane wall subjected to one dimensional conduction heat transfer. Let us take this element of thickness Δx in the plane wall at a distance of x from origin as shown. The heat transfer rate at two ends of element considered is Q_x and $Q_{x+\Delta x}$.

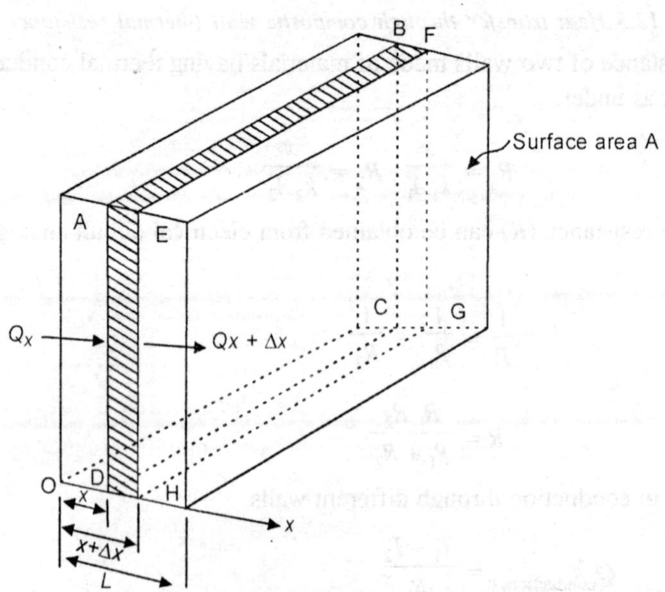

Fig. 12.6 One dimensional heat transfer by conduction in plane wall

Let us also consider that there occurs heat generation inside the element at the rate of Q_{gen}. For the thin element the energy balance may be written for the heat transfer taking place for some time Δt. Energy balance upon element yields.

(Heat entering from face ABCD due to conduction) + (Heat generated in element) – (Heat leaving from face EFGH due to conduction) = (Change in energy content of element in time Δt)

$$(Q_x \cdot \Delta t) + (Q_{gen} \cdot \Delta t) - (Q_{x+\Delta x} \cdot \Delta t) = \Delta E_{element}$$

$$Q_x + Q_{gen} - Q_{x+\Delta x} = \frac{\Delta E_{element}}{\Delta t} = \frac{\Delta E_{t+\Delta t} - E_t}{\Delta t}$$

Here the total heat generated within element can be estimated by knowing the rate of heat generation per unit volume. If the rate of heat generated per unit volume is q_{gen} then heat generation rate can be given as,

$$Q_{gen} = q_{gen} \cdot V_{element}$$
$$Q_{gen} = q_{gen} \cdot (A \cdot \Delta x)$$

Change in Energy content of element if its mass is m, specific heat is C and density of material is ρ can be given as;

$$\Delta E_{element} = E_{t+\Delta t} - E_t$$
$$= m \cdot C \cdot (T_{t+\Delta t} - T_t)$$
$$= (\rho \cdot A \cdot \Delta x) \cdot C \cdot (T_{t+\Delta t} - T_t)$$

Substituting in energy balance

$$Q_x + q_{gen}(A \cdot \Delta x) - Q_{x+\Delta x} = \frac{(\rho \cdot A \cdot \Delta x) \cdot C}{\Delta t}(T_{t+\Delta t} - T_1)$$

or,

$$q_{gen} + \frac{(Q_x - Q_{x+\Delta x})}{(A \cdot \Delta x)} = \frac{\rho \cdot C \cdot (T_{t+\Delta t} - T_t)}{\Delta t}$$

Considering thickness of element to be very small and duration of analysis also being very small we can write for limiting case when

$\Delta x \to 0$ & $\Delta t \to 0$;

$$q_{gen} + \lim_{\Delta x \to 0} \frac{(Q_x - Q_{x+\Delta x})}{\Delta x} = \lim_{\Delta t \to 0}\left\{\frac{\rho C \cdot (T_{t+\Delta t} - T_t)}{\Delta t}\right\}$$

Here, from the definition of derivative and Fourier's law of conduction following may be obtained as;

$$\lim_{\Delta x \to 0}\left(\frac{Q_{x+\Delta x} - Q_x}{\Delta x}\right) = \frac{\partial Q_{conduction}}{\partial x} = \frac{\partial}{\partial x}\left(-k \cdot A \cdot \frac{\partial T}{\partial x}\right)$$

and,

$$\lim_{\Delta t \to 0}\left\{\frac{\rho C \cdot (T_{t+\Delta t} - T_t)}{\Delta t}\right\} = \rho C \cdot \frac{\partial T}{\partial t}$$

Substituting these the one-dimensional conduction heat transfer equation is obtained as under.

$$\boxed{q_{gen} + \frac{\partial}{\partial x}\left(k \cdot \frac{\partial T}{\partial x}\right) = \rho \cdot C \cdot \frac{\partial T}{\partial t}}$$

Above is equation of one dimensional conduction heat transfer through plane wall with heat generation for unsteady state.

Case 1 : In special case when there is no heat generation it would be modified as $q_{gen} = 0$

$$\boxed{\frac{\partial}{\partial x}\left(k \cdot \frac{\partial T}{\partial x}\right) = \rho \cdot C \cdot \frac{\partial T}{\partial t}}$$

Above eauation is of one dimensional conduction heat transfer through plane wall without heat generation for unsteady state.

Case 2 : In case thermal conductivity of material remains constant then k can be taken out of derivative and conduction equation gets modified as under;

$$\boxed{q_{gen} + k\cdot\frac{\partial^2 T}{\partial x^2} = \rho C\cdot\frac{\partial T}{\partial t}}$$

Above eauation is of one dimensional conduction heat transfer through plane wall for constant thermal conductivity and heat generation for unsteady state.

or,
$$\frac{q_{gen}}{k} + \frac{\partial^2 T}{\partial x^2} = \frac{\rho C}{k}\cdot\frac{\partial T}{\partial t}$$

Here, $\frac{\rho C}{k} = \frac{1}{\propto}$ as thermal diffusivity $\propto = \frac{k}{\rho C}$. Thermal diffusivity of material indicates the capability of how fast heat diffuses through the material and is given by the ratio of heat conducted to heat stored per unit volume. Its unit is m^2/s. For higher rate of propagation of heat in material its thermal diffusivity should be large.

Therefore,

$$\boxed{\frac{q_{gen}}{k} + \frac{\partial^2 T}{\partial x^2} = \frac{1}{\propto}\cdot\frac{\partial T}{\partial t}}$$

Case 3 : In case of steady state heat transfer the conduction equation gets modified as shown ahead, considering constant thermal conductivity and heat generation.

Due to steady state there shall be no variation with time so, $T = f(x) \Rightarrow \frac{\partial T}{\partial t} = 0$ and partial derivative can be replaced by ordinary derivative.

Therefore,

$$\boxed{\frac{q_{gen}}{k} + \frac{d^2 T}{dx_2^2} = 0}$$

Above equation is of one dimensional steady state conduction heat transfer through plane wall for constant thermal conductivity and heat generation.

In the absence of heat generation, $q_{gen} = 0$ $\boxed{\frac{d^2 T}{dx^2} = 0}$

Thus, the one dimensional conduction heat transfer through plane wall can be suitably modeled by substituting the necessary conditions as applicable.

For steady state conduction heat transfer through plane wall without heat generation, considering temperature at two ends as T_1 and T_2 we can write;

$$\frac{d^2 T}{dx^2} = 0 \Rightarrow \frac{dT}{dx} = C_1 \Rightarrow C_1 x + C_2$$

Substituting boundary conditions to get integration constants C_1 & C_2.
at $x = 0$, $T = T_1$ and at $x = L$, $T = T_2$

$$C_2 = T_1 \text{ and } C_1 = \left(\frac{T_1 - T_2}{L}\right)$$

It yields, $\boxed{T = \left(\frac{T_1 - T_2}{L}\right) \cdot x + T_1}$

Thus, temperature distribution as given above in box is linear with x in plane wall.

Here, $$\frac{dT}{dx} = \left(\frac{T_1 - T_2}{L}\right)$$

Using Fourier's law of conduction the conduction heat transfer can be obtained as;

$$Q_{\text{conduction}} = kA\frac{dT}{dx}$$

$$Q_{\text{conduction}} = kA\left(\frac{T_1 - T_2}{L}\right) = \left(\frac{T_1 - T_2}{L/kA}\right)$$

Conduction heat transfer through wall

$$Q_{\text{conduction}} = \left(\frac{T_1 - T_2}{R}\right)$$, where R is thermal resistance and given as $R = \left(\frac{L}{kA}\right)$

12.2.1.2 Conduction in long cylinder

Let us consider a long cylinder in which the conduction heat transfer occurs across the thickness of cylinder wall. Heat transfer direction is perpendicular to the cylinder wall. Fig 12.7 shows the details of one dimensional heat transfer by conduction in long cylinder.

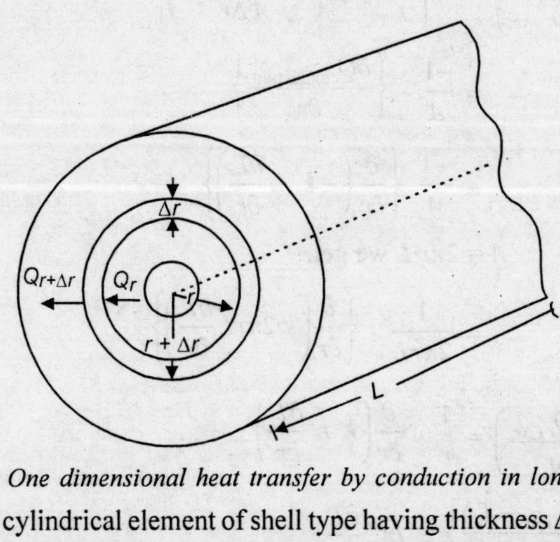

Fig. 12.7 One dimensional heat transfer by conduction in long cylinder.

Let us consider a thin cylindrical element of shell type having thickness Δr at distance of r from axis of cylinder. Let the length of cylinder be L, specific heat of cylinder material C and density of cylinder material as ρ. As shown the heat transfers at the rate of Q_r from inner surface towards outer surface and at the rate of $Q_{r+\Delta r}$ from outer surface towards periphery. Let us assume heat generation within element being at the rate of q_{gen} per unit volume. Writing down energy balance similar to that of plane wall case for time duration of Δt.

(Heat entering in element from inner surface of element) + (Heat generated) – (Heat leaving element from outer surface of element) = (Change in energy content of element)

$$(Q_r \cdot \Delta t) + (Q_{gen} \cdot \Delta t) - (Q_{r+\Delta r} \cdot \Delta t) = \Delta E_{element}$$

$$\Rightarrow \quad Q_r + Q_{gen} - Q_{r+\Delta r} = \frac{\Delta E_{element}}{\Delta t}$$

Here,

$$\Delta E_{element} = E_{t+\Delta t} - E_t$$

$$= m \cdot C \cdot (T_{t+\Delta t} - T_t)$$

$$\Delta E_{element} = (\rho \cdot A \cdot \Delta r) \cdot C \cdot (T_{t+\Delta t} - T_t), \text{ where } A = 2\pi r \cdot L$$

and,

$$Q_{gen} = q_{gen} \cdot V_{element}$$

$$Q_{gen} = q_{gen} \cdot (A \cdot \Delta r)$$

Substitution in energy balance equation yields,

$$Q_r + q_{gen} \cdot (A \cdot \Delta r) - Q_{r+\Delta r} = (\rho \cdot A \cdot \Delta r) \cdot C \cdot \frac{(T_{t+\Delta t} - T_t)}{\Delta t}$$

$$\text{or,} \quad q_{gen} + \frac{(Q_r - Q_{r+\Delta r})}{A \cdot \Delta r} = (\rho \cdot C) \cdot \frac{(T_{t+\Delta t} - T_t)}{\Delta t}$$

Taking the limiting case of very thin element i.e. $\Delta r \to 0$ and also $\Delta t \to 0$ it becomes

$$q_{gen} + \lim_{\Delta r \to 0} \frac{(Q_r - Q_{r+\Delta r})}{A \cdot \Delta r} = \lim_{\Delta t \to 0} \left\{ \frac{\rho C \cdot (T_{t+\Delta_t} - T_t)}{\Delta t} \right\}$$

Thus, from the definition of derivative and Fourier's law of conduction following is obtained;

$$\lim_{\Delta r \to 0} \left(\frac{Q_r - Q_{r+\Delta r}}{A \cdot \Delta r} \right) = -\left\{ \lim_{\Delta r \to 0} \left(\frac{Q_{r+\Delta r} - Q_r}{A \cdot \Delta r} \right) \right\}$$

$$= \frac{-1}{A} \cdot \left\{ \frac{\partial Q_{conduction}}{\partial r} \right\}$$

$$= \frac{-1}{A} \cdot \left\{ \frac{\partial}{\partial r} \left(-k.A.\frac{\partial T}{\partial r} \right) \right\}$$

Substituting $A = 2\pi r \cdot L$ we get,

$$= \frac{1}{2\pi r L} \cdot \left\{ \frac{\partial}{\partial r} \left(k \cdot 2\pi r L \frac{\partial T}{\partial r} \right) \right\}$$

$$\lim_{\Delta r \to 0} \left(\frac{Q_r - Q_{r+\Delta r}}{A \cdot \Delta r} \right) = \frac{1}{r} \cdot \frac{\partial}{\partial r} \left(k \cdot r \cdot \frac{\partial T}{\partial r} \right)$$

Also, $\lim_{\Delta t \to 0} \left\{ \frac{\rho C \cdot (T_{t+\Delta t} - T_t)}{\Delta t} \right\} = \rho C \cdot \frac{\partial T}{\partial t}$

Substitution yields one dimensional conduction heat transfer equation through a long cylinder as,

$$\boxed{q_{gen} + \frac{1}{r}\cdot\frac{\partial}{\partial r}\left(k\cdot r\cdot\frac{\partial T}{\partial r}\right) = \rho C\cdot\frac{\partial T}{\partial t}}$$

Equation for the one dimensional conduction heat transfer through long cylinder with heat generation for unsteady state.

Above equation can be modified for different special situations as described ahead.

Case 1 : In special case when there is no heat generation; $q_{gen} = 0$

$$\boxed{\frac{1}{r}\cdot\frac{\partial}{\partial r}\left(k\cdot r\cdot\frac{\partial T}{\partial r}\right) = \rho C\cdot\frac{\partial T}{\partial t}}$$

One dimensional conduction heat transfer equation for long cylinder having no heat generation and unsteady state.

Case 2 : In case of the cylinder material being such that its thermal conductivity '*k*' remains constant, conduction equation shall be,

$$q_{gen} + \frac{k}{r}\frac{\partial}{\partial r}\left(r\cdot\frac{\partial T}{\partial r}\right) = \rho C\cdot\frac{\partial T}{\partial t}$$

$$\frac{q_{gen}}{k} + \frac{1}{r}\frac{\partial}{\partial r}\left(r\cdot\frac{\partial T}{\partial r}\right) = \frac{\rho C}{k}\cdot\frac{\partial T}{\partial t}$$

$$\boxed{\frac{q_{gen}}{k} + \frac{1}{r}\frac{\partial}{\partial r}\left(r\cdot\frac{\partial T}{\partial r}\right) = \frac{1}{\propto}\cdot\frac{\partial T}{\partial t}}$$

One dimensional unsteady state equation for heat transfer in long cylinder when thermal conductivity is constant and heat generation is there.

In the absence of heat generation;

$$\boxed{\frac{1}{r}\frac{\partial}{\partial r}\left(r\cdot\frac{\partial T}{\partial r}\right) = \frac{1}{\propto}\cdot\frac{\partial T}{\partial t}}$$

Case 3 : In case of steady state during heat conduction $T = f(x)$ and so, $\frac{\partial T}{\partial t} = 0$ and $\frac{\partial T}{\partial r} = \frac{dT}{dr}$,

$$\boxed{\frac{q_{gen}}{k} + \frac{1}{r}\frac{d}{dr}\left(r\frac{dT}{dr}\right) = 0}$$

One dimensional steady state conduction equation for heat transfer in long cylinder when thermal conductivity is constant and heat generation is present.

In the absence of heat generation

$$\boxed{\frac{1}{r}\frac{d}{dr}\left(r\frac{dT}{dr}\right) = 0}$$

One dimensional steady state conduction equation for heat transfer in long cylinder when thermal conductivity is constant and heat generation is absent.

Thus, in case of cylinder having inner and outer radii as r_1 & r_2, length being L and temperature at inner and outer surface being T_1 & T_2 conduction heat transfer can be obtained as under ;

We know, $$\frac{d}{dr}\left(r\frac{dT}{dr}\right) = 0$$

or, $$r\frac{dT}{dr} = C_1$$

or, $$\frac{dT}{dr} = \frac{C_1}{r}$$

or, $$T = C_1\, ln\, r + C_2$$

Substituting boundary conditions the integration constants C_1 & C_2 may be obtained.

at $r = r_1$, $T = T_1$ and at $r = r_2$, $T = T_2$

$$T_1 = C_1 \ln r_1 + C_2$$

$$T_2 = C_1 \ln r_2 + C_2$$

It yields, $C_1 = \dfrac{(T_1 - T_2)}{\ln\left(r_1/r_2\right)}$ and $C_2 = T_1 - \dfrac{(T_1 - T_2)\cdot \ln r_1}{\ln\left(r_1/r_2\right)}$

Substituting C_1 & C_2 in expression for T at any location at radius r;

$$T = \left(\frac{(T_1 - T_2)}{\ln\left(r_1/r_2\right)}\right)\ln r + T_1 = \frac{(T_1 - T_2)\cdot \ln r_1}{\ln\left(r_1/r_2\right)}$$

$$\boxed{\left(\frac{T - T_1}{T_1 - T_2} = \frac{\ln\left(r/r_1\right)}{\ln\left(r_1/r_2\right)}\right)}$$

Heat transfer by conduction may be estimated using Fourier's equation and substituting $\dfrac{dT}{dr}$ in it.

$$Q_{\text{conduction}} = -k\cdot A\cdot\frac{dT}{dr} = -k\cdot 2\pi rL\cdot\left(\frac{T_1 - T_2}{r\cdot\ln\left(r_1/r_2\right)}\right)$$

$$Q_{\text{conduction}} = \left(\frac{-k\cdot 2\pi L(T_1 - T_2)}{\ln\left(r_1/r_2\right)}\right) = \left\{\frac{(T_2 - T_1)}{\ln(r_1/r_2)/(k2\pi L)}\right\}$$

or, $$Q_{\text{conduction}} = \frac{(T_2 - T_1)}{R}, \text{ where } R = \frac{\ln\left(r_1/r_2\right)}{(2\pi kL)} = \text{Thermal resistance}$$

Thus, thermal resistance for conduction heat transfer in long cylinder is $\left\{\dfrac{\ln\left(r_1/r_2\right)}{2\pi kL}\right\}$

12.2.1.3 Conduction in sphere

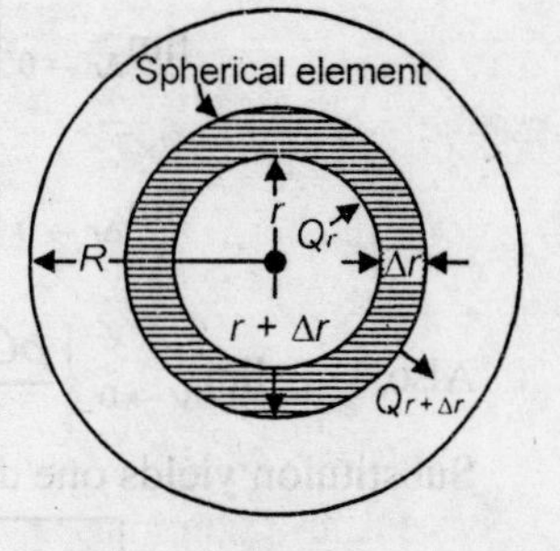

Fig. 12.8 One dimensional heat transfer by conduction in sphere.

Heat conduction through a sphere can also be quantified on the similar lines as done in case of long cylinder /wall by considering a thin spherical volume element in sphere of thickness Δr. Let us consider sphere element to have inner radius of r and outer radius of $(r + \Delta r)$. Sphere material may have density ρ and specific heat of C. Let us consider heat generation in sphere to be at the rate of q_{gen} per unit volume. Conduction heat transfer rate at inner surface and outer surface of spherical may be taken as Q_r and $Q_{r+\Delta r}$.

Energy balance can be written for the spherical element when considered for duration of Δt.

(Heat entering from inner surface of element) + (Heat generated) – (Heat leaving from outer surface of element) = (Change in energy content of spherical element in time Δt)

$$(Q_r \cdot \Delta t) + (Q_{gen} \cdot \Delta t) - (Q_{r+\Delta r} \cdot \Delta t) = \Delta E_{element}$$

$$Q_r + Q_{gen} - Q_{r+\Delta r} = \frac{\Delta E_{element}}{\Delta t}$$

Here, $$Q_{gen} = q_{gen} \cdot V_{element} = q_{gen} \cdot (A \cdot \Delta r)$$

For the element, $$A = 4\pi r^2$$

$$\Delta E_{element} = E_{t+\Delta t} - E_t$$
$$= m \cdot C \cdot (T_{t+\Delta t} - T_t)$$
$$\Delta E_{element} = (\rho \cdot A \cdot \Delta r) \cdot C \cdot (T_{t+\Delta t} - T_t)$$

Substitution in energy balance yields,

$$Q_r + q_{gen} \cdot (A \cdot \Delta r) - Q_{r+\Delta r} = (\rho \cdot A \cdot \Delta r) \cdot C \cdot \frac{(T_{t+\Delta t} - T_t)}{\Delta t}$$

Taking the limiting case of very thin spherical element i.e. $\Delta r \to 0$ and also $\Delta t \to 0$ the energy balance equation gets modified as.

$$q_{gen} + \lim_{\Delta r \to 0} \frac{(Q_r - Q_{r+\Delta r})}{A \cdot \Delta r} = \lim_{\Delta t \to 0} \left\{ \frac{\rho C \cdot (T_{t+\Delta t} - T_t)}{\Delta t} \right\}$$

From the definition of derivative and Fourier's law of conduction we get,

$$\lim_{\Delta r \to 0} \frac{(Q_r - Q_{r+\Delta r})}{A \cdot \Delta r} = -\left\{ \lim_{\Delta r \to 0} \left(\frac{(Q_{r+\Delta r} - Q_r)}{A \cdot \Delta r} \right) \right\}$$

$$= -\frac{1}{A} \left\{ \frac{\partial Q_{conduction}}{\partial r} \right\}$$

$$= \frac{1}{A} \left\{ \frac{\partial}{\partial r} \left(kA \frac{\partial T}{\partial r} \right) \right\}$$

Substituting $$A = 4\pi r^2$$

$$\lim_{\Delta r \to 0}\left(\frac{Q_r - Q_{r+\Delta r}}{A \cdot \Delta r}\right) = \frac{1}{4\pi r^2}\left\{\frac{\partial}{\partial r}\left(k \cdot 4\pi r^2 \frac{\partial T}{\partial r}\right)\right\}$$

$$\lim_{\Delta r \to 0}\left(\frac{Q_r - Q_{r+\Delta r}}{A \cdot \Delta r}\right) = \frac{1}{r^2}\left\{\frac{\partial}{\partial r}\left(k \cdot r^2 \frac{\partial T}{\partial r}\right)\right\}$$

Also, $$\lim_{\Delta t \to 0}\left\{\frac{\rho C\,(T_{t+\Delta t} - T_t)}{\Delta t}\right\} = \rho C \cdot \frac{\partial T}{\partial t}$$

Substituion yields one dimensional conduction heat transfer equation for sphere as,

$$\boxed{q_{gen} + \frac{1}{r^2}\left\{\frac{\partial}{\partial r}\left(k \cdot r^2 \frac{\partial T}{\partial r}\right)\right\} = \rho C \cdot \frac{\partial T}{\partial t}}$$ one dimensional conduction heat transfer equation for sphere.

Case 1 : In special case when there is no heat generation, then $q_{\text{gen}} = 0$

$$\boxed{\frac{1}{r^2}\left\{\frac{\partial}{\partial r}\left(k \cdot r^2 \frac{\partial T}{\partial r}\right)\right\} = \rho C \cdot \frac{\partial T}{\partial t}}$$ one dimensional conduction heat transfer equation for sphere in the absence of heat generation.

Case 2 : In special case when thermal conductivity of sphere material remains constant, one dimensional conduction equation shall be:

$$\boxed{q_{\text{gen}} + \frac{k}{r^2}\left\{\frac{\partial}{\partial r}\left(r^2 \frac{\partial T}{\partial r}\right)\right\} = \rho C \cdot \frac{\partial T}{\partial t}}$$

In the absence of heat generation the equation for one dimensional conduction heat transfer through sphere shall be:

$$\boxed{\frac{k}{r^2} \cdot \frac{\partial}{\partial r}\left(r^2 \frac{\partial T}{\partial r}\right) = \rho C \cdot \frac{\partial T}{\partial t}}$$

Above is one dimensional conduction heat transfer equation for sphere in absence of heat generation and constant thermal conductivity of material.

Case 3 : In case of steady state during conduction, $T = f(x)$ and

so, $$\frac{\partial T}{\partial t} = 0 \text{ and } \frac{\partial T}{\partial r} = \frac{dT}{dr}$$

$$q_{\text{gen}} + \frac{1}{r^2}\left\{\frac{\partial}{\partial r}\left(k \cdot r^2 \cdot \frac{\partial T}{\partial r}\right)\right\} = 0$$

or, $$\boxed{q_{\text{gen}} + \frac{1}{r^2}\frac{d}{dr}\left(k \cdot r^2 \frac{dT}{dr}\right) = 0}$$

Above is one dimensional steady state conduction equation for heat transfer in sphere with heat generation.

In the absence of heat generation, $q_{\text{gen}} = 0$,

so, $$\boxed{\frac{1}{r^2}\frac{d}{dr}\left(k \cdot r^2 \frac{dT}{dr}\right) = 0}$$ one dimensional steady state conduction equation for heat transfer in sphere in the absence of heat generation.

Let us consider a sphere of inner radius r_1, outer radius r_2, inner surface temperature T_1, outside surface temperature T_2 and constant thermal conductivity 'k' without any heat generation and steady state heat conduction.

One dimensional conduction equation for steady state heat transfer yields.

$$\frac{k}{r^2}\frac{d}{dr}\left(r^2\frac{dT}{dr}\right) = 0$$

$$\text{upon integrating, } T = \frac{-C_1}{r} + C_2$$

Using boundary conditions, at $r = r_1$, $T = T_1$ and at $r = r_2$, $T = T_2$

$$T_1 = \frac{-C_1}{r_1} + C_2$$

$$T_2 = \frac{-C_1}{r_2} + C_2$$

$$\text{Integration constants are, } C_1 = \frac{(T_1 - T_2)(r_1 \cdot r_2)}{(r_1 - r_2)}, \text{ and } C_2 = \frac{T_1 r_1 - T_2 r_2}{(r_1 - r_2)}$$

Thus, temperature variation can be given as,

$$\boxed{T = \left\{\frac{(T_2 - T_1)(r_1 r_2)}{r.(r_1 - r_2)}\right\} + \left\{\frac{T_1 r_1 - T_2 r_2}{(r_1 - r_2)}\right\}}$$

differentiating with respect to r,

$$\frac{dT}{dr} = \frac{(T_1 - T_2)(r_1\, r_2)}{(r_1 - r_2)r^2}$$

Rate of heat transfer by conduction in case of spherical wall by Fourier's law,

$$Q_{\text{conduction}} = -k \cdot A \cdot \frac{dT}{dr} = \frac{-k \cdot 4\pi r^2 \cdot (T_1 - T_2)(r_1\, r_2)}{(r_1 - r_2) \cdot r^2}$$

$$\boxed{Q_{\text{conduction}} = \frac{(T_1 - T_2)(k \cdot 4\pi r_1\, r_2)}{(r_2 - r_1)} = \frac{(T_1 - T_2)}{R}}$$

$$\text{where } R = \text{Thermal resistance in sphere} = \left(\frac{r_2 - r_1}{4\pi k \cdot r_1\, r_2}\right)$$

12.3 CONVECTION

Convection is that mechanism of heat transfer in fluid which involves the fluid motion in bulk. As described earlier, the heat transfer takes place due to mass movement of fluid i.e. fluid particles move from one place to other. This movement of fluid particles causing convection is realized either by external assistance or due to density difference which exists because of temperature difference. Due to the fluid motion the rate of heat transfer is more in convection as compared to conduction. Thus, convection is similar to conduction in one respect that both modes of heat transfer require medium but convection differs from conduction in another respect that it requires fluid motion in bulk which is not required in conduction.

Convection heat transfer which is caused due to fluid motion may be classified as 'natural convection' and 'forced convection'. When the heat transfer occurs due to fluid motion caused by density difference created because of temperature difference then it is called "natural convection". Fig 12.9 (*a*) shows natural convection heat transfer between the plate (maintained at high temperature than that of surrounding fluids) and surrounding fluid. Here the surrounding fluid in contact with plate gets lifted up upon heating due to buoyancy effect and the cold fluid at distance from plate moves to occupy the space created due to leaving of hot fluid. Thus, a convective current of circulation type is set up.

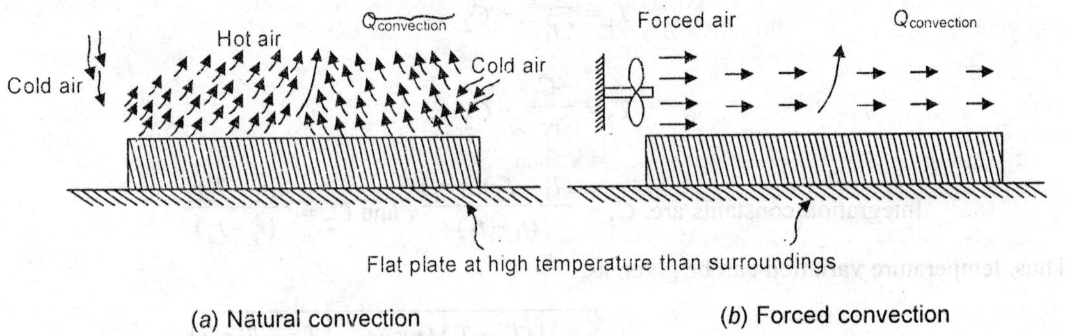

Fig. 12.9 Natural convection and forced convection over flat plate.

In case of "forced convection" the heat transfer due to fluid motion is realized by making fluid to move using external assistance such as forcing fluid to move using blower, compressor etc. Fig. 12.9 (*b*) shows the forced convection heat transfer where fluid moving along plate due to fan carries heat from hot surface to surroundings. Thus, it is obvious that in case of convection whether of natural or forced type there exists a layer of high temperature fluid in close vicinity of hot surface due to no slip condition between fluid particles and surface.

Heat is picked up by this fluid layer in contact with hot surface and subsequently transferred to surroundings. Here this thin layer may be termed as film in which temperature variation occurs from surface temperature (T_{hot}) to the surroundings temperature ($T_{surrounding}$). Heat transfer actually occurs through this fluid film between the temperature limits of T_{hot} and $T_{surrounding}$. It is seen that heat transfer in convection depends upon the fluid characteristics, type of fluid flow and geometry of hot surface etc. Different parameters upon which the convection heat transfer depends are 'dynamic viscosity of fluid', 'density of fluid', 'thermal conductivity of fluid', 'specific heat of fluid', 'fluid velocity', 'surface roughness', 'laminar or turbulent flow' etc. The rate of convection heat transfer can be estimated using 'Newton's law of cooling'. Newton's law of cooling states that heat transfer per unit area is directly proportional to temperature difference.

$$\left(\frac{Q_{convection}}{A}\right) \propto (T_{hot} - T_{surrounding})$$

or, $$\boxed{Q_{convection} = h \cdot A \cdot (T_{hot} - T_{surrounding})}$$

Here '*h*' is the convection heat transfer coefficient in $W/m^2 \cdot °C$, 'A' is surface area in m^2, T_{hot} & $T_{surrounding}$ are temperatures of hot surface and surroundings in °C, $Q_{convection}$ is convection heat transfer in *W*.

Convection heat transfer coefficient is defined as the rate of heat transfer per unit area and per unit temperature difference between surface and fluid. This could be mathematically given as below;

$$h = \frac{Q_{convection}}{A \cdot (T_{hot} - T_{surrounding})}$$

Taking; $A = 1\ m^2, (T_{hot} - T_{surrounding}) = 1°C$

$$h = Q_{convection}$$

Also, the convection could be understood to be realized due to conduction type heat transfer occurring within fluid film between temperatures of T_{hot} and $T_{surrounding}$. Let us consider thickness of fluid life as δ, thermal conductivity of fluid as *k*. By Fourier's law within the film;

$$Q = \frac{-k \cdot A \cdot (T_{surrounding} - T_{hot})}{\delta} = \frac{kA(T_{hot} - T_{surrounding})}{\delta}$$

If we look at convection heat transfer mechanism then it is obvious that this *Q* should be equal to $Q_{convection}$ and comparison with outcome from Newton's law of cooling shows heat transfer coefficient

$h = \frac{k}{\delta}$, W/m^2, °C

so, $$Q = Q_{convection} = h \cdot A \cdot (T_{hot} - T_{surrounding})$$

It indicates that heat transfer increases with increase in heat transfer coefficient. Heat transfer coefficient depends directly upon thermal conductivity of fluid in film and inversely upon thickness of film.

Also, $$Q_{convection} = \frac{(T_{hot} - T_{surrounding})}{(1/(h\,A))} = \frac{(T_{hot} - T_{surrounding})}{R}$$

so thermal resistance offered by fluid film, $\boxed{R = \frac{1}{(hA)}}$

Let us consider some examples involving convection heat transfer;

Example 1:

For the heat transfer occurring across a plate (wall) kept between hot fluid and cold fluid at temperatures T_{hot} and T_{cold} respectively. Fig. 12.10 gives further details.

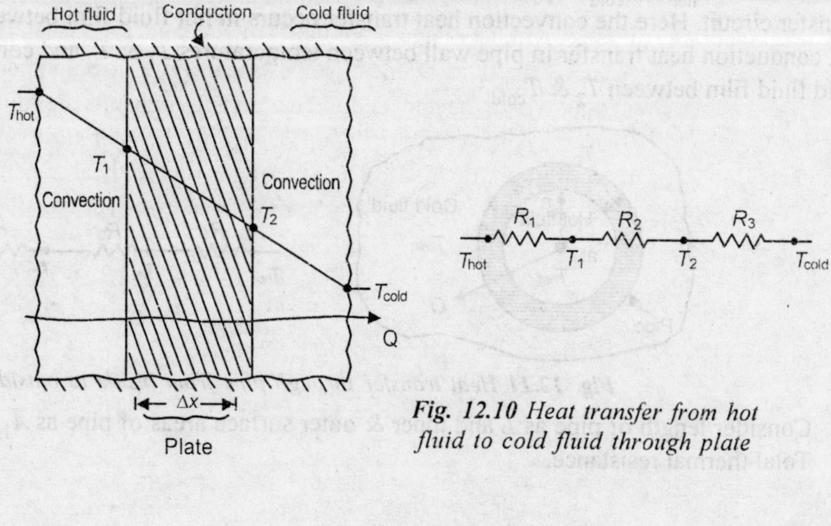

Fig. 12.10 Heat transfer from hot fluid to cold fluid through plate

Total thermal resistance may be obtained by considering it to be due to convection in hot fluid film between T_{hot} & T_1, conduction across plate (wall) between T_1 & T_2 and convection in cold fluid film between T_2 & T_{cold}. Thermal conductivity of plate may be taken as 'k', heat transfer coefficients for hot fluid film and cold fluid film may be taken as h_{hot} & h_{cold} respectively.

Thus,

$$R = R_{\text{convection, hot film}} + R_{\text{conduction, plate}} + R_{\text{convection, cold film.}}$$

$$R = R_1 + R_2 + R_3$$

Here, $R_1 = \left(\dfrac{1}{h_{hot} \cdot A}\right)$, $R_2 = \left(\dfrac{\Delta x}{k \cdot A}\right)$, $R_3 = \left(\dfrac{1}{h_{cold} \cdot A}\right)$

$$R = \left(\frac{1}{h_{hot} \cdot A}\right) + \left(\frac{\Delta x}{k \cdot A}\right) + \left(\frac{1}{h_{cold} \cdot A}\right)$$

Total heat transfer, $$Q = \frac{T_{hot} - T_{cold}}{\left\{\left(\dfrac{1}{h_{hot} \cdot A}\right) + \left(\dfrac{\Delta x}{k \cdot A}\right) + \left(\dfrac{1}{h_{cold} \cdot A}\right)\right\}}$$

$$Q = U \cdot A \cdot (T_{hot} - T_{cold})$$

Here U is overall heat transfer coefficient ($W/m^2 \cdot$ °C) and

$$\frac{1}{UA} = \left(\frac{1}{h_{hot} \cdot A}\right) + \left(\frac{\Delta x}{k \cdot A}\right) + \left(\frac{1}{h_{cold} \cdot A}\right)$$

or, $$\boxed{\frac{1}{U} = \frac{1}{h_{hot}} + \frac{\Delta x}{k} + \frac{1}{h_{cold}}}$$

Example 2:

Let us consider heat transfer from hot fluid inside pipe to cold fluid outside pipe at temperatures of T_{hot} & T_{cold} respectively. Let the temperatures at inside and outside pipe surface be T_1 & T_2 respectively. Let us also assume thermal conductivity of pipe material as k and heat transfer coefficients of hot & cold fluids as T_{hot} & T_{cold} respectively. Fig. 12.11 shows the heat transfer across the section and the heat transfer circuit. Here the convection heat transfer occurs in hot fluid film between temperatures T_{hot} & T_1, conduction heat transfer in pipe wall between temperatures T_1 & T_2 and convection heat transfer in cold fluid film between T_2 & T_{cold}.

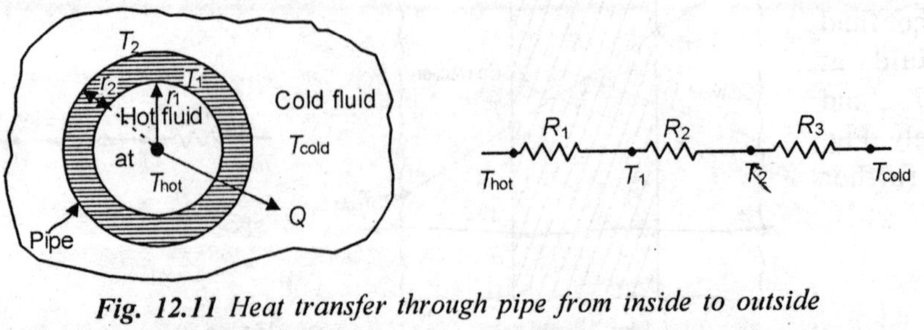

Fig. 12.11 Heat transfer through pipe from inside to outside

Consider length of pipe as L and inner & outer surface areas of pipe as A_1 & A_2 respectively. Total thermal resistance,

$$R = R_{\text{convection, hot fluid film}} + R_{\text{conduction, pipe}} + R_{\text{convection, cold fluid film}}$$

$$R = R_1 + R_2 + R_3$$

Here, $$R_1 = \frac{1}{h_{hot} \cdot A_1},\ R_2 = \frac{\ln(r_1/r_2)}{2\pi kL},\ R_3 = \frac{1}{h_{cold} \cdot A_2}$$

$$R_1 = \left(\frac{1}{h_{hot} \cdot 2\pi r_1 L}\right),\ R_2 = \left(\frac{\ln(r_1/r_2)}{2\pi kL}\right),\ R_3 = \left(\frac{1}{h_{cold} \cdot 2\pi r_2 \cdot L}\right)$$

Thus, $$Q = \left(\frac{T_{hot} - T_{cold}}{R_1 + R_2 + R_3}\right) = \frac{T_{hot} - T_{cold}}{\left\{\left(\frac{1}{h_{hot} \cdot 2\pi r_1 L}\right) + \left(\frac{\ln(r_1/r_2)}{2\pi kL}\right) + \left(\frac{1}{h_{cold} \cdot 2\pi r_2 \cdot L}\right)\right\}}$$

$Q = UA_2\,(T_{hot} - T_{cold})$, where U is overall heat transfer coefficient.

$$\boxed{\frac{1}{UA_2} = \left(\frac{1}{h_{hot} \cdot 2\pi r_1 L}\right) + \left(\frac{\ln(r_1/r_2)}{2\pi kL}\right) + \left(\frac{1}{h_{cold} \cdot 2\pi r_2 \cdot L}\right)}$$

Example 3 :

Analysis of heat transfer in insulated electric wire can be carried out similarly as described ahead.

CRITICAL THICKNESS OF INSULATION

In case of insulated wire the heat transfer analysis can be done by considering it similar to a long cylinder or tube having insulation covering it. Heat transfer from inside to outside is desired to be maximum in order to keep inside temperature minimum. It is seen that, as thickness of insulation is increased the resistance to conduction heat transfer shall get increased while the surface area for heat transfer also increases due to increase in diameter thereby causing increased heat transfer by convection. Thus, it is observed that there are two opposite consequences of change in thickness of insulation. This leads to quantification of optimum thickness of insulation. Let us consider wire as tube having covering insulation with inside and outside radius as r_1 & r_2 respectively. Thermal conductivity of

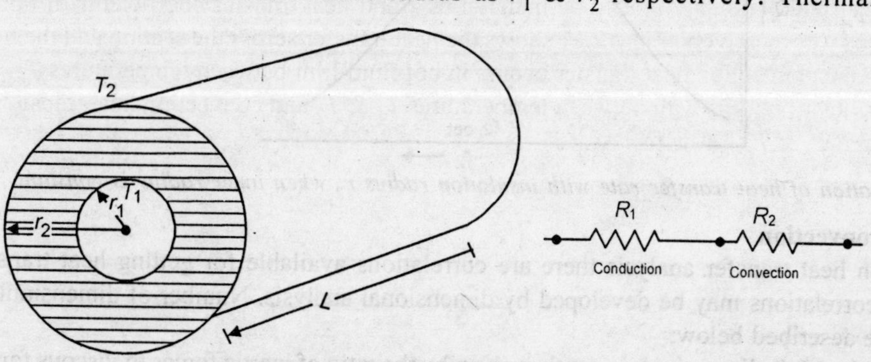

Fig. 12.12 One dimensional heat transfer in insulated wire

insulating material is 'k_1' and heat transfer coefficient responsible for convective heat transfer from outer surface to atmosphere be 'h'. Let temperature of wire be T_1 which shall be temperature at inner surface of insulation and temperature of atmosphere be T_2. Here total heat transfer shall comprise of heat transfer due to conduction across the insulation and convection between external surface and atmosphere. From electrical analogy it is obvious that there are two thermal resistances due to conduction and convection, say R_1 and R_2.

Thermal resistances are put in series so the total thermal resistance, $R = R_1 + R_2$, where

$$R_1 = \left(\frac{\ln(r_1/r_2)}{2\pi kL}\right) \text{ and } R_2 = \left(\frac{1}{2\pi r_2 L \cdot h}\right)$$

$$\text{Total heat transfer, } Q = \left(\frac{T_1 - T_2}{R_1 + R_2}\right) = \frac{(T_1 - T_2).(2\pi L)}{\frac{\ln(r_1/r_2)}{k} + \frac{1}{h.r_2}}$$

Here in case of insulated wire thickness of insulation is to be optimized as it is evident from above expression that with increase in thickness i.e. increase in r_2 for some fixed value of r_1 there occurs increase in $\left(\frac{\ln(r_1/r_2)}{k}\right)$ and decrease in $\left(\frac{1}{h \cdot r_2}\right)$. Therefore, Q should be optimized with respect to r_2.

$$\frac{dQ}{dr_2} = -2\pi L(T_1 - T_2)\left(\frac{1}{k \cdot r_2} - \frac{1}{h \cdot r_2^2}\right)$$

Equating differential $\left(\frac{dQ}{dr_2}\right)$ to zero; $\frac{dQ}{dr_2} = 0 \Rightarrow \boxed{r_{2,opt} = \frac{k}{h}}$

It is seen that Q is maximum for $\left(r_{2,opt} = \frac{k}{h}\right)$ and this outer radius yields critical thickness of insulation such that heat transfer is maximum. Graphical representation indicates how the heat transfer rate Q varies with outer radius r_2 for some fixed value of r_1 as shown in Fig. 12.13 ahead.

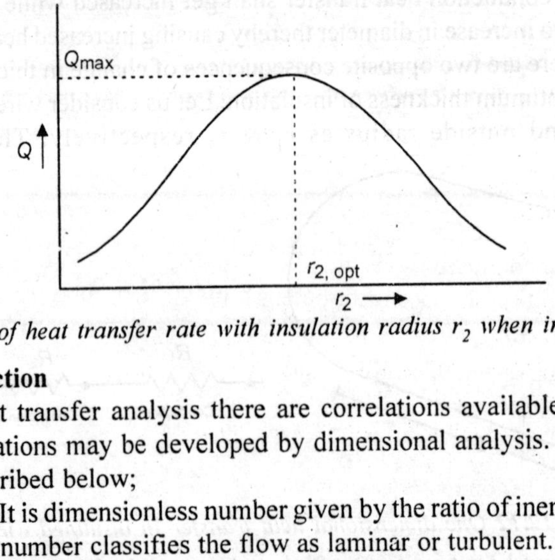

Fig. 12.13 Variation of heat transfer rate with insulation radius r_2 when inner radius is constant

Correlations for convection

For convection heat transfer analysis there are correlations available for getting heat transfer coefficient. These correlations may be developed by dimensional analysis. Number of dimensionless parameters used are described below;

(i) *Reynolds number:* It is dimensionless number given by the ratio of inertia forces to viscous forces in fluid. Reynolds number classifies the flow as laminar or turbulent. Reynolds number beyond which flow becomes turbulent is called critical Reynolds number. At small Reynolds number the flow is laminar as viscous forces are large and these viscous forces prevent fluid from becoming random and turbulent. Critical Reynolds number has different values for different geometry.

$$\text{Reynolds number} = \frac{\text{Inertia force}}{\text{Viscous force}}$$

Mathematically,

$$\boxed{Re = \frac{\delta V D}{\mu}}$$

where,

ρ : Density of fluid, kg/m^3
V : Mean velocity, m/s
D : Characteristic length, m
μ : Viscosity, kg/m.s
For flow over flat plate,
Laminar flow has Re < 2100 as
turbulent flow has Re > 2100
For flow through pipe, laminar flow has $Re < 5 \times 10^5$ and
turbulent flow has $Re > 5 \times 10^5$.

(*ii*) *Prandtl number*: It is dimensionless number given as the ratio of molecular diffusivity of momentum and molecular diffusivity of heat.

$$\text{Prandtl number, } P_r = \frac{\mu C_p}{k},$$

where

μ : Viscosity, kg/m·s
C_p : Specific heat, J/kg·k
k : Thermal conductivity, W/m·k

Prandtl number value depends upon fluid properties and has value lying between 0.01 to 10^5.

(*iii*) *Nusselt number*: Nusselt number is the dimensionless convection heat transfer coefficient.

$$\text{Nusselt number, } Nu = \frac{hD}{k}$$

where

h : heat transfer coefficient, W/m^2· K
D : characteristic length, m
k : thermal conductivity, W/m· K

For forced convection: In case of turbulent flow through tubes correlation is,
For fluid heating: $Nu = 0.023 \cdot Re^{0.8} \cdot pr^{0.4}$
For fluid cooling: $Nu = 0.023 \cdot Re^{0.8} \cdot pr^{0.4}$

12.4 RADIATION

Radiation is the mode of heat transfer which involves emission of energy in the form of electromagnetic waves or electromagnetic radiations. Radiation differs from conduction and convection as it does not require medium for heat transfer to take place. When the energy is emitted from a body due to its temperature then the radiations are termed as thermal radiations and these thermal radiations only cause radiation heat transfer. Rate of thermal radiation emission increases with increasing temperature as with the rise in temperature the activities at microscopic level increase. Generally electromagnetic radiations emitted lie in the range of wavelength varying between 10^{-10} μm to 10^{10} μm. In case of thermal radiations the electromagnetic radiations have wavelength lying between 0.1 to 100 μm and include the visible, infrared radiations and some ultraviolet radiations.

Radiations emitted by sun are called solar radiations and lie in the wavelength range of 0.3 - 0.3 μm. Radiations emitted by bodies at room temperature lie in infrared range of 0.76 – 100 μm. Light source emitting radiations in visible range lie in the wavelength range of 0.40 – 0.76 μm. Thermal radiations which may also lie in ultraviolet radiations range have wave length between 0.01 – 0.40 μm. It is seen that radiations are emitted by every body at temperature more than absolute zero. The radiation energy emitted shall depend upon its temperature. There shall be different amount of radiations per unit surface area depending upon its properties and state. There exists an ideal condition in which the emissions from the body are maximum at that temperature. Such idealized body has maximum emissions and also maximum absorption of radiations. This kind of body is called "black body". 'Black body' is defined as the one which absorbs all incident radiations. Black body is perfect emitter and absorber of radiations. Thus, at any temperature no surface can radiate energy more than that of black body. Bodies other than black body are called 'Gray body'. Generally, all bodies are gray body.

Black body emits radiation uniformly in all directions per unit area normal to the direction of emission and due to these uniform radiations in all directions it is called 'diffuse emitter'. Stefan & Boltzmann quantified the radiant energy emitted by a black body on per unit time and per unit surface area basis. Stefan-Boltzmann's law states that the rate of energy radiated by a black body at temperature (T in k) is given as;

$$\boxed{Q_{\text{radiation}} = \sigma \cdot A \cdot T^4}, \text{ where } \sigma = 5.67 \times 10^{-8} \text{ W/m}^2 \cdot \text{K}^4$$

Here $Q_{\text{radiation}}$ is rate of energy radiation in W, A is surface area radiating heat in m^2, T is temperature of body in K and σ is Stefan- Boltzmann constant with value of 5.67×10^{-8} $W/m^2 \cdot K^4$.

If some energy radiations ($Q_{\text{radiation}}$) fall on a surface then some part of it gets absorbed (Q_{absorbed}), some part of it gets reflected ($Q_{\text{reflected}}$) and remaining will be transmitted, ($Q_{\text{transmitted}}$) through the body. Here the fraction of energy absorbed is termed absorptivity, (α),fraction of energy reflected is called reflectivity (ρ) and fraction of energy transmitted is called transmissivity (τ). By energy balance.

Rate of energy radiations = $Q_{\text{radiation}} = Q_{\text{absorbed}} + Q_{\text{reflected}} + Q_{\text{transmitted}}$

$$1 = \frac{Q_{\text{absorbed}}}{Q_{\text{radiation}}} + \frac{Q_{\text{reflected}}}{Q_{\text{radiation}}} + \frac{Q_{\text{transmitted}}}{Q_{\text{radiation}}}$$

so, $$\boxed{1 = \alpha + \rho + \tau}$$

For an opaque body transmissivity is zero i.e. $\tau = 0$ and,

$$\boxed{\alpha + \rho = 1}$$

Similar to absorptivity, reflectivity and transmissivity the capability of a body to radiate energy may be quantified by emissivity (∈). Emissivity of gray body is defined as the fraction of actual radiations emitted in respect to the black body. Emissivity is given by the ratio of actual radiations of gray body at temperature, T (in K) and the radiations of black body at temperature, T (in K).

$$\boxed{\in = \frac{\text{Actual radiations from gray body at temperature } T}{\text{Radiations from black body at temperature } T}}$$

Thus, for gray body, $\boxed{Q_{\text{radiations}} = \in \cdot \sigma A T^4}$

Generally, emissivity of body is equal to absorptivity and this fact is given by "Kirchoff's law" Kirchoff's law states that the emissivity of a body at a particular temperature is equal to absorptivity for radiations from that body at same temperature.

It is seen that the amount of radiations from a body depend largely upon its temperature. The wavelength of radiations for maximum energy can be related to the temperature of body by "Wein's law". Wein's law states that the wavelength of radiations corresponding to maximum energy is inversely proportional to the absolute temperature of body i.e. $\boxed{\lambda_m \cdot T = \text{constant}}$ where λ_m is the wavelength of radiations corresponding to maximum energy.

Every body at temperature more than absolute zero emits radiations so if a body is kept at temperature equal to that of surroundings then the temperature of body does not change. This is due to the fact that the rate at which energy is radiated from body is equal to energy received from surroundings and so no change in its temperature. There is a theory called "Provost's theory of heat exchange" which states that in case of a hot body kept in low temperature surroundings the rate of energy radiated from body is more than the rate at which it receives energy from surroundings (being at low temperature) and so temperature of body drops down till thermal equilibrium is attained".

Radiation heat transfer between two gray bodies can be estimated using Stefan Boltzmann law. Let the temperature of two bodies be T_1 and T_2 with their surface areas as A_1 & A_2 respectively. The amount of radiation heat exchange between two bodies shall depend upon medium intervening two bodies and how the two bodies face each other (i.e. see each other) and also their emissivities and absorptivities. This consideration of how two bodies face each other can be quantified using 'shape factor', also called as 'view factor' or 'configuration factor' or 'geometry factor'. Shape factor quantifies the fraction of radiant energy leaving body 1 towards body 2.

If view factor for gray bodies is F_{1-2} and for the two similar black bodies view factor is F_{1-2} then the two can be related as,

$$F_{1-2} = \frac{1}{\left(\frac{1}{\epsilon_1} - 1\right) + \frac{1}{F_{1-2}} + \frac{A_1}{A_2}\left(\frac{1}{\epsilon_2} - 1\right)}$$

Here ϵ_1 & ϵ_2 are emissivities of two bodies, A_1 & A_2 are surface areas of two bodies facing each other.

Reciprocity theorem says that $F_{1-2} \cdot A_1 = A_2 F_{2-1}$

Amount of radiation energy heat exchange between two gray bodies shall be,

$$\boxed{Q_{1-2} = \sigma \cdot A_1 \cdot F_{1-2} \cdot \left(T_1^4 - T_2^4\right)}$$

Case I: In case of two infinite parallel gray plane surfaces, $A_1 = A_2$ and $F_{1-2} = 1$. Unity view factor means all radiations emitting from one surface fall on other.

So,

$$F_{1-2} = \frac{1}{\left\{\frac{1}{\epsilon_1} + \frac{1}{\epsilon_2} - 1\right\}}$$

Radiation heat exchange between two plane gray walls,

$$Q_{\text{radiation}} = \frac{\sigma A\left(T_1^4 - T_2^4\right)}{\left\{\frac{1}{\epsilon_1} + \frac{1}{\epsilon_2} - 1\right\}}$$

Case II: In case of two concentric cylinders,

$F_{1-2} = 1$, A_1 = Surface area of inner cylinder exposed to radiations

A_2 = Surface area of outer cylinder exposed to radiations

$$F_{1-2} = \frac{1}{\left\{\frac{1}{\epsilon_1} + \left(\frac{1}{\epsilon_2} - 1\right)\frac{A_1}{A_2}\right\}}$$

$$Q_{\text{radiation}} = \frac{\sigma \cdot A_1\left(T_1^4 - T_2^4\right)}{\left\{\frac{1}{\epsilon_1} + \left(\frac{1}{\epsilon_2} - 1\right)\frac{A_1}{A_2}\right\}}$$

Case III: In case of two concentric sphere; similar to the case of two concentric cylinders,

$F_{1-2} = 1$, A_1 = Surface area of inner sphere.

A_2 = Surface area of outer sphere.

$$F_{1-2} = \frac{1}{\frac{1}{\epsilon_1} + \left(\frac{1}{\epsilon_2} - 1\right)\frac{A_1}{A_2}}$$

$$Q_{\text{radiation}} = \frac{\sigma A_1\left(T_1^4 - T_2^4\right)}{\left\{\frac{1}{\epsilon_1} + \left(\frac{1}{\epsilon_2} - 1\right)\frac{A_1}{A_2}\right\}}$$

12.5 HEAT EXCHANGERS

Heat exchanger is the device in which heat exchange occurs between two fluids flowing in it. In heat exchanger the heat transfer between two fluids takes place across the metal (conducting) wall interface. Based on the relative flow of fluids the heat exchangers can be classified as;

(*i*) *Parallel flow heat exchanger* : In this both fluids flow parallel to each other in the same direction.

(*ii*) *Counter flow heat exchanger* : In this both fluids flow parallel to each other but in opposite direction.

(*iii*) *Cross flow heat exchanger* : In this heat exchanger two fluids flow normal to each other.

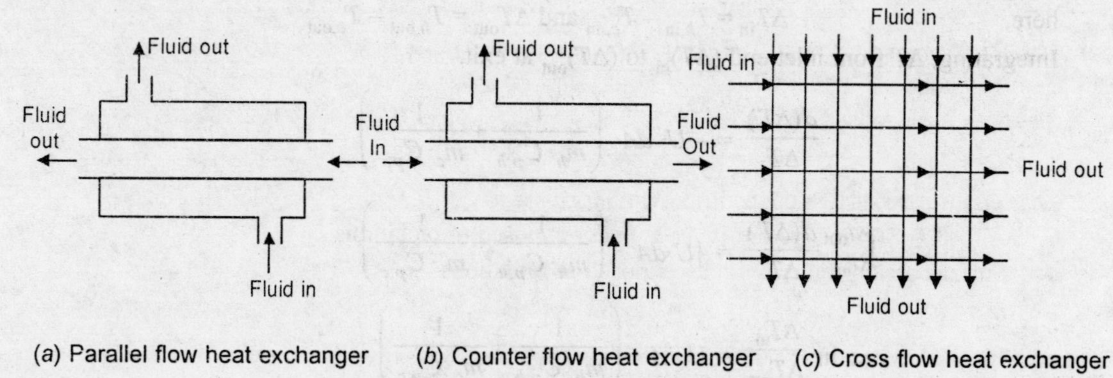

Fig. 12.14 Types of heat exchanger

Parallel flow heat exchanger

Let us assume cold fluid and hot fluid flowing through exchanger inside tube and annulus respectively.

$$\Delta T_{in} = T_{h,\,in} - T_{c,\,in}$$
$$\Delta T_{out} = T_{h,\,out} - T_{c,\,out}$$

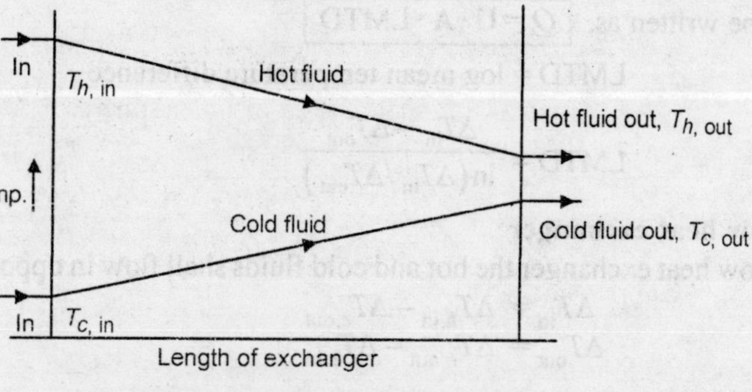

Fig. 12.15 Parallel flow heat exchanger

Let us consider small element with length dx in heat exchanger. Let the temperature of hot fluid be T_h and cold fluid be T_c in the element. Let the mass flow rate of hot & cold fluid be m_h & m_c and specific heat of hot & cold fluid be $C_{p,h}$ & $C_{p,c}$ respectively. If the over all heat transfer coefficient is U and surface area is dA and difference in temperature is ΔT between two fluids then amount of heat exchanged dQ shall be.

$$dQ = U \cdot dA \cdot \Delta T = m_c \cdot C_{p,\,c} \,.\, dT_c = -m_h \cdot C_{p,h} \cdot dT_h$$

Here, dT_h is negative as temperature of hot fluid decreases along length of exchanger.

Temperature difference, $\Delta T = T_h - T_c \Rightarrow d(\Delta T) = dT_h - dT_c$

$$d(\Delta T) = \frac{-dQ}{m_h \cdot C_{p,h}} - \frac{dQ}{m_c \cdot C_{p,c}} = -\,dQ\left(\frac{1}{m_h \cdot C_{p,h}} + \frac{1}{m_c \cdot C_{p,c}}\right)$$

$$\Rightarrow \int_{\Delta T_{in}}^{\Delta T_{out}} d(\Delta T) = -\int dQ \left(\frac{1}{m_h \cdot C_{p,h}} + \frac{1}{m_c \cdot C_{p,c}}\right) \Rightarrow \frac{\Delta T_{in} - \Delta T_{out}}{Q} = \left(\frac{1}{m_h \cdot C_{p,h}} + \frac{1}{m_c \cdot C_{p,c}}\right)$$

Integration yields,

here, $\Delta T_{in} = T_{h,in} - T_{c,\,in}$ and $\Delta T_{out} = T_{h,out} - T_{c,out}$

Integrating, ΔT from inlet end $(\Delta T)_{in}$ to $(\Delta T)_{out}$ at exit.

$$\frac{d(\Delta T)}{\Delta T} = -U\cdot dA\cdot\left(\frac{1}{m_h\cdot C_{p,h}} + \frac{1}{m_c\cdot C_{p,c}}\right)$$

$$\int_{\Delta T_{in}}^{\Delta T_{out}} \frac{d(\Delta T)}{\Delta T} = \int U\cdot dA\cdot\left(\frac{1}{m_h\cdot C_{p,h}} + \frac{1}{m_c\cdot C_{p,c}}\right)$$

$$\Rightarrow \ln\frac{\Delta T_{in}}{\Delta T_{out}} = U\cdot A\cdot\left(\frac{1}{m_h\cdot C_{p,h}} + \frac{1}{m_c\cdot C_{p,c}}\right)$$

Substitution yields, $\ln\dfrac{\Delta T_{in}}{\Delta T_{out}} = U\cdot A\cdot\dfrac{\Delta T_{in} - \Delta T_{out}}{Q}$

$$\Rightarrow \quad \boxed{Q = \frac{U\cdot A\cdot(\Delta T_{in} - \Delta T_{out})}{\ln(\Delta T_{in}/\Delta T_{out})}}$$

It may also be written as, $\boxed{Q = U\cdot A\cdot LMTD}$

where LMTD = log mean temperature difference

$$LMTD = \frac{\Delta T_{in} - \Delta T_{out}}{\ln(\Delta T_{in}/\Delta T_{out})}$$

Counter flow heat exchanger

In counter flow heat exchanger the hot and cold fluids shall flow in opposite direction to each other.

$$\Delta T_{in} = \Delta T_{h,in} - \Delta T_{c,out}$$
$$\Delta T_{out} = \Delta T_{h,out} - \Delta T_{c,in}$$

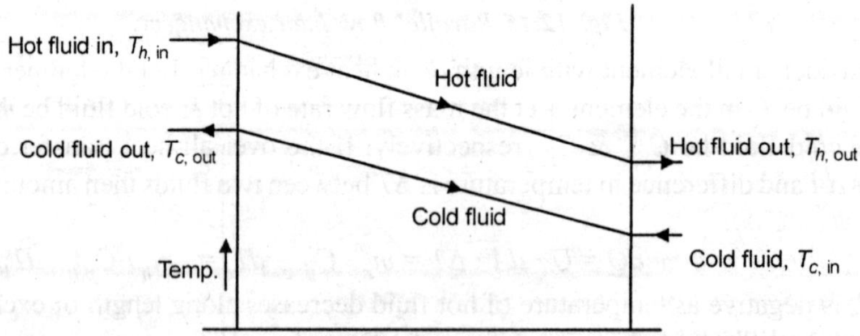

Fig. 12.16 Counter flow heat exchanger

Using similar nomenclature as in case of parallel flow heat exchanger;

$$dQ = U\cdot dA\cdot\Delta T = -m_c\cdot C_{p,c}\cdot dT_c = -m_h\cdot C_{p,h}\cdot dT_h$$

Temperature difference, $\Delta T = T_h - T_c \Rightarrow d(\Delta T) = dT_h - dT_c$

$$d(\Delta T) = \frac{-dQ}{m_h\cdot C_{p,h}} + \frac{dQ}{m_c\cdot C_{p,c}} = -dQ\left(\frac{1}{m_h\cdot C_{p,h}} - \frac{1}{m_c\cdot C_{p,c}}\right)$$

Integrating from inlet to outlet along +ve X direction

$$\int_{\Delta T_{in}}^{\Delta T_{out}} d(\Delta T) = -\int dQ\left(\frac{1}{m_h \cdot C_{p,h}} - \frac{1}{m_c \cdot C_{p,c}}\right) \Rightarrow \frac{\Delta T_{in} - \Delta T_{out}}{Q} = \left(\frac{1}{m_h \cdot C_{p,h}} - \frac{1}{m_c \cdot C_{p,c}}\right)$$

Also, $$d(\Delta T) = -U \cdot dA \cdot \Delta T\left(\frac{1}{m_h \cdot C_{p,h}} - \frac{1}{m_c \cdot C_{p,c}}\right)$$

$$\Rightarrow \frac{d(\Delta T)}{\Delta T} = -U \cdot dA\left(\frac{1}{m_h \cdot C_{p,h}} - \frac{1}{m_c \cdot C_{p,c}}\right)$$

$$\int_{\Delta T_{in}}^{\Delta T_{out}} \frac{d(\Delta T)}{\Delta T} = -\int U \cdot dA\left(\frac{1}{m_h \cdot C_{p,h}} - \frac{1}{m_c \cdot C_{p,c}}\right)$$

$$\Rightarrow \ln\left(\frac{\Delta T_{in}}{\Delta T_{out}}\right) = U \cdot A \cdot \left(\frac{1}{m_h \cdot C_{p,h}} - \frac{1}{m_c \cdot C_{p,c}}\right)$$

$$\ln\left(\frac{\Delta T_{in}}{\Delta T_{out}}\right) = U \cdot A \cdot \left(\frac{\Delta T_{in} - \Delta T_{out}}{Q}\right)$$

$$\Rightarrow Q = U \cdot A \cdot \frac{\Delta T_{in} - \Delta T_{out}}{\ln(\Delta T_{in}/\Delta T_{out})}$$

$$Q = U \cdot A \cdot \text{LMTD}$$

Here log mean temperature difference = LMTD = $\left\{\dfrac{\Delta T_{in} - \Delta T_{out}}{\ln(\Delta T_{in}/\Delta T_{out})}\right\}$

where $\Delta T_{in} = \Delta T_{h,in} - T_{c,out}$, $\Delta T_{out} = \Delta T_{h,out} - T_{c,in}$

EXAMPLES

***1.** A composite wall made of two different materials as shown has air temperatures of 50°C and 20°C on both ends of wall. Determine the rate of heat transfer across wall and temperature at interface. The wall has atmospheric air on both sides of wall. The heat transfer coefficients on the side of 50°C and 20°C and 30 and 10 W/m²·k. Surface area of heat transfer is 100 m².*

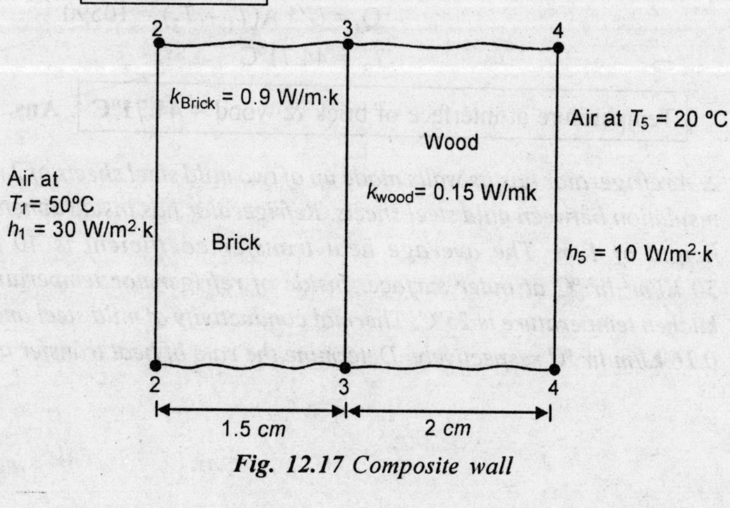

***Fig. 12.17** Composite wall*

Solution:

Here for one dimensional heat transfer across the wall the heat transfer circuit shall comprise of thermal resistance due to, 'convection between air & brick (R_1)', 'conduction in brick wall (R_2)', 'conduction in wood (R_3)', and 'convection between wood and air (R_4)'. Let temperature at outer brick wall be T_2°K, brick-wood interface be T_3°K, outer wood wall be T_4°K.

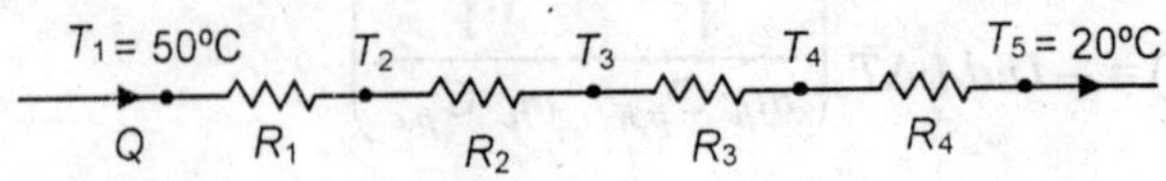

Overall heat transfer coefficient for steady state heat transfer

$$\frac{1}{U} = \frac{1}{h_1} + \frac{\Delta x_{\text{Brick}}}{k_{\text{Brick}}} + \frac{\Delta x_{\text{wood}}}{k_{\text{wood}}} + \frac{1}{h_5}$$

$$= \frac{1}{30} + \frac{1.5\times10^{-2}}{0.9} + \frac{2\times10^{-2}}{0.15} + \frac{1}{10}$$

$$\Rightarrow \qquad U = 3.53 \ W/m^2\cdot \text{k}$$

Rate of heat transfer, $Q = U{\cdot}A\,(T_1 - T_5) = 3.53 \times 100(50 - 20) = 10590\ W$

Rate of heat transfer = **10590 *W***

Ans.

Heat transfer across states 1 and 3 (at interface).
Overall heat transfer coefficient between 1 & 3

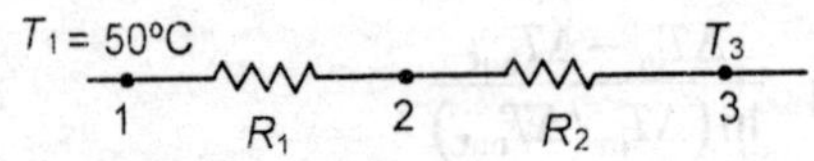

$$\frac{1}{U'} = \frac{1}{h_1} + \frac{\Delta x_{\text{Brick}}}{k_{\text{Brick}}}$$

$$= \frac{1}{30} + \frac{1.5\times10^{-2}}{0.9}$$

$$\Rightarrow \qquad U' = 20 \ W/m^2\cdot \text{k}$$

$$Q = U'\cdot A(T_1 - T_3) = 10590$$

$$\Rightarrow \qquad T_3 = 44.71°\text{C}$$

Temperature at interface of brick & wood = **44.71°C**

Ans.

2. *A refrigerator has its walls made up of two mild steel sheets of 2 mm thickness with 5 cm of glass wool insulation between mild steel sheets. Refrigerator has inside dimension of 0.5 × 0.5 m base surface and height of 1 m. The average heat transfer coefficient is 40 kJ/m²·hr·°C at inner surface and 50 kJ/m²·hr·°C at outer surface. Inside of refrigerator temperature is to be maintained at 5°C and kitchen temperature is 25°C. Thermal conductivity of mild steel and glass wool are 160 kJ/m·hr· °C and 0.16 kJ/m·hr·°C respectively. Determine the rate of heat transfer and temperature at wall surfaces.*

Solution:

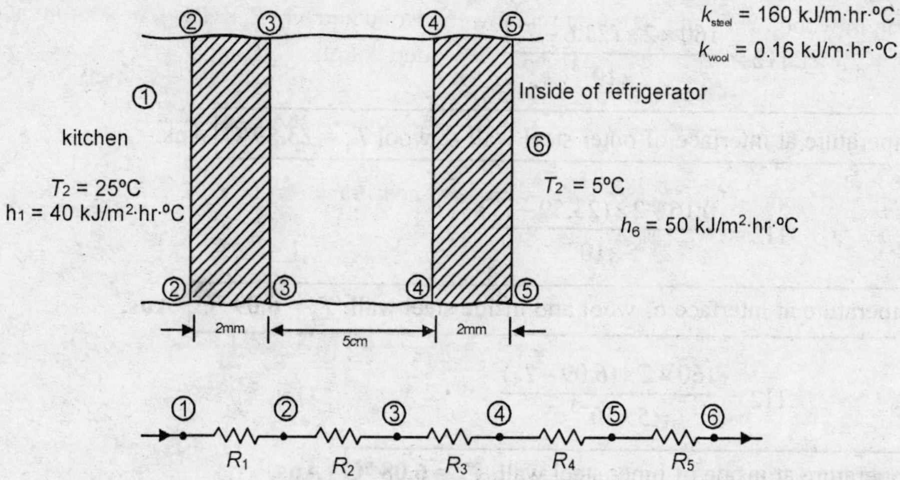

Fig. 12.18

Here thermal resistances as shown in figure are.

R_1 = Thermal resistance due to convection between kitchen air and outer surface of refrigerator wall (T_1 & T_2)

R_2 = Thermal resistance due to conduction across mild steel wall between 2 & 3 (T_2 & T_3)

R_3 = Thermal resistance due to conduction across glass wool between 3 & 4 (T_3 & T_4)

R_4 = Thermal resistance due to conduction across mild steel wall between 4 & 5 (T_4 & T_5)

R_5 = Thermal resistance due to convection between inside refrigerator wall and inside of refrigerator between 5 & 6(T_5 & T_6)

Overall heat transfer coefficient for one dimensional steady state heat transfer

$$\frac{1}{U} = \frac{1}{h_1} + \frac{\Delta x_{steel}}{k_{steel}} + \frac{\Delta x_{wool}}{k_{wool}} + \frac{\Delta x_{steel}}{k_{steel}} + \frac{1}{h_6}$$

$$= \frac{1}{40} + \frac{2\times10^{-3}}{160} + \frac{5\times10^{-2}}{0.16} + \frac{2\times10^{-3}}{160} + \frac{1}{50}$$

$\Rightarrow$ $U = 2.8$ kJ/m^2· hr·°C

Rate of heat transfer $Q = U{\cdot}A\,(T_1 - T_6)$

Wall surface area = 4(1× 0.5) = 2 m^2

$Q = 2.8 \times 2 \times (25 - 5)$ 112 kJ/m^2·hr.

Rate of heat transfer = **112 kJ/m²·hr**

Ans.

$$Q = A \cdot h_1\,(T_1 - T_2) = \frac{k_{steel} \times A \times (T_2 - T_3)}{\Delta x_{steel}} = \frac{k_{wool} \times A \times (T_3 - T_4)}{\Delta x_{wool}}$$

$$Q = \frac{k_{steel} \times A \times (T_4 - T_5)}{\Delta x_{steel}} = A \cdot h_6 \cdot (T_5 - T_6)$$

$\Rightarrow$ Substituting, $112 = 2 \times 40 \times (25 - T_2)$

Temperature of outer wall T_2 = **23.6 °C** **Ans.**

$$112 = \frac{160 \times 2 \times (23.6 - T_3)}{2 \times 10^{-3}}$$

Temperature at interface of outer steel wall & wool T_3 = **23.59°C** **Ans.**

$$112 = \frac{0.16 \times 2 \times (23.59 - T_4)}{5 \times 10^{-2}}$$

Temperature at interface of wool and inside steel wall. T_4 = **6.09 °C** **Ans.**

$$112 = \frac{160 \times 2 \times (6.09 - T_5)}{5 \times 10^{-3}}$$

Temperature at inside of inner steel wall. T_5 = **6.08 °C** **Ans.**

3. *A pipe carrying steam is covered with a layer of insulation of 3 cm thickness and thermal conductivities of pipe and insulation are 209 kJ/m·hr·°C and 0.3 kJ/m·hr·°C. Temperature of inner surface of steam pipe is 300°C and that of outer surface of insulation layer is 50°C. Steam pipe without insulation has inner and outer diameters as 15 cm and 16 cm respectively. Determine rate of heat loss per meter length of pipe and temperature at interface of pipe and insulation. Also determine state of steam coming out of pipe of 5 meter length if steam enters saturated at 300°C at the rate of 0.5 kg/min.*

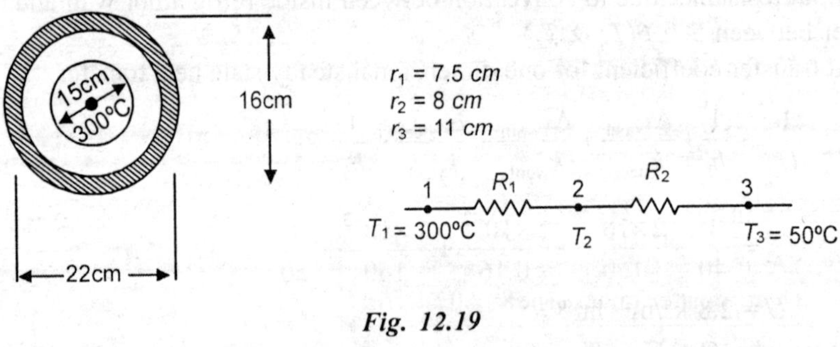

Fig. 12.19

Here, heat conduction is considered in pipe wall from 1 to 2 and conduction through insulation between 2 & 3 of one dimensional steady state type.

$$Q = \frac{(T_1 - T_3) \cdot 2\pi L}{\dfrac{1}{k_{pipe}} \cdot \ln\left(\dfrac{r_2}{r_1}\right) + \dfrac{1}{k_{insulation}} . \ln\left(\dfrac{r_3}{r_2}\right)}$$

$$= \frac{(300 - 50) \cdot 2\pi \cdot 1}{\left\{\dfrac{1}{209} \cdot \ln\left(\dfrac{8 \times 10^{-2}}{7.5 \times 10^{-2}}\right) + \dfrac{1}{0.3} \cdot \ln\left(\dfrac{11 \times 10^{-2}}{8 \times 10^{-2}}\right)\right\}}$$

$$Q = 1479.34 \text{ kJ/hr}$$

Heat loss per meter from pipe = **1479.34 kJ/hr** **Ans.**

$$\text{Heat lost from 5m length} = 5 \times 1479.34 = 7396.7 \text{ kJ/hr}$$

Enthalpy of saturated steam at 300°C $h_{\text{sat, }300°C} = 2749 \text{ kJ/kg} = h_{g \text{ at } 300°C}$

$$\text{Mass flow of steam} = 0.5 \times 60 = 30 \text{ kg per hr}$$

Final enthalpy of steam per kg at exit of 5 m pipe;

$$h = 2749 - \left(\frac{7396.7}{30}\right) = 2502.44 \text{ kJ/kg}$$

Let quality of steam at exit be x.

$$h = h_{f \text{ at } 300°C} + x \cdot h_{fg \text{ at } 300°C}$$

$$2502.44 = 1344 + x(1404.9)$$

$$x = 0.8245$$

⇒ Quality of steam at exit = **0.8245** **Ans.**

4. Determine the heat loss rate and heat flux from spherical shell having inner and outer diameters as 150 cm and 200 cm. Thermal conductivity of shell material is 28 kJ.m·hr·°C and inside surface temperature is 200°C while outer surface temperature is 40°C.

Solution:

Considering one dimensional heat transfer of steady state type

$$\text{For sphere } Q = \frac{(T_1 - T_2) \cdot 4\pi\, k \cdot r_1 r_2}{(r_2 - r_1)}$$

giving $r_1 = 75\text{cm}$, $r_2 = 100$ cm, k = 28 kJ/m·hr·°C

$T_1 = 200$ °C, $T_2 = 40$°C

$$\Rightarrow \quad \text{Heat transfer rate, } Q = \frac{(200-40) \times 4\pi \times 28 \times 0.75 \times 1}{(1-0.75)}$$

Heat transfer rate = **168892.02 kJ/hr.** **Ans.**

$$\text{Heat flux} = \frac{Q}{A} = \frac{168892.02}{4\pi \times (0.75)^2}$$

Heat flux = **23893.33 kJ/m²·hr.** **Ans.**

5. A room maintained at temperature of 25°C has a glass window of size 75cm × 100cm with glass thickness of 10 mm and thermal conductivity of 0.78 W/m·°C. During winter outside temperature is 2°C. Considering heat transfer coefficients on inner and outer window surfaces as 10 W/m²·°C and 30 W/m²·°C determine the steady state heat transfer through this glass window and temperature of inner surface of glass window.

Solution:

T_2 = Inner surface temperature of glass window.
T_3 = Outer surface temperature of glass window.

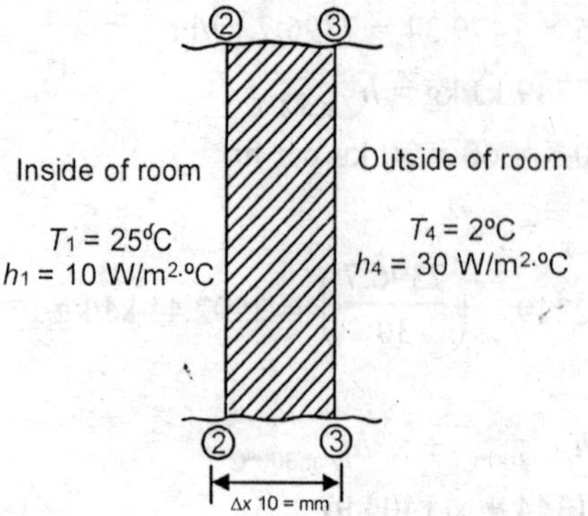

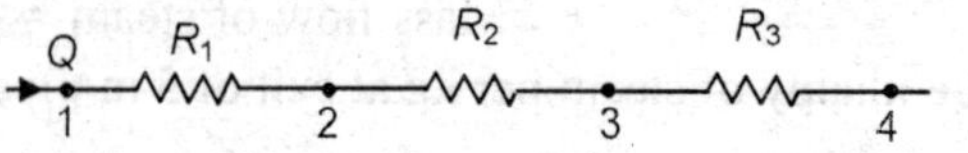

Fig. 12.20

R_1 = Thermal resistance for convection heat transfer between inside room (1) and inside surface of glass window (2) = $\frac{1}{h_1 A}$

R_2 = Thermal resistance for conduction through glass between inside of glass window (2) to outside surface of glass window (3) = $\frac{\Delta x}{k \cdot A}$

R_3 = Thermal resistance for convection heat transfer between outside surface of glass window (3) to outside atmosphere (4) = $\frac{1}{h_4 \cdot A}$

Total thermal resistance, $R_{\text{total}} = R_1 + R_2 + R_3$

$$= \left(\frac{1}{10 \times 0.75}\right) + \left(\frac{0.01}{0.78 \times 0.75}\right) + \left(\frac{1}{30 \times 0.75}\right)$$

$$R_{\text{total}} = 0.1949°\text{C/W}$$

$$\boxed{\text{Rate of heat transfer} = Q = \frac{(T_1 - T_4)}{R_{\text{total}}} = \left(\frac{25-2}{0.1949}\right) = \mathbf{118.01\ W}} \quad \textbf{Ans.}$$

Heat transfer rate from inside of room to inside surface of glass window.

$$Q = 118.01 = \frac{(T_1 - T_2)}{R_1} = \frac{(25 - T_2)}{(1/7.5)}$$

$$\Rightarrow \quad \boxed{T_2 = \mathbf{9.26°C}} \quad \textbf{Ans.}$$

Thus, inside surface of glass window will be at temperature of 9.26°C where as room inside temperature is 25°C.

6. Determine the rate of heat transfer in a pipe carrying water at mean temperature of 40°C whose inner diameter is 4 cm and length is 3m. Water flows with velocity of 1 m/s and the pipe wall temperature may be taken as 75°C. Take, for water at 40°C, k = 0.6 W/m K, P_r = 3 and ν = 0.478 × 10^{-6} m²/s. Use, Nu = 0.023 ·$(Re)^{0.8}$·$(Pr)^{0.4}$.

Solution:

$$\text{Reynolds number, Re} = \frac{VD}{\nu} = \frac{1 \times 4 \times 10^{-2}}{0.478 \times 10^{-6}} = 83682$$

Substituting in, $\text{Nu} = 0.023 \cdot (\text{Re})^{0.8} \cdot (\text{Pr})^{0.4}$

$$\left(\frac{h \times D}{k}\right) = 0.023 \times (83682)^{0.8} \cdot (3)^{0.4}$$

$$\Rightarrow \quad h = \frac{0.023 \times 0.6 \times (83682)^{0.8} \cdot (3)^{0.4}}{4 \times 10^{-2}} = 4642.73 \text{ W/m}^2\text{.k}$$

Rate of heat transfer due to convection,

$$Q = h \cdot A \cdot (75 - 40) = 4642{\cdot}73 \times (\pi \times 4 \times 10^{-2} \times 3)\,(75 - 40)$$

Heat transfer rate = **61259.38 W** or **61.26 kW** **Ans.**

7. A parallel flow tubular heat exchanger has hot gases flowing at the rate of 0.5 kg/s which are cooled from 500°C to 150°C using water. Specific heat of gas and water are 1.2 kJ/kg·K and 4.18 kJ/kg·K respectively and overall heat transfer coefficient from gases to water is 150W/m²·K. 1 kg/s of water enters the exchanger at 10°C. Determine surface area required for heat exchange.

Solution:

Let the temperature of water at exit be T.

Heat exchange, Q = Heat rejected by gases = Heat gained by water

$$Q = 0.5 \times 1.2 \times (500 - 150) = 1 \times 4.18 \times (T - 10)$$

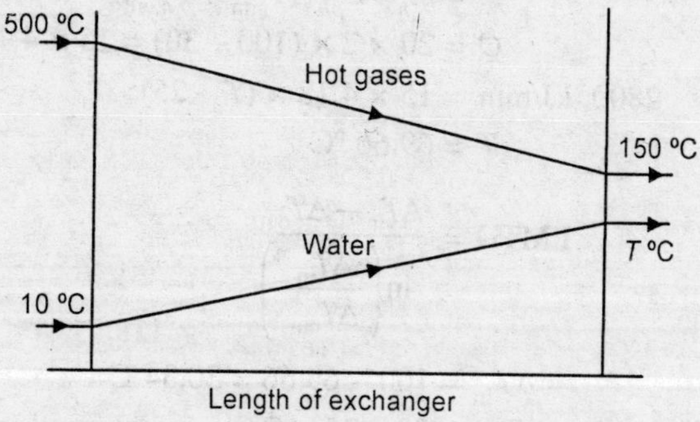

Fig. 12.21

$Q = 210$ kJ $\quad \Rightarrow T = 60.24°\text{C}$ $\quad$ so, $\Delta T_{in} = 500 - 10 = 490°\text{C}$

$\Delta T_{out} = 150 - 60.24 = 89.76°\text{C}$

For parallel flow heat exchanger,

$$\text{LMTD} = \frac{\Delta T_{in} - \Delta T_{out}}{\ln\left(\dfrac{\Delta T_{in}}{\Delta T_{out}}\right)} = 235.81\ °\text{C}$$

Also, $\qquad Q = \text{U} \cdot \text{A} \cdot \text{LMTD}$

$\Rightarrow \quad 10^3 \times 210 = 150 \times A \times 235.81$

$\Rightarrow \quad$ **Surface area, A = 5.936m² Ans.**

8. Determine surface area required for heat exchange in an oil cooler required to cool 20 kg/min oil from 100°C to 30°C using water available at 25°C. Water flows at the rate of 15 kg/min in direction opposite to that of oil. Take, overall heat transfer coefficient as 25 W/m²K, specific heat of oil as 2 kJ/kg·K, specific heat of water as 4.18 kJ/kg·K.

Solution:

This oil cooler has arrangement similar to a counter flow heat exchanger.

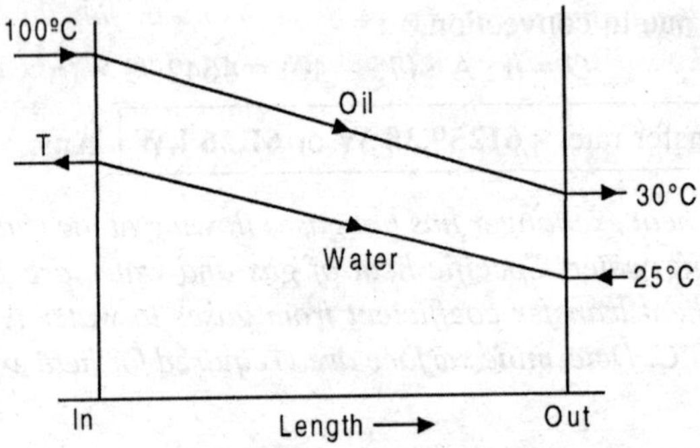

Fig. 12.22

By heat exchange, $Q = U \cdot A \cdot \text{LMTD} = m_c \cdot C_{pc} \cdot (T_{c,out} - T_{h,in})$

$= m_h \cdot C_{ph} \cdot (T_{c,in} - T_{h,out})$

$Q = 20 \times 2 \times (100 - 30) = 15 \times 4.18 \times (T - 25)$

$2800, \text{kJ/min} = 15 \times 4.18 \times (T - 25)$

$\Rightarrow \quad T = 69.66\ °C$

$$\text{LMTD} = \frac{\Delta T_{in} - \Delta T_{out}}{\ln\left(\dfrac{\Delta T_{in}}{\Delta T_{out}}\right)}$$

Here, $\Delta T_{in} = 100 - 69.66 = 30.34°C$

$\Delta T_{out} = 30 - 25 = 5°C$

$\Rightarrow \quad \text{LMTD} = 14.05$

Substituting in, $Q = U \cdot A \cdot \text{LMTD}$

$$\Rightarrow \quad \frac{2800 \times 10^3}{60} = 25 \times A \times 14.05$$

$\Rightarrow \quad A = 132.85\ m^2$

Surface area = 132.85 m² Ans.

9. A body at 1200°C is kept in black surroundings at 600°C. The emissivity of body is 0.4 at 1200°C and 0.7 at 600°C. Determine the rate of heat loss by radiation per m². Take $\sigma = 5.67 \times 10^{-8} W/m^2 \cdot K^4$

Solution:

$$T_1 = 1200 + 273 = 1473 \text{ K}$$
$$T_2 = 600 + 273 = 873 \text{ K}$$

$$\text{Rate of heat loss by radiation} = Q = \epsilon \cdot \sigma \cdot A \cdot \left(T_1^4 - T_2^4\right)$$

$$\text{Heat loss per unit area by radiation} = 0.4 \times 5.67 \times 10^{-8} \times \left\{(1473)^4 - (873)^4\right\}$$
$$= 93597.7 \text{ W}$$
$$\boxed{= 93.59 \text{ kW}} \quad \textbf{Ans.}$$

***10.** An electric cable of 8mm outer diameter has copper wire of 3mm diameter inside it with insulation covering it all around. The insulating plastic covering copper wire has thermal conductivity of 0.16 W/m·°C. Cable is exposed in the surroundings at temperature of 40°C and has heat transfer coefficient of 15W/m²·°C. Cable carries current of 5A when there is voltage drop of 16 volts along the 5 m length of cable. Determine the temperature at copper wire and plastic interface and also find out whether doubling the insulation thickness shall increase or decrease the temperature at copper wire and plastic interface.*

Solution:

Let us carry out one dimensional analysis for steady state. Due to flow of electricity the heat generated can be given as:

$$Q = VI = 5 \times 16 = 80 \text{ W}.$$

For steady state which means there should be no change in temperature of cable due to electricity flow, the heat generated should be transferred out to surroundings. Therefore, heat transfer across table should be 80 W.

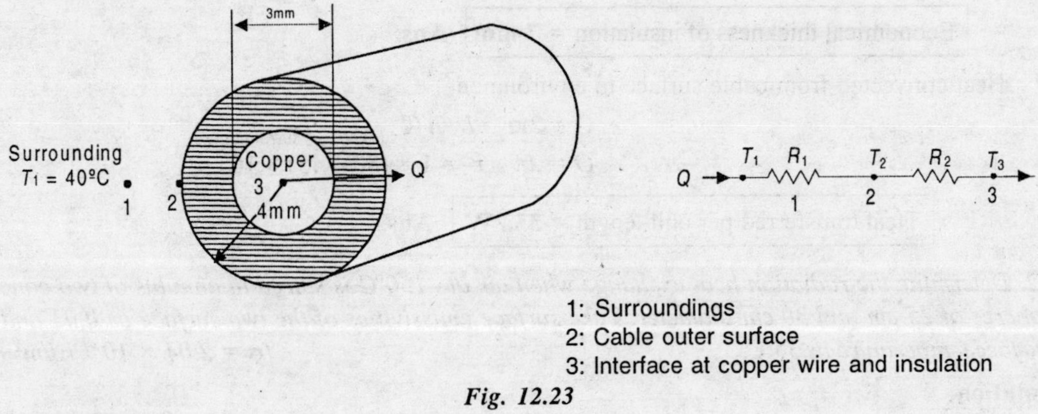

Fig. 12.23

Surface area for heat transfer, $A_2 = 2\pi \times 4 \times 10^{-3} \times 5 = 0.125 \text{ m}^2$

R_1 = Thermal resistance due to convection between surroundings and cable outer surface, (1 – 2)

$$= \frac{1}{h_1 A_2}$$

R_2 = Thermal resistance due to conduction across plastic insulation (2 – 3) $= \dfrac{\ln(r_2/r_3)}{(2\pi\, kL)}$

$$\text{Total resistance, } R_{\text{total}} = R_1 + R_2 = \frac{1}{(15 \times 0.125)} + \frac{\ln(4/1.5)}{(2\pi \times 0.16 \times 5)}$$

$$R_{\text{total}} = 1.064 \text{ °C/W}$$

$$Q = \frac{T_3 - T_1}{R_{\text{total}}} \Rightarrow T_3 = 40 + (80 \times 1.064) = 125.12\text{°C}$$

Temperature at interface = **125.12°C**	**Ans.**

$$\text{Critical radius of insulation, } r_c = \frac{k}{h} = \frac{0.16}{15}$$

$$r_c = 0.01066 \text{ m} = 10.66 \text{ mm}$$

This r_c is more than outer radius of cable so the increase in thickness of insulation upon $r_c = 10.66$ mm would increase rate of heat transfer. Doubling insulation thickness means new outer radius would be $r' = 1.5 + 5 = 6.5$ mm. Hence doubling (increase) of insulation thickness would increase heat transfer and thus temperature at interface would decrease if other parameters remain constant.

11. An electric wire of 3 mm diameter is to be insulated with plastic whose thermal conductivity is 0.16 W/m·°C. Cable is to be used in environment having temperature of 45°C. Find out the maximum heat which could be transferred per meter length through this cable and the economical thickness of insulation. Take heat transfer coefficient on cable surface as 16 W/m²·°C and cable surface temperature as 80°C.

Solution:

For maximum heat transfer the critical radius of insulation should be used.

$$\text{Critical radius of insulation, } r_c = \frac{k}{h} = \frac{0.16}{16} = 0.01 \text{ m} = 10 \text{ mm}$$

$$\text{Economical thickness of insulation} = r_c - r_{\text{wire}} = 10 - 3 = 7 \text{ mm}$$

Economical thickness of insulation = **7 mm**	**Ans.**

Heat convected from cable surface to environment,

$$Q = 2\pi r_c \cdot L \cdot h\,(T_{\text{surface}} - T_{\text{surrounding}})$$

$$Q = 2\pi \times r_c \times 1 \times (80 - 45) \times 16$$

Heat transferred per unit length = **35.2W**	**Ans.**

12. Determine the radiation heat exchange when air at –150°C is stored in annulus of two concentric spheres of 25 cm and 30 cm diameter. Take surface emissivities of the two surface as 0.03 and outer surface temperature as 35°C. *[$\sigma = 2.04 \times 10^{-4}$ kJ/m²·hr·K⁴]*

Solution:

$$\text{Temperature of air inside} = 123 \text{ K} = T_1$$

$$\text{Temperature of outer surface} = 308 \text{ K} = T_2$$

$$\text{Emissivity, } \epsilon_1 = \epsilon_2 = 0.03$$

Heat transfer through concentric sphere

$$Q = \frac{A_1 \cdot \sigma \cdot \left(T_1^4 - T_2^4\right)}{\frac{1}{\epsilon_1} + \frac{A_1}{A_2}\left(\frac{1}{\epsilon_2} - 1\right)}$$

$$= \frac{4\pi \times (0.125)^2 \times 2.04 \times 10^{-4} \times \{(123)^4 - (308)^4\}}{\left\{\frac{1}{0.03} + \left(\frac{0.125}{0.15}\right)^2 \left(\frac{1}{0.03} - 1\right)\right\}}$$

$$Q = -6297.1 \text{ kJ/hr}$$

Heat exchange = **6297.1 kJ/hr** **Ans.**

EXERCISE

12.1 Discuss the different modes of heat transfer. Briefly describe the mechanism for each one of them.

12.2 Define thermal conductivity and thermal diffusivity.

12.3 What are different types of convection heat transfer ?

12.4 Obtain the critical thickness of insulation for electric wire.

12.5 Obtain generalized one dimensional conduction heat transfer in case of plane wall having internal heat generation and without it.

12.6 State the Stefan-Boltzmann law for radiation.

12.7 Define emissivity, absorptivity, reflectivity and transmissivity.

12.8 Classify heat exchangers.

12.9 Obtain log mean temperature difference in case of parallel flow and counter flow heat exchangers.

12.10 Define Nusselt number, Reynolds number and Prandtl number.

12.11 Determine thickness of a wall having inside and outside temperatures as 60°C & 40°C to it the heat transferred through it is 2496 J/m^2·min and thermal conductivity of wall material is 0.52 W/m·k. [25 cm]

12.12 A composite wall made of steel and plastic has temperature difference across two exposed ends of wall as 250°C. If steel wall has thickness of 20 mm and plastic wall thickness is 5 mm then determine the rate of heat transfer in J/m^2·min. Take thermal conductivity of steel and plastic as 58W/m·k and 0.116W/m·k respectively. [3.48 × 10^5 J/m^2·min]

12.13 A composite plate comprising of two metals A and B is maintained such that the temperatures of outer surfaces of metals A & B are at boiling point and ice point of water. Thermal conductivities of two metals A and B are in ratio of 2:1. Show that temperature at interface is 66.7°C.

12.14 A refrigeration chamber is made up of mild steel sheet outside and wooden board inside with glass wool in between. Thickness of steel layer, wooden layer and glass wool layer are 1.5 mm, 10 mm and 20 mm respectively. Refrigeration unit is designed to extract heat at the rate of 6 kW/m^2. Determine the temperature of inside surface of wood if the steel surface outside is at 24°C. Take thermal conductivity of steel, glass wool and wood as 23.2 W/mk, 0.014W/mk and 0.052W/mk respectively. [–15°C]

12.15 A spherical shell of 120 cm diameter has inside surface at 450°C and outside shell surface as 250°C. Considering shell thickness as 10 cm and thermal conductivity of shell material as 0.083 W/m·°C determine the rate of heat loss from shell. [628.33W]

12.16 Determine the rate of heat loss per unit length of pipe carrying steam at 573 K. Pipe is kept in surroundings at temperature of 30°C. The heat transfer coefficients at inside and outside of pipe are 30 W/m^2·K and 5.8 W/m^{2}k. Steam carrying pipe is of mild steel and has two layers of insulating material as shown. [777.6 kJ/hr·m]

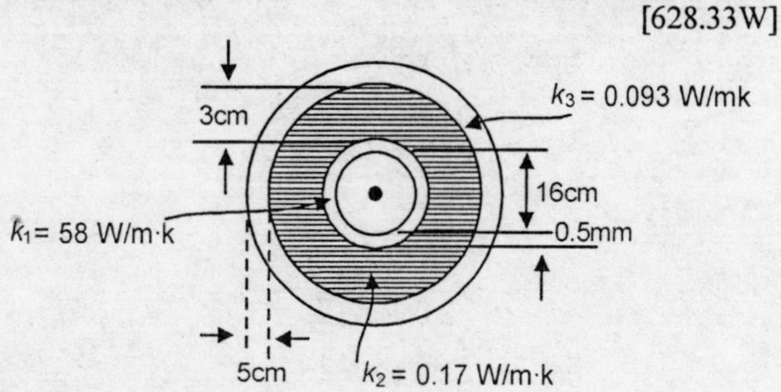

Fig. 12.24

12.17 Determine the heat transfer rate if water at 50°C flows inside tube having wall temperature of 70°C. Inside tube diameter is 5 cm and it is 3 m long carrying water at velocity of 0.8 m/s. Take; for water at 50°C, k = 0.66 W/m·k. $\nu = 0.478 \times 10^{-6}$ m²/s, Pr = 2.98. [38.4 kW]

12.18 Show that the total radiation energy exchanged between two large parallel plates at temperatures of 833 K and 300°C is 11.28 kW/m². Take emissivity of the two plates as 0.8 and 0.6 respectively.

12.19 An electric cable of 16 mm diameter is used for carrying electricity. Determine the optimum thickness of insulation of material having thermal conductivity as 0.17 W/m· °C. It is to be used in surroundings having temperature of 60°C and heat transfer coefficient on surface of insulated wire is 11.6 W/m² · °C. [6.65 mm]

12.20 An electric cable has copper wire of 0.6 mm radius insulated with 1.2 mm thick layer. Thermal conductivity of insulation and copper are 0.35 W/m·°C and 348.3 W/m·°C. The heat transfer coefficient on surface of insulated cable is 11.6 W/m·°C. Cable is put in a room having temperature of 30°C. Determine the heat transfer rate if the permissible cable outer surface temperature is 100°C. Also find the maximum possible heat transfer rate that could be realized by modifying the thickness of insulation. [8.59 W/m]

12.21 In a tubular heat exchanger gas flowing at the rate of 1440 kg/hr is cooled using 15°C cooling water flowing at the rate of 1800 kg/hr. Gas is cooled from 450°C to 150°C. Considering the specific heats of gas & water as 1.13 kJ/kg K and 4.18 kJ/kg. K respectively and overall heat transfer coefficient from gas to water as 140 W/m²·K obtain the ratio of surface areas required when it has parallel flow type and counter flow type arrangement. [1.17:1]

13

One Dimensional Compressible Fluid Flow

13.1 INTRODUCTION

The fluids whose density does not undergo variation with variation in other thermodynamic parameters are called incompressible fluids. If the density variation is seen in a fluid during change in other thermodynamic parameters, fluid is said to be a compressible fluid. The air a popular fluid normally behaves as incompressible fluid, but at very high speeds and temperatures the density variation is seen. Thus it becomes necessary for all the analysis pertaining to the high speed flow of a compressible fluid to incorporate the compressibility effect i.e. in mass and energy conservation, momentum change, entropy change etc. Such analysis invokes the equation of state of the gas and other thermodynamic relations.

The compressible fluids support the wave motion. The state variation in the flow of gases is obtained considering density to be varying like other thermodynamic properties. The study of such flows where the change in density with pressure change is appreciable is known as the science of GAS DYNAMICS. In the study of gas dynamics the fluid flow analysis is done with compressibility considerations, causing appreciable deviation from the analysis made using assumptions of incompressibility of fluid. For the analysis compressible fluids are classified in the different categories and the suitable type of a flow or their combination is selected for any given real fluid flow problem.

13.2 WAVE MOTION

One of the common modes of transfer of energy from one point to another is through the agency of waves, with which we are all familiar in a general way. This mode of transfer of energy is wave motion. The phenomenon of communicating the disturbance in the medium at any point to other point in the medium is known as wave motion. Thus, it may also be described as the movement of a disturbance in a medium. The nature and characteristics of the medium in which disturbance is created classifies the type of wave motion set up in the medium. The waves can be generally classified as:

1. Waves in elastic solid medium.
2. Infinitesimal pressure wave.
3. Non-steep finite pressure wave.
4. Steep finite pressure wave or shock wave.
5. Expansion or Rarefaction wave.

13.2.1 Wave propagation in solid elastic medium

The mechanism of wave motion setup in the solid elastic medium is explained by the Figure 2.1. Here a solid uniform rod of cross-section area 'A' and length 'l' is shown with a compressive stress 'f'

generated due to sudden force on end *X-X* of the rod. Let wave front moves with velocity "a" and the particle velocity of the medium be "c".

In the time of consideration "t" wave front reaches at section *Y-Y*, and the change in length of rod be "Δl".

Let the modulus of elasticity of rod be E.

A plane or a surface on which all particles of the medium are in an identical state of motion at a given instant is called a wave front.

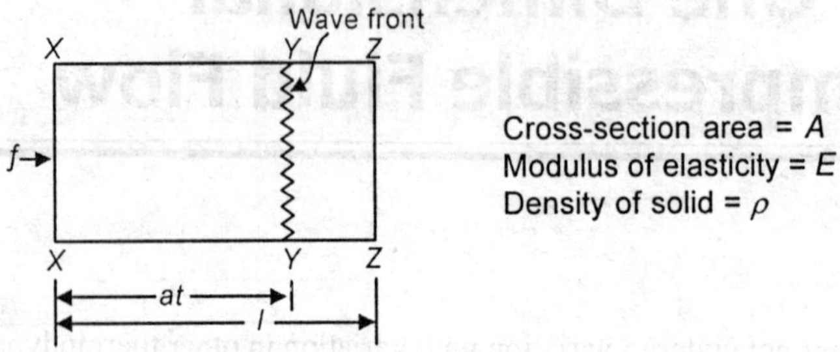

Fig. 13.1 Wave Propagation

Here the length of rod undergoing strain is the one through which wave has passed.

$$\text{Thus strain in the rod} = \frac{\Delta l}{at} \tag{13.1}$$

so

$$E = \frac{f}{(\Delta l / at)} \tag{13.2}$$

The particle velocity 'c' of the medium

$$c = \frac{\Delta l}{t} \tag{13.3}$$

or, from equations (13.2) & (13.3)

$$E = \frac{f}{(c/a)} \;;\; c = \frac{f}{(E/a)} \tag{13.4}$$

By Newton's law it can be written that the impulse force should be equal to the momentum change, i.e.

$$f \cdot A \cdot t = \rho . A . \text{ at } (c - o) \tag{13.5}$$

or

$$c = \frac{f}{\rho \cdot a} \tag{13.6}$$

The equations (13.4) and (13.6) yield

$$a = \sqrt{\frac{E}{\rho}} \tag{13.7}$$

i.e. the wave motion velocity $\qquad a = \sqrt{\dfrac{E}{\rho}}$

Thus, it is obvious from the above that the velocity of wave propagation is directly proportional to modulus of elasticity and inversely proportional to the density of material.

13.2.2 Infinitesimal pressure wave or Sound wave

Let us now look at the disturbance being generated in a gaseous medium. The complete mechanism of wave front propagation gets changed as compared with wave in a solid medium because of the molecular arrangement being different. The infinitesimal pressure wave is the wave with a very small (infinitely small) amplitude, which is also there in the sound wave. During the infinitesimal wave propagation in a gas the changes across the wave are very small and the speed of process corresponding to these changes is fast. Because of its speed the heat transfer during the wave front propagation is assumed to be negligible and the changes across an infinitesimal pressure wave can be considered to be reversible adiabatic or isentropic.

Let us consider an infinitesimal pressure wave set up in a constant area duct of stagnant gas, Figure 13.2 with observer at rest.

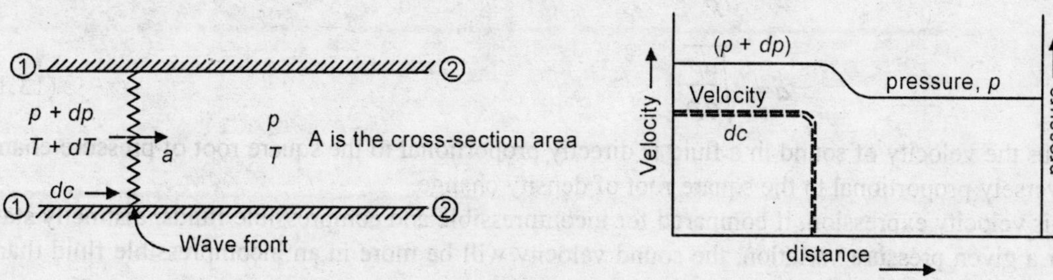

Fig. 13.2 Infinitesimal wave propagation in a constant area duct.

The wave moves with a velocity 'a' towards right as a wave front in stagnant gas at pressure 'p' and temperature 'T'. The pressure and temperatures of the gas due to the setting up of wave motion goes to ($p + dp$) and ($T + dT$) respectively and gas velocity is "dc".

Now if the observer is assumed to move with the wave and observe the various properties as pressure and also the velocity variation, the Figure 13.3 shows such arrangement. In this case the stagnant gas at pressure 'p' on right of wave front appears to move towards section 1-1. When the flow has passed across the wave to the left its pressure is raised to ($p + dp$) and the velocity becomes ($a - dc$), i.e. the relative velocity.

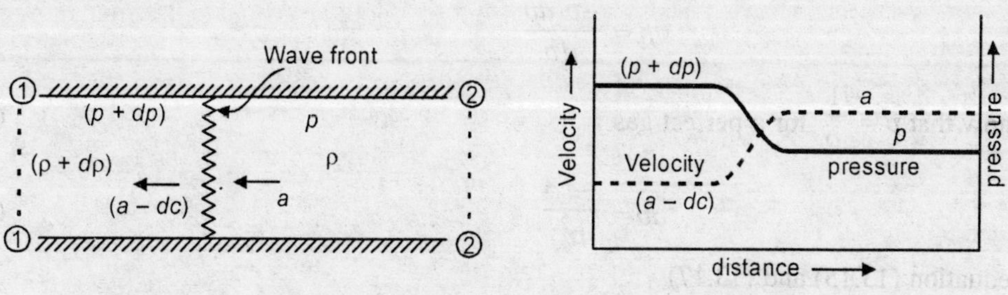

Fig. 13.3 Infinitesimal wave observation with observer moving along with wave

Let us write the momentum change equation, neglecting the shear force acting on the flow.

$$A\,[p - (p + dp)] = m\,[(a - dc) - a]$$

$$A \cdot dp = m \cdot dc \tag{13.8}$$

where 'm' is the mass flow rate of gas through duct.

From the continuity equation being applied on gas flow.

$$m = (\rho + d\rho) \cdot (a - dc) \cdot A \tag{13.9a}$$

$$m = \rho \cdot A \cdot a \tag{13.9b}$$

From equations (13.8) and (13.9*b*) we get

$$A \cdot dp = \rho \cdot A \cdot a \cdot dc$$

$$dp = \rho \cdot a \cdot dc \tag{13.10}$$

The equations (13.9*a*) and (13.9*b*) yield

$$\rho \cdot A \cdot a = (\rho + d\rho)\,(a - dc) \cdot A$$

$$\rho \cdot A \cdot a = \rho a A - \rho \cdot dc \cdot A + d\rho \cdot a \cdot A - d\rho \cdot dc \cdot A;$$ neglecting the higher order terms

$$\rho \cdot A \cdot dc = a \cdot A \cdot d\rho$$

$$\rho \cdot dc = a \cdot d\rho \tag{13.11}$$

Equations (13.10) and (13.11) finally yield

$$a^2 = \frac{dp}{d\rho} \tag{13.12}$$

or

$$a = \sqrt{\frac{dp}{d\rho}} \tag{13.13}$$

Thus the velocity of sound in a fluid is directly proportional to the square root of pressure change and inversely proportional to the square root of density change.

This velocity expression, if compared for incompressible and compressible fluids, distinctly states that for a given pressure variation, the sound velocity will be more in an incompressible fluid than a compressible fluid for setting up the wave.

Let us use the Bulk modulus of elasticity, K of the fluid. Since pressure rises from p to $(p + dp)$ the specific volume decreases from v to $(v - dv)$. For this control volume;

Hence

$$K = \frac{[(p + dp) - p]}{\dfrac{-dv}{v}} \tag{13.14}$$

$$K = \frac{-dp}{\dfrac{dv}{v}}$$

$$K = \frac{-v \cdot dp}{dv}$$

We know that $v = \dfrac{1}{\rho}$ for a perfect gas (13.16)

also

$$dv = \frac{-d\rho}{\rho^2} \tag{13.17}$$

from equation (13.15) and (13.17)

$$K = \frac{-1}{\rho} \cdot \frac{dp}{-d\rho} \cdot \rho^2$$

$$K = \frac{dp}{d\rho} \cdot \rho \tag{13.18}$$

This bulk modulus of elasticity will depend upon the gradient $\left(\dfrac{dp}{d\rho}\right)$.

For isentropic process

The governing equation is, $pv^{\gamma} =$ constant, for ideal gas partial differentiation results,

$$p \cdot \gamma \cdot v^{(\gamma-1)} dv + v^{\gamma} \cdot dp = 0 \tag{13.19}$$

or

$$\frac{-dp}{dv} = \frac{\gamma \cdot p}{v}$$

$$\frac{dp}{d\rho} = \frac{\gamma \cdot p}{\rho} \tag{13.20}$$

$$K = \gamma p \tag{13.21}$$

or $K = \gamma \cdot p$ for an isentropic process in an ideal gas.

For isothermal process;

The thermodynamic governing equation is

$$pv = \text{constant} \tag{13.22}$$

The partial differentiation yields,

$$p \cdot dv + v \cdot dp = 0$$

$$\frac{-dp}{dv} = \frac{p}{v} \tag{13.23}$$

or

$$\frac{dp}{d\rho} = \frac{p}{\rho}$$

$$K = p \tag{13.24}$$

$K = p$ for isothermal process in an ideal gas

Now combining the equations (13.13) and (13.20) we get

$$a = \sqrt{\frac{\gamma p}{\rho}} = \sqrt{\frac{K}{\rho}} \tag{13.25}$$

for a perfect gas $p = \rho RT$ hence $a = \sqrt{\gamma \cdot R \cdot T}$ (13.26)

Based on characteristic gas constant 'R', and universal gas constant $\bar{R}$, it can be given as;

$$R = \frac{\bar{R}}{M} \tag{13.27}$$

where M = molecular weight

The different forms of the expression for sound velocity are as under

$$\left.\begin{aligned} a &= \sqrt{\frac{K}{\rho}} \\ a &= \sqrt{\frac{\gamma p}{\rho}} \\ a &= \sqrt{\gamma \cdot R \cdot T} \\ a &= \sqrt{\gamma \cdot \frac{\bar{R}}{M} \cdot T} \\ a &= \sqrt{\frac{dp}{d\rho}} \end{aligned}\right\} \tag{13.28}$$

Based on set of expressions (13.28) the conclusions can be drawn such as:

- The velocity of sound is directly proportional to the temperature of the fluid.
- The velocity of sound is more for the fluids having high bulk modulus of elasticity.
- The velocity of sound is inversely proportional to the molecular weight of the gas in which it flows.
- The velocity of sound is directly proportional to the adiabatic index γ.
- The velocity of sound is more in a gas with low density than in a gas having higher density.

13.2.3 Mach Number and Mach Angle

The Mach number is in general defined by the ratio of the velocity of any moving object and the velocity of sound in the medium into which it is moving. This generalised definition is derived from the non-dimentionalized force expression i.e. the ratio of inertia force and elastic force. The Mach number is denoted by 'M' and plays a very important role in aeronautics for measuring the speed of aeroplane.

$$M^2 = \frac{\text{Inertia force}}{\text{Elastic force}} \qquad (13.29)$$

$$\text{Inertia force} = \rho \cdot A \cdot C^2$$

$$\text{Elastic force} = K \cdot A$$

Here 'C' is the velocity of object and K is the bulk modulus for the fluid in which object moves, A is the frontal area of the object.

$$M^2 = \frac{\rho \cdot A \cdot C^2}{\rho \cdot A \cdot a^2} = \left(\frac{C}{a}\right)^2$$

$$\boxed{M = \frac{C}{a}} \qquad (13.30)$$

$$\text{Mach number} = \frac{\text{Velocity of object}}{\text{Velocity of sound}}$$

Based on this Mach number the flow could be divided into different regimes as;

for $M < 1$ the flow is of Subsonic type.

for $M = 1$ the flow is of Sonic type.

for $M \approx 1$ the flow is of Transonic type.

for $M > 1$ the flow is of Supersonic type.

for $M > 5$ the flow is of Hypersonic type.

The aerodynamic effects differ significantly depending upon the type of flow regime. The Mach number reaches infinity when absolute temperature 'T' is reached to zero and under such circumstances another dimensionless velocity coefficient 'λ' is used. This new dimensionless velocity coefficient is the ratio of velocity of fluid to the velocity of sound at $M = 1$.

At $M = 1$, the velocity of sound is denoted by critical velocity i.e.'a^*'.

$$\text{Hence } \lambda = \frac{C}{a^*} \qquad (13.31)$$

Let us now analyze the different flow patterns with varied flow conditions.

(*a*) In case of an incompressible flow, the flow pattern is of the type shown in Figure 13.4. Here the velocity of object i.e. the source of disturbance, is negligibly small than the velocity of sound.

Thus the infinitesimal pressure wave (sound wave) keeps on spreading spherically as time passes, but the displacement of the source of disturbance is negligibly small.

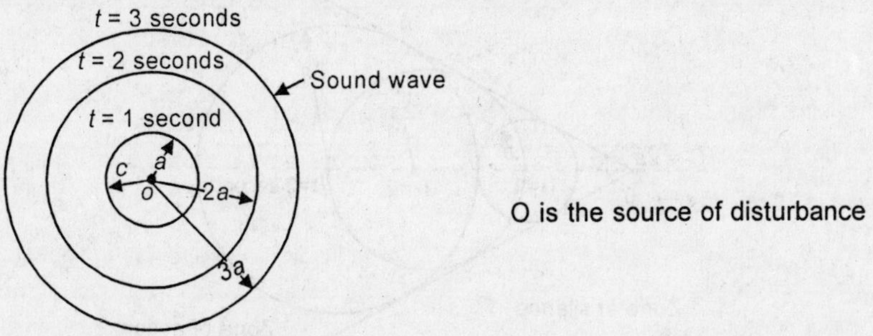

Fig. 13.4 Flow pattern in incompressible flow; $\{M = 0, C/a \approx 0\}$

(*b*) Let us now consider a compressible flow with the Mach number varying from subsonic to supersonic regimes. Wave propagation in subsonic flow, let M = 0.5, is shown in Figure 13.5. In the subsonic flow the source of disturbance moves slowly than the wave front. For M = 0.5, the source of disturbance moves at half of the speed of the wave front. The pattern is obtained for time intervals of 1 second, 2 seconds and 3 seconds. Here it could be seen that the wave front always leads the source of disturbance.

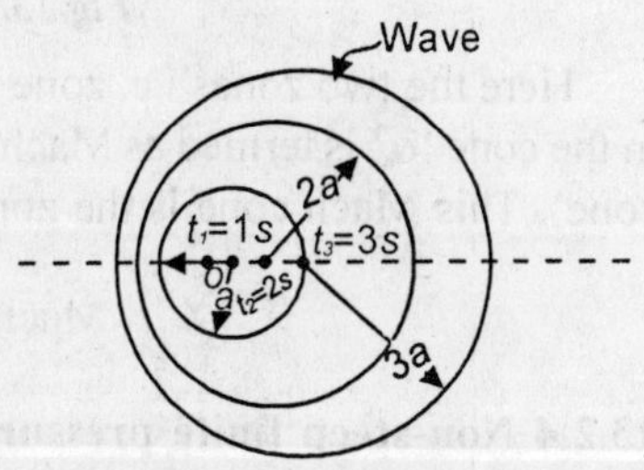

Fig. 13.5 Flow pattern for a compressible flow {M = 0.5, C = 0.5a}

The t_1, t_2, t_3 are three positions of centre of wave front at 1, 2 and 3 seconds.

(*c*) In the sonic flow region, i.e. M = 1 the point of disturbance '*O*' and the wave front both travel with same speed. The flow pattern shown in Figure 13.6 shows that inspite of time variation, the wave front does not move ahead of source nor it lags the source. Thus, the wave front moves in such a manner as to divide flow region into two zones i.e. the zone of silence and the zone of action. The zone of silence is the region lying unaffected by the wave front propagation, where as the zone of action is the region in which wave motion is there.

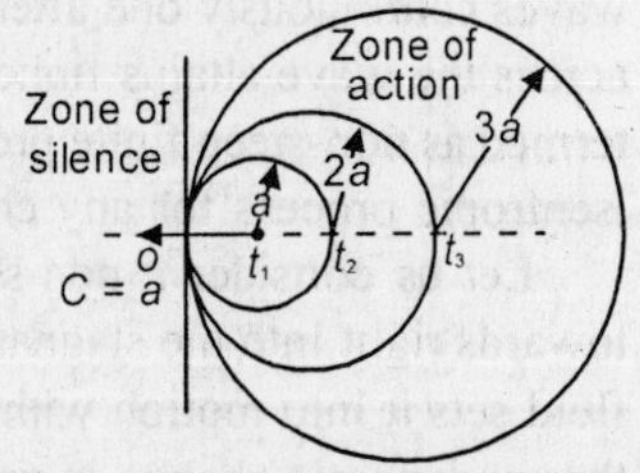

Fig. 13.6 Flow pattern for sonic flow {M=1, C=a}

In case of a supersonic flow, when M> 1, the point of disturbance moves ahead of the wave front and thus the flow pattern is as in Figure 13.7. At different time intervals of 1 second, 2 seconds, 3 seconds the position of wave front and source of disturbance are shown.

ΔOpt_3 shown that

$$\sin \alpha = \frac{a \cdot t}{C \cdot t} \Rightarrow \sin\alpha = \frac{1}{M}$$

or

$$\alpha = \sin^{-1}\left(\frac{1}{M}\right)$$

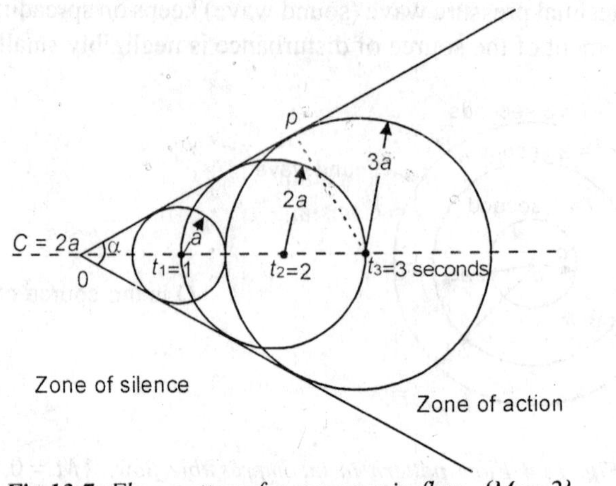

Fig.13.7. Flow pattern for supersonic flow. {M = 2}

Here the two zones i.e. zone of action and zone of silence are obvious and the semi angle included in the cone 'α' is termed as Mach angle. All the wave fronts remain confined to a cone known as 'Mach cone'. This Mach cone is the zone of action.

$$\text{Mach angle } \alpha = \sin^{-1}\left(\frac{1}{M}\right) \tag{13.32}$$

13.2.4 Non-steep finite pressure wave

The pressure wave discussed earlier has been the one having very small amplitude and small changes in the fluid properties with wave travel, and the thermodynamic behaviour as that of an isentropic process. Now let us think of the cluster of infinitesimal pressure waves travelling one after the other and cumulating the changes in thermodynamic properties by each of them e.g. say, we generate infinitesimal pressure waves continuously one after the other till the steady speed of wave is attained and the overall change across the wave attains finite magnitude. In these kind of waves since the waves are continuous, it is termed as non-steep finite pressure wave. Here it must be noted that the waves are still governed by the isentropic process for any change.

Let us consider a non-steep finite pressure wave as shown in Figure 13.8, where wave moves towards right into the stagnant gas at pressure 'p_1' and temperature 'T_1'. The passage of wave into the fluid sets it into motion with a velocity 'C' and increases its pressure to 'p_2' at section ② – ②. The thermodynamic change in properties from section 1 – 1 to 2 – 2 is governed by isentropic process.

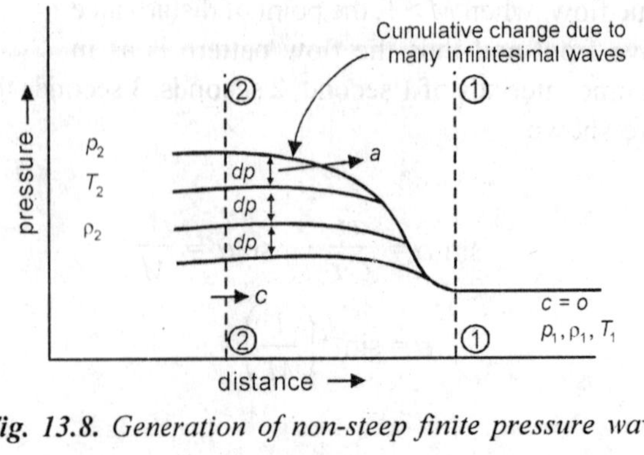

Fig. 13.8. Generation of non-steep finite pressure wave

Between section ① – ① and ② – ②, through isentropic relations,

$$\frac{a_2}{a_1} = \sqrt{\left(\frac{T_2}{T_1}\right)} = \left(\frac{p_2}{p_1}\right)^{\frac{\gamma-1}{2\gamma}} \tag{13.33}$$

We know $pv^{\gamma} = \text{Constant}$

or $p^{\gamma} = \text{Constant} \cdot \rho^{\gamma}$ (13.34)

Taking log of both sides;

$$ln\, p = ln \cdot \text{Constant} + \gamma \ln \rho \tag{13.35}$$

The partial differentiation yields as:

$$\frac{dp}{p} = \gamma \cdot \frac{d\rho}{\rho} \tag{13.36}$$

$$\frac{dp}{d\rho} = \frac{\gamma \cdot p}{\rho} = a^2 \tag{13.37}$$

or

$$\frac{dp}{p} = \frac{\gamma \cdot d\rho}{\rho}$$

$$\frac{d\rho}{\rho} = \frac{1}{\gamma} \cdot \frac{dp}{p} \tag{13.38}$$

For infinitesimal waves we know from eqn. (13.11)

$$\rho dc = a \cdot d\rho \tag{13.39}$$

$$dc = \frac{a \cdot d\rho}{\rho}$$

from eqns. (13.38) and (13.39)

$$dc = \frac{a}{\gamma} \cdot \frac{dp}{p} \tag{13.40}$$

From equation (13.33) generalised expression can be written as the function of stagnant gas properties,

$$a = a_1 \cdot \left(\frac{p}{p_1}\right)^{\frac{\gamma-1}{2\gamma}} \tag{13.41}$$

Eqns. (13.40) and (13.41) yield

$$dc = \frac{a_1}{\gamma} \cdot \left(\frac{p}{p_1}\right)^{\frac{\gamma-1}{2\gamma}} \cdot \frac{dp}{p}$$

partial integration gives

$$\int_{c=0}^{c} dc = \int_{p_1}^{p_2} \cdot \frac{a_1}{\gamma} \cdot \left(\frac{p}{p_1}\right)^{\frac{\gamma-1}{2\gamma}} \cdot \frac{dp}{p}$$

or

Velocity of gas
$$C = \frac{2a_1}{(\gamma-1)} \cdot \left[\left(\frac{p_2}{p_1} \right)^{\frac{\gamma-1}{2\gamma}} - 1 \right] \quad (13.42)$$

The expression can be generalised by putting $p_2 = p$.
The velocity of non-steep finite pressure wave propagation would be,
$$C_w = C + a \quad (13.43)$$

$$C_w = \frac{2a_1}{(\gamma-1)} \cdot \left[\left(\frac{p}{p_1} \right)^{\frac{\gamma-1}{2\gamma}} - 1 \right] + a_1 \left(\frac{p}{p_1} \right)^{\frac{\gamma-1}{2\gamma}}$$

$$C_w = \left[\left(\frac{\gamma+1}{\gamma-1} \right) \left(\frac{p}{p_1} \right)^{\frac{\gamma-1}{2\gamma}} - \frac{2}{(\gamma-1)} \right] a_1 \quad (13.44)$$

Thus the Mach no. of the wave.

$$\text{Wave Mach no. } M_w = \frac{C_w}{a_1} = \left\{ \left(\frac{\gamma+1}{\gamma-1} \right) \left(\frac{p}{p_1} \right)^{\frac{\gamma-1}{2\gamma}} - \frac{2}{(\gamma-1)} \right\} \quad (13.45)$$

Also, since non-steep finite pressure wave is the resultant of number of infinitesimal pressure waves put together, so.
$$p_2 = p = p_1 + \Delta p \quad (13.46)$$
From equations (13.45) & (13.46)

$$M_w = \left\{ \left(\frac{\gamma+1}{\gamma-1} \right) \cdot \left(\frac{p_1 + \Delta p}{p_1} \right)^{\frac{\gamma-1}{2\gamma}} - \frac{2}{(\gamma-1)} \right\} \quad (13.47)$$

Neglecting the higher order terms in binomial expansion of $\left(1 + \frac{\Delta p}{p_1}\right)$ we get;

$$C_w = \left\{ 1 + \left(\frac{\gamma+1}{2\gamma} \right) \left(\frac{\Delta p}{p_1} \right) \right\} \cdot a_1 \quad (13.48)$$

When ever the finite pressure wave behaves like the infinitesimal pressure wave;

$$\frac{\Delta p}{p} \to 0 \quad (13.49)$$

Hence
$C_w = a_1$ i.e. expression for wave velocity of infinitesimal pressure wave
The Mach no. of the flow induced by the non-steep pressure wave would be

$$M = \frac{C}{a} \tag{13.50}$$

Equations (13.42) and (13.41) give;

$$M = \frac{2}{(\gamma - 1)}\left[1 - \left(\frac{p_1}{p}\right)^{\frac{\gamma-1}{2\gamma}}\right] \tag{13.51}$$

Example:

Let us generate the non-steep finite pressure wave in a cylinder by continuous impact by the piston.

Due to the continuous passage of waves through the fluid its pressure and temperature increases, thus increasing the local velocity of sound for a wave generated. This continuous passage of pressure wave through the fluid also increases its velocity. Therefore, it is seen that the waves generated in later stages propagate at higher velocities of sound relative to the fluid and thus the propagation velocity of each subsequent wave is greater than that of the preceding wave.

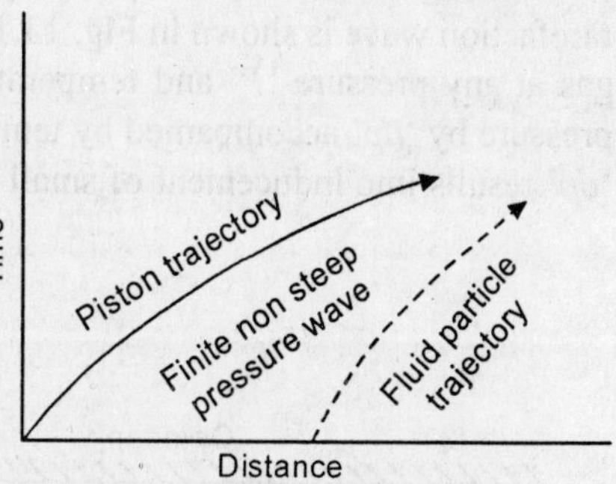

Fig. 13.9. *Development of finite non-steep pressure wave*

13.2.5 Steep finite pressure wave

These steep finite pressure waves are also known as the 'shock waves'. These are the special case of non-steep finite pressure wave. As we have already seen that the pressure waves generated by the continuous impulse application in a number of steps move at higher velocities than the preceding waves, these fast moving waves tend to overtake the slower waves generated earlier. Thus a situation is reached when the number of compression waves merge into one finite pressure wave and thus the variation in pressure and other flow properties due to cumulative effect becomes extremely large and abrupt across the final finite pressure wave. The large and abrupt change in flow properties across the pressure wave becomes the characteristics of "shock wave" i.e. the resultant of number of compression waves being merged into one.

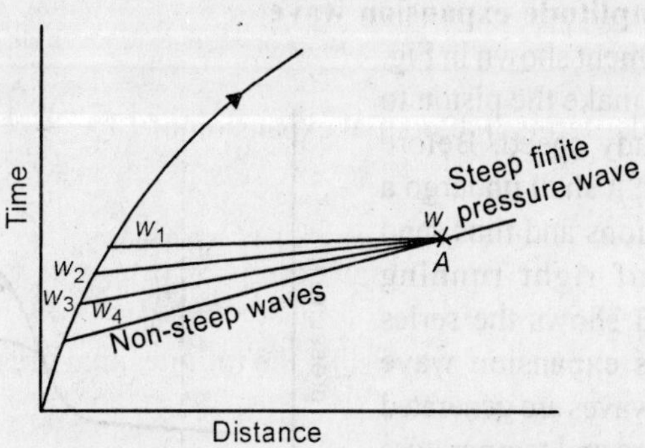

Fig. 13.10. *Development of steep finite pressure wave*

The Fig. 13.10, explains on time – distance axes that how the four different non-steep waves "w_1, w_2, w_3, w_4" are getting merged finally at point 'A' and result into steep finite pressure wave shown by 'w'.

13.2.6 Rare faction wave

The 'rare faction wave' is also called as 'expansion wave'. The expansion waves are generated with expansion process, i.e. all times rare faction wave generation has to be preceded by expansion or the drop in pressure. The expansion wave shall be generated if a piston undergoes sudden outward motion in a piston–cylinder arrangement. The sudden throttling process of any flowing stream shall also lead to generation of expansion wave. The expansion waves depending upon their amplitudes can be "infinitesimal expansion waves" and "finite amplitude expansion waves". The mechanism and brief analysis of each of the above expansion waves is discussed ahead.

(a) Mechanism of infinitesimal expansion wave

The Fig. 13.11 shows the piston cylinder arrangement with piston moving outwards. An infinitesimal rarefaction wave is shown in Fig. 13.12, which moves towards the right at sound velocity into stagnant gas at any pressure '*P*' and temperature '*T*'. Since it is of expansion type so the wave reduces the pressure by '*dp*' accompanied by temperature reduction of gas by '*dT*'. This small pressure difference '*dp*' results into inducement of small leftward velocity '*dc*'.

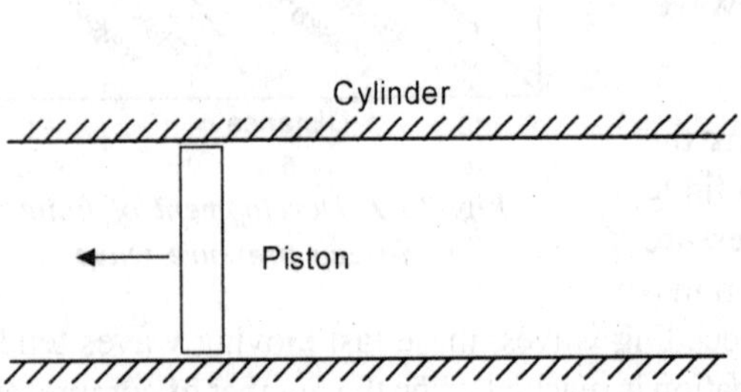

Fig. 13.11. Expansion of gas in a piston – cylinder arrangement

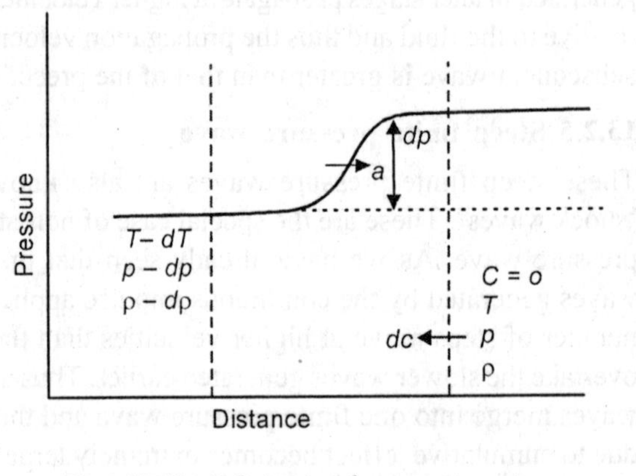

Fig. 13.12. Infinitesimal expansion wave

(b) Mechanism of finite amplitude expansion wave

Let us again take arrangement shown in Fig. 13.11 into consideration and make the piston to move towards left with steady speed. Before piston attaining steady speed, it shall undergo a series of impulsive accelerations and thus send simultaneously a series of right running expansion waves. Fig. 13.13 shows the series of expansion waves. In this expansion wave generation process the initial waves are generated in the gas having high pressure and temperature and thus initial waves contribute in high local sound velocity as compared to latter waves having their generation in gas with low pressure and temperature. The gas pressure and temperature continuously drop due to expansion

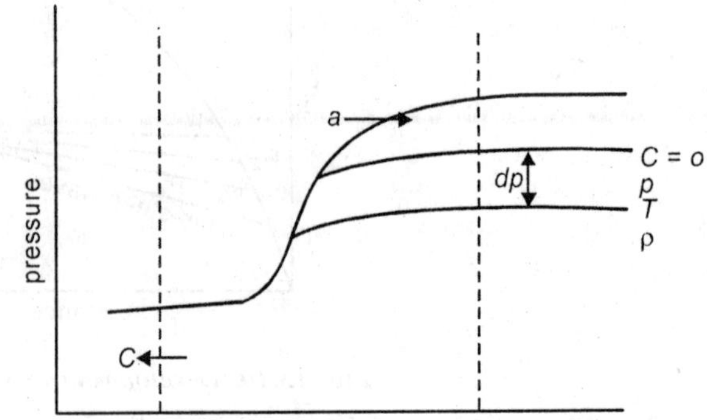

Fig. 13.13 Finite amplitude expansion wave

of gas. The movement of piston so as to cause expansion in gas shall induce fluid velocity "c" towards left. The velocity of wave propagation towards the right, with respect to local velocity of wave "a" shall be "$(a-c)$". The resultant velocity of wave propagation $(a-c)$, subsequently decreases as the expansion proceeds because the gas velocity "c" increases while "a" decreases.

The generation of finite amplitude expansion wave is shown in Fig. 13.13. It explains how the overall change of properties across wave is made up of a number of infinitesimal isentropic changes. If these are plotted on time–space diagram it shall be obvious that the paths of these waves are diverging and the merger of many expansion waves shall never result into shock wave generation, Fig. 13.14.

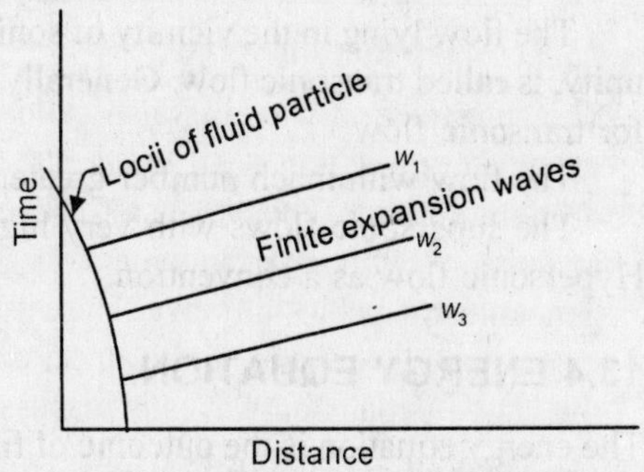

***Fig. 13.14** Expansion wave on time – space diagram*

13.3 CLASSIFICATION OF COMPRESSIBLE FLOW

The compressible flows can be classified, based on various criterion as:

(*a*) Based on nature of flow
1. One dimensional flow
2. Multi dimensional flow (Two dimensional or Three dimensional)

(*b*) Based on velocity of flow i.e. Mach number.
1. Subsonic flow; $M < 1$
2. Sonic flow; $M = 1$
3. Transonic flow; $0.9 > M < 1.1$; $M \approx 1$ but $M \neq 1$
4. Super sonic flow; $M > 1$
5. Hypersonic flow; $M > 5$

The compressible fluid flow when assumed to have the state variation in only one dimension is called uni-dimensional or one-dimensional flow. The uni-dimensional flow is sub-classified for the sake of analysis as:

(*i*) *Isentropic flow:* where the area change occurs treating the fluid to be inviscid and passage to be frictionless.

(*ii*) *Friction flow or Fanno flow:* where the frictional effects are considered assuming the heat transfer between the system and surrounding to be negligible.

(*iii*) *Raleigh flow:* where the heat transfer is considered but the fluid is considered to be inviscid.

The multidimensional flow analysis is performed when the analysis made considering uni-dimensional flow gives results differing very largely from actual situation. The uni-dimensional flow analysis is the simplification of real fluid flow situation to the maximum and hence inaccurate for complex problems. The multi-dimensional flow can be either of the two i.e. two dimensional flow and three dimensional flow.

The compressible fluid flows based on velocity are classified considering the mach number as basis. When the mach number which is the ratio of velocity of fluid to velocity of sound remains less than unity it is subsonic flow.

The flow having velocity equal to that of velocity of sound at that temperature and pressure is called sonic flow.

The flow lying in the vicinity of sonic flow i.e. mach number is of the order of unity but not exactly unity, is called transonic flow. Generally mach number lying between 0.9 and 1.1 is considered as range for transonic flow.

The flow with mach number greater than unity is called supersonic flow.

The supersonic flows with very high flow velocity i.e. mach number greater than 5 is termed as Hypersonic flow as a convention.

13.4 ENERGY EQUATION

The energy equation is the outcome of first law of thermodynamics i.e. the law of energy conservation applied to various processes.

Let us consider any fluid flowing. The energy at any point in the fluid shall be given by;

E = Internal energy + Kinetic energy + Potential energy

$$E = U + \frac{1}{2} m \cdot C^2 + m \cdot g \cdot z \quad (13.52)$$

For the energy change between any two states denoted by 1 and 2, $E_2 - E_1$ shall be given as;

$$E_2 - E_1 = (U_2 - U_1) + \frac{1}{2} m (C_2^2 - C_1^2) + mg \cdot (Z_2 - Z_1)$$

The first law of thermodynamic says

$$dQ = dU + dW. \quad (13.53)$$

$$\Delta Q = \Delta W + (E_2 - E_1)$$

$$\Delta Q = \Delta W + (U_2 - U_1) + \frac{1}{2} m (C_2^2 - C_1^2) + mg (Z_2 - Z_1)$$

$$q = w + (u_2 - u_1) + \frac{1}{2} (C_2^2 - C_1^2) + g (Z_2 - Z_1) \quad (13.54)$$

The equation (13.54) for a non flow process shall be

$$q = w + (u_2 - u_1) \quad (13.55)$$

or

$$\Delta Q = \Delta W + (U_2 - U_1) \quad (13.56)$$

$$\Delta Q = \int_1^2 pdv + mC_v (T_2 - T_1) \quad (13.57)$$

The equation (13.53) when applied to the flow processes has to be modified by introducing the component for flow work in the work.

$$dQ = dU + d(W + \text{Flow work}) \quad (13.58)$$

$$\Rightarrow \quad Q = W + (p_2 V_2 - p_1 V_1) + (U_2 - U_1) + mg (Z_2 - Z_1) + \frac{1}{2} m (C_2^2 - C_1^2)$$

$$Q = W + (U_2 + p_2 V_2) - (U_1 + p_1 V_1) + \frac{1}{2} m (C_2^2 - C_1^2) + mg(Z_2 - Z_1)$$

$$Q = W + (H_2 - H_1) + \frac{1}{2} m (C_2^2 - C_1^2) + mg(Z_2 - Z_1)$$

$$Q + H_1 + mgZ_1 + \frac{1}{2} m C_1^2 = W + H_2 + mgZ_2 + \frac{1}{2} m C_2^2 \quad (13.59)$$

or

$$q + h_1 + gz_1 + \frac{1}{2}C_1^2 = w + h_2 + gZ_2 + \frac{1}{2}C_2^2 \tag{13.60}$$

If during the gas flow change in elevation of gas at two sections is negligible then eqn. (13.60) gets modified as

$$q + h_1 + \frac{C_1^2}{2} = W + h_2 + \frac{C_2^2}{2} \tag{13.61}$$

If the heat transfer during gas flow is negligibly small or zero then; eqn. (13.61) reduces to adiabatic flow.

$$h_1 + \frac{C_1^2}{2} = h_2 + \frac{C_2^2}{2} + w \tag{13.62}$$

If in a gas flow situation the work interaction is absent then eqn. (13.62) becomes Energy equation.

$$h_1 + \frac{C_1^2}{2} = h_2 + \frac{C_2^2}{2} \tag{13.63}$$

or

$$h + \frac{C^2}{2} = \text{Constant} \tag{13.64}$$

If the gas is being decelerated under isentropic conditions to zero velocity i.e. the gas is brought to stagnation state, then, it is obtained as

$$h_1 + \frac{C_1^2}{2} = h_2 \text{ ; as } C_2 = 0 \tag{13.65}$$

also

$$h_0 = h + \frac{C^2}{2}\text{; where } h_0 \text{ is stagnation enthalphy} \tag{13.66}$$

The temperature at this state of gas is known as stagnation temperature and can be given as;

$$C_p T_0 = C_p T + \frac{C^2}{2} \tag{13.67}$$

$$T_0 = T + \frac{C^2}{2C_p} \tag{13.68}$$

When flow of an ideal gas is brought to rest isentropically the kinetic energy of gas changes into heat energy. The stagnation temperature and stagnation enthalpy thus may have fixed values, but values of stagnation pressure p_0 and stagnation density ρ_0 may change in such a way that their ratio $\frac{p_0}{\rho_0}$ remains constant.

Rewriting the energy equation (13.64) we see

$$h + \frac{C^2}{2} = \text{Constant}$$

or
$$dh + C \cdot dC = 0$$
$$dh = -C \cdot dc \quad (13.69)$$

Equation (13.69) shows that increase in kinetic energy is possible with decrease in enthalpy or vice-versa.

13.5 ISENTROPIC FLOW

When during the gas flow thermodynamic changes occur in such a way that there is no change in its entropy, process is called isentropic and it is reversible in nature. For above conditions to prevail it is essential to have no heat transfer from the fluid to surroundings and vice-versa during the process.

Let us assume a flow of isentropic type in the pressure differential of 'dp' and density variation of '$d\rho$' in gas.

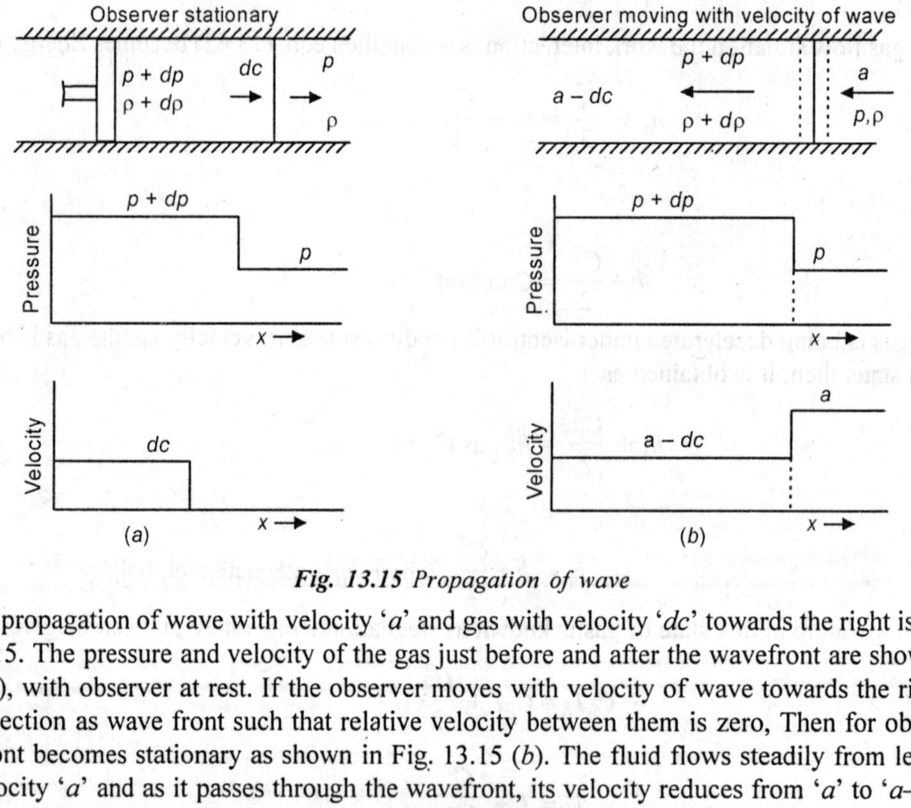

Fig. 13.15 Propagation of wave

The propagation of wave with velocity 'a' and gas with velocity 'dc' towards the right is shown in Fig. 13.15. The pressure and velocity of the gas just before and after the wavefront are shown in Fig. 13.15 (*a*), with observer at rest. If the observer moves with velocity of wave towards the right in the same direction as wave front such that relative velocity between them is zero, Then for observer the wave front becomes stationary as shown in Fig. 13.15 (*b*). The fluid flows steadily from left to right with velocity 'a' and as it passes through the wavefront, its velocity reduces from 'a' to '$a-dc$' along with density rise from ρ to $\rho + d\rho$ and pressure rise from 'p' to '$p + dp$'.

If the momentum change equations are applied then;
$$A[p - (p + dp)] = m'[(a - dc) - a] \quad (13.70)$$
$$A \cdot dp = m' \cdot dc \text{ where } m' \text{ is mass flow rate.}$$
$$A \cdot dp = (\rho \cdot A \cdot a) \cdot dc$$
$$dp = \frac{\rho \cdot A \cdot a \cdot dc}{A}$$
$$dp = \rho \cdot a \cdot dc \quad (13.71)$$

The continuity equations applied to control surface on both sides of wave front gives.

$$\rho \cdot A \cdot a = (\rho + d\rho)\,(a - dc) \cdot A \tag{13.72}$$

$$\rho \cdot A \cdot a = \rho \cdot a \cdot A - \rho \cdot dc \cdot A + d\rho \cdot a \cdot A - d\rho \cdot dc \cdot A$$

Neglecting the higher order terms.

$$\rho \cdot dc \cdot A = d\rho \cdot a \cdot A$$

$$\rho \cdot dc = a \cdot d\rho \tag{13.73}$$

Equations (13.71) and (13.73) yield

$$dp = d\rho \cdot a^2 \tag{13.74}$$

$$a = \sqrt{\frac{dp}{d\rho}} \tag{13.75}$$

The velocity of wave propagation is given by the ratio of change in pressure and change in density as given in eqn. 13.75 for isentropic flow.

Let us write the energy equation for isentropic flow across the wave front, assuming that enthalpy on one end be (h) and at other it is ($h + dh$); eqn. (13.64)

$$h + \frac{a^2}{2} = (h + dh) + \frac{(a - dc)^2}{2} \tag{13.76}$$

Neglecting the higher order terms

$$dh = a \cdot dc \tag{13.77}$$

From eqns. (13.77) and (13.71)

$$dh = \frac{dp}{\rho} \tag{13.78}$$

By the thermodynamics of isentropic flow;

$$\frac{p}{\rho^{\gamma}} = \text{Constant} \tag{13.79}$$

$$p \cdot \rho^{-\gamma} = \text{Constant} \tag{13.80}$$

For a perfect gas ; $p = \rho RT$

Eqn. (13.80) yields; taking log of both sides

$$p \cdot \delta^{-\gamma} = \text{Constant}$$

$$\ln p - \gamma \ln \rho = \text{Constant} \tag{13.81}$$

Differentiating $\dfrac{dp}{p} - \dfrac{\gamma \cdot d\rho}{\rho} = 0$

$$\frac{dp}{p} = \frac{\gamma \cdot d\rho}{\rho}$$

$$\frac{dp}{d\rho} = \frac{p \cdot \gamma}{\rho} = \gamma \cdot RT \tag{13.82}$$

Now using eqn. (13.75) we get

$$a = \sqrt{\gamma \cdot RT} = \sqrt{\frac{\gamma \cdot p}{\rho}} = \sqrt{\frac{dp}{d\rho}} \tag{13.83}$$

The equation (13.64) yields

$$h + \frac{C^2}{2} = \text{Constant}$$

$$C_pT + \frac{C^2}{2} = \text{Constant} \tag{13.84}$$

We know $$C_p = \frac{\gamma R}{\gamma - 1}$$

substituting

$$\frac{\gamma RT}{(\gamma - 1)} + \frac{C^2}{2} = \text{Constant} \tag{13.85}$$

$$C^2 + \frac{2\gamma RT}{(\gamma - 1)} = \text{Constant} \tag{13.86}$$

From eqn. (13.83) & (13.86)

$$C^2 + \frac{2a^2}{(\gamma - 1)} = \text{Constant} \tag{13.87}$$

or $$C^2 + \frac{2\gamma p}{\rho(\gamma - 1)} = \text{Constant}$$

$$C^2 + \frac{2 \cdot \gamma}{(\gamma - 1)} \cdot \frac{p}{\rho} = \text{Constant} \tag{13.88}$$

Equations (13.86) and (13.88) are the energy equations for isentropic flow.

The boundary conditions will be ; At $C = 0$, when the gas is stationary, the wave propagation velocity is a_0. At $T = 0$, when absolute zero temperature is achieved, the wave propagation shall not be there i.e $a = 0$ and $C = C_{max}$

From equation (13.87)

$$\frac{2a_0^2}{(\gamma - 1)} = \text{Constant.}$$

Energy equation: $$\boxed{C^2 + \frac{2a^2}{(\gamma - 1)} = \frac{2a_0^2}{(\gamma - 1)} = C_{max}^2 = \text{Constant}}$$

The velocity a_0 is also known as stagnation velocity of sound.

$$\frac{C^2}{C_{max}^2} + \frac{2a^2}{(\gamma - 1)C_{max}^2} = 1$$

$$\frac{C^2}{C_{max}^2} + \frac{a^2}{\frac{(\gamma - 1)}{2} \cdot C_{max}^2} = 1 \tag{13.89}$$

Equation (13.89) when plotted for wave propagation velocity 'a' and velocity of particle 'C' gives

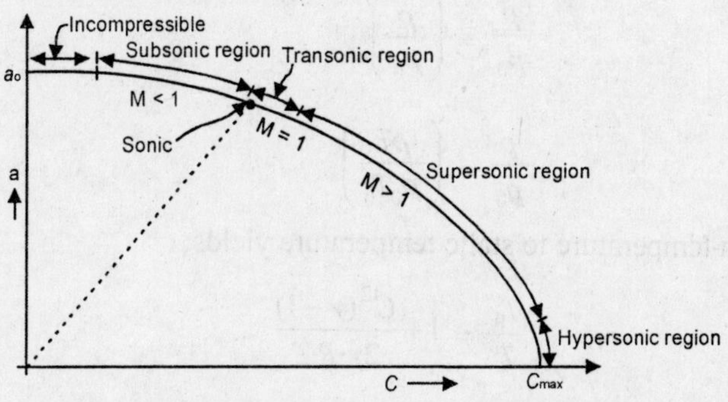

Fig. 13.16 Various regions of flow in adiabatic gas flow

The steady flow ellipse shows the various regions of flow. The regions can be categorically shown by the use of Mach number. As the fluid particle velocity C increases starting from region close to Y-axis i.e. near a_0, the fluid behaves as incompressible fluid. Beyond some value of C, fluid shows compressible fluid nature and flow changes from subsonic to transonic, sonic, supersonic and hypersonic flow subsequently. The half lengths of this adiabatic ellipse are; given by major axis equal to C_{max} and minor axis equal to $\left(C_{max}\cdot\sqrt{\frac{\gamma-1}{2}}\right)$ or a_0 respectively.

13.6 STAGNATION PROPERTIES

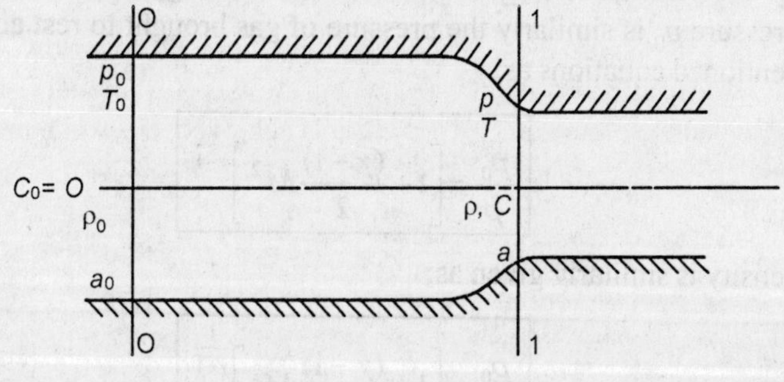

Fig. 13.17 Flow between stagnation and any other section of duct

The isentropic deceleration process which leads to the stagnation state and the thermodynamic properties as pressure, temperature and density can be given using the equation 13.68.

Isentropic flow ; $\frac{p}{\rho^r}$ = Constt.

Perfect gas ; $p = \delta RT$

Stagnation temperature ; $T_0 = T + \frac{C^2}{2C_p}$ or $\frac{T_0}{T} = 1 + \frac{C^2}{2C_p \cdot T}$

at the two sections 0 – 0 and 1 – 1 using the above,

$$\frac{p}{p_0} = \left(\frac{\rho}{\rho_0}\right)^{\gamma}$$

and
$$\frac{p}{p_0} = \left(\frac{\rho T}{\rho_0 T_0}\right)$$

Ratio of stagnation temperature to static temperature yields;

$$\frac{T_0}{T} = 1 + \frac{C^2(\gamma-1)}{2\gamma \cdot R \cdot T}$$

or
$$\frac{T_0}{T} = 1 + \frac{C^2 \cdot (\gamma-1)}{2a^2}$$

$$\boxed{\frac{T_0}{T} = 1 + \frac{(\gamma-1)}{2} \cdot M^2}$$

The stagnation temperature T_0 defined is equation (13.68) is algebraic sum of the dynamic temperature, $\frac{C^2}{2C_p}$ and static temperature T. It is thus obvious that the stagnation temperature shall remain constant, if there is no heat or work transfer. The use of stagnation temperatures is preferred over static temperatures as it eliminates the reference to kinetic energy term which is contained in stagnation temperature. The measurement of stagnation temperature is easier than static temperature. Also it gives actual energy in the form of heat as by deceleration to stagnation state kinetic energy is restored.

The stagnation pressure p_0 is similarly the pressure of gas brought to rest adiabatically and can be given using earlier mentioned equations as;

$$\boxed{\frac{p_0}{p} = \left[1 + \frac{(\gamma-1)}{2} M^2\right]^{\frac{\gamma}{\gamma-1}}}$$

The stagnation density is similarly given as;

$$\boxed{\frac{\rho_0}{\rho} = \left[1 + \frac{(\gamma-1)}{2} M^2\right]^{\frac{1}{\gamma-1}}}$$

13.7 SPECIAL RELATIONS

The some of useful and important relations based on energy equation, gas properties, isentropic flow condition and stagnation properties can be obtained as explained below.

By the energy equation

When $T = 0,\ C = C_{max} = \sqrt{\frac{2\gamma R \cdot T_0}{(\gamma-1)}}$ (as $a_0 = \sqrt{\gamma R T_0}$)

When $T = T_0,\ C = 0\ \&\ a_0 = \sqrt{\gamma R T_0}$

When $T = T^*$ = Temperature when sonic flow conditions are achieved.

$$C = C^* = a^* = \sqrt{\gamma R T^*} = \sqrt{\frac{2\gamma R\left(T_0 - T^*\right)}{(\gamma - 1)}}$$

i.e.

$$M = 1$$

$$T^* = \frac{2}{(\gamma - 1)}\,(T_0 - T^*)$$

or

$$\frac{T^*}{T_0} = \left(\frac{2}{\gamma + 1}\right)$$

Hence at sonic flow condition, $M = 1$

$$\boxed{C^* = a^* = \sqrt{\left(\frac{2\gamma R T_0}{\gamma + 1}\right)}}$$

The various velocity ratios can be expressed as a function of temperature, T_0 as following using the earlier formulation

$$\boxed{\frac{C_{\max}}{a_0} = \sqrt{\frac{2}{(\gamma - 1)}}}$$

Also we may write

$$\boxed{\frac{C_{\max}}{C^*} = \frac{C_{\max}}{a^*} = \sqrt{\frac{(\gamma + 1)}{(\gamma - 1)}}}$$

$$\boxed{\frac{C^*}{a_0} = \frac{a^*}{a_0} = \sqrt{\frac{2}{(\gamma + 1)}}}$$

Substituting in the energy equation we get;

$$C^2 + \frac{2a^2}{(\gamma - 1)} = C_{\max}^2 = \left(\frac{\gamma + 1}{\gamma - 1}\right)\cdot C^{*2} = \left(\frac{\gamma + 1}{\gamma - 1}\right)\cdot a^{*2} \qquad (13.90)$$

The mach number M which is given by the ratio of the velocity of object to velocity of sound can also be defined for the velocity of sound or wave taken at sonic flow conditions. This special mach number is denoted by λ, as

$$M = \frac{C}{a}$$

$$\lambda = \frac{C}{a^*}$$

$$\lambda = \frac{C}{a}\cdot\frac{a}{a^*}$$

$$\lambda = M \cdot \frac{a}{a^*} \qquad \{a^* = C^*\}$$

or

$$\lambda = M \cdot \frac{a}{C^*}$$

$$\frac{\lambda}{M} = \frac{a}{C^*}$$

squaring both sides

$$\frac{\lambda^2}{M^2} = \frac{a^2}{C^{*2}}$$

Combining equations (13.90) & (13.91) it yields;

$$\frac{C^2}{a^{*2}} + \frac{2}{(\gamma-1)} \cdot \frac{a^2}{C^{*2}} = \left(\frac{\gamma+1}{\gamma-1}\right)$$

$$\frac{C^2}{a^{*2}} + \frac{2}{(\gamma-1)} \cdot \frac{\lambda^2}{M^2} = \left(\frac{\gamma+1}{\gamma-1}\right)$$

or

$$\lambda^2 + \frac{\lambda^2}{M^2} \cdot \left(\frac{2}{\gamma-1}\right) = \left(\frac{\gamma+1}{\gamma-1}\right)$$

$$\boxed{\lambda^2 = \frac{\left(\frac{\gamma+1}{2}\right) \cdot M^2}{1 + \left(\frac{\gamma-1}{2}\right) \cdot M^2}}$$

or

$$\boxed{M^2 = \frac{\frac{2}{(\gamma+1)} \cdot \lambda^2}{1 - \left(\frac{\gamma-1}{\gamma+1}\right)\lambda^2}}$$

The above equations relate the mach number with special mach number and it can be seen that special mach number, λ behaves similar to mach number M.

As $M < 1$; $\lambda < 1$

As $M > 1$; $\lambda > 1$

As $M = 1$; $\lambda = 1$

As $M = 0$; $\lambda = 0$

As $M = \infty$; $\lambda = \sqrt{\left(\frac{\gamma+1}{\gamma-1}\right)}$ i.e. special mach number is limited as Mach no. tends to infinite.

For air having $\gamma = 1.4$, λ is limited to 2.45 as $M = \infty$.

Similar to special mach number, the non dimensional number can also be defined using maximum velocity of fluid, C_{max}. This ratio of velocity of fluid to maximum velocity of fluid is called Crocco number. Thus, Crocco number mon-dimensionalises the fluid velocity by maximum fluid velocity.

$$Cr = \frac{C}{C_{max}} \tag{13.92}$$

$$Cr = \frac{C}{C^*} \cdot \frac{C^*}{C_{max}}$$

$$Cr = \frac{C}{a^*} \cdot \frac{C^*}{C_{max}}$$

$$Cr = \lambda \cdot \frac{C^*}{C_{max}}$$

Substitution for $\left(\frac{C^*}{C_{max}}\right)$ yields

$$\boxed{Cr = \lambda \cdot \sqrt{\left(\frac{\gamma - 1}{\gamma + 1}\right)}}$$

or, substituting λ yields,

$$\boxed{Cr = \sqrt{\frac{(\gamma - 1) \cdot M^2}{2 + (\gamma - 1) \cdot M^2}}}$$

or

$$\boxed{M = \sqrt{\frac{2 \cdot C^2 r}{(\gamma - 1)\ (1 - C^2 r)}}}$$

From above equations the stagnation temperature can be defined in terms

$$\boxed{\frac{T_0}{T} = \frac{1}{1 - C^2 r}}$$

13.8 COMPRESSIBILITY CORRECTION FACTOR

The effect of compressibility in a fluid flow can be quantified based on stagnation pressure, pressure and mach number. The compressibility correction factor gives the error when the gas is assumed incompressible. For isentropic compressible flow the ratio of the stagnation and static pressures is given by.

$$\frac{p_0}{p} = \left[1 + \frac{\gamma - 1}{2} M^2\right]^{\frac{\gamma}{(\gamma - 1)}}$$

Expanding the right hand side using Binomial theorem;

$$\frac{p_0}{p} = 1 + \frac{(\gamma - 1)}{2} M^2 \cdot \frac{\gamma}{(\gamma - 1)} + \left(\frac{\gamma}{Y - 1}\right) \frac{\left(\frac{\gamma}{\gamma - 1} - 1\right)}{L^2} \cdot \left(\frac{\gamma - 1}{2}\right)^2 M^4 + ...$$

$$\frac{p_0}{p} = 1 + \frac{\gamma}{2}M^2 + \frac{\gamma}{8}M^4 + \left(\frac{\gamma(2-\gamma)}{48}\right)M^6 + ...$$

$$\frac{p_0}{p} = 1 + \frac{\gamma}{2}M^2 + \frac{\gamma}{8}M^4 + ...$$

$$(p_0 - p) = \frac{\gamma p M^2}{2} + \frac{\gamma p M^4}{8} + \frac{\gamma(2-\gamma)}{48}M^6 \cdot p$$

also $\gamma p M^2$ can be given as $\rho \cdot C^2$

$$(p_0 - p) = \frac{\rho C^2}{2} + \frac{\rho C^2 M^2}{8} + \frac{(2-\gamma)\rho \cdot C^2 M^4}{48}$$

$$= \frac{\rho C^2}{2}\left[1 + \frac{M^2}{4} + \frac{M^4 \cdot (2-\gamma)}{24} + ...\right]$$

or

$$\frac{(p_0 - p)}{\frac{\rho C^2}{2}} = \left[1 + \frac{M^2}{4} + \frac{M^4 \cdot (2-\gamma)}{24} + ...\right]$$

or,

$$\boxed{\frac{(p_0 - p)}{(\rho C^2 / 2)} = (1 + \xi)}$$

Where ξ = Compressibility correction factor

$$\boxed{\xi = \frac{M^2}{4} + \frac{(2-\gamma) \cdot M^4}{24} + ...}$$

The compressibility correction factor value increases with the increase in mach number.

13.9 VARIABLE CROSS SECTIONAL AREA DUCT FLOW

Here the one dimensional isentropic type flow in a variable area duct is studied. The popular examples of the variable cross-section area duct are the flow through nozzles and diffusers. The significance of such flow analysis can be understood by the interchangeable nozzle and diffuser behaviour shown by the same convergent/divergent duct or a combination thereoff, depending upon the flow conditions. The flow conditions can be expressed based on Mach no.

The analysis of flows through a duct can be done with certain assumptions such as steady flow, one dimensionality of the flow etc. Here the flow conditions are assumed to be isentropic and these act as reference process for comparing the actual processes with it. Some of significant analysis such as Mach number variation in duct, mass flow rate variation, choking flow etc. are presented here.

(a) Mach number variation in duct flows

Considering the flow through a duct to be isentropic the energy equation can be written as;

$$h + \frac{C^2}{2} = \text{Constant}$$

The continuity equation yields

Mass flow rate $m = \rho \cdot A \cdot C$

Taking log and differentiating continuity equation we get.

$$ln\, \rho + ln\, A + ln\, C = 0$$

$$\frac{d\rho}{\rho} + \frac{dA}{A} + \frac{dC}{C} = 0$$

$$dC = -C\left(\frac{d\rho}{\rho} + \frac{dA}{A}\right)$$

Partial differentiation of energy equation yields.

$$dh + C \cdot dC = 0$$

also it is known $dh = \dfrac{dp}{\rho}$ and $(dh = v\, dp)$

Substituting dh into differential form of energy equation;

$$\frac{dp}{\rho} = -C \cdot dC$$

Combining above equations,

$$\frac{dp}{\rho C^2} = \left(\frac{d\rho}{\rho} + \frac{dA}{A}\right)$$

$$\frac{dA}{A} = \frac{dp}{\rho C^2}\left(1 - \frac{d\rho}{dp} \cdot C^2\right)$$

Also for an isentropic process the velocity of sound 'a' is given as;

$$\frac{dp}{d\rho} = a^2$$

Hence we get

$$\frac{dA}{A} = \frac{dp}{\rho C^2}\left(1 - \frac{C^2}{a^2}\right)$$

$$\boxed{\frac{dA}{A} = \frac{dp}{\rho C^2}\left(1 - M^2\right)} \qquad (13.93)$$

The equation (13.93) involving duct cross-section area, pressure variation, density, gas particle velocity and mach no. is the governing equation for variable cross-section area duct flow.

Let us now analyse the question (13.93) for various ducts;

(*a*) Converging duct shall have the cross sectional area gradient $\dfrac{dA}{A}$ as (–*ve*).

If now the flow is subsonic then $M < 1$, for which the '*dp*' shall be –*ve* i.e. the '*dc*' shall be +*ve*.

If now the flow is supersonic then $M > 1$ for which the '*dp*' shall be +*ve* i.e. the '*dc*' shall be –*ve*.

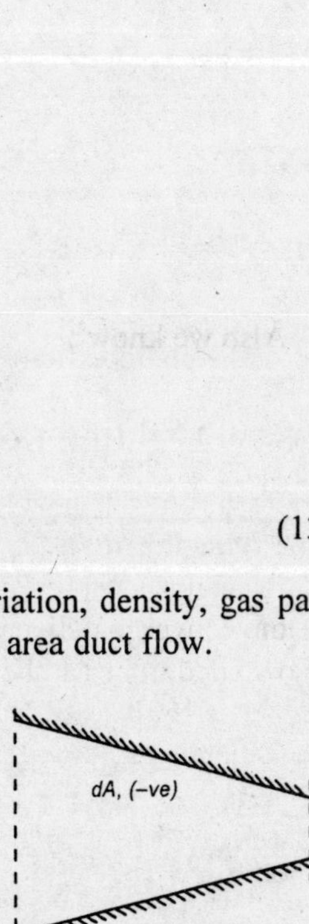

for $M < 1$; dC, +*ve*

for $M > 1$; dC, –*ve*

for $M = 1$; no change

(*b*) Diverging duct shall have the cross-sectional area gradient $\frac{dA}{A}$ as (+*ve*).

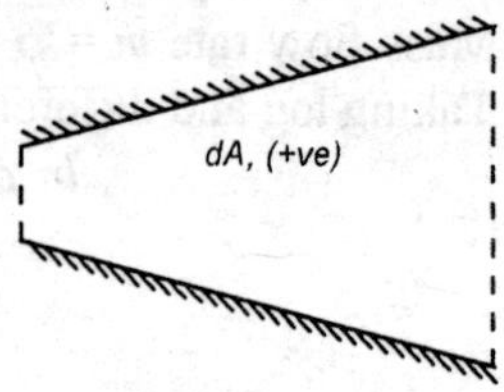

If now the flow is subsonic then $M < 1$, for which the '*dp*' shall be +*ve* i.e. the '*dc*' shall be –*ve*.

If now the flow is supersonic then $M > 1$, for which the '*dp*' shall be –*ve*, i.e. the '*dc*' shall be +*ve*.

for $M < 1$; dC, –*ve*
for $M > 1$; dC, +*ve*
for $M = 1$; no change

Thus the above interesting results show that the same duct may behave in different ways depending upon the flow conditions and compressibility characteristics of the gas.

(b) Mass Flow Rate Variation in Duct Flow

The mass flow rate can be obtained using the isentropic flow conditions and continuity equation.

Mass flow rate $m = \rho \cdot A \cdot C$

$$\frac{m}{A} = \rho \cdot C = \frac{P}{RT} \cdot C$$

$$\frac{m}{A} = \frac{P \cdot C}{\sqrt{RT} \cdot \sqrt{RT}}$$

$$\frac{m}{A} = \frac{P \cdot C}{\sqrt{\gamma \cdot RT}} \cdot \sqrt{\frac{\gamma}{R}} \cdot \sqrt{\frac{T_0}{T}} \cdot \frac{1}{\sqrt{T_0}}$$

$$\frac{m}{A} = \sqrt{\frac{\gamma}{R}} \cdot M \cdot \frac{P}{\sqrt{T_0}} \sqrt{\left(1 + \left(\frac{\gamma - 1}{2}\right) \cdot M^2\right)}$$

Also we know ;

$$\frac{p_0}{p} = \left(1 + \frac{\gamma - 1}{2} M^2\right)^{\frac{\gamma}{(\gamma - 1)}}$$

$$\frac{m}{A} = \sqrt{\frac{\gamma}{R}} \cdot \frac{M}{\sqrt{T_0}} \cdot \frac{P_0}{\left(1 + \frac{\gamma - 1}{2} M^2\right)^{\frac{\gamma}{\gamma - 1}}} \cdot \sqrt{1 + \left(\frac{\gamma - 1}{\gamma}\right) M^2}$$

or

$$\frac{m}{A} = \sqrt{\frac{\gamma}{R}} \cdot \frac{M \cdot P_0}{\sqrt{T_0}} \cdot \frac{1}{\left(1 + \frac{\gamma - 1}{2} M^2\right)^{\frac{(\gamma + 1)}{2(\gamma - 1)}}}$$

or

$$\frac{m}{A} \cdot \sqrt{\frac{R}{\gamma}} \cdot \frac{\sqrt{T_0}}{P_0} = \frac{M}{\left(1 + \left(\frac{\gamma - 1}{2}\right) M^2\right)^{\frac{(\gamma + 1)}{2(\gamma - 1)}}}$$

$$\frac{m}{A}\cdot\sqrt{\frac{C_p\cdot(\gamma-1)}{\gamma^2}}\cdot\frac{\sqrt{T_0}}{P_0}=\frac{M}{\left(1+\left(\frac{\gamma-1}{2}\right)M^2\right)^{\frac{\gamma+1}{2(\gamma-1)}}}$$

or
$$\frac{m\cdot\sqrt{C_p\cdot T_0}}{AP_0}=\sqrt{\frac{\gamma^2}{(\gamma-1)}}\frac{M}{\left(1+\frac{\gamma-1}{2}M^2\right)^{\frac{\gamma+1}{2(\gamma-1)}}}$$

Also the mass flow rate can be given in terms of pressure ratio;
Mass flow rate can be rewritten as;

$$m=\rho A C$$

from isentropic flow conditions
$$\frac{p}{p_0}=\left(\frac{\rho}{\rho_0}\right)^{\gamma}$$

thus
$$\frac{m}{A}=\left(\frac{p}{p_0}\right)^{\frac{1}{\gamma}}\cdot\rho_0\cdot C$$

Using the definition of stagnation temperature.

$$T_0=T+\frac{C^2}{2C_p}$$

or
$$C^2=\frac{2\gamma}{(\gamma-1)}\cdot RT_0\left(1-\frac{T}{T_0}\right)$$

$$C=\sqrt{\frac{2\gamma}{(\gamma-1)}\cdot RT_0\left\{1-\left(\frac{p}{p_0}\right)^{\frac{\gamma-1}{\gamma}}\right\}}$$

Substitution of C in above mass flow equation yields

$$\frac{m}{A}=\rho_0\left(\frac{p}{p_0}\right)^{\frac{1}{\gamma}}\sqrt{2\left(\frac{\gamma}{\gamma-1}\right)\cdot RT_0\left\{1-\left(\frac{p}{p_0}\right)^{\frac{\gamma-1}{\gamma}}\right\}}$$

or

$$\frac{m}{A}=\frac{p_0}{RT_0}\cdot\left(\frac{p}{p_0}\right)^{\frac{1}{\gamma}}\sqrt{2\left(\frac{\gamma}{\gamma-1}\right)\cdot RT_0\left\{1-\left(\frac{p}{p_0}\right)^{\frac{\gamma-1}{\gamma}}\right\}}$$

In non-dimensional form;

$$\frac{m\sqrt{T_0}}{A.\,p_0}\cdot\sqrt{\frac{R}{\gamma}}=\sqrt{\frac{2}{(\gamma-1)}\left\{\left(\frac{p}{p_0}\right)^{2/\gamma}-\left(\frac{p}{p_0}\right)^{(\gamma+1)/\gamma}\right\}}$$

The maximum possible mass flow through this variable area duct can be obtained as a function of pressure ratio, by partial differentiation of above equation involving mass flow rate;

$$\left(\frac{p}{p_0}\right)_{max} = \left(\frac{2}{\gamma+1}\right)^{\frac{\gamma}{(\gamma-1)}}$$

Corresponding to maximum flow condition : $A = A^*$
Substitution gives

$$\boxed{\frac{m_{max}\cdot\sqrt{T_0}}{A^*\cdot P_0}\sqrt{\frac{R}{\gamma}} = \left(\frac{2}{\gamma+1}\right)^{\frac{\gamma+1}{2(\gamma-1)}}} \qquad (13.94)$$

Using the above equation the maximum mass flow rate can be calculated, if the stagnation pressure, temperature and A^* is known.

(c) Choking Flow

The choked flow is the flow condition, when the mass flow rate through the duct is maximum and irrespective of any change mass flow rate can't be increased beyond this maximum limit.

The equation (13.94) obtained above is the mathematical expression of choked flow operation of duct.

$$\left(\frac{m}{A}\right)_{max} = \sqrt{\frac{\gamma}{R}}\cdot\frac{P_0}{\sqrt{T_0}}\cdot\left(\frac{2}{\gamma+1}\right)^{\frac{(\gamma+1)}{2(\gamma-1)}}$$

or

$$\left(\frac{m}{A^*}\right)_{max} = \sqrt{\frac{\gamma}{R}\cdot\left(\frac{2}{\gamma+1}\right)^{\frac{(\gamma+1)}{(\gamma-1)}}}\cdot\frac{P_0}{\sqrt{T_0}}$$

Analysing above equation it could be seen that the maximum flow per unit area depends only on P_o and T_o i.e. stagnation properties. The maximum flow per unit area also depends upon R, which supports the conclusion that the flow per unit area shall be large for high molecular weight gases and small for low molecular weight gases.

(d) Area Ratio

The area ratio is non-dimensional parameter which quantifies the area at any section relative to choked section area. This ratio is also readily available from gas tables for isentropic flow conditions.

Let the area at any section be A, where as at choked section be A^* then the ratio $\left(\frac{A}{A^*}\right)$ can be obtained from earlier equations.

$$\frac{A}{A^*} = \frac{m/A^*}{m/A} = \frac{\sqrt{\frac{\gamma}{R}}\cdot\frac{p_0}{\sqrt{T_0}}\cdot\sqrt{\left(\frac{2}{\gamma+1}\right)^{\left(\frac{\gamma+1}{\gamma-1}\right)}}}{\sqrt{\frac{\gamma}{R}}\cdot\frac{p_0}{\sqrt{T_0}}\cdot M\cdot\frac{1}{\sqrt{\left(1+\frac{\gamma-1}{2}M^2\right)^{\left(\frac{\gamma+1}{\gamma-1}\right)}}}}$$

or

$$\frac{A}{A^*} = \frac{1}{M}\sqrt{\left\{\frac{2}{(\gamma+1)}\left(1+\frac{\gamma-1}{2}M^2\right)\right\}^{\left(\frac{\gamma+1}{\gamma-1}\right)}}$$

The ratio $\frac{A}{A^*}$ shall have the value always greater than '1' since A^* is the smallest possible cross-section in duct.

For any value of $\frac{A}{A^*}$ the mach number 'M' shall have two values, out of which one shall be in subsonic and other in supersonic flow regime. The graphical representation of $\left(\frac{A}{A^*}\right)$ area ratio with respect to mach number (M) shall appear as shown in Fig 13.18.

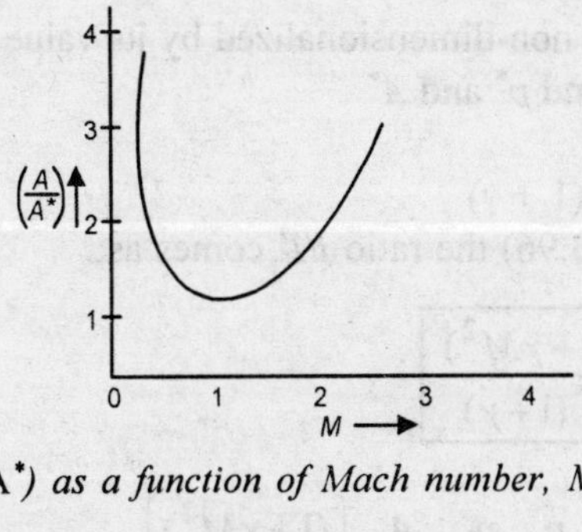

Fig. 13.18 Area ratio (A/A) as a function of Mach number, M in an isentropic flow*

(e) Impulse Function

Impulse function is defined as a parameter to quantify the net thrust forces acting on the duct. It shall include the force due to pressure and due to momentum change. It acts as a very important gas dynamics parameter as it serves as wall force function in a compressible flow.

Let us consider a one-dimensional flow through a control surface as shown in Fig. 13.19 for a perfect gas;

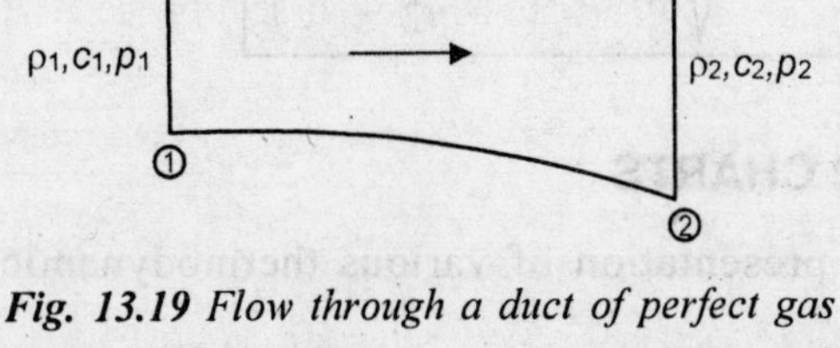

Fig. 13.19 Flow through a duct of perfect gas

The effective thrust force experienced by the duct wall shall be due to the changes in the momentum and pressure forces. Writing the force balance the thrust can be given as.

$$\text{Thrust} = (p_2 A_2 + \rho_2 \cdot A_2 \cdot C_2^2) - (p_1 A_1 + \rho_1 A_1 C_1^2)$$

$$= \left(p_2A_2 + \frac{p_2}{RT_2}\cdot A_2\cdot C_2^2\right) - \left(p_1A_1 + \frac{p_1}{RT_1}\cdot A_1\cdot C_1^2\right)$$

$$= \left(p_2A_2 + \frac{\gamma\, p_2}{\gamma RT_2}\cdot C_2^2\cdot A_2\right) - \left(p_1A_1 + \frac{\gamma\, p_1}{\gamma RT_1}\cdot A_1\cdot C_1^2\right)$$

$$= \left(p_2A_2 + \gamma\cdot p_2M_2^2\cdot A_2\right) - \left(p_1A_1 + \gamma\cdot p_1M_1^2\cdot A_1\right)$$

$$= \{pA + \gamma pM^2\cdot A\}_{\text{section } 2-2} - \{pA + \gamma pM^2\cdot A\}_{\text{section } 1-1}$$

$$= F_{2-2} - F_{1-1}$$

$$\text{Thrust} = \left[pA\{1+\gamma M^2\}\right]_{\text{section } 2-2} - \left[pA\{1+\gamma M^2\}\right]_{\text{section } 1-1}$$

The F_{2-2} and F_{1-1} can be termed as impulse function and are denoted by 'I'

so $\quad \text{Thrust} = I_2 - I_1$

or
$$\boxed{I = pA(1+\gamma M^2)} \quad (13.95)$$

The impulse function can be non-dimensionalized by its value at sonic flow conditions i.e. M = 1 which shall give the value of I^* and p^* and A^*

so

$$I^* = p^*A^*(1+\gamma) \quad (13.96)$$

By equations (13.95) and (13.96) the ratio I/I^* comes as;

$$\boxed{\frac{I}{I^*} = \frac{p\cdot A\cdot(1+\gamma M^2)}{p^*\cdot A^*(1+\gamma)}}$$

or
$$\frac{I}{I^*} = \frac{p}{p_0}\cdot\frac{p_0}{p^*}\cdot\frac{A}{A^*}\cdot\left\{\frac{(1+\gamma M^2)}{(1+\gamma)}\right\}$$

Substituting for the ratios $\left(\frac{p}{p_0}\right)\left(\frac{p_o}{p^*}\right)\cdot\left(\frac{A}{A^*}\right)$, it results into;

$$\boxed{\frac{I}{I^*} = \frac{1+\gamma M^2}{M\sqrt{2(1+\gamma)\left(1+\frac{\gamma-1}{2}M^2\right)}}}$$

13.10 GAS TABLES AND CHARTS

The gas tables are tabular presentation of various thermodynamic parameter ratios such as $\frac{T}{T^*}, \frac{p}{p^*}, \frac{\rho}{\rho^*}, \frac{p_0}{p_0^*}, \frac{I}{I^*}, \frac{A}{A^*}$ etc. with respect to mach number 'M' for isentropic flow. The ratios have been readily calculated and tabulated based on the basic equations discussed earlier. Using these tables and charts calculations become easy, avoiding the unnecessary calculations of basic thermodynamic parametric ratios. With the nodal values available the desired states can be located by interpolation or extrapolation during the engineering calculations. Air tables are given in appendix.

EXAMPLES

1. Determine Mach no., velocity and stagnation pressure at same section in the duct where stagnation temperature is 27ºC and static pressure is 0.4 bar. Duct has air flowing at the rate of 3000 kg/hr and the diameter of duct is 80 mm.

Solution:

Stagnation temperature = 27 + 273 = 300 K

$$\text{Mass flow rate, } m = \rho AC \Rightarrow \frac{3000}{3600} = \rho \times \frac{\pi}{4}(80\times10^{-3})^2 \times C$$

$$\rho C = 165.79$$

$$\Rightarrow \rho = \frac{165.79}{C}$$

Using perfect gas equation

$$p = \rho RT$$

$$\Rightarrow \quad T = \frac{p}{\rho R} = \frac{0.4\times10^5 \times C}{165.79\times0.287\times10^3}$$

$$\Rightarrow \quad C = (1.19 \cdot T)$$

$$\text{We know, } C^2 = \frac{2\gamma R}{\gamma-1}(T_0 - T)$$

$$C^2 = \frac{2\times1.4\times0.287\times10^3}{(1.4-1)}(300 - T)$$

$$C^2 = 602.7 \times 10^3 - 2009\,T$$

$$C^2 + 1688.23C - 602.7 \times 10^3 = 0$$

Solving $\Rightarrow C = 302.72$ m/s & $T = 254.39$ K

Using stagnation property relation,

$$\frac{T_0}{T} = 1 + \left(\frac{\gamma-1}{2}\right)M^2 \Rightarrow \frac{300}{254.39} = 1 + \left(\frac{1.4-1}{2}\right)\cdot M^2$$

$$M = 0.947$$

$$\text{Stagnation pressure} \Rightarrow P_0 = P\left(1 + \frac{\gamma-1}{2}M^2\right)$$

$$= 0.4\left(1 + \frac{(1.4-1)}{2}\cdot(0.947)^2\right)$$

$$P_0 = 0.472 \text{ bar}$$

Mach number = 0.947, Stagnation pressure = 0.472 bar, Velocity = **302.72 m/s**	**Ans.**

2. Determine the ratio of average kinetic energy at exit to exit kinetic energy corresponding to isentropic expansion upto measured value of 0.25 bar static pressure for the air flowing through converging –

diverging-nozzle. It is seen from Schlieren photograph that average Mach angle over exit cross - section is 45° and pressure at upstream side of nozzle is 1.01 bar and velocity is very-very small here. Take measured value of stagnation temperature as 1100°C and air having $\gamma = 1.4$.

Solution:

Given, $T_0 = 1100 + 273 = 1373$ K

Mach number, $M' = \dfrac{1}{\sin\alpha} = \dfrac{1}{\sin 45} = 1.414$

Also, it is given that stagnation pressure $p_0 = 1.01$ bar

so, $\dfrac{p}{p_0} = \dfrac{0.25}{1.01} = 0.2475$. Corresponding to this $\dfrac{p}{p_0}$ ratio the Mach number & $\dfrac{T}{T_0}$ can be seen from air table as $M = 1.564$ & $\dfrac{T}{T_0} = 0.6717 \Rightarrow T = 922.244$ K $\Rightarrow C_{max} = M \cdot \sqrt{\gamma RT}$

$$C_{max} = 1.564 \cdot \sqrt{1.4 \times 0.287 \times 10^3 \times 922.244}$$

$$C_{max} = 952.06 \text{ m/s.}$$

Corresponding to Mach number ($M' = 1.414$) as obtained from experimental observation, the $\dfrac{T}{T_0}$ can be seen from air table and it comes out as $\left(\dfrac{T}{T_0}\right) = 0.7145$

$$\Rightarrow \qquad T = 0.7145 \times 1373 = 981 \text{ K}$$

$$C_{av} = M'\sqrt{\gamma RT} = 1.414\sqrt{1.4 \times 0.287 \times 10^3 \times 981} = 887.75 \text{ m/s}$$

$$\text{Ratio of kinetic energy} = \frac{\frac{1}{2} \times (887.75)^2}{\frac{1}{2} \times (952.06)^2} = \mathbf{0.869} \quad \textbf{Ans.}$$

3. *Determine the speed indicator reading in Mach nos. for an aircraft flying with speed of 300 m/s at some altitude where pressure and density are 0.472 bar and 0.659 kg/m³ respectively. Also determine the compressibility correction factor.*

Solution:

Velocity, $C = 300$ m/s, $P = 0.472$ bar, $\rho = 0.659$ kg/m^3

From Bernoulli's equation $p_0 - p = \dfrac{1}{2}\rho C^2$

$$p_0 = (0.472 \times 10^5) + \frac{1}{2} \times 0.659 \times (300)^2$$

$$p_0 = 76855 \text{ N/m}^2$$

Speed indicator reading shall be given by Mach nos.

$$\text{Mach no.} = \frac{C}{a} = \frac{300}{\sqrt{\gamma RT}}$$

Using perfect gas equation,

$$p = \rho RT \Rightarrow T = \frac{0.472 \times 10^5}{287 \times 0.659}$$

$$T = 249.56 \text{ K}$$

$$\text{Mach no.} = \frac{300}{\sqrt{(1.4 \times 287 \times 249.56)}} = \mathbf{0.947} \quad \textbf{Ans.}$$

Considering compressibility effect

$$\frac{p_0}{p} = \left(1 + \frac{\gamma - 1}{2} M^2\right)^{\frac{\gamma}{\gamma - 1}}$$

$$\text{Stagnation pressure } p_0 = 0.472 \times 10^5 \times \left[1 + \left(\frac{1.4 - 1}{2}\right) \times (0.947)^2\right]^{\frac{1.4}{1.4-1}}$$

$$p_0 = 84082.71 \text{ N/m}^2$$

$$\text{Also } p_0 - p = (1 + \xi)\ \frac{1}{2}\ \rho C^2$$

Substitution yields,

$$(84082.71 - 47200) = (1 + \xi)\ (\frac{1}{2} \times 0.659 \times (300)^2)$$

$$\text{Compressibility correction factor } \xi = \mathbf{0.2437} \quad \textbf{Ans.}$$

4. Determine the Mach number of a stream of supersonic air flow for which Pitot static tube shows total pressure of 2 bar and static pressure of 0.3 bar.

Solution:

Given, $p_0 = 2$ bar, $p = 0.3$ bar

$$\text{We know, } \frac{p_0}{p} = \left(1 + \frac{(\gamma - 1)}{2} M^2\right)^{\left(\frac{\gamma}{(\gamma - 1)}\right)}$$

$$\frac{2}{0.3} = \left(1 + \frac{(1.4 - 1)}{2} \times M^2\right)^{\left(\frac{1.4}{0.4}\right)}$$

$$\text{Mach number, } M = \mathbf{1.89} \quad \textbf{Ans.}$$

5. In a Pitot static tube the static pressure shown is 0.3 bar. Barometer reading is 1.01 bar and stagnation temperature of air stream is 305 K. Difference between total pressure and static pressure is 0.6 bar. Determine air stream velocity considering air to be incompressible and also compressible.

Solution:

Actual Static pressure = 1 + 0.3 = 1.3 bar = P

It is also given that $P_0 - P = 06 \Rightarrow P_0 = 1.9$ bar, $T_0 = 305$ K

$$\text{Air velocity, } a_0 = \sqrt{\gamma R T_0} = \sqrt{1.4 \times 287 \times 305} = 350.06 \text{ m/s}$$

$$\text{Density of air, } \rho_0 = \frac{p_0}{RT_0} = \frac{1.9\times10^5}{287\times305} = 2.17 \text{ kg/m}^3$$

Considering air to be in-compressible,

$$\boxed{P_0 = P + \frac{\rho_0 C^2}{2} \Rightarrow C = \mathbf{235.16\ m/s}} \quad \textbf{Ans.}$$

$\Rightarrow$ For compressible fluid,

$$\frac{P_0}{P} = \left\{1+\left(\frac{\gamma-1}{2}\right)M^2\right\}^{\left(\frac{\gamma}{(\gamma-1)}\right)}$$

$$\Rightarrow \quad \frac{1.9}{1.3} = \left\{1+\left(\frac{0.4}{2}\right)\cdot M^2\right\}^{3.5}$$

$$\Rightarrow \quad M = 0.7567$$

Compressibility correction factor,

$$\xi = \frac{M^2}{4} + \left(\frac{2-\gamma}{24}\right)M^4$$

$$\xi = 0.1513$$

Stagnation temperature,

$$\frac{T_0}{T} = 1+\left(\frac{\gamma-1}{2}\right)M^2$$

$$\Rightarrow \quad T = 273.66 \text{ K}$$

$$\text{Density } \rho = \frac{P}{RT} = \frac{1.3\times10^5}{287\times273.66} = 1.655 \text{ kg/m}^3$$

Substituting in

$$P_0 - P = \frac{1}{2}\delta C^2 (1 + \xi)$$

$$\Rightarrow \quad 10^5 \times (1.9 - 1.3) = \frac{1}{2} \times 1.655 \times C^2 \times 1.1513$$

$$\Rightarrow \quad \boxed{C = \mathbf{250.95\ m/s}} \quad \textbf{Ans.}$$

EXERCISE

13.1 What do you understand by compressible fluid flow?

13.2 Describe the different flow regimes based on Mach no.

13.3 What do you understand by energy equation? Derive it and discuss its relevance.

13.4 What is meant by zone of silence and zone of action?

13.5 Differentiate between static and stagnation properties.

13.6 Show that converging duct shall act as nozzle instead of diffuser if the type of flow is changed from subsonic to supersonic.

13.7 What is significance of compressibility correction factor? Derive it.

13.8 What is meant by choked flow?

13.9 Differentiate between steep finite pressure wave and non steep finite pressure wave.

13.10 Explain the rarefaction or expansion wave.

APPENDIX

Table 1 Ideal-gas specific heats of various common gases at 300 K

Gas	*Formula*	*Gas constant R* kJ/(kg.K)	c_p kJ/(kg.K)	c_v kJ/(kg.K)	γ
Air	–	0.2870	1.005	0.718	1.400
Argon	Ar	0.2081	0.5203	0.3122	1.667
Butane	C_4H_{10}	0.1433	1.7164	1.5734	1.091
Carbon dioxide	CO_2	0.1889	0.846	0.657	1.289
Carbon monoxide	CO	0.2968	1.040	0.744	1.400
Ethane	C_2H_6	0.2765	1.7662	1.4897	1.186
Ethylene	C_2H_4	0.2964	1.5482	1.2518	1.237
Helium	He	2.0769	5.1926	3.1156	1.667
Hydrogen	H_2	4.1240	14.307	10.183	1.405
Methane	CH_4	0.5182	2.2537	1.7354	1.299
Neon	Ne	0.4119	1.0299	0.6179	1.667
Nitrogen	N_2	0.2968	1.039	0.743	1.400
Octane	C_8H_{18}	0.0729	1.7113	1.6385	1.044
Oxygen	O_2	0.2598	0.918	0.658	1.395
Propane	C_3H_8	0.1885	1.6794	1.4909	1.126
Steam	H_2O	0.4615	1.8723	1.4108	1.327

Table 2 Saturated Steam (temperature) table

Temp. °C	Sat. press. kPa	*Specific volume* m^3/kg		*Internal energy* kJ/kg			*Enthalpy* kJ/kg			*Entropy* kJ/(kg.K)		
T	P_{sat}	Sat. liquid v_f	Sat. vapour v_g	Sat. liquid u_f	Evap. u_{fg}	Sat. vapour u_g	Sat. liquid h_f	Evap. h_{fg}	Sat. vapour h_g	Sat. liquid s_f	Evap. s_{fg}	Sat. vapour s_g
0.01	0.6113	0.001000	206.14	0.0	2375.3	2375.3	0.01	2501.4	2501.4	0.000	9.1562	9.1562
5	0.8721	0.001000	147.12	20.97	2361.3	2382.3	20.98	2489.6	2510.6	0.0761	8.9496	9.0257
10	1.2276	0.001000	106.38	42.00	2347.2	2389.2	42.01	2477.7	2519.8	0.1510	8.7498	8.9008
15	1.7051	0.001001	77.93	62.99	2333.1	2396.1	62.99	2465.9	2528.9	0.2245	8.5569	8.7814
20	2.339	0.001002	57.79	83.95	2319.0	2402.9	83.96	2454.1	2538.1	0.2966	8.3706	8.6672
25	3.169	0.001003	43.36	104.88	2409.8	2309.8	104.89	2442.3	2547.2	0.3674	8.1905	8.5580
30	4.246	0.001004	32.89	125.78	2290.8	2416.6	125.79	2430.5	2556.3	0.4369	8.0164	8.4533
35	5.628	0.001006	25.22	146.67	2276.7	2423.4	146.68	2418.6	2565.3	0.5053	7.8478	8.3531
40	7.384	0.001008	19.52	167.56	2262.6	2430.1	167.57	2406.7	2574.3	0.5725	7.6845	8.2570
45	9.593	0.001010	15.26	188.44	2248.4	2436.8	188.45	2394.8	2583.2	0.6387	7.5261	8.1648
50	12.349	0.001012	12.03	209.32	2234.2	2443.5	209.33	2382.7	2592.1	0.7038	7.3725	8.0763
55	15.758	0.001015	9.568	230.21	2219.9	2450.1	230.23	2370.7	2600.9	0.7679	7.2234	7.9913
60	19.940	0.001017	7.671	251.11	2205.5	2456.6	251.13	2358.5	2609.6	0.8312	7.0784	7.9096
65	25.03	0.001020	6.197	272.02	2191.1	2463.1	272.06	2346.2	2618.3	0.8935	6.9375	7.8310
70	31.19	0.001023	5.042	292.95	2176.6	2469.6	292.98	2333.8	2626.8	0.9549	6.8004	7.7553
75	38.58	0.001026	4.131	313.90	2162.0	2475.9	313.93	2321.4	2635.3	1.0155	6.6669	7.6824
80	47.39	0.001029	3.407	334.86	2147.4	2482.2	334.91	2308.8	2643.7	1.0753	6.5369	7.6122
85	57.83	0.001033	2.828	355.84	2132.6	2488.4	355.90	2296.0	2651.9	1.1343	6.4102	7.5445
90	70.14	0.001036	2.361	376.85	2117.7	2494.5	376.92	2283.2	2660.1	1.1925	6.2866	7.4791
95	84.55	0.001040	1.982	397.88	2102.7	2500.6	397.96	2270.2	2668.1	1.2500	6.1659	7.4159

Table 2 Saturated Steam (temperature) table

Temp. °C	Sat. press. kPa	*Specific volume* m^3/kg		*Internal energy* kJ/kg			*Enthalpy* kJ/kg			*Entropy* $kJ/(kg.K)$		
		Sat. liquid	Sat. vapour	Sat. liquid	Evap.	Sat. vapour	Sat. liquid	Evap.	Sat. vapour	Sat. liquid	Evap.	Sat. vapour
T	P_{sat}	v_f	v_g	u_f	u_{fg}	u_g	h_f	h_{fg}	h_g	s_f	s_{fg}	s_g
100	0.10135	0.001044	1.6729	418.94	2087.6	2506.5	419.04	2257.0	2676.1	1.3069	6.0480	7.3549
105	0.12082	0.001048	1.4194	440.02	2072.3	2512.4	440.15	2243.7	2683.8	1.3630	5.9328	7.2958
110	0.14327	0.001052	1.2102	461.14	2057.0	2518.1	461.30	2230.2	2691.5	1.4185	5.8202	7.2387
115	0.16906	0.001056	1.0366	482.30	2041.4	2523.7	482.48	2216.5	2699.0	1.4734	5.7100	7.1833
120	0.19853	0.001060	0.8919	503.50	2025.8	2529.3	503.71	2202.6	2706.3	1.5276	5.6020	7.1296
125	0.2321	0.001065	0.7706	524.74	2009.9	2534.6	524.99	2188.5	2713.5	1.5813	5.4962	7.0775
130	0.2701	0.001070	0.6685	546.02	1993.9	2539.9	546.31	2174.2	2720.5	1.6344	5.3925	7.0269
135	0.3130	0.001075	0.5822	567.35	1977.7	2545.0	567.69	2159.6	2727.3	1.6870	5.2907	6.9777
140	0.3613	0.001080	0.5089	588.74	1961.3	2550.0	589.13	2144.7	2733.9	1.7391	5.1908	6.9299
145	0.4154	0.001085	0.4463	610.18	1944.7	2554.9	610.63	2129.6	2740.3	1.7907	5.0926	6.8833
150	0.4758	0.001091	0.3928	631.68	1927.9	2559.5	632.20	2114.3	2746.5	1.8418	4.9960	6.8379
155	0.5431	0.001096	0.3468	653.24	1910.8	2564.1	653.84	2098.6	2752.4	1.8925	4.9010	6.7935
160	0.6178	0.001102	0.3071	674.87	1893.5	2568.4	675.55	2082.6	2758.1	1.9427	4.8075	6.7502
165	0.7005	0.001108	0.2727	696.56	1876.0	2572.5	697.34	2066.2	2763.5	1.9925	4.7153	6.7078
170	0.7917	0.001114	0.2428	718.33	1858.1	2576.5	719.21	2049.5	2768.7	2.0419	4.6244	6.6663
175	0.8920	0.001121	0.2168	740.17	1840.0	2580.2	741.17	2032.4	2773.6	2.0909	4.5347	6.6256
180	1.0021	0.001127	0.19405	762.09	1821.6	2583.7	763.22	2015.0	2778.2	2.1396	4.4461	6.5857
185	1.1227	0.001134	0.17409	784.10	1802.9	2587.0	785.37	1997.1	2782.4	2.1879	4.3586	6.5465
190	1.2544	0.001141	0.15654	806.19	1783.8	2590.0	807.62	1978.8	2786.4	2.2359	4.2720	6.5079
195	1.3978	0.001149	0.14105	828.37	1764.4	2592.8	829.98	1960.0	2790.0	2.2835	4.1863	6.4698
200	1.5538	0.001157	0.12736	850.65	1744.7	2595.3	852.45	1940.7	2793.2	2.3309	4.1014	6.4323
205	1.7230	0.001164	0.11521	873.04	1724.5	2597.5	875.04	1921.0	2796.0	2.3780	4.0172	6.3952
210	1.9062	0.001173	0.10441	895.53	1703.9	2599.5	897.76	1900.7	2798.5	2.4248	3.9337	6.3585
215	2.104	0.001181	0.09479	918.14	1682.9	2601.1	920.62	1879.9	2800.5	2.4714	3.8507	6.3221
220	2.318	0.001190	0.08619	940.87	1661.5	2602.4	943.62	1858.5	2802.1	2.5178	3.7683	6.2861
225	2.548	0.001199	0.07849	963.73	1639.6	2603.3	966.78	1836.5	2803.3	2.5639	3.6863	6.2503
230	2.795	0.001209	0.07158	986.74	1617.2	2603.9	990.12	1813.8	2804.0	2.6099	3.6047	6.2146
235	3.060	0.001219	0.06537	1009.89	1594.2	2604.1	1013.62	1790.5	2804.2	2.6558	3.5233	6.1791
240	3.344	0.001229	0.05976	1033.21	1570.8	2604.0	1037.32	1766.5	2803.8	2.7015	3.4422	6.1437
245	3.648	0.001240	0.05471	1056.71	1546.7	2603.4	1061.23	1741.7	2803.0	2.7472	3.3612	6.1083
250	3.973	0.001251	0.05013	1080.39	1522.0	2602.4	1085.36	1716.2	2801.5	2.7927	3.2802	6.0730
255	4.319	0.001263	0.04598	1104.28	1596.7	2600.9	1109.73	1689.8	2799.5	2.8383	3.1992	6.0375
260	4.688	0.001276	0.04221	1128.39	1470.6	2599.0	1134.37	1662.5	2796.9	2.8838	3.1181	6.0019
265	5.081	0.001289	0.03877	1152.74	1443.9	2596.6	1159.28	1634.4	2793.6	2.9294	3.0368	5.9662
270	5.499	0.001302	0.03564	1177.36	1416.3	2593.7	1184.51	1605.2	2789.7	2.9751	2.9551	5.9301
275	5.942	0.001317	0.03279	1202.25	1387.9	2590.2	1210.07	1574.9	2785.0	3.0208	2.8730	5.8938
280	6.412	0.001332	0.03017	1227.46	1358.7	2586.1	1235.99	1543.6	2779.6	3.0668	2.7903	5.8571
285	6.909	0.001348	0.02777	1253.00	1328.4	2581.4	1262.31	1511.0	2773.3	3.1130	2.7070	5.8199
290	7.436	0.001366	0.02557	1278.92	1297.1	2576.0	1289.07	1477.1	2766.2	3.1594	2.6227	5.7821
295	7.993	0.001384	0.02354	1305.2	1264.7	2569.9	1316.3	1441.8	2758.1	3.2062	2.5375	5.7437
300	8.581	0.001404	0.02167	1332.0	1231.0	2563.0	1344.0	1404.9	2749.0	3.2534	2.4511	5.7045
305	9.202	0.001425	0.019948	1359.3	1195.9	2555.2	1372.4	1366.4	2738.7	3.3010	2.3633	5.6643
310	9.856	0.001447	0.018350	1387.1	1159.4	2546.4	1401.3	1326.0	2727.3	3.3493	2.2737	5.6230

Table 2 Saturated Steam (temperature) table

Temp. °C T	Sat. press. MPa P_{sat}	Specific volume m^3/kg Sat. liquid v_f	Sat. vapour v_g	Internal energy kJ/kg Sat. liquid u_f	Evap. u_{fg}	Sat. vapour u_g	Enthalpy kJ/kg Sat. liquid h_f	Evap. h_{fg}	Sat. vapour h_g	Entropy kJ/(kg.K) Sat. liquid s_f	Evap. s_{fg}	Sat. vapour s_g
315	10.547	0.001472	0.016867	1415.5	1121.1	2536.6	1431.0	1283.5	2714.5	3.3982	2.1821	5.5804
320	11.274	0.001499	0.015488	1444.6	1080.9	2525.5	1461.5	1238.6	2700.1	3.4480	2.0882	5.5362
330	12.845	0.001561	0.012996	1505.3	993.7	2498.9	1525.3	1140.6	2665.9	3.5507	1.8909	5.4417
340	14.586	0.001638	0.010797	1570.3	894.3	2464.6	1594.2	1027.9	2622.0	3.6594	1.6763	5.3357
350	16.513	0.001740	0.008813	1641.9	776.6	2418.4	1670.6	893.4	2563.9	3.7777	1.4335	5.2112
360	18.651	0.001893	0.006945	1725.2	626.3	2351.5	1760.5	720.3	2481.0	3.9147	1.1379	5.0526
370	21.03	0.002213	0.004925	1844.0	384.5	2228.5	1890.5	441.6	2332.1	4.1106	0.6865	4.7971
374.14	22.09	0.003155	0.003155	2029.6	0	2029.6	2099.3	0	2099.3	4.4298	0	4.4298

Table 3 Saturated steam (pressure) table

Press. kPa P	Sat. Temp. °C T_{sat}	Specific volume m^3/kg Sat. liquid v_f	Sat. vapour v_g	Internal energy kJ/kg Sat. liquid u_f	Evap. u_{fg}	Sat. vapour u_g	Enthalpy kJ/kg Sat. liquid h_f	Evap. h_{fg}	Sat. vapour h_g	Entropy kJ/(kg.K) Sat. liquid s_f	Evap. s_{fg}	Sat. vapour s_g
0.6113	0.01	0.001000	206.14	0.00	2375.3	2375.3	0.01	2501.3	2501.4	0.0000	9.1562	9.1562
1.0	6.98	0.001000	129.21	29.30	2355.7	2385.0	29.30	2484.9	2514.2	0.1059	8.8697	8.9756
1.5	13.03	0.001001	87.98	54.71	2338.6	2393.3	54.71	2470.6	2525.3	0.1957	8.6322	8.8279
2.0	17.50	0.001001	67.00	73.48	2326.0	2399.5	73.48	2460.0	2533.5	0.2607	8.4629	8.7237
2.5	21.08	0.001002	54.25	88.48	2315.9	2404.4	88.49	2451.6	2540.0	0.3120	8.3311	8.6432
3.0	24.08	0.001003	45.67	101.04	2307.5	2408.5	101.05	2444.5	2545.5	0.3545	8.2231	8.5776
4.0	28.96	0.001004	34.80	121.45	2293.7	2415.2	121.46	2432.9	2554.4	0.4226	8.0520	8.4746
5.0	32.88	0.001005	28.19	137.81	2282.7	2420.5	137.82	2423.7	2561.5	0.4764	7.9187	8.3951
7.5	40.29	0.001008	19.24	168.78	2261.7	2430.5	168.79	2406.0	2576.8	0.5764	7.6750	8.2515
10	45.81	0.001010	14.67	191.82	2246.1	2437.9	191.83	2392.8	2584.7	0.6493	7.5009	8.1502
15	53.97	0.001014	10.02	225.92	2222.8	2448.7	225.94	2373.1	2599.1	0.7549	7.2536	8.0085
20	60.06	0.001017	7.649	251.38	2205.4	2456.7	251.40	2358.3	2609.7	0.8320	7.0766	7.9085
25	64.97	0.001020	6.204	271.90	2191.2	2463.1	271.93	2346.3	2618.2	0.8931	6.9383	7.8314
30	69.10	0.001022	5.229	289.20	2179.2	2468.4	289.23	2336.1	2625.3	0.9439	6.8247	7.7686
40	75.87	0.001027	3.993	317.53	2159.5	2477.0	317.58	2319.2	2636.8	1.0259	6.6441	7.6700
50	81.33	0.001030	3.240	340.44	2143.4	2483.9	340.49	2305.4	2645.9	1.0910	6.5029	7.5939
75	91.78	0.001037	2.217	384.31	2112.4	2496.7	384.39	2278.6	2663.0	1.2130	6.2434	7.4564

Table 3 Saturated steam (pressure) table

Press. MPa P	Sat. Temp. °C T_{sat}	*Specific volume* m^3/kg Sat. liquid v_f	Sat. vapour v_g	*Internal energy* kJ/kg Sat. liquid u_f	Evap. u_{fg}	Sat. vapour u_g	*Enthalpy* kJ/kg Sat. liquid h_f	Evap. h_{fg}	Sat. vapour h_g	*Entropy* $kJ/(kg.K)$ Sat. liquid s_f	Evap. s_{fg}	Sat. vapour s_g
0.100	99.63	0.001043	1.6940	417.36	2088.7	2506.1	417.46	2258.0	2675.5	1.3026	6.0568	7.3594
0.125	105.99	0.001048	1.3749	444.19	2069.3	2513.5	444.32	2241.0	2685.4	1.3740	5.9104	7.2844
0.150	111.37	0.001053	1.1593	466.94	2052.7	2519.7	467.11	2226.5	2693.6	1.4336	5.7897	7.2233
0.175	116.06	0.001057	1.0036	486.80	2038.1	2524.9	486.99	2213.6	2700.6	1.4849	5.6868	7.1717
0.200	120.23	0.001061	0.8857	504.49	2025.0	2529.5	504.70	2201.9	2706.7	1.5301	5.5970	7.1271
0.225	124.00	0.001064	0.7933	520.47	2013.1	2533.6	520.72	2191.3	2712.1	1.5706	5.5173	7.0878
0.250	127.44	0.001067	0.7187	535.10	2002.1	2537.2	535.37	2181.5	2716.9	1.6072	5.4455	7.0527
0.275	130.60	0.001070	0.6573	548.59	1991.9	2540.5	548.89	2172.3	2721.3	1.6408	5.3801	7.0209
0.300	133.55	0.001073	0.6058	561.15	1982.4	2543.6	561.47	2163.8	2725.3	1.6718	5.3201	6.9919
0.325	136.30	0.001076	0.5620	572.90	1973.5	2546.4	573.25	2155.8	2729.0	1.7006	5.2646	6.9652
0.350	138.88	0.001079	0.5243	583.95	1965.0	2548.9	584.33	2148.1	2732.4	1.7275	5.2130	6.9405
0.375	141.32	0.001081	0.4914	594.40	1956.9	2551.3	594.81	2140.8	2735.6	1.7528	5.1647	6.9175
0.40	143.63	0.001084	0.4625	604.31	1949.3	2553.6	604.74	2133.8	2738.6	1.7766	5.1193	6.8959
0.45	147.93	0.001088	0.4140	622.77	1934.9	2557.6	623.25	2120.7	2743.9	1.8207	5.0359	6.8565
0.50	151.86	0.001093	0.3749	639.68	1921.6	2561.2	640.23	2108.5	2748.7	1.8607	4.9606	6.8213
0.55	155.48	0.001097	0.3427	655.32	1909.2	2564.5	665.93	2097.0	2753.0	1.8973	4.8920	6.7893
0.60	158.85	0.001101	0.3157	669.90	1897.5	2567.4	670.56	2086.3	2756.8	1.9312	4.8288	6.7600
0.65	162.01	0.001104	0.2927	683.56	1886.5	2570.1	684.28	2076.0	2760.3	1.9627	4.7703	6.7331
0.70	164.97	0.001008	0.2729	696.44	1876.1	2572.5	697.22	2066.3	2763.5	1.9922	4.7158	6.7080
0.75	167.78	0.001112	0.2556	708.64	1866.1	2574.7	709.47	2057.0	2766.4	2.0200	4.6647	6.6847
0.80	170.43	0.001115	0.2404	720.22	1856.6	2576.8	721.11	2048.0	2769.1	2.0462	4.6166	6.6628
0.85	172.96	0.001118	0.2270	731.27	1847.4	2578.7	732.22	2039.4	2771.6	2.0710	4.5711	6.6421
0.90	175.38	0.001121	0.2150	741.83	1838.6	2580.5	742.83	2031.0	2773.9	2.0946	4.5280	6.6226
0.95	177.69	0.001124	0.2042	751.95	1830.2	2582.1	753.02	2023.1	2776.1	2.1172	4.4869	6.6041
1.00	179.91	0.001127	0.19444	761.68	1822.0	2583.6	762.81	2015.3	2778.1	2.1387	4.4478	6.5865
1.10	184.09	0.001133	017753	780.09	1806.3	2586.4	781.34	2000.4	2781.7	2.1792	4.3744	6.5536
1.20	187.99	0.001139	0.16333	797.29	1791.5	2588.8	798.65	1986.2	2784.8	2.2166	4.3067	6.5233
1.30	191.64	0.001144	0.15125	813.44	1777.5	2591.0	814.93	1972.7	2787.6	2.2515	4.2438	6.4953
1.40	195.07	0.001149	0.14084	828.70	1764.1	2592.8	830.30	1959.7	2790.0	2.2842	4.1850	6.4693
1.50	198.32	0.001154	0.13177	843.16	1751.3	2594.5	844.89	1947.3	2792.2	2.3150	4.1298	6.4448
1.75	205.76	0.001166	0.11349	876.46	1721.4	2597.8	878.50	1917.9	2796.4	2.3851	4.0044	6.3896
2.00	212.42	0.001177	0.09963	906.44	1693.8	2600.3	908.79	1890.7	2799.5	2.4474	3.8935	6.3409
2.25	218.45	0.001187	0.08875	933.83	1668.2	2602.0	936.49	1865.2	2801.7	2.5035	3.7937	6.2972
2.5	223.99	0.001197	0.07998	959.11	1644.0	2603.1	962.11	1841.0	2803.1	2.5547	3.7028	6.2575
3.0	233.90	0.001217	0.06668	1004.78	1599.3	2604.1	1008.42	1795.7	2804.2	2.6457	3.5412	6.1869
3.5	242.60	0.001235	0.05707	1045.43	1558.3	2603.7	1049.75	1753.7	2803.4	2.7253	3.4000	6.1253
4	250.40	0.001252	0.04978	1082.31	1520.0	2602.3	1087.31	1741.1	2801.4	2.7964	3.2737	6.0701
5	263.99	0.001286	0.03944	1147.81	1449.3	2597.1	1154.23	1640.1	2794.3	2.9202	3.0532	5.9734
6	275.64	0.001319	0.03244	1205.44	1384.3	2589.7	1213.35	1571.0	2784.3	3.0267	2.8625	5.8892
7	285.88	0.001351	0.02737	1257.55	1323.0	2580.5	1267.00	1505.1	2772.1	3.1211	2.6922	5.8133
8	295.06	0.001384	0.02352	1305.57	1264.2	2569.8	1316.64	1441.3	2758.0	3.2068	2.5364	5.7432
9	303.40	0.001418	0.02048	1350.51	1207.3	2557.8	1363.26	1378.9	2742.1	3.2858	2.3915	5.6722
10	311.06	0.001452	0.018026	1393.04	1151.4	2544.4	1407.56	1317.1	2724.7	3.3596	2.2544	5.6141

Contd.

Table 3 Saturated steam (pressure) table

Press. MPa P	Sat. Temp. °C T_{sat}	Specific volume m^3/kg Sat. liquid v_f	Sat. vapour v_g	Internal energy kJ/kg Sat. liquid u_f	Evap. u_{fg}	Sat. vapour u_g	Enthalpy kJ/kg Sat. liquid h_f	Evap. h_{fg}	Sat. vapour h_g	Entropy $kJ/(kg.K)$ Sat. liquid s_f	Evap. s_{fg}	Sat. vapour s_g
11	318.15	0.001489	0.015987	1433.7	1096.0	2529.8	1450.1	1255.5	2705.6	3.4295	2.1233	5.5527
12	324.75	0.001527	0.014263	1473.0	1040.7	2513.7	1491.3	1193.3	2684.9	3.4962	1.9962	5.4924
13	330.93	0.001567	0.012780	1511.1	985.0	2496.1	1531.5	1130.7	2662.2	3.5606	1.8718	5.4323
14	336.75	0.001611	0.011485	1548.6	928.2	2476.8	1571.1	1066.5	2637.6	3.6232	1.7485	5.3717
15	342.24	0.001658	0.010337	1585.6	869.8	2455.5	1610.5	1000.0	2610.5	3.6848	1.6249	5.3098
16	347.44	0.001711	0.009306	1622.7	809.0	2431.7	1650.1	930.6	2580.6	3.7461	1.4994	5.2455
17	352.37	0.001770	0.008364	1660.2	744.8	2405.0	1690.3	856.9	2547.2	3.8079	1.3698	5.1777
18	357.06	0.001840	0.007489	1698.9	675.4	2374.3	1732.0	777.1	2509.1	3.8715	1.2329	5.1044
19	361.54	0.001924	0.006657	1739.9	598.1	2338.1	1776.5	688.0	2464.5	3.9388	1.0839	5.0228
20	365.81	0.002036	0.005834	1785.6	507.5	2293.0	1826.3	583.4	2409.7	4.0139	0.9130	4.9269
21	369.89	0.002207	0.004952	1842.1	388.5	2230.6	1888.4	446.2	2334.6	4.1075	0.6938	4.8013
22	373.80	0.002742	0.003568	1961.9	125.2	2087.1	2022.2	143.4	2165.6	4.3110	0.2216	4.5327
22.09	374.12	0.003155	0.003155	2029.6	0	2029.6	2099.3	0	2099.3	4.4298	0	4.4298

Table 4 Superheated steam table

T °C	v m^3/kg	u kJ/kg	h kJ/kg	s kJ/(kg·K)	v m^3/kg	u kJ/kg	h kJ/kg	s kJ/(kg·K)	v m^3/kg	u kJ/kg	h kJ/kg	s kJ/(kg·K)
	P = 0.01 MPa (45.81 °C)				P = 0.05 MPa (81.33 °C)				P = 0.10 MPa (99.63 °C)			
Sat.	14.674	2437.9	2584.7	8.1502	3.240	2483.9	2645.9	7.5939	1.6940	2506.1	2675.5	7.3594
50	14.869	2483.9	2592.6	8.1749								
100	17.196	2515.5	2687.5	8.4479	3.418	2511.6	2682.5	7.6947	1.6958	2506.7	2676.2	7.3614
150	19.512	2587.9	2783.0	8.6882	3.889	2585.6	2780.1	7.9401	1.9364	2582.8	2776.4	7.6134
200	21.825	2661.3	2879.5	8.9038	4.356	2659.9	2877.7	8.1580	2.172	2658.1	2875.3	7.8343
250	24.136	2736.0	2977.3	9.1002	4.820	2735.0	2976.0	8.3556	2.406	2733.7	2974.3	8.0333
300	26.445	2812.1	3075.5	9.2813	5.284	2811.3	3075.5	8.5373	2.639	2810.4	3074.3	8.2158
400	31.063	2968.9	3279.6	9.6077	6.209	2968.5	3278.9	8.8642	3.103	2967.9	3278.2	8.5435
500	35.679	3132.3	3489.1	9.8978	7.134	3132.0	3488.7	9.1546	3.565	3131.6	3488.1	8.8342
600	40.295	3302.5	3705.4	10.1608	8.057	3302.2	3705.1	9.4178	4.028	3301.9	3704.4	9.0976
700	44.911	3479.6	3928.7	10.4028	8.981	3479.4	3928.5	9.6599	4.490	3479.2	3928.2	9.3398
800	49.526	3663.8	4159.0	10.6281	9.904	3663.6	4158.0	9.8852	4.952	3663.5	4158.6	9.5652
900	54.141	3855.0	4396.4	10.8396	10.828	3854.9	4396.3	10.0967	5.414	3854.8	4396.1	9.7767
1000	58.757	4053.0	4640.6	11.0393	11.751	4052.9	4640.5	10.2964	5.875	4052.8	4640.3	9.9764
1100	63.372	4257.5	4891.2	11.2287	12.674	4257.4	4891.1	10.4859	6.337	4257.3	4891.0	10.1659
1200	67.987	4467.9	5147.8	11.4091	13.597	4467.8	5147.7	10.6662	6.799	4467.7	5147.6	10.3463
1300	72.602	4683.7	5409.7	11.5811	14.521	4683.6	5409.6	10.8382	7.260	4683.5	5409.5	10.5183

Table 4 Superheated steam table

T °C	*v* m³/kg	*u* kJ/kg	*h* kJ/kg	*s* kJ/(kg·K)	*v* m³/kg	*u* kJ/kg	*h* kJ/kg	*s* kJ/(kg·K)	*v* m³/kg	*u* kJ/kg	*h* kJ/kg	*s* kJ/(kg·K)
	P = 0.20 MPa (120.23 °C)				*P = 0.30 MPa (133.55 °C)*				*P = 0.40 MPa (143.63 °C)*			
Sat.	0.8857	2529.5	2706.7	7.1272	0.6058	2543.6	2725.3	6.9919	0.4625	2553.6	2738.6	6.8959
150	0.9596	2576.9	2768.8	7.2795	0.6339	2570.8	2761.0	7.0778	0.4708	2564.5	2752.8	6.9299
200	1.0803	2654.4	2870.5	7.5066	0.7163	2650.7	2865.6	7.3115	0.5342	2646.8	2860.5	7.1706
250	1.1988	2731.2	2971.0	7.7086	0.7964	2728.7	2967.6	7.5166	0.5951	2726.1	2964.2	7.3789
300	1.3162	2808.6	3071.8	7.8926	0.8753	2806.7	3069.3	7.7022	0.6548	2804.8	3066.8	7.5662
400	1.5493	2966.7	3276.6	8.2218	1.0315	2965.6	3275.0	8.0330	0.7726	2964.4	3273.4	7.8985
500	1.7814	3130.8	3487.1	8.5133	1.1867	3130.0	3486.0	8.3251	0.8893	3129.2	3484.9	8.1913
600	2.013	3301.4	3704.0	8.7770	1.3414	3300.8	3703.2	8.5892	1.0055	3300.2	3702.4	8.4558
700	2.244	3478.8	3927.6	9.0194	1.4957	3478.4	3927.1	8.8319	1.1215	3477.9	3926.5	8.6987
800	2.475	3663.1	4158.2	9.2449	1.6499	3662.9	4157.8	9.0576	1.2312	3662.4	4157.3	8.9244
900	2.705	3854.5	4395.8	9.4566	1.8041	3854.2	4395.4	9.2692	1.3529	3853.9	4395.1	9.1362
1000	2.937	4052.5	4640.0	9.5663	1.9581	4052.3	4639.7	9.4690	1.4685	4052.0	4639.4	9.3360
1100	3.168	4257.0	4890.7	9.8458	2.1121	4256.8	4890.4	9.6585	1.5840	4256.5	4890.2	9.5256
1200	3.399	4467.5	5147.5	10.0262	2.2661	4467.2	5147.1	9.8389	1.6996	4467.0	5146.8	9.7060
1300	3.630	4683.2	5409.3	10.1982	2.4201	4683.0	5409.0	10.0110	1.8151	4682.8	5408.8	9.8780

Table 4 Superheated steam table

T °C	*v* m³/kg	*u* kJ/kg	*h* kJ/kg	*s* kJ/(kg·K)	*v* m³/kg	*u* kJ/kg	*h* kJ/kg	*s* kJ/(kg·K)	*v* m³/kg	*u* kJ/kg	*h* kJ/kg	*s* kJ/(kg·K)
	P = 0.50 MPa (151.86 °C)				*P = 0.60 MPa (158.85 °C)*				*P = 0.80 MPa (170.43 °C)*			
Sat.	0.3749	2561.2	2748.7	6.8213	0.3157	2567.4	2756.8	6.7600	0.2404	2576.8	2769.1	6.6628
200	0.4249	2642.9	2855.4	7.0592	0.3520	2638.9	2850.1	6.9665	0.2608	2630.6	2839.3	6.8158
250	0.4744	2723.5	2960.7	7.2709	0.3938	2720.9	2957.2	7.1816	0.2931	2715.5	2950.0	7.0384
300	0.5226	2802.9	3064.2	7.4599	0.4344	2801.0	3061.6	7.3724	0.3241	2797.2	3056.5	7.2328
350	0.5701	2882.6	3167.7	7.6329	0.4742	2881.2	3165.7	7.5464	0.3544	2878.2	3161.7	7.4089
400	0.6173	2963.2	3271.9	7.7938	0.5137	2962.1	3270.3	7.7079	0.3843	2959.7	3267.1	7.5716
500	0.7109	3128.4	3483.9	8.0873	0.5920	3127.6	3482.8	8.0021	0.4433	3126.0	3480.6	7.8673
600	0.8041	3299.6	3701.7	7.3522	0.6697	3299.1	3700.9	8.2674	0.5018	3297.9	3699.4	8.1333
700	0.8969	3477.5	3925.9	8.5952	0.7472	3477.0	3925.3	8.5107	0.5601	3476.2	3924.2	8.3770
800	0.9896	3662.1	4156.9	8.8211	0.8215	3661.8	4156.5	8.7367	0.6181	3661.1	4155.6	8.6033
900	1.0822	3853.6	4394.7	9.0329	0.9017	3853.4	4394.4	8.9486	0.6761	3852.8	4393.7	8.8153
1000	1.1747	4051.8	4639.1	9.2328	0.9788	4051.5	4638.8	9.1485	0.7340	4051.0	4638.2	9.0153
1100	1.2672	4256.3	4889.9	9.4224	1.0559	4256.1	4889.6	9.3381	0.7919	4255.6	4889.1	9.2050
1200	1.3596	4466.8	5146.6	9.6029	1.1330	4466.5	5146.3	9.5185	0.8497	4466.1	5145.9	9.3855
1300	1.4521	4682.5	5408.6	9.7749	1.2101	4682.3	5408.3	9.6906	0.9076	4681.8	5407.9	9.5575

Table 4 Superheated steam table

T °C	v m³/kg	u kJ/kg	h kJ/kg	s kJ/(kg·K)	v m³/kg	u kJ/kg	h kJ/kg	s kJ/(kg·K)	v m³/kg	u kJ/kg	h kJ/kg	s kJ/(kg·K)
	P = 1.00 MPa (179.91 °C)				*P = 1.20 MPa (187.99 °C)*				*P = 1.40 MPa (195.70 °C)*			
Sat.	0.19444	2583.6	2778.1	6.5865	0.16333	2588.8	2784.8	6.5233	0.14084	2592.8	2790.0	6.4693
200	0.2060	2621.9	2827.9	6.6940	0.16930	2612.8	2815.9	6.5898	0.14302	2603.1	2803.3	6.4975
250	0.2327	2709.9	2942.6	6.9247	0.19234	2704.2	2935.0	6.8294	0.16350	2698.3	2927.2	6.7467
300	0.2579	2793.2	3051.2	7.1229	0.2138	2789.2	3045.8	7.0317	0.18228	2785.2	3040.4	6.9534
350	0.2825	2875.2	3157.7	7.3011	0.2345	2872.2	3153.6	7.2121	0.2003	2869.2	3149.5	7.1360
400	0.3066	2957.3	3263.9	7.4651	0.2548	2954.9	3260.7	7.3774	0.2178	2952.5	3257.5	7.3026
500	0.3541	3124.4	3478.5	7.7622	0.2946	3122.8	3476.3	7.6759	0.2521	3121.1	3474.1	7.6027
600	0.4011	3296.8	3697.9	8.0290	0.3339	3295.6	3696.3	7.9435	0.2860	3294.4	3694.8	7.8710
700	0.4478	3475.3	3923.1	8.2731	0.3729	3474.4	3922.0	8.1881	0.3195	3473.6	3920.8	8.1160
800	0.4943	3660.4	4154.7	8.4996	0.4118	3659.7	4153.8	8.4148	0.3528	3659.0	4153.0	8.3431
900	0.5407	3852.2	4392.9	8.7118	0.4505	3851.6	4392.2	8.6272	0.3861	3851.1	4391.5	8.5556
1000	0.5871	4050.5	4637.6	8.9119	0.4892	4050.0	4637.0	8.8274	0.4192	4049.5	4636.4	8.7559
1100	0.6335	4255.1	4888.6	9.1017	0.5278	4254.6	4888.0	9.0172	0.4524	4254.1	4887.5	8.9457
1200	0.6798	4465.6	5145.4	9.2822	0.5665	4465.1	5144.9	9.1977	0.4855	4464.7	5144.4	9.1262
1300	0.7261	4681.3	5407.4	9.4543	0.6051	4680.9	5407.0	9.3698	0.5186	4680.4	5406.5	9.2984

Table 4 Superheated steam table

T °C	v m³/kg	u kJ/kg	h kJ/kg	s kJ/(kg·K)	v m³/kg	u kJ/kg	h kJ/kg	s kJ/(kg·K)	v m³/kg	u kJ/kg	h kJ/kg	s kJ/(kg·K)
	P = 1.60 MPa (201.41 °C)				*P = 1.80 MPa (207.15 °C)*				*p = 2.00 MPa (212.42 °C)*			
Sat	0.12380	2596.0	2794.0	6.4218	0.11042	2598.4	2797.1	6.3794	0.09963	2600.3	2799.5	6.3409
225	0.13287	2644.7	2857.3	6.5518	0.11673	2636.6	2846.7	6.4808	0.10377	2628.3	2835.8	6.4147
250	0.14184	2692.3	2919.2	6.6732	0.12497	2686.0	2911.0	6.6066	0.11144	2679.6	2902.5	6.5453
300	0.15862	2781.1	3034.8	6.8844	0.14021	2776.9	3029.2	6.8226	0.12547	2772.6	3023.5	6.7664
350	0.17456	2866.1	3145.4	7.0694	0.15457	2863.0	3141.2	7.0100	0.13857	2859.8	3137.0	6.9563
400	0.19005	2950.1	3254.2	7.2374	0.16847	2947.7	3250.9	7.1794	0.15120	2945.2	3247.6	7.1271
500	0.2203	3119.5	3472.0	7.5390	0.19550	3117.9	3469.8	7.4825	0.17568	3116.2	3467.6	7.4317
600	0.2500	3293.3	3693.2	7.8080	0.2220	3292.1	3691.7	7.7523	0.19960	3290.9	3690.1	7.7024
700	0.2794	3472.7	3919.7	8.0535	0.2482	3471.8	3971.8	7.9983	0.2232	3470.9	3917.4	7.9487
800	0.3086	3658.3	4152.1	8.2808	0.2742	3657.6	4151.2	8.2258	0.2467	3657.0	4150.3	8.1765
900	0.3377	3850.5	4390.8	8.4935	0.3001	3849.9	4390.1	8.4386	0.2700	3849.3	4389.4	8.3895
1000	0.3668	4049.0	4635.8	8.6938	0.3260	4048.5	4635.2	8.6391	0.2933	4048.0	4634.6	8.5901
1100	0.3958	4253.7	4887.0	8.8837	0.3518	4253.2	4886.4	8.8290	0.3166	4252.7	4885.9	8.7800
1200	0.4248	4464.2	5143.9	9.0643	0.3776	4463.7	5143.4	9.0096	0.3398	4463.3	5142.9	8.9607
1300	0.4538	4679.9	5406.0	9.2364	0.4034	4679.5	5405.6	9.1818	0.3631	4679.0	5405.1	9.1329

Table 4 Superheated steam table

T °C	v m³/kg	u kJ/kg	h kJ/kg	s kJ/(kg·K)	v m³/kg	u kJ/kg	h kJ/kg	s kJ/(kg·K)	v m³/kg	u kJ/kg	h kJ/kg	s kJ/(kg·K)
	P = 2.50 MPa (223.99 °C)				P = 3.00 MPa (233.90 °C)				P = 3.50 MPa (242.60 °C)			
Sat.	0.07998	2603.1	2803.1	6.2575	0.06668	2604.1	2804.2	6.1869	0.05707	2603.7	2803.4	6.1253
225	0.08027	2605.6	2806.3	6.2639								
250	0.08700	2662.6	2880.1	6.4085	0.07058	2644.0	2855.8	6.2872	0.05872	2623.7	2829.2	6.1749
300	0.09890	2761.6	3008.8	6.6438	0.08114	2750.1	2993.5	6.5390	0.06842	2738.0	2977.5	6.4461
350	0.10976	2851.9	3126.3	6.8403	0.09053	2843.7	3115.3	6.7428	0.07678	2835.3	3104.0	6.6579
400	0.12010	2939.1	3239.3	7.0148	0.09936	2932.8	3230.9	6.9212	0.08453	2926.4	3222.3	6.8405
450	0.13014	3025.5	3350.8	7.1746	0.10787	3020.4	3344.0	7.0834	0.09196	3015.3	3337.2	7.0052
500	0.13993	3112.1	3462.1	7.3234	0.11619	3108.0	3456.5	7.2338	0.09918	3103.0	3450.9	7.1572
600	0.15930	3288.0	3686.3	7.5960	0.13243	3285.0	3682.3	7.5085	0.11324	3282.1	3678.4	7.4339
700	0.17832	3468.7	3914.5	7.8435	0.14838	3466.5	3911.7	7.7571	0.12699	3464.3	3908.8	7.6837
800	0.19716	3655.3	4148.2	8.0720	0.16414	3653.5	4145.9	7.9862	0.14056	3651.8	4143.7	7.9134
900	0.21590	3847.9	4387.6	8.2853	0.17980	3846.5	4385.9	8.1999	0.15402	3845.0	4384.1	8.1276
1000	0.2346	4046.7	4633.1	8.4861	0.19541	4045.4	4631.6	8.4009	0.16743	4044.1	4630.1	8.3288
1100	0.2532	4251.5	4884.6	8.6762	0.21098	4250.3	4883.3	8.5912	0.18080	4249.2	4881.9	8.5192
1200	0.2718	4462.1	5141.7	8.8569	0.22652	4460.9	5140.5	8.7720	0.19415	4459.8	5139.3	8.7000
1300	0.2905	4677.8	5404.0	9.0291	0.24206	4676.6	5402.8	8.9442	0.20749	4675.5	5401.7	8.8723

Table 4 Superheated steam table

T °C	v m³/kg	u kJ/kg	h kJ/kg	s kJ/(kg·K)	v m³/kg	u kJ/kg	h kJ/kg	s kJ/(kg·K)	v m³/kg	u kJ/kg	h kJ/kg	s kJ/(kg·K)
	P = 4.0 MPa (250.40 °C)				P = 4.5 MPa (257.49 °C)				P = 5.0 MPa (263.99 °C)			
Sat.	0.04978	2602.3	2801.4	6.0701	0.04406	2600.1	2798.3	6.0198	0.03944	2597.1	2794.3	5.9734
275	0.05457	2667.9	2886.2	6.2285	0.04730	2650.3	2863.2	6.1401	0.04141	2631.3	2838.3	6.0544
300	0.05884	2725.3	2960.7	6.3615	0.05135	2712.0	2943.1	6.2828	0.04532	2698.0	2924.5	6.2084
350	0.06645	2826.7	3092.5	6.5821	0.05840	2817.8	3080.6	6.5131	0.05194	2808.7	3068.4	6.4493
400	0.07341	2919.9	3213.6	6.7690	0.06475	2913.3	3204.7	6.7047	0.05781	2906.6	3195.7	6.6459
450	0.08002	3010.2	3330.3	6.9363	0.07074	3005.0	3323.3	6.8746	0.06330	2999.7	3316.2	6.8186
500	0.08643	3099.5	3445.3	7.0901	0.07651	3095.3	3439.6	7.0301	0.06857	3091.0	3433.8	6.9759
600	0.09885	3279.1	3674.4	7.3688	0.08765	3276.0	3670.5	7.3110	0.07869	3273.0	3666.5	7.2589
700	0.11095	3462.1	3905.9	7.6198	0.09847	3459.9	3903.0	7.5631	0.08849	3457.6	3900.1	7.5122
800	0.12287	3650.0	4141.5	7.8502	0.10911	3648.3	4139.3	7.7942	0.09811	3646.6	4137.1	7.7440
900	0.13469	3843.6	4382.3	8.0647	0.11965	3842.2	4380.6	8.0091	0.10762	3840.7	4378.8	7.9593
1000	0.14645	4042.9	4628.7	8.2662	0.13013	4041.6	4627.2	8.2108	0.11707	4040.4	4625.7	8.1612
1100	0.15817	4248.0	4880.6	8.4567	0.14056	4246.8	4879.3	8.4015	0.12648	4245.6	4878.0	8.3520
1200	0.16987	4458.6	5138.1	8.6376	0.15098	4457.5	5136.9	8.5825	0.13587	4456.3	5135.7	8.5331
1300	0.18156	4674.3	5400.5	8.8100	0.16139	4673.1	5399.4	8.7549	0.14526	4672.0	5398.2	8.7055

Table 4 Superheated steam table

T °C	v m³/kg	u kJ/kg	h kJ/kg	s kJ/(kg·K)	v m³/kg	u kJ/kg	h kJ/kg	s kJ/(kg·K)	v m³/kg	u kJ/kg	h kJ/kg	s kJ/(kg·K)
	P = 6.0 MPa (275.64 °C)				P = 7.0 MPa (285.88 °C)				P = 8.0 MPa (295.06 °C)			
Sat.	0.03244	2589.7	2784.3	5.8892	0.02737	2580.5	2772.1	5.8133	0.02352	2569.0	2758.0	5.7432
300	0.03616	2667.2	2884.2	6.0674	0.02947	2632.2	2838.4	5.9305	0.02426	2590.9	2785.0	5.7906
350	0.04223	2789.6	3043.0	6.3335	0.03524	2769.4	3016.0	6.2283	0.02995	2747.7	2987.3	6.1301
400	0.04739	2892.9	3177.2	6.5408	0.03993	2878.6	3158.1	6.4478	0.03432	2863.8	3138.3	6.3634
450	0.05214	2988.9	3301.8	6.7193	0.04416	2978.0	3287.1	6.6327	0.03817	2966.7	3272.0	6.5551
500	0.05665	3082.2	3422.2	6.8803	0.04814	3073.4	3410.3	6.7975	0.04175	3064.3	3398.3	6.7240
550	0.06101	3174.6	3540.6	7.0288	0.05195	3167.2	3530.9	6.9486	0.04516	3159.8	3521.0	6.8778
600	0.06525	3266.9	3658.4	7.1677	0.05565	3260.7	3650.3	7.0894	0.04845	3254.4	3642.0	7.0206
700	0.07352	3453.1	3894.2	7.4234	0.06283	3448.5	3888.3	7.3476	0.05481	3443.9	3882.4	7.2812
800	0.08160	3643.1	4132.7	7.6566	0.06981	3639.5	4128.2	7.5822	0.06097	3636.0	4123.8	7.5173
900	0.08958	3837.8	4375.3	7.8727	0.07669	3835.0	4371.8	7.7991	0.06702	3832.1	4368.3	7.7351
1000	0.09749	4037.8	4622.7	8.0751	0.08350	4035.3	4619.8	8.0020	0.07301	4032.8	4616.9	7.9384
1100	0.10536	4243.3	4875.4	8.2661	0.09027	4240.9	4872.8	8.1933	0.07896	4238.6	4870.3	8.1300
1200	0.11321	4454.0	5133.3	8.4474	0.09703	4451.7	5130.9	8.3747	0.08489	4449.5	5128.5	8.3115
1300	0.12106	4669.6	5396.0	8.6199	0.10377	4667.3	5393.7	8.5475	0.09080	4665.0	5391.5	8.4842

Table 4 Superheated steam table

T °C	v m³/kg	u kJ/kg	h kJ/kg	s kJ/(kg·K)	v m³/kg	u kJ/kg	h kJ/kg	s kJ/(kg·K)	v m³/kg	u kJ/kg	h kJ/kg	s kJ/(kg·K)
	P = 9.0 MPa (303.40 °C)				P = 10.0 MPa (311.06 °C)				P = 12.5 MPa (327.89 °C)			
Sat.	0.02048	2557.8	2742.1	5.6772	0.018026	2544.4	2724.7	5.6141	0.013495	2505.1	2673.8	5.4624
325	0.02327	2646.6	2856.0	5.8712	0.019861	2610.4	2809.1	5.7568				
350	0.02580	2724.4	2956.6	6.0361	0.02242	2699.2	2923.4	5.9443	0.016126	2624.6	2826.2	5.7118
400	0.02993	2848.4	3117.8	6.2854	0.02641	2832.4	3096.5	6.2120	0.02000	2789.3	3039.3	6.0417
450	0.03350	2955.2	3256.6	6.4844	0.02975	2943.4	3240.9	6.4190	0.02299	2912.5	3199.8	6.2719
500	0.03677	3055.2	3386.1	6.6576	0.03279	3045.8	3373.7	6.5966	0.02560	3021.7	3341.8	6.4618
550	0.03987	3152.2	3511.0	6.8142	0.03564	3144.1	3500.9	6.7561	0.02801	3125.0	3475.2	6.6290
600	0.04285	3248.1	3633.7	6.9589	0.03837	3241.7	3625.3	6.9029	0.03029	3225.4	3604.0	6.7810
650	0.04574	3343.6	3755.3	7.0943	0.04101	3338.2	3748.2	7.0398	0.03248	3324.4	3730.4	6.9218
700	0.04857	3439.3	3876.5	7.2221	0.04358	3434.7	3870.5	7.1687	0.03460	3422.9	3855.3	7.0536
800	0.05409	3632.5	4119.3	7.4596	0.04859	3628.9	4114.8	7.4077	0.03869	3620.0	4103.6	7.2965
900	0.05950	3829.2	4364.8	7.6783	0.05349	3826.3	4361.2	7.6272	0.04267	3819.1	4352.5	7.5182
1000	0.06485	4030.3	4614.0	7.8821	0.05832	4027.8	4611.0	7.8315	0.04658	4021.6	4603.8	7.7237
1100	0.07016	4236.3	4867.7	8.0740	0.06312	4234.0	4865.1	8.0237	0.05045	4228.2	4858.8	7.9165
1200	0.07544	4447.2	5126.2	8.2556	0.06789	4444.9	5123.8	8.2055	0.05430	4439.3	5118.0	8.0937
1300	0.08072	4662.7	5389.2	8.4284	0.07265	4460.5	5387.0	8.3783	0.05813	4654.8	5381.4	8.2717

Table 4 Superheated steam table

T °C	v m³/kg	u kJ/kg	h kJ/kg	s kJ/(kg·K)	v m³/kg	u kJ/kg	h kJ/kg	s kJ/(kg·K)	v m³/kg	u kJ/kg	h kJ/kg	s kJ/(kg·K)
	P = 15.0 MPa (342.24 °C)				P = 17.5 MPa (354.75 °C)				P = 20.0 MPa (365.81 °C)			
Sat.	0.010337	2455.5	2610.5	5.3098	0.007920	2390.2	2528.8	5.1419	0.005834	2293.0	2409.7	4.9269
350	0.011470	2520.4	2692.4	5.4421								
400	0.015649	2740.7	2975.5	5.8811	0.012447	2685.0	2902.9	5.7213	0.009942	2619.3	2818.1	5.5540
450	0.018445	2879.5	3156.2	6.1404	0.015174	2844.2	3109.7	6.0184	0.012695	2806.2	3060.1	5.9017
500	0.02080	2996.6	3308.6	6.3443	0.017358	2970.3	3274.1	6.2383	0.014768	2942.9	3238.2	6.1401
550	0.02293	3104.7	3448.6	6.5199	0.019288	3083.9	3421.4	6.4230	0.016555	3062.4	3393.5	6.3348
600	0.02491	3208.6	3582.3	6.6776	0.02106	3191.5	3560.1	6.5866	0.018178	3174.0	3537.6	6.5048
650	0.02680	3310.3	3712.3	6.8224	0.02274	3296.0	3693.9	6.7357	0.019693	3281.4	3675.3	6.6582
700	0.02861	3410.9	3840.1	6.9572	0.02434	3324.6	398.7	6.8736	0.02113	3386.4	3809.0	6.7993
800	0.03210	3610.9	4092.4	7.2040	0.02738	3601.8	4081.1	7.1244	0.02385	3592.7	4069.7	7.0544
900	0.03546	3811.9	4343.8	7.4279	0.03031	3804.7	4335.1	7.3507	0.02645	3797.5	4326.4	7.2830
1000	0.03875	4015.4	4596.6	7.6348	0.03316	4009.3	4589.5	7.5589	0.02897	4003.1	4582.5	7.4925
1100	0.04200	4222.6	4852.6	7.8283	0.03597	4216.9	4846.4	7.7531	0.03145	4211.3	4840.2	7.6874
1200	0.04523	4433.8	5112.3	8.0108	0.03876	4428.3	5106.6	7.9360	0.03391	4422.8	5101.0	7.8707
1300	0.04845	4649.1	5376.0	8.1840	0.04154	4643.5	5370.5	8.1093	0.03636	4638.0	5365.1	8.0442

Table 4 Superheated steam table

T °C	v m³/kg	u kJ/kg	h kJ/kg	s kJ/(kg·K)	v m³/kg	u kJ/kg	h kJ/kg	s kJ/(kg·K)	v m³/kg	u kJ/kg	h kJ/kg	s kJ/(kg·K)
	P = 25.0 MPa				P = 30.0 MPa				P = 35.0 MPa			
375	0.0019731	1798.7	1848.0	4.0320	0.0017892	1737.8	1791.5	3.9305	0.0017003	1702.9	1762.4	3.8722
400	0.006004	2430.1	2580.2	5.1418	0.002790	2067.4	2151.1	4.4728	0.002100	1914.1	1987.6	4.2126
425	0.007881	2609.2	2806.3	5.4723	0.005303	2455.1	2614.2	5.1504	0.003428	2253.4	2373.4	4.7747
450	0.009162	2720.7	2949.7	5.6744	0.006735	2619.3	2821.4	5.4424	0.004961	2498.7	2672.4	5.1962
500	0.011123	2884.3	3162.4	5.9592	0.008678	2820.7	3081.1	5.7905	0.006927	2751.9	2994.4	5.6282
550	0.012724	3017.5	3335.6	6.1765	0.010168	2970.3	3275.4	6.0342	0.008345	2921.0	3213.0	5.9026
600	0.014137	3137.9	3491.4	6.3602	0.011446	3100.5	3443.9	6.2331	0.009527	3062.0	3395.5	6.1179
650	0.015433	3251.6	3637.4	6.5229	0.012596	3221.0	3598.9	6.4058	0.010575	3189.8	3559.9	6.3010
700	0.016646	3361.3	3777.5	6.6707	0.013661	3335.8	3745.6	6.5606	0.011533	3309.8	3713.5	6.4631
800	0.018912	3574.3	4047.1	6.9345	0.015623	3555.5	4024.2	6.8332	0.013278	3536.7	4001.5	6.7450
900	0.021045	3783.0	4309.1	7.1680	0.017448	3768.5	4291.9	7.0718	0.014883	3754.0	4274.9	6.9386
1000	0.02310	3990.9	4568.5	7.3802	0.019196	3978.8	4554.7	7.2867	0.016410	3966.7	4541.1	7.2064
1100	0.02512	4200.2	4828.2	7.5765	0.020903	4189.2	4816.3	7.4845	0.017895	4178.3	4804.6	7.4037
1200	0.02711	4412.0	5089.9	7.7605	0.022589	4401.3	5079.0	7.6692	0.019360	4390.7	5068.3	7.5910
1300	0.02910	4626.9	5354.4	7.9342	0.024266	4616.0	5344.0	7.8432	0.020815	4605.1	5333.6	7.7653

Table 4 Superheated steam table

T °C	v m³/kg	u kJ/kg	h kJ/kg	s kJ/(kg·K)	v m³/kg	u kJ/kg	h kJ/kg	s kJ/(kg·K)	v m³/kg	u kJ/kg	h kJ/kg	s kJ/(kg·K)
	P = 40.0 MPa				P = 50.0 MPa				P = 60.0 MPa			
375	0.0016407	1677.1	1742.8	3.8290	0.0015594	1638.6	1716.6	3.7639	0.0015028	1609.4	1699.5	3.7141
400	0.0019077	1854.6	1930.9	4.1135	0.0017309	1788.1	1874.6	4.0031	0.0016335	1745.4	1843.4	3.9318
425	0.002532	2096.9	2198.1	4.5029	0.002007	1959.7	2060.0	4.2734	0.0018165	1892.7	2001.7	4.1626
450	0.003693	2365.1	2512.8	4.9459	0.002486	2159.6	2284.0	4.5884	0.002085	2053.9	2179.0	4.4121
500	0.005622	2678.4	2903.3	5.4700	0.003892	2525.5	2720.1	5.1726	0.002956	2390.6	2567.9	4.9321
550	0.006984	2869.7	3149.1	5.7785	0.005118	2763.6	3019.5	5.5485	0.003956	2658.8	2896.2	5.3441
600	0.008094	3022.6	3346.4	6.0144	0.006112	2942.0	3247.6	5.8178	0.004834	2861.1	3151.2	5.6452
650	0.009063	3158.0	3520.6	6.2054	0.006966	3093.5	3441.8	6.0342	0.005595	3028.8	3364.5	5.8829
700	0.009941	3283.6	3681.2	6.3750	0.007727	3230.5	3616.8	6.2189	0.006272	3177.2	3553.5	6.0824
800	0.011523	3517.8	3978.7	6.6662	0.009076	3479.8	3933.6	6.5290	0.007459	3441.5	3889.1	6.4109
900	0.012962	3739.4	4257.9	6.9150	0.010283	3710.3	4224.4	6.7882	0.008508	3681.0	4191.5	6.6805
1000	0.014324	3954.6	4527.6	7.1356	0.011411	3930.5	4501.1	7.0146	0.009480	3906.4	4475.2	6.9121
1100	0.015642	4167.4	4793.1	7.3364	0.012496	4145.7	4770.5	7.2184	0.010409	4124.1	4748.6	7.1195
1200	0.016940	4380.1	5057.7	7.5224	0.013561	4359.1	5037.2	7.4058	0.011317	4338.2	5017.2	7.3083
1300	0.018229	4594.3	5323.5	7.6969	0.014616	4572.8	5303.6	7.5808	0.012215	4551.4	5284.3	7.4837

Table 5 Compressed liquid water table

T °C	v m³/kg	u kJ/kg	h kJ/kg	s kJ/(kg·K)	v m³/kg	u kJ/kg	h kJ/kg	s kJ/(kg·K)	v m³/kg	u kJ/kg	h kJ/kg	s kJ/(kg·K)
	P = 5 MPa (263.99 °C)				P = 10 MPa (311.06 °C)				P = 15 MPa (342.24 °C)			
Sat.	0.0012859	1147.8	1154.2	2.9202	0.0014524	1393.0	1407.6	3.3596	0.0016581	1585.6	1610.5	3.6848
0	0.0009977	0.04	5.04	0.0001	0.0009952	0.09	10.04	0.0002	0.0009928	0.15	15.05	0.0004
20	0.0009995	83.65	83.65	0.2956	0.0009972	83.36	93.33	0.2945	0.0009950	83.06	97.99	0.2934
40	0.0010056	166.95	171.97	0.5705	0.0010034	166.35	176.38	0.5686	0.0010013	165.76	180.78	0.5666
60	0.0010149	250.23	255.30	0.8285	0.0010127	249.36	259.49	0.8258	0.0010105	248.51	263.67	0.8232
80	0.0010268	333.72	338.85	1.0720	0.0010245	332.59	342.83	1.0688	0.0010222	331.48	346.81	1.0656
100	0.0010410	417.52	422.72	1.3030	0.0010385	416.12	426.50	1.2992	0.0010361	414.74	430.28	1.2955
120	0.0010576	501.80	507.09	1.5233	0.0010549	500.08	510.64	1.5189	0.0010522	498.40	514.19	1.5145
140	0.0010768	586.76	592.15	1.7343	0.0010737	584.68	595.42	1.7292	0.0010707	582.66	598.72	1.7242
160	0.0010988	672.62	678.12	1.9375	0.0010953	670.13	681.08	1.9317	0.0010918	667.71	684.09	1.9260
180	0.0011240	759.63	765.25	2.1341	0.0011199	756.65	767.84	2.1275	0.0011159	753.76	770.50	2.1210
200	0.0011530	848.1	853.9	2.3255	0.0011480	844.5	856.0	2.3178	0.0011433	841.0	858.2	2.3104
220	0.0011866	938.4	944.4	2.5128	0.0011805	934.1	945.9	2.5039	0.0011748	929.9	947.5	2.4953
240	0.0012264	1031.4	1037.5	2.6979	0.0012187	1026.0	1038.1	2.6872	0.0012114	1020.8	1039.0	2.6771
260	0.0012749	1127.9	1134.3	2.8830	0.0012645	1121.1	1133.7	2.8699	0.0012550	1114.6	1133.4	2.8576
280					0.0013216	1220.9	1234.1	3.0548	0.0013084	1212.5	1232.1	3.0393
300					0.0013972	1328.4	1342.3	3.2469	0.0013770	1316.6	1337.3	3.2260
320									0.0014724	1431.1	1453.2	3.4247
340									0.0016311	1567.5	1591.9	3.6546

Table 5 Compressed liquid water table

T °C	v m³/kg	u kJ/kg	h kJ/kg	s kJ/(kg·K)	v m³/kg	u kJ/kg	h kJ/kg	s kJ/(kg·K)	v m³/kg	u kJ/kg	h kJ/kg	s kJ/(kg·K)
	P = 20.0 MPa (365.81 °C)				P = 30 MPa				P = 50 MPa			
Sat.	0.002036	1785.6	1826.3	4.0139								
0	0.0009904	0.19	20.01	0.0004	0.0009856	0.25	29.82	0.0001	0.0009766	0.20	49.03	0.0014
20	0.0009928	82.77	102.62	0.2923	0.0009886	82.17	111.84	0.2899	0.0009804	81.00	130.02	0.2848
40	0.0009992	165.17	185.16	0.5646	0.0009951	164.04	193.89	0.5607	0.0009872	161.86	211.21	0.5527
60	0.0010084	247.68	267.85	0.8206	0.0010042	246.06	276.19	0.8154	0.0009962	242.98	292.79	0.8052
80	0.0010199	330.40	350.80	1.0624	0.0010156	328.30	358.77	1.0561	0.0010073	324.34	374.70	1.0440
100	0.0010337	413.39	434.06	1.2917	0.0010290	410.78	441.66	1.2844	0.0010201	405.88	456.89	1.2703
120	0.0010496	496.76	517.76	1.5102	0.0010445	493.59	524.93	1.5018	0.0010348	487.65	539.39	1.4857
140	0.0010678	580.69	602.04	1.7193	0.0010621	576.88	608.75	1.7098	0.0010515	569.77	622.35	1.6915
160	0.0010885	665.35	687.12	1.9204	0.0010821	660.82	693.28	1.9096	0.0010703	652.41	705.92	1.8891
180	0.0011120	750.95	773.20	2.1147	0.0011047	745.59	778.73	2.1024	0.0010912	735.69	790.25	2.0794
200	0.0011388	837.7	860.5	2.3031	0.0011302	831.4	865.3	2.2893	0.0011146	819.7	875.5	2.2634
220	0.0011695	925.9	949.3	2.4870	0.0011590	918.3	953.1	2.4711	0.0011408	904.7	961.7	2.4419
240	0.0012046	1016.0	1040.0	2.6674	0.0011920	1006.9	1042.6	2.6490	0.0011702	990.7	1049.2	2.6158
260	0.0012462	1108.6	1133.5	2.8459	0.0012303	1097.4	1134.3	2.8243	0.0012034	1078.1	1138.2	2.7860
280	0.0012965	1204.7	1230.6	3.0248	0.0012755	1190.7	1229.0	2.9986	0.0012415	1167.2	1229.3	2.9537
300	0.0013596	1306.1	1333.3	3.2071	0.0013304	1287.9	1327.8	3.1741	0.0012860	1258.7	1323.0	3.1200
320	0.0014437	1415.7	1444.6	3.3979	0.0013997	1390.7	1432.7	3.3539	0.0013388	1353.3	1420.2	3.2868
340	0.0015684	1539.7	1571.0	3.6075	0.0014920	1501.7	1546.5	3.5426	0.0014032	1452.0	1522.1	3.4557
360	0.0018226	1702.8	1739.3	3.8772	0.0016265	1626.6	1675.4	3.7494	0.0014838	1556.0	1630.2	3.6291
380					0.0018691	1781.4	1837.5	4.0012	0.0015884	1667.2	1746.6	3.8101

Table 6 Saturated ice-steam (temperature) table

Temp. °C	Sat. Press. kPa	Specific volume m³/kg Sat. ice	Sat. vapour	Internal energy kJ/kg Sat. ice	Subl.	Sat. vapour	Enthalpy kJ/kg Sat. ice	Subl.	Sat. vapour	Entropy kJ/(kg.K) Sat. ice	Subl.	Sat. vapour
T	T_{sat}	$v_i \times 10^3$	v_g	u_i	u_{ig}	u_g	h_i	h_{ig}	h_g	s_i	s_{ig}	s_g
0.01	0.6113	1.0908	206.1	–333.40	2708.7	2375.3	–333.40	2834.8	2501.4	–1.221	10.378	9.156
0	0.6108	1.0908	206.3	–333.43	2708.8	2375.3	–333.43	2834.8	2501.3	–1.221	10.378	9.157
–2	0.5176	1.0904	241.7	–337.62	2710.2	2372.6	–337.62	2835.3	2497.7	–1.237	10.456	9.219
–4	0.4375	1.0901	283.8	–341.78	2711.6	2369.8	–341.78	2835.7	2494.0	–1.253	10.536	9.283
–6	0.3689	1.0898	334.2	–345.91	2712.9	2367.0	–345.91	2836.2	2490.3	–1.268	10.616	9.348
–8	0.3102	1.0894	394.4	–350.02	2714.2	2364.2	–350.02	2836.6	2486.6	–1.284	10.698	9.414
–10	0.2602	1.0891	466.7	–354.09	2715.5	2361.4	–354.09	2837.0	2482.9	–1.299	10.781	9.481
–12	0.2176	1.0888	553.7	–358.14	2716.8	2358.7	–358.14	2837.3	2479.2	–1.315	10.865	9.550
–14	0.1815	1.0884	658.8	–362.15	2718.0	2355.9	–362.15	2837.6	2475.5	–1.331	10.950	9.619
–16	0.1510	1.0881	786.0	–366.14	2719.2	2353.1	–366.14	2837.9	2471.8	–1.346	11.036	9.690
–18	0.1252	1.0878	940.5	–370.10	2720.4	2350.3	–370.10	2838.2	2468.1	–1.362	11.123	9.762
–20	0.1035	1.0874	1128.6	–374.03	2721.6	2347.5	–374.03	2838.4	2464.3	–1.377	11.212	9.835
–22	0.0853	1.0871	1358.4	–377.93	2722.7	2344.7	–377.93	2838.6	2460.6	–1.393	11.302	9.909
–24	0.0701	1.0868	1640.1	–381.80	2723.7	2342.0	–381.80	2838.7	2456.9	–1.408	11.394	9.985
–26	0.0574	1.0864	1986.4	–385.64	2724.8	2339.2	–385.64	2838.9	2453.2	–1.424	11.486	10.062

Contd.

Temp. °C	Sat. Press. kPa	Specific volume m³/kg		Internal energy kJ/kg			Enthalpy kJ/kg			Entropy kJ/(kg.K)		
		Sat. ice	Sat. vapour	Sat. ice	Subl.	Sat. vapour	Sat. ice	Subl.	Sat. vapour	Sat. ice	Subl.	Sat. vapour
T	T_{sat}	$v_i \times 10^3$	v_g	u_i	u_{ig}	u_g	h_i	h_{ig}	h_g	s_i	s_{ig}	s_g
–28	0.0469	1.0861	2413.7	–389.45	2725.8	2336.4	–389.45	2839.0	2449.5	–1.439	11.580	10.141
–30	0.0381	1.0858	2943	–393.23	2726.8	2333.6	–393.23	2839.0	2445.8	–1.455	11.676	10.221
–32	0.0309	1.0854	3600	–396.98	2727.8	2330.8	–396.98	2839.1	2442.1	–1.471	11.773	10.303
–34	0.0250	1.0851	4419	–400.71	2728.7	2328.0	–400.71	2839.1	2438.4	–1.486	11.872	10.386
–36	0.0201	1.0848	5444	–404.40	2729.6	2325.2	–404.40	2839.1	2434.7	–1.501	11.972	10.470
–38	0.0161	1.0844	6731	–408.06	2730.5	2322.4	–408.06	2839.0	2430.9	–1.517	12.073	10.556
–40	0.0129	1.0841	8354	–411.70	2731.3	2319.6	–411.70	2839.9	2427.2	–1.532	12.176	10.644

Table 7 Critical Point Data For Some Substances

Substance	*Formula*	*Molecular Weight*	*Temperature* T_c.*K*	*Pressure* p_c. *MPa*	*Volume* *m³/kmol*
Ammonia	NH_3	17.03	405.5	11.28	.0724
Argon	Ar	39.948	151	4.86	.0749
Bromine	Br_2	159.808	584	10.34	.1355
Carbon Dioxide	CO_2	44.01	304.2	7.39	.0943
Carbon Monoxide	CO	28.001	133	3.50	0.930
Chlorine	Cl_2	70.906	417	7.71	.1242
Deuterium	D_2	4.00	38.4	1.66	
Helium	He	4.003	5.3	0.23	0.578
Helium^3	He	3.00	3.3	0.12	—
Hydrogen	H_2	2.016	33.3	1.30	.0649
Krypton	Kr	83.80	209.4	5.50	.0924
Neon	Ne	20.183	44.5	2.73	.0417
Nitrogen	N_2	28.013	126.2	3.39	.0899
Nitrous Oxide	N_2O	44.013	309.7	7.27	.0961
Oxygen	O_2	31.999	154.8	5.08	.0780
Sulfur Dioxide	SO_2	64.063	430.7	7.88	.1217
Water	H_2O	18.015	647.3	22.09	.0568
Xenon	Xe	131.30	289.8	5.88	.1186
Benzene	X_2H_6	78.115	562	4.92	.2603
n-Butane	C_4H_{10}	58.124	425.5	3.80	.2547
Carbon Tetrachloride	CCl_4	153.82	556.4	4.56	.2759
Chloroform	$CHCl_2$	119.38	536.6	5.47	.2403
Dichlorodifluoromethane	CCl_2F_2	120.91	384.7	4.01	.2179
Dichlorofluoromethane	$CHCl_2F$	102.92	451.7	5.17	.1973
Ethane	C_2H_6	30.070	305.5	4.88	.1480
Ethyl Alcohol	C_2H_5OH	46.07	516	6.38	.1673
Ethylene	C_2H_{14}	28.054	282.4	5.12	.1242
n-Hexane	C_6H_4	86.178	507.9	3.03	.3677
Methane	CH_4	16.043	191.1	4.64	.993
Methyl Alcohol	CH_3OH	32.042	513.2	7.95	.1180
Methyl Chloride	CH_3Cl	50.488	416.3	6.68	.1430
Propane	C_3H_8	44.097	370	4.26	.1998
Propene	C_3H_6	42.081	365	4.62	.1810
Propyne	C_3H_4	40.065	401	5.35	—
Trichlorofluoromethane	CCl_3F	137.37	471.2	4.38	.2478

Table 8 Saturated Ammonia Table

		Specific volume m^3/kg			*Enthalpy* kJ/kg			*Entropy* $kJ/(kg.K)$		
Te-mp. °C	*Abs. Press. kpa* P	*Sat. Liquid* v_f	*Evap.* v_{fg}	*Sat. Vapor* v_g	*Sat. Liquid* h_f	*Evap.* h_{fg}	*Sat. Vapor* h_g	*Sat. Liquid* s_f	*Evap.* s_{fg}	*Sat. Vapor* s_g
–50	40.88	0.001 424	2.6239	2.6254	–44.3	1416.7	1372.4	–0.1942	6.3502	6.1561
–48	45.96	0.001 429	2.3518	2.3533	–35.5	1411.3	1375.8	–0.1547	6.2696	6.1149
–46	51.55	0.001 434	2.1126	2.1124	–26.6	1405.8	1379.2	–0.1156	6.1902	6.0746
–44	57.69	0.001 439	1.9018	1.9032	–17.8	1400.3	1382.5	–0.0768	6.1120	6.0352
–42	64.42	0.001 444	1.7155	1.7170	–8.9	1394.7	1385.8	–0.0382	6.0349	5.9967
–40	71.77	0.001 449	1.5506	1.5521	0.0	1389.0	1389.0	0.0000	5.9589	5.9589
–38	79.80	0.001 454	1.4043	1.4058	8.9	1383.3	1392.2	0.0380	5.8840	5.9220
–36	88.54	0.001 460	1.2742	1.2757	17.8	1377.6	1395.4	0.0757	5.8101	5.8858
–34	98.05	0.001 465	1.1582	1.1597	26.8	1371.8	1398.5	0.1132	5.7372	5.8504
–32	108.37	0.001 470	1.0547	1.0562	35.7	1365.9	1401.6	0.1504	5.6652	5.8156
–30	119.55	0.001 476	0.9621	0.9635	44.7	1360.0	1404.6	0.1873	5.5942	5.7815
–28	131.64	0.001 481	0.8790	0.8805	53.6	1354.0	1407.6	0.2240	5.5241	5.7481
–26	144.70	0.001 487	0.8044	0.8059	62.6	1347.9	1410.5	0.2605	5.4548	5.7153
– 24	158.78	0.001 492	0.7373	0.7388	71.6	1341.8	1413.4	0.2967	5.3864	5.6831
–22	173.93	0.001 498	0.6768	0.6783	80.7	1335.6	1416.2	0.3327	5.3188	5.6515
–20	190.22	0.001 504	0.6222	0.6237	89.7	1329.3	1419.0	0.3684	5.2520	5.6205
–18	207.71	0.001 510	0.5728	0.5743	98.8	1322.9	1421.7	0.4040	5.1860	5.5900
–16	226.45	0.001 515	0.5280	0.5296	107.8	1316.5	1424.4	0.4393	5.1207	5.5600
–14	246.51	0.001 521	0.4874	0.4889	116.9	1310.0	1427.0	0.4744	5.0561	5.5305
–12	267.95	0.001 528	0.4505	0.4520	126.0	1303.5	1429.5	0.5093	4.9922	5.5015
–10	290.85	0.001 534	0.4169	0.4185	135.2	1296.8	1432.0	0.5440	4.9290	5.4730
–8	315.25	0.001 540	0.3863	0.3878	144.3	1290.1	1434.4	0.5785	4.8664	5.4449
–6	341.25	0.001 546	0.3583	0.3599	153.5	1283.3	1436.8	0.6128	4.8045	5.4173
–4	368.90	0.001 553	0.3328	0.3343	162.7	1276.4	1439.1	0.6469	4.7432	5.3901
–2	398.27	0.001 559	0.3094	0.3109	171.9	1269.4	1441.3	0.6808	4.6825	5.3633
0	429.44	0.001 566	0.2879	0.2895	181.1	1262.4	1443.5	0.7145	4.6223	5.3369
2	462.49	0.001 573	0.2683	0.2698	190.4	1255.2	1445.6	0.7481	4.5627	5.3108
4	497.49	0.001 580	0.2502	0.2517	199.6	1248.0	1447.6	0.7815	4.5037	5.2852
6	534.51	0.001 587	0.2335	0.2351	208.9	1240.6	1449.6	0.8148	4.4451	5.2599
8	573.64	0.001 594	0.2182	0.2198	218.3	1233.2	1451.5	0.8479	4.3871	5.2350
10	614.95	0.001 601	0.2040	0.2056	227.6	1225.7	1453.3	0.8808	4.3295	5.2104
12	658.52	0.001 608	0.1910	0.1926	237.0	1218.1	1455.1	0.9136	4.2725	5.1861
14	407.44	0.001 616	0.1789	0.1805	246.4	1210.4	1456.8	0.9463	4.2159	5.1621
16	752.79	0.001 623	0.1677	0.1693	255.9	1202.6	1458.5	0.9788	4.1597	5.1385
18	803.66	0.001 631	0.1574	0.1590	265.4	1194.7	1460.0	1.0112	4.1039	5.1151
20	857.12	0.001 639	0.1477	0.1494	274.9	1186.7	1461.5	1.0434	4.0486	5.0920
22	913.27	0.001 647	0.1388	0.1405	284.4	1178.5	1462.9	1.0755	3.9937	5.0692
24	972.19	0.001 655	0.1305	0.1322	294.0	1170.3	1464.3	1.1075	3.9392	5.0467
26	1033.97	0.001 663	0.1228	0.1245	303.6	1162.0	1465.6	1.1394	3.8850	5.0244
28	1098.71	0.001 671	0.1156	0.1173	313.6	1153.6	1466.8	1.1711	3.8312	5.0023
30	1166.49	0.001 680	1.1089	0.1106	322.9	1145.0	1467.9	1.2028	3.7777	4.9805
32	1237.41	0.001 689	0.1027	0.1044	332.6	1136.4	1469.0	1.2343	3.7246	4.9589
34	1311.55	0.001 698	0.0969	0.0986	342.3	1127.6	1469.9	1.2656	3.6718	4.9374
36	1389.03	0.001 707	0.0914	0.0931	352.1	1118.7	1470.8	1.2969	3.6192	4.9161
38	1469.92	0.001 716	0.0863	0.0880	361.9	1109.7	1471.5	1.3281	3.5669	4.8950
40	1554.33	0.001 726	0.0815	0.0833	371.7	1100.5	1472.2	1.3591	3.5148	4.8740
42	1442.35	0.001 735	0.0771	0.0788	381.6	1091.2	1472.8	1.3901	3.4630	4.8530
44	1734.09	0.001 745	0.0728	0.0746	391.5	1081.7	1473.2	1.4209	3.4112	4.8322
46	1829.65	0.001 756	0.0689	0.0707	401.5	1072.0	1473.5	1.4518	3.3595	4.8113
48	1929.13	0.001 766	0.0652	0.0669	411.5	1062.2	1473.7	1.4826	3.3079	4.7905
50	2032.62	0.001 777	0.0617	0.0635	421.7	1052.0	1473.7	1.5135	3.2561	4.7696

Table 9 Superheated Ammonia Table

Abs. Press. kPa		*Temperature, °C*											
		–20	–10	0	10	20	30	40	50	60	70	80	100
	v	2.4474	–2.5481	2.6482	2.7479	2.8473	2.9464	3.0453	3.1441	3.2427	3.3413	3.4397	
50	*h*	1435.8	1457.0	1478.1	1499.2	1520.4	1541.7	1563.0	1584.5	1606.1	1627.8	1649.7	
	s	6.3256	6.4077	6.4865	6.5625	6.6360	6.7073	6.7766	6.8441	6.9099	6.9743	7.0372	
	v	1.6233	1.6915	1.7591	1.8263	1.8932	1.9597	2.0261	2.0923	2.1584	2.2244	2.2903	
75	*h*	1433.0	1454.7	1476.1	1497.5	1518.9	1540.3	1561.8	1583.4	1605.1	1626.9	1648.9	
	s	6.1190	6.2028	6.2828	6.3597	6.4339	6.5058	6.5756	6.6434	6.7096	6.7742	6.8373	
	v	1.2110	1.2631	1.3145	1.3654	1.4160	1.4664	1.5165	1.5664	1.6163	1.6659	1.7155	1.8145
100	*h*	1430.1	1452.2	1474.1	1495.7	1517.3	1538.9	1560.5	1582.2	1604.1	1626.0	1648.0	1692.6
	s	5.9695	6.0552	6.1366	6.2144	6.2894	6.3618	6.4321	6.5003	6.5668	6.6316	6.6950	6.8177
	v	0.9635	1.0059	1.0476	1.0889	1.1297	1.1703	1.2107	1.2509	1.2909	1.3309	1.3707	1.4501
125	*h*	1427.2	1449.8	1472.0	1493.9	1515.7	1537.5	1559.3	1581.1	1603.0	1625.0	1647.2	1691.8
	s	5.8512	5.9389	6.0217	6.1006	6.1763	6.2494	6.3201	6.3887	6.4555	6.5206	6.5842	6.7072
	v	0.7984	0.8344	0.8697	0.9045	0.9388	0.9729	1.0068	1.0405	1.0740	1.1074	1.1408	1.2072
150	*h*	1424.1	1447.3	1469.8	1492.1	1514.1	1536.1	1558.0	1580.0	1602.0	1624.1	1646.3	1691.1
	s	5.7526	5.8424	5.9266	6.0066	6.0831	6.1568	6.2280	6.2970	6.3641	6.4295	6.4933	6.6167
	v		0.6199	0.6471	0.6738	0.7001	0.7261	0.7519	0.7774	0.8029	0.8282	0.8533	0.9035
200	*h*		1442.0	1465.5	1488.4	1510.9	1533.2	1555.5	1577.7	1599.9	1622.2	1644.6	1689.6
	s		5.6863	5.7737	5.8559	5.9342	6.0091	6.0813	6.1512	6.2189	6.2849	6.3491	6.4732
	v		0.4910	0.5135	0.5354	0.5568	0.5780	0.5989	0.6196	0.6401	0.6605	0.6809	0.7212
250	*h*		1436.6	1461.0	1484.5	1507.6	1530.3	1552.9	1575.4	1597.8	1620.3	1642.8	1688.2
	s		5.5609	5.6517	5.7365	5.8165	5.8928	5.9661	6.0368	6.1052	6.1717	6.2365	6.3613
	v			0.4243	0.4430	0.4613	0.4792	0.4968	0.5143	0.5316	0.5488	0.5658	0.5997
	h			1456.3	1480.6	1504.2	1527.4	1550.3	1573.0	1595.7	1618.4	1641.1	1686.7
300	*s*			5.5493	5.6366	5.7186	5.7963	5.8707	5.9423	6.0114	6.0785	6.1437	6.2693
	v			0.3605	0.3770	0.3929	0.4086	0.4239	0.4391	0.4541	0.4689	0.4837	0.5129
350	*h*			1451.5	1476.5	1500.7	1524.4	1547.6	1570.7	1593.6	1616.5	1639.3	1685.2
	s			5.4600	5.5502	5.6342	5.7135	5.7890	5.8615	5.9314	5.9990	6.0647	6.1910
	v			0.3125	0.3274	0.3417	0.3556	0.3692	0.3826	0.3959	0.4090	0.4220	0.4478
400	*h*			1446.5	1472.4	1497.2	1521.3	1544.9	1568.3	1591.5	1614.5	1637.6	1683.7
	s			5.3803	5.4735	5.5597	5.6405	5.7173	5.7907	5.8613	5.9296	5.9957	6.1228
	v			0.2752	0.2887	0.3017	0.3143	0.3266	0.3387	0.3506	0.3624	0.3740	0.3971
450	*h*			1441.3	1468.1	1493.6	1518.2	1542.2	1565.9	1589.3	1612.6	1635.8	1682.2
	s			5.3078	5.4042	5.4926	5.5752	5.6532	5.7275	5.7989	5.8678	5.9345	6.0623
		20	30	40	50	60	70	80	100	120	140	160	180
	v	0.2698	0.2813	0.2926	0.3036	0.3144	0.3251	0.3357	0.3565	0.3771	0.3975		
500	*h*	1489.9	1515.0	1539.5	1563.4	1587.1	1610.6	1634.0	1680.7	1727.5	1774.7		
	s	5.4314	5.5157	5.5950	5.6704	5.7425	5.8120	5.8793	6.0079	6.1301	6.2472		
	v	0.2217	0.2317	0.2414	0.2508	0.2600	0.2691	0.2781	0.2957	0.3130	0.3302		
600	*h*	1482.4	1508.6	1533.8	1558.5	1582.7	1606.6	1630.4	1677.7	1724.9	1772.4		

	s	5.3222	5.4102	5.4923	5.5697	5.6436	5.7144	5.7826	5.9129	6.0363	6.1541		
	v	0.1874	0.1963	0.2048	0.2131	0.2212	0.2291	0.2369	0.2522	0.2672	0.2821		
700	h	1474.5	1501.9	1528.1	1553.4	1578.2	1602.6	1626.8	1674.6	1722.4	1770.2		
	s	5.2259	5.3179	5.4029	5.4826	5.5582	5.6303	5.6997	5.8316	5.9562	6.0749		
	v	0.1615	0.1696	0.1773	0.1848	0.1920	0.1991	0.2060	0.2196	0.2329	0.2459	0.2589	
800	h	1466.3	1495.0	1522.2	1548.3	1573.7	1598.6	1623.1	1671.6	1719.8	1768.0	1816.4	
	s	5.1387	5.2351	5.3232	5.4053	5.4827	5.5562	5.6268	5.7603	5.8861	6.0057	6.1202	
	v		0.1488	0.1559	0.1627	0.1693	0.1757	0.1820	0.1942	0.2061	0.2178	0.2294	
900	h		1488.0	1516.2	1543.0	1569.1	1594.4	1619.4	1668.5	1717.1	1765.7	1814.4	
	s		5.1593	5.2508	5.3354	5.4147	5.4897	5.5614	5.6968	5.8237	5.9442	6.0594	
	v		0.1321	0.1388	0.1450	0.1511	0.1570	0.1627	0.1739	0.1847	0.1954	0.2058	0.2162
1000	h		1480.6	1510.0	1537.7	1564.4	1590.3	1615.6	1665.4	1714.5	1763.4	1812.4	1861.7
	s		5.0889	5.1840	5.2713	5.3525	5.4292	5.5021	5.6392	5.7674	5.8888	6.0047	6.1159
	v			0.1129	0.1185	0.1238	0.1289	0.1338	0.1434	0.1526	0.1616	0.1705	0.1792
1200	h			1497.1	1526.6	1554.7	1581.7	1608.0	1659.2	1709.2	1758.9	1808.5	1858.2
	s			5.0629	5.1560	5.2416	5.3215	5.3970	5.5379	5.6687	5.7919	5.9091	6.0214
	v			0.0944	0.0995	0.1042	0.1088	0.1132	0.1216	0.1297	0.1376	0.1452	0.1528
1400	h			1483.4	1515.1	1544.7	1573.0	1600.2	1652.8	1703.9	1754.3	1804.5	1854.7
	s			4.9534	5.0530	5.1434	5.2270	5.3053	5.4501	5.5836	5.7087	5.8273	5.9406
	v				0.0851	0.0895	0.0937	0.0977	0.1053	0.1125	0.1195	0.1263	0.1330
1600	h				1502.9	1534.4	1564.0	1592.3	1646.4	1698.5	1749.7	1800.5	1851.2
	s				4.9584	5.0543	5.1419	5.2232	5.3722	5.5084	5.6355	5.7555	5.8699
	v				0.0739	0.0781	0.0820	0.0856	0.0926	0.0992	0.1055	0.1116	0.1177
1800	h				1490.0	1523.5	1554.6	1584.1	1639.8	1693.1	1745.1	1796.5	1847.7
	s				4.8693	4.9715	5.0635	5.14825	3018	5.4409	5.5699	5.6914	5.8069
	v				0.0648	0.0688	0.0725	0.0760	0.0824	0.0885	0.0943	0.0999	0.1054
2000	h				1476.1	1512.0	1544.9	1575.6	1633.2	1687.6	1740.4	1792.4	1844.1
	s				4.7834	4.8930	4.9902	5.0786	5.2371	5.3793	5.5104	5.6333	5.7499

Table 10 Saturated Freon - 12 Table

		Specific volume m^3/kg			Enthalpy kJ/kg			Entropy $kJ/(kg.K)$		
Temp. °C	Abs. Press. MPa P	Sat. Liquid v_f	Evap. v_{fg}	Sat. Vapor v_g	Sat. Liquid h_f	Evap. h_{fg}	Sat. Vapor h_g	Sat. Liquid s_f	Evap. s_{fg}	Sat. Vapor s_g
–90	0.0028	0.000 608	4.414 937	4.415 545	–43.243	189.618	146.375	–0.2084	1.0352	0.8268
–85	0.0042	0.000 612	3.036 704	3.037 316	–38.968	187.608	148.640	–0.1854	0.9970	0.8116
–80	0.0062	0.000 617	2.137 728	2.138 345	–34.688	185.612	150.924	–0.1630	0.9609	0.7979
–75	0.0088	0.000 622	1.537 030	1.537 651	–30.401	183.625	153.224	–0.1411	0.9266	0.7855
–70	0.0123	0.000 627	1.126 654	1.127 280	–26.103	181.640	155.536	–0.1197	0.8940	0.7744
–65	0.0168	0.000 632	0.840 534	0.841 166	–21.793	179.651	157.857	–0.0987	0.8630	0.7643
–60	0.0226	0.000 637	0.637 274	0.637 910	–17.469	177.653	160.184	–0.0782	0.8334	0.7552
–55	0.0300	0.000 642	0.490 358	0.491 000	–13.129	175.641	162.512	–0.0581	0.8051	0.7470
–50	0.0391	0.000 648	0.382 457	0.383 105	–8.772	173.611	164.840	–0.0384	0.7779	0.7396
–45	0.0504	0.000 654	0.302 029	0.302 682	–4.396	171.558	167.163	–0.0190	0.7519	0.7329
–40	0.0642	0.000 659	0.241 251	0.241 910	–0.000	169.479	169.479	–0.0000	0.7269	0.7269
–35	0.0807	0.000 666	0.194 732	0.195 398	4.416	167.368	171.784	0.0187	0.7027	0.7214
–30	0.1004	0.000 672	0.158 703	0.159 375	8.854	165.222	174.076	0.0371	0.6795	0.7165
–25	0.1237	0.000 679	0.130 487	0.131 166	13.315	163.037	176.352	0.0552	0.6570	0.7121
–20	0.1509	0.000 685	0.108 162	0.108 847	17.800	160.810	178.610	0.0730	0.6352	0.7082
–15	0.1826	0.000 693	0.090 326	0.091 018	22.312	158.534	180.846	0.0906	0.6141	0.7046
–10	0.2191	0.000 700	0.075 946	0.076 646	26.851	156.207	183.058	0.1079	0.5936	0.7014
–5	0.2610	0.000 708	0.064 255	0.064 963	31.420	153.823	185.243	0.1250	0.5736	0.6986
0	0.3086	0.000 716	0.054 673	0.055 389	36.022	151.376	187.397	0.1418	0.5542	0.6960
5	0.3626	0.000 724	0.046 761	0.047 485	40.659	148.859	189.518	0.1585	0.5351	0.6937
10	0.4233	0.000 733	0.040 180	0.040 914	45.337	146.265	191.602	0.1750	0.5165	0.6916
15	0.4914	0.000 743	0.034 671	0.035 413	50.058	143.586	193.644	0.1914	0.4983	0.6897
20	0.5673	0.000 752	0.030 028	0.030 780	54.828	140.812	195.641	0.2076	0.4803	0.6879
25	0.6516	0.000 763	0.026 091	0.026 854	59.653	137.933	197.586	0.2237	0.4626	0.6863
30	0.7449	0.000 774	0.022 734	0.023 508	64.539	134.936	199.475	0.2397	0.4451	0.6848
35	0.8477	0.000 786	0.019 855	0.020 641	69.494	131.805	201.299	0.2557	0.4277	0.6834
40	0.9607	0.000 798	0.017 373	0.018 171	74.527	128.525	203.051	0.2716	0.4104	0.6820
45	1.0843	0.000 811	0.015 220	0.016 032	79.647	125.074	204.722	0.2875	0.3931	0.6806
50	1.2193	0.000 826	0.013 344	0.014 170	84.868	121.430	206.298	0.3034	0.3758	0.6792
55	1.3663	0.000 841	0.011 701	0.012 542	90.201	117.565	207.766	0.3194	0.3582	0.6777
60	1.5259	0.000 858	0.010 253	0.011 111	95.665	113.443	209.109	0.3355	0.3405	0.6760
65	1.6988	0.000 877	0.008 971	0.009 847	101.279	109.024	210.303	0.3518	0.3224	0.6742
70	1.8858	0.000 897	0.007 828	0.008 725	107.067	104.255	211.321	0.3683	0.3038	0.6721
75	2.0874	0.000 920	0.006 802	0.007 723	113.058	99.068	212.126	0.3851	0.2845	0.6697
80	2.3046	0.000 946	0.005 875	0.006 821	119.291	93.373	212.665	0.4023	0.2644	0.6667
85	2.5380	0.000 976	0.005 029	0.006 005	125.818	87.047	212.865	0.4201	0.2430	0.6631
90	2.7885	0.001 012	0.004 246	0.005 258	132.708	79.907	212.614	0.4385	0.2200	0.6585
95	3.0569	0.001 056	0.003 508	0.004 563	140.068	71.658	211.726	0.4579	0.1946	0.6526
100	3.3440	0.001 113	0.002 790	0.003 903	148.076	61.768	209.843	0.4788	0.1655	0.6444
105	3.6509	0.001 197	0.002 045	0.003 242	157.085	49.014	206.099	0.5023	0.1296	0.6319
110	3.9784	0.001 364	0.001 098	0.002 462	168.059	28.425	196.484	0.5322	0.0742	0.6064
112	4.1155	0.001 792	0.000 005	0.001 797	174.920	0.151	175.071	0.5651	0.0004	0.5655

Table 11 Superheated Freon - 12 Table

Temp. °C	*v* m^3/kg	*h* kJ/kg	*s* kJ/kg K	*v* m^3/kg	*h* kJ/kg	*s* kJ/kg K	*v* m^3/kg	*h* kJ/kg	*s* kJ/kg K
	0.05 *MPa*			0. 10 *MPa*			0.15 *MPa*		
–20.0	0.341 857	181.042	0.7912	0.167 701	179.861	0.7401			
–10.0	0.356 227	186.757	0.8133	0.175 222	185.707	0.7628	0.114 716	184.619	0.7318
0.0	0.370 508	102.567	0.8350	0.182 647	191.628	0.7849	0.119 866	190.660	0.7543
10.0	0.384 716	198.471	0.8562	0.189 994	197.628	0.8064	0.124 932	196.762	0.7763
20.0	0.398 863	204.469	0.8770	0.197 277	203.707	0.8275	0.129 930	202.927	0.7977
30.0	0.412 959	210.557	0.8974	0.204 506	209.866	0.8482	0.134 873	209.160	0.8186
40.0	0.427 012	216.733	0.9175	0.211 691	216.104	0.8684	0.139 768	215.463	0.8390
50.0	0.441 030	222.997	0.9372	0.218 839	222.421	0.8883	0.144 625	221.835	0.8591
60.0	0.455 017	229.344	0.9565	0.225 955	228.815	0.9078	0.149 450	228.277	0.8787
70.0	0.468 978	235.774	0.9755	0.233 044	235.285	0.9269	0.154 247	234.789	0.8980
80.0	0.482 917	242.282	0.9942	0.240 111	241.829	0.9457	0.159 020	241.371	0.9169
90.0	0.496 838	248.868	1.0126	0.247 159	248.446	0.9642	0.163 774	248.020	0.9354
	0.20 *MPa*			0.25 *MPa*			0.30 *MPa*		
0.0	0.088 608	189.669	0.7320	0.069 752	188.644	0.7139	0.057 150	187.583	0.6984
10.0	0.092 550	195.878	0.7543	0.073 024	194.969	0.7366	0.059 984	194.034	0.7216
20.0	0.096 418	202.135	0.7760	0.076 218	201.322	0.7587	0.062 734	200.490	0.7440
30.0	0.100 228	208.446	0.7972	0.079 350	207.715	0.7801	0.065 418	206.969	0.7658
40.0	0.103 989	214.814	0.8178	0.082 431	214.153	0.8010	0.068 049	213.480	0.7869
50.0	0.107 710	221.243	0.8381	0.085 470	220.642	0.8214	0.070 635	220.030	0.8075
60.0	0.111 397	227.735	0.8578	0.088 474	227.185	0.8413	0.073 185	226.627	0.8276
70.0	0.115 055	234.291	0.8772	0.091 449	233.785	0.8608	0.075 705	233.273	0.8473
80.0	0.118 690	240.910	0.8962	0.094 398	240.443	0.8800	0.078 200	239.971	0.8665
90.0	0.122 304	247.593	0.9149	0.097 327	247.160	0.8987	0.080 673	246.723	0.8853
100.0	0.125 901	254.339	0.9332	0.100 238	253.936	0.9171	0.083 127	253.530	0.9038
110.0	0.129 483	261.147	0.9512	0.103 114	260.770	0.9352	0.085 566	260.391	0.9220
	0.40 *MPa*			0.50 *MPa*			0.60 *MPa*		
120.0	0.045 836	198.762	0.7199	0.035 646	196.935	0.6999			
130.0	0.047 971	205.428	0.7423	0.037 464	203.814	0.7230	0.030 422	202.116	0.7063
140.0	0.050 046	212.095	0.7639	0.039 214	210.656	0.7452	0.031 966	209.154	0.7291
150.0	0.052 072	218.779	0.7849	0.040 911	217.484	0.7667	0.033 450	216.141	0.7511
160.0	0.054 059	225.488	0.8054	0.042 565	224.315	0.7875	0.034 887	223.104	0.7723
170.0	0.056 014	232.230	0.8253	0.044 184	231.161	0.8077	0.036 285	230.062	0.7929
180.0	0.057 941	239.012	0.8448	0.045 774	238.031	0.8275	0.037 653	237.027	0.8129
190.0	0.059 846	245.837	0.8638	0.047 340	244.932	0.8467	0.038 995	244.009	0.8324
200.0	0.061 731	252.707	0.8825	0.048 886	251.869	0.8656	0.040 316	251.016	0.8514
210.0	0.063 600	259.624	0.9008	0.050 415	258.845	0.8840	0.041 619	258.053	0.8700
220.0	0.065 455	266.590	0.9187	0.051 929	265.862	0.9021	0.042 907	265.124	0.8882
230.0	0.067 298	273.605	0.9364	0.053 430	772.923	0.9198	0.044 181	272.231	0.9061

	0.70 *MPa*			0.80 *MPa*			0.90 *MPa*		
0.0	0.026 761	207.580	0.7148	0.022 830	205.924	0.7016	0.019 744	204.170	0.6982
0.0	0.028 100	214.745	0.7373	0.024 068	213.290	0.7248	0.020 912	211.765	0.7131
0.0	0.029 387	221.854	0.7590	0.025 247	220.558	0.7469	0.022 012	219.212	0.7358
0.0	0.030 632	228.931	0.7799	0.026 380	277.766	0.7682	0.023 062	226.564	0.7575
0.0	0.031 843	235.997	0.8002	0.027 477	234.941	0.7888	0.024 072	233.856	0.7785
0.0	0.033 027	243.066	0.8199	0.028 545	242.101	0.8088	0.025 051	241.113	0.7987
0.0	0.034 189	250.146	0.8392	0.029 588	249.260	0.8283	0.026 005	248.355	0.8184
0.0	0.035 332	257.247	0.8579	0.030 612	256.428	0.8472	0.026 937	255.593	0.8376
0.0	0.036 458	264.374	0.8763	0.031 619	263.613	0.8657	0.027 851	262.839	0.862
0.0	0.037 572	271.531	0.8943	0.032 612	270.820	0.8838	0.028 751	270.100	0.8745
0.0	0.038 673	278.720	0.9119	0.033 592	278.055	0.9016	0.029 639	277.381	0.8923
0.0	0.039 764	285.946	0.9292	0.034 563	285.320	0.9189	0.030 515	284.687	0.9098
	1.00 *MPa*			1.20 *MPa*			1.40 *MPa*		
50.0	0.018 366	210.162	0.7021	0.014 483	206.661	0.6812			
60.0	0.019 410	217.810	0.7254	0.015 463	214.805	0.7060	0.012 579	211.457	0.6876
70.0	0.020 397	225.319	0.7476	0.016 368	222.687	0.7293	0.013 448	219.822	0.7123
80.0	0.021 341	232.739	0.7689	0.017 221	230.398	0.7514	0.014 247	227.891	0.7355
90.0	0.022 251	240.101	0.7895	0.018 032	237.995	0.7727	0.014 997	235.766	0.7575
100.0	0.023 133	247.430	0.8094	0.018 812	245.518	0.7931	0.015 710	243.512	0.7785
110.0	0.023 993	254.743	0.8287	0.019 567	252.993	0.8129	0.016 393	251.170	0.7988
120.0	0.024 835	262.053	0.8475	0.020 301	260.441	0.8320	0.017 053	258.770	0.8183
130.0	0.025 661	269.369	0.8659	0.021 018	267.875	0.8507	0.017 695	266.334	0.8373
140.0	0.026 474	276.699	0.8839	0.021 721	275.307	0.8689	0.018 321	273.877	0.8558
150.0	0.027 275	284.047	0.9015	0.022 412	282.745	0.8867	0.018 934	281.411	0.8738
160.0	0.028 068	291.419	0.9187	0.023 093	290.195	0.9041	0.019 535	288.946	0.8914
	1.60 *MPa*			1.80 *MPa*			2.00 *MPa*		
70.0	0.011 208	216.650	0.6959	0.009 406	213.049	0.6794			
80.0	0.011 984	225.177	0.7204	0.010 187	222.198	0.7057	0.008 704	218.859	0.6909
90.0	0.012 698	233.390	0.7433	0.010 884	230.835	0.7298	0.009 406	228.056	0.7166
100.0	0.013 366	241.397	0.7651	0.011 526	239.155	0.7524	0.010 035	236.760	0.7402
110.0	0.014 000	249.264	0.7859	0.012 126	247.264	0.7739	0.010 615	245.154	0.7624
120.0	0.014 608	257.035	0.8059	0.012 697	255.228	0.7944	0.011 159	253.341	0.7835
130.0	0.015 195	264.742	0.8253	0.013 244	263.094	0.8141	0.011 676	261.384	0.8037
140.0	0.015 765	272.406	0.8440	0.013 772	270.891	0.8332	0.012 172	269.327	0.8232
150.0	0.016 320	280.044	0.8623	0.014 284	278.642	0.8518	0.012 651	277.201	0.8420
160.0	0.016 864	287.669	0.8801	0.014 784	286.364	0.8698	0.013 116	285.027	0.8603
170.0	0.017 398	295.290	0.8975	0.015 272	294.069	0.8874	0.013 570	292.822	0.8781
180.0	0.017 923	302.914	0.9145	0.015 752	301.767	0.9046	0.014 013	300.598	0.8955

Table 11 Superheated Freon - 12 Table

Temp. °C	v m^3/kg	h kJ/kg	s kJ/kg K	v m^3/kg	h kJ/kg	s kJ/kg K	v m^3/kg	h kJ/kg	s kJ/kg K
	2.50 MPa			3.00 MPa			3.50 MPa		
90.0	0.006 595	219.562	0.6823						
100.0	0.007 264	229.852	0.7103	0.005 231	220.529	0.6770			
110.0	0.007 837	239.271	0.7352	0.005 886	232.068	0.7075	0.004 324	222.121	0.6750
120.0	0.008 351	248.192	0.7582	0.006 419	242.208	0.7336	0.004 959	234.875	0.7078
130.0	0.008 827	256.794	0.7798	0.006 887	251.632	0.7573	0.005 456	245.661	0.7349
140.0	0.009 273	265.180	0.8003	0.007 313	260.620	0.7793	0.005 884	255.524	0.7591
150.0	0.009 697	273.414	0.8200	0.007 709	269.319	0.8001	0.006 270	264.846	0.7814
160.0	0.010 104	281.540	0.8390	0.008 083	277.817	0.8200	0.006 626	273.817	0.8023
170.0	0.010 497	289.589	0.8574	0.008 439	286.171	0.8391	0.006 961	282.545	0.8222
180.0	0.010 879	297.583	0.8752	0.008 782	294.422	0.8575	0.007 279	291.100	0.8413
190.0	0.011 250	305.540	0.8926	0.009 114	302.597	0.8753	0.007 584	299.528	0.8597
200.0	0.011 614	313.472	0.9095	0.009 436	310.718	0.8927	0.007 878	307.864	0.8775
	4.00 MPa								
120.0	0.003 736	224.863	0.6771						
130.0	0.004 325	238.443	0.7111						
140.0	0.004 781	249.703	0.7386						
150.0	0.005 172	259.904	0.7630						
160.0	0.005 522	269.492	0.7854						
170.0	0.005 845	278.684	0.8063						
180.0	0.006 147	287.602	0.8262						
190.0	0.006 434	296.326	0.8453						
200.0	0.006 708	304.906	0.8636						
210.0	0.006 972	313.380	0.8813						
220.0	0.007 228	321.774	0.8985						
230.0	0.007 477	330.108	0.9152						

Table 12 Enthalpies of Formation. Gibbs Function of Formation and Absolute Entropy at 25°C and 1 atm Pressure

Substance	*M*	*Enthalpy of formation kJ/kgmol*	*Gibbs function of formation kJ/kgmol*	*Absolute Entropy kJ/kgmol.K*
Acetylene $C_2H_2(g)$	26.038	226 866	209 290	200.98
Ammonia $NH_3(g)$	17.032	–45 926	–16 390	192.72
Benzene $C_6H_6(g)$	78.108	82 976	129 732	269.38
Butane $C_4H_{10}(g)$	58.124	–126 223	–17 164	310.32
Carbon C graphite	12.011	0	0	5.69
Carbon Dioxide $CO_2(g)$	44.01	–393 757	–344 631	213.83
Carbon Monoxide CO(g)	28.001	–110 596	–137 242	197.68
Ethane $C_2H_6(g)$	30.07	–84 718	–32 905	229.64
Ethene $C_2H_4(g)$	28.054	52 315	681 59	219.58
Hydrazine $N_2H_4(g)$	32.048	95 410	159 260	238.77
Methane $CH_4(g)$	16.043	–74 917	–50 844	186.27
Octane $C_8H_{18}(g)$	114.23	–208 581	16 536	467.04
Octane $C_8H_{18}(l)$	114.23	–250 102	6 614	361.03
Propane $C_3H_8(g)$	44.097	–103 909	–23 502	270.09
Water $H_2O(g)$	18.015	–241 971	–228 729	188.85
Water $H_2O(l)$	18.015	–286 010	–237 327	69.98
Dodecane $C_{12}H_{26}(g)$	170.328	–290 971		
Dodecane $C_{12}H_{26}(l)$	170.328	–394 199		

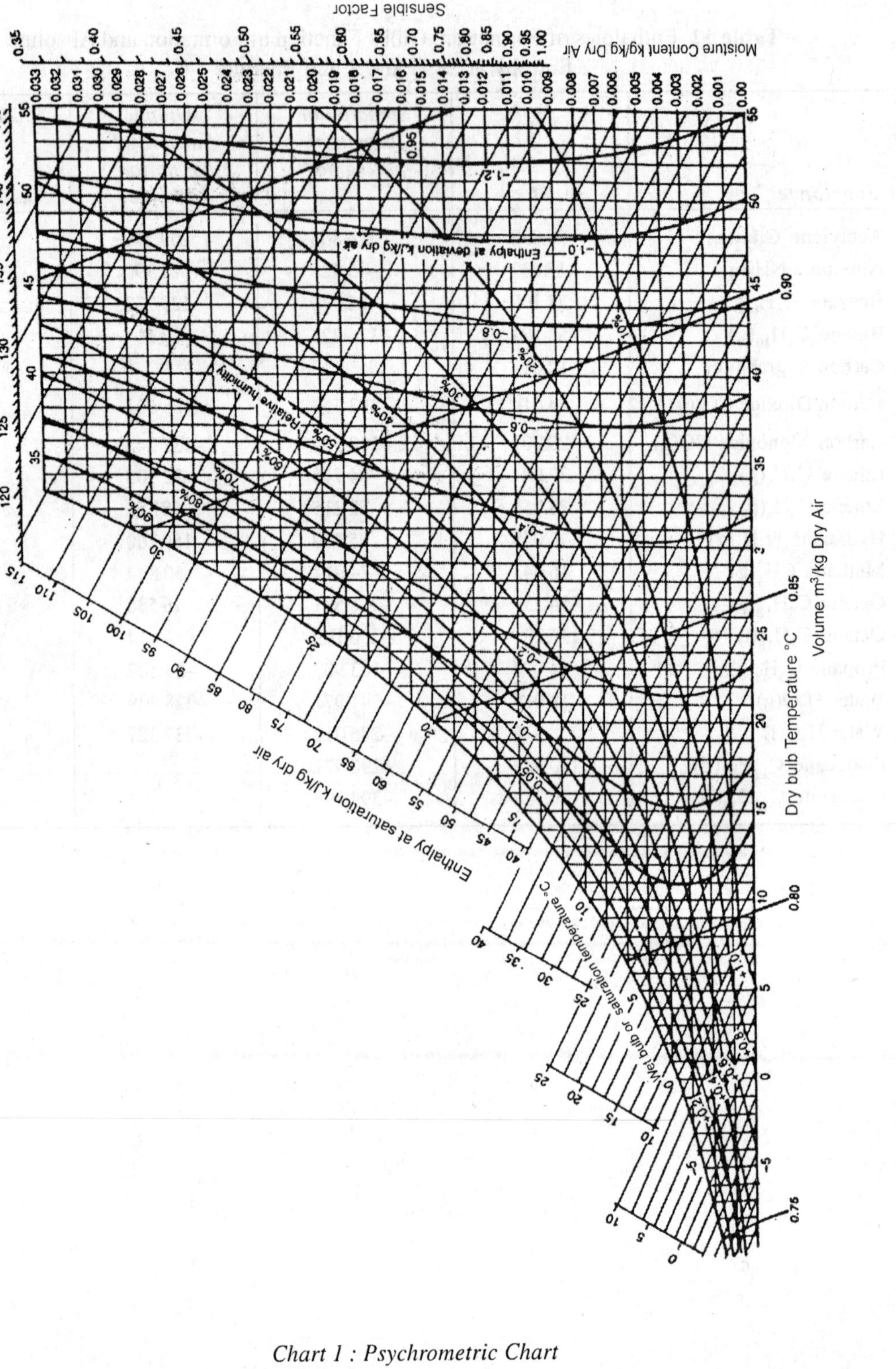

Chart 1 : Psychrometric Chart

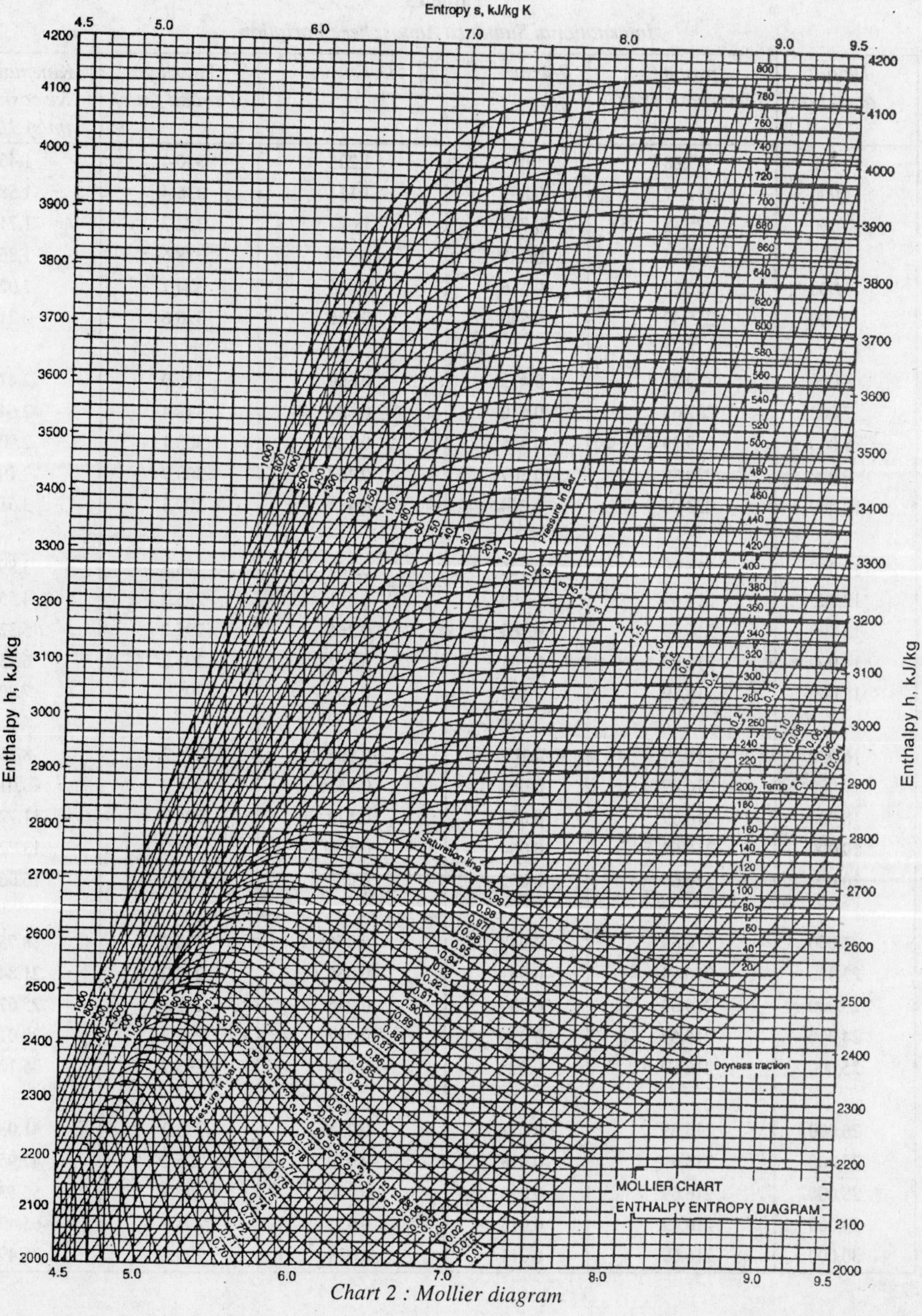

Chart 2 : Mollier diagram

Table 13

International Standard Atmosphere Variation

Height (meters)	Absolute Temperature (K)	Pressure (bar)	Density (kg/m^3)	Velocity of sound (m/s)	Kinematic viscosity $(m/s)\cdot 10^{-5}$
0	288	1.013	1.225	340.4	1.46
1,000	281.37	0.899	1.111	336.4	1.58
2,000	275.18	0.794	1.007	332.7	1.71
3,000	268.56	0.701	0.909	328.6	1.86
4,000	262.22	0.616	0.820	324.7	2.02
5,000	255.59	0.540	0.736	320.6	2.21
6,000	249.25	0.472	0.659	316.0	2.41
7,000	242.62	0.410	0.589	312.4	2.64
8,000	235.99	0.356	0.526	308.1	2.90
9,000	229.66	0.307	0.467	303.9	3.19
10,000	222.90	0.264	0.413	299.5	3.51
11,000	216.69	0.226	0.364	295.2	3.89
12,000	216.69	0.193	0.311	295.2	4.55
13,000	216.69	0.165	0.266	295.2	5.32
14,000	216.69	0.141	0.227	295.2	6.23
15,000	216.69	0.121	0.194	295.2	7.30
16,000	216.69	0.102	0.165	295.2	8.55
17,000	216.69	0.088	0.141	295.2	10.01
18,000	216.69	0.075	0.120	295.2	11.72
19,000	216.69	0.064	0.103	295.2	13.72
20,000	216.69	0.055	0.088	295.2	16.06
21,000	216.69	0.048	0.076	295.2	18.75
22,000	216.69	0.041	0.065	295.2	21.84
23,000	216.69	0.034	0.055	295.2	25.67
24,000	216.69	0.029	0.047	295.2	30.01
25,000	216.69	0.025	0.040	295.2	35.10
26,000	216.69	0.021	0.034	295.2	41.03
27,000	216.69	0.018	0.029	295.2	47.95
28,000	216.69	0.015	0.026	295.2	56.53
29,000	216.69	0.013	0.022	295.2	65.47
30,000	216.69	0.011	0.018	295.2	76.49

Table 14

Air table for isentropic flow of air ($\gamma = 1.4$)

M	λ	$\frac{T}{T_0}$	$\frac{P}{P_0}$	$\frac{A}{A^*}$	$\frac{I}{I^*}$	$\frac{A}{A^*}\frac{P}{P_0}$	$\frac{m\sqrt{C_P T_0}}{A \cdot P_0}$
0.00	0.000	1.000	1.000				0
0.02	0.022	0.999	0.999	28.242	22.834	28.934	0.043
0.04	0.044	0.999	0.998	14.482	11.435	14.465	0.088
0.06	0.066	0.999	0.997	9.666	7.643	9.642	0.133
0.08	0.088	0.998	0.995	7.262	5.753	7.229	0.176
0.10	0.109	0.998	0.993	5.822	4.624	5.781	0.220
0.12	0.131	0.997	0.989	4.864	3.875	4.815	0.263
0.14	0.153	0.996	0.986	4.182	3.343	4.126	0.306
0.16	0.175	0.995	0.982	3.673	2.947	3.607	0.349
0.18	0.197	0.994	0.976	3.278	2.642	3.205	0.391
0.20	0.218	0.992	0.973	2.964	2.400	2.882	0.432
0.22	0.239	0.990	0.967	2.707	2.205	2.618	0.473
0.24	0.261	0.988	0.961	2.495	2.043	2.398	0.513
0.26	0.283	0.986	0.954	2.317	1.909	2.211	0.553
0.28	0.304	0.985	0.947	2.166	1.795	2.051	0.592
0.30	0.325	0.982	0.939	2.035	1.698	1.912	0.629
0.32	0.347	0.979	0.932	1.922	1.614	1.790	0.667
0.34	0.368	0.977	0.923	1.823	1.542	1.683	0.703
0.36	0.389	0.975	0.914	1.735	1.479	1.587	0.738
0.38	0.410	0.972	0.905	1.658	1.424	1.501	0.772
0.40	0.431	0.969	0.895	1.590	1.375	1.424	0.806
0.42	0.452	0.966	0.886	1.529	1.332	1.354	0.838
0.44	0.473	0.963	0.876	1.474	1.294	1.290	0.869
0.46	0.494	0.959	0.865	1.425	1.259	1.232	0.899
0.48	0.514	0.956	0.854	1.380	1.230	1.178	0.928
0.50	0.534	0.952	0.843	1.340	1.203	1.129	0.956
0.52	0.555	0.948	0.832	1.303	1.178	1.084	0.983
0.54	0.575	0.945	0.820	1.270	1.157	1.042	1.008
0.56	0.595	0.941	0.808	1.240	1.138	1.002	1.033
0.58	0.615	0.937	0.796	1.213	1.121	0.966	1.056

Contd.

Table 14 *Air table for isentropic flow of air, ($\gamma = 1.4$)*

M	λ	$\frac{T}{T_0}$	$\frac{P}{P_0}$	$\frac{A}{A^*}$	$\frac{I}{I^*}$	$\frac{A}{A^*}\frac{P}{P_0}$	$\frac{m\sqrt{C_P T_0}}{A \cdot P_0}$
0.60	0.635	0.933	0.784	1.188	1.105	0.932	1.078
0.62	0.655	0.928	0.772	1.166	1.091	0.900	1.099
0.64	0.674	0.924	0.759	1.145	1.078	0.869	1.119
0.66	0.693	0.919	0.746	1.126	1.068	0.841	1.137
0.68	0.713	0.915	0.734	1.109	1.058	0.814	1.154
0.70	0.732	0.911	0.721	1.094	1.049	0.789	1.171
0.72	0.751	0.906	0.708	1.081	1.041	0.765	1.185
0.74	0.769	0.901	0.695	1.068	1.035	0.742	1.199
0.76	0.788	0.896	0.682	1.057	1.028	0.721	1.212
0.78	0.807	0.892	0.669	1.047	1.023	0.700	1.223
0.80	0.825	0.886	0.656	1.038	1.018	0.681	1.234
0.82	0.843	0.881	0.643	1.031	1.015	0.663	1.243
0.84	0.861	0.876	0.630	1.024	1.011	0.645	1.251
0.86	0.879	0.871	0.617	1.018	1.008	0.628	1.259
0.88	0.897	0.869	0.604	1.013	1.006	0.612	1.265
0.90	0.915	0.861	0.591	1.009	1.004	0.596	1.270
0.92	0.932	0.855	0.578	1.006	1.002	0.582	1.274
0.94	0.949	0.849	0.566	1.003	1.001	0.567	1.277
0.96	0.966	0.844	0.553	1.001	1.0006	0.554	1.279
0.98	0.983	0.839	0.541	1.0003	1.0001	0.541	1.281
1.00	1.000	0.834	0.528	1.000	1.000	0.528	1.281
1.02	1.017	0.828	0.516	1.0003	1.0001	0.516	1.281
1.04	1.033	0.822	0.510	1.0007	1.0003	0.510	1.279
1.06	1.049	0.816	0.492	1.003	1.001	0.493	1.277
1.08	1.065	0.811	0.480	1.005	1.002	0.483	1.274
1.10	1.081	0.805	0.468	1.008	1.003	0.472	1.271
1.12	1.097	0.799	0.457	1.011	1.004	0.462	1.267
1.14	1.113	0.794	0.445	1.015	1.005	0.452	1.262
1.16	1.128	0.788	0.434	1.019	1.007	0.443	1.256
1.18	1.143	0.782	0.423	1.025	1.009	0.434	1.250

Contd.

Table 14 *Air table for isentropic flow of air, ($\gamma = 1.4$)*

M	λ	$\frac{T}{T_0}$	$\frac{P}{P_0}$	$\frac{A}{A^*}$	$\frac{I}{I^*}$	$\frac{A}{A^*}\frac{P}{P_0}$	$\frac{m\sqrt{C_P T_0}}{A \cdot P_0}$
1.20	1.158	0.776	0.412	1.030	1.010	0.425	1.243
1.22	1.173	0.771	0.402	1.037	1.013	0.416	1.236
1.24	1.188	0.765	0.391	1.043	1.015	0.408	1.228
1.26	1.203	0.759	0.381	1.050	1.017	0.400	1.220
1.28	1.217	0.753	0.371	1.058	1.019	0.392	1.211
1.30	1.231	0.747	0.361	1.066	1.022	0.385	1.201
1.32	1.245	0.742	0.351	1.075	1.024	0.378	1.192
1.34	1.259	0.736	0.342	1.084	1.027	0.37	1.181
1.36	1.273	0.730	0.332	1.094	1.029	0.364	1.171
1.38	1.286	0.724	0.323	1.104	1.032	0.357	1.160
1.40	1.300	0.718	0.314	1.115	1.035	0.350	1.149
1.42	1.313	0.713	0.305	1.126	1.037	0.344	1.138
1.44	1.326	0.706	0.297	1.138	1.040	0.338	1.126
1.46	1.330	0.701	0.288	1.150	1.043	0.332	1.114
1.48	1.352	0.695	0.280	1.163	1.046	0.326	1.102
1.50	1.365	0.689	0.272	1.176	1.048	0.320	1.089
1.52	1.377	0.684	0.265	1.189	1.052	0.315	1.077
1.54	1.389	0.678	0.257	1.204	1.055	0.309	1.064
1.56	1.402	0.673	0.249	1.219	1.057	0.304	1.051
1.58	1.414	0.667	0.242	1.234	1.061	0.300	1.038
1.60	1.425	0.661	0.235	1.250	1.063	0.294	1.025
1.62	1.437	0.656	0.228	1.267	1.067	0.289	1.011
1.64	1.448	0.650	0.222	1.284	1.069	0.285	0.998
1.66	1.460	0.645	0.215	1.301	1.073	0.280	0.985
1.68	1.471	0.639	0.208	1.319	1.076	0.275	0.971
1.70	1.483	0.634	0.203	1.337	1.078	0.271	0.958
1.72	1.494	0.628	0.196	1.357	1.082	0.267	0.944
1.74	1.504	0.623	0.191	1.376	1.085	0.262	0.931
1.76	1.515	0.617	0.185	1.397	1.087	0.258	0.917
1.78	1.526	0.612	0.179	1.418	1.091	0.254	0.904

Contd.

Table 14 *Air table for isentropic flow of air,* ($\gamma = 1.4$)

M	λ	$\frac{T}{T_0}$	$\frac{P}{P_0}$	$\frac{A}{A^*}$	$\frac{I}{I^*}$	$\frac{A}{A^*}\frac{P}{P_0}$	$\frac{m\sqrt{C_P T_0}}{A \cdot P_0}$
1.80	1.536	0.607	0.174	1.439	1.094	0.250	0.890
1.82	1.546	0.602	0.168	1.461	1.096	0.247	0.877
1.84	1.556	0.596	0.164	1.484	1.099	0.243	0.863
1.86	1.566	0.591	0.158	1.507	1.102	0.239	0.850
1.88	1.576	0.585	0.154	1.531	1.105	0.235	0.837
1.90	1.586	0.581	0.149	1.555	1.108	0.232	0.824
1.92	1.596	0.575	0.145	1.580	1.111	0.228	0.811
1.94	1.605	0.570	0.140	1.606	1.114	0.225	0.798
1.96	1.615	0.566	0.136	1.633	1.117	0.222	0.785
1.98	1.624	0.561	0.132	1.659	1.120	0.219	0.772
2.00	1.633	0.555	0.128	1.687	1.123	0.216	0.759
2.02	1.642	0.551	0.124	1.716	1.126	0.213	0.747
2.04	1.651	0.546	0.120	1.745	1.128	0.209	0.734
2.06	1.660	0.541	0.116	1.775	1.131	0.207	0.722
2.08	1.668	0.536	0.113	1.805	1.134	0.204	0.709
2.10	1.677	0.531	0.109	1.837	1.137	0.200	0.697
2.12	1.685	0.527	0.106	1.869	1.139	0.198	0.685
2.14	1.694	0.522	0.103	1.902	1.142	0.195	0.674
2.16	1.702	0.517	0.099	1.935	1.145	0.193	0.662
2.18	1.710	0.513	0.097	1.969	1.147	0.190	0.650
2.20	1.718	0.508	0.094	2.005	1.150	0.187	0.639
2.22	1.726	0.504	0.091	2.041	1.153	0.185	0.628
2.24	1.734	0.499	0.088	2.078	1.155	0.183	0.617
2.26	1.741	0.495	0.085	2.115	1.158	0.180	0.606
2.28	1.749	0.490	0.082	2.154	1.160	0.178	0.595
2.30	1.756	0.486	0.080	2.193	2.163	0.175	0.584
2.32	1.764	0.482	0.078	2.233	1.165	0.173	0.574
2.34	1.771	0.477	0.075	2.274	1.168	0.171	0.563
2.36	1.778	0.473	0.073	2.316	1.170	0.163	0.553
2.38	1.785	0.469	0.071	2.359	1.173	0.166	0.543

Contd.

Table 14 *Air table for isentropic flow of air, ($\gamma = 1.4$)*

M	λ	$\frac{T}{T_0}$	$\frac{P}{P_0}$	$\frac{A}{A^*}$	$\frac{I}{I^*}$	$\frac{A}{A^*}\frac{P}{P_0}$	$\frac{m\sqrt{C_P T_0}}{A \cdot P_0}$
2.40	1.792	0.464	0.068	2.403	1.179	0.164	0.533
2.42	1.799	0.461	0.066	2.448	1.178	0.162	0.523
2.44	1.806	0.456	0.064	2.494	1.180	0.160	0.514
2.46	1.813	0.452	0.062	2.540	1.182	0.158	0.504
2.48	1.819	0.448	0.060	2.588	1.184	0.156	0.495
2.50	1.826	0.444	0.059	2.637	1.187	0.154	0.486
2.60	1.857	0.425	0.050	2.896	1.198	0.145	0.442
2.70	1.886	0.407	0.043	3.183	1.208	0.137	0.402
2.80	1.914	0.389	0.037	3.500	1.218	0.129	0.366
2.90	1.940	0.373	0.031	3.850	1.228	0.122	0.333
3.00	1.964	0.357	0.027	4.235	1.237	0.115	0.303
3.50	2.046	0.286	0.013	6.789	1.274	0.089	0.189
4.00	2.138	0.238	6.6×10^{-3}	10.719	1.303	0.071	0.1200
4.50	2.194	0.198	3.5×10^{-3}	16.562	1.325	0.057	0.77×10^{-1}
5.00	2.236	0.167	1.89×10^{-3}	25.000	1.341	0.047	0.51×10^{-1}
5.50	2.269	0.142	1.07×10^{-3}	36.869	1.355	0.039	0.34×10^{-1}
6.00	2.295	0.122	0.63×10^{-3}	53.180	1.365	0.033	0.24×10^{-1}
6.50	2.316	0.106	0.38×10^{-3}	75.134	1.374	0.028	0.17×10^{-1}
7.00	2.333	0.093	0.24×10^{-3}	104.143	1.381	0.025	0.12×10^{-1}
7.50	2.347	0.082	0.15×10^{-3}	141.842	1.387	0.021	0.9×10^{-2}
8.00	2.359	0.072	0.1×10^{-3}	190.11	1.392	0.019	0.67×10^{-2}
8.50	2.369	0.065	0.07×10^{-3}	251.087	1.396	0.018	0.51×10^{-2}
9.00	2.377	0.058	0.04×10^{-3}	327.189	1.399	0.016	0.39×10^{-2}
9.50	2.384	0.052	0.03×10^{-3}	421.132	1.402	0.013	0.30×10^{-2}
10.00	2.390	0.047	0.02×10^{-3}	536.00	1.404	0.010	0.24×10^{-2}
∞	2.45	0	0	∞	1.429	0	0

INDEX

D

E

F

N

O

P

Q

R